Toxic Legacy

Toxic Legacy

Synthetic Toxins in the Food, Water, and Air of American Cities

Patrick J. Sullivan
James J. J. Clark
Franklin J. Agardy
Paul E. Rosenfeld

ELSEVIER
ACADEMIC
PRESS

AMSTERDAM • BOSTON • HEIDELBERG • LONDON • NEW YORK • OXFORD • PARIS
SAN DIEGO • SAN FRANCISCO • SINGAPORE • SYDNEY • TOKYO

Academic Press is an imprint of Elsevier
30 Corporate Drive, Suite 400, Burlington, MA 01803, USA
525 B Street, Suite 1900, San Diego, California 92101-4495, USA
84 Theobald's Road, London WC1X 8RR, UK

This book is printed on acid-free paper. ∞

Library of Congress Cataloging-in-Publication Data
Application Submitted

British Library Cataloguing-in-Publication Data
A catalogue record for this book is available from the British Library.

ISBN 13: 978-0-12-370640-9
ISBN 10: 0-12-370640-8

For information on all Academic Press publications
visit our Web site at www.books.elsevier.com

Printed in the United States of America
06 07 08 09 10 9 8 7 6 5 4 3 2 1

CONTENTS

CHAPTER 9

Toxicity and Synthetic Chemical Mixtures

CHAPTER 10

The Body's Synthetic Chemical Burden

Appendixes

PREFACE

It should be no surprise that synthetic toxic chemicals released into the environment contaminate the human body. After all, the food we eat, the water we drink, and the air we breathe are all contaminated with these man-made chemicals. This is a known condition of our daily lives that the vast majority of us passively accept as a reality of the modern world. Our acceptance of this reality is buffered by the assurances of both the United States Environmental Protection Agency (EPA) and the Food and Drug Administration (FDA) that there is little health risk from being exposed to low levels of these individual environmental contaminants.

When considering the reality of our health and the health of our children and families, however, can we confidently accept the assurances of the EPA and the FDA that some level of allowable pollution does us no harm? The honest answer is that we really do not know. This is because the EPA, the FDA, or any other scientific institution in the United States has the ability to calculate the potential health impact of consuming two or more different chemicals. In other words, in the real world where we are all exposed to mixtures of chemical residues in our food, water, and air, the health risk from these mixtures is totally unknown.

An article in *Scientific American* (May 2006) stated that "With tens of thousands of chemicals in regular use worldwide, assessing which combinations might prove harmful is a gargantuan task. . . . The potential effects of contaminant mixtures on people, aquatic life, and fish-eating wildlife are still poorly understood" (p. 16).

Given these uncertainties, most of us would prefer to have no daily intake of chemical residues. Yet, depending on where we live in the United States, the amount and diversity of chemical residues in food, water, and air can vary dramatically. Because of the unknown health risks associated with our exposure to residual chemical mixtures, the purpose of this book is to (1) provide extensive information on the occurrence of chemical mixtures in the food, water, and air in cities around the United States, (2) show how consumer preferences for specific food and water products, as well as particular diets, influence exposure to chemical mixtures, (3) discuss the toxicity and risks associated with our exposure to synthetic chemicals, and (4) offer some guidance as to what steps we can all take to limit our exposure to complex chemical mixtures.

FOREWORD

Since the start of recorded human history, pollution has been understood to be a significant cause of human suffering and disease. Yet today, we have set up a system in which polluters are allowed to discharge massive quantities of chemicals that pollute our food, water, and air. Ancient cultures knew to keep waste away from homes and food sources yet we license polluters to adulterate our food, water, and air. Lead poisoning from food, water, and in the air of Roman cities may have contributed to the ultimate fall of the Roman Empire. Romans preferred wine that was distilled in lead lined pots and kettles, resulting in a sweet flavor that was thought to complement foods, even though they knew that lead was poisoning the populace. Two thousand years later, we have failed to learn the lessons of other civilizations and the network of laws meant to protect us from pollution in our food, water, and air has failed to prevent the daily assault on our bodies. With more than 75,000 regulated synthetic chemicals in use in the United States, the long term impacts from the constant exposure to these chemicals is understood for less than 1% of these chemicals. The synthetic chemicals released by polluters into the environment not only affects our health today, but the health and well being of future generations, including newborns who are born with a toxic soup of chemicals passed on by their mothers. The time has come to carefully evaluate how we are impacting our children's world with this toxic legacy.

Erin Brockovich

1 CONFRONTING THE UNKNOWN

As a university professor, you are frequently receiving fliers from professional associations and interest groups. So it was not unusual when one day in my mail I found, addressed to Dr. Patrick Sullivan, a glossy, colored postcard from the nuclear power industry. The postcard had a photograph of a nuclear power plant with a small black pellet enclosed in its plastic window.

When concerns over the safe disposal of nuclear waste were first raised, a representative of a public utility commented with a straight face that "if every man, woman and child in the United States just ate one gram of nuclear waste there would not be a disposal problem." The postcard was actually designed to illustrate how much electrical power could be generated by this small pellet, but for a moment I thought I had received my personal gram.

I think it is safe to assume that anyone asked to consume a gram of nuclear waste would not believe it was a serious request. But what if we were asked to eat only a milligram (one thousandth of a gram)[1] or a microgram (one millionth of a gram) of nuclear waste? Would you then give your consent? Although we are not specifically confronted with this decision, many people do make similar decisions when they choose to drink bottled water, eat a salad with organic lettuce, or not live in an industrialized area.

Individuals make these choices based more on hearsay than on fact. Why? Because, in reality, medical scientists cannot yet predict the toxic effects to individual humans from consuming low-level chemical mixtures that permeate our food, water, and air. Being unable to predict if low-level chemical mixtures do us harm, each individual's health faces an uncertain future.

[1] As a point of reference, a common aspirin tablet is 325 milligrams or 0.325 grams.

That is why individual choices that influence the type and amount of synthetic chemicals that a person allows into his or her body need to be understood.[2] This means that everyone should have at least some rudimentary knowledge of the synthetic chemicals in our food, water, and air, and the potential threat they may pose if we are to confront the unknown.

1.1 SYNTHETIC CHEMICALS IN THE HUMAN HABITAT

In 1983, I was teaching environmental and soil science courses at Ball State University, Indiana. During the summer months, however, I was always faced with having to find other sources of income. Fortunately, along with five other professors from across the nation, I had been selected as an environmental and engineering fellow by the American Association for the Advancement of Science to work at the Environmental Protection Agency (EPA) in Washington, D.C. As part of this program, we attended various seminars that discussed current and future environmental problems that needed to be studied.

The speaker at our first seminar asked the six of us if we knew the three most common chemical pollutants in water. We all sat there mute, pondering the question. I remember thinking that they had to be either pesticides or other common industrial chemicals. Before we could reply, our speaker preempted any response by correctly assuming we wouldn't know the answer anyway and blurted out, "Valium, estrogen, and diuretics."

Valium (diazepam) was introduced by Roche Labs in 1963 to relieve anxiety, nervousness, and tension. It received a vast amount of media coverage and was popularized as a lifestyle drug. Given the ever-increasing stress load of modern society, it is no wonder that Valium became the first billion-dollar drug. In May 1960, the Food and Drug Administration (FDA) approved Searle to sell oral contraceptives (estrogen-type compounds). By 1962, 1.2 million American women were on the "Pill." Sales slowed during the early 1970s, due to safety concerns and Senate hearings but surged to 10 million users by 1973. During this same time period, diuretic compounds (which promote water loss and sodium excretion) were becoming routinely prescribed to treat high blood pressure.

Given the popularity and use of these drugs, why wouldn't they be present in water? I wrongly assumed that if pharmaceuticals found their way into one's urine or feces and were flushed to a sewage treatment plant, then these drugs would be destroyed during the treatment process. Based on the speaker's information, this was simply not the case. Clearly, wastewater containing pharmaceuticals was being discharged into creeks, rivers, lakes, and groundwater, and these same waters would eventually become someone's drinking water. Nothing

[2] A synthetic chemical is an organic chemical (i.e., a carbon-based compound) that is generated by human activity.

has changed since 1983, except for the type and number of different drugs that are currently being found in our water resources. In today's water, we also find drugs like Prozac, aspirin, ibuprofen, cholesterol-lowering drugs, and even chemotherapy chemicals.

Thousands of synthetic chemicals could be found in our drinking water sources if we would only analyze for them. However, the water agencies that provide drinking water to our communities routinely check for only 65 known toxic chemicals. This trend also exists for the food we eat and the air we breathe. There are currently 975 different active ingredients (chemicals used for their toxic effects) registered with the EPA, and the EPA allows another 1918 inert ingredients (chemicals not used for their toxic effects) to be mixed into pesticide products. However, the Food and Drug Administration (FDA) monitoring programs only look for approximately 10% of the synthetic chemicals that could be found in food products. Just as with water, air resources can contain thousands of different synthetic chemicals, but the EPA requires that only 188 toxic chemicals be tracked and their emissions reported.

Our bodies are currently being bombarded with a vast mixture of both known and unknown chemicals, and it is only going to get worse. Our nation's manufacturers produce approximately 87,000 different chemicals with about 2000 new chemicals being added each year. In recognition of our exposure to this mixture of chemicals, in 2001 the Centers for Disease Control and Prevention (CDC) began looking for synthetic chemicals in the human body. The chemical data collected by the CDC should not be surprising.

The January 2003 CDC study analyzed for 49 synthetic chemicals,[3] from a selected group of pesticides and one group of industrial chemicals, known as phthalates,[4] in both blood and urine. The CDC reported the following total average concentrations (parts per billion, or ppb) of either pesticides or phthalates in human blood or urine.

Age group	Insecticide organophosphates	Insecticide organochlorine	Herbicides	Phthalates
6–11 years old	8.4	No data	2.6	177
12–19 years old	6.3	122	2.4	276
20–59 years old	4.9	332	2.0	227

In 2005, the Environmental Working Group released a report on chemical contaminants in newborns.[5] In this study, an average of 200 industrial chemicals and pollutants were identified in umbilical cord blood from 10 babies born in U.S. hospitals. The information collected by the CDC and the Environmental Working Group confirms what we all should suspect. Since we are exposed

[3] Second National Report on Human Exposure to Environmental Chemicals.

[4] Phthalates occur in plastics, detergents, soaps, shampoo, pesticides, and pharmaceuticals.

[5] The Environmental Working Group report (July 14, 2005) on "Body Burden, the Pollution in Newborns," noted 287 chemicals in umbilical cord blood.

to a mixture of chemicals in our water, food, and air, some combination of these same chemicals will be in our bodies, and this combination of chemicals will be passed on to our children.

To many, the verification of synthetic chemicals in our bodies is old news and is of little or no concern since state and federal governments regulate how much of a specific chemical is "allowed" in drinking water, in each food product, or in the air. For example, there is a 5.0 ppb limit for the chemical trichloroethylene (a common industrial chemical) in drinking water, yet there is no set limit for trichloroethylene in food or air. Although regulations that limit the concentration of individual chemicals are a good idea, they do not provide the consumer with comprehensive protection. For example, a report by the Government Accountability Office in June 2005 concluded that "EPA's reviews of new chemicals provide limited assurance that health and environmental risks are identified before the chemicals enter commerce." In fact, when you consider that we do not know what chemicals we actually consume, coupled with the fact that all of us are exposed to a massive number of chemicals and chemical mixtures that have unknown toxic effects, we really do not know the potential health risks of this exposure.

1.2 DEFINING THE RISK

In 1988, I chose to leave academics and research to work in the emerging field of environmental forensics.[6] Of all the potential environmental forensic problems that can occur, the most difficult to prove is the "toxic tort" lawsuit. In this type of lawsuit, individuals who have been exposed to toxic chemicals must prove that the chemicals to which they were exposed actually resulted in damaging their health. This type of lawsuit was portrayed in the movies *A Civil Action* and *Erin Brockovitch*.

I became involved in a similar case in 1991. I had been retained by the Spence Law Firm in Jackson Hole, Wyoming, and was asked to determine which industries in Casper, Wyoming, were responsible for polluting the residents' drinking water wells, within the Brookhurst subdivision, with a mixture of petroleum products and chlorinated industrial chemicals.

Based on the maps and chemical data, I was confident that my expert opinion given at trial would establish which companies were responsible for polluting the residents' wells. My opinion, however, was not enough to prove that the residents had been harmed by drinking and bathing in the chemical soup that had become their tap water. It was also necessary that the medical experts (toxicologist and physician) be able to show a direct connection between drink-

[6] The field of environmental forensics is described in *Practical Environmental Forensics, Process and Case Histories*, by Patrick Sullivan, Franklin Agardy, and Richard Traub, John Wiley & Sons (2001).

ing chemically polluted water and the residents' existing medical problems. Unlike my opinion, which was fairly straightforward, the medical experts had a difficult time establishing this connection since they had no actual data on human health effects from exposure to low levels of the chemicals that were found in the residents' well water. There were animal studies showing toxic effects for individual chemicals, but no human health studies. Because of this major shortcoming, it was practically impossible for medical experts, in this specific case, to technically support the opinion that it was a scientific certainty the residents were harmed or would be harmed. I found this result deeply disturbing because if the basic science used to establish a chemical standard, which was supposed to be protective of human health, was at the same time not necessarily suitable for establishing harm in a court of law, what protection do chemical standards actually provide and what are the real risks associated with consuming low levels of synthetic chemicals?

1.3 DEFINING THE UNKNOWN

Everyone should have a reasonable assurance that food, water, and air containing synthetic chemical contaminants do not pose any health effects, but I am not convinced that today's science is capable of providing a reasonable answer any time soon. A case in point is the infamous history of DDT. In the mid-1940s, DDT was heralded as one of the modern miracles of science. There was finally a pesticide that killed a wide range of insects, was relatively cheap, was persistent, and was virtually nontoxic to humans. As a result, DDTs popularity grew rapidly throughout the United States, but by the early 1950s an increasing number of massive fish kills were associated with its use. In 1954, the FDA allowed the use of DDT on food products with a tolerance level of 7000 ppb. Due to continued incidents of DDT poisoning of wildlife in the 1960s, the most widely used pesticide of the time fell under intensive scrutiny. The EPA (formed in 1969) canceled the use of DDT on food crops and its use around beef, sheep, and hogs in 1970. From approximately 1945 through 1970, 1,350,000,000 pounds of DDT were used domestically.

Today, DDT and its metabolites (DDE and TDE) are classified as poisonous or deleterious substances in human foods and have established tolerance levels that range from 50 to 500 ppb, depending on the commodity. Even though DDT has not been used on food crops for 35 years, DDT, DDE, and TDE still occur in food crops, milk products, and protein products because these chemicals either persist in the soil or are used on food products imported to the United States. Thus, it should not be surprising that even organic foods may contain these chemicals for decades or centuries to come.

When looking at the history of DDT, it should be remembered that it took 35 years before this toxin was banned from food production. Given such a dramatic shift in the understanding of DDT toxicity to humans, what assurances do we have that scientists have any better ability to timely predict human health effects for the massive number of pesticide mixtures used in agriculture today? Furthermore, a food tolerance level is only an estimate of a pesticide's concentration that will "unavoidably" occur in a crop during agricultural production using best management practices, instead of a value selected deliberately for its ability to protect the public health. These conditions emphasize the unknown nature of the potential harm synthetic chemicals may present in our food and water, as well as in the air we breathe.

As a naive chemistry student in the early 1970s, it was not uncommon to use benzene to clean glassware in the laboratory. Because of this practice, I and other students would routinely breathe concentrated benzene vapors while unaware of their potential harm. By the time I stopped working in the laboratory, around the mid-1980s, benzene was finally recognized in the United States as a carcinogen. In work environments today, individuals are not supposed to be exposed to more than 1600 micrograms (ug) of benzene in one cubic meter of air (ug/m^3) during an 8-hour time period. Smokers, however, inhale around 1800 ug of benzene a day. In 1987, the state of California determined that the maximum level to which an individual could be exposed to benzene in air was 13 ug/day or approximately 0.75 ug/m^3 for an adult.[7] For most nonsmokers, a person's main source of exposure to benzene is from gasoline and automobile exhaust. Outdoor levels of benzene have been reported to range from about 0.2 ug/m^3 in rural regions to 350 ug/m^3 in urban areas associated with a high density of automobile traffic. While standing at the gasoline pump, motorists have been known to breathe benzene at levels up to about 20 ug/m^3. Most of us are being exposed to benzene every day, yet there is no national air exposure limit for benzene outside the work environment.

In fact, there are no federal exposure limits for any toxic chemical in the air that occurs outside the work environment. Instead, facilities that release air toxins into the atmosphere are required to limit the amount of their emissions. Motor vehicles, however, are exempt from any limitations on their release of toxic chemicals to the atmosphere. Thus, the quality of the air is affected by toxic emissions in both localized neighborhoods (e.g., as a result of automobile traffic, a landfill, or a chemical plant) and within metropolitan areas. To date, air quality standards have focused on the "criteria pollutants" which the U.S. EPA has recognized as regional problems that contribute to the decline in the health of the environment and reduce the aesthetic value of the environment (i.e., ozone, carbon monoxide, lead, nitrogen dioxide, particulate matter, sulfur dioxide). However, air quality standards for toxins have yet to be derived. Given the regional nature of pollution, it is important to recognize that we are exposed

---

[7] The State of California's Safe Drinking Water and Toxic Enforcement Act of 1986 (Proposition 65) requires the state to develop safe harbor levels — no significant risk levels for carcinogens and maximum allowable dose levels for chemicals that cause reproductive toxicity. These standards were developed for both indoor and outdoor environments.

to the same chemical from multiple sources (i.e., food, water, and air) as well as from exposure to chemical mixtures.

For example, benzene is not only a common air pollutant, but it is routinely found in many different food products as well as in water. However, our exposure to multiple sources of benzene is not being considered by regulatory agencies, nor is it likely to be considered any time soon. There is no food tolerance level nor an ambient air quality standard for benzene,[8] yet drinking water has a maximum contaminant level of 0.005 milligrams of benzene per liter of water (mg/L). *In fact, there is not a single combined exposure limit for any of the thousands of synthetic chemicals used in the United States, let alone standards for complex chemical mixtures.*

This lack of exposure limits for both multiple exposures to the same chemical and chemical mixtures is understandable given that toxicologists do not have the necessary information to actually predict the health effects of a known toxic chemical found in the human body at very small concentrations.[9] Furthermore, when humans are exposed to a mixture of known chemical toxins, it is not known whether a potential health effect could be additive (for example, 1 + 1 = 2) or potentially worse (for example, 1 + 1 = 20). Because of all these unknowns, can we be certain that drinking water standards, food tolerance levels, and air emission controls truly protect our health?

1.4 CONFRONTING THE UNKNOWN

Toxicologists have just begun to address the influence on human health of individual synthetic chemicals at low levels. Yet this knowledge represents only the tip of the iceberg, since a true understanding of the real health effects posed by our exposure to unknown chemicals, chemicals with unknown toxicities, and chemical mixtures remains hidden from our view. Although toxicologists know virtually nothing about the toxic effects of chemical mixtures, toxicologists do know that there can be synergistic effects from multiple chemical exposures. Therefore, it is appropriate and realistic *to assume that the greater the number of synthetic chemicals to which a person is exposed, the greater the probability an individual may experience either a synergistic or additive health effect.*[10]

[8] "Ambient" refers to the air that surrounds us outdoors, as opposed to indoor air in the home or workplace.

[9] For example, an acute high-level exposure to benzene in air has been known to cause convulsions, coma, and sudden death from an irregular heartbeat, while chronic low-level workplace exposures to benzene in air are known to cause leukemia, when low levels of benzene occur in the environment (e.g., in food, water, or air) and in the human body, benzene is considered a suspected carcinogen.

[10] Such an interaction would be similar to eating two different foods that in combination result in an upset stomach but eaten individually have no effect.

Given these possibilities, our exposure to complex mixtures of synthetic chemicals could be responsible for a portion of the background health effects measured in our society. It is this realization that forms the foundation of this book.

Because chemical exposures come from multiple sources, any attempt to understand potential health effects must be based on the actual occurrence of synthetic chemicals found in food products, drinking water, and air resources. This necessitates that information be collected in a way that is representative of our exposure to known toxic synthetic chemicals across the country. Given these requirements and the availability of regional data collected by federal, state, and local agencies, this book uses information on the actual synthetic chemicals that are known to be toxic and found in food products, drinking water, and the air resources around the nation.

Unfortunately, it is not possible to characterize the occurrence of synthetic chemicals in the food, water, and air for every city in the United States. This is because the availability of information on synthetic chemical contaminants in food is limited. Based on this limitation, the following cities were evaluated:

Bakersfield, California
Baton Rouge, Louisiana
Binghamton, New York
Birmingham, Alabama
Boise, Idaho
Charleston, South Carolina
Columbus, Georgia
Dayton, Ohio
Denver, Colorado
Fresno, California
Grand Rapids, Michigan
Harrisburg, Pennsylvania
Jackson, Missouri
Lafayette, Louisiana
Long Beach, New Jersey
Los Angeles, California
Milwaukee, Wisconsin
New York, New York
Ogden, Utah
Oklahoma City, Oklahoma
Orlando, Florida
Peoria, Illinois
Phoenix, Arizona
Portland, Oregon
Providence, Rhode Island

Pueblo, Colorado
Reno, Nevada
San Antonio, Texas
Savannah, Georgia
Sioux Falls, South Dakota
Springfield, Illinois
Tacoma, Washington
Tulsa, Oklahoma
Washington, D.C.
West Palm Beach, Florida
Worcester, Massachusetts

These cities represent a cross section of the types of communities present in the United States (small versus large, urban versus rural, high pollution versus low pollution, etc.).

The information on synthetic chemicals found in the food, water, and air from each city is used to (1) show which synthetic chemical mixtures in the food, water, and air can be found in each community; (2) evaluate consumer choices relative to food preferences, diets (for example, there will be different chemical mixtures in low-carbohydrate, low-fat, vegetarian, and balanced diets) and menus, drinking tap water versus purified water or bottle water, and where one chooses to live (which can influence the type and amount of chemicals an individual actually consumes by city, region, and nationally); and (3) discuss the potential health effects that may be associated with actual chemical exposures.

This chemical information is also summarized to identify those cities and regions of the nation that have the greatest number and concentration of chemical contaminants in food products, drinking water, and air resources.[11] Finally, the available methods by which communities and individuals can limit their exposure to chemicals in food, water, and air are also evaluated and discussed.

Based on my personal experiences, and in cooperation with my coauthors, the intent of this book is to encourage the student or interested individual to explore the known and the unknown realm of synthetic chemicals in our natural resources, the potential health effects we all face, and the methods available to each individual and society for reducing our exposure to synthetic chemical contaminants. It is also hoped that access to this information will (1) allow each person to make informed decisions about which "synthetic chemicals" they elect to consume and (2) provide toxicologists with current data on the most common chemical mixtures found in our food, water, and air resources with the hope that meaningful multichemical toxicity studies will become a reality.

[11] Comparisons have been made of rural versus urban as well as general geographic regions.

The need for this information was addressed in the May 2006 issue of *Scientific America*. This article (titled "Mixing It Up," p. 16) makes the following conclusions:

> One chemical alone may do no harm in low doses, but in conjunction with a few of its peers, even in doses that are individually safe, it can inflict serious harm. . . . The potential effects of contaminant mixtures on people, aquatic life and fish-eating wildlife are still poorly understood. . . . Our results indicate, however, that studies of mixtures should be a high priority.

2 AN INTRODUCTION TO SYNTHETIC CHEMICAL CONTAMINANTS IN FOOD

The curative powers of chicken soup have been known since it was prescribed as a remedy for the common cold in ancient Egypt. In America today, chicken noodle soup is still a popular comfort food for treating the symptoms of the common cold and the flu. Given its healing powers, we should all be consoled by the fact that of the 240 food products tested by the Food and Drug Administration (FDA), chicken noodle soup was one of only five food products that were *not* contaminated with synthetic chemicals.[1] Other popular comfort foods faired much worse.

After an exhaustive evaluation of the available FDA data on contaminants in American foods, three things are certain. Every person who eats a balanced diet of fruits, vegetables, grains, dairy, and protein products potentially consumes approximately 32 different types of synthetic pesticides and industrial chemicals. Second, the actual number of different synthetic chemicals a person will consume depends largely on one's diet, the city and region of the country where one lives, and the amount of organic food the person elects to eat.

Finally, even those foods defined by the United States Department of Agriculture (USDA) as "organic" will contain multiple pesticides and industrial chemicals unless they have been carefully screened and produced. This is because a food product can be classified as a USDA-certified organic food if the food was produced on land that has not use pesticides for at least 3 years.

[1] The other products were canned pineapple, corn, kidney beans, and pinto beans.

Unfortunately, this length of time is not sufficient to eliminate the occurrence of persistent pesticides that were banned from agricultural use in the late 1970s. Furthermore, the handling and processing of organic foods can and do introduce industrial chemicals into "organic" products. Because the definition of an "organic" food is so important, it is critical for the purposes of this book to understand what a real "organic food" product should be. A true "organic food" should meet the following definition:

> "Organic Food" is any food product that is (1) produced without pesticides *nor* grown on land that used DDT and other banned pesticides in the last 51 years, (2) not processed (e.g., handling, sorting, mixing, or packaging) so that known toxic industrial chemicals contaminate a product, and (3) an animal product that has been shown to be free of synthetic chemicals that are known to accumulate in the edible parts of that product (i.e., by chemical analysis or is the progeny of an animal already tested as being free of synthetic chemicals and is given only "organic" feed).

In other words, USDA-certified organic foods are not necessarily free of synthetic chemicals. By understanding the difference in definitions, it is possible to conclude from the information presented in the market baskets that

> Approximately 98% of the foods tested by the FDA and consumed in the United States are contaminated with pesticides and industrial chemicals. Chemical mixtures in food is the norm rather than the exception. Furthermore, synthetic chemical mixtures also occur in USDA-certified "organic" foods.

In attempting to meaningfully describe and interpret the massive extent of this contamination, the synthetic chemical data provided in this book have been summarized to (1) determine what is a "safe" food (i.e., in terms of synthetic chemical contamination), (2) show how selected food choices will influence the type and number of synthetic chemicals an individual will consume, (3) provide a picture of how a city or region of the country can influence the magnitude of an individual's exposure to chemical mixtures, and (4) discuss the impact of "organic" foods and other techniques employed to reduce synthetic chemical exposures. The process of understanding these data and their importance and relevance to human health begins with a most critical first step. We must know how these chemical data were collected.

2.1 UNDERSTANDING THE FDA DATA COLLECTION PROCESS

The data presented in this chapter were recompiled from the original data sets published in the FDA's Total Diet Studies to focus on synthetic chemical mixtures in foods. The original Total Diet Studies are available online at www.fda.gov under Foods. In these studies, the FDA presented its data in a "market basket." The market basket process uses the following approach:

- Samples of specific foods (e.g., pizza, chicken, potatoes) are purchased in three different cities. These three cities and the purchased foods make up the market basket for that sampling period. No purchased food products have been identified in the market baskets as being produced organically.
- Each food is prepared so that it is ready to eat. For example, a frozen pizza is baked, chicken is roasted or fried, and potatoes are fried, baked, or mashed. This is significant since the cooking process does influence what contaminants may be in the cooked product (e.g., fried chicken will contain cooking oils and flour compared to a roasted chicken).
- After each food is prepared, the three individual food samples are mixed together into a composite sample. For example, if oranges were purchased, the three orange samples would be peeled and the flesh combined into one mixture.
- It is this composite mixture that is sent to an FDA laboratory to determine the concentration of selected synthetic chemicals.

Thus, the FDA sampling method shows that the market basket-composited foods are representative of the selected cities and the region spanned by their location.

Market baskets have been collected quarterly since 1982 from the west, north central, south, and northeast geographic areas of the nation. The market baskets available from the FDA begin in the third quarter of 1991 (MB91-3) and currently end with the fourth quarter of 2001 (MB01-4). During this time period, there have been changes in the types of chemicals that were looked for in the market basket foods. The most significant change occurred in 1996 when industrial chemicals (as opposed to pesticides) were added to the list of compounds looked for in foods. The complete list of chemicals looked for in the market basket food products is given in Appendix 2-1.

At this point, it needs to be stressed that a knowledge of chemistry is *not* necessary to understand the toxicological implications of the presented data. It is more important to recognize the unique differences of the chemicals to which we are all exposed, since the toxicological diversity of a synthetic chemical mixture provides an added opportunity that an *individual may experience either a synergistic or additive health effect*. However, the toxicity and chemistry (i.e., when necessary to understand a toxicological concept) of the synthetic compounds that occur in the market basket foods will be addressed in greater detail in the discussions on chemical toxicity.

Synthetic pesticide chemicals are toxic by design. We do not need to know the individual chemical properties that make pesticides toxic, but we do need to recognize that different chemical properties cause different toxic effects. That is why there are different classes of pesticides as well as individual pesticides within each class. This is well illustrated in Appendix 2-1. The FDA looks for the following main groups of pesticides in food:

- *The organochlorine pesticides.* This class of organic chemicals (a carbon-based compound) contains chlorine. This class includes 82 chemically unique compounds. For example, the pesticides from this class that most commonly occur in food are DDE (by-product of DDT) and endosulfans.
- *Organophosphorus pesticides.* This class of organic chemicals contains phosphorus but may also contain chlorine in addition to the phosphorus. This class includes 64 chemically unique compounds. For example, the pesticides from this class that most commonly occur in food are chlorpyrifos, and chlorpyrifos by-products, and malathion.
- *Methylcarbamate pesticides.* This class of organic chemicals contains an amide compound. This class includes 15 chemically uniquc compounds. For example, the pesticide from this class that most commonly occurs in food is carbaryl.
- *Chlorophenoxy acid pesticides.* This class of compounds contains a phenol-type compound with chlorine. This class includes 30 chemically unique compounds. For example, the pesticide from this class that most commonly occurs in food is dicamba.
- *Pyrethroid pesticides.* This class of compounds contains benzene-type compounds with either chlorine or bromine. This class includes 14 chemically unique compounds. For example, the pesticide from this class that most commonly occurs in food is permethrin.
- *Substituted urea pesticides.* This class of compounds contains a phenyl compound with chlorine. This class includes 10 chemically unique compounds. These compounds were not commonly found in the market basket foods.

The FDA also expects that, in addition to pesticides, other known toxic organic industrial chemicals will be found in food products. The primary chemical sources that may contact food products include the following:

- Inert chemicals that are combined into pesticide formulations contain a complex mixture of petroleum-based organic chemicals.[2] For example, the most common are toluene and xylene.

[2] Inert chemicals are *not* the active toxic ingredient in a pesticide formulation.

- Many of these same chemicals can occur in petroleum-based cleaning solvents (with and without chlorine) or lubricants that could be used on equipment for handling of food products.[3] The most common solvents are trichloroethylene (TCE) and tetrachloroethylene (PCE). In some cases, these chemicals may be associated with the direct contact with packaging (e.g., TCE with sytrene).
- Water that is contaminated with petroleum-based organic chemicals provides direct contact with aquatic food products (fish, shrimp, crab, etc.). The polychlorinated biphenyls (PCBs) that were banned in 1979 are a good example.
- Disinfection by-products contaminate food because chlorine or bromine (i.e., halogens) are used to disinfect food processing equipment. These halogens react with water-soluble organic matter to produce compounds like chloroform.

The FDA looks for the following specific classes of petroleum-based organic chemicals in food:

- *Aromatic hydrocarbons.* These are benzene ring-type compounds (e.g., benzene). This class includes 13 chemically unique compounds.
- *Halogenated aromatic hydrocarbons.* These are benzene ring-type compounds with chlorine or bromine. This class includes seven chemically unique compounds (e.g:, dichlorobenzene).
- *Halogenated aliphatic hydrocarbons.* These are single and branched chain-type compounds (no rings) with chlorine or bromine (e.g., TCE). This class includes 22 chemically unique compounds.

In summary, the FDA looks for nine different chemical classes of synthetic compounds and a total of 257 chemically unique compounds in our food products. *This means that some combination of these chemicals will occur in most foods. It should be remembered, however, that hundreds of other pesticide and industrial chemicals used in food production are not tested for by the FDA. This omission demonstrates that a true or total picture of our exposure to chemical mixtures in food is still not available.* Even with this omission, however, the data from the market basket studies provide critical baseline information on the principal chemical mixtures that are found in food today.

[3] The occurrence of known toxic petroleum chemicals in FDA-approved petroleum-based food-grade grease is not reported.

2.3 THE SELECTED MARKET BASKETS

The market baskets discussed in this chapter were selected to (1) provide information across the major geographic regions of the United States with no individual market basket using the same city and (2) provide chemical information on both pesticides and organic industrial compounds in food products. Based on these selection criteria, the market baskets that were chosen are given in Table 2-1. These 12 market baskets span the time period from the first quarter of 1996 through the third quarter of 2001.

The selected market baskets tested virtually all of the same food products each year for the same chemicals. A total of 240 food products and 12 beverages were evaluated for each of the 12 selected market baskets. The synthetic chemicals that were detected in any food or beverage of these market baskets are listed in Appendix 2-2, along with their abbreviations (i.e., symbols used in this chapter's data tables), specific chemical class, and chemical formula. In order to present a meaningful summary of the data, each food product was placed into one of the following categories:

- Child/infant products (produced/processed specifically for children and infants)
- Dairy products
- Fruit products
- Grain and nut products (the majority of an individual product is a grain or nut)
- Mixed products (combination of dairy, fruit, grain, protein, or vegetable products)
- Protein products
- Vegetable products (includes some products that are technically fruits but are usually consumed in salads)

Using these categories, the market basket data have been summarized to illustrate the extent of multiple chemical contamination within (1) food product categories at the city, regional, and national levels and (2) individual food products within each category. By using this information, the consumer can evaluate and choose safer foods.

2.4 CHOOSING SAFER FOODS

A safe food contains no detectable synthetic chemical residue. However, given standard agricultural production and processing practices, foods are contaminated with synthetic pesticides and synthetic industrial chemicals. As discussed previously, this also includes foods that receive the USDA "organic" certification. Because a safer food contains no synthetic chemical residues, the more

TABLE 2-1 List of Market Basket Cities

Market Baskets	City	State	Region
96-1	Bakersfield	California	West
	Ogden	Utah	
	Tacoma	Washington	
96-2	Charleston	South Carolina	Southeast
	Lafayette	Louisiana	
	West Palm Beach	Florida	
97-1	Phoenix	Arizona	Southwest
	Fresno	California	
	San Antonio	Texas	
98-1	Boise	Idaho	West
	Los Angeles	California	
	Pueblo	Colorado	
98-2	Baton Rouge	Louisiana	South Central
	Savannah	Georgia	
	Tulsa	Oklahoma	
98-3	Long Beach/Asbury Park	New Jersey	Northeast
	New York	New York	
	Providence	Rhode Island	
98-4	Dayton	Ohio	North Central
	Peoria	Illinois	
	Sioux Falls	South Dakota	
99-1	Denver	Colorado	West
	Portland	Oregon	
	Reno	Nevada	
99-3	Binghamton	New York	Northeast
	Harrisburg	Pennsylvania	
	Worcester	Massachusetts	
00-1	Grand Rapids	Michigan	North Central
	Milwaukee	Wisconsin	
	Springfield	Illinois	
00-3	Birmingham	Alabama	Southeast
	Columbus	Georgia	
	Jackson	Mississippi	
03-1	Oklahoma City	Oklahoma	East
	Orlando	Florida	
	Washington, D.C.		

chemicals found in a food product, the less safe the product. Therefore, the information provided in this book is designed to do the following:

- Illustrate the degree to which individual food products are contaminated
- Identify those *organic* foods that can be expected to contain synthetic chemicals
- Identify those chemicals that can be expected to occur in a balanced diet

- Show how low-fat or high-fat, vegetarian, and children's diets will influence the number and type of chemicals consumed

Given these objectives, the information provided in the market baskets can be used to choose foods for a specific diet or recipe that have the least number of chemicals. If a food contains moderate, high, or very high levels of contaminants, it is recommended that an organic product be substituted. When selecting an organic food, that specific food should be checked against the food category data to identify the "organic" foods that are the most likely to be contaminated with banned pesticides. This is particularly true for dairy, protein, and vegetable products. If the food product contains banned pesticides, the users should consider substituting a generally equivalent product. For example, substitute a potato without its skin for a baked potato, skim milk for whole milk, or ground pork or turkey for ground beef. Finally, choosing safer foods means that the total diet can be modified to minimize daily chemical exposures. Thus, the market basket data can be used to design a diet that minimizes the number and diversity of chemicals in a specific recipe or a daily diet.

3 THE NUMBER OF SYNTHETIC CHEMICALS IN FOOD

The total number of chemical contaminants varies widely depending on the food product category. For example, Table 3-1 shows the total number of chemicals detected in each food category for every market basket. The averages at the bottom of the table represent the average number of chemicals per food product in each category. Thus, this average makes it clear that the least contaminated food category is the child/infant category, while mixed foods contain the greatest average number of contaminants per food. This would be expected, since child/infant food producers actively strive to reduce the number of pesticides in their products, as opposed to mixed foods that combine multiple sources of pesticides and industrial chemicals from the different food categories. *These data also suggest that vegetarians eat food groups (i.e., fruits, grains, nuts, and vegetables) with the greatest number of contaminants.*

As most of us would anticipate, there should be variations in the number of different chemicals that contaminate the individual food products within each food category. This variation is illustrated for each of the food categories.

3.1 DAIRY PRODUCTS

The number of different synthetic chemicals (either pesticides or industrial chemicals) detected in each dairy product for the 12 market baskets is given in Table 3-2. When looking at these data, it would appear that dairy products with high-fat content are more contaminated, with butter being significantly more contaminated than the other dairy products. Because pesticides are known to be soluble in fat and oil, it should be expected that high-fat and oil foods should

TABLE 3-1 Market Basket Product Category Comparison

MB	Child Number	Protein Number	Dairy Number	Fruit Number	Grain Number	Vegetable Number	Mixed Number	Total Number
96-1	40	51	47	90	121	149	197	695
96-2	43	36	29	83	68	121	171	551
97-1	37	39	52	80	81	118	197	604
98-1	23	35	46	71	106	142	175	598
98-2	24	37	37	59	76	114	160	507
98-3	28	31	40	57	82	135	173	546
98-4	43	48	53	67	98	168	207	684
99-1	32	54	67	84	107	150	217	711
99-3	43	75	67	89	138	143	268	823
00-1	31	79	85	70	123	135	283	806
00-3	46	66	68	85	140	146	284	835
01-3	39	76	64	82	153	163	311	888
Totals	429	627	655	917	1293	1684	2643	8248
Average	1.23	2.18	2.73	2.83	2.91	3.12	3.73	

have more pesticides. Using the grams of fat per serving and the grams per serving listed on these dairy products, the data presented in Table 3-3 compare the total number of pesticides detected in each product (i.e., for all market baskets) with the grams of fat in each gram of product.

These data generally illustrate that as the fat content of a dairy product increases, the number of pesticides increases. The cheeses tested contain a significant number of pesticides. Thus, consumers should consider selecting low-fat cheeses. Although the consumer is not provided grams of fat per serving information on protein products, we believe that it is safe to conclude that high-fat meat products will also be more contaminated.

3.2 PROTEIN PRODUCTS

The number of different synthetic chemicals (either pesticides or industrial chemicals) detected in each protein product for the 12 market baskets is given in Table 3-4. These data suggest that processed meats (e.g., hot dogs, bologna, and salami) contain more chemicals than muscle products (e.g., steak or chicken breast). This could be a function of the fat content of these processed meats. In general, beef products are more contaminated than pork products, whereas poultry have very few contaminants. Clearly, there appears to be some correlation between the amount of fat in protein products and the occurrence of synthetic chemicals in a given product. The number of chemical contaminants in fish may also be correlated to the occurrence of fatty acids. For example, salmon and tuna are known to have more fatty acids than haddock. Therefore, it is expected that salmon would contain more contaminants than haddock.

TABLE 3-2 Number of Chemical Contaminants in the Dairy Category by Individual Food Product

	Market Baskets												
Dairy Product	**96-2**	**98-2**	**98-3**	**98-1**	**96-1**	**97-1**	**98-4**	**01-3**	**99-1**	**99-3**	**00-3**	**00-1**	**Totals**
Skim milk	0	0	0	0	1	0	0	0	0	0	0	0	1
Chocolate milk	0	0	0	0	1	1	1	0	1	0	0	0	4
Plain yogurt (low fat)	0	1	0	1	0	1	0	0	1	0	0	2	6
Cottage cheese (4%)	0	0	0	1	1	1	0	0	2	0	0	2	7
Chocolate milk shake	0	0	0	1	1	0	1	2	1	1	0	1	8
Fruit-flavored yogurt	0	2	0	2	1	1	2	0	2	0	1	0	11
Evaporated milk	1	1	1	1	2	1	1	0	1	1	0	2	12
Milk (2%)	0	0	0	1	2	3	3	0	1	1	0	2	13
Half-and-half cream	1	1	1	2	1	1	3	2	3	1	2	2	20
Ice cream (light)	2	1	1	2	1	2	2	4	1	2	2	2	22
Whole milk	0	1	0	1	1	4	5	4	2	2	1	3	24
Sherbet	3	2	2	2	1	3	1	3	1	2	5	3	28
Vanilla ice cream	1	2	2	1	1	3	4	5	5	5	7	5	41
Sour cream	2	2	4	4	5	5	4	4	5	6	6	5	52
Swiss cheese	5	3	4	4	2	5	4	6	6	6	8	8	61
Cheddar cheese	3	4	5	4	5	5	3	5	10	6	6	8	64
Cream cheese	2	4	5	4	6	3	5	6	7	7	6	10	65
American cheese	3	3	6	4	5	5	4	8	5	9	10	10	72
Butter	6	10	9	11	10	8	10	15	13	18	14	20	144
Totals	29	37	40	46	47	52	53	64	67	67	68	85	

TABLE 3-3 Dairy Fat and Pesticide Comparison

Dairy Products	Total Number of Pesticides	Grams of Fat per Gram of Dairy Product
Butter	67	0.79
Cream cheese	37	0.36
Swiss cheese	31	0.29
American cheese	29	0.26
Sour cream	28	0.2
Cheddar cheese	24	0.32
Half-and-half cream	20	0.1
Vanilla ice cream	18	0.14
Whole milk	15	0.03
Milk (2%)	13	0.02
Evaporated milk	11	0.06
Ice cream (light)	7	0.06
Cottage cheese (4%)	7	0.05
Plain yogurt (low fat)	6	0.02
Skim milk	1	0

Both hard-boiled and fried eggs have on average one or fewer contaminants. Scrambled eggs, however, have on average 2.1 contaminants. Obviously, how these eggs were prepared (i.e., use of butter or margarine) influenced the number of chemicals detected.

3.3 FRUIT PRODUCTS

The number of different synthetic chemicals (either pesticides or industrial chemicals) detected in each fruit product for the 12 market baskets is given in Table 3-5. When reviewing the fruit data, we see several distinctive trends. First, the juice of a fruit generally has fewer contaminants than the parent fruit (e.g., orange juice versus oranges). The only exception seems to be for pineapples versus pineapple juice. In other words, the contaminants are in the flesh or pulp of the fruit. Second, a peeled fruit product has fewer contaminants than the whole fruit (e.g., applesauce and apples). Finally, citrus fruits such as oranges and grapefruits tend to have fewer contaminants than noncitrus tree fruits. For example, apples, cherries, and peaches have more contaminants than oranges and grapefruit, whereas pears and apricots have more contaminants than grapefruit.

3.4 GRAIN AND NUT PRODUCTS

The number of different synthetic chemicals (either pesticides or industrial chemicals) detected in each grain and nut product for the 12 market baskets is

TABLE 3-4 Number of Chemical Contaminants in the Protein Category by Individual Food Product

	Market Baskets												
Protein Products	**98-3**	**98-1**	**96-2**	**98-2**	**97-1**	**98-4**	**96-1**	**99-1**	**00-3**	**99-3**	**01-3**	**00-1**	**Totals**
Ham (luncheon meat)	0	0	0	0	0	0	0	0	0	0	0	0	0
Fish (haddock)	ND	ND	0	ND	1	ND	2	ND	ND	ND	ND	ND	3
Chicken breast	1	0	0	0	1	0	1	0	0	0	0	0	3
Veal cutlet	0	1	0	1	0	0	0	0	0	1	0	1	4
Ham	0	0	1	0	0	0	1	0	0	2	0	0	4
Eggs (fried)	0	1	0	0	1	0	1	0	1	0	0	1	5
Turkey breast	0	1	0	0	0	1	3	0	0	0	0	0	5
Pork roast	0	0	0	0	0	1	4	0	0	0	0	0	5
Shrimp	0	0	0	0	1	1	2	0	0	0	2	0	6
Pork chop	0	0	0	0	0	2	2	0	1	0	1	0	6
Eggs (boiled)	0	2	0	0	1	1	2	1	1	0	0	1	9
Beef liver	0	0	0	0	1	1	3	0	4	2	1	0	12
Beef steak	0	0	1	3	1	1	2	4	0	0	2	0	14
Lamb chop	1	2	3	2	1	2	3	3	1	3	1	0	22
Eggs (scrambled)	0	2	0	1	1	3	1	0	6	4	3	5	26
Tuna (canned)	1	1	2	0	3	1	2	1	3	6	4	4	28
Pork sausage	0	0	1	4	3	4	2	5	2	3	3	1	28
Roast beef	2	3	2	2	6	4	3	4	7	5	3	7	48
Pork bacon	3	5	3	2	5	2	3	3	6	6	7	9	54
Salami	2	2	5	3	5	2	3	5	6	8	8	6	55
Bologna	3	4	9	6	1	1	5	4	8	8	9	10	68
Ground beef	4	2	2	5	4	6	3	8	5	9	10	13	71
Fish (salmon)	10	7	ND	6	ND	8	ND	9	4	9	9	10	72
Hot dogs (beef)	4	2	7	2	3	7	3	7	11	9	13	11	79
Totals	31	35	36	37	39	48	51	54	66	75	76	79	

ND = no data on product.

TABLE 3-5 Number of Chemical Contaminants in the Fruit Category by Individual Food Product

	Market Baskets												
Fruit Products	**98-3**	**98-2**	**98-4**	**00-1**	**98-1**	**97-1**	**01-3**	**96-2**	**99-1**	**00-3**	**99-3**	**96-1**	**Totals**
Pineapple (canned)	0	0	0	0	0	0	0	0	0	0	0	0	0
Pear (canned)	0	0	1	0	0	0	0	0	0	0	0	1	2
Lemonade	0	0	0	0	1	0	0	0	0	0	0	1	2
Grapefruit juice	0	0	1	0	1	0	0	1	0	0	0	3	6
Watermelon	0	0	0	0	2	1	1	2	0	0	0	1	7
Pineapple juice	0	0	1	2	1	0	1	1	0	0	0	1	7
Prune juice	0	0	1	0	0	2	2	0	1	0	0	3	9
Peach (canned)	0	3	1	1	2	1	1	0	1	1	1	1	13
Grape juice	1	1	2	2	0	1	1	1	4	1	1	2	17
Grapefruit	2	1	1	1	2	2	1	2	1	2	1	2	18
Fruit cocktail (canned)	0	0	3	1	2	1	1	4	2	2	0	3	19
Apple juice	1	2	3	2	2	1	3	2	2	1	3	2	24
Orange juice	0	0	2	3	1	1	5	0	1	6	4	2	25
Applesauce	3	3	0	1	2	3	2	3	3	2	1	4	27
Banana	2	0	3	5	2	2	4	3	3	4	2	1	31
Apricot	2	2	2	4	4	8	0	3	4	3	8	ND	40
Prunes	1	1	1	0	2	6	4	7	8	3	4	4	41
Pear	4	1	4	5	3	3	3	4	2	5	3	6	43
Orange	3	3	1	4	2	3	3	1	2	6	10	8	46
Grapes	9	5	4	3	2	1	5	5	1	6	2	4	47
Cantaloupe	2	3	4	0	6	4	5	6	8	6	3	4	51
Plums	4	6	3	5	4	6	4	4	4	3	3	5	51
Raisins	7	3	4	5	2	4	5	5	4	9	7	9	64
Peach	4	6	3	5	8	8	4	7	6	4	7	9	71
Cherries	2	4	9	7	5	5	11	5	10	8	8	ND	74
Strawberries	5	8	7	7	7	12	5	9	9	4	11	5	89
Apple	5	7	6	7	8	5	11	8	8	9	10	9	93
Totals	57	59	67	70	71	80	82	83	84	85	89	90	

ND = no data on product.

given in Table 3-6. When reviewing the grain and nut product table, note that many of the products may have been prepared with milk, eggs, margarine, butter, or oils (or cooked in oil). However, the composition of these products is dominated by a grain. As a result, these products do not accurately represent what is specifically in a given pure corn, oat, wheat, or rye product. Furthermore, all of the food products listed were commercially produced or prepared unless they are marked as homemade.

Even with the ingredients added to many of these products, the FDA data do suggest that wheat-based products consistently have more chemical contaminants than corn, rice, and oat-based products. As a whole, corn-based products are the least contaminated. Furthermore, products containing oils (e.g., nut products) or cooked in oils (e.g., popcorn and chips) tend to be more contaminated. These data show that the nut products are the most contaminated foods (particularly peanut butter). Once again, this is probably due to the high content of peanut oil. The least contaminated products were dried kidney and pinto beans. It is also important to note that all wheat breads contained at least three synthetic chemicals.

3.5 VEGETABLE PRODUCTS

The number of different synthetic chemicals (either pesticides or industrial chemicals) detected in each vegetable product for the 12 market baskets is given in Table 3-7. In general, the dark green and yellow vegetables tend to contain elevated contaminant levels along with salad vegetables (tomato, cucumber, celery, sweet peppers, lettuce). It is also important to note that vegetable oils, margarine, and olives contain elevated contaminant levels (i.e., oils tend to accumulate synthetic chemicals). Furthermore, some of the highest contaminant levels are associated with vegetables cooked in oils (e.g., potato chips and french fries from fast-food restaurants).

3.6 MIXED FOOD PRODUCTS

The number of different synthetic chemicals (either pesticides or industrial chemicals) detected in each mixed product for the 12 market baskets is given in Table 3-8. In general, fast-food and carryout products are more contaminated than homemade products (e.g., hamburgers and pizza). It is also clear that foods containing high-fat protein and dairy products tend to have a greater number of contaminants.

TABLE 3-6 Number of Chemical Contaminants in the Grain and Nut Category by Individual Food Product

	Market Baskets												
Grain Products	**96-2**	**98-2**	**97-1**	**98-3**	**98-4**	**98-1**	**99-1**	**96-1**	**00-1**	**99-3**	**00-3**	**01-3**	**Totals**
Corn	0	0	0	0	0	0	0	0	0	0	0	0	0
Kidney beans	0	0	0	0	0	0	0	0	0	0	0	0	0
Pinto beans	0	0	0	0	0	0	0	0	0	0	0	0	0
Corn grits	0	0	0	0	0	0	1	0	0	0	0	0	1
Wheat cereal	0	0	0	0	0	0	2	0	0	0	0	0	2
Cornflakes	0	0	0	0	0	0	0	1	0	0	0	2	3
Crisped rice cereal	0	0	0	1	0	0	0	0	0	1	1	1	4
Oatmeal	1	0	0	0	0	0	0	2	0	2	0	0	5
Creamed corn	0	0	1	0	0	0	0	2	1	1	0	0	5
White rice	1	0	0	2	1	1	1	0	0	1	1	1	9
Macaroni	0	1	0	0	1	2	1	2	1	2	2	2	14
Biscuit	0	1	1	2	2	2	2	2	0	2	2	1	17
Oat ring cereal	1	1	1	1	1	1	2	1	2	2	2	2	17
Noodles	1	2	0	2	0	1	2	2	0	2	2	4	18
Corn bread (HM)	0	1	3	0	2	2	2	2	2	2	2	3	21
English muffin	2	2	2	0	2	3	3	2	2	2	2	2	24
Granola	3	1	2	2	2	3	1	4	1	2	1	4	26
Pancake (mix)	1	2	3	2	3	3	2	5	0	2	2	2	27
Pretzels (hard)	3	3	2	3	3	2	3	1	2	3	3	2	30
Shredded wheat	0	2	3	2	2	3	3	5	2	4	2	2	30
Bagel	2	1	5	0	4	4	4	2	2	3	2	2	31
Rye bread	2	2	1	2	2	3	4	5	2	3	2	3	31
Saltine crackers	1	2	2	3	3	2	2	6	2	2	3	5	33
Flour tortilla	3	3	2	3	4	3	2	4	1	4	3	4	36
White roll	3	3	2	4	2	5	3	3	3	3	3	5	39
Whole-wheat bread	3	3	3	4	3	5	3	5	3	3	4	4	43
Cracked-wheat bread	2	4	3	4	6	4	2	5	3	4	3	6	46
Corn chips	1	2	5	0	2	2	1	5	9	7	8	6	48
Popcorn (in oil)	1	2	3	3	3	2	6	3	12	5	12	8	60
Fruit-flavored cereal	4	0	1	6	5	6	6	6	7	5	7	8	61
Butter crackers	2	2	2	3	4	3	4	3	10	7	11	13	64
Sweet roll	3	5	3	4	6	5	5	4	8	8	9	9	69
White bread	3	3	2	4	5	7	4	5	7	9	10	12	71
Graham crackers	5	3	4	3	6	5	4	5	11	10	9	10	75
Dry-roasted peanuts	7	8	7	6	6	8	8	11	7	7	7	8	90
Mixed nuts (no peanuts)	4	7	7	6	6	7	9	6	10	11	11	8	92
Peanut butter	9	10	11	10	12	12	15	12	13	19	14	14	151
Totals	68	76	81	82	98	106	107	121	123	138	140	153	

HM = homemade.

TABLE 3-7 Number of Chemical Contaminants in the Vegetable Category by Individual Food Product

	Market Baskets												
Vegetable Products	**98-2**	**97-1**	**96-2**	**98-3**	**00-1**	**98-1**	**99-3**	**00-3**	**96-1**	**99-1**	**01-3**	**98-4**	**Totals**
Onion	0	0	0	0	0	0	0	0	1	1	0	0	2
Cauliflower	0	0	0	0	0	0	0	0	0	1	1	0	2
Cabbage	1	0	1	0	1	0	0	0	0	1	0	0	4
Sauerkraut	0	0	0	0	0	0	1	0	1	1	1	0	4
Green peas	1	1	1	1	0	0	0	0	1	1	0	0	6
Asparagus	0	0	0	0	0	0	1	2	1	1	0	2	7
Okra	0	1	0	2	0	0	0	4	0	0	0	1	8
Eggplant	1	2	1	1	1	0	0	1	0	0	2	0	9
Peas (dried)	0	0	0	0	1	3	0	0	1	1	2	1	9
Tomato juice	0	0	1	0	3	2	0	3	1	0	3	0	13
Stewed tomato	0	0	0	0	0	2	2	0	1	2	7	2	16
Carrot	2	1	1	1	0	3	3	0	2	1	0	3	17
Beets	2	3	3	0	2	1	2	0	1	2	1	2	19
Sweet potato	1	0	1	2	1	2	2	1	4	0	2	3	19
Mixed vegetables (F)	1	0	2	3	0	4	1	1	2	2	2	3	21
Mushrooms	1	3	1	2	1	2	1	2	2	4	1	1	21
Tomato sauce	0	1	0	2	2	2	2	3	5	0	0	6	23
Tomato catsup	3	0	2	3	2	4	0	2	3	1	3	3	26
Avocado	3	1	3	1	2	1	4	4	4	0	4	2	29
Turnip	0	5	3	1	2	3	1	2	5	3	0	5	30
Broccoli	3	4	3	4	2	1	1	1	0	8	1	3	31
Brussels sprouts	2	3	1	1	3	3	7	3	3	2	3	1	32
White potato wo/s	1	1	5	3	3	2	2	3	3	1	3	5	32
Radish	1	4	1	3	4	3	2	1	5	3	0	6	33
Winter squash	2	3	2	2	3	4	1	2	2	5	2	5	33

TABLE 3-7 *Continued*

	Market Baskets												
Vegetable Products	**98-2**	**97-1**	**96-2**	**98-3**	**00-1**	**98-1**	**99-3**	**00-3**	**96-1**	**99-1**	**01-3**	**98-4**	**Totals**
Lima beans	3	6	2	2	4	4	0	0	2	5	3	3	34
Mashed potatoes	2	3	2	3	3	2	2	3	2	2	5	5	34
Green beans	0	3	3	4	2	6	3	9	0	2	9	3	44
French fries (F)	4	4	3	2	4	4	6	4	3	5	4	3	46
Iceberg lettuce	8	5	5	5	0	5	2	2	4	5	3	3	47
Black olives	5	6	4	5	3	3	2	2	6	5	3	4	48
Olive/safflower oil	4	0	7	1	10	1	10	3	1	3	10	9	59
Tomato	5	0	5	7	7	5	3	4	8	7	5	5	61
Cucumber	4	4	6	4	4	4	8	5	4	8	3	10	64
White potato w/s	7	4	7	5	3	5	4	6	6	4	6	7	64
Margarine	2	1	4	7	11	1	9	12	2	2	9	4	64
Summer squash	5	3	6	3	3	8	7	5	5	8	5	7	65
Celery	5	6	1	6	4	8	6	4	7	5	5	8	65
Dill cucumber pickles	4	5	4	9	7	6	6	5	6	6	4	7	69
Sweet cucumber pickles	5	5	5	10	3	5	7	9	8	5	8	3	73
Green pepper	6	6	9	8	4	6	3	9	10	3	6	4	74
Potato chips	4	3	5	4	12	6	9	10	2	4	10	5	74
Collards	5	4	5	5	3	8	6	3	8	14	9	11	81
French fries (FF)	6	4	2	6	9	3	12	10	9	7	11	6	85
Spinach	5	13	4	7	6	10	5	6	8	9	7	7	87
Totals	114	118	121	135	135	142	143	146	149	150	163	168	

F = frozen.
FF = fast food or carry out.
wo/s = without skin.
w/s = with skin.

TABLE 3-8 Number of Chemical Contaminants in the Mixed Category by Individual Food Product

	Market Baskets												
Mixed Products	**98-2**	**96-2**	**98-3**	**98-1**	**97-1**	**96-1**	**98-4**	**99-1**	**99-3**	**00-1**	**00-3**	**03-1**	**Totals**
Chicken noodle soup	0	0	0	0	0	0	0	0	0	0	0	0	0
Bean and bacon soup	0	0	0	0	0	0	0	1	0	0	0	0	1
Tomato soup	0	0	0	0	0	0	1	0	0	0	1	0	2
Spaghetti and tomato	2	0	0	0	0	1	0	0	0	0	0	1	4
Chocolate pudding	0	0	0	1	1	1	0	1	0	0	0	0	4
Fried chicken (HM)	0	0	0	0	1	3	0	2	0	0	0	0	6
Mushroom soup	0	0	1	2	1	0	1	0	0	1	0	0	6
Suckers	0	1	0	1	2	1	0	0	0	0	1	0	6
Pork and beans	0	0	0	0	0	1	2	1	0	1	0	1	6
Veg/beef soup	1	1	1	1	0	0	3	0	1	1	1	1	11
Brown gravy (HM)	2	0	0	1	0	2	2	1	2	0	2	1	13
Spaghetti and meatballs	1	0	0	0	1	4	0	5	1	1	0	3	16
Mayonnaise	0	2	2	1	1	0	0	3	3	1	1	2	16
Popsicle	1	1	0	2	2	2	1	0	0	3	2	3	17
Jelly	1	2	0	2	5	1	1	2	0	0	1	3	18
Clam chowder	1	1	1	1	1	2	1	2	1	1	2	4	18
Yellow cake	2	1	2	2	1	3	2	1	2	2	2	0	20
Lasagna and meat (HM)	0	1	2	1	1	3	1	1	1	3	2	5	21
Beef stew	1	1	1	2	1	2	2	1	1	2	4	3	21
Italian salad dressing (low calorie)	0	0	0	2	1	1	3	3	3	1	4	4	22
White sauce (HM)	1	1	3	1	2	1	3	2	2	2	1	3	22
Beef chow mein (FF)	1	0	3	4	3	2	1	4	2	0	2	2	24

TABLE 3-8 *Continued*

	Market Baskets												
Mixed Products	**98-2**	**96-2**	**98-3**	**98-1**	**97-1**	**96-1**	**98-4**	**99-1**	**99-3**	**00-1**	**00-3**	**03-1**	**Totals**
Macaroni and cheese	1	2	1	1	3	5	2	3	2	2	3	1	26
Turkey dinner	2	0	1	1	0	3	2	4	4	5	3	4	29
Beef Stroganoff (HM)	2	1	1	1	3	3	3	3	3	3	3	5	31
Scalloped potatoes (HM)	3	2	1	3	2	3	3	2	1	3	3	5	31
Caramel candy	3	2	2	3	2	2	2	0	3	3	4	8	34
Chili and beans (HM)	0	0	4	3	3	3	5	6	5	2	5	3	39
Stuffed peppers w/rice and beef (HM)	6	3	4	4	4	3	5	1	2	2	5	1	40
Muffin w/egg/cheese/ham (FF)	4	3	3	2	1	4	7	2	3	4	4	4	41
Chicken potpie	2	3	2	3	3	3	3	4	3	9	2	5	42
Salisbury steak and gravy	5	3	3	1	4	5	2	3	6	4	3	4	43
Raisin bran	3	6	4	4	3	5	5	4	2	3	2	3	44
French salad dressing	0	1	1	5	4	4	5	3	5	2	11	4	45
Chocolate snack cake	3	2	1	1	4	2	4	6	5	6	6	6	46
Fried chicken (FF)	1	5	2	3	3	2	3	3	6	10	6	5	49
Coleslaw with dressing (HM)	3	10	1	2	6	2	3	7	4	5	3	7	53
Mustard	8	2	4	6	6	6	3	5	2	5	3	3	53
Fish and bun (FF)	4	6	5	3	3	5	7	4	5	3	3	6	54
Sandwich cookies w/crème filling	4	3	2	5	3	4	4	4	4	6	8	8	55

Tuna noodle casserole (HM)	8	4	6	3	2	3	8	4	3	6	4	6	57
Hot dog and bun (HM)	4	5	4	4	6	4	5	6	8	3	3	7	59
Apple pie	2	2	2	4	7	4	3	4	10	8	8	8	62
Sugar cookies	3	4	5	6	4	4	6	6	6	9	11	8	72
Chicken nuggets (FF)	2	4	3	4	3	2	2	5	14	11	14	9	73
Chocolate cake	4	5	4	4	8	5	5	4	12	6	10	6	73
Doughnut—cake and icing	5	3	6	5	7	3	5	6	7	8	10	11	76
Meatloaf (HM)	5	4	7	2	4	3	5	10	9	13	6	9	77
Brownies	3	4	8	5	5	7	4	4	8	9	11	9	77
Blueberry muffin	4	5	3	6	6	6	5	4	8	9	9	12	77
Pumpkin pie	6	5	9	6	6	10	8	8	8	6	8	6	86
Fish sticks (F)	5	5	5	7	5	7	6	4	11	10	12	10	87
Chocolate chip cookies	6	8	5	6	6	5	6	6	10	9	11	10	88
Cheese pizza (FF)	5	11	7	4	6	8	8	9	9	12	10	14	103
Milk chocolate candy bar	3	8	9	7	6	8	8	11	11	9	11	12	103
1/4-lb hamburger and bun (FF)	5	7	6	6	8	7	7	9	14	17	8	13	107
Taco/tostada (FF)	8	4	8	9	10	4	8	4	11	14	14	14	108
1/4-lb cheeseburger and bun (FF)	8	8	5	4	8	8	6	10	14	17	9	15	112
Cheese and pepperoni pizza (FF)	6	9	13	8	9	5	10	9	11	11	12	14	117
Totals	160	171	173	175	197	197	207	217	268	283	284	311	

F = frozen
FF = fast food or carry out
HM = homemade

3.7 CHILD AND INFANT PRODUCTS

The number of different synthetic chemicals (either pesticides or industrial chemicals) detected in each child and infant product for the 12 market baskets is given in Table 3-9. Most of the child and infant food products have either no chemicals or very low chemical levels, yet 50% of all the child/infant products contain multiple chemical contaminants. The food products with low to moderate chemical levels in this category tend to have moderate to very high levels in the other food categories (i.e., the protein, fruit, and vegetable categories). In other words, it would seem that adult foods with high numbers of chemical contaminants will occur in children's food at relatively higher levels.

If parents are not satisfied with the level of contaminants in some of the foods tested by the FDA, they have two basic choices. They can either select organic child and infant products or they can prepare child and infant food from organic products.

3.8 SUMMARY OF THE FOOD PRODUCT CATEGORY DATA

If our goal is to eat safer foods, then we should choose foods that contain, on average, one or fewer synthetic chemicals and minimize our consumption of foods with more synthetic chemicals. To aid in the selection of safer foods, the following contaminant intervals were defined to illustrate the relative magnitude of the number of synthetic chemicals in individual food products:

Average number of contaminants	Contaminant level
<1.0	Very low
1.0–2.99	Low
3.0–4.99	Moderate
5.0–6.99	High
>7.0	Very high

These contaminant levels were used to provide the relative food product rankings illustrated in Figures 3-1 through 3-7. Based on this classification, only those food products that average fewer than 1.0 contaminant (i.e., No chemicals or Very Low) have the lowest risk of exposing the consumer to multiple synthetic chemicals. All of the food products in the other contaminant levels have an increasing risk of exposing the consumer to multiple synthetic chemicals.

When reviewing all of the previous data, the FDA detected synthetic chemicals in virtually all the foods tested. Only corn, ham (luncheon), pineapple

TABLE 3-9 Number of Chemical Contaminants in the Child/Infant Category by Individual Food Product

	Market Baskets												
Child/Infant Products	**98-1**	**98-2**	**98-3**	**00-1**	**99-1**	**97-1**	**03-1**	**96-1**	**99-3**	**98-4**	**96-2**	**00-3**	**Totals**
Bananas and tapioca	0	0	0	0	0	0	0	0	0	0	0	0	0
Creamed corn (strained)	0	0	0	0	0	0	ND	0	0	0	0	0	0
Custard pudding (strained)	0	0	0	0	0	0	0	0	0	0	0	0	0
Peas (strained)	0	0	0	0	0	0	0	1	0	0	0	0	1
Orange juice (strained)	0	0	0	0	0	0	0	2	0	0	0	0	2
Milk-based formula	0	0	0	0	0	0	1	0	1	0	0	1	3
Chicken (strained)	0	0	0	0	0	1	0	1	0	1	0	0	3
Macaroni/tomato/beef	0	0	0	1	0	1	1	0	0	0	0	1	4
Chicken noodle dinner	0	1	0	0	0	0	0	1	1	0	1	1	5
Rice cereal (infant)	1	0	1	0	1	1	1	0	0	0	0	0	5
Soy-based formula	0	0	0	2	0	0	4	0	0	0	1	1	8
Mixed vegetables (strained)	1	1	1	0	0	1	0	2	1	1	1	0	9
Turkey and rice	0	0	1	0	1	1	1	4	0	3	0	0	11
Vegetables and beef	1	0	1	2	0	1	1	1	0	2	1	1	11
Sweet potatoes (strained)	1	1	1	1	1	1	1	1	1	1	1	0	11
Carrots (strained)	0	0	0	3	0	0	2	0	5	0	0	2	12
Spinach (strained)	2	2	1	1	1	2	0	1	1	2	0	1	14
Green beans (strained)	0	0	2	3	0	2	2	1	1	0	1	3	15
Squash (strained)	1	2	1	1	4	0	3	1	2	0	0	1	16
Vegetables and ham	0	1	1	1	1	1	1	1	2	3	1	3	16
Vegetables and chicken	0	0	2	1	0	1	3	2	1	3	1	2	16
Apple juice (strained)	2	2	2	3	2	2	1	3	4	1	3	2	27
Teething biscuits	3	2	2	2	2	4	2	3	2	2	2	3	29
Fruit pudding (strained)	1	0	1	1	1	4	0	3	2	5	6	6	30
Beef (strained)	4	2	1	0	2	1	7	1	4	2	3	5	32
Pears (strained)	1	2	0	2	3	3	1	3	4	6	9	1	35
Rice w/apple cereal (strained)	ND	3	3	2	5	3	2	0	4	4	4	6	36
Applesauce (strained)	2	4	3	2	4	5	2	4	1	3	4	4	38
Peaches (strained)	3	1	4	3	4	2	3	4	6	4	4	2	40
Totals	23	24	28	31	32	37	39	40	43	43	43	46	

ND = no data on product.

No Chemicals	Very Low	Low	Moderate	High	Very High
None	Skim milk	Evaporated milk	Vanilla ice cream	Swiss cheese	Butter
	Chocolate milk	Milk (2%)	Sour cream	Cheddar cheese	
	Plain yogurt (lowfat)	Half & Half cream		Cream cheese	
	Cottage cheese (4%)	Ice cream (light)		American cheese	
	Chocolate milk shake	Whole milk			
	Fruit flavored yogurt	Sherbet			

FIGURE 3-1 Food contamination summary for the dairy category. All columns (except "No Chemicals") show increasing contamination from top to bottom.

No Chemicals	Very Low	Low	Moderate	High	Very High
Ham (luncheon meat)	Chicken breast	Fish (Haddock)	Roast beef	Bologna	Fish (Salmon)
	Veal cutlet	Beef liver	Pork bacon	Ground beef	
	Ham	Beef steak	Salami	Hot dogs (beef)	
	Eggs (fried)	Lamb chop			
	Turkey breast	Eggs (scrambled)			
	Pork roast	Tuna (canned)			
	Shrimp	Pork sausage			
	Pork chop				
	Eggs (boiled)				

FIGURE 3-2 Food contamination summary for the protein category. All columns (except "No Chemicals") show increasing contamination from top to bottom.

No Chemicals	Very Low	Low	Moderate	High	Very High
Pineapple (canned)	Pear (canned)	Peaches (canned)	Prunes	Raisins	Strawberries
	Lemonade	Grape juice	Pears	Peaches	Apple
	Grapefruit juice	Grapefruit	Apricots	Cherries	
	Watermelon	Fruite cocktail (canned)	Oranges		
	Pineapple juice	Apple juice	Grapes		
	Prune juice	Orange juice	Cantaloupe		
		Applesauce	Plums		
		Banana			

FIGURE 3-3 Food contamination summary for the fruit category. All columns (except "No Chemicals") show increasing contamination from top to bottom.

No Chemicals	Very Low	Low	Moderate	High	Very High
Corn	Corn grits	Macaroni	Flour tortilla	Popcorn (in oil)	Dry roasted peanuts
Kidney beans	Wheat cereal	Biscuit	White roll	Fruit-flavored cereal	Mixed nuts (no peanuts)
Pinto beans	Corn flakes	Oat ring cereal	Whole wheat bread	Butter crackers	Peanut butter
	Crisped rice cereal	Noodles	Cracked wheat bread	Sweet roll	
	Oatmeal	Cornbread (HM)	Corn chips	White bread	
	Creamed corn	English muffin		Graham crackers	
	White rice	Granola			
		Pancake (mix)			
		Pretzels (hard)			
		Shredded wheat			
		Bagel			
		Rye bread			
		Saltine crackers			

FIGURE 3-4 Food contamination summary for the grain and nut category. All columns (except "No Chemicals") show increasing contamination from top to bottom.

No Chemicals	Very Low	Low	Moderate	High	Very High
None	Onion	Tomato juice	Green beans	Tomato	French fries (FF)
	Cauliflower	Stewed tomato	French fries (F)	Cucumber	Spinach
	Cabbage	Carrot	Iceberg lettuce	White potato with skin	
	Sauerkraut	Beets	Black olives	Margarine	
	Green peas	Sweet potato	Olive/safflower oil	Summer squash	
	Asparagus	Mixed vegetables (F)		Celery	
	Okra	Mushrooms		Dill cucumber pickles	
	Eggplant	Tomato sauce		Sweet cucumber pickles	
	Peas (dried)	Tomato catsup		Green pepper	
		Avocado		Potato chips	
		Turnip		Collards	
		Broccoli			
		Brussels sprouts			
		White potato - no skin			
		Radish			
		Winter squash			
		Lima beans			
		Mashed potatoes			

FIGURE 3-5 Food contamination summary for the vegetable category. All columns (except "No Chemicals") show increasing contamination from top to bottom.
Note: FF = fast food, F = frozen.

No Chemicals	Very Low	Low	Moderate	High	Very High
Chicken noodle soup	Bean & bacon soup	Brown gravy (HM)	Chili & beans (HM)	Apple pie	Pumpkin pie
	Tomato soup	Spaghetti & meatballs	Stuffed peppers w/rice & beef (HM)	Sugar cookies	Fish sticks (F)
	Spaghetti & tomato	Mayonnaise	Muffin w/egg/cheese/ham (FF)	Chicken nuggets (FF)	Chocolate chip cookies
	Chocolate pudding	Popsicle	Chicken potpie	Chocolate cake	Cheese pizza (FF)
	Fried Chicken (HM)	Jelly	Salisbury steak & gravy	Doughnut-cake & icing	Milk chocolate candy bar
	Mushroom soup	Clam chowder	Raisin bran	Meatloaf (HM)	1/4lb hamburger & bun (FF)
	Suckers	Yellow cake	French salad dressing	Brownies	Taco/tostada (FF)
	Pork and Beans	Lasagna & meat (HM)	Chocolate snack cake	Blueberry muffin	1/4lb cheeseburger & bun (FF)
	Veg/beef soup	Beef stew	Fried chicken (FF)		Cheese & pepperoni pizza (FF)
		Italian salad dressing	Coleslaw with dressing (HM)		
		White sauce (HM)	Mustard		
		Beef chow mein (FF)	Fish & bun (FF)		
		Macaroni & cheese	Sandwich cookies w/crème filling		
		Turkey dinner	Tuna noodle casserole (HM)		
		Beef stroganoff (HM)	Hot dog & bun (HM)		
		Scalloped potatoes (HM)			
		Caramel candy			

FIGURE 3-6 Food contamination summary for the mixed food category. All columns (except "No Chemicals") show increasing contamination from top to bottom.
Note: FF = fast food, F = frozen, HM = home made.

No Chemicals	Very Low	Low	Moderate	High	Very High
Bananas & tapioca	Peas (strained)	Carrots (strained)	Applesauce (strained)	None	None
Creamed corn (strained)	Orange juice (strained)	Spinach (strained)	Rice w/apple cereal (strained)		
Custard pudding (strained)	Milk-based formula	Green beans (strained)	Peaches (strained)		
	Chicken (strained)	Squash (strained)			
	Macaroni/tomato/beef	Vegetables and ham			
	Chicken noodle dinner	Vegetables and chicken			
	Rice cereal (infant)	Apple juice (strained)			
	Soy-based formula	Teething biscuits			
	Mixed vegetables (strained)	Fruit pudding (strained)			
	Turkey and rice	Beef (strained)			
	Vegetables and beef	Pears (strained)			
	Sweet potatoes (strained)				

FIGURE 3-7 Food contamination summary for the child/infant category. All columns (except "No Chemicals") show increasing contamination from top to bottom.

(canned), pinto beans, kidney beans, and chicken noodle soup were shown not to be contaminated. The fact that virtually every food we eat has been shown to be contaminated raises the following question: What synthetic chemicals are we most likely going to be exposed to if we consume a balanced diet of dairy, fruits, grains, protein, and vegetable products?

4 SYNTHETIC CHEMICALS IN A BALANCED DIET

The U.S. Department of Agriculture's Food Guide Pyramid (www.usda.gov/cnpp/pyramid) recommends that "no one food group is more important than another — for good health you need them all." Most people are well aware of the USDA recommendations, but individual choices obviously dictate the degree to which these recommendations are actually followed. Individuals who attempt to maintain a balanced diet must also realize that they will consume a mixture of chemicals that is unique to each food group. Therefore, the chemical diversity of an individual's diet is dictated by the foods that person consumes.

To evaluate an individual's exposure to chemical mixtures, it was necessary to determine how frequently pesticide and industrial chemicals were found in each food product. The results of this determination are discussed for the basic food categories (i.e., dairy, protein, vegetable, fruit, and grain) as well as for the mixed-food and child/infant food categories.

4.1 SYNTHETIC CHEMICALS IN DAIRY PRODUCTS

The pesticides found in dairy products are given in Table 4-1. To highlight the occurrence of any banned pesticide in a food, all detected occurrences have been shaded gray. This will be done in all the food groups. The industrial chemicals in dairy products are given in Table 4-2. Each table provides the following information:

1. The number of times a specific chemical was found in a market basket (e.g., there were 12 market baskets). For example in Exhibit 1, DDE was

TABLE 4-1 Pesticide Frequency in Dairy Products (All Market Baskets)

Product	DDE	Die	EnS	HCB	HE	Oct	Mxy	Lin	Cum	P	Ipr	Mal	Dip	Nac	Vin	Azi	BHC
Butter	12	10	7	11	6	5	1	5	4	2			1	2			1
American cheese	12	8	4	2	1			1					1				
Cream cheese	12	12	3	1	3	1	2		1	2							
Cheddar cheese	8	9	1		3	1	1									1	
Swiss cheese	10	11		3	3	1	1			1							
Sour cream	11	9	2	4	2												
Vanilla ice cream	10	6	1	1													
Sherbet	6		1									3					
Whole milk	7	5	1		1		1										
Ice cream (light)	7																
Half-and-half	11	6		1	2												
Milk (2%)	6	4	2				1										
Evaporated milk	10	1															
Fruit yogurt	5										4				2		
Chocolate milk shake	7	1															
Cottage cheese (4%)	5	2															
Plain yogurt	5		1														
Chocolate milk	4																
Skim milk	1																
Totals	149	84	23	23	21	8	7	6	5	5	4	3	2	2	2	1	1
Frequency (top 10)	65.4	36.8	10.1	10.1	9.2												

Banned pesticides

Total number of analyses = 228

TABLE 4-2 Industrial Chemical Frequency in Dairy Products (All Market Baskets)

Product	Cfo	Tol	Xyl	Sty	BZ	TBZ	PCE	EBZ	BDM	DBZ	TCA	CBZ	DCE	TCE	PCB	Pyl
Butter	12	12	11	10	3	7	5	6		3	3			1	2	2
American cheese	9	10	10	4	1	2	1	2	1	1	1			1		
Cream cheese	10	8	3	2	3		1				1					
Cheddar cheese	10	10	8	2	2		1	1		1	1	1	3			
Swiss cheese	7	11	3	2	2	2	1		1			1		1		
Sour cream	11	2	2	1	2		1		2	1	1	1				
Vanilla ice cream	9	4	4	1		1	1		2	1						
Sherbet	5	7	4		1				1							
Whole milk	6	1		1	1											
Ice cream (light)	9	1	4		1											
Half-and-half																
Milk (2%)																
Evaporated milk		1														
Fruit yogurt																
Chocolate milk shake																
Cottage cheese (4%)																
Plain yogurt																
Chocolate milk																
Skim milk																
Totals	88	67	49	23	16	12	11	9	7	7	7	3	3	3	2	2
Frequency (top 10)	38.5	29.4	21.5	10.1	7											
Banned chemical																
Total number of analyses = 228																

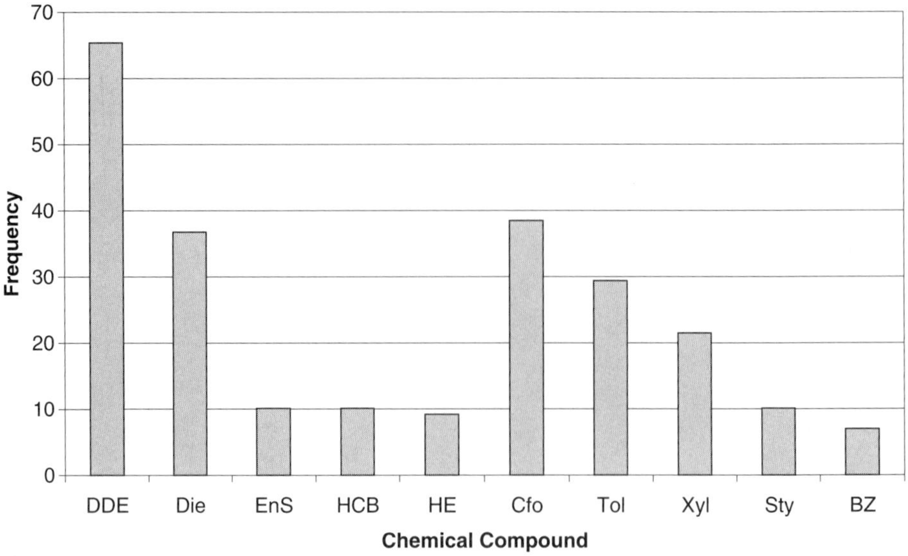

FIGURE 4-1 Frequency of top-10 chemicals in dairy products.

found all 12 times or 100% of the time, whereas HE was found in butter 6 out of 12 times or 50% of the time.

2. The top-10 chemicals (or most frequently found chemicals in a food category) are also identified. For example, DDE was found in dairy products 65.4% of the time.

The top-10 synthetic chemicals in dairy products are summarized in Figure 4-1. When reviewing Table 4-1, observe that every dairy product had detectable levels of at least one banned pesticide, whereas 13 dairy products had at least two or more detected banned pesticides. Clearly, the higher the fat content of the dairy products, the greater the number of the banned pesticides. Also common in dairy products are the endosulfan pesticides. Because endosulfan pesticides are common contaminants in most agricultural products, their occurrence in dairy products (i.e., found in 10 out of 19 products) is expected.

Of all the food categories, dairy products also contain one of the higher frequencies of industrial chemicals (see Table 4-2). This is expected because these industrial chemicals tend to accumulate in the fat of dairy products. The diversity of the chemicals found in dairy products, however, occurs for several reasons. Because chloroform is a by-product of using a chlorine disinfectant, it would appear that a significant number of dairy producers in the United States either do not rinse or do not completely rinse their equipment after disinfection.[1] Another source of disinfection by-products is from water that may be used in a dairy. Benzene, toluene, ethyl benzene, xylenes, and the other detected

[1] Also note that at least five dairy products were contaminated with both chloroform and bromodichloromethane (i.e., associated with both bromine and chlorine disinfection).

chlorinated petroleum solvents (e.g., CBZ, DCE, PCE, TCA, and TCE) occur in dairy foods because (1) these chemicals were in products used to lubricate or clean machinery that contacted the dairy products or (2) these chemicals were in the ambient air of the dairy.

Plastic packaging is another source of chemical contaminants in dairy products. Styrene is a basic building block of various plastics. However, some of the styrene that is in plastic is not linked or bound into the plastic but is free to leach out of this seemingly solid product. Furthermore, styrene commonly carries its own chemical contaminants, such as ethyl benzene, benzene, and TCE. These data suggest that the occurrence of styrene, ethyl benzene, benzene, and TCE is not unexpected in dairy products that come into contact with plastic containers.[2] This is consistent with the chemical contaminants detected in dairy products (e.g., styrene is detected when benzene and TCE are detected). Another unique industrial chemical that has only been reported in butter is polychlorinated biphenyl (PCB). This industrial chemical was banned in 1979 but is still a widespread contaminant in soil, sediment, and water resources.

In addition to the previously summarized data, specific synthetic chemical concentrations in dairy products for all market baskets are given in Appendixes 4-1 through 4-7.

4.2 SYNTHETIC CHEMICALS IN PROTEIN PRODUCTS

The pesticides found in protein products are given in Table 4-3, and the industrial chemicals are given in Table 4-4. Relative to the distribution of chemicals in protein products, the top-10 synthetic chemicals are summarized in Figure 4-2. As the tables show, protein products have contaminant patterns similar to those of dairy products. In other words, protein products with a higher fat content contain the greatest number of pesticides and industrial chemicals. Thus, the food products containing the greatest number of pesticides and industrial chemicals are hot dogs, salmon, ground beef, bologna, salami, bacon, roast beef, and pork sausage. Another similarity with dairy products is the occurrence of the same four banned pesticides (i.e., DDE, dieldrin, hexachlorobenzene, and heptachlor epoxide) in almost all of the protein. Only ham luncheon meat contained none of these chemicals. Furthermore, the vast majority of beef, veal, pork, and lamb is not imported. This demonstrates that the occurrence of banned pesticides in protein products is from past agricultural practices in the United States.

Once again, these data demonstrate that USDA-certified organic protein products do contain multiple organochlorine pesticides and industrial

[2] The occurrence of toluene and xylene contaminants in food products could also be associated with styrene plastics.

TABLE 4-3 Pesticide Frequency for Protein Products (All Market Baskets)

Product	DDE	Die	HCB	HE	BHC	Cno	Cpa	DDT	Oct	TDE	Lin	Atz	Cam	Cmy	Dip	Cd	Cpy	EnS	Etn	Mal	Nac	Cum	Dan	Diz	P
Hot dogs	10	5	2	4		3	3	1						2											
Fish (salmon)	9	8	9	6	8			1	5	6	2		1			2					2				
Ground beef	10	8	3	5					1						2		1								
Bologna	8	3				2	1	1												1					
Salami	9	3				1	2							1											
Bacon	6							2																	
Roast beef	7	1		1									1												
Pork sausage	7				2	6	4	2		1	2								2						
Tuna (canned)	1																								
Egg (scrambled)	4	2																				1			
Lamb chop	10	2	5						1								1	1						1	
Beef steak	6	4		2					1																
Beef liver	3	4		1														1							
Egg (hard boiled)	5	3																							
Pork chop	2	1		1							1														
Shrimp	3				2							1													
Pork roast		1																					1		
Turkey	1														1										
Egg (fried)	3	2																							
Ham	2							1																	
Veal cutlet	1										1		1												1
Chicken	1																			1					
Fish (haddock)		1	2																						
Ham (luncheon meat)																									
Totals	108	48	21	20	12	12	10	8	8	7	6	1	3	3	3	2	2	2	2	2	2	1	1	1	1
Frequency (top 10)	39.1	17.4	7.6	7.2																					
Banned pesticides																									
Total number of analyses = 276																									

TABLE 4-4 Industrial Chemistry Frequency for Protein Products (All Market Baskets)

Product	Tol	BZ	Xyl	Sty	EBZ	PCE	TCE	Cfo	TBZ	DBZ	PCB	TCA	BDM	BBZ	Ply	CTC	PBZ	RBZ	CBZ
Hot dogs	12	5	5	4	2	4	6	4					3	2	1	1			
Fish (salmon)									1		8						2	2	
Ground beef	11	8	5	4	2	2	3	1	1	3		1							
Bologna	9	5	5	6	4	7	3	5	1			2	2	2					1
Salami	12	5	7	5	3	2		1					1		2	1			
Bacon	12	4	7	7	4	2		1	5	2		2							
Roast beef	11	11	3	6	2		2			1		2							
Pork sausage	1		1																
Tuna (canned)	10	7	4				2	1	1	1						1			
Egg (scrambled)	6	2	3	5					1	1					1				
Lamb chop	1																		
Beef steak	1																		
Beef liver	1		1		1														
Egg (hard boiled)	1																		
Pork chop	1																		
Shrimp																			
Pork roast	1	1												1					
Turkey	1	1		1															
Egg (fried)																			
Ham	1																		
Veal cutlet																			
Chicken	1																		
Fish (haddock)																			
Ham (luncheon meat)																			
Totals	93	49	41	38	18	17	16	13	10	8	8	7	6	5	4	3	2	2	1
Frequency (top 10)	33.7	17.8	14.9	13.8	6.5	6.2													
Banned chemical																			
Total number of analyses = 276																			

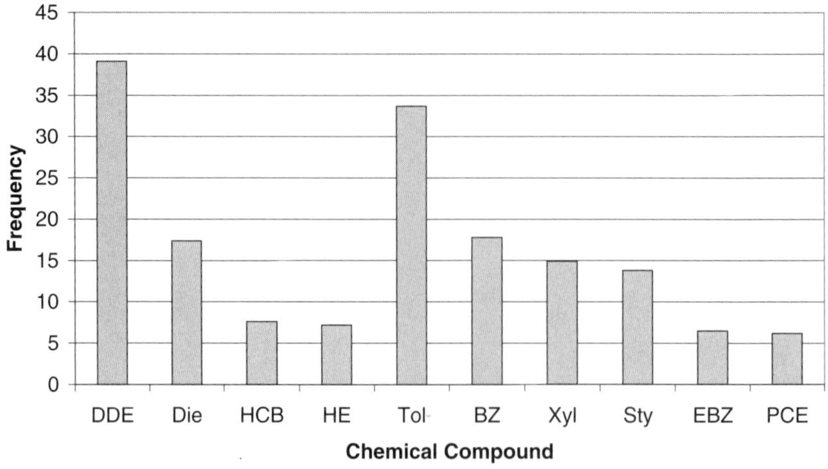

FIGURE 4-2 Frequency of top-10 chemicals in protein products.

chemicals and that only low-fat protein products are less likely to contain multiple synthetic chemicals, yet low-fat products will most likely contain DDE.

There are yet more similarities between protein and dairy products. For example, highly processed protein products (e.g., ground, mixed, and canned) also contain disinfection by-products. Thus, hot dogs, ground beef, bologna, salami, and canned tuna are contaminated with chloroform and sometimes with bromodichloromethane. Benzene, toluene, ethyl benzene, and xylenes, as well as the other detected chlorinated petroleum solvents (e.g., PCE, TCA, and TCE), also occur in protein products because these chemicals were in products used to lubricate or clean machinery that contacted the dairy products or because these chemicals were in the ambient air of the meatpacking plant. Protein products also contain styrene from contact with plastics. Finally, PCB (a banned industrial chemical) was found to occur in every market basket sample of salmon.

In addition to the previously summarized data, specific synthetic chemical concentrations in protein products for all market baskets are given in Appendixes 4-8 through 4-12.

4.3 SYNTHETIC CHEMICALS IN FRUIT PRODUCTS

The pesticides found in fruit products are given in Table 4-5, and the industrial chemicals are given in Table 4-6. Relative to the distribution of synthetic chemicals in fruit products, the top-10 synthetic chemicals are summarized in Figure 4-3. Fruit products are contaminated with a significant number of diverse insecticides and fungicides. However, several of these pesticides are responsible for the majority of the contamination. Iprodione and its metabolite make up 44.6% of the pesticides detected, whereas endosulfan and its by-product

TABLE 4-5 Pesticide Frequency for Fruit Products (All Market Baskets)

Product	Ipr	Car	Thi	Ipm	EnS	Azi	Cpy	Dic	Ell	Pho	Enl	Dan	Ben	Cap	DDE	Prg	Dat	Par	Mal	Etn	Die	Mxy	P
Apples		2	12		5	11	6	1	7	3	7	1	3	3		1						8	
Strawberries	10	8		3	4	2		2	3		3		4	8	3	1		1	5		4		
Cherries	9	9		9	5	7			5	2	3		2	1	1		1		5			1	3
Peaches	11	6		9	1	3	2	1	1	9	1	7	5	1	1	2		7			1		1
Raisins	2	5		1	1		2	7	1	2	1				10	12		1					
Plums	12	2		12	1		5	2				10	4			1		1					
Cantaloupe		1	3		10		3		6		6	2							1		4		1
Grapes	8	2		7	1		3	2	1		1			5	3		4						
Oranges		2	9		1		7	2				1					1						
Pears	2	2	11		2	7		2	1	6		1	2	2				3					1
Prunes	10			6	4	1	2	6	3		2	1			1	2		1					1
Apricots	9	2	1	2	2	3	1		2	7	1	1	2	2				1					
Bananas			10																				
Applesauce		7	3		1		2						1	1			3						
Orange juice		2						1												6			
Apple juice		11	7														5						
Fruit cocktail	6	1		5				4															1
Grapefruit			12					2											1	2			
Grape juice	3	11						1									1						1
Peaches (canned)	4	6		2																			
Prune juice	4			2	1						1												
Pineapple juice					6																		
Watermelon					4				1			1											
Grapefruit juice			1																	3			
Lemonade							1																
Pears (canned)					1																		
Pineapple (canned)																							
Totals	90	79	69	58	50	34	34	33	31	29	26	25	23	23	19	19	15	15	12	11	9	9	9
Frequency (top 10)	27.1	23.8	20.8	17.5	15.1	10.2	10.2	10	9.3														
Banned pesticides																							
Total number of analyses = 332																							

TABLE 4-5 *Continued*

Product	Far	Fen	Tbp	Tox	Vin	Mos	Ome	Diz	Dip	Eth	Mat	Bif	DCP	Efv	Fex	Fol	Met	Pal	Pne	Ace	Cd	Cil	DDT	Edr	Pon
Apples	1																		2						1
Strawberries				1	5			1	1		1	2			1	1	2				1				
Cherries	4	2			1		1		1					1						1					
Peaches		1				1																			
Raisins	2	1						1															1		
Plums																									
Cantaloupe				6		4							2					1						1	
Grapes	1				1	1	3	1							1	1						1			
Oranges							1				2														
Pears																									
Prunes																									
Apricots		3												1											
Bananas																									
Applesauce			6							2															
Orange juice																									
Apple juice										1															
Fruit cocktail																									
Grapefruit								1																	
Grape juice																									
Peaches (canned)			1																						
Prune juice																									
Pineapple juice									1																
Watermelon																		1							
Grapefruit juice																									
Lemonade																									
Pears (canned)																									
Pineapple (canned)																									
Totals	8	7	7	7	7	6	5	4	3	3	3	2	2	2	2	2	2	2	2	1	1	1	1	1	1
Frequency (top 10)																									
Banned pesticides																									
Total number of analyses = 332																									

TABLE 4-6 Industrial Chemistry Frequency for Fruit Products (All Market Baskets)

Product	Tol	Xyl	BZ	Sty	Cfo	TCE	EBZ	PCE	DBZ	DNB
Apples	3	9	4		2				1	
Strawberries	1	1	1	8				1		
Cherries										
Peaches										
Raisins	4	3	2	1	2	1		1		
Plums										1
Cantaloupe										
Grapes										
Oranges	8	5	2	2	2	1				
Pears		1								
Prunes	1									
Apricots										
Bananas	5	3	9	1	1	1	1			
Applesauce		1								
Orange juice	5	2	3	2	3		1			
Apple juice										
Fruit cocktail	1	1								
Grapefruit										
Grape juice										
Peaches (canned)										
Prune juice		1								
Pineapple juice										
Watermelon										
Grapefruit juice			1							
Lemonade	1									
Pears (canned)	1									
Pineapple (canned)										
Totals	30	27	22	14	10	3	2	2	1	1
Frequency (top 10)	9									
Total number of analyses = 332										

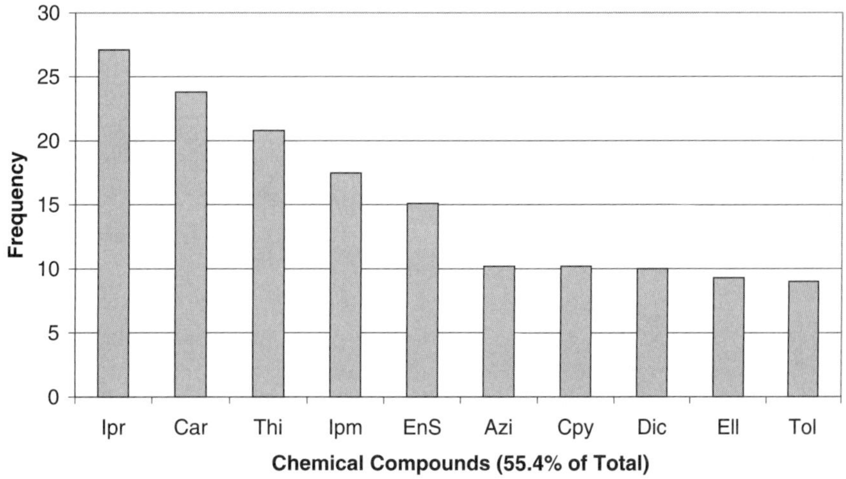

FIGURE 4-3 Frequency of top-10 chemicals in fruit products.

(endosulfan sulfate) make up an additional 24.4% of the detected pesticides, for a total of 69%. The remaining five pesticides are distributed throughout the fruit products.

Toluene and the other industrial chemicals are associated primarily with apples, oranges and orange juice, bananas, and raisins. It is assumed that the occurrence of benzene, ethylbenzene, toluene, and xylene compounds in these fruit products may be associated with the inactive portion of pesticide formulations or is the result of contacting process equipment. The most frequent occurrence of styrene is in strawberries. This occurrence would be consistent with the ubiquitous plastic wrap or baskets associated with packaging strawberries.

Note that several fruit products contain banned pesticides. These data suggest that raisins, cantaloupe, and strawberries have the greatest probability of being contaminated with banned pesticides, while some grape, peach, cherry, and prune samples have reported DDE contamination. The vast majority of these fruits was also produced domestically. This means that these pesticides occurred in these products because of past agricultural practices.

Citrus fruits do not tend to be as contaminated with pesticides as are noncitrus fruits. Specifically, all of the citrus fruits and fruit juices are essentially free of endosulfan and iprodione contaminants, yet they are contaminated with ethion, which does not occur in noncitrus fruits.

Another interesting characteristic of the fruit category is that juices tend to be far less contaminated than their parent fruit (except for pineapple). This suggests that synthetic chemicals remain in the rind or peel of the parent fruit. This is also consistent with canned peaches and pears that have far fewer contaminants than the whole fruit.

In addition to the previously summarized data, specific synthetic chemical concentrations in fruit products for all market baskets are given in Appendixes 4-13 through 4-23.

4.4 SYNTHETIC CHEMICALS IN GRAIN AND NUT PRODUCTS

The pesticides found in grain and nut products are given in Table 4-7, and the industrial chemicals are given in Table 4-8. Relative to the distribution of synthetic chemicals in grain and nut products, the top-10 synthetic chemicals are summarized in Figure 4-4. Unlike the previous food categories, many grain and nut food products are mixtures of other food category products and were produced commercially (except corn bread, which was homemade). The grain and nut food products in this category were selected for this group if the dominant ingredient was a grain or nut product; otherwise the food product was placed in the mixed food category. Because of these mixtures, some products may exhibit chemical contaminant patterns associated with dairy, protein (i.e., eggs), or fruit products.

TABLE 4-7 Pesticide Frequency for Grain Products (All Market Baskets)

Product	Mal	Cpm	Cpy	Mxy	Dip	Pri	Dca	Pal	Tox	DDE	Die	24-D	EnS	Pme	BHC
Peanut butter	4		11	1				12	12	8	11	1	7	6	4
Mixed nuts (no peanuts)			8					2		3	2	1			5
Roasted peanuts	3		8					11	12	8	9		7	8	3
Graham crackers	12	12	3	4		1									
White bread	10	10	1		11	1	4					6			
Sweet roll	11	11	1	2						1					
Butter crackers	12	12		4	1								1		
Fruit-flavored cereal	3	2	4				6								
Popcorn	2			7		7			1		1				
Corn chips	2		1	1		5									
Cracked-wheat bread	11	11	2		3	2	6					8			
Whole-wheat bread	12	12	3		9	4									
White roll	11	9	1	1	4	3									
Flour tortilla	11	11	2	3	7										
Saltine crackers	12	11	3	4		1									
Rye bread	11	12			3	1									
Bagel	10	11	3	1		5									
Shredded wheat	10	10	1	3											
Pretzels	11	12		4		2									
Pancake mix	10	11		1						3					
Granola	5	9	3	1											
English muffin	11	11				2									
Corn bread (HM)	9	8				2				1					
Noodles	7	7											1		
Oat ring cereal							12					1			
Biscuit	5	9		1		1									
Macaroni	6	6		1											
White rice				1	1										
Creamed corn															
Oatmeal	1												1		
Crisped rice cereal															
Cornflakes				1											
Wheat cereal	1	1													
Corn grits	1														
Pinto beans															
Kidney beans															
Corn															
Totals	214	208	55	41	39	37	28	25	25	24	23	17	17	14	12
Frequency (top 10)	48.2	46.8	12.3	9.2	8.8	8.3									

Banned pesticides

Total number of analyses = 444

TABLE 4-7 *Continued*

Product	Cam	Cum	Qac	Diz	Edi	Mat	Qui	Clo	Ipr	Lin	Etn	Pms	Cno	Dic	Ell	Mec	P
Peanut butter							4					2			1		
Mixed nuts (no peanuts)		4								2							
Roasted peanuts							1					1	2				
Graham crackers		1															
White bread	3	1							1								
Sweet roll		2				1											
Butter crackers		1															
Fruit-flavored cereal		1			6	4					3					2	
Popcorn									1								
Corn chips																	
Cracked-wheat bread				1													1
Whole-wheat bread																	
White roll	7																
Flour tortilla																	
Saltine crackers																	
Rye bread										1							1
Bagel																	
Shredded wheat				2													
Pretzels				1													
Pancake mix																	
Granola				1					1					2			
English muffin																	
Corn bread (HM)																	
Noodles				1													
Oat ring cereal								4									
Biscuit																	
Macaroni										1							
White rice			6						1								
Creamed corn																	
Oatmeal															1		
Crisped rice cereal			4														
Cornflakes																	
Wheat cereal																	
Corn grits																	
Pinto beans																	
Kidney beans																	
Corn																	
Totals	10	10	10	6	6	5	5	4	4	4	3	3	2	2	2	2	2
Frequency (top 10)																	
Total number of analyses = 444																	

Product	Par	Tbp	Cpa	Cte	Dat	EDC	Eth	Fit	HCB	Mxo	Pcy	Ton
Peanut butter												
Mixed nuts (no peanuts)												
Roasted peanuts			1	1								
Graham crackers												
White bread							1					
Sweet roll												
Butter crackers									1			
Fruit-flavored cereal						1						1
Popcorn												
Corn chips										1		
Cracked-wheat bread												
Whole-wheat bread	1											
White roll	1											
Flour tortilla												
Saltine crackers												
Rye bread												
Bagel								1				
Shredded wheat		1										
Pretzels												
Pancake mix												
Granola											1	
English muffin												
Corn bread (HM)												
Noodles												
Oat ring cereal												
Biscuit												
Macaroni												
White rice												
Creamed corn					1							
Oatmeal												
Crisped rice cereal												
Cornflakes		1										
Wheat cereal												
Corn grits												
Pinto beans												
Kidney beans												
Corn												
Totals	2	2	1	1	1	1	1	1	1	1	1	1
Frequency (top 10)												
Banned pesticides												
Total number of analyses = 444												

Note: HM = homemade.

TABLE 4-8 Industrial Chemical Frequency for Grain Products (All Market Baskets)

Product	Tol	Xyl	Sty	TBZ	BZ	EBZ	PBZ	DBZ	TCE	Cfo
Peanut butter	11	6	10	5	6	5	11	1	1	2
Mixed nuts (no peanuts)	12	10	12	9	2	8	1	3	3	1
Roasted peanuts	1	1					12			
Graham crackers	9	6	4	7	3	4		3		1
White bread	4	4	3	4	1	1		3	1	
Sweet roll	9	6	11	4	3	3		1	1	
Butter crackers	9	5	3	3	1	3		1	2	1
Fruit-flavored cereal	7	5	3	1	5	1			1	4
Popcorn	10	7	3	3	5	2		1	2	
Corn chips	10	5	4	4	5	2		1	2	2
Cracked-wheat bread										
Whole-wheat bread		1		1						
White roll				1						
Flour tortilla	1	1								
Saltine crackers		1		1						
Rye bread	1	1								
Bagel										
Shredded wheat	1	1		1						
Pretzels										
Pancake mix	1	1								
Granola	1	1	1							
English muffin										
Corn bread (HM)		1								
Noodles	1	1								
Oat ring cereal										
Biscuit										1
Macaroni										
White rice										
Creamed corn	3				1					
Oatmeal		1		1						
Crisped rice cereal										
Cornflakes		1								
Wheat cereal										
Corn grits										
Pinto beans										
Kidney beans										
Corn										
Totals	91	66	54	45	32	29	24	14	13	12
Frequency (top 10)	20.5	14.9	12.2	10.1						
Total number of analyses = 444										

Note: HM = homemade.

The first six organophosphorus and chlorinated organophosphorus pesticides dominate wheat-based products. Table 4-7 shows that virtually all wheat-based products are, at a minimum, contaminated with malathion and chlorpyrifos-methyl. These data also suggest that although corn-based products can contain malathion and chlorpyrifos-methyl, they are much less likely to be contami-

Product	TCA	PCB	Pyl	BBZ	PCE	Tpp	BDM
Peanut butter	5	1	2	1			
Mixed nuts (no peanuts)	1		1		2		
Roasted peanuts		1					
Graham crackers			2	2	1		
White bread			1				
Sweet roll		2					
Butter crackers		3	1				
Fruit-flavored cereal			1				
Popcorn	3	2		2			1
Corn chips	2			1			
Cracked-wheat bread						1	
Whole-wheat bread							
White roll						1	
Flour tortilla							
Saltine crackers							
Rye bread							
Bagel							
Shredded wheat							
Pretzels							
Pancake mix							
Granola							
English muffin							
Corn bread (HM)							
Noodles							
Oat ring cereal							
Biscuit							
Macaroni							
White rice							
Creamed corn							
Oatmeal							
Crisped rice cereal							
Cornflakes							
Wheat cereal							
Corn grits							
Pinto beans							
Kidney beans							
Corn							
Totals	11	9	8	6	3	2	1
Frequency (top 10)							
Banned chemical							
Total number of analyses = 444							

nated by the first six organophosphorus and chlorinated organophosphorus pesticides.

Theoretically, the organophosphorus pesticides are not persistent. This means that USDA-certified organic wheat-based and corn-based products should be relatively free of pesticides. Just like the previous categories, some

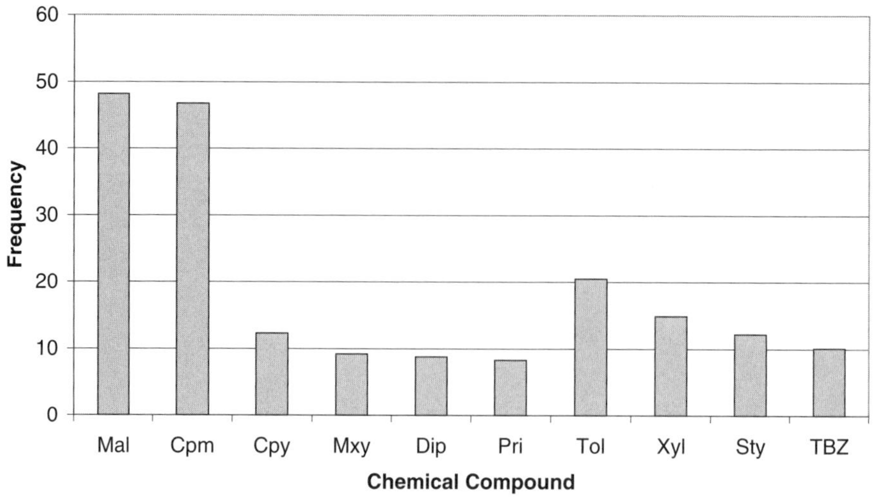

FIGURE 4-4 Frequency of top-10 chemicals in grain products.

USDA-certified organic grain and nut products will, however, be contaminated with banned pesticides.

These data show that peanut products (peanut butter and dry-roasted peanuts) are contaminated with at least three banned pesticides, while mixed nuts, butter crackers, popcorn, sweet roll, pancake mix, and corn bread were reported to contain banned pesticides. Given the low percentage of imports for these products, the occurrence of these pesticides can be said to be the result of past U.S. agricultural practices. Once again, this suggests that even USDA-certified organic peanut products will contain a mixture of banned pesticides. The occurrence of banned pesticides in the other grain-based products may be associated with the occurrence of butter or oils in these mixed products.

Grain products with a high oil content or prepared with butter or margarine may have contacted processing equipment and plastics containing industrial chemicals. As a result, several of the grain products were contaminated with a significant number of industrial chemicals. These products were graham crackers, white bread, sweet rolls, butter crackers, and fruit-flavored cereal.

In addition to the previously summarized data, specific synthetic chemical concentrations in grain and nut products for all market baskets are given in Appendixes 4-24 through 4-33.

4.5 SYNTHETIC CHEMICALS IN VEGETABLE PRODUCTS

The pesticides found in vegetable products are given in Table 4-9, and the industrial chemicals are given in Table 4-10. Relative to the distribution of synthetic chemicals in vegetable products, the top-10 synthetic chemicals are summarized in Figure 4-5. Compared to both dairy and protein products, the vegetable foods contain far fewer industrial chemicals except for those foods that are (1) oils (i.e., olive/safflower oil and margarine) or fried in oils (i.e., french fries and potato chips) and (2) mechanically handled or sorted (i.e., tomatoes and avocados). Vegetables are, however, contaminated with a significant number of diverse insecticides and herbicides.

The chlorinated endosulfan insecticides contaminate 75% of all the vegetable products and account for 70% of the detected pesticides. Since 2002, the use of endosulfan insecticides on succulent beans and peas, spinach, and grapes has been discontinued. Even with this discontinuation, however, the majority of vegetable products are still expected to be contaminated with endosulfan pesticides. Also note that the endosulfan pesticides are moderately persistent in soil. As a result, these pesticides could possibly still be detected in succulent beans and peas, spinach, and grapes today.

Table 4-9 shows that many of the vegetable products contain banned pesticides. These data show that 33 vegetable products are contaminated with banned pesticides, whereas 21% or 47% of the vegetable products are contaminated with multiple banned pesticides. This widespread contamination transcends the five recommended vegetable groups that are included in a balanced diet:

- Dark green vegetables: spinach, collards, and broccoli
- Deep yellow vegetables: carrots, sweet potatoes, and winter squash
- Beans and peas: lima beans
- Starchy vegetables: potatoes (white potatoes, french fries, potato chips, and mashed potatoes)

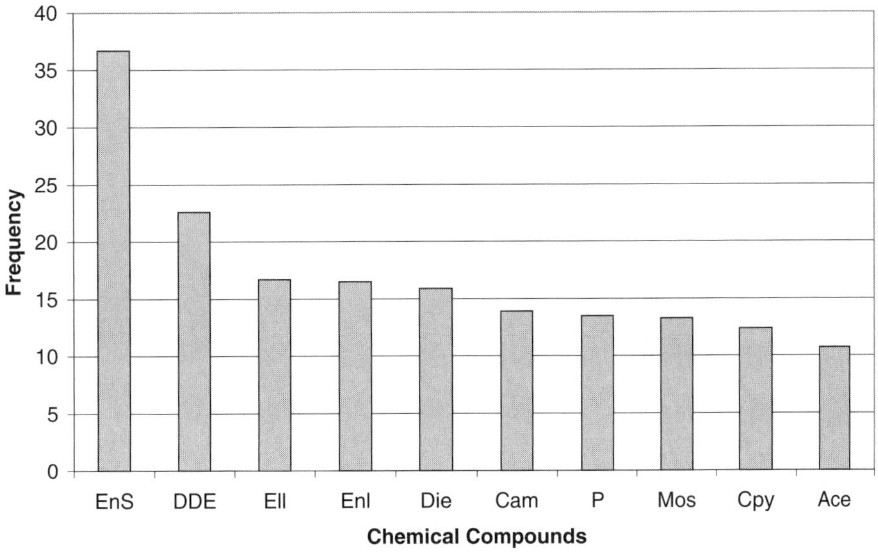

FIGURE 4-5 Frequency of top-10 chemicals in vegetable products.

TABLE 4-9 Pesticide Frequency for Vegetable Products (All Market Baskets)

Product	EnS	DDE	Ell	Enl	Die	Cam	P	Mos	Cpy	Ace	DCP	DDT	Dan	Cd	Tox	Ipr	Dat	Dip	Eth	Icm
Spinach	7	12	3	1	6		12		6	1	4	11	1	2	4		2		1	
French fries (FF)	11	3			4	8			2			1	1							2
Collards	6	11	1	2	5		12		4		10	4		2	3		1		1	
Potato chips	6	2			2	11						1				1				2
Green peppers	6	3	6	6		2	4	11	3	10	1	1					3			
Sweet pickles	12	2	11	10	10				3			1		6	3					
Dill pickles	12	2	11	10	10		1					1		4	6				1	
Celery	3	7	3	2		1	11	5		9		2	9					1		
Summer squash	12	3	8	11	9							1		2	7					
Margarine																		6		
White potato with skin	11	7	1	2		12		1	2			6				3			5	4
Cucumber	12		12	12	4			4					1	4	2					
Tomato	7		8	6		1	7	6	6				2							
Olive/safflower oil	8	5	3	2					1											
Black olives	11	6		5					10		6									
Iceberg lettuce	7	8	6	6	1		4	3		6	1		1				3			
French fries (F)	11	2			2	12							1			1		1	6	4
Green beans	3	2	2	2				7		8	1		2			3	1		2	
Mashed potatoes	8	3				12												6	1	3
Lima beans		2			1			8		9	1					2	5			
Winter squash	7	2			6	1			1				1	5	2					
Radish	3	9			5				3		4	4		1						
White potato without skin	8	2	1			11		1											3	4
Brussels sprouts	4	3	1	2	2		5	1	9							2		1		

Broccoli	3	9	3	3	2		5		1		4									
Turnip	2	7	1		5				7		5	1					1			
Avocado																				
Tomato catsup	1	2	1			1	2	3	2	2				1						
Tomato sauce	1	2	2	2			2	4	1	1										
Mushrooms							3													
Mixed vegetables (F)		1			1		1	7		9										
Sweet potatoes					2	3			1				9							
Beets	3	4			6						3					1				
Carrots	1	4														9				
Stewed tomatoes	1			1			1	1		1										
Tomato juice	3		3	2				3												
Peas (dried)								2		1								4		
Okra	3		3	1													1	1		
Eggplant	1			1				5		1										
Asparagus		2							5											
Green peas																	5			
Sauerkraut					2		1						1							
Cabbage	4																			
Cauliflower							2													
Onion					1											1				
Totals	198	122	90	89	86	75	73	72	67	58	40	34	29	27	24	23	22	20	20	19
Frequency (top 10)	36.7	22.6	16.7	16.5	15.9	13.9	13.5	13.3	12.4	10.7										

Banned pesticides

Total number of analyses = 540

TABLE 4-9 *Continued*

Product	Thi	HE	Lin	Nac	Ome	Cno	Dic	Pal	Cin	Cpa	Car	Vin	Diz	TDE	Tpp	Met	Fen	Ben	Cum	BHC
Spinach			1	1	4			1					1	5						
French fries (FF)		1											1						2	
Collards				1	1			1	7					2			5			
Potato chips		1																		
Green peppers					3		2				2					4	1			
Sweet pickles		1	4	4					1											2
Dill pickles		1	4	1							2				1					1
Celery													1			1				
Summer squash		3		1				4												
Margarine															6				4	
White potato with skin	8			1				1												
Cucumber		3		3								1	1			1				2
Tomato					2		3								1		1			
Olive/safflower oil			4										1							
Black olives						1	6			1										
Iceberg lettuce					1				2			1				2				
French fries (F)		1				1				1										
Green beans					1						2	6								
Mashed potatoes															1					
Lima beans			2		3		1													
Winter squash		5		2																
Radish								1						1						
White potato without skin	1			1																
Brussels sprouts								1			1									
Broccoli													1							

Turnip														1						
Avocado																				
Tomato catsup						5				3										
Tomato sauce						3			1	1										
Mushrooms	10												2					6		
Mixed vegetables (F)												2								
Sweet potatoes								1												
Beets								2												
Carrots													1							
Stewed tomatoes						4				5	1									
Tomato juice											2									
Peas (dried)			1																	
Okra																				
Eggplant																				
Asparagus																				
Green peas																				
Sauerkraut																				
Cabbage																				
Cauliflower																				
Onion																				
Totals	19	16	16	15	15	14	12	12	11	11	10	10	9	9	9	8	7	6	6	5
Frequency (top 10)																				
Banned pesticides																				
Total number of analyses = 540																				

TABLE 4-9 *Continued*

Product	Efv	Oct	Cil	Cyf	Edr	Mal	Tri	Bif	Cte	Lac	Qui	Cau	HCB	Lrn	Mat	Mxy	Neb	Oxa	Sim	Tos
Spinach																				
French fries (FF)						1														
Collards	1										1									
Potato chips							3													1
Green peppers	2			2				1		1										
Sweet pickles		2																		
Dill pickles																				
Celery			3			1												1		
Summer squash		1			1						1		1							
Margarine																				
White potato with skin																				
Cucumber		1			1															
Tomato	1																			
Olive/safflower oil															1				1	
Black olives																				
Iceberg lettuce																				
French fries (F)						1														
Green beans								1									1			
Mashed potatoes																				
Lima beans																				
Winter squash		1																		
Radish				1	1															
White potato without skin																				
Brussels sprouts																				

Broccoli																				
Turnip																				
Avocado																				
Tomato catsup	1								1			1								
Tomato sauce									1	1										
Mushrooms																				
Mixed vegetables (F)																				
Sweet potatoes																				
Beets																				
Carrots														1						
Stewed tomatoes																				
Tomato juice																				
Peas (dried)																				
Okra																				
Eggplant																				
Asparagus																				
Green peas																1				
Sauerkraut																				
Cabbage																				
Cauliflower																				
Onion																				
Totals	5	5	3	3	3	3	3	2	2	2	2	1	1	1	1	1	1	1	1	1
Frequency (top 10)																				
Banned pesticides																				
Total number of analyses = 540																				

Note: FF = fast food, F = frozen.

TABLE 4-10 Industrial Chemical Frequency for Vegetable Products (All Market Baskets)

Product	Tol	Sty	BZ	Xyl	EBZ	Cfo	TBZ	DBZ	PCE	TCE	DNB	PBZ	BBZ	BDM	CBZ	Pyl	BRB	RBZ
Spinach						1												
French fries (FF)	11	10	2	6	5	1	3	2	1	1		1	1	1	2	1		
Collards																		
Potato chips	10	5	3	6	5	2	5	3	2	2			1					
Green peppers																		
Sweet pickles			1															
Dill pickles	1																	
Celery	1			1				1			1	1						
Summer squash																		
Margarine	12	5	3	7	4	2	4	4	4					1		1	1	
White potato with skin																		
Cucumber																		
Tomato	2	1	3	1	1	2												
Olive/safflower oil	7	6	6	5	4	2	2		1									
Black olives	1					1												
Iceberg lettuce																		
French fries (F)												1						1
Green beans																		
Mashed potatoes																		
Lima beans																		
Winter squash																		
Radish																		
White potato without skin																		

Brussels sprouts																		
Broccoli																		
Turnip																		
Avocado	2	5	8	1		6		1	3	3								
Tomato catsup																		
Tomato sauce	1																	
Mushrooms																		
Mixed vegetables (F)																		
Sweet potatoes	1										2							
Beets																		
Carrots																		
Stewed tomatoes			1															
Tomato juice																		
Peas (dried)	1																	
Okra																		
Eggplant																		
Asparagus																		
Green peas																		
Sauerkraut																		
Cabbage																		
Cauliflower																		
Onion																		
Totals	50	32	27	27	19	17	14	11	11	6	3	3	2	2	2	2	1	1

Note: FF = fast food, F = frozen.

- Other vegetables: summer squash, pickles, turnips, beets, cucumbers, celery, olives, oils, brussels sprouts, iceberg lettuce, green peppers, mixed vegetables, asparagus, green beans, sauerkraut, and onions

When a fairly high percentage of a vegetable are imported (e.g., squash, cucumbers, olives), that vegetable may contain banned pesticides from foreign sources. The remaining vegetables were grown mostly in the United States. These data demonstrate that the majority of both USDA-certified organic vegetables and nonorganic products may contain at least one banned pesticide, while at least 47% are likely to be contaminated with multiple banned organochlorine pesticides. Clearly, the vegetables with the greatest probability of being contaminated with multiple banned pesticides are spinach, collards, squash, pickles, cucumber, radish, turnip, potatoes with skin, broccoli, and beets.

These data suggest that those vegetable products most likely to be contaminated with a mix of pesticides are green peppers, celery, tomato, spinach, mixed vegetables, lima beans, collards, green beans, brussels sprouts, potatoes with skin, and iceberg lettuce. All of these products purchased as USDA-certified organic foods would not be expected to contain these pesticides.

In addition to the previously summarized data, specific synthetic chemical concentrations in vegetable products for all market baskets are given in Appendixes 4-34 through 4 42.

4.6 SYNTHETIC CHEMICALS IN MIXED PRODUCTS

The pesticides found in mixed products are given in Table 4-11, and the industrial chemicals are given in Table 4-12. Relative to the distribution of synthetic chemicals in mixed products, the top-10 synthetic chemicals are summarized in Figure 4-6. A review of the mixed food products in these two exhibits shows

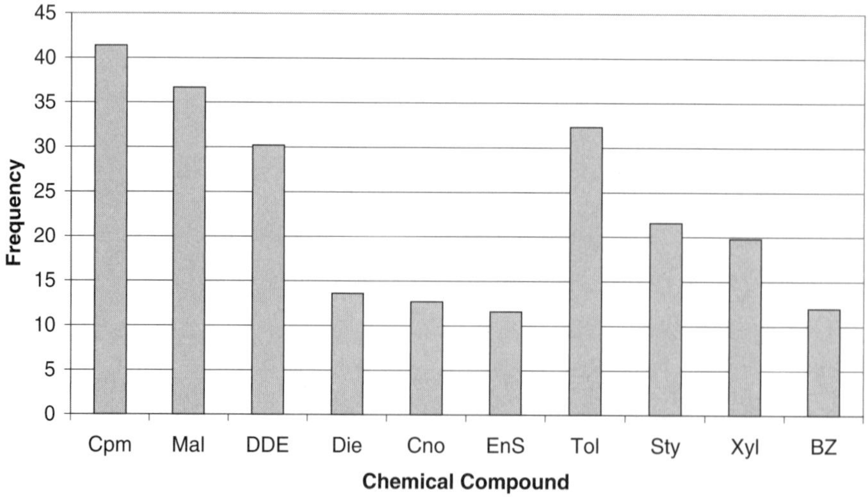

FIGURE 4-6 Frequency of top-10 chemicals in mixed products.

TABLE 4-11 Pesticide Frequency for Mixed Products (All Market Baskets)

Product	Cpm	Mal	DDE	Die	Cno	EnS	Cam	Cpa	Cpy	Lin	Cmy	BHC	Cum	Dip	P	Ell	DDT	Enl	Dan	Cau	Diz	Ipr	Etn	HE	Pri
Cheese and pepperoni pizza (FF)	12	12	11	2	9	1		6	5		1	1	1		2	3		3	1		1		2		
1/4-lb cheeseburger and bun (FF)	11	11	12	12		7			2	2		1	1			3	1	2						2	1
Taco/tostada (FF)	2	4	12	2	11	5		8	2		4	1	4		2	3		3		1					1
1/4- lb hamburger and bun (FF)	11	12	12	8		5	1		2	1		1	2			1	3	1						3	1
Milk chocolate candy bar			9			3			3	12		3					5	1							1
Cheese pizza (FF)	12	11	11	2	6	2		4	5		1		2		2	1		1			1				
Chocolate chip cookies	11	12							1	10		3	1	1											
Fish sticks	11	12	1										1				1								
Pumpkin pie	8	11	2	11	11		2	10			12	1			6					7					
Blueberry muffin	10	10											3												2
Brownies	12	12								1			1								1				
Meatloaf			12	7	1		1	1	3		1	1	1	1	2						2	2		4	
Doughnut — cake and icing	11	11							2		2		1								1				1
Chocolate cake	8	6	2	1						1			1												
Chicken nuggets (FF)	6	6	1		2			2					3				1						1		
Sugar cookies	11	10	2							1	1									1					
Apple pie	10	12											2												
Hot dog and bun (FF)	11	12	10	4	3	1			6	1		2					3								1
Tuna noodle casserole (HM)	8	3	12	2	1	7			1						5	2	1	2	10						
Sandwich cookies	11	11																							
Fish and bun (FF)	12	12	11	8		1	2		2			1					2	1			1				

TABLE 4-11 *Continued*

Product	Cpm	Mal	DDE	Die	Cno	EnS	Cam	Cpa	Cpy	Lin	Cmy	BHC	Cum	Dip	P	Ell	DDT	Enl	Dan	Cau	Diz	Ipr	Etn	HE	Pri
Mustard	1		1		9	1		8	2	11	4	8				1	3	1							
Coleslaw and dressing			4	4	2	6	1	2	1	2		7			1		1								
Fried chicken (FF)	3	3	1	1	2			1					2												
Chocolate snack cake	8	6								1															
French salad dressing			1	2	10			9	1	1	8	1					1			2			4		
Raisin bran	12	7	1			2			2										1			1			
Salisbury steak and gravy	3	2	6	3	2	6	12												1						
Chicken potpie	10	12		1	1	1	8	1	2	1											1				1
Muffin with egg/cheese/ham (FF)	10	10	12	4					1												1				1
Peppers with rice and beef (HM)	3		5	1		5			3						4	5	1	4							
Chili and beans (HM)			7	1	8			8			5			1						5		1			
Caramel candy			1											10											
Scalloped potatoes			4	1		6	10							5					1						
Beef Stroganoff (HM)	8	6	6	1		1	1				1			1			1				2	2		1	
Turkey dinner	2	2	2	1		5	10		2									1							
Macaroni and cheese	10	6	5						3					1											
Beef chow mein (FF)	3	1	1			3			3						3	1			5		1				
White sauce (HM)	6	3	4	1										7											
Italian salad dressing (low fat)					8			5			4		1							1			2		

Beef stew	3		4	2		5	4															2			
Lasagna and meat (HM)	4		9	1	1	1		1	1							1									
Yellow cake	9	9		1																					
Clam chowder	1		3	1		2	9												1						
Jelly						1								1		1		1				5			
Popsicle													3												
Mayonnaise				9		4								1											
Spaghetti and meatballs	2	1	6						2	1						1									
Brown gravy (HM)	5	1	3	1					1			1													
Vegetable beef soup			1				8																		
Pork and beans					3			1	1																
Suckers		1							1					1									2		
Mushroom soup	1		1						1												1				
Fried chicken (HM)			3							1															
Chocolate pudding			3																						
Spaghetti and tomato	1															2									
Tomato soup						1										1									
Bean and bacon soup				1																					
Chicken noodle soup																									
Totals	293	260	214	96	90	82	69	67	61	47	44	32	30	30	27	26	24	21	20	17	13	13	11	10	10
Frequency (top 10)	41.4	36.7	30.2	13.6	12.7	11.6																			

Banned pesticides

Total number of analyses = 708

TABLE 4-11 *Continued*

Product	Cte	HCB	Mxy	Prg	Car	Vin	Ccp	Dic	Azi	24-D	DCP	Icm	TDE	Cap	Cyf	Fen	Oct	Thi	Cin	Dat	Dca	Efv	Ipm	Lac	Pho	Pme
Cheese and pepperoni pizza (FF)	2						3																			
1/4-lb cheeseburger and bun (FF)		2															1									
Taco/tostada (FF)	3						2																			
1/4-lb hamburger and bun (FF)		1																								
Milk chocolate candy bar													3													1
Cheese pizza (FF)																										
Chocolate chip cookies			1																							
Fish sticks																										
Pumpkin pie	2																									
Blueberry muffin			1						1																1	
Brownies																										
Meatloaf		3									1						1									
Doughnut — cake and icing			1																							
Chocolate cake																										
Chicken nuggets (FF)																										
Sugar cookies																										
Apple pie																										
Hot dog and bun (FF)		3																								
Tuna noodle casserole (HM)						1	1																			

Sandwich cookies																
Fish and bun (FF)																
Mustard						1										
Coleslaw and dressing																
Fried chicken (FF)																
Chocolate snack cake																
French salad dressing	1				1	1							1			
Raisin bran			8		2		4			2				1		
Salisbury steak and gravy				4		1										
Chicken potpie		2														
Muffin with egg/ cheese/ham (FF)		1														
Peppers with rice and beef (HM)					2	1		1			1	2			1	1
Chili and beans (HM)	1															
Caramel candy																
Scalloped potatoes									2							
Beef Stroganoff (HM)																
Turkey dinner				2												
Macaroni and cheese																
Beef chow mein (FF)											1					
White sauce (HM)																
Italian salad dressing (low fat)								1								
Beef stew									1							

TABLE 4-11 *Continued*

Product	Cte	HCB	Mxy	Prg	Car	Vin	Ccp	Dic	Azi	24-D	DCP	Icm	TDE	Cap	Cyf	Fen	Oct	Thi	Cin	Dat	Dca	Efv	Ipm	Lac	Pho	Pme
Lasagna and meat (HM)																										
Yellow cake																										
Clam chowder																										
Jelly					6															1			1			
Popsicle								1																		
Mayonnaise			2																							
Spaghetti and meatballs																										
Brown gravy (HM)																										
Vegetable beef soup						1																				
Pork and beans																										
Suckers																										
Mushroom soup																		2								
Fried chicken (HM)																										
Chocolate pudding																										
Spaghetti and tomato																										
Tomato soup																										
Bean and bacon soup																										
Chicken noodle soup																										
Totals	9	9	8	8	6	8	6	6	5	4	3	3	3	2	2	2	2	2	1	1	1	1	1	1	1	1
Frequency (top 10)																										
Banned pesticides																										
Total number of analyses = 708																										

Note: FF = fast food, HM = homemade.

TABLE 4-12 Industrial Chemical Frequency for Mixed Products (All Market Baskets)

Product	Tol	Sty	Xyl	BZ	TBZ	EBZ	Cfo	PCE	TCE	DBZ	BBZ	Pyl	TCA	Tpp	CBZ	BDM	CTC
Cheese and pepperoni pizza (FF)	11	7	10	1		4	1	3	2								
1/4-lb cheeseburger and bun (FF)	12	5	5	6		3	3	1	1	1	2	1			1		
Taco/tostada (FF)	10	7	6	2		4	3	1	1		1	1			1	1	
1/4-lb hamburger and bun (FF)	11	7	6	6	1	3	3	1	1		1	1	1				
Milk chocolate candy bar	11	10	9	2	7	6	1	8	2	2	1		1		1	1	
Cheese pizza (FF)	12	7	7	3	1	3	3	4	1				1				
Chocolate chip cookies	12	12	5		7	2	2	3	1	4							
Fish sticks	11	4	6	3	10	3	7	4		4	3	1	2		2		1
Pumpkin pie	1	1	1														
Blueberry muffin	12	8	7	4	3	5	1	4	1	2		1	1				
Brownies	11	6	7	4	8	6	2	2	1	1	1				1		
Meatloaf	9	5	4	7		1	2	2	2				1				
Doughnut — cake and icing	12	9	8	3	4	3	1	1	1		2	2					
Chocolate cake	12	10	8	2	6	4	2	4	1	3	1				1		
Chicken nuggets (FF)	12	11	6	6		4	2	1	2		1	2	2			2	
Sugar cookies	11	12	6	1	2	4	4	3	1	1		1					
Apple pie	5	8	5	4	8	3	1	1		1	2						
Hot dog and bun (FF)	1		1														
Tuna noodle casserole (HM)	1																

TABLE 4-12 *Continued*

Product	Tol	Sty	Xyl	BZ	TBZ	EBZ	Cfo	PCE	TCE	DBZ	BBZ	Pyl	TCA	Tpp	CBZ	BDM	CTC
Sandwich cookies	7	11	6	3	2		2	1							1		
Fish and bun (FF)			1														
Mustard	2																
Coleslaw and dressing	4	1	2	12			1			1							1
Fried chicken (FF)	12	5	5	5		3	1		2			1	2				
Chocolate snack cake	11	5	6	2	4		2	1									
French salad dressing			1														
Raisin bran					1												
Salisbury steak and gravy	1		1				1										
Chicken potpie			1														
Muffin with egg/ cheese/ham (FF)			1														
Peppers with rice and beef (HM)																	
Chili and beans (HM)	1		1														
Caramel candy	1	1	1	2		1	2	2	2					9			2
Scalloped potatoes	1													1			
Beef Stroganoff (HM)																	
Turkey dinner	1						1										
Macaroni and cheese			1														
Beef chow mein (FF)	1	1															
White sauce (HM)														1			
Italian salad dressing (low fat)																	
Beef stew																	

Lasagna and meat (HM)			1	1													
Yellow cake			1														
Clam chowder				1													
Jelly				1													
Popsicle	5			3			3									2	
Mayonnaise																	
Spaghetti and meatballs	1		1	1													
Brown gravy (HM)	1																
Vegetable beef soup																	
Pork and beans			1														
Suckers	1																
Mushroom soup																	
Fried chicken (HM)	1		1														
Chocolate pudding	1																
Spaghetti and tomato			1														
Tomato soup																	
Bean and bacon soup																	
Chicken noodle soup																	
Totals	229	153	140	85	64	62	51	47	22	20	15	11	11		8	6	4
Frequency (top 10)	32.3	21.6	19.8	12													
Total number of analyses = 708																	

Note: FF = fast food, HM = homemade.

that the composition of these food products is dominated by grain, dairy, and meat products. Therefore, it is not unexpected that the most common pesticides and industrial chemicals in the mixed foods are essentially the same ones that dominate the grain, dairy, and meat product chemicals. In general, these data show that those food products that contain cheese, butter or margarine, milk, high-fat meat products, and oils (or cooked in oils) are contaminated with the greatest number of pesticides and industrial chemicals. Furthermore, the majority of the mixed food products is contaminated with a mixture of banned pesticides. Once again, those products containing dairy or protein products are contaminated with the greatest number of banned pesticides.

It also appears that fewer industrial chemicals are found in homemade mixed products than in commercially prepared mixed products. These data also show a consistent lack of styrene in homemade products, suggesting that homemade products have less contact with plastics during their preparation (as opposed to for storage).

In addition to the previously summarized data, specific synthetic chemical concentrations in mixed products for all market baskets are given in Appendixes 4-43 through 4-47.

4.7 SYNTHETIC CHEMICALS IN CHILD AND INFANT PRODUCTS

The pesticides found in child and infant products are given in Table 4-13, and the industrial chemicals are given in Table 4-14. Relative to the distribution of synthetic chemicals in child and infant products, the top-10 synthetic chemicals are summarized in Figure 4-7. In general, the type and distribution of pesticides and industrial chemicals in child/infant products are somewhat similar to the basic food categories (i.e., dairy, protein, fruit, grain, and vegetable), but

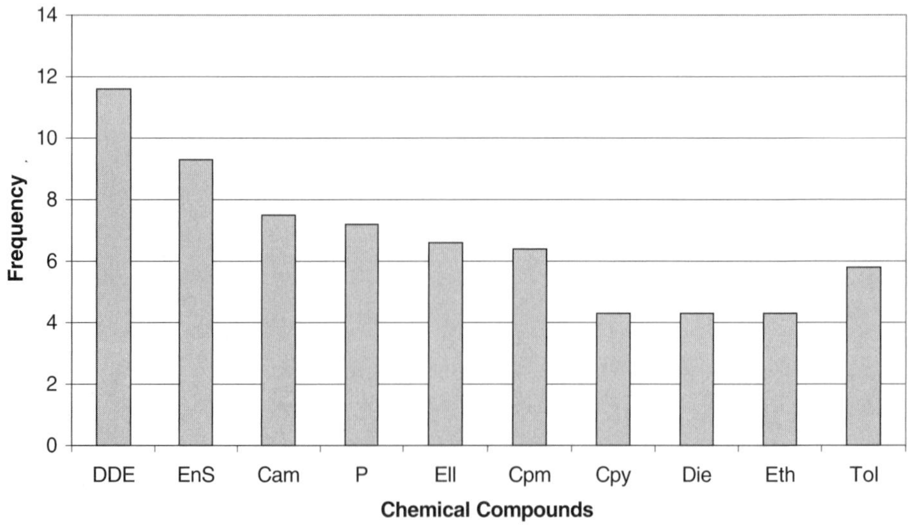

FIGURE 4-7 Frequency of top-10 chemicals in child/infant products.

TABLE 4-13 Pesticide Frequency in Child/Infants Products (All Market Baskets)

	Pesticides																			
Product	**DDE**	**EnS**	**Cam**	**P**	**Ell**	**Cpm**	**Cpy**	**Die**	**Eth**	**Dat**	**Ipr**	**Enl**	**Dan**	**Mal**	**Car**	**Thi**	**Ipm**	**Ace**	**Mos**	**Ome**
Peaches		4		9	2	1	1				5	1			4		3			
Applesauce (strained)		6			5		6		6	4		1			3					3
Rice with apple cereal (strained)		10			8		7			1		6			2					
Pears (strained)		4		2	4				9	1		3				1		1		
Beef (strained)	9																			
Fruit pudding (strained)		3		8	3		1			1	7	1	2				3			
Teething biscuits						12								12						
Apple juice (strained)										7					2	8				1
Vegetables and chicken	3					8		4			1									
Vegetables and ham	3		10					1			1						1			
Squash (strained)	1	5		1				5												
Green beans (strained)								1										5	5	
Spinach (strained)	9			4																
Carrots (strained)																				
Sweet potatoes (strained)	1												10							

TABLE 4-13 *Continued*

	Pesticides																			
Product	**DDE**	**EnS**	**Cam**	**P**	**Ell**	**Cpm**	**Cpy**	**Die**	**Eth**	**Dat**	**Ipr**	**Enl**	**Dan**	**Mal**	**Car**	**Thi**	**Ipm**	**Ace**	**Mos**	**Ome**
Vegetables and beef	6		4																	
Turkey and rice			4		1			2				1								
Mixed vegetables (strained)			6																1	
Soy-based formula						1														
Rice cereal (infant)	2																			
Chicken noodle dinner	1		2					2												
Macaroni/tomato/beef	4																			
Chicken (strained)	1			1																
Milk-based formula																				
Orange juice (strained)																				
Peas (strained)																				
Custard pudding (strained)																				
Creamed corn (strained)																				
Bananas and tapioca																				
Totals	40	32	26	25	23	22	15	15	15	14	14	13	12	12	11	9	7	6	6	4
Frequency (top 10)	9.3	7.5	6.1	5.8	5.4	5.1	3.5	3.5	3.5											

Banned pesticides

Product	Ben	Par	Pho	Qac	Azi	Bif	Efv	HCB	Pzo	Vin	Cap	Cum	Dic	Dip	Etn	Fen	Mxy	Pal	Prg	Tca	TDE
Peaches		1					1		2							1			1		
Applesauce (strained)	2																				
Rice with apple cereal (strained)	1										1										
Pears (strained)		2	3		2								1								
Beef (strained)																					
Fruit pudding (strained)							1														
Teething biscuits														1			1				
Apple juice (strained)																					
Vegetables and chicken																					
Vegetables and ham																					
Squash (strained)								2										1			1
Green beans (strained)						2				2											
Spinach (strained)																					
Carrots (strained)												1									
Sweet potatoes (strained)																					
Vegetables and beef																					

TABLE 4-13 *Continued*

Product	Ben	Par	Pho	Qac	Azi	Bif	Efv	HCB	Pzo	Vin	Cap	Cum	Dic	Dip	Etn	Fen	Mxy	Pal	Prg	Tca	TDE
Turkey and rice																					
Mixed vegetables (strained)																				1	
Soy-based formula																					
Rice cereal (infant)				3																	
Chicken noodle dinner																					
Macaroni/tomato/ beef																					
Chicken (strained)																					
Milk-based formula																					
Orange juice (strained)															1						
Peas (strained)																					
Custard pudding (strained)																					
Creamed corn (strained)																					
Bananas and tapioca																					
Totals	3	3	3	3	2	2	2	2	2	2	1	1	1	1	1	1	1	1	1	1	1
Frequency (top 10)																					
Banned pesticides																					

TABLE 4-14 Industrial Chemical Frequency in Child/Infants Products (All Market Baskets)

Product	Tol	Cfo	BZ	Xyl	TBZ	Sty	BBZ	BDM	CTC	PCE	Ply	TCA
Peaches	1	1		1								1
Applesauce (strained)	1			1								
Rice with apple cereal (strained)												
Pears (strained)	1			1								
Beef (strained)	10	3	3	2	3			1			1	
Fruit pudding (strained)												
Teething biscuits				1	1		1					
Apple juice (strained)		3	3			1			1	1		
Vegetables and chicken												
Vegetables and ham												
Squash (strained)												
Green beans (strained)												
Spinach (strained)				1								
Carrots (strained)	4	3	2	1		1						
Sweet potatoes (strained)												
Vegetables and beef				1								
Turkey and rice		1	1	1								
Mixed vegetables (strained)				1								
Soy-based formula	1	2	3			1						
Rice cereal (infant)												
Chicken noodle dinner												
Macaroni/tomato/beef												
Chicken (strained)	1											
Milk-based formula	1	1	1									
Orange juice (strained)				1								
Peas (strained)				1								
Custard pudding (strained)												
Creamed corn (strained)												
Bananas and tapioca												
Totals	20	14	13	13	4	3	1	1	1	1	1	1
Frequency (top 10)	4.7											
Banned pesticides												

TABLE 4-15 Average Number of Contaminants for All Foods with Low to Very High Ranking

Food Category	Average Number of Contaminants
Child/infant products	2.1
Protein products	3.6
Fruit products	3.7
Vegetable products	3.8
Grain and nut products	3.9
Dairy products	4
Mixed products	4.4

their concentrations are significantly lower. This is illustrated in a comparison of child and infant products with all foods, by category, with a low to very high contaminant ranking (see Table 4-15). On average, 45% of the child/infant products contain multiple contaminants as well as banned or discontinued pesticides.

These data also show that banned pesticides also occur in child/infant products. These data are consistent with the occurrence of banned pesticides in protein and vegetable products. The only apparent inconsistency is that milk-based formula did not contain any of these pesticides. In reality, however, there should be little or no banned or discontinued pesticides in milk-based formulas since commercially available products have removed the butterfat.

In addition to the previously summarized data, specific synthetic chemical concentrations in child and infant products for all market baskets are given in Appendixes 4-48 through 4-58.

4.8 SYNTHETIC CHEMICALS IN A BALANCED DIET

When we examine all of the chemical contaminants in each food category (except the child/infant category since it represents a specialized group), we see that a balanced diet will expose an individual to a complex mixture of synthetic chemicals. Although each food category had multiple contaminants, only the top-10 most frequently detected synthetic chemicals (i.e., see Figures 4-1 through 4-6) were used to illustrate the chemical diversity of a balanced diet.[3] This chemical exposure is summarized in Table 4-16.

[3] This percentage was determined by dividing the number of times a chemical was detected by the total number of chemical analyses in a food category. For example, in the dairy category, DDE was detected in 149 samples of the 228 samples tested, or 65.4% of the samples were contaminated with DDE.

TABLE 4-16 Chemical Frequency for All Market Baskets — Balanced Diet Exposure as a Percentage

Pesticides	Dairy %	Fruits %	Grains %	Protein %	Vegetables %	Mixed %	Average* %
Acephate (Ace)		0.3			10.7		
Azinphos-methyl (Azi)	0.4	10.2				0.7	
Carbaryl (Car)		23.8			1.9	1.1	
Chloroethyl linoleate (Cno)			0.5	4.3	2.6	12.7	
Chlorpropham (Cam)			2.2	1.2	13.9	9.7	
Chlorpyrifos (Cpy)		10.2	12.3	0.7	12.4	8.6	7.4
Chlorpyrifos-methyl (Cpm)			46.8			41.4	14.7
DDE	65.4	5.7	5.4	39.1	22.6	30.2	28.1
Dicofol (Dic)		10				0.9	
Dieldrin (Die)	36.8	2.7	5.1	17.4	15.9	13.6	15.3
Diphenyl-2-ethylhexyl phosphate (Dip)	0.9	0.9	8.8	1.1	3.7	4.2	3.3
Endosulfan I (Enl)		7.8			16.5	2.9	
Endosulfan II (Ell)		9.3	0.5		16.7	3.7	
Endosulfan sulfate (EnS)	10.1	15.1	3.8	0.7	36.7	11.6	13
Heptachlor epoxide (HE)	10.1			7.2	3	1.4	
Hexachlorobenzene (HCB)	9.2			7.6		1.3	
Iprodione (Ipr)	0.4	27.1	0.9		4.3	1.8	5.8
Iprodione metabolite (Ipm)		17.5					
Malathion (Mal)	1.3	3.6	48.2	0.7	0.6	36.7	15.2
Methamidophos (Mos)		1.8			13.3		
Methoxychlor (Mxy)	3.1	2.7	9.2		0.2	1.1	2.7
Permethrin (P)	2.2	2.7	0.5	0.4	13.5	3.8	3.9
Pirimiphos-methyl (Pri)			8.3				
Thiabendazole (Thi)		20.8			3.5		
Industrial Chemicals							
Benzene (BZ)	7	6.6	7.2	17.8	5	12	9.3
Chloroform (Cfo)	38.5	3	2.7	4.7	3.1	7.2	9.9
Ethyl benzene (EBZ)	3.9	0.6	6.5	6.5	3.5	8.8	5
Perchloroethylene (PCE)	4.8	0.6	0.7	6.2	2	6.6	3.5
Styrene (Sty)	10.1	4.2	12.2	13.8	5.9	21.6	11.3
Trimethylbenzene (TBZ)	5.3		10.1	3.6	2.6	11.7	5.6
Toluene (Tol)	29.4	9	20.5	33.7	9.3	32.3	22.3
Xylenes (Xyl)	21.5	8.1	14.9	14.9	5	19.8	14
Total number of analyses	228	322	444	276	540	708	

* Average if chemical detected in five or more food categories.

The frequency of each top-10 chemical within a food category is given in prior figures.[4] The frequency of all top-10 chemicals is also displayed in each food category. For example, acephate was a top-10 chemical in the vegetable category (10.7%) but was only detected in the fruit category (0.3%, this com-

[4] The chemical class of each compound is given in Appendix 2-2.

pound is not a top-10 chemical in the fruit category). Thus, if you look at acephate from left to right, the dairy column is blank (i.e., no chemicals detected), the fruit column shows 0.3, followed by the grain and protein columns, which are also blank, the vegetable column shows 10.7, and the mixed column is blank. The overall frequency of chemical occurrence in all of the food products is summarized in Figure 4-8.

When looking at the five basic food categories (i.e., dairy, fruit, grain, protein, and vegetable), anyone who eats a balanced diet can be exposed to a complex chemical mixture of 24 pesticides and 8 industrial chemicals. This exposure can be reduced, however, by limiting the intake of high-fat foods or eating USDA-certified organic products. For example, substituting margarine and oil for butter provides a significant reduction in the number of chemicals. This comparison is given in Table 4-17. For example, 27 different chemicals have been found in butter as compared to 14 in margarine. Furthermore, it is important to remember that USDA-certified organic butter will provide little reduction in the number of contaminants since at least five banned or discontinued pesticides will most likely occur in butter. Margarine, however, had no detectable levels of banned pesticides.

The information in Table 4-16 has also been used to determine the makeup of the most common chemical mixture associated with eating a balanced diet. If a chemical occurs in at least five of the six food categories, it was selected as a common dietary contaminant, and its average frequency in food products was determined. This complex chemical mixture is represented by two banned organochlorine pesticides, two chlorinated organophosphorus pesticides, four organochlorine pesticides, two organophosphorus pesticides, six aromatic petroleum hydrocarbons, and two chlorinated petroleum hydrocarbons (see Figure 4-8).

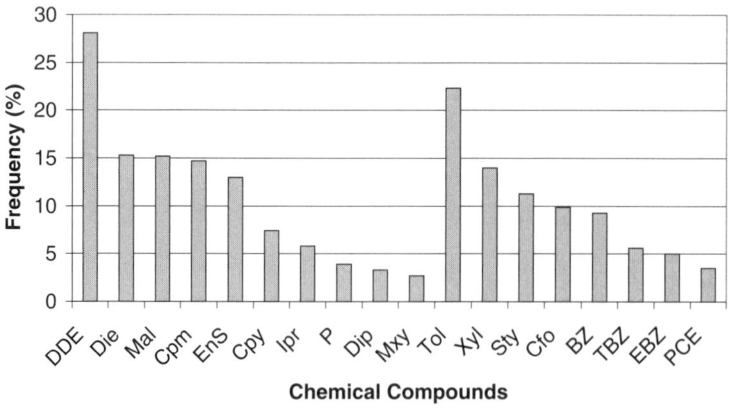

FIGURE 4-8 Average frequency of most common chemicals in all food products (see Table 2-26).

TABLE 4-17 A Butter Comparison

Product	Average Number of Contaminants for All Market Baskets	Chemicals That Probably Occur in Product*								
Butter (Appendix 4-10)	12	DDE	Dieldrin	Endosulfan sulfate	Hexachloro-benzene	Chloroform	Styrene	Trimethyl-benzene	Toluene	Xylene
Margarine (Appendix 4-42)	5.3								Toluene	Xylene
Olive/safflower oil (Appendix 4-42)	4			Endosulfan sulfate					Toluene	

* Indicates chemicals detected in seven or more market baskets (i.e., greater than 50% of the time).

TABLE 4-18 Basic Foods with Banned Pesticides

	Number of Market Baskets with Detected Pesticides						
Food Products	**DDE**	**DDT**	**Dieldrin**	**HCB**	**PCB**	**Toxaphene**	**Perecentage Imported***
Dairy							
Milk (2%)	6						2.5
Sherbet	6						2.5
Chocolate milk shake	7						2.5
Evaporated milk	10						2.5
Half-and-half	11						2.5
Whole milk	7						2.5
Ice cream — light	7						2.5
Vanilla ice cream	10						2.5
Swiss cheese	10		11				2.5
Cheddar cheese	8		9				2.5
American cheese	12		8				2.5
Butter	12		10	11			2.5
Protein							
Bacon	6						3.9
Beef steak	6						9.2
Lamb chop	10						29.5
Pork sausage	7						3.9
Roast beef	7						9.2
Salami	9						
Bologna	8						
Ground beef	10		8				9.2
Fish — salmon	9		8	9	8		42.7
Hot dogs	10						
Fruit							
Raisins	10						5
Grain and Nuts							
Peanut butter	8		11			12	2.7
Dry-roasted peanuts	8		9			12	3.4
Vegetables							
Turnip	7						
Broccoli	9						4.8
Radish	9						21.8
Winter squash			6				33.5
Black olives	6						43.6
White potato with skin	7	6					6.2
Summer squash			9			7	33.5
Celery	7						3.7
Dill pickles			10			6	40.1
Sweet pickles			10				40.1
Collards	11						0.6
Spinach	12	11	7				2.2
Child and Infant							
Beef (strained)	9						
Vegetables and beef	6						
Spinach (strained)	9						

* The lower the percentage of imports, the greater the probability that the pesticides in these products originated from past agricultural practices.

TABLE 4-19 Foods with Multiple Industrial Chemicals

	Number of Market Baskets with Detected Industrial Chemicals								
Food Product	**Benzene**	**Ethyl Benzene**	**Styrene**	**TBZ**	**Toluene**	**Xylene**	**Chloroform**	**PCE**	**TCE**
Dairy									
Swiss cheese					11		7		
Cheddar cheese					10	8	10		
American cheese					10	10	9		
Butter		6	10	7	12	11	12		
Grain and Nuts									
Popcorn					10	7			
Graham crackers				7	9	6			
Sweet roll			11		9	6			
Peanut butter	6		10		11	6			
Mixed nuts (no peanuts)			12	9	12	10			
Protein									
Tuna (canned)	7				10				
Salami					12	7			
Hot dogs					12				6
Ground beef	8				11				
Bacon			7		12	7			
Roast beef	11		6		11				
Bologna			6		9			7	

TABLE 4-19 *Continued*

Vegetable								
Avocado	8						6	
Margarine					12	7		
Potato chips					10	6		
Olive/safflower oil	6		6		7			
French fries (FF)			10		11	7		
Mixed								
Fried chicken (FF)					12			
Apple pie			8	8				
Meatloaf (HM)	7				9			
1/4-lb cheeseburger and bun (FF)	6				12			
Chocolate snack cake					11	6		
Sandwich cookies			11		7	6		
Chocolate chip cookies			12	7	12			
Sugar cookies			12		11	6		
Blueberry muffin			8		12	7		
Doughnut — cake and icing			9		12	8		
Taco/tostada (FF)			7		10	7		
Cheese pizza (FF)			7		12	7		
Cheese and pepperoni pizza (FF)			7		11	10		
1/4-lb hamburger and bun (FF)	6		7		11	6		
Chicken nuggets (FF)	6		11		12	6		
Fish sticks				10	11	6	7	
Chocolate cake			10	6	12	8		
Brownies		6	6	8	11	7		
Milk chocolate candy bar		6	10	7	11	9		8

Note: FF = fast food, HM = homemade.

There are also two fairly special chemical categories that need to be summarized to illustrate their common distribution in a large number of our food products. The first category is composed of the banned pesticides. Their distribution is given in Table 4-18. This table also includes a summary of the average percentage of imported food products collected by the FDA Residue Monitoring program from 1996 through 2001 (the same period as that used for the Total Diet Study data presented in this book). These percentages demonstrate that the vast majority of foods consumed in the United States is grown in the United States. Thus, the occurrence of banned pesticides is from past agricultural practices. The second category of compounds are the industrial chemicals. The distribution of these common chemicals in the basic food groups is given in Table 4-19. These data demonstrate that although pesticides are a primary concern in foods, industrial chemicals also contaminate a significant number of foods.

4.9 OPTIONS FOR CHOOSING SAFER FOODS

Because so many foods are contaminated with multiple chemicals, the information in this book can be used to choose foods with the minimum number of different synthetic chemicals or to identify those foods that should be purchased as USDA-certified organic products. However, even an organic product can contain banned pesticides and industrial chemicals unless it is carefully produced. Therefore, it is important to remember that organic food producers do not analyze their products and inform the consumer as to the chemical mixtures in their products (both pesticides and industrial chemicals). Because the consumer is not provided with this information, individuals have two basic choices:

1. *Purchase an organic product and be confident that this product will most likely contain fewer pesticides, but not necessarily fewer industrial chemicals, than nonorganic products.*
2. *Limit or eliminate specific high-fat dairy and protein products and DDE-contaminated vegetable products from one's diet as well as purchase organic products.*

These choices are not always easy to make nor are they convenient. They are also expensive. This means that individuals with a limited income need to select foods carefully if they are unable to purchase organic products.

The previous discussion has focused on an individual's exposure to synthetic chemicals in a balanced diet, yet many individuals choose to eat only specific foods that fit their lifestyle preferences. Thus, it is important to look at how specific diets can affect an individual's exposure to chemical mixtures.

5 DIETS AND CHEMICAL MIXTURES

All of the previous data show that an individual's diet will dictate the number of different chemicals he or she will consume on a daily basis. This chapter gives examples of how individual food choices influence a person's chemical exposure. With these examples, the consumer can utilize this guide to determine the chemical mixtures that may occur in a specific meal or daily diet. For example, using the chemical exposure data given in the appendixes, examples of how specific food choices can influence a person's exposure to synthetic chemicals are shown for low-fat and high-fat diets, vegetarian diets, children's lunches, and comfort foods. Specific synthetic chemicals are listed for a food product if a chemical occurred in at least 7 out of 12 market baskets (i.e., approximately 60% of the time). Otherwise, the chemical was not listed even if was detected in that food.

The first example compares the daily chemical exposure of a balanced high-fat and low-fat diet using individual food products (i.e., no mixed products) that are nonorganic. The selected high-fat and low-fat menus with the most probable chemical exposures are given in Table 5-1. This comparison shows that the low-fat diet has 19 different chemicals (22 total detected chemicals), whereas the high-fat diet has only 17 different chemicals (39 total detected chemicals);[1] yet the high-fat diet has a greater number of banned chemicals that occur in multiple foods. For example, DDE occurs in butter, bologna, American cheese, ground beef, baked potato, and sour cream. These data also show that both diets

[1] It is also important to note that a low-carbohydrate diet has the potential to have similar contaminant characteristics as a high-fat diet.

TABLE 5-1 Low-Fat and High-Fat Diet Comparison

Low-Fat Example														
Breakfast				Lunch						Dinner				
Oatmeal	Raisins	Skim Milk	Grapefruit	Turkey	Lettuce	Tomato	Mustard	Cracked-Wheat Bread	Apple Juice	Ham	Kidney Beans	Carrots	Celery	Chemical Summary
								2,4-D						2,4-D
													Ace	Ace
							BHC							BHC
							Cno							Cno
							Cpa							Cpa
								Cpm						Cpm
						Cpy								Cpy
									Car					Car
	DDE												DDE	DDE (2)
													Dan	Dan
	Dic													Dic
								Dip						Dip
						EII								EII
					EnS	EnS								EnS (2)
							Lin							Lin
								Mal						Mal
						P							P	P (2)
	Prg													Prg
									Thi					Thi
Total different chemicals													19	
Total number of detected chemicals														22

High-Fat Example														
Breakfast				Lunch						Dinner				
Bacon	Eggs (Fried)	Biscuit	Butter	Applesauce	Bologna	American Cheese	Mayonnaise	Wheat Bread	Banana	Ground Beef	Baked Potato	Sour Cream	Asparagus	Chemical Summary
					BZ				BZ	BZ				BZ (3)
											Cam			Cam
				Car										Car
			Cfo			Cfo						Cfo		Cfo (3)
								Cpm						Cpm
			DDE		DDE	DDE				DDE	DDE	DDE		DDE (6)
			Die		Die	Die	Die			Die		Die		Die (6)
								Dip						Dip
			EnS								EnS			EnS (2)
			HCB											HCB
								Mal						Mal
		P												P
Sty			Sty											Sty
			TBZ											TBZ
									Thi		Thi			Thi (2)
Tol			Tol		Tol	Tol				Tol				Tol (5)
Xly			Xyl			Xyl								Xyl (3)
Total different chemicals													17	
Total number of detected chemicals													39	
					Banned pesticides									

Note: ( ) indicates the number of multiple occurrences.

share eight pesticides (carbaryl, chlorpyrifos-methyl, DDE, diphenyl-2-ethylhexyl phosphate, endosulfan sulfate, malathion, permethrin, and thiabendazole), but that the low-fat diet has no industrial chemicals (i.e., benzene, chloroform, styrene, toluene, trimethlybenzene, and xylene).

This comparison shows that (1) both diets have a significant number of individual chemical contaminants and (2) the high-fat diet has a greater number and occurrence of banned pesticides. The number of chemicals in either diet can be significantly reduced by purchasing organic products. However, the low-fat diet (i.e., fewer banned pesticides) will ultimately contain fewer multiple sources of specific chemicals as well as fewer synthetic chemicals.

The second example looks at chemical exposures that can occur in a vegetarian diet. In this example, a strict vegetarian diet (i.e., no dairy, meat, eggs, or fish) is compared to a vegetarian diet that does include eggs and dairy products. The selected vegetarian menus with the most probable chemical exposures are given in Table 5-2. This comparison shows that the strict vegetarian diet has 27 different chemicals (39 total detected chemicals), while the vegetarian diet with dairy and eggs has only 21 different chemicals (36 total detected chemicals). Furthermore, just like the high-fat and low-fat diets, the vegetarian diets contain multiple exposures to the same chemical from different foods.

Clearly, vegetarian diets that do not substitute organic food products for nonorganic foods will probably contain at least 20 to 30 different synthetic chemicals per day. Vegetarian diets will generally contain much fewer banned pesticides than diets that include protein and dairy products. This suggests that substituting organic foods for nonorganic foods will significantly reduce the number of chemicals in a vegetarian diet (e.g., this assumes that vegetables like spinach or collards are avoided or minimized unless the consumer can find spinach or collards that do not contain these chemicals).

Another good example of a chemical-exposure comparison is for foods that go into a child's lunch box. Given the morning or evening rush to put together a child's lunch, many of the foods are prepackaged or easy to assemble. For example, simple lunches often include individually packaged snacks (e.g., cookies, cakes, chips, chocolate candy, nuts, pudding) and individual fruits (e.g., box of raisins, orange, banana, apple, pear). Many of these products have a substantial number of chemical contaminants. In addition to snacks and fruits, common sandwiches may also contain a significant number of synthetic chemicals. A comparison of the chemicals that may be expected to be in different sandwiches is given in Table 5-3.

Because so many lunch box foods contain numerous synthetic chemicals, it is important to at least estimate the number of different synthetic chemicals that can occur under worst- and best-case lunch box scenarios. The worst-case lunch box example (Table 5-4) includes a peanut butter and jelly sandwich, potato chips, an apple, and chocolate chip cookies. This single meal would contain 21 different synthetic chemicals. The best-case lunch box example (Table 5-5) includes a turkey sandwich (on rye with mayonnaise), corn chips, a banana,

TABLE 5-2 Vegetarian Diet Comparison

Strict Vegetarian Diet															
Breakfast				Lunch				Dinner							
Granola	Raisins	Soy Milk (No Data)	Apple Juice	Peanut Butter	Banana	Cracked-Wheat Bread	Corn Chips	Lettuce	Tomato	Green Peppers	Avocado	Italian Dressing	Mixed Nuts	White Roll	Chemical Summary
						2,4-D									2,4-D
										Ace					Ace
											BZ				BZ
														Cam	Cam
			Car												Car
												Cno			Cno
Cpm						Cpm								Cpm	Cpm (3)
				Cpy					Cpy				Cpy		Cpy (3)
	DDE			DDE											DDE (2)
	Dic														Dic
				Die											Die
						Dip									Dip
													EBZ		EBZ
									Ell						Ell
				EnS				EnS	EnS						EnS (3)
						Mal								Mal	Mal (2)
										Mos					Mos
									P						P
				Pal											Pal
				PBZ											PBZ
	Prg														Prg
													TBZ		TBZ
			Thi		Thi										Thi (2)
				Tol			Tol						Tol		Tol (3)
				Tox											Tox
				Sty									Sty		Sty (2)
													Xyl		Xyl
Total different chemicals														27	
Total number of detected chemicals														39	

Banned pesticides

TABLE 5-2 *Continued*

Vegetarian Plus Eggs and Dairy															
Breakfast				Lunch				Dinner							
Plain Yogurt	Strawberries	Bagel	Jelly	Eggs (Hard)	Mayonnaise	Whole-Wheat Bread	Potato Chips	Apple	Macaroni	Cheddar Cheese	Lettuce	Tomato	Cucumber	Olive Oil	Chemical Summary
								Azi							Azi
							Cam								Cam
	Car														Car
	Cap														Cap
										Cfo					Cfo
		Cpm				Cpm									Cpm (2)
												Cpy			Cpy
										DDE					DDE
					Die					Die					Die (2)
						Dip									Dip
								Enl					Enl		Enl (2)
								Ell				Ell	Ell		Ell (3)
								EnS			EnS	EnS	EnS	EnS	EnS (5)
	Ipr														Ipr
		Mal				Mal									Mal (2)
								Mxy							Mxy
												P			P
	Sty														Sty
			Thi					Thi							Thi (2)
							Tol			Tol				Tol	Tol (3)
								Xyl		Xyl					Xyl (2)
Total different chemicals														21	
Total number of detected chemicals														35	

Banned pesticides

Note: ( ) indicates the number of multiple occurrences.

TABLE 5-3 Sandwich Comparison

Bologna and American Cheese on White with Mustard	Peanut Butter and Jelly on White	American Cheese on Wheat with Mustard	Ham and Swiss Cheese on Rye with Mustard	Tuna Salad on Wheat with Lettuce and Tomato	Egg Salad on Wheat	Ham on Rye with Mustard	Tuna Salad on Wheat	Turkey on Rye with Mayonnaise
BHC	Cpm	BHC	BHC	BZ	Cpm	BHC	BZ	Cpm
Cpa	Cpy	Cpa	DDE	Cpm	Cpy	Cpm	Cpm	Die
Cno	DDE	Cno	Die	Cpy	DDE	Cpa	Die	Mal
Cfo	Die	Cfo	Cpa	Die	Die (2)	Cno	Dip	
Cpm	Dip	Cpm	Cno	Dip	Dip	Lin	Mal	
DDE (2)	EnS	DDE	Cfo	Ell	Mal	Mal		
Die	Mal	Die	Cpm	EnS (2)	Pir			
Dip	Pal	Dip	Lin	Mal				
Lin	PBZ	Lin	Mal	P				
Mal	Sty	Mal	Tol					
Tol	Tol	Tol						
PCE	Tox	Xyl						
Xyl								
Banned pesticides								

Note: ( ) indicates the number of multiple occurrences.

TABLE 5-4 A Lunch Box with Worst-Case Food Products

Peanut Butter and Jelly on White Bread	Potato Chips	Apple	Chocolate Chip Cookies	Lunch Box Chemical Summary
		Azi		Azinphos-methyl
	Cam			Chlorpropham
Cpy				Chlorpyrifos
Cpm			Cpm	Chlorpyrifos-methyl (2)
DDE				DDE
Die				Dieldrin
Dip				Diphenyl-ethylhexyl
		Enl		Endosulfan I
		Ell		Endosulfan II
EnS				Endosulfan sulfate
			Lin	Lindane
Mal			Mal	Malathion (2)
		Mxy		Methoxychlor
Pal				Pentachloroaniline
PBZ				Pentachlorobenzene
Sty			Sty	Styrene (2)
		Thi		Thiabendazole
Tol	Tol		Tol	Toluene (3)
Tox				Toxaphene
			TBZ	Trimethylbenzene
		Xyl		Xylene
				Banned pesticides

Note: ( ) indicates the number of food products with the same chemical.

TABLE 5-5 Lunch Box with Best-Case Food Products

Turkey on Rye with Mayonnaise	Corn Chips	Banana	Sandwich Cookies	Lunch Box Chemical Summary
Cpm			Cpm	Chlorpyrifos-methyl (2)
Die				Dieldrin
Mal			Mal	Malathion (2)
			Sty	Styrene
		Thi		Thiabendazole
	Tol		Tol	Toluene (2)
				Banned pesticides

Note: ( ) Indicates the number of food products with the same chemical.

and sandwich cookies. This single meal would contain six different synthetic chemicals.

A child's school lunch may include both homemade and commercially prepared products. When considering the number of chemicals that occurs in the mixed food category (i.e., commercially available and fast-food products), parents should seriously consider using fewer prepackaged products and switch to homemade lunch box foods made from organic products.

The previous examples illustrate the chemical diversity associated with individual food preferences. Because the examples cannot be comprehensive, individuals who wish to regulate the number and type of chemicals they consume can use the chemical information from the appendixes to evaluate their own food choices.

5.1 SUMMARY

Based on the data collected by the Food and Drug Administration, no diet is free of synthetic chemical mixtures and banned pesticides. This even includes *organic*-based diets. Therefore, the data in this book are meant to provide chemical information to those individuals who wish to make informed decisions about the chemical mixtures they choose to allow in their own daily diets and the diets of their children.

All of the data provided in this book have been summarized to evaluate, at a national level, the trends of pesticide and industrial chemical contamination in food products. In addition to this type of analysis, these data also reflect contaminant trends in different cities and regions around the United States.

6 FOOD CONTAMINATION AT THE CITY AND REGIONAL LEVELS

Based on the market basket sampling method, it is not possible to identify which city within a specific market basket is the most or least contaminated. This shortcoming aside, the total number of chemical contaminants found in all of the food products within each market basket was determined.

As Table 6-1 shows, it is possible to identify those cities with the highest and lowest number of chemical contaminants in their food products that occurs within a given market basket. The greater the number of detected chemical contaminants, the greater one's potential exposure to chemical mixtures. *Table 6-1 shows that the greater the market basket ranking (i.e., number 12), the poorer the quality of the food products.*

Obviously, these rankings also reflect the total contaminant contributions from each food category. For example, in a given market basket, the greatest number of synthetic chemicals may originate from the grain and vegetable products categories as opposed to the dairy and mixed products categories. The differences within each market basket are presented in Figures 6-1 through 6-12 by a relative contaminant rank (as given in Table 6-1). These figures illustrate which food categories have the greatest number of contaminants in a region's market basket.

This means that the product category figures can be used to identify those categories of food that contain elevated numbers of synthetic chemicals. For example, Figure 6-2, market basket 98-3 for the Long Beach/New York/Providence region, shows that consumers should be aware that vegetables are the only food category with elevated numbers of synthetic chemicals.

In addition to the total number of contaminants found in a market basket, each market basket also has a unique number of chemical compounds or

TABLE 6-1 Market Basket and City Ranking by the Total Number of Chemical Contaminants

Total Number of Detected Chemicals	Market Baskets	Market Basket Cities	Total Number of Different Chemicals in a Market Basket
888	01-3	Oklahoma City — Orlando — Washington	84
835	00-3	Birmingham — Columbus — Jackson	83
823	99-3	Binghamton — Harrisburg — Worcester	82
806	00-1	Grand Rapids — Milwaukee — Springfield	78
711	99-1	Denver — Portland — Reno	76
695	96-1	Bakersfield — Ogden — Tacoma	76
684	98-4	Dayton — Peoria — Sioux Falls	75
604	97-1	Phoenix — Fresno — San Antonio	72
598	98-1	Boise — Los Angeles — Pueblo	77
551	96-2	Charleston — Lafayette — Palm Beach	73
546	98-3	Long Beach — New York — Providence	80
507	98-2	Baton Rouge — Savannah — Tulsa	66

Contaminant Levels	Protein Rank	Child Rank	Dairy Rank	Fruit Rank	Grain Rank	Mixed Rank	Vegetable Rank
Very High							
Very High							
High							
High							
Moderately High							
Moderate							
Moderate							
Moderately Low							
Low	9						
Low							
Very Low		11	11	11	11		
Very Low						12	12

FIGURE 6-1 Market basket 98-2 for Baton Rouge, LA, Savannah, GA, and Tulsa, OK.

Contaminant Levels	Vegetable Rank	Grain Rank	Child Rank	Dairy Rank	Mixed Rank	Fruit Rank	Protein Rank
Very High							
Very High							
High							
High							
Moderately High							
Moderate							
Moderate							
Moderately Low	8						
Low		9					
Low			10	10	10		
Very Low							
Very Low						12	12

FIGURE 6-2 Market basket 98-3 for Long Beach, NJ, New York, NY, and Providence, RI.

Increasing Number of Chemicals ↑

Contaminant Levels	Child Rank	Fruit Rank	Protein Rank	Vegetable Rank	Mixed Rank	Dairy Rank	Grain Rank
Very High							
Very High	2						
High							
High							
Moderately High		5					
Moderate							
Moderate							
Moderately Low							
Low							
Low			10	10			
Very Low					11		
Very Low						12	12

FIGURE 6-3 Market basket 96-2 for Charleston, SC, Lafayette, LA, and West Palm Beach, FL.

Increasing Number of Chemicals ↑

Contaminant Levels	Vegetable Rank	Grain Rank	Fruit Rank	Dairy Rank	Mixed Rank	Protein Rank	Child Rank
Very High							
Very High							
High							
High							
Moderately High							
Moderate							
Moderate	7	7					
Moderately Low			8				
Low				9	9		
Low							
Very Low						11	
Very Low							12

FIGURE 6-4 Market basket 98-1 for Boise, ID, Los Angeles, CA, and Pueblo, CO.

Increasing Number of Chemicals ↑

Contaminant Levels	Child Rank	Dairy Rank	Fruit Rank	Mixed Rank	Protein Rank	Grain Rank	Vegetable Rank
Very High							
Very High							
High							
High							
Moderately High							
Moderate							
Moderate	7	7	7				
Moderately Low				8	8		
Low							
Low						10	
Very Low							11
Very Low							

FIGURE 6-5 Market basket 97-1 for Phoenix, AZ, Fresno, CA, and San Antonio, TX.

Contaminant Levels	Vegetable Rank	Child Rank	Dairy Rank	Mixed Rank	Protein Rank	Grain Rank	Fruit Rank
Very High	1						
Very High							
High		3					
High							
Moderately High							
Moderate			6	6			
Moderate					7		
Moderately Low						8	
Low							
Low							10
Very Low							
Very Low							

FIGURE 6-6 Market basket 98-4 for Dayton, OH, Peoria, IL, and Sioux Falls, SD.

Contaminant Levels	Fruit Rank	Vegetable Rank	Child Rank	Grain Rank	Protein Rank	Mixed Rank	Dairy Rank
Very High	1						
Very High							
High							
High		4					
Moderately High			5	5			
Moderate					6		
Moderate						7	
Moderately Low							8
Low							
Low							
Very Low							
Very Low							

FIGURE 6-7 Market basket 96-1 for Bakersfield, CA, Ogden, UT, and Tacoma, WA.

Contaminant Levels	Vegetable Rank	Dairy Rank	Fruit Rank	Mixed Rank	Protein Rank	Grain Rank	Child Rank
Very High							
Very High							
High	3	3					
High			4				
Moderately High				5	5		
Moderate						6	
Moderate							
Moderately Low							8
Low							
Low							
Very Low							
Very Low							

FIGURE 6-8 Market basket 99-1 for Denver, CO, Portland, OR, and Reno, NV.

Contaminant Levels	Dairy Rank	Protein Rank	Mixed Rank	Grain Rank	Fruit Rank	Vegetable Rank	Child Rank
Very High	1	1					
Very High							
High			3				
High				4			
Moderately High							
Moderate							
Moderate							
Moderately Low							
Low					9	9	9
Low							
Very Low							
Very Low							

FIGURE 6-9 Market basket 00-1 for Grand Rapids, MI, Milwaukee, WI, and Springfield, IL.

Contaminant Levels	Fruit Rank	Grain Rank	Protein Rank	Child Rank	Dairy Rank	Mixed Rank	Vegetable Rank
Very High							
Very High	2						
High		3	3				
High				4	4	4	
Moderately High							
Moderate							6
Moderate							
Moderately Low							
Low							
Low							
Very Low							
Very Low							

FIGURE 6-10 Market basket 99-3 for Binghamton, NY, Harrisburg, PA, and Worcester, MA.

Contaminant Levels	Child Rank	Dairy Rank	Grain Rank	Mixed Rank	Fruit Rank	Protein Rank	Vegetable Rank
Very High	1						
Very High		2	2	2			
High					3		
High						4	
Moderately High							5
Moderate							
Moderate							
Moderately Low							
Low							
Low							
Very Low							
Very Low							

FIGURE 6-11 Market basket 00-3 for Birmingham, AL, Columbus, GA, and Jackson, MS.

Contaminant Levels	Grain Rank	Mixed Rank	Protein Rank	Vegetable Rank	Dairy Rank	Child Rank	Fruit Rank
Very High	1	1					
Very High			2	2			
High							
High							
Moderately High					5		
Moderate						6	6
Moderate							
Moderately Low							
Low							
Low							
Very Low							
Very Low							

FIGURE 6-12 Market basket 01-3 for Oklahoma City, OK, Orlando, FL, and Washington, D.C.

TABLE 6-2 Food Quality Ranking

Market Baskets	Rank Total Number of Chemicals	Rank Chemical Diversity	Totals		Combined Rank
01-3	12	12	24	Oklahoma City — Orlando — Washington	12
00-3	11	11	22	Birmingham — Columbus — Jackson	11
99-3	10	10	20	Binghamton — Harrisburg — Worcester	10
00-1	9	8	17	Grand Rapids — Milwaukee — Springfield	9
99-1	8	6	14	Denver — Portland — Reno	8
96-1	7	5	12	Bakersfield — Ogden — Tacoma	7
98-3	2	9	11	Long Beach — New York — Providence	6
98-1	4	7	11	Boise — Los Angeles — Pueblo	5
98-4	6	4	10	Dayton — Peoria — Sioux Falls	4
97-1	5	2	7	Phoenix — Fresno — San Antonio	3
96-2	3	3	6	Charleston — Lafayette — Palm Beach	2
98-2	1	1	2	Baton Rouge — Savannah — Tulsa	1

chemical diversity. In other words, the chemical diversity within a market basket can be expressed by determining the number of different pesticides and industrial chemicals that actually contaminate foods within a given market basket. The number of these chemicals is also given in Table 2-35. Given both the total number of chemicals and the chemical diversity, each market basket was ranked to determine the best food quality (i.e., a number 1 ranking) versus the worst food quality of the 12 market baskets evaluated. This combined ranking is illustrated in Table 6-2. *Clearly, a market basket with a high total number of contaminants and chemical diversity will expose the consumer in that market basket region to a more complex chemical mixture than will a market basket with fewer chemical contaminants with less chemical diversity.*

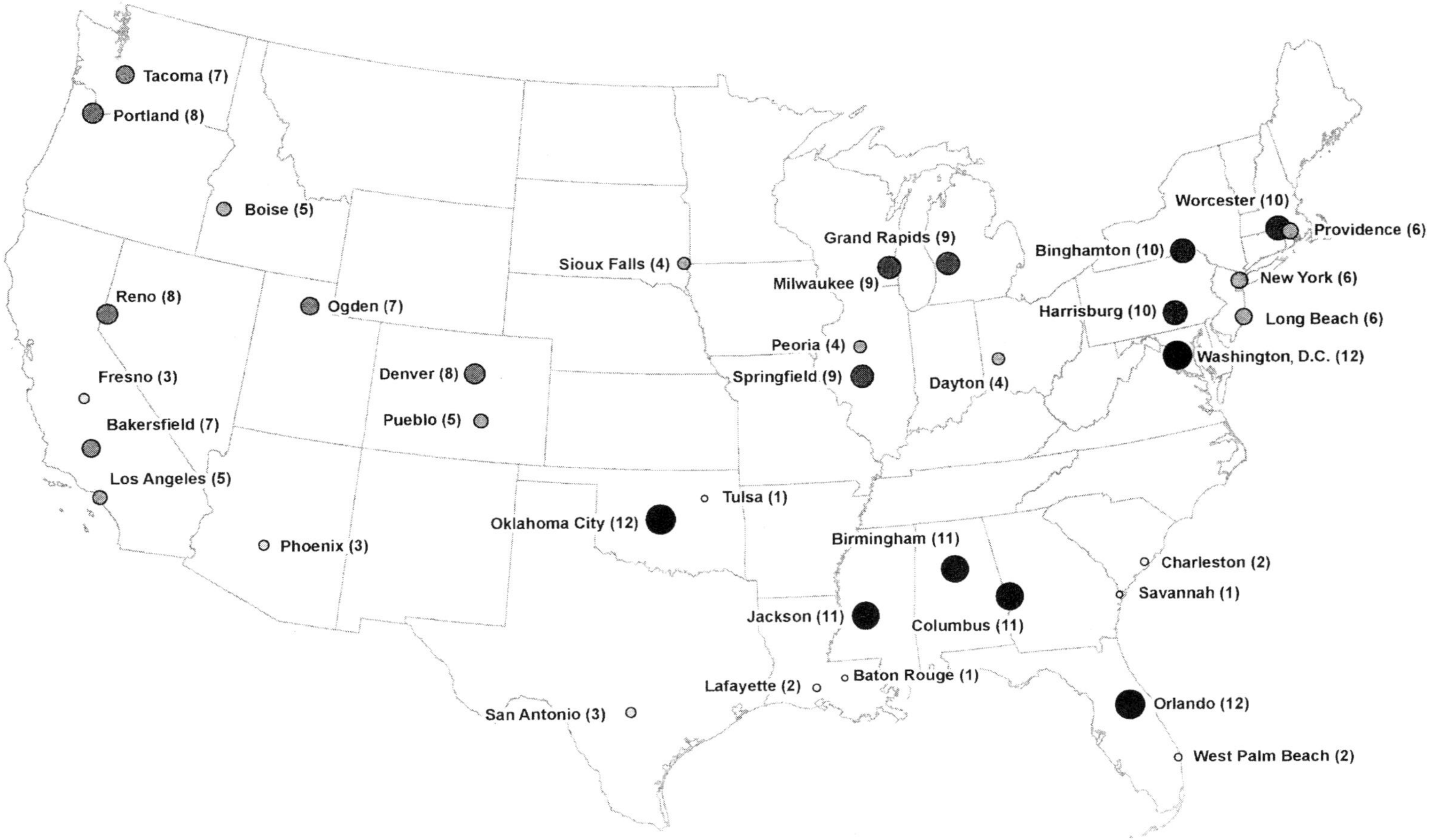

FIGURE 6-13 Food quality ranking.

TABLE 6-3 Reported Chemicals in Beverages

	Number of Market Baskets with Detected Chemicals															
Beverage	**BZ**	**Cfo**	**Car**	**Dan**	**Dat**	**Ipr**	**Ipm**	**Tol**	**BDM**	**CTC**	**CBZ**	**Ome**	**Pcy**	**Pyl**	**Tpp**	**Xyl**	**Totals**
Powdered fruit drink															1		1
Cola	1	6							1								8
Low-calorie cola	9	5							1					1			16
Carbonated fruit-flavored beverage	5	6	1					2	1	1	1					1	18
White wine			10	4	4	10	4					1	1				36

A map illustrating this ranking is given in Figure 6-13. The previous sections of this text have only focused on regional food products and not beverages. The FDA Total Diet Studies also evaluated a limited number of beverages, as well as tap water, around the nation. These data are presented in the following section.

6.1 BEVERAGES AND DRINKING WATER

Based on all the market basket data, the following beverages reported no detected chemicals: coffee, tea, canned fruit drinks, martinis, beer, and whiskey. A summary of the market basket data on beverages that did contain synthetic chemicals is given in Table 6-3. Of the beverages that are contaminated, the dominant chemicals are petroleum hydrocarbons and disinfection by-products. The exception to this trend was white wine, which was contaminated by a number of pesticides. The data in Table 6-3 show that white wine can be expected to contain at least two pesticides (carbaryl and iprodione).

For the most part, drinking water samples were not contaminated. The samples taken in the Total Diet Study, however, do not represent the broad range of water resources available to consumers in the cities sampled. This deficiency, however, is addressed in Chapter 7 on water quality where the contamination of drinking water for each market basket city is critically evaluated.

7 SYNTHETIC CHEMICAL CONTAMINANTS IN DRINKING WATER

Is drinking water in the United States safe? This simple question does not have a simple answer. However, the behavior of the American public is a good indicator of their perception of our nation's drinking water quality. For example, an article published by the American Water Works Association characterized customer expectations and behavior:[1]

1. Eighty-six percent of Americans are concerned about their tap water.
2. Forty-eight percent of the American public either treat/filter their tap water or use bottled water.
3. In 1993, the per capita consumption of bottled water in the United States was a little over 10 gallons. By 2003, it had increased to about 22 gallons.
4. Safety was the primary motivator for those who drank home-filtered water, while those who drank bottled water cited taste, safety, or healthiness as motivation for drinking this product.

Clearly, the American public is concerned for their safety, but is this concern justified?

[1] Means, Spina, and R. Patrick. 2005. The primary trends and their implications for water utilities. Journal of the American Water Works Association 97 (7).

7.1 WHAT IS SAFE DRINKING WATER?

Water at the earth's surface comes into contact with biological organisms, soil, rock, and the works of humans. As a result, water can be polluted by hazardous microorganisms, dissolved toxic metals, or radioactive elements from rock and pick up poisonous synthetic chemicals from human activities. Of all these hazards, a microbial infection can result in illness or death within days of contact. This means that safe drinking water must be free of hazardous microorganisms or treated (see Appendix 7-1) to ensure that microbial levels pose no risk to human health. Water that contains no biological threat, however, can still be polluted with chemical toxins.

When dealing with chemical toxins, the dose makes the poison. In other words, a high concentration of a specific chemical can result in an immediate illness or death, while a long-term chemical exposure to low-level concentrations may also cause an illness, death, or have no discernable effect. Because dose is important, both federal and state governments establish maximum contaminant levels (MCL) for drinking water (Table 7-1). Drinking water that does not exceed these dose levels is not anticipated to have an adverse health effect.

Thus, according to the federal government, drinking water that meets all of the microbial treatment criteria given in Appendix 7-1 and does not exceed any of the chemical concentrations listed in Table 7-1 is safe to drink. The vast majority of the tap water in the United States meets these criteria. Yet the American public is still concerned with the safety of this product. There just seems to be an uneasiness or basic mistrust that drinking water is safe. Although consumers may not be able to define their uneasiness, there is reason why we should all be concerned.

This is because the vast majority of Americans consumes a mixture of synthetic chemicals in their drinking water. This mixture of chemicals is made up of compounds with concentrations that are usually below established MCLs (i.e., "safe" levels) and with chemicals that do not have defined MCLs (i.e., unregulated chemicals[2]). Consuming these mixtures should give Americans pause for concern because, as discussed in Chapter 1, scientists do not know the impact on human health from consuming chemical mixtures.[3] Because the true impact of consuming mixtures of synthetic chemicals is unknown, individuals should have some understanding of the chemical mixtures in drinking water so they can choose safer water products.

---

[2] A chemical is unregulated because it is known to be nontoxic or the chemical's toxic characteristics or impact on individual human health is unknown.

[3] Approximately 70,000 synthetic organic chemicals are produced or used in the United States. Yet the federal government has established MCLs for only 63 of these synthetic chemicals. For a complete evaluation of this subject, refer to "The Environmental Science of Drinking Water," by Sullivan, Agardy, and Clark (2005).

TABLE 7-1 Federal Drinking Water Standards

National Primary Drinking Water Standards for Chemicals. These are the legally enforceable standards that apply to public water systems (as of June, 2001).

Contaminant	Maximum Contaminant Level (MCL) (Micrograms per Liter)
Inorganic Compounds	
Antimony	6.0
Arsenic	50.0
Barium	2000.0
Beryllium	4.0
Bromate	10.0
Cadmium	5.0
Chlorine gas (as Cl_2)	4000.0
Chlorine dioxide (as ClO_2)	800.0
Chlorite	1000.0
Chromium (total)	100.0
Copper	1300.0
Cyanide	200.0
Fluoride	4000.0
Lead	15.0
Mercury	2.0
Nitrate	10000.0
Nitrite	1000.0
Selenium	50.0
Thallium	2.0
Organic Compounds	
Acrylamide	1000.0
Alachlor	2.0
Atrazine	3.0
Benzene	5.0
Benzo(a)pyrene	0.2
Carbofuran	40.0
Carbon tetrachloride	5.0
Chloramines (as Cl_2)	4000.0
Chlordane	2.0
Chlorobenzene	100.0
Dichlorophenoxyacetic acid (2,4-D)	70.0
Dalapon	200.0
DBCP	0.2
o-Dichlorobenzene	600.0
p-Dichlorobenzene	75.0
1,2-Dichloroethane	5.0
1,1-Dichloroethylene	7.0
cis-1,2-Dichloroethylene	70.0
Trans-1,2-dichloroethylene	100.0
Dichloromethane	5.0
1,2-Dichloropropane	5.0
Di(2-ethylhexyl)adipate	400.0
Di(2-ethylhexyl)phthalate	6.0
Dinoseb	7.0
2,3,7,8-dioxin(TCDD)	0.00003
Diaquat	20.0

TABLE 7-1 *Continued*

Contaminant	Maximum Contaminant Level (MCL) (Micrograms per Liter)
Organic Compounds	
Endothall	100.0
Endrin	2.0
Epichlorohydrin	20,000.0
Ethylbenzene	700.0
Ethylene dibromide	0.05
Glyphosate	700.0
Haloacetic acids (HAA5)*	60.0
Heptachlor	0.4
Heptachlor epoxide	0.2
Hexachlorobenzene	1.0
Hexachlorocyclopentadiene	50.0
Lindane	0.2
Methoxychlor	40.0
Oxamyl(Vydate)	200.0
Polychlorinated biphenyls	0.5
Pentachlorophenol	1.0
Picloram	500.0
Simazine	4.0
Styrene	100.0
Tetrachloroethylene	5.0
Toluene	1000.0
Total Trihalomethanes (TTHM)**	100.0
Toxaphene	0.003
2,4,5-TP (Silvex)	50.0
1,2,4-Trichlorobenzene	70.0
1,1,1-Trichloroethane	200.0
1,1,2-Trichloroethane	5.0
Trichloroethylene	5.0
Vinyl chloride	2.0
Xylenes (total)	10,000.0

National Secondary Drinking Water Regulations

These standards are nonenforceable guidelines regulating contaminants that may cause cosmetic effects or aesthetic effects in drinking water. The EPA recommends secondary standards to water systems but does not require systems to comply. However, individual states may choose to adopt them as enforceable standards.

Contaminant	Secondary Standard (Micrograms per Liter)
Aluminum	0.05
Chloride	250
Copper	1.0
Fluoride	2.0
Iron	0.3
Manganese	0.05
Silver	0.1
Sulfate	250
Total dissolved solids	500
Zinc	5.0

* The HAA5 compounds are dibromoacetic acid, dichloroacetic acid, monobromacetic acid, monochloroacetic acid, and trichloroacetic acid.

** The TTHM compounds are bromodichloromethane, bromoform, chloroform, and dibromochloromethane.

In reality, a safer drinking water product would meet all federal or state MCLs for biological, radioactive, inorganic, and organic constituents but also contain no more than one detected synthetic organic chemical (i.e., no synthetic chemical mixtures). For some drinking water resources in the United States, this is in fact a reality. Whether or not this is your reality largely depends on where you live and the degree to which your source of drinking water is protected from synthetic chemical pollution. Given these major variables, the ability to make informed decisions about the water you choose to drink requires three basic types of information.

1. If the provider of the water product (e.g., a water utility or bottled water company) has identified and eliminated all potential sources of synthetic chemicals in the watershed[4] and has protected its equipment from leaching synthetic chemicals into its product, the consumer can be confident that their water will most likely not contain synthetic chemicals. The process of eliminating or limiting all chemicals (i.e., both organic and inorganic) in drinking water is called source water protection. As expected, the degree of protection offered by a provider varies greatly depending on the watershed and groundwater resource and the ability to control pollutant sources.
2. If the provider uses chlorine or bromine (i.e., halogens), your drinking water has an extremely high probability of being contaminated with halogenated organic compounds that are the by-product of the disinfection process.
3. If the provider of the water product analyzes for all anticipated synthetic chemicals (based on its watershed assessment) that might occur in its product and publishes this information on a regular basis, the consumer can also be confident that the product will be most likely not contain synthetic chemicals. This process is called comprehensive water quality monitoring.

Unfortunately, the providers of drinking water furnish only a fraction of this important information. Furthermore, source water protection is a process that is just beginning to be investigated and applied to drinking water resources in the United States, and water quality monitoring is far from comprehensive. So what is the consumer to do? The first step is to understand how drinking water resources are polluted by synthetic chemicals so that consumers can generally

[4] A watershed is usually defined as a geographic landform that accumulates and transports surface water resources within a valley basin. This definition can be applied to groundwater resources as well (i.e., a defined aquifer as opposed to a valley).

assess the potential threats to their primary source of drinking water (e.g., from a water utility or private well). Second, consumers who are serviced by a water utility should review the utility's annual water quality report. With this information, each consumer can determine if he or she feels in-home treatment or bottled water to be a safer alternative.

With this goal in mind, the objectives of this chapter are to (1) provide an explanation and examples of how drinking water sources are polluted, (2) explain the importance and shortcomings of chemical monitoring, (3) discuss the synthetic chemical monitoring data collected for the cities evaluated in Chapter 2 to illustrate how regional environmental characteristics influence drinking water quality (to help consumers understand their own drinking water pollution issues), (4) determine which of the 36 cities have the safest drinking water, and (5) discuss what options the consumer has for choosing safer drinking water.

7.3 SOURCES OF DRINKING WATER

Supplies of drinking water are obtained from bodies of surface water such as creeks, rivers, lakes, and reservoirs and groundwater aquifers. It is important to remember that surface water percolates into groundwater aquifers and that groundwater moving in aquifers can emerge naturally at the earth's surface as a spring or discharge unseen into a river, lake, or ocean.

The diversion of surface water resources by individuals and communities can be a very convenient source of drinking water. This assumes, however, that the surface water occurs throughout the year in sufficient quantities to serve the users' needs and that they also have a legal right to the amount of water they are diverting. This is usually not a problem in the humid climatic regions of the world. In the United States, these regions generally lie east of the Mississippi River and in the Pacific Northwest. When the demand for water resources exceeds surface water supplies, groundwater supplies can usually be obtained by installing groundwater supply wells. In today's industrialized societies, the users must usually have a permit to install a groundwater extraction well. Depending on an aquifer's capability to produce a usable amount of water, a user may need multiple wells. Therefore, any given community may utilize surface water, groundwater, or a mixture of both.

7.4 SYNTHETIC CHEMICAL POLLUTION OF DRINKING WATER

The synthetic chemicals that are known to pollute water resources are pesticides, organic industrial chemicals and petroleum products, pharmaceuticals, and disinfection by-products. These chemicals can occur in both urban and rural environments and are released into both surface water and groundwater in the following ways.

7.4.1 Pesticides

Because pesticides are applied at and above the soil surface, rainfall can transport pesticide residues into ponds, lakes, creeks, and rivers. Pesticides are also injected into the root zone. Thus, pesticides at the soil surface can be leached downward into groundwater resources. Once in groundwater, pesticides can be sorbed onto the porous media, and, as a result, groundwater can be purified as it travels through underground aquifers. In general, this means that when dealing with diffuse pollution from pesticide applications (as opposed to pollution from a leaking underground tank), surface water has a higher probability of being polluted than groundwater. This is true even for remote wilderness areas where airborne pesticides have polluted surface water.

Although pesticide use in urban environments is ubiquitous, its magnitude and concentrations in agricultural areas are much greater. Therefore, both surface water and groundwater have a much greater probability of being polluted in rural agricultural areas. This means that surface waters taken for drinking water supplies in rural agricultural areas have the greatest probability of being polluted with pesticides than other water resources (i.e., groundwater and surface water in urban areas).

7.4.2 Organic Industrial Chemicals and Petroleum Products

These chemicals are used in virtually every home and business in America. This means that the only area that would not be expected to have potential pollution from organic industrial chemicals and petroleum products would be a designated wilderness. In general, this means that any of a vast number of synthetic chemicals that have been leaked, dripped, or spilled on soil and impervious surfaces can be picked up in rainfall runoff and transported to bodies of surface water or percolate into the groundwater. The major threat posed by organic industrial chemicals and petroleum products is usually associated with their release from storage facilities and their disposal.

For example, organic chemicals or petroleum fuels that leak from aboveground or underground storage tanks at industrial facilities and gas stations can pollute vast areas of groundwater. Industrial sites, large commercial parks (e.g., metal finishing and auto repair), and dry cleaning facilities have been shown to dispose of chemicals onto the bare ground or into lagoons, ponds, and sewers. In all of these cases, groundwater resources can be polluted. Solid waste and recycling facilities can also be a source of groundwater pollution. In other words, urban and industrialized areas tend to have some degree of groundwater contamination from industrial chemicals and petroleum products.

7.4.3 Pharmaceuticals

The vast majority of over-the-counter and prescription drugs are not metabolized and pass through the body into urine and feces that discharge to sewage treatment plants or a rural septic system. In some cases, unused pharmaceuticals are dumped down a drain or into a toilet. These synthetic chemicals are not completely destroyed in municipal sewage treatment plants and remain in the discharged wastewater. When this wastewater is released into an adjacent river or lake, it can become the drinking water source of another community. It should also be remembered that sewage discharges also contain industrial chemicals.

Clearly, wastewater containing pharmaceuticals that discharge to a septic system or leaking cesspool will pollute groundwater. This is a particular threat to adjacent home owners who obtain the same groundwater for their drinking water. In some communities, treated wastewater that still contains dissolved pharmaceuticals or other unregulated industrial chemicals is injected into groundwater resources. When wastewater is injected into groundwater aquifers, it is assumed that the porous media of the aquifer will "purify" the water so that trace chemicals will not pollute drinking water sources.

In summary, surface water resources that are used for drinking water should be considered as being polluted until demonstrated otherwise. Groundwater resources should contain less pollution than surface water if there are no sources of pollution into the groundwater (e.g., leaking sewer lines above or in groundwater; chemical spills and leaks from industrial sites or commercial facilities like dry cleaners and fueling facilities, leaching of pesticides through soil in agricultural areas, leakage from land disposal facilities, pollution from septic systems and cesspools, pollution from mining and metallurgy operations), there is no direct recharge of treated wastewater into the groundwater, and there is no polluted surface water that is recharging back into groundwater.

Because biological pollution of our water resources is likely, most communities in the industrial nations provide basic water treatment for drinking water. For the most part, the collection, treatment, and distribution of drinking water are provided by local governmental agencies (city or regional governments or private utilities). However, the disinfection process creates its own synthetic chemical pollution.

7.4.4 Disinfection By-Products

Water in contact with the earth's surface will dissolve natural organic matter. When water containing dissolved organic matter is disinfected with chlorine or bromine, some of these organic compounds are chemically altered. The common disinfection by-products are given in Table 7-1 (e.g., trihalomethanes and haloacetic acids). Surface water tends to have higher concentrations of dissolved organic matter than groundwater. Therefore, surface water tends to have high concentrations of disinfection by-products. Virtually all utilities that provide drinking water that is disinfected with chlorine or bromine are polluted with disinfection by-products.

Water products that do not contain disinfection by-products are bottled waters (which are not required to be disinfected) or waters that are disinfected with ultraviolet light or ozone. The vast majority of drinking water provided by a community or private water utilities in the United States should be assumed to polluted with disinfection by-products.

This is a great concern given the following statements from the World Health Organization:[5]

> Disinfected drinking water is a very complex mixture of chemicals, most of which have not been identified. Studies on individual DBPs [disinfection by-products] may not represent the risk posed by the mixture. Studies of simple combinations of chemicals provided positive results, but only at concentrations so much greater than those that occur in drinking water as to be irrelevant. Studies utilizing complex mixtures of chemicals as they could be isolated from water or produced by chlorinating high concentrations of humic or fulvic acids produce little convincing evidence of adverse effect. A variety of methodological issues prevent our being too comfortable with that conclusion.

In other words, the risk to human health is unknown.

7.4.5 Consumer Unknowns

Numerous cities or communities in the United States obtain their water from protected watersheds (i.e., minimal human impact), for example, Seattle and San Francisco. Cities and communities that have a protected watershed tend to advertize these characteristics in their federally mandated annual water quality reports. It is highly recommended that consumers obtain a current copy of their water provider's annual water quality report and determine the degree to which their source water is protected (both surface water and groundwater resources). When a well-thought-out and implemented source water protection plan is in

[5] United Nations environment program, Environmental Health Criteria 216 on disinfectants and disinfectant by-products (2000).

place, the only contaminants that are anticipated to be in drinking water are disinfection by-products.

Because most water utilities do not have (1) a truly protected source water (e.g., water originates in a wilderness area with no human impacts), (2) a fully implemented comprehensive source water protection plan, or (3) monitoring programs that look for all potential contaminants (e.g., pharmaceuticals), the consumer is really left in the dark as to degree of synthetic chemical pollution that may actually be impacting the drinking water.

7.5 MONITORING FOR SYNTHETIC CHEMICALS

Water utilities in the United States consistently analyze their product for the primary[6] and secondary drinking water chemicals listed in Table 7-1. They occasionally monitor for a handful of chemicals that may eventually find their way onto the primary drinking water list (i.e., unregulated chemicals). The results of this monitoring are reported in the water utility's annual water quality report. However, the vast majority of industrial chemicals and pesticides as well as all of the pharmaceuticals is not looked for in drinking water. This means that consumers have no real assurance that their drinking water does not contain complex chemical mixtures. With this major shortcoming in mind, chemical monitoring data have been collected for the 36 cities that were evaluated in Chapter 2 and are summarized in the following sections.

In reviewing the monitoring data, it is important to note that relative to the disinfection by-products that occur in treated drinking water, two general parameters reflect the sum of several individual contaminants:

- The concentration of the total trihalomethanes (TTHM) are the sum of the concentrations of bromodichloromethane, bromoform, chloroform, and dibromochloromethane.
- The concentration of the haloacetic acids (HAA) is the sum of the concentrations of dibromoacetic acid, dichloroacetic acid, monobromacetic acid, monochloroacetic acid, and trichloroacetic acid.

This means that for some water quality reports, the individual disinfection by-products may be reported along with the total parameters (TTHMs and HAAs). Some utilities also report the detection of the following infrequently monitored disinfection by-products: haloketones, halonitriles, and total halogenated organic compounds.

[6] These chemicals are usually referred to as "regulated" compounds.

7.6 COLLECTION AND INTERPRETATION OF THE WATER DATA

The drinking water data summarized for each city was obtained by (1) a Freedom of Information Act request to each state for all available data (2000 to 2005) on synthetic organic chemicals detected in community water systems or (2) from the most recent annual water quality reports available from the individual cities. Because each state administers its own collection program under the Clean Water Act, the data provided by each state varied widely. Thus, some city data are much more extensive than others.[7] For example, since it is against Colorado's policy to release its data, the information for both Denver and Pueblo can only come from annual water quality reports, which means that the data collected for each city are not directly comparable. Therefore, the objective of the data analysis is to provide a snapshot of the synthetic chemical contaminant mixtures found in community water systems that have reported detected synthetic chemicals in a region's drinking water. Using this information, the water quality for each city is given in the following sections. The vast majority of the reported concentrations (part per billion, or ppb) is averages, but in some cases the value reported is a single sample.

7.7 BAKERSFIELD, CALIFORNIA

The city of Bakersfield is located in the southern portion of the San Joaquin Valley of California and has a population of approximately 406,000. The region surrounding the city is dominated by irrigated agriculture, while industry in the area is dominated by food processing and oil production. Drinking water resources for the city come from both groundwater and surface water.

For the city of Bakersfield, the state of California provided data on 85 active community water systems that derive their water from groundwater resources. Both groundwater and surface water systems serve approximately the same number of customers. The data for these community water systems that had detected synthetic chemicals are given in Appendix 7-2. A total of 26 different synthetic compounds were found in the groundwater (see Table 7-2).

Given that irrigated agricultural in this region of the United States requires soils to be constantly leached in order to minimize salt accumulations, it is understandable that groundwater contains pesticides (e.g., 2,4-D, aldicarb, aldrin). This is particularly a problem with pesticides injected into a below-the-root zone, which would include dibromochloropropane (DBCP), ethylene dibromide, and 1,2,3-trichloropropane (e.g., a contaminant in DBCP). These

[7] In some case sampling, frequency is so great that many chemical compounds were tested for hundreds of times. Thus, when more than 100 samples were tested for the same chemical, only the average will be reported along with the sample frequency.

TABLE 7-2 Bakersfield Water Data

Chemical Compound	Groundwater Average Concentration (ppb)	Groundwater Raw Water Average Concentration (ppb)	Surface Water Average Concentration (ppb)	Type of Chemical Compound
2,4-D	65			Pesticide
Acenaphthene	99.1			Petroleum
Acenaphthylene	98.8			Petroleum
Acetone	2			Industrial chemical
Aldicarb sulfone	86			Pesticide
Aldrin	100.6			Pesticide
Atrazine			1.4	Pesticide
Benzene	0.67			Petroleum
Bromobenzene	83			Industrial chemical
Bromomethane			11.1	Industrial chemical*
Chloromethane	0.6		0.6	Industrial chemical*
Dichloromethane	0.81			Industrial chemical
Dichlorobenzene	1.1			Industrial chemical
Dibromochloropropane (DBCP)	0.08	1.03	0.02	Pesticide
Diuron			3	Pesticide
Ethylbenzene	0.83			Petroleum
Ethylene dibromide	0.04	0.08		Pesticide
Tetrachloroethylene	3.9			Industrial chemical
Trichloroethylene	0.6			Industrial chemical
Trichloropropane	0.09		0.01	Pesticide
Xylenes	13.1			Industrial chemical
Total reported individual chemicals — 30				
Haloacetic acids	3.2		26	Disinfection by-product
Dibromoacetic acid	1.9		6.28	Disinfection by-product
Dichloroacetic acid	1.4		10.5	Disinfection by-product
Monobromoacetic acid	1.1		2.4	Disinfection by-product
Monochloroacetic acid			5	Disinfection by-product
Trichloroacetic acid	1		12.7	Disinfection by-product
Total trihalomethanes	5.4		47.3	Disinfection by-product
Bromodichlormethane	1.91		7.17	Disinfection by-product
Bromoform	2.35		6.92	Disinfection by-product
Chloroform	1.49		37.9	Disinfection by-product
Dibromochloromethane	2.14		7.35	Disinfection by-product

* Possible disinfection by-product.

data also show that even treating the groundwater to remove dibromochloropropane and ethylene dibromide does not completely remove these pesticides from drinking water. In addition, water treatment did not remove a range of petroleum compounds and industrial chemicals. The largest number of detected synthetic chemicals is the disinfection by-products.

The water quality data for surface water sources of drinking water are given in Appendix 7-3. A total of 15 different synthetic compounds were found in the surface water resources (see Table 7-2). The surface water quality data show that the major contaminants are disinfection by-products with several pesticides. When comparing the groundwater with the surface water contamination, it is clear that the concentration of disinfection products is significantly higher in the surface water. This concentration difference was expected since surface water usually has higher concentrations of dissolved organic matter. When combining both groundwater and surface water data, drinking water resources in Bakersfield have a total of 29 different synthetic chemicals as well as the combined disinfection by-product constituents measured as total trihalomethanes and haloacetic acids.

7.8 BATON ROUGE, LOUISIANA

The city of Baton Rouge is located along the Mississippi River and has a population of approximately 230,000 with the greater Baton Rouge area having a population of about 600,000. The city is a mixture of industrial, commercial, and residential areas surrounded by agricultural land use. The state of Louisiana provided no data for the city of Baton Rouge.

As a result, the available 2004 annual water quality monitoring reports for the three Baton Rouge community water systems (Baton Rouge, Parish, and Ascension) were sumerized. The source waters come from groundwater resources. The water quality summary for these systems is given in Table 7-3. All of the detected contaminants were disinfection by-products.

TABLE 7-3 Baton Rouge Water Data

Chemical Compound	Groundwater Average Concentration (ppb)	Type of Chemical Compound
Haloacetic acids*	4	Disinfection by-product
Total trihalomethanes*	3.23	Disinfection by-product

* Concentrations of individual compounds not reported.

7.9 BINGHAMTON, NEW YORK

The city of Binghamton is located in a wooded river valley at the confluence of the Chenango and Susquehanna rivers and has a population of approximately 47,000. The region surrounding the city is dominated by recreational and agriculture land uses and some light industry. Drinking water resources for the city come from both groundwater and surface water.

For the city of Binghamton, the state of New York provided data on six groundwater systems and one surface water system. The surface water system serves the vast majority of the city's customers. All of these systems reported at least one contaminant during the record of sampling. The synthetic chemical data for the groundwater community water systems are given in Appendix 7-4, while data for the surface water system are given in Appendix 7-5.

No annual water quality reports were available for the city. Using the state data alone, six different synthetic compounds were found in all of the community water systems (see Table 7-4). The detected chemicals in the city's drinking water are fairly evenly distributed between petroleum compounds (toluene), chlorinated solvents (trichloroethylene), and the remaining disinfection by-products.

TABLE 7-4 Binghamton Water Data

Chemical Compound	Groundwater Average Concentration (ppb)	Surface Water Average Concentration (ppb)	Type of Chemical Compound
Bromomethane	2.46		Industrial chemical*
Dichloroethane		0.6	Industrial chemical*
Dichloromethane	1.55		Industrial chemical*
Methyl tert butyl ether (MTBE)	1	1.4	Petroleum
Toluene	1.65		Industrial chemical
Trichloroethylene		0.7	Industrial chemical
Total reported individual chemicals — 6			
No haloacetic acid or total trihalomethane data avilable			

* Possible disinfection by-product.

7.10 BIRMINGHAM, ALABAMA

The city of Birmingham is located in a valley of Appalachian foothills and is drained by several small creeks that flow into the Black Warrior River. Birmingham's population is approximately 240,000 with the surrounding metropolitan area exceeding 1 million. The urban sprawl contains a mixture of light industry and commercial centers and is bisected by a major railroad corridor. There are also several water supply reservoirs in the region.

For the city of Birmingham, the state of Alabama provided data on one groundwater systems and one surface water system. The surface water system serves the majority of the city. The synthetic chemical data for the groundwater community water systems are given in Appendix 7-6, while the data for the surface water system are given in Appendix 7-7. Seven different disinfection by-products and one industrial chemical were found in all of the community water systems (see Table 7-5). As expected, the concentration of disinfection by-products was, in general, greater in surface water sources (which serve most of the population).

TABLE 7-5 Birmingham Water Data

Chemical Compound	Surface Water Average Concentration (ppb)	Groundwater Average Concentration (ppb)	Type of Chemical Compound
Chlorodibromomethane	1.05		Industrial chemical*
Dichloromethane	1.59		Industrial chemical*
Dichlorodifluoromethane	0.89		Industrial chemical*
Total reported individual chemicals — 3			
Total haloacetic acids**	24.1	0.7	Disinfection by-product
Total trihalomethanes	34.1	3.45	Disinfection by-product
Chloroform	19.5	1.79	Disinfection by-product
Bromodichloromethane	3.93	1.45	Disinfection by-product
Bromoform		1.18	Disinfection by-product

* Possible disinfection by-product.

** Concentrations of individual compounds not reported.

7.11 BOISE, IDAHO

The city of Boise is located against the Sawtooth Mountain Range in a broad valley dissected by the Boise River and has a population of approximately 185,000. The region is surrounded by ranching/agricultural and recreational areas with some light industry and commercial centers.

For the city of Boise, the state of Idaho provided data on 10 groundwater systems and one surface water system. The surface water system serves the majority of the city. The synthetic chemical data for the groundwater community water systems are given in Appendix 7-8, while data for the surface water system are given in Appendix 7-9. Most of the synthetic chemicals found in both surface water and groundwater community water systems (see Table 7-6) were industrial chemicals. Given the type of industrial chemicals found in both resources, the groundwater was expected to contain more and higher concentrations of these chemicals.

Surprisingly, the state data showed the lack of disinfection by-products. Thus, the 2004 and 2005 Annual Water Quality Reports for United Water of Idaho were reviewed for additional compounds not reported by the state of Idaho. These data are also summarized in Table 7-5. This example demonstrates the variability in reporting and administering state data.

TABLE 7-6 Boise Water Data

Chemical Compound	Surface Water Average Concentration (ppb)	Groundwater Average Concentration (ppb)	Type of Chemical Compound
Atrazine	0.54		Pesticide
Carbon tetrachloride		1.6	Industrial chemical
Dichloroethylene		7.49	Industrial chemical
Cis-dichloroethylene		3.48	Industrial chemical
Di(2-ethylhexyl)-phthalate	9.8	1.71	Industrial chemical
Hexachlorocyclopentadiene	0.09		Industrial chemical
Methyl tert butyl ether (MTBE)	0.5		Petroleum
Tetrachloroethlyene	3.12	1.62	Industrial chemical
Trichloroethane		14.9	Industrial chemical
Trichloroethylene	0.5	27.3	Industrial chemical
Xylenes		3.45	Petroleum
Total reported individual chemicals — 11			
Haloacetic acids*	6.7		Disinfection by-product
Total trihalomethanes	12.2		Disinfection by-product
Bromodichloromethane	6.9		Disinfection by-product
Bromoform	4.1		Disinfection by-product
Chloroform	16.9		Disinfection by-product
Dibromochloromethane	3.8		Disinfection by-product

* Concentrations of individual compounds not reported.

7.12 CHARLESTON, SOUTH CAROLINA

Charleston is a harbor city located in the coastal plain. This developed urban area of business complexes and industry is surrounded by agricultural and forested regions. The drinking water supply comes from the Edisto River and the Bushy Park Reservoir and serves approximately 350,000 people.

The state of South Carolina provided no water quality data for this city. Therefore, the water quality data were taken from the available annual reports (2004–2001). These averaged data are given in Table 7-7. The regulated chemicals of concern were disinfection by-products with several unregulated chemicals.

7.13 COLUMBUS, GEORGIA

The Columbus metropolitan area is surrounded by agricultural and recreational areas (numerous lakes) and has a population of approximately 187,000. For the city of Columbus, the state of Georgia provided data on one surface water system. Water from this system is from Lake Oliver, which is fed by the

TABLE 7-7 Charleston Water Data

Chemical Compound	Surface Water Average Concentration (ppb)	Type of Chemical Compound
Bromoacetic acid	6.4	Disinfection by-product
Chloropicrin	0.6	Disinfection by-product
Haloacetonitriles	3.7	Disinfection by-product
Haloketones	2.1	Disinfection by-product
Total reported individual chemicals — 4		
Haloacetic acids*	35.6	Disinfection by-product
Total organic halides	203	Disinfection by-product
Total trihalomethanes*	59.3	Disinfection by-product

* Concentrations of individual compounds not reported.

TABLE 7-8 Columbus Water Data

Chemical Compound	Surface Water Average Concentration (ppb)	Type of Chemical Compound
Haloacetic acids	36	Disinfection by-product
Dibromoacetic acid	1.65	Disinfection by-product
Dichloroacetic acid	18.4	Disinfection by-product
Monochloroacetic acid	4.35	Disinfection by-product
Trichloroacetic acid	11.1	Disinfection by-product
Total trihalomethanes	65	Disinfection by-product
Bromodichloromethane	7.65	Disinfection by-product
Chloroform	31.2	Disinfection by-product

Chattahoochee River. The synthetic chemical data for this community water system are given in Appendix 7-10.

Because no summary was given for haloacetic acids or total trihalomethanes, the 2004 annual water quality report was reviewed. All of the synthetic chemicals detected in the surface water are given in Table 7-8. All of these chemicals are disinfection by-products.

7.14 DAYTON, OHIO

The city of Dayton is located in the Miami Valley at the confluence of the Miami, Stillwater, and Mad rivers. The greater Dayton area encompasses a number of contiguous communities and has a population of approximately 850,000. Outside the metropolitan area the region is primarily agricultural.

For the city of Dayton, the state of Ohio provided data on two groundwater systems and one surface water system. Both surface water and groundwater systems serve approximately the same number of consumers. The synthetic chemical data for the groundwater community water systems are given in Appendix 7-11, while data for the surface water system are given in Appendix 7-12. Because there was a lack of reported haloacetic acids and total trihalomethanes, the 2005 annual water quality report was also reviewed. The annual report specifically indicates that the groundwater is influenced by surface water recharge.

The majority of the synthetic chemicals found in both surface water and groundwater community water systems (see Table 7-9) was disinfection by-products with several industrial chemicals and pesticides in the groundwater resources.

TABLE 7-9 Dayton Water Data

Chemical Compound	Groundwater Average Concentration (ppb)	Surface Water Average Concentration (ppb)	Type of Chemical Compound
Atrazine	1.21		Pesticide
Carbon tetrachloride	2.26		Industrial chemical
Chloromethane	0.65		Industrial chemical*
Cis-dichloroethene	0.45		Industrial chemical
Metolachlor	0.9		Pesticide
Naphthalene	0.28		Petroleum
Simazine	0.84		Disinfection by-product
Trichloroethene	0.32		Industrial chemical
Trichlorofluoromethane	0.75		Industrial chemical
Xylenes	0.5		Petroleum
Total reported individual chemicals — 10			
Haloacetic acids	6		Disinfection by-product
Dibromoacetic acid	2.25	2	Disinfection by-product
Dichloroacetic acid	2.27		Disinfection by-product
Monochloroacetic acid	4.09	4.4	Disinfection by-product
Trichloroacetic acid	1.42	1.25	Disinfection by-product
Total trihalomethanes	22		Disinfection by-product
Bromodichloromethane	2.17		Disinfection by-product
Bromoform	2.5		Disinfection by-product
Chloroform	7.9		Disinfection by-product
Dibromochloromethane	10		Disinfection by-product

* Possible disinfection by-product.

7.15 DENVER, COLORADO

The city of Denver is located along the front range of the Rocky Mountains and has a population of approximately 560,000. The downtown area is surrounded by business centers, industrial parks, and commercial and residential land use. The state of Colorado did not provide any water quality data. Therefore, the data from Denver's annual water quality reports of 2004, 2003, and 2002 were averaged and are listed in Table 7-10. Denver's drinking water comes from rivers, lakes, streams, and reservoirs and springs fed by mountain runoff. As expected, drinking water contaminants are dominated by disinfection by-products.

TABLE 7-10 Denver Water Data

Chemical Compound	Surface Water Average Concentration (ppb)	Type of Chemical Compound
Chlorodibromomethane	2.4	Industrial chemical*
Chloropicrin	0.66	Disinfection by-product
Haloacetonitriles	2.4	Disinfection by-product
Haloketones	172	Disinfection by-product
Total reported individual chemicals — 4		
Haloacetic acids**	2.6	Disinfection by-product
Total trihalomethanes	22	Disinfection by-product
Bromodichloromethane	7.6	Disinfection by-product
Bromoform	1.9	Disinfection by-product
Chloroform	23.4	Disinfection by-product
Total organic halides	26.3	Disinfection by-product

* Possible disinfection by-product.

** Concentrations of individual compounds not reported.

7.16 FRESNO, CALIFORNIA

The city of Fresno is located in the central San Joaquin Valley of California with a population of approximately 470,000. The region surrounding the city is dominated by irrigated agriculture. Drinking water resources for the city come from both groundwater and surface water.

For the city of Fresno, the state of California provided data on four community water systems that derive their water from groundwater resources, which serve the majority of the population, and four community water systems that derive their product from surface water resources. The groundwater data are provided in Appendix 7-13. It should be noted that for some compounds (e.g., DBCP) there were so many analyses that only the average concentration was provided. The surface water data are provided in Appendix 7-14. Because only data on trihalomethanes were provided, the 2004 annual water quality report was also reviewed. The summary of detected synthetic compounds found in community drinking water is given in Table 7-11.

Given that irrigated agriculture in this region of the United States requires soils to be constantly leached in order to minimize salt accumulations, it is understandable that groundwater contains pesticides (e.g., DBCP). These data also show that even treating the groundwater to remove DBCP, dichloroethylene, dichlorodifluoromethane, ethylene dibromide, trichloroethylene, and trichloropropane does not completely remove these synthetic chemicals from drinking water. Once again, the concentration of the disinfection by-products was greater in surface water sources.

7.17 GRAND RAPIDS, MICHIGAN

The city of Grand Rapids sits on the banks of the Grand River and has a population of approximately 200,000. The city's mix of commercial and residential land use also incorporates major furniture and automobile manufacturing. For the city of Grand Rapids, the state of Michigan provided data on four community water systems that derive their water from groundwater resources and one community water system that derives its water from surface water resources. The surface water system serves the majority of the city's population. The groundwater data are provided in Appendix 7-15, while the surface water data are given in Appendix 7-16.

Because there was a lack of reported haloacetic acids and trihalomethanes, the 2002 through 2004 annual water quality reports were also reviewed. The detected synthetic compounds found in community drinking water are given in Table 7-12. All of the industrial and petroleum chemicals were found in the groundwater sources. Both systems reported disinfection by-products with synthetic chemical concentrations being higher in the surface water sources.

7.18 HARRISBURG, PENNSYLVANIA

The city of Harrisburg sits against the Susquehanna River north of the Blue Ridge chain of the Appalachian Mountains and has a population of approximately 620,000. The city's metropolitan area is a mixture of commercial, industrial, and residential land use. For the city of Harrisburg, the state of Pennsylvania provided data on four community water systems that derive their water from groundwater resources and one community water system that derives its water from surface water resources. The surface water system serves

TABLE 7-11 Fresno Water Data

Chemical Compound	Groundwater Average Concentration (ppb)	Treated Groundwater Average Concentration (ppb)	Surface Water Average Concentration (ppb)	Type of Chemical Compound
Bromomethane	1			Industrial chemical*
Butylbenzene	0.9			Industrial chemical
Chloroethane	0.63			Industrial chemical*
Chloromethane	1.46		1.17	Industrial chemical*
DCPA	4.7			Pesticide
Diazinon	507			Pesticide
Dibromochloropropane (DBCP)	0.34	0.08		Pesticide
Dichloroethylene	6.31			Industrial chemical
Cis-dichloroethylene	1.78	0.56		Industrial chemical
Dichlorodifluoromethane	9.8	1.48		Industrial chemical
Dichloromethane	0.66		2.8	Industrial chemical*
1,2-Dichloropropane	1.28			Industrial chemical
2,2-Dichloropropane	35			Industrial chemical
Ethylene dibromide	0.08	0.04		Pesticide
Methyl tert butyl ether (MTBE)	5			Petroleum
Molinate	5.7			Pesticide
Monochlorobenzene	0.52			Industrial chemical
Styrene	0.52			Industrial chemical
Tetrachloroethylene	1.43			Industrial chemical
Thiobencarb	7.2			Pesticide
Toluene	0.88			Industrial chemical
Trichloroethane	0.88			Industrial chemical
Trichloroethylene	15.2	2.06		Industrial chemical
Trichloropropane	0.09	0.12		Pesticide
Trimethylbenzene	1			Industrial chemical
Xylenes	0.89			Industrial chemical
Total reported individual chemicals — 26				
Haloacetic acids			20.3	Disinfection by-product
Dibromacetic acid			7.26	Disinfection by-product
Dichloroacetic acid			7.7	Disinfection by-product
Trichloroacetic acid	5.3			Disinfection by-product
Total trihalomethanes			12	Disinfection by-product
Bromodichloromethane	2.4			Disinfection by-product
Bromoform	4.5			Disinfection by-product
Chloroform	2.6			Disinfection by-product
Dibromochloromethane	3.2			Disinfection by-product

* Possible disinfection by-product.

TABLE 7-12 Grand Rapids Water Data

Chemical Compound	Groundwater Average Concentration (ppb)	Surface Water Average Concentration (ppb)	Type of Chemical Compound
Bromochloroacetic acid	1		Disinfection by-product
Chlorodibromomethane		2.56	Industrial chemical*
1,2-Dichlorobenzene	0.78		Industrial chemical
1,3-Dichlorobenzene	0.55		Industrial chemical
1,4-Dichlorobenzene	0.61		Industrial chemical
Dichlorobromomethane	1.55	5.92	Industrial chemical*
Tetrahydrofuran	0.59		Petroleum
Total reported individual chemicals — 7			
Haloacetic acids		26.5	Disinfection by-product
Dichloroacetic acid	11.8		Disinfection by-product
Total trihalomethanes		40.3	Disinfection by-product
Bromoform		2.1	Disinfection by-product
Chloroform	5.18	8.48	Disinfection by-product
Total organic halogens		110	Disinfection by-product

* Possible disinfection by-product.

the majority of the city's population. The groundwater data are provided in Appendix 7-17, while the surface water data are given in Appendix 7-18.

Because the surface water data reported no disinfection by-products, the available annual reports for 2003 and 2004 were also reviewed. All of the monitoring data are summarized in Table 7-13. These data show that the detection of industrial chemicals was in the groundwater sources, while the primary con-

TABLE 7-13 Harrisburg Water Data

Chemical Compound	Groundwater Average Concentration (ppb)	Surface Water Average Concentration (ppb)	Type of Chemical Compound
Atrazine	0.2	0.2	Pesticides
Dichloromethane	1.2		Industrial chemical*
Styrene	1.98		Industrial chemical
Trichloroethylene	0.65		Industrial chemical
Toluene	0.5		Industrial chemical
Xylenes	0.62		Industrial chemical
Total reported individual chemicals — 6			
Haloacetic acids**		50.1	Disinfection by-product
Total trihalomethanes**		35.2	Disinfection by-product

* Possible disinfection by-product.

** Concentration of individual compounds not reported.

taminants in the surface water sources were disinfection by-products. Given the nature of the data omissions, the groundwater systems are also assumed to contain disinfection by-products.

7.19 JACKSON, MISSISSIPPI

The city of Jackson is located on the Pearl River and has a population of approximately 185,000. The city's metropolitan area is a mixture of commercial and residential land use surrounded by agricultural areas. The state of Mississippi provided no data for the city of Jackson. The city of Jackson has one groundwater and one surface water community water system. The monitoring data for these systems (2004) are given in Table 7-14. For both systems, the only reported contaminants are disinfection by-products. These data are unusual in that the treated groundwater has higher concentrations of disinfection by-products. These values also exceed the primary drinking water standards (Table 7-1).

TABLE 7-14 Jackson Water Data

Chemical Compound	Groundwater Average Concentration (ppb)	Surface Water Average Concentration (ppb)	Type of Chemical Compound
Haloacetic acids*	85.8	53.3	Disinfection by-product
Total trihalomethanes*	65.5	62	Disinfection by-product

* Concentrations of individual compounds not reported.

7.20 LAFAYETTE, LOUISIANA

The city of Lafayette is located on the Vermillion River and has a population of approximately 110,000. The city's metropolitan area is a mixture of commercial and residential land use surrounded by agricultural areas. The state of Louisiana provided no data for the city of Lafayette. The city of Lafayette has one groundwater community water system that serves the majority of the city. The average chemical concentrations for the monitoring data from the annual water quality reports (2000 through 2004) are given in Table 7-15. The only reported contaminants are disinfection by-products.

7.21 LONG BEACH, NEW JERSEY

Long Beach is located on one of New Jersey's outer islands (Long Beach Island) and has a population of approximately 3300. The township is made up

TABLE 7-15 Lafayette Water Data

Chemical Compound	Groundwater Average Concentration (ppb)	Type of Chemical Compound
Haloacetilenitriles	1.2	Disinfection by-product
Total number of individual chemicals — 1		
Haloacetic acids*	5.5	Disinfection by-product
Total trihalomethanes*	10.5	Disinfection by-product

* Concentrations of individual compounds not reported.

of commercial and residential areas. The state of New Jersey provided no water quality data for the township. Long Beach Island has five groundwater community water systems that serve the majority of the island. The monitoring data from the annual water quality reports (2002 through 2004) indicate that there are no detected synthetic chemicals (including disinfection by-products).

7.22 LOS ANGELES, CALIFORNIA

The city of Los Angeles and associated urban sprawl is located between the San Gabriel Mountains and the Pacific Ocean. The city of Los Angeles has an approximate population of 3.6 million, while the Los Angeles metropolitan area has a population of about 18 million. The entire area is dominated by residential, commercial, and industrial land use.

For the city of Los Angeles, the state of California provided data on a community water system that utilized both surface water (California, Los Angeles, Colorado aqueducts, and reservoirs) and groundwater sources. Surface water sources provide approximately 85% of the drinking water. The chemical data are provided in Appendix 7-19. If the number of samples was more than a hundred, only the average has been reported. Because of the lack of monitoring data on disinfection by-products, the annual water quality reports from 2003 and 2004 were used to determine the detection of these types of chemicals. A total of 22 different synthetic compounds was detected in the water system (see Table 7-16). These chemicals include pesticides, industrial chemicals, petroleum products, and disinfection by-products.

7.23 MILWAUKEE, WISCONSIN

The city of Milwaukee sits at the shore of Lake Michigan at the confluence of the Menomonee, Kinnickinnic, and Milwaukee rivers and has a population of

TABLE 7-16 Los Angeles Water Data

Chemical Compound	Average Concentration (ppb)	Type of Chemical Compound
1,1-Dichloroethane	0.92	Industrial chemical*
1,2-Dichloroethane	0.83	Industrial chemical*
1,1-Dichloroethylene	2.77	Industrial chemical
Cis-dichloroethylene	2.2	Industrial chemical
Dichlorodifluoromethane	2.72	Industrial chemical*
Dichloromethane	0.87	Industrial chemical*
1,2-Dichloropropane	1.18	Industrial chemical
Diquat	4	Pesticide
Ethylbenzene	0.87	Petroleum
Ethylene dibromide	0.02	Pesticide
Indeno(1,2,3-CD)pyrene	0.05	Petroleum
Methyl isobutyl ketone	2.2	Industrial chemical
Methyl tert butyl ether (MTBE)	2.5	Petroleum
Nitrosodimethylamine	0.004	Disinfection by-product
Tetrachloroethylene	4.59	Industrial chemical
Toluene	2.56	Petroleum
Trichloroethylene	10.7	Industrial chemical
1,1,2-Trichloro-1,2,2-trifluoroethane	1.73	Industrial chemical*
Trichlorofluoromethane	8.76	Industrial chemical*
Xylenes	17.4	Industrial chemical
Total reported individual chemicals — 20		
Haloacetic acids**	52	Disinfection by-product
Total trihalomethanes	52.5	Disinfection by-product
Bromodichloromethane	19.4	Disinfection by-product
Bromoform	7.2	Disinfection by-product
Chloroform	17.4	Disinfection by-product
Dibromochloromethane	23.3	Disinfection by-product

* Possible disinfection by-product.

** Concentrations of individual compounds not reported.

approximately 592,000 (1.7 million in the Milwaukee metropolitan area). Outside the urban sprawl, agricultural land uses dominate the region.

For the city of Milwaukee, the state of Wisconsin provided data on 24 groundwater systems and four surface water systems. The surface water systems serve the majority of the community. The synthetic chemical data for the groundwater community water systems are given in Appendix 7-20, while data for the surface water systems are given in Appendix 7-21. The majority of the synthetic chemicals was found in the groundwater systems (see Table 7-17). However, both groundwater and surface water systems detected a large number of industrial chemicals and disinfection by-products.

TABLE 7-17 Milwaukee Water Data

Chemical Compound	Groundwater Average Concentration (ppb)	Surface Water Average Concentration (ppb)	Type of Chemical Compound
1,1,1-Trichloroethane	0.13	0.63	Industrial chemical
1,1-Dichloroethane	0.23		Industrial chemical
1,2-Dichloroethane	0.43		Industrial chemical
1,2-Dichloropropane	0.12	0.3	Industrial chemical
2,2-Dichloropropane	0.56		Industrial chemical
Bromomethane	0.48		Industrial chemical*
Carbon tetrachloride	0.72	0.1	Industrial chemical
Methylchloride	0.29	0.63	Industrial chemical
Di(2-ethylhexyl)phthalate	4.48		Industrial chemical
Dichloromethane	0.83	1.02	Industrial chemical*
Dichlorobenzene	0.27		Industrial chemical
Ethylbenzene		0.4	Petroleum
Styrene	0.33		Industrial chemical
Tetrachloroethylene	0.08	0.3	Industrial chemical*
Toluene	4.54	1.1	Petroleum
Triazine	0.1		Pesticide
Xylenes	37.3	1.03	Petroleum
Total reported individual chemicals — 17			
Haloacetic acids	10.4	4.69	Disinfection by-product
Dibromoacetic acid		1.05	Disinfection by-product
Dichloroacetic acid	3.85	2.51	Disinfection by-product
Monochloroacetic acid	15.6		Disinfection by-product
Trichloroacetic acid	1.33		Disinfection by-product
Total trihalomethanes	2.03	14.10	Disinfection by-product
Bromodichloromethane	0.56		Disinfection by-product
Bromoform	0.5	0.43	Disinfection by-product
Chloroform	1.66	7.82	Disinfection by-product
Dibromochloromethane	0.69	2.44	Disinfection by-product

* Possible disinfection by-product.

7.24 NEW YORK, NEW YORK

The city of New York is situated on the three major islands of Manhattan, Staten Island, and western Long Island while the Bronx is the only borough that is part of the mainland United States. The city is dominated by commercial and residential land use and has a population of approximately 8.1 million, while the metropolitan area has a population of 22 million.

For the city of New York, the state of New York provided data on two small groundwater systems. These data are given in Appendix 7-22. The only contaminant reported by either system was methyl tert butyl ether (MTBE), which is derived from gasoline. Because no other data were provided, the New York water quality monitoring data were taken from the annual water quality reports. The annual reports show that 99% of New York's drinking water comes from surface water sources (i.e., the Catskill/Delaware System and the Croton System in Putnam, Westchester, and Dutchess counties). The data from the annual water quality reports are given in Table 7-18.

TABLE 7-18 New York Water Data

Compound	Year of Annual Report					Type of Chemical Compound
	2004	2003	2002	2001	2000	
Acetone			22	12		Industrial chemical
Bromochloroacetic acid	1.33	1.66	1.66	1.9	1.1	Disinfection by-product
Bromomethane			0.6	0.9		Industrial chemical*
Butylbenzene			0.6			Industrial chemical
Carbon tetrachloride	0.6	0.5	0.6			Industrial chemical
Chloromethane				0.5		Industrial chemical*
Chloropicrin	0.55	0.35	0.45	0.43	0.4	Disinfection by-product
Dalapon		1.9				Pesticide
DCPA			0.77		0.95	Pesticide
Dibromochloropropane		0.06				Industrial chemical*
Dichlorofluoromethane	1.8	2.3	2.2	0.9	1	Industrial chemical*
Dieldrin			0.7			Pesticide
Haloacetonitriles	4.45	3.9	5	3.4	2.95	Disinfection by-product
Halogenated ketones	4.1	2.45	3.6	2.87	2.3	Disinfection by-product
Hexachlorobutadiene					1	Petroleum
Hexachlorocylopentadiene				0.08		Petroleum
Methyl tert butyl ether	1	1	1.3	14	10.4	Petroleum
Phenanthrene		0.02				Petroleum
Propylbenzene				0.6		Industrial chemical
Simazine				0.05		Industrial chemical
Styrene			0.6			Industrial chemical
Tetrachloroethylene	2.6	17.3	4.7	3.4	3	Industrial chemical
Toluene			1.9	0.8	1	Petroleum
Trichlorobenzene		0.8	0.7	0.6		Industrial chemical
Trichloroethylene	1.1	1.2	1.7	1.3	2	Industrial chemical
Xylenes			4.3	2.2		Petroleum
Total reported individual chemicals — 26						
Haloacetic acids**	49.7	42	32.3	36.7	24.6	Disinfection by-product
Total trihalomethanes**	42	36	32.3	37.3	33	Disinfection by-product
Total organic halogens	176	145	136	154	161	Disinfection by-product

* Possible disinfection by-product.

** Concentrations of individual compounds not reported.

Because various unregulated chemicals were not consistently monitored over the 2000–2004 period, the data for each year are summarized in Table 7-18. The data for the disinfection by-products are averages for all the community water systems, while the data reported for the pesticides, industrial chemicals, and petroleum products are the highest reported values. Most of the chemical analysis detected disinfection by-products, while the pesticides, petroleum products, and industrial chemicals occurred less frequently.

7.25 OGDEN, UTAH

The city of Ogden is located against the western slope of the Wasatch Mountains and east of the Great Salt Lake and has a population of approximately 77,000. The city is a mixture of commercial and residential areas surrounded by recreational and ranching land uses. The state of Utah provided no data for the city of Ogden.

As a result, the available annual water quality monitoring reports for Ogden and North Ogden (2003 and 2002) were reviewed. Ogden obtains its water from both surface and groundwater sources, while North Ogden only utilizes groundwater. For both systems, the only reported contaminants were disinfection by-products as measured by total trihalomethanes (15.9 ppb).

7.26 OKLAHOMA CITY, OKLAHOMA

Oklahoma City is located approximately in the center of the state with an estimated population of 530,000. The city is a complex mixture of residential, commercial, and industrial land uses. The state of Oklahoma provided data on three groundwater systems and one surface water system. The surface water system serves the vast majority of the city. The surface water resources come from the North Canadian River (i.e., northwestern and north central Oklahoma) to feed two reservoirs.

The synthetic chemical data for the groundwater community water systems are given in Appendix 7-23, while data for the surface water system are given in Appendix 7-24. The majority of the synthetic chemicals found in both surface water and groundwater community water systems (see Table 7-19) was disinfection by-products. The 2003 annual water quality report data are also included in Table 7-19. These data show that the number and concentration of disinfection by-products were significantly greater in the surface water source.

7.27 ORLANDO, FLORIDA

The city of Orlando occupies a generally low-lying area with numerous natural lakes and sand hills and has a population of approximately 186,000 (1.6 million in the Orlando metropolitan area). The urban areas are generally a mix of commercial and residential land use.

TABLE 7-19 Oklahoma City Water Data

Chemical Compound	Groundwater Average Concentration (ppb)	Surface Water Average Concentration (ppb)	Type of Chemical Compound
Benzene	1.2		Industrial chemical
Chlorodibromomethane	2.75	14.5	Industrial chemical*
Toluene	0.7		Industrial chemical
Total reported individual chemicals — 3			
Haloacetic acids**		43.2	Disinfection by-product
Total trihalomethanes		41.9	Disinfection by-product
Bromodichloromethane	3.8	21.9	Disinfection by-product
Chloroform	2.49	45.6	Disinfection by-product

* Possible disinfection by-product.

** Concentrations of individual compounds not reported.

For the city of Orlando, the state of Florida provided data on two groundwater systems. The synthetic chemical data for the groundwater community water systems are given in Appendix 7-25. Because of the lack of detected disinfection by-products, the annual water quality reports from 2000 through 2004 were also reviewed and summarized. The source waters in the annual reports were from groundwater resources. All of the summarized data on detected synthetic organic chemicals are given in Table 7-20. The majority of the detected synthetic chemicals was disinfection by-products. This result is in spite of the fact that Orlando also uses ozone as a disinfectant.

7.28 PEORIA, ILLINOIS

The city of Peoria is located against the Illinois River with a population of approximately 112,000. The city is a mixture of commercial and residential areas surrounded by agricultural land use. The state of Illinois provided no data for the city of Peoria.

As a result, the available annual water quality monitoring report (2003) for Peoria was reviewed. The source water comes from both groundwater wells and surface resources. The water quality summary is given in Table 7-21. All of the detected contaminants were disinfection by-products.

7.29 PHOENIX, ARIZONA

The city of Phoenix is located in the Phoenix Valley and surrounded by the McDowell, White Tank, and Superstition mountains and the Sierra Estrella. The

TABLE 7-20 Orlando Water Data

Chemical Compound	Groundwater Average Concentration (ppb)	Type of Chemical Compound
Bromochloroacetic acid	5.1	Disinfection by-product
Bromochloroacetonitrile	1.43	Disinfection by-product
Chlorodibromomethane	3	Disinfection by-product
Chloropicrin	0.07	Disinfection by-product
Dibromochloroacetic acid	0.24	Disinfection by-product
Dibromoacetonitrile	0.44	Disinfection by-product
Dichlorobromomethane	8.1	Disinfection by-product
Dichloroacetonitrile	2.29	Disinfection by-product
Cis-dichloroethylene	0.38	Industrial chemical
Dichloropropanone	0.18	Disinfection by-product
Di(2-ethylhexyl)phthalate	0.49	Industrial chemical
Trichloropropanone	1.75	Disinfection by-product
Total reported individual chemicals — 12		
Haloacetic acids	32	Disinfection by-product
Dichloracetic acid	8.4	Disinfection by-product
Monobromoacetic acid	0.49	Disinfection by-product
Monochloracetic acid	3.82	Disinfection by-product
Trichloracetic acid	4.9	Disinfection by-product
Total trihalomethanes	55.6	Disinfection by-product
Chloroform	21.8	Disinfection by-product
Total organic halides	92	Disinfection by-product

TABLE 7-21 Peoria Water Data

Chemical Compound	Average Concentration (ppb)	Type of Chemical Compound
Chlorodibromomethane	8.2	Industrial chemical*
Total reported individual chemicals — 1		
Haloacetic acids	10.1	Disinfection by-product
Dibromochloroacetic acid	1.9	Disinfection by-product
Dichloracetic acid	3.4	Disinfection by-product
Trichloracetic acid	4.4	Disinfection by-product
Total trihalomethanes	31.6	Disinfection by-product
Bromodichloromethane	10.7	Disinfection by-product
Bromoform	2.1	Disinfection by-product
Chloroform	10.7	Disinfection by-product

* Possible disinfection by-product.

Salt River runs through the city and is normally dry except during rainfall, when water releases from the upriver dams. The city is a mixture of commercial, residential, and industrial areas with a population of approximately 3.2 million. For the city of Phoenix, the state of Arizona provided data on six groundwater systems and one surface water system. The synthetic chemical data for the groundwater community water systems are given in Appendix 7-26, and the surface water community water system data are given in Appendix 7-27. Because of the lack of detected disinfection by-products, the annual water quality reports from 2000 through 2004 were also reviewed and summarized. The source waters in the annual reports indicated that 98% of the drinking water came from surface water resources. All of the summarized data on detected synthetic organic chemicals are given in Table 7-22. For the surface water system, the majority of the detected synthetic chemicals was disinfection by-products. The drinking water from groundwater sources contained detected levels of industrial chemicals and petroleum products.

TABLE 7-22 Phoenix Water Data

Chemical Compound	Groundwater Average Concentration (ppb)	Surface Water Average Concentration (ppb)	Type of Chemical Compound
2,4-D		0.2	Pesticide
Chlorodibromomethane		14.4	Industrial chemical*
Chloromethane		0.1	Industrial chemical*
Di(2-ethylhexyl)phthalate		1.1	Industrial chemical
Dichloroethylene	3.3		Industrial chemical
Ethylbenzene	1.7		Petroleum
Trichlorobenzene	0.94		Industrial chemical
Toluene	1.2		Industrial chemical
Xylenes	2.96		Petroleum
Total reported individual chemicals — 9			
Haloacetic acids**		24	Disinfection by-product
Total trihalomethanes		54	Disinfection by-product
Bromodichloromethane		20.3	Disinfection by-product
Bromoform		2.1	Disinfection by-product
Chloroform		14.7	Disinfection by-product

* Possible disinfection by-product.

** Concentrations of individual compounds not reported.

7.30 PORTLAND, OREGON

The city of Portland straddles the Willamette River south of its confluence with the Columbia River. The city is a mixture of commercial, residential, and industrial areas and has a population of approximately 550,000 with a metropolitan population of 2.1 million.

For the city of Portland, the state of Oregon provided limited data on two groundwater systems and one surface water system. However, the surface water system appears to be associated with groundwater sources. Therefore, all the synthetic chemical data for all the community water systems are given in Appendix 7-28. Because of the lack of detected disinfection by-products, the annual water quality reports from 2004 and 2005 were also reviewed and summarized.

Based on the annual reports, the majority of the city's water comes from the protected Bull Run watershed in the Mount Hood National Forest. All of the summarized data on detected synthetic organic chemicals are given in Table 7-23. For the surface water system, the majority of the detected synthetic chemicals was disinfection by-products. The drinking water from groundwater sources contained detected levels of industrial chemicals and a pesticide.

TABLE 7-23 Portland Water Data

Chemical Compound	Groundwater Average Concentration (ppb)	Surface Water Average Concentration (ppb)	Type of Chemical Compound
Dichlorobenzene	0.8		Industrial chemical
Dichloroethylene	0.63		Industrial chemical
Diquat	0.4		Pesticide
Tetrachloroethylene	4.19		Industrial chemical
Total reported individual chemicals — 4			
Haloacetic acids*		35	Disinfection by-product
Total trihalomethanes*		25	Disinfection by-product

* Concentrations of individual compounds not reported.

7.31 PROVIDENCE, RHODE ISLAND

The city of Providence is located at the head of Narragansett Bay, with the Providence River running into the bay through the center of the city. Providence has a population of approximately 180,000 with a metropolitan population of about 1.2 million. The city is a mixture of commercial, residential, and industrial areas. The state of Rhode Island provided no data for the city of Providence.

As a result, the available annual water quality monitoring reports (2004 and 2005) for Providence were reviewed. The source water comes from the Scituate Reservoir, which obtains runoff from a predominately forested area, but there is also a considerable amount of urban development in the watershed. The water quality summary is given in Table 7-24. All of the detected contaminants were disinfection by-products.

TABLE 7-24 Providence Water Data

Chemical Compound	Surface Water Average Concentration (ppb)	Type of Chemical Compound
Haloacetic acids*	16	Disinfection by-product
Total trihalomethanes*	35.4	Disinfection by-product

* Concentrations of individual compounds not reported.

7.32 PUEBLO, COLORADO

The city of Pueblo is situated along the front range of the Colorado Rockies on the confluence of the Arkansas River and Fountain Creek with a population of approximately 100,000. The city is a mixture of commercial and residential areas surrounded by ranching and agriculture. The state of Colorado provided no data for the city of Pueblo.

As a result, the available annual water quality monitoring reports (2003, 2004, and 2005) for Pueblo were reviewed. The source water comes from the Arkansas River, which flows into the Pueblo Reservoir. The reservoir water quality summary is given in Table 7-25. All of the detected contaminants were disinfection by-products.

TABLE 7-25 Pueblo Water Data

Chemical Compound	Surface Water Average Concentration (ppb)	Type of Chemical Compound
Bromochloroacetic acid	1.51	Disinfection by-product
Bromochloroacetonitrile	0.41	Disinfection by-product
Chlorodibromomethane	1.5	Industrial chemical*
Dichloroacetonitrile	0.49	Disinfection by-product
Dichloropropanone	1.2	Disinfection by-product
Trichloropropanone	0.26	Disinfection by-product
Total reported individual chemicals — 6		
Haloacetic acids	15	Disinfection by-product
Dibromoacetic acid	0.46	Disinfection by-product
Dichloroacetic acid	9.3	Disinfection by-product
Monochloroacetic acid	1.97	Disinfection by-product
Trichloroacetic acid	2.25	Disinfection by-product
Total trihalomethanes	7.5	Disinfection by-product
Bromodichloromethane	1.66	Disinfection by-product
Chloroform	4.23	Disinfection by-product
Dibromochloromethane	0.4	Disinfection by-product

* Possible disinfection by-product.

7.33 RENO, NEVADA

The city of Reno lies near the eastern slope of the Sierra Nevada along the Truckee River with an approximate population of 180,000. For the city of Reno, the state of Nevada provided data on 18 groundwater systems and two surface water system. The vast majority of the city's drinking water comes from the surface water systems.

The synthetic chemical data for the groundwater community water systems are given in Appendix 7-29, and the surface water community water system data are given in Appendix 7-30. Given the lack of other disinfection by-products reported in the state data, the available chemical data from the 2004 annual report were also reviewed. All of the summarized data on detected synthetic organic chemicals are given in Table 7-26. Besides the anticipated disinfection by-products, industrial chlorinated solvents were detected in this surface water.

TABLE 7-26 Reno Water Data

Chemical Compound	Groundwater Average Concentration (ppb)	Surface Water Average Concentration (ppb)	Type of Chemical Compound
Chlorodibromomethane	2.28	2.51	Industrial chemical*
Tetrachloroethylene		0.99	Industrial chemical
Trichloroethylene		0.84	Industrial chemical
Total reported individual chemicals — 3			
Haloacetic acids*		28.9	Disinfection by-product
Total trihalomethanes		32.2	Disinfection by-product
Bromodichloromethane	3.45	6.91	Disinfection by-product

* Possible disinfection by-product.

** Concentrations of individual compounds not reported.

7.34 SAN ANTONIO, TEXAS

The city of San Antonio is situated along the San Antonio River and has a population of approximately 1.1 million. The city is a mixture of commercial, residential, and industrial areas. For the city of San Antonio, the state of Texas provided data on 40 groundwater systems and one surface water system. The majority of the city's drinking water is provided from groundwater resources.

The synthetic chemical data for the groundwater community water systems are given in Appendix 7-31, and the surface water community system data are given in Appendix 7-32. Given the lack of trihalomethane and haloacetic acid data reported by the state, the available chemical data from the 2004 and 2005 annual reports were also reviewed. All of the summarized data on detected synthetic organic chemicals are given in Table 7-27. The detected compounds included pesticides, petroleum products, industrial chemicals, and disinfection by-products.

7.35 SAVANNAH, GEORGIA

The city of Savannah is situated along the Savannah River and the U.S. Intracoastal Waterway and has a population of approximately 130,000. The city is a mixture of commercial, residential, and industrial areas. For the city of Savannah, the state of Georgia provided data on 70 groundwater systems. All of the city's drinking water is provided from groundwater resources.

The synthetic chemical data for the groundwater community water systems are given in Appendix 7-33. Given the lack of specific trihalomethane and haloacetic acid summaries reported by the state, the available chemical data from the 2004 annual reports (i.e., six different systems) were also reviewed. All of the summarized data on detected synthetic organic chemicals are given in Table 7-28. The groundwater contaminants were dominated by disinfection by-products and petroleum compounds.

7.36 SIOUX FALLS, SOUTH DAKOTA

The city of Sioux Falls is situated along the banks of the Big Sioux River and has a population of approximately 140,000. The city is a mixture of commercial and residential areas surrounded by ranching and agricultural land uses. For the city of Sioux Falls, the state of North Dakota provided data on surface water sampling, but the community water system derives its water from both surface and groundwater resources.

The synthetic chemical data for this system are given in Appendix 7-34. Given the lack of specific trihalomethane and haloacetic acid summaries reported by the state, the available chemical data from the 2004 annual report were also reviewed. All of the summarized data on detected synthetic organic chemicals are given in Table 7-29. The detected chemical contaminants were dominated by the disinfection by-products.

7.37 SPRINGFIELD, ILLINOIS

The city of Springfield is located in the Sangamon River Valley and has a population of approximately 120,000. The city is a mixture of commercial and residential areas surrounded by agricultural land uses. For the city of Springfield,

TABLE 7-27 San Antonio Water Data

Chemical Compound	Groundwater Average Concentration (ppb)	Surface Water Average Concentration (ppb)	Type of Chemical Compound
Acetone	29.3		Industrial chemical
Atrazine	0.59	3.3	Pesticide
Bromochloroacetic acid	5.3		Disinfection by-product
Butanone (MEK)	6		Industrial chemical
Butylbenzylphthalate	26.4		Industrial chemical
Carbon disulfide	2.25		Industrial chemical
Carbon tetrachloride	0.59		Industrial chemical
Chloroacetic acid	7		Disinfection by-product
Chlorobenzene	0.7		Industrial chemical
Chlorodimethylbenzene	2.75		Industrial chemical
Dibromomethane	2.1		Industrial chemical*
Dichlorobenzene	0.5		Industrial chemical
Dichlorobromomethane	2.9		Industrial chemical*
Dichloromethane	2.16		Industrial chemical*
Di(2-ethylhexyl)adipate	2.59		Industrial chemical
Di(2-ethylhexyl)phthalate	4.19		Industrial chemical
Diethylphthalate	2.2		Industrial chemical
Ethylbenzene	1.46		Petroleum
Fluorene	0.13		Petroleum
Hexachlorobutadiene	1.3		Industrial chemical
Methyl methacrylate	4.4		Industrial chemical
Methylpentanone (MIBK)	9.4		Industrial chemical
Phenanthrene	0.16		Petroleum
Styrene	0.6		Industrial chemical
Tetrachloroethylene	0.94		Industrial chemical
Tetrahydrofuran	21.6		Petroleum
Toluene	4.8		Industrial chemical
Trichloroethylene	0.6		Industrial chemical
Trimethylbenzene	1.6		Industrial chemical
Xylenes	3.58	0.5	Petroleum
Total reported individual chemicals — 30			
Haloacetic acids	16.2		Disinfection by-product
Dibromoacetic acid	8.7		Disinfection by-product
Dichloroacetic acid	7.4		Disinfection by-product
Total trihalomethanes	18.3		Disinfection by-product
Bromodichloromethane	4.52	8.46	Disinfection by-product
Bromoform	1.1		Disinfection by-product
Chloroform	5.7		Disinfection by-product
Dibromochloromethane	4.23	7	Disinfection by-product

* Possible disinfection by-product.

TABLE 7-28 Savannah Water Data

Chemical Compound	Groundwater Average Concentration (ppb)	Type of Chemical Compound
Chlorodibromomethane	1.47	Industrial chemical*
Dichloromethane	1.18	Industrial chemical*
Ethylbenzene	1.2	Industrial chemical
Methyl ethyl ketone	2.4	Industrial chemical
Toluene	1.63	Industrial chemical
Trimethylbenzene	2.18	Industrial chemical
Xylenes	5.16	Industrial chemical
Total reported individual chemicals — 7		
Haloacetic acids		
Dichloroacetic acid	2.07	Disinfection by-product
Monochloroacetic acid	3.78	Disinfection by-product
Trichloroacetic acid	1.95	Disinfection by-product
Total trihalomethanes	1.8	Disinfection by-product
Bromodichloromethane	1.46	Disinfection by-product
Bromoform	1.28	Disinfection by-product
Chloroform	1.53	Disinfection by-product

* Possible disinfection by-product.

TABLE 7-29 Sioux Falls Water Data

Chemical Compound	Groundwater Average Concentration (ppb)	Type of Chemical Compound
2,4-D	0.6	Pesticide
Total reported individual chemicals — 1		
Haloacetic acids*	10.8	Disinfection by-product
Total trihalomethanes	33.1	Disinfection by-product
Bromodichloromethane	12	Disinfection by-product
Bromoform	3	Disinfection by-product
Chloroform	8	Disinfection by-product
Dibromochloromethane	13	Disinfection by-product

* Concentrations of individual compounds not reported.

the state of Illinois provided data on one surface water system that derives its water from Lake Springfield.

The synthetic chemical data for this system are given in Appendix 7-35. Given the lack of specific trihalomethane and haloacetic acid summaries reported by the state, the available chemical data from the 2004 annual report were also reviewed. All of the summarized data on detected synthetic organic chemicals are given in Table 7-30. The detected chemical contaminants were dominated by pesticides and disinfection by-products. Because agricultural runoff discharges to Lake Springfield, the raw water is treated with granulated activated carbon, to specifically remove atrazine. The atrazine MCL of three parts-per-billion has not been exceeded, yet the contaminant remains in the water supply.

TABLE 7-30 Springfield Water Data

Chemical Compound	Groundwater Average Concentration (ppb)	Type of Chemical Compound
Atrazine	0.63	Pesticide
Dichloroacetic acid	17.2	Disinfection by-product
Simazine	1.2	Pesticide
Total reported individual chemicals — 3		
Haloacetic acids	21	Disinfection by-product
Monochloroacetic acid	0.7	Disinfection by-product
Trichloroacetic acid	2.63	Disinfection by-product
Total trihalomethanes*	49	Disinfection by-product

* Concentrations of individual compounds not reported.

7.38 TACOMA, WASHINGTON

The city of Tacoma is situated on Puget Sound's Commencement Bay, Tacoma Narrows, and the estuary of the Puyallup River and has a population of approximately 197,000 (the metropolitan population is about 750,000). The city is a mixture of commercial, residential, and industrial areas. The state of Washington provided no data for the city of Tacoma.

As a result, the available annual water quality monitoring report of 2004 was reviewed. The source water for the community system is derived from both surface water (a protected forested watershed) and groundwater resources. The water quality summary is given in Table 7-31. The groundwater supplies reported numerous industrial chemical contaminants with surface water supplies only reporting disinfection by-products.

7.39 TULSA, OKLAHOMA

The city of Tulsa is situated along the Arkansas River and has a population of approximately 390,000 (the metropolitan population is about 930,000). The city

TABLE 7-31 Tacoma Water Data

Chemical Compound	Average Concentration (ppb)	Groundwater Average Concentration (ppb)	Type of Chemical Compound
Dichlorodifluoromethane		1	Industrial chemical*
Dichloromethane		0.88	Industrial chemical*
Fluorotrichloromethane		2.5	Industrial chemical*
Total reported individual chemicals — 3			
Haloacetic acids**	46.5		Disinfection by-product
Total trihalomethanes	67.4		Disinfection by-product
Chloroform		23.5	Disinfection by-product
Bromodichloromethane		2.09	Disinfection by-product

* Possible disinfection by-product.

** Concentrations of individual compounds not reported.

TABLE 7-32 Tulsa Water Data

Chemical Compound	Surface Water Average Concentration (ppb)	Type of Chemical Compound
Atrazine	0.16	Pesticide
Chlorodibromomethane	4.48	Industrial chemical*
Xylenes	1.82	Petroleum
Total reported individual chemicals — 5		
Haloacetic acids**	24	Disinfection by-product
Total trihalomethanes	32.0	
Bromodichloromethane	6.74	Disinfection by-product
Chloroform	11.1	Disinfection by-product

* Possible disinfection by-product.

** Concentrations of individual compounds not reported.

is a mixture of commercial, residential, and industrial areas. For the city of Tulsa, the state of Oklahoma provided data on one community surface water system.

The source water for the city comes from Oologah Lake. The synthetic chemical data for the community water system are given in Appendix 7-36. Given the lack of specific data on haloacetic acids, the available chemical data from the 2004 annual report of cities that purchase water from the Tulsa system were also reviewed. The water quality summary is given in Table 7-32. The contaminants reported in the drinking water supplies were dominated by disinfection by-products.

7.40 WASHINGTON, D.C.

The city of Washington is situated on the Potomac River and has a population of approximately 550,000 (the metropolitan population is about 4.8 million). The city is a mixture of commercial/governmental and residential areas. The District of Columbia provided no data for the city of Washington.

As a result, the available annual water quality monitoring reports (2000 through 2004) were reviewed. The source water for the community system is derived solely from the Potomac River. The water quality summary is given in Table 7-33. The contaminants detected in the water supply were dominated by pesticides and disinfection by-products.

TABLE 7-33 Washington Water Data

Chemical Compound	Surface Water Average Concentration (ppb)	Type of Chemical Compound
2,4-D	0.3	Pesticide
Atrazine	0.29	Pesticide
Chloropicrin	0.8	Disinfection by-product
Di(2-ethylhexyl)adipate	0.68	Industrial chemical
Di(2-ethylhexyl)phthalate	3.5	Industrial chemical
Dalpon	3.2	Pesticide
Haloacetonitriles	9.25	Disinfection by-product
Haloketones	3.65	Disinfection by-product
Metolachlor	0.06	Pesticide
Phenanthrene	0.03	Petroleum
Simazine	0.1	Pesticide
Total reported individual chemicals — 11		
Haloacetic acids*	38.1	Disinfection by-product
Total organic halides	283	Disinfection by-product
Total trihalomethanes*	56.6	Disinfection by-product

* Concentrations of individual compounds not reported.

7.41 WEST PALM BEACH, FLORIDA

The city of West Palm Beach is situated along the Atlantic-Intracoalstal Waterway and has a population of approximately 82,000. The city is a mixture of commercial and residential areas. For the city of West Palm Beach, the state of Florida provided data on one surface water system and one groundwater system. The synthetic chemical data for the groundwater system are given in Appendix 7-37, and the data for the surface water system are given in Appendix 7-38. Given the lack of specific data on trihalomethanes and haloacetic acids, the

TABLE 7-34 West Palm Beach Water Data

Chemical Compound	Average Concentration (ppb)	Type of Chemical Compound
Atrazine	0.18	Pesticide
Dalapon	2.6	Pesticide
Total reported individual chemicals — 2		
Haloacetic acids*	53.1	Disinfection by-product
Total trihalomethanes*	62.3	Disinfection by-product

* Concentrations of individual compounds not reported.

available chemical data from the 2004 annual report for Jupiter (the city's water utility, which serves the local region) was reviewed.

The vast majority of the drinking water in the region is from groundwater resources and is treated using reverse osmosis to remove excess salts. The water quality summary is given in Table 7-34. The contaminants detected in the water supply were dominated by pesticides and disinfection by-products.

7.42 WORCESTER, MASSACHUSETTS

The city of Worcester is split by the Blackstone River and has a population of approximately 175,000. The city is a mixture of commercial, residential, and industrial areas. For the city of Worcester, the state of Massachusetts provided data on one surface water system and one groundwater system. The synthetic chemical data for the groundwater system are given in Appendix 7-39, and the data for the surface water system are given in Appendix 7-40. Given the lack of specific data on trihalomethanes and haloacetic acids, the available chemical data from the 2004 annual report was reviewed.

The vast majority of the drinking water in the region is from surface water resources that come from a network of reservoirs. The water quality summary is given in Table 7-35. The contaminants detected in the water supply were dominated by disinfection by-products.

7.43 DATA SUMMARY

The synthetic chemical data reported by each state and the water utility annual water quality reports establish that consumers in individual cities are exposed to a unique chemical mixture. This exposure may be continuous or fluctuate with seasonal pollution patterns and resource utilization (e.g., supplementing a primary surface water source with groundwater). Ultimately, this means that

TABLE 7-35 Worcester Water Data

Chemical Compound	Groundwater Average Concentration (ppb)	Surface Water Average Concentration (ppb)	Type of Chemical Compound
Chlorodibromomethane		0.7	Industrial chemical*
Methyl tertiary butyl ether (MTBE)	2.8		Petroleum
Total reported individual chemicals — 2			
Haloacetic acids**		23.7	Disinfection by-product
Total trihalomethanes		36.8	Disinfection by-product
Bromodichloromethane		7.78	Disinfection by-product
Chloroform		17.2	Disinfection by-product

* Possible disinfection by-product.

** Concentrations of individual compounds not reported.

the consumer has the potential to be exposed to the chemical mixture that has been characterized for each city.

From approximately 2000 through 2005, each community water system monitored for both regulated and unregulated chemical contaminants. However, it should be remembered that the regulated chemicals are looked for much more frequently then unregulated chemicals. Therefore, it should be no surprise that the vast majority of the databases reports more frequently on disinfection by-products.

7.43.1 Disinfection By-Products

When reviewing the water quality for each city, it is abundantly clear that anyone who drinks water from a community water system in the United States will most likely consume a mixture of disinfection by-products. *In other words, we are all exposed to a baseline number of synthetic chemicals in our drinking water independent of any other environmental contaminants. This complex mixture of specifically identified chlorinated, bromated, and fluoridated methanes, ethanes, ketones, nitriles, and acetic acids are almost always in our drinking water. This does not even include a whole range of other "unidentified" halogenated (i.e., primarily chlorine and bromine) compounds that are measured as "total organic halogens."*

For example, the New York City data (Table 7-18) shows that in 2004 the combined concentration of total trihalomethanes and haloacetic acids was 91.7 ppb; there was an additional 10.4 ppb of other disinfection by-products and 6.1 ppb of chlorinated industrial chemicals for a total of 108.1 ppb while the concentration for "total organic halogens" was 176 ppb. This means that there were an additional 67.8 ppb of and unknown number of "unidentified" halogenated chemicals in New York City's drinking water. Should the consumer be concerned? You be the judge.

According to the New York City 2004 Drinking Water Supply and Quality Report, which was discussing a system violation for exceeding haloacetic acids, "*Some studies of people who drink chlorinated drinking water for 20 to 30 years show that long-term exposure to disinfection by-products (possibly including haloacetic acids) is associated with an increased risk for certain types of cancer. However, how long and how frequently people actually drank the water as well as how much haloacetic acids the water contained is not known for certain.*" Given that the mass consumption of bottled water (which is usually not chlorinated or bromated) is a relatively recent occurrence and that community water systems in the United States have been chlorinating water since before the 1900s, who hasn't frequently drank water containing disinfection by-products?

Thus, when looking at drinking water quality in each city, it is necessary to segregate the regularly monitored disinfection by-products (total trihalomethanes and haloacetic acids) from the other chemical monitoring data to determine the extent of other environmental contaminants in drinking water. Thus, those chemicals that occur in addition to the regularly monitored disinfection by-products define the diversity of the synthetic chemical contamination due to pollutant sources. Clearly, the chemical contaminant diversity in a drinking water resource is largely dependent on the makeup of the source waters.

7.43.2 Groundwater and Surface Water Resources

When reviewing the drinking water pollution of the 36 cities that were evaluated, it is clear that both groundwater resources and surface water resources can contain a significant number chemicals. For example, Fresno and Bakersfield are located in a physical environment that is conducive to the chemical transport of contaminants through the soil and into groundwater. Thus, when chemicals from commercial/industrial facilities and agricultural lands in the Central Valley of California are released into soil, the contamination of groundwater resources is not surprising. In a similar way, large watersheds that include both urban and agricultural land use can contribute a significant number of chemicals to surface water resources. This is well illustrated by the drinking water quality of both New York City and Washington, D.C.

In all cases, the degree of pollution found in each drinking water supply is highly dependent on region-specific releases of synthetic chemicals to the surface water and groundwater resources. This emphasizes the need for consumers to review chemical data from their own water utility to determine the extent to which their drinking water is polluted. It is also important to note that even when drinking water resources were specifically treated to remove compounds like DBCP, atrazine, ethylene dibromide, tetrachloroethylene, and trichloroethylene, not all of the contaminants were removed (see Fresno, Bakersfield, and Springfield). This means that once polluted, most drinking water will still contain low levels of contaminants. This is particularly true for groundwater resources. Yet large watersheds can have fairly continuous sources of pollution that are being discharged to drinking water resources.

7.43.3 Watershed Protection

A truly protected watershed would not allow any man-made activities. Large watersheds that are all located in designated wilderness areas are a rare occurrence. Because the human habitat is polluted, there are multiple sources of synthetic chemicals discharged into both surface and groundwater resources. For example, surface runoff from both urban and rural areas will routinely be polluted with petroleum chemicals, industrial chemicals, and pesticides. In many rural areas, runoff from farm/dairy/feedlot areas can also be contaminated with pharmaceuticals and hormones from animal waste. The discharge of treated wastewater from municipal sewage treatment plants will contain pharmaceuticals as well as low levels of petroleum compounds, pesticides, and industrial chemicals that are below state discharge standards. Septic systems, sewers, landfills, and underground storage tanks can and do pollute groundwater resources.

Given these varied sources of chemical pollution, water utilities attempt to protect their water sources by diverting urban surface runoff into recharge basins (it is assumed that polluted surface water recharged to groundwater will be purified by the soil media) instead of letting the runoff into a river or lake (this is sometimes also attempted for runoff from agricultural lands and feedlots), reminding people not to dump waste into storm drains, imposing lower and lower limits on chemical concentrations allowed to be discharged by an industry or municipal sewage treatment plant, and by purchasing land in a watershed and eliminating all human activities on this land.

Beginning in 2003, many communities began the process of conducting and publishing source water assessments to determine those steps that would be necessary to protect their source waters. The assessment process will identify those sources of chemical pollution in a watershed, and consumers should be aware of the chemical discharges to their drinking water resources. This is particularly true if sewage treatment plants discharge to a drinking water resource, since pharmaceuticals that occur in wastewater are almost never removed prior to discharge. Therefore if a water utility provides water from a watershed that has human activities, it can generally be assumed that some synthetic chemicals (i.e., pharmaceuticals, unregulated industrial chemicals, pesticides, and disinfection by-products) will be in the drinking water.

Every drinking water in this study was contaminated with disinfection by-products except the drinking water in Long Beach, New Jersey. The drinking water in the Long Beach area also did not report the detection of any routinely monitored chemical.[8] However, it should be remembered that there could be chemical contaminants in this drinking water, but the monitoring program did not look for them. Because the existing monitoring program did not detect any contaminants, drinking water from this area is ranked number 1 (i.e., differentiate quality of the 36 cities).

With Long Beach, New Jersey, ranked number 1, the water quality of the remaining 35 cities was ranked solely on the number of different synthetic chemicals that was detected in the water of any community water system. This ranking, however, does not include the occurrence of disinfection by-products, trihalomethanes, and haloacetic acids, since they occur in the drinking water of all of these 35 cities. The final ranking is presented in Table 7-36 and Figure 7-1. As this table shows, each city was ranked from 1 to 12 with the 12th-ranked cities reporting the greatest number of detected chemicals.

The drinking water contaminants found in the 36 cities do not represent a statistical presentation of water quality for all cities across the United States. These data do, however, represent trends in water quality for cities that obtain their drinking water from similar geographic regions and levels of industrial development. When looking at the synthetic chemicals that were detected most frequently, the chemicals that occurred in at least 25% or more of the drinking water of all the cities (see Table 7-37 for the complete list) were as follows:

Haloacetic acids (disinfection by-product)	97.2%
Total trihalomethanes (disinfection by-product)	97.2%
Chlorodibromomethane (disinfection by-product)	33.3%
Xylenes (petroleum chemcial)	33.3%
Atrazine (pesticide)	27.8%
Dichloromethane (disinfection by-product)	27.8%
Trichloroethylene (industrial chemical)	27.8%
Tetrachloroethylene (industrial chemical)	25%
Toluene (petroleum chemical)	25%

Clearly, drinking water across the United States is contaminated with disinfection by-products. Furthermore, the high frequency of detection of atrazine (one of the most extensively used agricultural chemicals in the United States),

[8] Depending on the size of the water utility, the number of different chemicals monitored and the frequency of chemical testing will vary from city to city.

TABLE 7-36 Water Quality Ranking

Cities	Baseline* Disinfection By-Products	Number of Chemicals above Baseline	Relative** Contaminant Ranking
Long Beach, New Jersey	No	None	1
Ogden, Utah	Yes	None	2
Baton Rouge, Louisiana	Yes	None	2
Columbus, Georgia	Yes	None	2
Jackson, Mississippi	Yes	None	2
Providence, Rhode Island	Yes	None	2
Peoria, Illinois	Yes	1	3
Lafayette, Louisiana	Yes	1	3
Sioux Falls, South Dakota	Yes	1	3
West Palm Beach, Florida	Yes	2	3
Worcester, Massachusetts	Yes	2	3
Birmingham, Alabama	Yes	3	4
Oklahoma City, Oklahoma	Yes	3	4
Reno, Nevada	Yes	3	4
Springfield, Illinois	Yes	3	4
Tacoma, Washington	Yes	3	4
Charleston, South Carolina	Yes	4	4
Denver, Colorado	Yes	4	4
Portland, Oregon	Yes	4	4
Tulsa, Oklahoma	Yes	5	5
Binghamton, New York	Yes	6	5
Harrisburg, Pennsylvania	Yes	6	5
Pueblo, Colorado	Yes	6	5
Grand Rapids, Michigan	Yes	7	6
Savannah, Georgia	Yes	7	6
Phoenix, Arizona	Yes	9	7
Dayton, Ohio	Yes	10	7
Boise, Idaho	Yes	11	8
Washington, D.C.	Yes	11	8
Orlando, Florida	Yes	12	8
Milwaukee, Wisconsin	Yes	17	9
Los Angeles, California	Yes	20	10
Fresno, California	Yes	26	11
New York, New York	Yes	26	11
Bakersfield, California	Yes	30	12
San Antonio, Texas	Yes	30	12

* Reported concentration of total trihalomethanes and haloacetic acids.

** As the rank increases, there is an increasing risk of exposure to a more diverse group of chemicals.

xylenes, and toluene (water-soluble components of gasoline), and tetrachloroethylene and trichloroethylene (two of the most widely used chlorinated solvents used in the United States) is not unexpected.

These data show that almost everyone in the United States who drinks water provided by water utilities consumes a very large number of disinfection

FIGURE 7-1 Water quality ranking.

TABLE 7-37 Chemicals Detected Most Frequently in Drinking Water

Synthetic Chemicals	Frequency of Detection (%)*
Haloacetic acids	97.2
Total trihalomethanes	97.2
Chlorodibromomethane	33.3
Xylenes	33.3
Atrazine	27.8
Dichloromethane	27.8
Trichloroethylene	27.8
Tetrachloroethylene	25
Toluene	25
Dichloroethylene	19.4
Chloromethane	16.7
Di(2-ethylhexyl)phthalate	16.7
Ethylbenzene	16.7
Methyl tertiary butyl ether (MTBE)	16.7
Bromochloroacetic acid	13.9
Bromomethane	13.9
Carbon tetrachloride	13.9
Chloropicrin	13.9
Dichlorobenzene	13.9
Dichlorodifluoromethane	13.9
Haloacetonitriles	13.9
Styrene	13.9
2,4-D	11.1
Haloketones	11.1
Simazine	11.1
Trimethylbenzene	11.1
Acetone	8.3
Dalapon	8.3
Dichlorobromomethane	8.3
Dichloroethane	8.3
Dichloropropane	8.3
Ethylene dibromide	8.3
Dibromochloropropane	8.3
Phenanthrene	8.3
Trichloroethane	8.3
Benzene	5.6
Butylbenzene	5.6
Dcpa	5.6
Di(2-ethylhexyl)adipate	5.6
Dichloroacetonitrile	5.6
Dichloropropanone	5.6
Diquat	5.6
Hexachlorobutadiene	5.6
Hexachlorocylopentadiene	5.6
Methyl ethyl ketone	5.6
Methylpentanone (MIBK)	5.6
Metolachlor	5.6
Tetrahydrofuran	5.6
Trichloropropane	5.6

TABLE 7-37 *Continued*

Synthetic Chemicals	Frequency of Detection (%)*
Trichloropropanone	5.6
Trichlorotrifluoroethane	5.6
Bromochloroacetonitrile	5.6
Acenaphthene	2.8
Acenaphthylene	2.8
Aldicarb sulfone	2.8
Aldrin	2.8
Bromoacetic acid	2.8
Butylbenzylphthalate	2.8
Bromobenzene	2.8
Chlorobenzene	2.8
Propylbenzene	2.8
Carbon disulfide	2.8
Chloroacetic acid	2.8
Chlorodimethylbenzene	2.8
Chloroethane	2.8
Diazinon	2.8
Dibromoacetonitrile	2.8
Dibromochloroacetic acid	2.8
Dibromomethane	2.8
Dieldrin	2.8
Diethylphthalate	2.8
Diuron	2.8
Fluorene	2.8
Fluorotrichloromethane	2.8
Indeno(1,2,3-CD)pyrene	2.8
Methyl methacrylate	2.8
Molinate	2.8
Monochlorobenzene	2.8
Naphthalene	2.8
Nitrosodimethylamine	2.8
Thiobencarb	2.8
Trichlorobenzene	2.8
Trichlorofluoromethane	2.8

* Percentage is based on the number of cities that detected each chemical in their drinking water (1 city = 2.8%).

by-products.[9] Many others also have the potential to consume various industrial chemicals, pesticides, and petroleum chemicals. Because of the extent of the chemical contamination illustrated by the data collected for the 36 cities, what alternatives do consumers have if they do not want synthetic chemicals in the water they drink?

[9] The EPA estimates that there can be in excess of 100 different disinfection by-products in drinking water. Only nine are routinely monitored for in drinking water.

7.45 CONSUMER ALTERNATIVES

Some water utilities that exclusively derive their drinking water from groundwater resources do not disinfect their product with chlorine or bromine. This is also true for many individuals who produce drinking water from their own private wells. For most consumers, however, the water they drink has been disinfected with chlorine or bromine. As a result, most drinking water in the United States is contaminated with a complex mixture of disinfection by-products (as demonstrated by all but one city's monitoring data). This contamination is independent of the same drinking water potentially containing other industrial chemicals, pesticides, petroleum products, and hormones/pharmaceuticals. For consumers who wish not to partake of this complex mixture of synthetic chemicals, there are several alternatives.

7.45.1 Drink Bottled Water Derived from a Protected Source

The first and best alternative is to drink a bottled water that is derived from a truly protected source. Many providers of bottled water advertise a protected source, yet there is no governmental oversight that regulates these claims. In general, consumers should just avoid any bottled water that either (1) does not make any claims of a protected source or (2) simply claims to be spring, artisan, or mountain water without a detailed explanation (this information is usually available from the provider's website).

The reason this is the first and best alternative is that drinking water can and does contain unregulated chemicals, hormones, and pharmaceuticals (i.e., water utilities only test for a minuscule number of synthetic chemicals of the tens of thousands of synthetic organic chemicals used in the United States today), and treating drinking water does not remove all the chemicals of concern (e.g., see the Fresno and Bakersfield data). Bottled water is, however, not always free of contaminants.

If a bottled water does not come from a protected source (either a high-mountain/wilderness area or a deep protected groundwater aquifer), it may contain contaminants. It is also possible that the handling of the product may also introduce contaminants. For example, a study by the Natural Resources Defense Counsel found that one third of the 103 brands of bottled water tested contained elevated levels of bacteria, inorganic chemicals, or organic chemicals. In this study, the specific chemicals that were tested for varied widely between laboratories. However, in general these researchers looked for chemicals with established primary drinking water standards. The chemicals found in bottled water included acetone, butylbenzene, chlorotoluene, dichloroethane, ethylbenzene, isopropyltoluene, methylene chloride, styrene, trichloroethylene, toluene, and xylene. Most of these compounds were also found in the drinking water of the cities we evaluated. The Natural Resources Defense Counsel report concluded that bottled water was not necessarily safer than tap water. Clearly, one reason for contaminants in bottled water is that both government and industry estimate that between 25% and as high as 40% of all bottled water is actually tap water.

This means that the buyer must do some minimal homework to determine which bottled water offered in their region of the United States provides the best possible product at a cost they are willing to pay. It is also important to note that the bottled water industry has not established the concentration of water-soluble contaminants that occur in their products from plastic bottles. Currently, FDA regulations require that these plastics do not produce a chloroform-soluble extract value greater than 0.5 mg/in^2 of plastic for distilled water with a contact time of 2 hours at 250 °F. Thus, a prudent consumer would purchase water in glass bottles; if plastic containers cannot be avoided, the consumer should use a disposal carbon filter that is specifically designed for plastic bottles.[10]

Here are some general guidelines for selecting a bottled water whose product has the least probability of containing chemical pollutants:

- Do not drink bottled water whose source is from a "community" or "municipal" water utility.
- Do not drink bottled water whose source is from an artesian well, spring, or groundwater well unless it can be determined from the provider that the source water originates from a deep aquifer and that the landscape of the recharge zone is dominated by wilderness or a land use having a minimum of human development.[11] An example of a water source that originates from a deep aquifer and wilderness landscape would be the bottled water sold by Mountain Valley in Arkansas. Another example would be Voss, which comes from glacial meltwater and is sold in a glass bottle.

Drinking bottled water that is not contaminated is one step toward limiting a person's exposure to pollutants. However, for those individuals who wish to minimize their exposure and their family's exposure to chemical pollutants in their household water, the only real alternative is to treat the water prior to its consumption in the home.

---

[10] If a carbon filter is used, the consumer should make sure that the manufacturer's recommendations on the volume of water that can be treated per filter are followed.

[11] The recharge zone is the area at the earth's surface where surface water (i.e., rivers, lakes, rainfall) infiltrates through the soil and percolates downward until it flows into or "recharges" an aquifer. The longer the travel time of the water (i.e., the time between recharge and the water's exit at the spring), the better.

7.45.2 Treat Household Water

The only remaining alternative to is to treat household water to remove disinfection by-products and any additional chemicals that my be in a source of drinking water. When treating drinking water, consumers have two basic choices. Home purification systems fall into two broad categories: point-of-use systems and point-of-entry systems. A point-of-use system is installed at the location of the water's use, while a point-of-entry system is installed to treat all water entering the house. Since point-of-use systems only provide a means of supplying chemically pure water to one location in a household, they are not universally recommended as they do not treat shower or bath water, dishwasher water, or water used for the washing machine. However, if space limitations or economic considerations do not allow the use of a point-of-entry system, a point-of-use system can be used.

Although there are various methods of removing pollutants from water, the two most commonly used techniques employed by in-home water treatment systems are reverse osmosis (RO) and granulated activated carbon (GAC). RO technologies use fine, porous membranes to separate inorganic and organic chemicals from water. These units are effective at removing dissolved salts, suspended matter, and a wide variety of dissolved organic chemicals as well as bacteria and viruses. GAC filters can reduce chlorine, many man-made synthetic chemicals, some radiological constituents, fluoride, radon, and some metals. Depending on the odor, taste, bacterial levels, chlorine, or chloramine levels, and the amount of suspended sediment in a water, there are additional treatment methods that can be employed.

The number and type of water treatment methods needed to provide a pure, good-tasting water depend largely on the source of the water supply. Water from a community water supply will usually require less treatment than water from a private well. This means that each consumer should consult with a local provider of water treatment equipment and services to determine what basic water treatment methods need to be used to maximize contaminant removal. Those who want a simple solution with the lowest economic cost (and who realize that this simple solution may be inadequate but is better than nothing) can install a point-of-use GAC filter at their primary drinking water tap. It is important to follow the manufacturer's recommended schedule for changing GAC filters.

Consumers who install point-of-entry systems also assume the responsibility for maintaining them. Failure to do so will result in either decreased pollutant removal efficiency or no pollutant removal at all. Home owners who do not want to assume these very important responsibilities should engage a water purification service company to provide routine monitoring surveillance and maintenance activities.

8 SYNTHETIC CHEMICAL CONTAMINANTS IN AIR

Air pollution does not recognize political boundaries, nor does exposure to toxic chemicals necessarily occur near their source of release. The inherent complexity of the atmosphere and the difficulty of monitoring and assessing chemical exposures require the close cooperation of federal, state, and regional air pollution agencies to control air toxins. Because of these complexities, emissions of toxic compounds to the atmosphere from stationary sources (e.g., industries and disposal sites) are controlled by placing limits on the amount of specific chemicals that can be released to the atmosphere and then requiring the discharger to report their releases. Thus, toxic emissions are controlled but they are not eliminated.

This means that although the air that surrounds us appears to be "crystal clear" (except in those instances when smog is apparent), our air resources actually contains chemical mixtures. Considering that each individual takes in approximately 3400 gallons of air every day, the pollutant content and chemical diversity of the hazardous compounds in the air we breathe should be a primary concern.

8.1 HAZARDOUS AIR POLLUTANTS

As discussed in previous chapters, the United States uses or manufactures about 72,000 chemicals. The Environmental Protection Agency (EPA) has identified 188 of these chemicals as hazardous (Appendix 8-1). The EPA has placed emission limits on these 188 hazardous chemicals and requires that industry report the amount each facility releases. Some of these chemicals can also be released from mobile sources (automobiles, boats, lawnmowers, aircraft, etc.) as well as from small businesses (e.g., vapors from service stations, exhaust from wood-burning heaters, vapors from dry cleaning or spray painting operations). Given these multiple sources, the EPA monitors and tracks toxic air emissions under the National Air Toxics Control Program, which was implemented in 1990.

Under this program, state air pollution control agencies actually determine the concentration of selected hazardous chemicals in the environment through 264 state-maintained monitoring stations. This program focuses on monitoring 33 urban hazardous air pollutants that represent the most hazardous threat to public health in urban environments. In addition to air monitoring, this program also tracks the amount of hazardous air emissions released by industrial sources using the Toxic Release Inventory (TRI).

The TRI program compiles air pollution discharges specifically from industrial sources. The TRI program dates back to 1988 and has been refined over time to include more chemicals, bringing today's total to 188. It should also be noted that, just like for food and water, numerous other chemicals occur in air emissions that are not reported. Since 1994, federal facilities have had to report their releases as part of the program. Industry and federal facilities present these data to the EPA on an annual basis (this information is often based on industry estimates regarding both the composition of and the amount of each chemical released). Once compiled, these data are made available to the public on the EPA TRI website.

In any given community, there can be multiple sources of specific chemicals being discharged from stationary industrial facilities. This is illustrated using data collected by the EPA on four common air contaminants. For example, four common chemical sources are given for tetrachloroethylene, ethylene oxide, benzene, and acetaldehyde (Tables 8-1, 8-2, 8-3, and 8-4). This type of information is available for all 188 hazardous chemicals. Clearly, these data demonstrate that the greater the number of source areas in a community, the greater the diversity of chemicals discharged into the air.

TABLE 8-1 Tetrachloroethylene (PCE) Emissions (1999)

Source	Percentage
Halogenated solvent cleaners	11
Dry cleaning	63
Consumer and commercial products	12
Municipal landfills	2
Other categories	12

TABLE 8-2 Ethylene Oxide Emissions (1999)

Source	Percentage
Consumer and commercial product use	34
Hospital sterilizers	26
Commercial sterilization facilities	11
Organic chemical production	4
Pharmaceuticals production	4
Cellulose products manufacturing	2
Industrial gases	2
Miscellaneous organic chemicals manufacturing	2
Other categories	15

TABLE 8-3 Benzene Emissions (1999)

Source	Percentage
Mobile on road	49
Mobile nonroad	19
Open burning	14
Residential wood burning	6
Oil and natural gas production	2
Gasoline distribution	1
Other categories	9

TABLE 8-4 Acetaldehyde Emissions (1999)

Source	Percentage
Mobile on road	32
Mobile nonroad	26
Open burning	19
Residential wood heating	3
Pulp and paper production	7
Other categories	13

8.2 DATA SOURCE AND LIMITATIONS

The purpose in this study of evaluating toxic releases to air is to define the chemical mixtures to which a community may be exposed. Because of this objective, the actual chemical monitoring data on the 33 hazardous air pollutants collected by various states around the nation were not used because the number of chemicals included in this program is so limited. Thus, to evaluate the chemical diversity of air contaminants in a given city, the data available from the TRI were used. This database, however, also has some limitations.

First of all, virtually all nongovernment releases are most often based on estimates rather than on direct measurements. Furthermore, there is no way to verify the quality of self-reported industrial discharges. The problem with industrial discharge estimates is typical of the problems encountered by Thermatrix Corporation located in Santa Clara County, California. During the 1990s, Thermatrix manufactured thermal oxidizers that were specifically designed to destroy complex organic chemical emissions from industrial sources. Unfortunately for Thermatrix, units that were designed using plant-specific information failed to operate properly because the industries' reported emissions were underestimated.

It should also be remembered that the TRI database does not have any data on mobile source emissions or emissions from small business. Given these limitations, the TRI data should only be considered qualitative and viewed with caution. For example, the EPA explains the basis for its data reporting through the Toxics Release Inventory (TRI) program as follows:

> Users of TRI information should be aware that TRI data reflect releases and other waste management activities of chemicals, not whether (or to what degree) the public has been exposed to those chemicals. Release estimates alone are not sufficient to determine exposure or to calculate potential adverse effects on human health and the environment. TRI data, in conjunction with other information, can be used as a starting point in evaluating exposures that may result from releases and other waste management activities which involve toxic chemicals. The determination of potential risk depends upon many factors, including the toxicity of the chemical, the fate of the chemical, and the amount and duration of human or other exposure to the chemical after it is released.

These data, however, do indicate the chemical mixtures to which a community is exposed.

8.3 TOXIC CHEMICAL RELEASES BY CITY

The toxic chemical emissions for each of the 36 cities were compiled for 1996 through 2001. All facilities reporting toxic emissions within a 5-mile radius of each city were included in this evaluation. Of the 36 cities, only one had no reported toxic air emissions. That city was Long Beach, New Jersey. This result was not surprising because this is a resort city located on a small island off the New Jersey coast. Other cities were not as fortunate. A summary of toxic air emissions for each city can be found in the following appendixes:

Appendix 8-2 for Bakersfield, California
Appendix 8-3 for Baton Rouge, Louisiana
Appendix 8-4 for Binghamton, New York

Appendix 8-5 for Birmingham, Alabama
Appendix 8-6 for Boise, Idaho
Appendix 8-7 for Charleston, South Carolina
Appendix 8-8 for Columbus, Georgia
Appendix 8-9 for Dayton, Ohio
Appendix 8-10 for Denver, Colorado
Appendix 8-11 for Fresno, California
Appendix 8-12 for Grand Rapids, Michigan
Appendix 8-13 for Harrisburg, Pennsylvania
Appendix 8-14 for Jackson, Missouri
Appendix 8-15 for Lafayette, Louisiana
Appendix 8-16 for Los Angeles, California
Appendix 8-17 for Milwaukee, Wisconsin
Appendix 8-18 for New York, New York
Appendix 8-19 for Ogden, Utah
Appendix 8-20 for Oklahoma City, Oklahoma
Appendix 8-21 for Orlando, Florida
Appendix 8-22 for Peoria, Illinois
Appendix 8-23 for Phoenix, Arizona
Appendix 8-24 for Portland, Oregon
Appendix 8-25 for Providence, Rhode Island
Appendix 8-26 for Pueblo, Colorado
Appendix 8-27 for Reno, Nevada
Appendix 8-28 for San Antonio, Texas
Appendix 8-29 for Savannah, Georgia
Appendix 8-30 for Sioux Falls, South Dakota
Appendix 8-31 for Springfield, Illinois
Appendix 8-32 for Tacoma, Washington
Appendix 8-33 for Tulsa, Oklahoma
Appendix 8-34 for Washington, D.C.
Appendix 8-35 for West Palm Beach, Florida
Appendix 8-36 for Worcester, Massachusetts

By reviewing the data in these appendixes, the extent to which each community may be impacted by complex mixtures is directly related to the number of individual chemicals released. Thus, each city has been ranked according to the number of individual chemicals reported for each city. This ranking is provided in Table 8-5 and illustrated in Figure 8-1.

Based on the TRI database, Washington, D.C., ranks as having the least number of reported chemicals in its air for those cities having industries reporting emissions. The other top tier (or best cities) having the least number of chemical mixtures are Binghamton, New York City, and Lafayette. Among the cities with the greatest number of different chemicals are Baton Rouge, Los Angeles, Birmingham, Portland, and Denver.

TABLE 8-5 Air Contaminant Ranking

City	Number of Different Chemicals	Relative Rank
Long Beach	0	1
Washington	1	2
Binghamton	2	2
New York	3	2
Lafayette	4	2
Harrisburg	8	3
Reno	9	3
Sioux Falls	9	3
Worcester	12	4
Fresno	13	4
West Palm Beach	13	4
Boise	14	4
Peoria	14	4
Pueblo	15	4
Ogden	18	5
Bakersfield	22	6
Orlando	23	6
Jackson	24	6
Springfield	26	7
Oklahoma City	27	7
Providence	29	7
San Antonio	29	7
Grand Rapids	33	8
Savannah	34	8
Tacoma	37	9
Columbus	39	9
Tulsa	40	9
Milwaukee	41	9
Phoenix	41	9
Charleston	42	9
Dayton	43	9
Denver	45	10
Portland	45	10
Birmingham	47	10
Los Angeles	51	11
Baton Rouge	108	12

Although this might come as a surprise, these results parallel the lack of manufacturing in the core areas of the best cities. Generally, one does not find concentrated manufacturing activity in the center of major cities composed of dense residential and white-collar commercial land use (e.g., Washington, D.C., and New York City). On the other hand, cities with dense urban areas like Los Angeles do have large pockets of industrial areas surrounded by residential and commercial land use. Baton Rouge, on the other hand, simply has a large number of chemical and petroleum facilities located within the area.

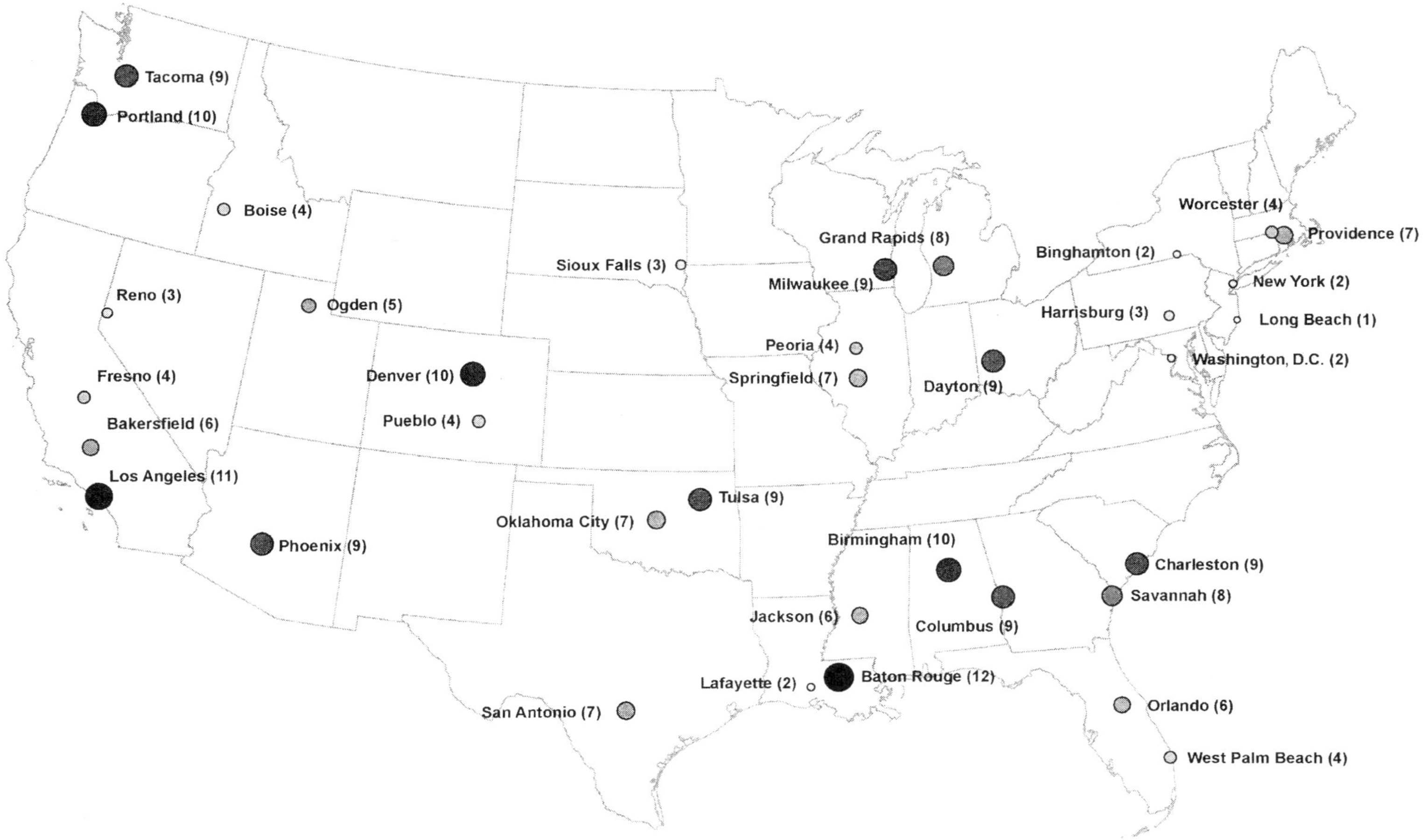

FIGURE 8-1 Air quality ranking.

Looking at each city in terms of just how much total organic pollutant emissions (in pounds per year) are released to the atmosphere reveals patterns similar to the number of pollutants identified. Release rankings by city are found in Table 8-6. These rankings show the total pounds of pollutants released between 1996 and 2001. On a weight basis it should come as no surprise that Washington shows the least pounds of chemicals released. New York and Binghamton also have low releases. However, a different pattern appears when

TABLE 8-6 Ranking of Toxic Chemical Releases to the Atmosphere

City	Pounds Released 1996–2001	Ranked Minimum to Maximum
Long Beach/Asbury Park, NJ	0	1
Washington, D.C.	0.2	2
New York, NY	8	3
Binghamton, NY	90	4
Lafayette, LA	4556	5
West Palm Beach, FL	117,401	6
Harrisburg, PA	179,541	7
Pueblo, CO	183,246	8
Ogden, UT	309,813	9
Fresno, CA	580,224	10
Bakersfield, CA	662,469	11
Worcester, MA	886,540	12
Boise, ID	1,027,187	13
Columbus, GA	1,379,532	14
Jackson, MS	1,536,094	15
Sioux Falls, SD	1,773,877	16
Los Angeles, CA	2,444,398	17
Denver, CO	2,555,455	18
Reno, NV	2,593,630	19
Providence, RI	2,883,357	20
San Antonio, TX	2,996,834	21
Springfield, MO	3,294,011	22
Dayton, OH	3,539,301	23
Peoria, IL	3,566,512	24
Tulsa, OK	5,173,647	25
Orlando, FL	5,824,689	26
Milwaukee, WI	6,545,885	27
Oklahoma City, OK	7,070,523	28
Phoenix, AZ	7,179,764	29
Portland, OR	7,467,609	30
Tacoma, WA	8,093,819	31
Birmingham, AL	11,270,988	32
Grand Rapids, MI	14,835,516	33
Savannah, GA	17,878,197	34
Charleston, SC	19,943,561	35
Baton Rouge, LA	32,051,286	36

reviewing the largest contributors of chemicals to the atmosphere This list includes Baton Rouge, Charleston, Savannah, Grand Rapids, Birmingham, and Tacoma. One can conclude that although there is a close correlation between the number of chemicals and pounds of chemicals among the least polluted cities, such a correlation does not necessarily exist within the dirtiest cities. The only cities with the largest releases and largest number of chemicals are Baton Rouge and Birmingham. By way of prospective, Table 8-7 shows the total pounds of toxic air pollutants discharged by year and in total during 1996 through 2001 by each city. At the "quality" end of the scale, Washington released only 0.2 pounds of toxic pollutants in the 6-year period, while Baton Rouge released more than 32 million pounds of organic toxic pollutants.

When looking across all the cities in the United States, the most frequently detected chemicals (i.e., the number of cities where the chemical was discharged) are summarized in Table 8-8. Chemicals such as xylene, glycol ethers, ethyl benzene, methanol, methyl ethyl ketone, styrene, and toluene are most prominent. The frequency of occurrence is found in Table 8-8. The highest reported frequency values were for xylenes (86%), toluene (81%), and glycol ethers (81%). Based on the TRI data, this would be the most common mixture of chemicals individuals could be exposed to due to industrial releases.

When looking at the actual monitoring data collected around the United States, many of the most frequently reported chemicals that were released (Table 8-8) are also detected by over a quarter of the urban monitoring locations (Table 8-9). For example, the most common compounds are the single-ring aromatics like toluene, xylene, benzene, and styrene. This is significant, because this includes all sources that would be monitored. Therefore, this suggests that in many urban areas, individuals are exposed primarily to the single-ring aromatic compounds. The concentration, however, would appear to be decreasing.

8.4 CONCENTRATION TRENDS

Based on a review of the total amount of organic toxic pollutants released by all the cities in the study, on a year-to-year basis, one can conclude that significant gains have been made in reducing toxic emissions to the atmosphere (Table 8-10). This table shows that each year there has been a reduction in total emissions. As a result, there was an overall reduction of 42%.

When looking at all of the available data and based on the fact that both benzene and toluene appear prominently in the TRI and monitoring databases, a review of the reported data was made to determine if there were reductions in the discharges of these chemicals over the 6-year study period. The results of this analysis are summarized in Table 8-11 for benzene and Table 8-12 for toluene. In both cases there have been considerable reductions.

TABLE 8-7 Total Pounds of Toxic Air Pollutants Discharged

City	1996	1997	1998	1999	2000	2001	Total
Bakersfield, CA	90,913	100,054	272,591	67,422	68,067	63,422	662,469
Baton Rouge, LA	5,535,130	5,177,948	5,965,773	5,423,043	5,390,553	4,558,839	32,051,286
Binghamton, NY	10	0	0	20	20	10	60
Birmingham, AL	1,882,540	2,102,089	2,073,151	1,963,761	1,914,295	1,335,152	11,270,988
Boise, ID	18,508	75,303	257,556	242,266	243,434	190,120	1,027,187
Charleston, SC	4,394,446	3,723,656	3,114,007	3,245,741	2,952,239	2,513,472	19,943,561
Columbus, GA	564,117	326,465	166,683	124,289	120,028	78,273	1,379,855
Dayton, OH	1,344,321	609,517	579,138	487,775	270,658	247,892	3,539,301
Denver, CO	638,323	539,395	467,011	375,988	271,085	263,654	2,555,456
Fresno, CA	63,623	61,785	206,012	130,612	70,095	48,097	580,224
Grand Rapids, MI	3,411,470	3,069,578	3,423,973	2,194,909	2,006,523	729,063	14,835,516
Harrisburg, PA	59,710	30,181	79,025	3747	3501	3377	179,541
Jackson, MS	301,786	143,966	70,696	541,671	210,638	267,337	1,536,094
Lafayette, LA	2053	2280	223	0	0	0	4556
Long Beach/ Asbury Park, NJ	0	0	0	0	0	0	0
Los Angeles, CA	795,466	565,160	395,551	228,100	238,652	221,469	2,444,398
Milwaukee, WI	1,014,831	1,037,842	1,121,479	968,997	661,729	592,672	5,397,550
New York, NY	0	0	0	0	1	6	7
Ogden, UT	45,046	60,705	59,449	52,321	43,232	49,060	309,813
Oklahoma City, OK	1,321,135	1,329,945	1,328,225	1,245,288	1,417,046	428,884	7,070,523
Orlando, FL	972,063	879,420	832,059	1,208,543	1,146,609	785,995	5,824,689
Peoria, IL	116,047	172,301	158,058	1,156,127	1,269,796	694,184	3,566,513
Phoenix, AZ	1,059,090	1,713,149	1,179,588	1,146,309	1,127,504	954,124	7,179,764
Portland, OR	1,804,543	1,467,299	1,290,996	1,314,142	920,127	670,502	7,467,609
Providence, RI	948,924	700,987	324,105	351,714	328,833	228,794	2,883,357
Pueblo, CO	75,900	45,000	21,188	15,745	10,654	14,758	183,245
Reno, NV	324,288	649,197	576,430	478,460	329,906	235,349	2,593,630
San Antonio, TX	738,217	914,562	428,141	358,133	287,030	270,751	2,996,834
Savannah, GA	2,967,969	3,649,565	3,330,116	2,599,317	2,883,264	2,447,955	17,878,186
Sioux Falls, SD	346,783	303,341	373,487	273,061	265,083	212,122	1,773,877
Springfield, MO	933,582	1,023,915	218,775	475,468	373,992	268,279	3,294,011
Tacoma, WA	1,498,447	1,737,870	1,461,725	1,248,572	1,079,504	1,067,701	8,093,819
Tulsa, OK	1,059,164	1,072,419	1,091,760	592,118	925,806	432,381	5,173,648
Washington, D.C.	0	0	0	0	0.1	0.1	0.2
West Palm Beach, FL	28,718	24,567	13,398	14,603	13,214	22,901	117,401
Worcester, MA	148,543	174,634	180,455	182,161	152,476	48,281	886,550
Total	34,505,706	33,484,095	31,060,824	28,710,423	26,995,594	19,944,876	174,701,518

TABLE 8-8 Frequency of Occurence of Air Contaminants

Air Contaminant	Number of Cities with Detected Contaminant	Frequency of Occurrence
Xylenes	31	86
Toluene	29	81
Glycol ethers	29	81
Ethylbenzene	27	35
Methyl ethyl ketone	27	75
Methanol	25	69
Styrene	25	69
Ethylene glycol	24	66
Methyl isobutyl ketone	24	66
PAHs	24	66
Trimethylbenzene	24	66
Hexane	22	61
Benzene	21	58
Butyl alcohol	21	58
Dioxins	21	58
Dichloromethane	20	55
Trichloroethylene	20	55
Napthalene	19	53
Methyl pyrrolidone	19	53
Diisocyanates	18	50

TABLE 8-9 Organic Pollutant Frequency of Occurrence in Urban Areas

Pollutant	Percentage*
Benzene	31
Toluene	30
Perchloroethylene	28
Chloroform	28
Methylene chloride	27
Trichloroethylene	25
1,3-Butadiene	23
Styrene	22
Ethylene dichloride	21
Carbon tetrachloride	21
Vinyl chloride	20

* Percentage of monitoring sites with detected levels of pollutant.

TABLE 8-10 Comparison of Total Pollutants Discharged Year to Year

Year	Total Pounds	Delta	Percentage Change	Overall Reduction
1996	34,505,706			
1997	33,484,095	1,154,007	3.34	
1998	31,060,824	2,423,271	7.24	
1999	28,710,423	2,350,401	7.57	
2000	26,995,594	1,714,829	5.97	
2001	19,944,876	7,050,718	26.1	42.2%

TABLE 8-11 Total Pounds of Benzene Discharged

City	1996	1997	1998	1999	2000	2001	Gain/loss 2001–1996
Bakersfield, CA	4969	0	0	0	0	2760	
Baton Rouge, LA	167,329	0	0	0	0	81,947	
Binghamton, NY	0	0	0	0	0	0	
Birmingham, AL	26,412	0	0	0	0	63,419	
Boise, ID	0	0	5976	0	0	12,706	
Charleston, SC	117,000	0	0	0	0	17,233	
Columbus, GA	0	54,414	0	0	33,210	0	
Dayton, OH	260	0	0	0	0	1770	
Denver, CO	15,421	0	0	0	0	7793	
Fresno, CA	0	0	0	0	0	0	
Grand Rapids, MI	0	0	0	0	0	0	
Harrisburg, PA	0	0	845	0	0	608	
Jackson, MS	0	0	0	0	0	0	
Lafayette, LA	0	0	0	0	0	0	
Long Beach/Asbury Park, NJ	0	0	0	0	0	0	
Los Angeles, CA	0	0	0	0	0	0	
Milwaukee, WI	0	0	714	0	0	330	
New York, NY	0	0	0	0	0	0	
Ogden, UT	0	0	0	0	0	0	
Oklahoma City, OK	296	0	0	0	500	0	
Orlando, FL	0	0	0	0	0	0	
Peoria, IL	0	0	0	2569	0	790	
Phoenix, AZ	180	0	0	0	0	2199	
Portland, OR	6354	0	0	0	0	23,206	
Providence, RI	0	0	2319	0	0	3670	
Pueblo, CO	0	0	3579	0	0	3122	
Reno, NV	0	0	0	0	0	0	
San Antonio, TX	310	0	0	0	0	3966	
Savannah, GA	1262	0	604	0	0	0	
Sioux Falls, SD	0	0	0	0	0	0	
Springfield, MO	0	0	0	0	0	0	
Tacoma, WA	9132	0	0	0	0	3301	
Tulsa, OK	22,034	0	0	0	0	20,210	
Washington, D.C.	0	0	0	0	0	0	
West Palm Beach, FL	500	0	0	500	0	0	
Worcester, MA	0	0	0	0	0	0	
Total	371,459	54,414	14,037	3069	33,710	249,030	−122,429
Percentage reduction							32.60%

TABLE 8-12 Total Pounds of Toluene Discharged

City	1996	1997	1998	1999	2000	2001	Gain/Loss 2001–1996
Bakersfield, CA	7097	6319	4087	3410	3847	4464	
Baton Rouge, LA	214,972	156,351	183,385	135,021	147,976	166,446	
Binghamton, NY	0	0	0	0	0	0	
Birmingham, AL	133,356	208,739	324,172	153,448	180,527	229,521	
Boise, ID			8616	7874	5299	15,840	
Charleston, SC	50,577	49,626	45,583	66,715	56,142	13,986	
Columbus, GA					138	137	
Dayton, OH	18,695	13,268	27,734	32,320	23,615	26,388	
Denver, CO	166,106	114,103	106,039	92,085	79,824	73,296	
Fresno, CA	0	0	0	0	0	0	
Grand Rapids, MI	684,479	710,915	673,151	554,538	334,713	67,188	
Harrisburg, PA			1276	1173	1097	1022	
Jackson, MS	46,307	3557	4791	4902	5548	7728	
Lafayette, LA	571	850	0	0	0	0	
Long Beach/Asbury Park, NJ	0	0	0	0	0	0	
Los Angeles, CA	131,115	108,564	61,525	49,823	66,901	49,307	
Milwaukee, WI	14,604	13,077	73,976	79,875	76,726	68,812	
New York, NY	0	0	0	0	0	0	
Ogden, UT	20,893	22,032	0	96	0	0	
Oklahoma City, OK	52,695	46,198	52,473	0	81,403	26,097	
Orlando, FL	6330	3030	21,131	18,176	3426	2310	
Peoria, IL	10,700	12,350	8856	5076	0	0	
Phoenix, AZ	66,285	78,135	59,045	50,102	47,547	46,892	
Portland, OR	316,916	235,601	210,703	207,564	88,852	53,781	
Providence, RI	647,565	438,851	51,756	65,830	86,815	47,175	
Pueblo, CO	12,000	12,000	284	284	282	0	
Reno, NV	299,324	619,855	555,859	471,286	319,721	235,337	
San Antonio, TX	249,387	278,627	186,937	116,380	73,339	51,916	
Savannah, GA	274,912	331,320	308,047	290,965	294,885	282,785	
Sioux Falls, SD	27,476	0	0	0	17,967	25,994	
Springfield, MO	324,558	371,245	44,240	139,110	137,280	129,892	
Tacoma, WA	105,410	114,875	43,880	42,071	75,741	72,148	
Tulsa, OK	174,008	225,655	265,315	134,009	198,747	86,103	
Washington, D.C.	0	0	0	0	0	0	
West Palm Beach, FL	500	0	500	500	0	0	
Worcester, MA	18,062	23,727	20,655	21,277	15,085	3474	
Total	4,074,900	4,198,870	3,344,016	2,743,910	2,423,443	1,788,039	−2,286,861
Percentage reduction							56.10%

8.5 THE OCCURRENCE OF SINGLE-RING AROMATICS

Virtually all databases support the observation that toxic chemicals such as the single-ring aromatics (i.e., benzene, toluene, ethylbenzene, styrene, and xylene) and chlorinated solvents (i.e., trichloroethylene and tetrachloroethylene) are pervasive across the 35 reporting cities. There is merit in examining just why there is such of a preponderance of these chemicals being released into the atmosphere. Among the most common chemicals are the BTEX (benzene toluene, ethylbenzene, and xylene) compounds. The sources, manufacturing, and uses of these chemicals shed light on just why they appear in so many air emissions.

Benzene is a colorless liquid that evaporates very quickly. Benzene ranks in the top 20 chemicals produced for production of other chemicals. Benzene is employed in the manufacture of plastics, resins, synthetic fibers, rubber, lubricants, dyes, detergents, drugs, styrene, phenol, and pesticides. Some of the specific chemicals produced from benzene include ethylbenzene, cumene cyclohexane, chlorobenzene, and nitrobenzene. The annual production of benzene is in excess of 3 billion gallons. Not to be overlooked, benzene also occurs naturally with sources being volcanoes, forest fires, and cigarette smoke. It is therefore no surprise that benzene is so often found in the air.

Similar to benzene, toluene is also a liquid and has found many uses including the production of polyurethanes, solvents, gasoline, and explosives. The annual production capacity approaches 2 billion gallons. To add further prospective, it is estimated that there are in excess of 3000 facilities that produce or process toluene in the United States. Every state where we have identified one of our study cities has at least two (Nevada) or more (208 in Texas) plants producing or processing toluene.

Produced from benzene, ethylbenzene has many uses including being a precursor in the production of styrene. In fact, the major use of ethylbenzene is for the manufacture of styrene, taking up about 90% of the total ethylbenzene production. Additional uses of this chemical include incorporation into asphalt and naphtha. Ethylbenzene is also found in fuels. The chemical is also produced in each of the states where we have identified study cities. The estimated annual production volume is in excess of 12 billion pounds.

Xylene, another of the BTEX compounds, is also produced in high volume; is incorporated into most pesticides, weed killers, and insecticides; and is a common component of petroleum fuels. Xylenes are also used in the manufacture of dyes as well as other organic chemicals.

8.6 CONCLUSIONS

The massive amount of available data relative to each study city has been tabulated, discussed, and analyzed. Based on this summary, the following conclusions can be made:

1. The EPA has been reporting information, city by city, in its TRI program; however, as stated earlier, the information is qualitative not quantitative.
2. The greater the industrial base in any given city, the greater the number and quantity of chemicals released to the air environment.
3. On an overall basis, chemicals such as the single-ring aromatics group (e.g., the BTEX compounds) as well as trichloroethylene and perchloroethylene are common to most if not all cities.
4. Multiple chemical releases are reported for most of the cities.
5. The actual exposure to the TRI chemicals cannot be directly correlated to the reported releases.
6. The ranking of the study cities, best to worst, does represent a reasonable "safety guide."
7. The comparison of the amount of chemicals released between 1996 and 2001 demonstrates significant gains in the reduction of air toxic releases across the spectrum of the cities in this study. Wider-reaching studies demonstrate that this trend (reduction in air pollutant discharges) has been ongoing since the 1970s and continued through 2004.

It is clear that many different chemicals are routinely discharged to the atmosphere and that in certain locations the amounts discharged appear to be large. However, these data do show that in the face of population increases and continued economic growth, there has been a substantial improvement in air quality across the 36-city study as well as across the nation. This demonstrates that federal and state air quality control programs are having a positive effect on reducing air pollution.

9 TOXICITY AND SYNTHETIC CHEMICAL MIXTURES

So why are we worried about the presence of synthetic chemical mixtures in our food, water, and air? In short, the reasons have to do with our current health, our future health, and the health of future generations. While the types of reactions that are possible from exposure to chemicals are varied and complex, the primary reasons for evaluating chemicals in food, water, and air are simple:

1. Toxicological and epidemiological data have clearly demonstrated that the presence of synthetic chemicals in food, water, and air result in adverse health effects and will be a significant driver in health care costs.
2. Federal and state legislation requiring that food and water be unadulterated.
3. Exposures in the real world are not "neat," therefore the basis for the previous toxicological studies (and the predicted public health outcomes) are limited in their ability to predict disease outcomes in the general population.
4. The toxic outcome (e.g., cancer in a specific organ) becomes less specific with exposure to more toxins (i.e., a greater number of cancers will occur that are less specific to any one contaminant).
5. Exposure to toxins early in the development of a biological system causes irreversible damage that is expressed later.

Given these reasons, it is important to understand how individual synthetic chemicals and mixtures of synthetic chemicals in the environment can have such a far-reaching impact on our lives.

For example, for individuals with compromised immune systems, exposure to chemicals in our environment can lead to serious consequences (e.g., increased incidence of allergic reactions, immuno-suppression, and autoimmunity). For women of childbearing age or those who will be of childbearing age exposure to synthetic compounds in the environment can impact their immediate and future health as well as affect their potential to have children or to alter the development of their children. Exposure to trihalomethanes (disinfection by-products in drinking water) can result in fetotoxicity including miscarriage/spontaneous abortion, and/or stillbirths. Males are also not immune to the reproductive disorders. Exposure of men to dibromochloropropane (DBCP) (involved in applying or formulating the pesticide) resulted in sterility. Children may develop hyperactivity, Attention Deficit Disorder, Attention Deficit Hyperactivity Disorder, decreased IQ, mental retardation, and developmental delays following exposure to polychlorinated biphenyls (PCBs) as well as chloracne, cirrhosis of the liver, and cancer.

The cumulative effects of exposure to all of these compounds simultaneously have yet to be comprehensively studied, nor do toxicologists know which chemical mixtures that contaminate the environment should be studied. This major shortcoming emphasizes the need to appreciate how single chemical toxicity influences human health.

9.1 SINGLE CHEMICAL TOXICOLOGICAL EFFECTS

While Paracelsus (1493–1541) is generally credited with the adage "All substances are poisons; there is none that is not a poison. The right dose differentiates a poison and a remedy," it is Mathieu Orfila who is generally considered to be the father of modern toxicology, since he was the first to write about it in 1813 in his *Trait des Poisons*, also called *Toxicologie Generate*. Orfila, who was renowned in courts across Europe for his expertise in the toxicity of arsenic, the poison of choice during the 1800s, moved toxicology as a forensic science forward. Modern toxicology is an applied science that is coupled with biological sciences (physiology, biochemistry, pathology, pharmacology, and chemistry), medical sciences (general medicine, occupational medicine, and pathology), and public health (environmental sciences, epidemiology, and biostatistics) to determine a cause-and-effect relationship between exposure to chemicals and health outcomes. Since Orfila's time, toxicology has moved away from death as the only adverse health outcome to an understanding that the "sentinel" events at a cellular level ultimately lead to adverse health outcomes.

For example, halogenated hydrocarbons (e.g., methyl chloride, vinyl chloride, trichloroethylene, and perchloroethylene), commonly used as solvents, aerosol propellants, chemical intermediates, refrigerants, and dry cleaning agents, suppress the central nervous system resulting in anesthetic properties. Exposure to halogenated hydrocarbons can lead to altered mental states

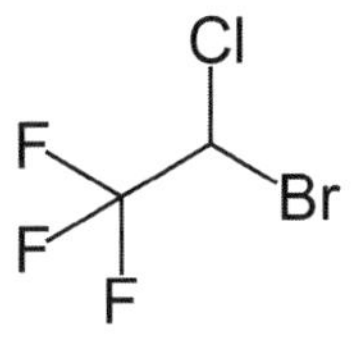

FIGURE 9-1 Chemical structure of halothane.

(vapor/fume intoxication), tremors, blurred vision, nausea, and speech difficulties. Longer-term health effects such as Central Nervous System depression, liver and kidney damage, cardiac arrythmias, cardiac arrest, cancer, or death are possible since the detoxification process for halogenated hydrocarbons may result in the activation of toxic by-products.

An example of the signficant toxicity of halogenated hydrocarbons is demonstrated in 2-bromo-2-chloro-1,1,1-trifluoroethane. Halothane (2-bromo-2-chloro-1,1,1-trifluoroethane) is a general anaesthetic delivered via inhalation that is still used in veterinary medicine and outside the United States because its is relatively inexpensive. Once widely used as a general anaesthetic, it has largely been replaced by sevoflurane and desflurane because of its toxicity. Figure 9-1 provides a diagram of the structure of halothane.

Multiple exposures to halothane increase the risk of developing two types of liver damage. Type I hepatotoxicity is benign and self-limiting. Type I hepatotoxicity is relatively common (up to 25% to 30% incidence) and characterized by mild, transient increases in serum transaminase and glutathione *S*-transferase concentrations and by altered postoperative drug metabolism. It is assumed that Type I results from reductive (anaerobic) biotransformation of halothane rather than the normal oxidative pathway.

Type II hepatotoxicity is characterized by massive centrilobular liver cell necrosis that leads to fulminant liver failure. Clinical features of Type II hepatotoxicity include fever, jaundice, and grossly elevated serum transaminase levels. Type II hepatotoxicity is immune mediated and is initiated by oxidative halothane metabolism to an intermediate compound. This intermediate compound then binds to trifluoroacetylate proteins in the hepatic endoplasmic reticulum. Halothane can also cause malignant hyperthermia.

The understanding that the cellular necrosis is the sentinel event leading to liver failure, and ultimately the death of the individual with exposure to halothane, has resulted in the minimization of the use of the compound. Other halogenated compounds, such as PCE, TCE, and vinyl chloride, are more ubiquitous in the environment and are likely to follow similar detoxification pathways. Figure 9-2 provides a clear example of how dexotification may end up leading to activation of a chemical in the body.

Occupational medicine predates modern toxicology and is a significant contributor to our understanding of the adverse health effects from exposure to chemicals. Bernardino Ramazzini first described systematically occupational

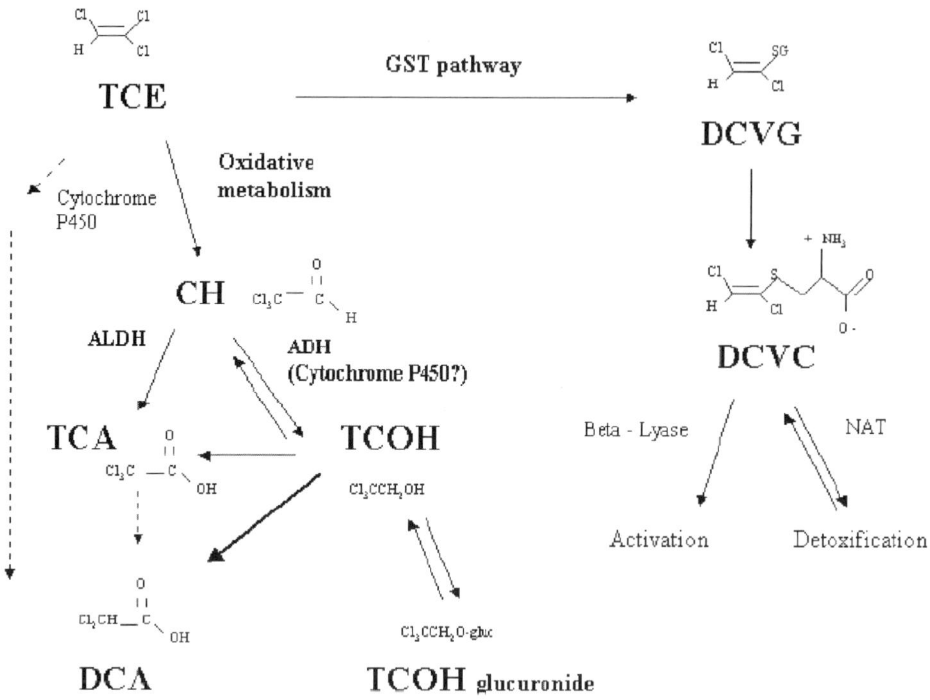

FIGURE 9-2 Metabolic pathways for detoxification and activation of TCE.

diseases according to their causes in the 17th century.[1] In 1700, Ramazzini published the first edition of his most famous book, *De Morbis Artificum Diatriba* (Diseases of Workers), the first comprehensive work on occupational diseases, outlining the health hazards of irritating chemicals, dust, metals, and other abrasive agents encountered by workers in 52 occupations. *De Morbis* is based on Ramizzini's personal interviews of workers to discuss their illnesses, observations of their workplaces and activities, and in-depth analyses of the problems. Ramizzini was the first to report blindness in sewage workers and postulated that the effects were due to the continual exposure of workers to ammonia vapors present in the sewers. Ramizzini was unique in his assessments in that he also offered remedies for the health hazards.

9.2 CAUSATION OF DISEASES: EPIDEMIOLOGICAL EVIDENCE

Assigning a causal relationship to chemical exposures is a complex process that is fraught with uncertainty. Epidemiological data from occupational exposures and public health studies of affected communities (affected by synthetic chem-

[1] Ramazzini B. 1700. *De Morbis Artificum Diatribe*. (Mutinae: Typis Antonii Capponi.)

ical releases) provide clear evidence that adverse health effects do occur following exposure to synthetic chemicals. What is less clear is the number of diseases/adverse health effects that may be expressed in a community. Communities such as Woburn, Massachusetts, Seviso, Italy, Minamata, Japan, and Times Beach, Missouri, provide clear evidence that multiple disease expressions are likely from chemical mixtures.

Table 9-1 shows the known health effects of exposure to "neat," or pure, doses of three chemicals commonly found in the market basket. Exposure to each of the compounds is heavily regulated in the work environment, as well

TABLE 9-1 Toxicity Characteristics of Benzene, Chlorpyrifos, and Trichloroethylene

Chemical	Carcinogen	Organs/Systems Affected	Health Effect
Benzene Aromatic hydrocarbon found in gasoline	A: Known human carcinogen	• Immune • Nervous • Hematological • Hepatic • Cardiovascular • Autoimmune antibodies • Developmental disorders • Renal	• Adult-onset leukemias • Aplastic anemia • Childhood leukemias • Multiple myeloma • Myelodysplastic syndrome • Porphyria • Arrhythmias • Non-Hodgkin's lymphomas • Preterm deliveries • Renal cancer • Scleroderma
Chlorpyrifos Primary agent in flea bombs and most frequently detected chemical in human blood	C: Possbile human carcinogen	• Immune • Nervous	• Immune suppression • Psychiatric disturbances • Attention deficit disorder • Hyperactivity • Autoimmune antibodies • Developmental delays
Trichloroethylene Agent used in degreasing operations; less toxic substitute for Perchloroethylene; one of the primary chemicals found at superfund sites	No classification at this time (under reevalation by USEPA); Likely a human carcinogen	• Hematological • Nervous • Hepatic • Renal • Immune	• Acute hepatocellular injury (hepatitis) • Cardiac congenital malformations • Childhood leukemias • Cirrhosis • Decreased coordination/ dysequilibrium • Fetotoxicity • Hepatocellular cancer • Non-Hodgkin's lymphoma • Psychiatric disturbances • Renal cancer • Scleroderma

as in air and drinking water supplies, yet significant exposures still occur among the general population.

9.2.1 The Case for Leukemia

The following section details the expression of diseases from exposure to synthetic chemicals. Hematopoietic diseases (leukemias) represent one of the most frequently expressed diseases following exposure to synthetic chemicals. Two chemicals frequently associated with leukemias are PCE and benzene. The epidemiology showing the variety of disease expression by these compounds is summarized in the following sections.

Leukemia is a type of cancer involving the hematopoietic system. Leukemia is cancer of the blood cells. Each year, nearly 27,000 adults and more than 2000 children in the United States learn that they have leukemia. The blood is made up of a fluid called plasma and three types of cells. Each type has special functions:

- White blood cells (also called WBCs or leukocytes) help the body fight infections and other diseases.
- Red blood cells (also called RBCs or erythrocytes) carry oxygen from the lungs to the body's tissues and take carbon dioxide from the tissues back to the lungs. The red blood cells give blood its color.
- Platelets (also called thrombocytes) help form blood clots that control bleeding.

Blood cells are formed in the bone marrow, the soft, spongy center of bones. New (immature) blood cells are called blasts. Some blasts stay in the marrow to mature. Some travel to other parts of the body to mature. Normally, blood cells are produced in an orderly, controlled way as the body needs them.

There are several types of leukemia. They are grouped in two ways. One way is by how quickly the disease develops and gets worse. The other way is by the type of blood cell that is affected.

Leukemia is either acute or chronic. In acute leukemia, the abnormal blood cells are blasts that remain very immature and cannot carry out their normal functions. The number of blasts increases rapidly, and the disease becomes worse quickly. In chronic leukemia, some blast cells are present, but in general, these cells are more mature and can carry out some of their normal functions. Also, the number of blasts increases less rapidly than in acute leukemia. As a result, chronic leukemia worsens gradually.

Leukemia can arise in either of the two main types of white blood cells: lymphoid cells or myeloid cells. When leukemia affects lymphoid cells, it is called lymphocytic leukemia. When myeloid cells are affected, the disease is called myeloid or myelogenous leukemia.

These are the most common types of leukemia:

1. Acute lymphocytic leukemia (ALL) is the most common type of leukemia in young children. This disease also affects adults, especially those age 65 and older.
2. Acute myeloid leukemia (AML) occurs in both adults and children. This type of leukemia is sometimes called acute nonlymphocytic leukemia (ANLL).
3. Chronic lymphocytic leukemia (CLL) most often affects adults over the age of 55. It sometimes occurs in younger adults, but it almost never affects children.
4. Chronic myeloid leukemia (CML) occurs mainly in adults. A very small number of children also develop this disease.
5. Hairy cell leukemia is an uncommon type of chronic leukemia.

9.2.2 PCE

A number of occupational cohort studies have examined the relationship between exposure to PCE and various cancer types.[2] The most significant tumor types that emerged from the epidemiological studies are lymphohematopoietic, esophageal, and female genital cancers. Cancers of the bladder, liver, kidney, lung, and breast have also been associated with PCE exposure in at least one study. The majority of occupational studies with relevance to PCE exposure concern the dry cleaning industry in the United States (most recent updates[3,4]) or laundry and dry cleaning occupations in Nordic countries.[5] A cohort of chemical industry workers,[6] air force base employees,[7] and a study of Finnish workers in a variety of occupations have also provided important information.[8]

---

[2] OEHHA. 2001. Public Health Goal for Tetrachloroethylene in Drinking Water, pp 31–34.

[3] Blair A, Stewart PA, Tolbert PE, Grauman D. 1990. Cancer and other causes of death among a cohort of dry cleaners. Brit J Indust Med 47:162–168.

[4] Ruder AM, Ward EM, Brown DP. 1994. Cancer mortality in female and male dry-cleaning workers. J Occup Med 36(8):867–874.

[5] Lynge E, Thygesen L. 1990. Primary liver cancer among women in laundry and dry-cleaning work in Denmark. Scand J Work Environ Health 16(2):108–112.

[6] Olsen GW, Hearn S, Cook RR, Currier MF, Allen S. 1989. Mortality experience of a cohort of Louisiana U.S.A. chemical workers, 31:1.

[7] Spirtas R, Stewart PA, Lee JS, Marano DE, Forbes CD, Grauman DJ, et al. 1991. Retrospective cohort mortality study of workers at an aircraft maintenance facility. I. Epidemiological results. Brit J Indust Med 49:515–530.

[8] Anttila A, Pukkala E, Sallmen M, Hernberg S, Hemminki K. 1995. Cancer incidence among Finnish workers exposed to halogenated hydrocarbons. J Occup Environ Med 37(7):797–806.

An increased risk of lymphohematopoietic cancers in humans has been observed for individuals exposed to PCE. This class of cancers is of particular interest in light of the increased leukemia rate in rats exposed to PCE. Olsen and colleagues[9] reported an elevated standard mortality ratio (SMR) for leukemia/aleukemia in a cohort of chemical workers; the SMR was statistically significant when local cancer rates were used for comparison. However, there were only 11 total cancer deaths in the cohort of 2,610 subjects, and exposure was not limited to PCE. A drinking water study also reported increased relative risk of leukemia in the group with the highest exposure to PCE (90th percentile of exposure); however, the result was based on only two cases.[10] Elevated risks of non-Hodgkin's lymphoma (NHL) have been reported in three cohort studies.[11,12,13] In contrast, a case-control study and a mortality study did not find any association between PCE and NHL.[14,15] A second case-control study of NHL found an increased odds ratio (OR) of 2.0 for employment in laundry or dry cleaning establishments, after adjustment for smoking, although the association was not statistically significant. The New Jersey Department of Health conducted a study of 18 organic chemicals in drinking water in 75 towns.[16] Incidence rates of high-grade lymphoma were elevated in females from the towns with the highest concentrations of PCE in drinking water. After stratifying towns into 0, 0.1–5.0, and >5.0 ppb exposure levels, the relative risk of high-grade NHL was significantly higher in the highest exposure stratum compared to controls. However, PCE levels in drinking water were also highly correlated with trichloroethylene levels, so whether the lymphoma risk was due to PCE is not clear.

---

[9] Olsen GW, Hearn S, Cook RR, Currier MF, Allen S. 1989. Mortality experience of a cohort of Louisiana U.S.A. chemical workers, 31:1.

[10] Aschengrau A, Ozonoff D, Paulu C, Coogan P, Vezina R, Heeren T, et al. 1993. Cancer risk and tetrachloroethylene-contaminated drinking water in Massachusetts. Arch Environ Health 48(5):284–292.

[11] Blair A, Stewart PA, Tolbert PE, Grauman D. 1990. Cancer and other causes of death among a cohort of dry cleaners. Brit J Indust Med 47:162–168.

[12] Anttila A, Pukkala E, Sallmen M, Hernberg S, Hemminki K. 1995. Cancer incidence among Finnish workers exposed to halogenated hydrocarbons. J Occup Environ Med 37(7):797–806.

[13] Spirtas R, Stewart PA, Lee JS, Marano DE, Forbes CD, Grauman DJ, et al. 1991. Retrospective cohort mortality study of workers at an aircraft maintenance facility. I. Epidemiological results. Brit J Indust Med 49:515–530.

[14] Siemiatycki J, ed. 1991. Risk Factors for Cancer in the Workplace. Boca Raton, FL: CRC Press.

[15] Walker J, Burnett C, Lalich N, Sestito J, Halperin W. 1997. Cancer mortality among laundry and dry cleaning workers. Amer J Indust Med 32(6):614–619.

[16] Cohn P, Bove F, Klotz J, Berkowitz M, Fagliano J. 1994. Drinking water contamination and the incidence of leukemia and non-Hodgkin's lymphoma. Environ Health Perspect 102(6–7):556–561.

Both of the major U.S. dry cleaner cohorts[17,18] reported an increased risk of bladder cancer. One study defined a subcohort of workers whose exposures were primarily to PCE; the bladder cancer risk was not elevated in the subcohort. Two Nordic laundry/dry cleaning cohorts did not experience excess bladder cancer.[19] Malker and Weiner,[20] reported in two proportionate mortality studies of laundry and dry cleaning workers some increases in bladder cancer risk,[21] as did a study of a population in Massachusetts exposed via drinking water.[22] A significant increased risk of bladder cancer was reported in non-whites employed 6 months or longer in dry cleaning establishments.[23] An occupational surveillance study in Canada concluded that the elevated odds ratio for bladder cancer observed in laundry/dry cleaning workers (after adjustment for smoking, coffee drinking, and other possible risk factors) merited further surveillance follow-up.[24] Whether smoking or exposure to other solvents could play a key role in the bladder cancer observed in these studies is not clear.[25] IARC concluded that the risk for cancer of the bladder "may be increased by employment in dry-cleaning," but also that "little or no information was available" about the specific relation to PCE from the majority of bladder cancer studies.[26]

Epidemiological studies provide evidence that PCE is possibly carcinogenic in humans. Especially important tumor sites are the esophagus and lymphatic system, but available data are quite limited in terms of the ability to quantify these cancer risks. This is particularly the case for risks associated with oral

---

[17] Ruder AM, Ward EM, Brown DP. 1994. Cancer mortality in female and male dry-cleaning workers. J Occup Med 36(8):867–874.

[18] Blair A, Stewart PA, Tolbert PE, Grauman D. 1990. Cancer and other causes of death among a cohort of dry cleaners. Brit J Indust Med 47:162–168.

[19] Lynge E, Thygesen L. 1990. Primary liver cancer among women in laundry and dry-cleaning work in Denmark. Scand J Work Environ Health 16(2):108–112.

[20] Malker HS, Weiner JA, McLaughlin JK. 1990. Register epidemiology studies of recent cancer trends in selected workers. Ann NY Acad Sci 609:322–332.

[21] Katz RM, Jowett D. 1981. Female laundry and dry cleaning workers in Wisconsin: A mortality analysis. Amer J Pub Health 71(3):305–307.

[22] Aschengrau A, Ozonoff D, Paulu C, Coogan P, Vezina R, Heeren T, et al. 1993. Cancer risk and tetrachloroethylene-contaminated drinking water in Massachusetts. Arch Environ Health 48(5):284–292.

[23] Silverman DT, Levin LI, Hoover RN. 1989. Occupational risks of bladder cancer in the United States: II. Nonwhite men. J Natl Cancer Inst 81:1480–1483.

[24] Teschke K, Morgan MS, Checkoway H, Franklin G, Spinelli JJ, Van Belle G, et al. 1997. Surveillance of nasal and bladder cancer to locate sources of exposure to occupational carcinogens. Occup Environ Med 54(6):443–451.

[25] Weiss NS. 1995. Cancer in relation to occupational exposure to perchloroethylene. Cancer Causes Control 6(3):257–266.

[26] IARC. 1995. International Agency for Research on Cancer. Tetrachloroethylene. IARC Monographs on the Evaluation of Carcinogenic Risks to Humans 63:159–221.

exposure to PCE, since the majority of the human data, and the most quantitatively informative studies, concerns occupational exposure by the inhalation route. Most of the evidence for human carcinogenicity is based on solvent exposures in the dry cleaning industry, of which PCE is a major but not exclusive component. Studies of solvent exposures in other industries also implicate PCE as a possible risk factor for human cancer, but these typically involve exposure to multiple solvents and other potentially carcinogenic agents.

Epidemiology studies of dry cleaning and laundry workers have demonstrated excesses in mortality due to various types of cancer, including liver cancer, but the data are regarded as inconclusive because of various confounding factors.[27,28] The chemical also induced mononuclear cell leukemia and renal tubular cell tumors in rats. PCE was negative for tumor initiation in a dermal study and for tumor induction in a pulmonary tumor assay.[29,30]

9.2.3 Benzene

There is general agreement within the scientific community that benzene causes a host of diseases including aplastic anemia, adult-onset leukemias, aplastic anemia, childhood leukemias, multiple myeloma, myelodysplastic syndrome, porphyria, arrhythmias, non-Hodgkin's lymphomas, preterm deliveries, renal cancer, and scleroderma. Human epidemiological and toxicological study data have characterized minimal exposures that result in significant increased incident rates of aplastic anemia and AML. There is strong evidence that exposure to benzene significantly increases the risk of developing aplastic anemia and AML.

Yin et al. examined more than 500,000 workers in China who had been exposed to benzene or benzene mixtures.[31] Benzene toxicity was found in workers exposed to concentrations above approximately 29 ppm. In shoe factory workers exposed to benzene, the incidence of aplastic anemia was 5.8 times greater than in the general population.

---

[27] Lynge E, Thygesen L. 1990. Primary liver cancer among women in laundry and dry-cleaning work in Denmark. Scand J Work Environ Health 16(2):108–112.

[28] U.S. EPA. 1988. Updated Health Effects Assessment for Tetrachloroethylene. Final draft. Prepared for Office of Solid Waste and Emergency Response. Environmental Criteria and Assessment Office, U.S. EPA, Cincinnati, OH. ECAO-CIN-H009a.

[29] Van Duuren BL, Goldschmidt BM, Loewengart G, Smith AC, Melchionne S, Seidman I. 1979. Carcinogenicity of halogenated olefinic and aliphatic hydrocarbons in mice. J Nat Cancer Inst 63:1433–1439.

[30] Theiss JC, Stoner GD, Shimkin MB, Weisberger EK. 1977. Tests for carcinogenicity of organic contaminants of United States drinking waters by pulmonary tumor response in Strain A mice. Cancer Res 37:2717–2720.

[31] Yin S-N, Li Q, Liu Y, Tian F, Du C, Jin C. 1987. Occupational exposure to benzene in China. Brit J Ind Med 44:192–195.

Another study by Yin et al.[32] used an expanded cohort study design to evaluate cancer in more than 100,000 workers (approximately 75% had been exposed to benzene). The incidence of aplastic anemia in this study was significantly increased in exposed individuals. Other studies that established a link between benzene exposure and aplastic anemia include those by IARC,[33] Aksoy,[34,35] Young and Alter,[36] Aksoy et al.,[37] Yin et al.,[38,39] Aksoy and Erdem,[40] and Paci et al.[41]

AML risk from benzene exposure is documented in numerous studies, including two that related increasing levels of benzene exposure to increased risk of AML, strengthening the conclusion that the observed effects were due to the benzene exposure. Rinsky et al.[42] examined 1,165 workers who were exposed from 1935 to 1965 to benzene at a natural rubber manufacturing facility in the United States. The risk of death from AML in the workers was increased significantly in exposed workers (SMR = 3.37) and followed a dose-dependent pattern. Wong et al.[43] examined the Pliofilm data and estimated the workers' exposure, finding extremely significant increases in risk for AML among workers. Workers with 200 to 400 and >400 ppm-years of exposure

---

[32] Yin SN, Hayes RB, Linet MS, Li GL, Dosemeci M, Travis LB, Shang ZN, Li DG, Chow WH, Wacholder S, Blot WJ. 1996. An expanded cohort study of cancer among benzene exposed workers in China. Environ Health Perspect 104(Supp 6):1339–1341.

[33] IARC. 1982. Benzene. IARC monographs on the evaluation of the carcinogenic risk of chemicals to humans: some industrial chemicals and dyestuffs. Lyon: International Agency for Research on Cancer, pp 93–148.

[34] Aksoy M. 1988. Benzene Hematotoxicity. In: Aksoy M, Editor. *Benzene Carcinogenicity*. Boca Raton: CRC Press, pp 59–104.

[35] Aksoy M. 1989. Hematotoxicity and carcinogenicity of benzene. Environ Health Perspect 82:193–197.

[36] Young N, Alter BP. 1994. Aplastic Anemia, Acquired and Inherited. Philadelphia: W.B. Saunders.

[37] Aksoy M, Dincol K, Akgun T, Erdem S, Dincol G. 1971. Haematological effects of chronic benzene poisoning in 217 workers. Brit J Ind Med 28:296–302.

[38] Yin SN, Li G, Tain F, Fu Z, Jin C, Chen Y, Luo S, Ye P, Zhang J, Wang G, Zhang X, Wu H, Zhong Q. 1987. Leukaemia in benzene workers: a retrospective cohort study. Brit J Ind Med 44:124–128.

[39] Yin SN, Hayes RB, Linet MS, Li GL, Dosemeci M, Travis LB, Shang ZN, Li DG, Chow WH, Wacholder S, Blot WJ. 1996. An expanded cohort study of cancer among benzene exposed workers in China. Environ Health Perspect 104(Supp 6):1339–1341.

[40] Aksoy M, Erdem S. 1978. Followup study on the mortality and the development of leukemia in 44 pancytopenic patients with chronic exposure to benzene. Blood 52:285–292.

[41] Paci E, Buiatti E, Costantini AS, Miligi L, Pucci N, Scarpelli A, Petrioli G, Simonato L, Winkelmann R, Kaldor JM. 1989. Aplastic anemia, leukemia and other cancer mortality in a cohort of shoe workers exposed to benzene. Scand J Work Environ Health 15:313–318.

[42] Rinsky RA, Smith AB, Hornung R, Filloon TG, Young RJ, Okun AH, Landrigan PJ. 1987. Benzene and leukemia: an epidemiologic risk assessment. N Engl J Med 316:1044–1050.

[43] Wong O. 1995. Risk of acute myeloid leukaemia and multiple myeloma in workers exposed to benzene. Occup Environ Med 52:380–384.

were, respectively, 27 and 98 times more likely to develop AML than unexposed individuals.

In a study by Hayes et al.,[44] more than 74,000 benzene-exposed Chinese workers were examined for toxicity. The relative risk for AML in exposed workers was 3.0 based on 21 exposed cases. This study demonstrated an increase in AML risk at exposure levels below 200 ppm-years.

Additional studies that demonstrate risk of AML in benzene exposed individuals include those by IARC,[45] Aksoy,[46] McCraw et al.,[47] Bond et al.,[48] Rinsky et al.,[49] Wong,[50,51] Yin et al.,[52] Hayes et al.,[53] Rushton and Romaniuk,[54] Austin et al.,[55] and Swaen and Meijers.[56]

Peer-reviewed studies by Glass, et al.[57] have demonstrated by logistic regression analysis increased rates of leukemia among benzene-exposed workers with as low as 1 to 2 ppm-year exposures. The current National Institute for Occu-

---

[44] Hayes RB, Yin SN, Dosemeci M, Li GL, Wacholder S, Travis LB, Li CY, Rothman N, Hoover RN, Linet MS. 1997. Benzene and the dose related incidence of hematologic neoplasms in China. J Natl Cancer Inst 89:1065–1071.

[45] IARC. 1982. Benzene. IARC monographs on the evaluation of the carcinogenic risk of chemicals to humans: some industrial chemicals and dyestuffs. Lyon: International Agency for Research on Cancer, pp 93–148.

[46] Aksoy M. 1985. Benzene as a leukemogenic and carcinogenic agent. Am J Ind Med 8:9–20.

[47] McCraw DS, Joyner RE, Cole P. 1985. Excess leukemia in a refinery population. J Occup Environ Med 27:220–222.

[48] Bond GG, McLaren EA, Baldwin CL, Cook RR. 1986. An update of mortality among chemical workers exposed to benzene. Brit J Ind Med 43:685–691.

[49] Rinsky RA, Smith AB, Hornung R, Filloon TG, Young RJ, Okun AH, Landrigan PJ. 1987. Benzene and leukemia: an epidemiologic risk assessment. N Engl J Med 316:1044–1050.

[50] Wong O. 1987. An industry wide mortality study of chemical workers occupationally exposed to benzene. II. Dose response analyses. Brit J Ind Med 44:382–395.

[51] Wong O. 1987. An industry wide mortality study of chemical workers occupationally exposed to benzene. I. General results. Brit J Ind Med 44:365–381.

[52] Yin SN, Hayes RB, Linet MS, Li GL, Dosemeci M, Travis LB, Shang ZN, Li DG, Chow WH, Wacholder S, Blot WJ. 1996. An expanded cohort study of cancer among benzene exposed workers in China. Environ Health Perspect 104(Supp 6):1339–1341.

[53] Hayes RB, Yin SN, Dosemeci M, Li GL, Wacholder S, Travis LB, Li CY, Rothman N, Hoover RN, Linet MS. 1997. Benzene and the dose related incidence of hematologic neoplasms in China. J Natl Cancer Inst 89:1065–1071.

[54] Rushton L, Romaniuk H. 1997. A case-control study to investigate the risk of leukaemia associated with exposure to benzene in petroleum marketing and distribution workers in the United Kingdom. Occup Environ Med 54:152–166.

[55] Austin H, Delzell E, Cole P. 1988. Benzene and leukemia. A review of the literature and a risk assessment. Am J Epidemiol 127:419–439.

[56] Swaen GMH, Meijers JMM. 1989. Risk assessment of leukemia and occupational exposure to benzene. Brit J Ind Med 46:826–830.

[57] Glass DC, Gray CN, Jolley DJ, Gibbons C, Sim MR, Fritschi L, Adams GC, Bisby JA, Manuell R. 2003. Leukemia risk associated with low-level benzene exposure. Epidemiology 14(5):569–577.

pational Safety and Health (NIOSH) recommended exposure limit (REL) is 0.1 ppm, while the legally enforceable OSHA standard is 1.0 ppm.

The diagnosis of aplastic anemia is highly consistent with benzene exposures. Lower-level benzene exposures (1 ppm) are associated with a significantly elevated risk of leukemia and chromosomal breakage, while an elevated risk of leukopenia has been associated with levels of 3 ppm and less. Several additional hematological indicators, including severe pancytopenia, are highly consistent with previous benzene exposure occurring several months prior to the acute onset. Dozens of human epidemiologic studies have confirmed the causal relationship between benzene exposure and induction of AML as diagnosed.

Data of benzene-induced leukemia indicate the onset of leukemia as brief — 0.8 years from initial exposure to benzene and up to 49 years after initial exposure. The NIOSH cohort mortality study of benzene-exposed workers[58] based on 15 leukemia deaths indicates an average latency period of 24.9 years with a range of 2 years (a death from CML) to 45 years (a death from ALL). Aksoy[59] revealed an average latency period of 11 years between initial exposure and clinical detection of leukemia. Vigliani[60] revealed a median latency of 9 years between initial exposure and diagnosis with leukemia. In the Vigliani case series, the latency period ranged from 3 to 24 years. The study by Yin et al.[61] revealed an average latency period (start of exposure to diagnosis of leukemia) of 11.4 years (minimum 0.8 years, maximum 49.5 years) for cases of leukemia.

9.3 FEDERAL AND STATE LEGISLATION REQUIRING THAT FOOD AND WATER BE UNADULTERATED

Laws governing water quality and public health date back to the 1800s. The first public health service (PHS), known as the Marine Hospital Service, was created following the passage of an act in 1798 that provided for the care and relief of sick and injured merchant seamen.[62] The scope of activities of the Marine Hospital Service also began to expand well beyond the care of merchant seamen beginning with the control of infectious disease and expanding to regulating the food and drug supply.

[58] Paxon et al. 1994. Leukemia risk associated with benzene exposure in the Pliofilm cohort: mortality update and exposure distribution. Risk Anal 14:147–154.

[59] Aksoy. 1977. Leukemia in workers due to occupational exposure to benzene. New Istabbul Contrib Clin Sci 12:3–14.

[60] Vigliani. 1976. Leukemia associated with benzene exposure. NY Acad Sci 271:143–151.

[61] Yin et al. 1989. A retrospective cohort study of leukemia and other cancers in benzene workers. Environ Health Perspect 207–213.

[62] USPHSCC. 2003. The History of the Commissioned Corps. United States Public Health Service Commissioned Corps, www.usphs.gov/html/history.html.

By 1904, water pollution laws in the United States prohibited pollution.[63] As a consequence, there appeared to be no immediate need for water pollution standards. National water quality standards were first published in 1925.[64] The objective of the U.S. Public Health Service (USPHS) in publishing drinking water standards[65] was to "safeguard the health of the public" against the most seriously recognized dangers, namely "that of drinking water contamination by disease such as typhoid fever and other illnesses of similar origin and transmission." The standards did, however, warn that "to state that a water supply is 'safe' does not necessarily signify that absolutely no risk is ever incurred in drinking it."

Few federal laws dealing with the health and safety or directly with the environment were promulgated prior to the passage of the National Environmental Protection Act (NEPA) in 1970. The Pure Food and Drugs Act (1906) and the Meat Inspection Act (1906) covered food safety and tolerances for pesticides in food. The Federal Insecticide, Fungicide, and Rodenticide Act (FIFRA) of 1947, set guidelines for pesticide registration, classification (general versus restricted), and use, as well as applicator certification. The Federal Hazardous Substances Labeling Act (FHLSA) and the Delaney Act (1959) were intended to prevent the introduction of carcinogens into foodstuffs. Nearly 40 years later, the Food Quality Protection Act (FQPA) amended the Food, Drug and Cosmetic Act (FDCA) and the FIFRA by doing the following:

- Replacing the Delaney Clause with the standard of "reasonable certainty that no harm will result from aggregate exposure to pesticide residue"
- Providing uniformity of state regulations on pesticide tolerances unless a state has successfully petitioned the Environmental Protection Agency (EPA) for a different tolerance level or can demonstrate that pesticide residues pose unreasonable dietary risks to the state's residents
- Imposing civil penalties of up to $50,000 for individuals and up to $250,000 for companies that introduce pesticide-adultered food into the marketplace

The act changed FIFRA by doing the followng:

- Defining the term "nitrogen stabilizer"
- Providing for uses of existing stocks of suspended or canceled pesticides
- Providing for a review of pesticide registrations at least every 15 years

---

[63] Goodell, EB. 1904. A review of the laws forbidding pollution of inland waters in the United States, U.S. Geological Survey, Water-Supply and Irrigation Paper No. 103.

[64] Given that surface and groundwater resources cross state boundaries, a national water quality policy was necessary.

[65] U.S. Public Health Service. 1925. Report of advisory committee on official water standards, Public Health Rep 40(15):April 10.

- Adding definitions and minimum requirements for training of maintenance applicators and service technicians
- Providing for coordination between federal and state officials on pesticide issues
- Addressing pesticide labeling issues
- Expediting registration of reduced-risk pesticides
- Increasing data collection to ensure the health of infants and children
- Promoting the use of integrated pest-management practices

Following the passage of the NEPA in 1970, a safety net of legislation was promulgated dealing with safety or the environment including the following:

- Clean Air Act (CAA) of 1955, reauthorized in 1963 and 1982
- Clean Air Act Amendments (CAAA) of 1966, 1970, 1982, and 1990
- Occupational Safety Health Act (OSHA)
- Poison Prevention Packaging Act (PPPA)
- Federal Environmental Pesticide Control Act (FEPCA)
- Consumer Product Safety Act (CPSA)
- Marine Protection, Research, and Sanctuaries Act (MPRSA)
- Clean Water Act (CWA)
- Ports and Waterways Safety Act (PWA)
- Endangered Species Act (ESA)
- Hazardous Materials Transportation Act (HMTA)
- Safe Drinking Water Act (SDWA)
- Resource Conservation and Recovery Act (RCRA)
- Toxic Substances Control Act (TSCA)
- Surface Mining Control and Reclamation Act (SMCRA)
- Comprehensive Environmental Response, Compensation, and Liability Act (CERCLA)
- Hazard Communication Standard Act (HCSA)
- Superfund Amendment and Reauthorization Act (SARA)

To supplement federal regulations, states have created a host of regulations, typically driven by federal statues (such as the CAA or SDWA), requiring polluters to meet discharge standards, notify the public, and to minimize toxic impacts. In 1986, California approved the Safe Drinking Water and Toxic Enforcement Act or Proposition 65. Proposition 65 requires the state to publish a list of chemicals known to cause cancer or birth defects or other reproductive harm. This list, which must be updated at least once a year by the Office of Environmental Health and Hazard Assessment (OEHHA), has grown to include approximately 750 chemicals since it was first published in 1987.

Since numerous chemicals identified by the FDA in food products and by EPA-mandated monitoring in drinking water are on the Proposition 65 lists, it is important to consider labeling food and bottled water products that meet Proposition 65 labeling requirements.

9.4 THE NEED FOR WARNING LABELS IN THE SUPERMARKET

The Proposition 65 list contains a wide range of naturally occurring and synthetic chemicals that are known to cause cancer or birth defects or other reproductive harm. These chemicals include additives or ingredients in pesticides, common household products, food, drugs, dyes, or solvents. Listed chemicals may also be used in manufacturing and construction, or they may be byproducts of chemical processes, such as motor vehicle exhaust.

If a warning is placed on a product label or posted or distributed at the workplace, a business, or in rental housing, the business issuing the warning is aware or believes that one or more listed chemicals are present. By law, a warning must be given for listed chemicals unless exposure is low enough to pose no significant risk of cancer or is significantly below levels observed to cause birth defects or other reproductive harm.

For a chemical that causes cancer, the "no significant risk level," or NSRL, is defined as the level of exposure that would result in not more than one excess case of cancer in 100,000 individuals exposed to the chemical over a 70-year lifetime. In other words, a person exposed to the chemical at the "no significant risk level" for 70 years would not have more than a "one in 100,000" chance of developing cancer as a result of that exposure.

For chemicals that are listed as causing birth defects or reproductive harm, the "no observable effect level," or NOEL, is determined by identifying the level of exposure that has been shown to not pose any harm to humans or laboratory animals. Proposition 65 then requires this no observable effect level to be divided by 1000 in order to provide an ample margin of safety. Businesses subject to Proposition 65 are required to provide a warning if they cause exposures to chemicals listed as causing birth defects or reproductive harm that exceed 1/1000th of the no observable effect level.

OEHHA has developed guidance levels, known as "safe harbor numbers" for determining whether a warning is necessary or whether discharges of a chemical into drinking water sources are prohibited. The safe harbor numbers consist of no significant risk levels for chemicals listed as causing cancer and maximum allowable dose levels for chemicals listed as causing birth defects or other reproductive harm. OEHHA has established safe harbor numbers for the chemicals given in Appendix 9-1.

Although the objective of this text is to illustrate the complexity and diversity of chemical mixtures in the environment independent of chemical concentrations, the concentrations listed for synthetic chemicals found in both food and drinking water are a great concern. Thus, it is important to discuss the toxicity and occurrence of these compounds in our food and water. Although acrylamide (this compound is also used to treat drinking water) is not monitored by the FDA in its Total Diet Studies, it is an important synthetic chemical that occurs in the environment.

9.4.1 Acrylamide

Acrylamide was listed as a chemical known to cause cancer under Proposition 65 on January 1, 1990. At the time that acrylamide was initially listed under Proposition 65, the state of California's Environmental Health and Hazard Assessment (OEHHA) determined that the NSRL for exposure to acrylamide was 0.2 micrograms per day (ug/day). The NSRL is typically derived assuming a human body weight of 70 kilograms. Sensitive subpopulations such as infants (10 kg body weight), children (20 kg body weight), and adolescents (40 kg body weight) may be considered when appropriate. For the acrylamide NSRL, these subpopulations have not been specifically considered.

In 2002, it was widely reported that acrylamide is formed during high-temperature cooking of starchy foods.[66] Many plant-based foods that are baked or fried at high temperatures were found to contain acrylamide. Acrylamide forms in starchy foods through Maillard reaction products,[67] and it is influenced by the raw food's storage temperature, sugar content, moisture content, cooking temperature, pH, and the presence of asparaginase. A study by Svensson found acrylamide in potato products, bread, breakfast cereals, biscuits, cookies, snacks, and coffee in Sweden and estimated daily intake rates of 9.1, 27, and 62 ug/day for the 5th, 50th, and 95th percentiles, respectively.[68] Food agencies in Europe and the United States have also established the presence of acrylamide in foodstuffs.[69] Reports of acrylamide in prepared foodstuffs led to an evaluation of the potential health impacts from the consumption of acrylamide in average diets.[70]

---

[66] Biedermann M, Noti A, Biedermann-Brem S, Mozzetti V, Grob K. 2002. Experiments on acrylamide formation and possibilities to decrease the potential of acrylamide formation in potatoes. (Mitteilungen aus Lebensmitteluntersuchung und Hygiene) 93:668–687, and Biedermann M, Biedermann-Brem S, Noti A, Grob K. 2002. Methods for determining the potential of acrylamide formation and its elimination in raw materials for food preparation, such as potatoes. (Mitteilungen aus Lebensmitteluntersuchung und Hygiene) 93:653–667.

[67] Stadler RH, Blank I, Varga N, Robert F, Hau J, Guy PA, Robery MC, Riediker S. 2002. Acrylamide from Maillard reaction products. Nature 419:448–449.

[68] Svennson K, Abramsson L, Becker W, Glynn A, Hellenas KE, Lind Y, Rosen J. 2003. Dietary intake of acrylamide in Sweden. Food Chem Toxicol 41:1581–1586.

[69] Svennson K, Abramsson L, Becker W, Glynn A, Hellenas KE, Lind Y, Rosen J. 2003. Dietary intake of acrylamide in Sweden. Food Chem Toxicol 41:1581–1586.

[70] European Commission. 2003. Note of the Meeting of Experts on Industrial Contaminants in Food: Acrylamide Workshop. In: Information on Ways to Lower the Levels of Acrylamide Formed in Food European Commission (EU) Acrylamide Workshop October 20–21, 2003.

After lobbying by industry groups, in August 2003, OEHHA drafted a work plan to address regulatory options for acrylamide in the environment. The final work plan, released March 2005, proposed an NSRL of 1.0 ug/day. The new NSRL took into consideration data showing the dose-response relationship of acrylamide and the formation of cancer in animals. The NSRL decision used data from key animal cancer studies and took into consideration differences between humans and animals in absorbing, distributing, and metabolizing acrylamide. The revised NSRL for acrylamide has been widely criticized by industry groups as being overly protective and flawed. Written comments from industry groups have advocated NSRLs in the range of 50 to 226 ug/day.

The U.S. EPA considers acrylamide a B-2 carcinogen or probable human carcinogen in light of animal data. In a 2-year carcinogenesis bioassay, acrylamide (>98% purity) was administered in drinking water to F344 rats (60/sex/dose) at doses of 0, 0.01, 0.1, 0.5, or 2.0 mg/kg/day.[71] A maximum tolerated dose (MTD) was determined based on decreased body weight gain, decreased survival, and the observance of several toxic effects in the high-dose group. There were transient symptoms of a viral infection (sialodacryoadenitis virus) in some rats beginning on day 210 of the study; however, all animal groups were equally affected. This viral infection did not significantly affect the body weight, survival, or tumor incidences of F344 rats.[72] Acrylamide induced statistically significant increases in the incidences of several tumor types in test rats of both sexes when compared with control animals. In males, significantly increased incidences of tumors included the following: scrotal mesotheliomas in the two highest doses (11/53 and 10/54 at the two highest doses versus 3/57 in the control), adrenal pheochromocytomas in the high dose (10/54 versus 3/57), and thyroid adenomas in the high dose (7/54 versus 1/57). In high-dose females, gliomas and astrocytomas of the central nervous system (9/61 high dose versus 1/60 control), adenomas and adenocarcinomas of the mammary gland (28/60 versus 10/60), adenomas and adenocarcinomas of the thyroid gland (5/50 versus 1/54), adenocarcinomas of the uterus (5/49 versus 1/56), and papillomas and carcinomas of the oral cavity (8/60 versus 0/60) were significantly increased.

A series of mouse skin papilloma and lung adenoma assays showed that acrylamide initiated skin tumorigenesis in both SENCAR and Swiss-ICR mice and induced lung tumors in SENCAR, Swiss-ICR, and A/J mouse

---

[71] Johnson KA, Gorzinski SJ, Bodner KM, Campbell RA, Wolf CH, Friedman MA, et al. 1986. Chronic toxicity and oncogenicity study on acrylamide incorporated in the drinking water of Fischer 344 rats. Toxicol Appl Pharmacol 85(2):154–168.

[72] Rao GN, Edmondson J, Haseman J. 1988. Influence of viral infection on tumor incidences, body weight and survival of Fischer 344 rats. Toxicologist 8:166.

strains.[73] A 1984 study by Bull[74] administered a total of 0, 75, 150, or 300 mg acrylamide/kg body weight during six applications over a 2-week period via gavage, intraperitoneal injection (i.p.), or dermal application to groups of female SENCAR mice, followed by triweekly applications of 1 ug TPA (12-o-tetradecanoyl-phorbal-13-acetate) for 20 weeks. Acrylamide initiated skin tumorigenesis in Swiss-ICR mice (by gavage) and induced lung neoplasms in Swiss-ICR mice (by gavage) and A/J mice (by gavage and i.p.). A dose-response increase in skin tumors was observed in the mice. Other studies by Bull[75] found that skin tumor development was dependent on promotion by TPA whereas lung tumor induction was not. Robinson[76] noted significant increases of skin and lung tumors in SENCAR mice administered 50 mg/kg of acrylamide by a single i.p. injection followed by treatment with TPA.

Acrylamide has also been shown to be a clastogenic agent, inducing chromosomal aberrations, dominant lethality, sister-chromatid exchanges, and unscheduled DNA synthesis in various *in vivo* and *in vitro* systems. Acrylamide produces cell transformation *in vitro* and causes amplification of SV40 DNA inserts of SV40-transformed Chinese hamster cells. Furthermore, there is evidence that C^{14}-acrylamide binds covalently to DNA and protein in rodents.[77] It should be noted that acrylamide is structurally analogous to vinyl carbamate and acrylonitrile; these compounds contain a vinyl group, which may interact with cellular macromolecules via activation to an epoxide.

The upper-bound cancer potency estimate of 2.8 $(mg/kg\text{-}d)^{-1}$, derived from the occupational cohort study for exposure of the general population to acrylamide over a lifetime,[78] is higher than the upper-bound cancer potencies

[73] Bull R, Robinson M, Laurie R, Stoner G, Greisiger E, Meier J, Stober J. 1984. Carcinogenic effects of acrylamide in Sencar and A/J mice. Cancer Res 44:107–111; Bull R, Robinson M, Stober J. 1984. Carcinogenic activity of acrylamide in the skin and lung of Swiss-ICR mice. Cancer Lett 24:209–212; and Robinson M, Bull RJ, Knutsen GL, Shields RP, Stober J. 1986. A combined carcinogen bioassay utilizing both the lung adenoma and skin papilloma protocols. Environ Health Perspect 68:141–145.

[74] Bull R, Robinson M, Laurie R, Stoner G, Greisiger E, Meier J, Stober J. 1984. Carcinogenic effects of acrylamide in Sencar and A/J mice. Cancer Res 44:107–111.

[75] Bull R, Robinson M, Laurie R, Stoner G, Greisiger E, Meier J, Stober J. 1984. Carcinogenic effects of acrylamide in Sencar and A/J mice. Cancer Res 44:107–111, and Bull R, Robinson M, Stober J. 1984. Carcinogenic activity of acrylamide in the skin and lung of Swiss-ICR mice. Cancer Lett 24:209–212.

[76] Robinson M, Bull RJ, Knutsen GL, Shields RP, Stober J. 1986. A combined carcinogen bioassay utilizing both the lung adenoma and skin papilloma protocols. Environ Health Perspect 68:141–145.

[77] Dearfield KL, Abernathy CO, Ottley MS, Brantner JH, Hayes PF. 1988. Acrylamide: its metabolism, developmental and reproductive effects, genotoxicity and carcinogenicity. Mutat Res 195:45–77.

[78] Marsh GM, Lucas LJ, Youk AO, Schall LC. 1999. Mortality patterns among workers exposed to acrylamide: 1994 follow up. Occup Environ Med 56(3):181–190.

estimated from the long-term rat drinking water studies, which ranged from 0.69 to 1.5 $(mg/kg\text{-}d)^{-1}$. This comparison illustrates that, if humans are equally or less sensitive to acrylamide-induced cancer as rats, then the expected relative risk estimates in the worker study would have been small. If the animal cancer studies are predictive of human risk, one would not expect to observe a statistically significant increase in cancer rates from the exposures experienced by the workers, a conclusion also reached by other researchers.[79] Thus, the available human cancer data on acrylamide do not reduce OEHHA's concern regarding the potential induction of cancers by acrylamide.

A cancer potency estimate of 0.70 $(mg/kg\text{-}day)^{-1}$ was derived from the geometric means of combined distributions of cancer potency estimates for all acrylamide-related tumor sites across four studies in rats. This is associated with a dose[80] of 0.014 ug/kg-day at a lifetime cancer risk of 10^{-5}.

Neurotoxicity is a sensitive noncancer endpoint of acrylamide toxicity where primate studies (oral and subcutaneous) and a human occupational study support a NOEL for acrylamide neuropathy of 0.5 mg/kg body weight per day.[81]

In addition, acrylamide has been shown to adversely affect reproduction and fetal development and induce heritable DNA damage. In a draft report by the Expert Panel on Acrylamide of the National Toxicology Program's Center for the Evaluation of Risks to Human Reproduction, the lowest observed effect level (LOEL) for both developmental and male reproductive toxicity was reported to be 5 mg/kg-day. Dividing this level by a standard uncertainty factory of 10 for LOEL to no observable effect level (NOEL) extrapolation results in a NOEL estimate of 0.5 mg/kg-day, approximating the NOEL determined by the World Health Organization (WHO) for neurotoxic effects. The U.S. EPA established a NOEL of 0.2 mg/kg-day (200 ug/kg-day) and a corresponding reference dose (RfD) of 0.2 ug/kg-day for noncancer effects based on experimental neurotoxicity data.

The daily dose level posing a 10^{-5} lifetime risk of cancer of 0.014 ug/kg-day (= 1.0 ug/day ÷ 70 kg) is more than 10,000 times lower than the NOELs of 200 or 500 ug/kg-day for noncancer (reproductive, developmental, or neurotoxic)

---

[79] Calleman C. 1996. The metabolism and pharmacokinetics of acrylamide: implications for mechanisms of toxicity and human risk estimation. Drug Metab Rev 28(4):527–590; Dybing E, Sanner T. 2003. Risk assessment of acrylamide in foods. Toxicol Sci 75(1):7–15; Erdreich LS, Friedman MA. 2004. Epidemiologic evidence for assessing the carcinogenicity of acrylamide. Reg Toxicol Pharmacol 39(2):150–157; and Granath F, Ehrenberg L, Paulsson B, Tornqvist M. 2001. Cancer risk from exposure to occupational acrylamide. Occup Environ Med 58(9):608–609.

[80] OEHHA, 2005. Characterization of Acrylamide Intake from Certain Foods. Sacramento, CA: California Office of Environmental Health Hazard Assessment.

[81] OEHHA, 2005. Characterization of Acrylamide Intake from Certain Foods. Sacramento, CA: California Office of Environmental Health Hazard Assessment.

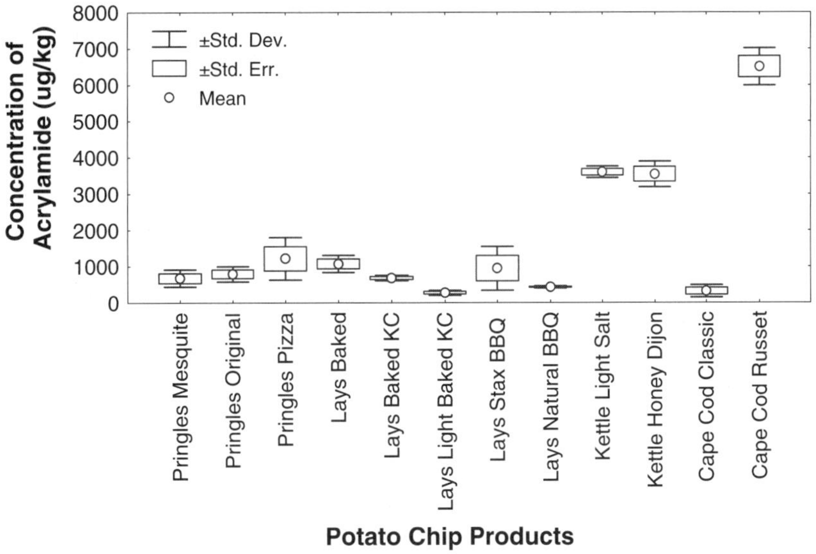

FIGURE 9-3 Measured concentrations of acrylamide in commercially available potato chips.

endpoints and more than 10 times lower than the U.S. EPA RfD. This suggests that cancer is the most sensitive health endpoint for acrylamide.[82]

The NSRL for Proposition 65 is the intake associated with a lifetime cancer risk of 10^{-5}. The combined cancer potency estimate for all acrylamide-related tumor sites, 0.70 $(mg/kg\text{-}day)^{-1}$, derived earlier, was used to calculate the 1.0 ug/day NSRL for acrylamide. In an experiment by Rosenfeld and Clark, a total of 12 different potato chip products from four brands were purchased for acrylamide analysis. The results of the laboratory analyses demonstrated that the acrylamide content in the chips dramatically exceeded the proposed Proposition 65 NSRL (Figure 9-3). The variance in the products could be due to a variety of factors, but it is apparent that the concentration of acrylamide in potato chips can be influenced by handling processes.

Based on the health impacts on humans for acrylamide consumption via potato chips, the daily dose of acrylamide was calculated using several studies estimating various daily food consumption rates, including potato chips.[83,84] The estimated daily consumption rates for potato chips ranged from 4 to 41 grams

[82] OEHHA, 2005. Characterization of Acrylamide Intake from Certain Foods. Sacramento, CA: California Office of Environmental Health Hazard Assessment.

[83] Smiciklas-Wright H, Mitchell DC, Mickle SJ, Goldman JD, Cook A. 2002. Foods Commonly Eaten in the United States, Quantities Consumed per Eating Occasion in a Day, 1994–96. U.S. Department of Agriculture (USDA) NFS Report Number: 96-5.

[84] Robie D, DiNovi M. 2003. The Exposure Assessment for Acrylamide. U.S. Food and Drug Administration, Office of Food Additive Safety presentation to the JIFSAN workshop. Chicago, IL.

per day. In this study, the potato chip daily consumption rates were multiplied by the average acrylamide concentration determined by the laboratory to calculate the amount of acrylamide consumed per day. This was used to determine which potato chips, if consumed at the estimated rates, would exceed the proposed 1.0 ug/day NSRL. All 12 varieties of potato chip products contained enough acrylamide to exceed the NSRL for all consumption rate estimates.

The manufacturers' recommended serving sizes were also considered when calculating the potential acrylamide exposure from the potato chips. Ten of the 12 products have a recommended serving size of 28 grams; the Pringles Original Snack Stack and the Pringles Pizzalicious Snack Stack have recommended serving sizes of 23 grams. These manufacturers' recommended serving sizes are similar to the 26.6 grams used by Robie and DiNovi[85] in their calculation to determine daily acrylamide consumption. When considering serving size, all 12 products contain enough acrylamide to exceed the 1.0 ug/day NSRL.

Consumption of 12 different commercially available potato chip products at the manufacturers' recommended serving size will result in exposure to acrylamide above the NSRL for adults. Based on the body weights of sensitive subpopulations, equivalent servings of potato chips would result in doses that are 1.5 to 7 times higher in adolescents, children, and infants. Therefore, the lifetime risk from exposure to acrylamide in food products is likely to be much higher than initially estimated.

Under the guidance established by Proposition 65 potato chips therefore require warning labels informing consumers of the potential hazards of ingesting the products. Potato chips are not the only food requiring labeling though.

9.4.2 Market Basket Foods

When reviewing the market basket data and the listed Proposition 65 chemicals, the synthetic compounds that most frequently contaminate our food are given in Table 9-2. This means that food products that are contaminated with these chemicals may require a warning label. By using the pesticide and industrial chemical concentration data for the various food product categories (i.e., dairy, protein, fruit, grain and nuts, vegetables, and child/infant products) that are given in Appendixes 4-1 through 4-58, those foods with multiple contaminants that could exceed Proposition 65's Safe Harbor concentrations are listed in Table 9-3. Given that an individual daily diet can contain a mixture of the foods listed in Table 9-3, all of these foods could contribute to an individual exceeding safe harbor levels. On an individual basis, 16 adult food products

[85] Robie D, DiNovi M. 2003. The Exposure Assessment for Acrylamide. U.S. Food and Drug Administration, Office of Food Additive Safety presentation to the JIFSAN workshop. Chicago, IL.

TABLE 9-2 Reproductive Toxins and Carcinogens in Market Basket Foods

Toxicological Hazard	Food
Chemicals known to cause cancer	Benzene Chlordane Chloroform Dieldrin DDE DDT Heptachlor epoxide Hexachlorobenzene Iprodine PCB Propargite Tetrachloroethlyene Trichloroethylene Toxaphene
Chemicals known to cause reproductive toxicity	Benomyl Benzene 2,4-D DDT PCB Toluene Vinclozolin

and one child/infant product exceed recommended safe harbor levels for dieldrin, toxaphene, and/or PCBs. A list of these foods, their serving sizes, and calculated daily exposure based on one serving are given in Table 9-4. The data used for this analysis were based on mean chemical concentrations determined for all food analyses by the FDA.[86] Based on the serving size, all of these foods should have a warning label. This includes butter since an individual's daily exposures can exceed one tablespoon. Furthermore, when considering that these products are just a small portion of a balanced diet, a significant number of individuals will exceed safe harbor levels for dieldrin considering its distribution throughout numerous food products.

Furthermore, the data provided in Appendixes 4-1 through 4-58 show that food products contain DDE, dieldrin, and toxaphene. This means that even if they are classified as "organic" foods they will still contain pesticides. This means that diets high in butter, cheese, and certain vegetables (see Table 9-4) that substitute organic products will most probably exceed Proposition 65's Safe Harbor numbers.

[86] Food and Drug Administration Total Diet Study, Summary of Residues Found Ordered by Pesticide, Market Baskets 91-3–01-4, June 2003.

TABLE 9-3 Market Basket Foods That May Exceed Proposition 65 Safe Harbor Numbers in a Daily Diet

	Food Product Average Concentration (micrograms/kg of product)							
Food Product	**Benzene (7)**	**Chloroform (20)**	**DDE (2)**	**Dieldrin (0.04)**	**Hexachloro-benzene (0.4)**	**PCB (0.09)**	**Styrene* (4)**	**Toxaphene (0.6)**
American cheese		14.7	9.9	0.4				
Butter		54.3	20.5	1.6	1.4		13.3	
Cheddar cheese		28	2.2	0.7				
Cream cheese		38.2	5.3	0.63				
Dry-roasted nuts			1.2	0.7				32.6
Ground beef	10.5		1.2	0.3				
Sour cream		34	2	0.3				
Salmon			8.9	1.9	0.8	26.6		
Swiss cheese		14.2	1.2	0.6				
Peanut butter			1.7	0.8			19.6	29
Spinach			18.4	1.5				

* As styrene oxide, which is a metabolite of styrene in the human body.

Note: ( ) Indicates safe harbor number in micrograms per day.

TABLE 9-4 Foods and Serving Sizes That Exceed Proposition 65 Safe Harbor Numbers

Product	Container	Serving	Dieldrin μg/kg — Mean	Calculated ug per Day of Dieldrin	Toxaphene μg/kg — Mean	Calculated ug per Day of Toxaphene	PCB μg/kg — Mean	Calculated μg per Day of PCB
Peanut butter	16 oz jar	2 Tbs 35 g	1.1	0.0385	28.3	0.991		
Dry-roasted nuts	16 oz jar	1 oz 39 pieces	1.1	0.031	34.8	0.989		
Beets	15 oz can	1/2 cup 123 g	1.1	0.135				
Cucumber (1 lb)	gourd	1 oz	5.1	0.144	21.3	0.605		
Dill pickles	16 oz jar	1/2 spear 28 g	5.5	0.154	19.5	0.546		
Sweet pickles	16 oz jar	1 spear 1 oz	3.3	0.094	19.4	0.551		
Spinach	10 oz (frozen)	1/3 cup 83 g	2	0.166				
Summer squash	16 oz (frozen)	2/3 cup 85 g	12	1.02	21	1.785		
Winter squash	10 oz (frozen)	1/2 cup 130 g	9.4	1.222				
Strained squash	4 oz jar	4 oz	2.6	0.295				
Cantaloupe	melon (3.5 lb)	2 oz	0.9	0.051	17.2	0.977		
Butter	16 oz package	1 Tbs 0.5 oz	2.3	0.033				
Ice cream	1 pint	1/2 cup 62 g	0.7	0.043				
Hot dogs	1 lb (8/pack)	1 link 2 oz	0.8	0.045				
Beef steak		3 oz	0.7	0.6				
Ground beef		3 oz	0.5	0.043				
Salmon steak		3 oz	1.9	0.162			26.3	2.241
Safe harbor numbers				0.04 ug/day		0.6 ug/day		0.09 ug/day

9.5 DRINKING WATER AND WARNING LABELS

Two of the most common disinfection by-products measured in drinking water in the 36 cities studied were bromodichloromethane and chloroform. However, only a third of the cities routinely monitored for these compounds since most cities only reported total trihalomethanes (both bromodichloromethane and chloroform are counted as part of the trihalomethane measurement). For those cities that reported these compounds, the average concentration of bromodichloromethane in drinking water was 7.6, ug/L while the average concentration of chloroform was 16.1 ug/L.

It is recommended that a person should drinking 64 fluid ounces of water a day. Since a liter is approximately 33.8 fluid ounces, a person following this recommendation would consume about 2 liters of water a day. Given the averages noted previously, this means that a significant number of people in the 36 cities studied were drinking tap water that would exceed Proposition 65's safe harbor number of 5 ug per day for bromodichloromethane and 20 ug per day for chloroform.

For the three cities in California, the drinking water taps in the cities of Bakersfield (see Table 7-2) and Los Angeles (see Table 7-16) should have a Proposition 65 warning label. Furthermore, based on the Los Angeles data, a resident only needs a little more than one 8-ounce glass of water to exceed the bromodichloromethane Safe Harbor number.

Clearly, California's Proposition 65 is not the law of the land. However, California's reliance on the numbers presented in Proposition 65 as they relate to protecting human health should give everyone cause for concern. This is especially true when you consider our total daily exposure to the Proposition 65 chemicals that can be found in our food and water. This problem is further expressed in the occurrence of chemical mixtures in human breast milk.

9.6 BREAST MILK AS AN INDICATOR OF WHAT IS WRONG IN THE ENVIRONMENT

Bioaccumulation is the process by which chemicals are stored in the bodies of exposed individuals, increasing in concentration over time. The effects of bioaccumulation in the environment have been reported repeatedly with the cases of mercury levels in fish and the well-demonstrated impacts of DDT exposure to raptor populations (egg shell thinning accelerated the rapid decline in bald eagle populations in the United States). In most biological systems (including humans), under periods of stress, pregnancy, and fasting, for example, chemicals stored in the body are released into the bloodstream when fat stores are tapped, increasing the amount of toxins circulating in the body. The continued accumulation of synthetic chemicals in the body can therefore be very significant in determining the overall health of the individual or the person's offspring.

Breast milk contains essential nutrients that are specifically designed for developing babies and provide essential building blocks for the immune system and growth.[87] Proteins, white blood cells, nucleotides, enzymes, and antibodies are each important components of breast milk. Breast milk also contains docosahexaenoic acid, which is responsible for growth of the brain and retina and myelinization of nervous tissue; cholesterol, which enhances myelinization of nervous tissue; taurine, which is important for bile acid function; choline, which enhances memory; lactoferrin, which binds to iron transporting it to the baby; inositol, which enhances synthesis of surfactant in immature lung tissue; and polysaccharides, which inhibit bacterial binding to intestinal surfaces.

While breast milk is the most important food for newborns and infants during critical periods of development, it may also be the primary route of exposure to many unsafe environmental contaminants. Fat-soluble chemicals such as polychlorinated biphenyls (PCBs), polybrominated diphenyl ethers (PBDEs), dioxins, and DDT are commonly found in maternal adipose tissue and are excreted via lipid-rich breast milk. Although mothers can limit contaminants in their breast milk by avoiding foods that may contain high concentrations of these compounds, many of these chemicals are already stored in the body from prior exposures and will be excreted in maternal milk.[88]

Babies and children are much more susceptible to the effects of contaminants than are adults. The rapid growth and development of sensitive systems make chemical exposure particularly hazardous to babies.[89] The immature and porous blood-brain barrier allows chemicals to gain access to the brain.[90] In addition, systems that detoxify compounds in adults are not fully developed in infants.[91] The risk associated with children's chemical exposures is greater, pound for pound, than the risk for adults.[92]

The synthetic chemicals most commonly found in breast milk are PCBs, DDT and its metabolites, dioxins, dibenzofurans, and PBDEs.[93] Other

---

[87] Bates CJ, Prentice A. 1994. Breast milk as a source of vitamins, essential minerals and trace elements. Pharmacol Ther 62(1–2):193–220.

[88] Solomon GM, Weiss PM. 2002. Chemical contaminants in breast milk: time trends and regional variability. Environ Health Perspect 110(6):A339–A347.

[89] Landrigan PJ, Sonawane B, Mattison D, McCally M, Garg A. 2002. Chemical contaminants in breast milk and their impacts on children's health: an overview. Environ Health Perspect 110(6):A313–A315.

[90] Landrigan PJ. 1999. Risk assessment for children and other sensitive populations. Ann NY Acad Sci 895:1–9.

[91] Landrigan PJ, Garg A. 2002. Chronic effects of toxic environmental exposures on children's Health. J Toxicol 40(4):449–456.

[92] EWG (Environmental Working Group). 2005. Body burden: the pollution in newborns. Available at www.ewg.org/reports/bodyburden2/execsumm.php [accessed March 23, 2006].

[93] Landrigan PJ, Sonawane B, Mattison D, McCally M, Garg A. 2002. Chemical contaminants in breast milk and their impacts on children's health: an overview. Environ Health Perspect 110(6):A313–A315.

synthetic chemicals detected include chlordane, dieldrin and aldrin, heptachlor, hexachlorobenzene (HCB), and hexachlorocyclohexane (HCH).[94] The levels of these compounds in breast milk can even exceed the drinking water standards set by the U.S. Environmental Protection Agency. Contamination of breast milk is not unique to the United States, and studies of breast milk in China, Spain, the Netherlands, Ukraine, Canada, and Sweden revealed concentrations of pesticides, heavy metals, industrial compounds, and other harmful compounds.[95–99] The characteristics of some of the most common compounds are described next. Since there are no standards for many of these compounds, comparisons will be made to the U.S. EPA's preliminary remediation goals (PRGs) for drinking water. PRGs are risk-based tools that standardize the cleanup of contaminant sites.

9.6.1 PBDEs (Polybrominated Diphenyl Ethers)

PDBE's are one of the most prevalent chemicals found in maternal milk.[100,101] They are used in flame retardants in furniture foams, computers, and televisions.[102] PBDEs are persistent in the environment, resisting breakdown and bioaccumulating in animals. As a result, PBDEs build up in their tissues. These chemicals accumulate in fatty tissues and fluids like breast milk because they are lipophilic or "fat loving." Consequences of high exposure to PBDEs

[94] Solomon GM, Weiss PM. 2002. Chemical contaminants in breast milk: time trends and regional variability. Environ Health Perspect 110(6):A339–A347.

[95] Ip HMH. 1983. Breast milk contaminants in Hong Kong. (The Bulletin of the Hong Kong Medical Association) 35:1–16.

[96] Hernández LM, Fernández MA, Hoyas E, González MJ, García JF. 1993. Organochlorine insecticide and polychlorinated biphenyl residues in human breast milk in Madrid (Spain). Bull Environ Contam Toxicol 50(2):308–315.

[97] Soechitram D, Chan SM, Nelson EAS, Brouwer A, Sauer PJJ. 2003. Comparison of dioxin and PCB concentrations in human breast milk samples from Hong Kong and the Netherlands. Food Addit Contam 20(1):65–69.

[98] Gladen BC, Monaghan SC, Lukyanova EM, Hulchiy OP, Shkyryak-Nyzhnyk ZA, Sericano JL, Little RE. 1999. Organochlorines in breast milk from two cities in Ukraine. Environ Health Perspect 107(6):459–462.

[99] Craan AG, Haines DA. Twenty-five years of surveillance for contaminants in human breast milk. Arch Environ Contam Toxicol 355:702–710.

[100] Hooper K, McDonald TA. 2000. The PBDEs: an emerging environmental challenge and another reason for breast-milk monitoring programs. Environ Health Perspect 108:387–392.

[101] Schecter A, Pavuk M, Papke O, Ryan JJ, Birnbaum L, Rosen R. 2003. Polybrominated diphenyl ethers (PBDEs) in U.S. mother's milk. Environ Health Perspect 111(14):1723–1729.

[102] Solomon GM, Weiss PM. 2002. Chemical contaminants in breast milk: time trends and regional variability. Environ Health Perspect 110(6):A339–A347.

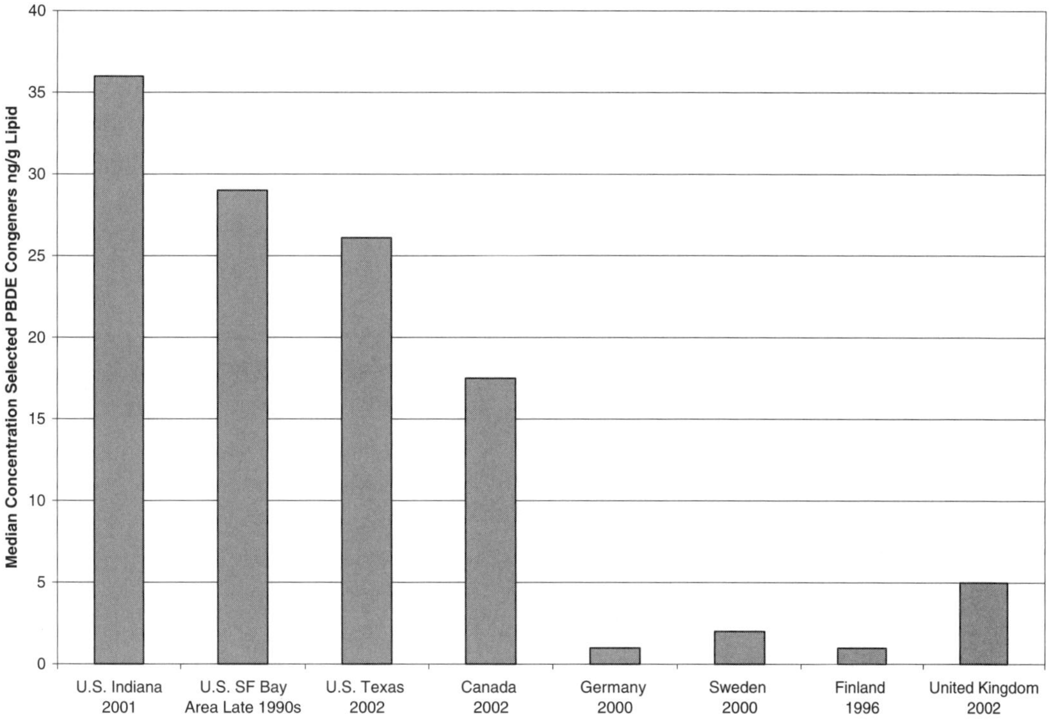

FIGURE 9-4 PBDEs in breast milk and fat samples around the world. From NRDC (Natural Resources Defense Council). 2005. Healthy Milk, Healthy Baby. Available at www.nrdc.org/breastmilk [accessed March 23, 2006].

can include impaired development of the brain and thyroid, hearing deficits, delayed puberty, decreased sperm count, fetal malformations, and possibly cancer.[103] Unlike levels of PCBs, dioxins, and DDT, PBDE concentrations in breast milk are actually increasing over time.[104,105] Figures 9-4 and 9-5, respectively, show concentrations of PBDEs around the world and in Swedish women, where the concentration is increasing. The United States has the highest levels of PBDE at 73.9 ppb,[106] which is approximately 10,000 times higher than the U.S. EPA's Region 9 2005 PBDE preliminary remediation goal (PRG) of 7.6E-03 ppb for polybrominated phenols in drinking water.

[103] Schecter A, Pavuk M, Papke O, Ryan JJ, Birnbaum L, Rosen R. 2003. Polybrominated diphenyl ethers (PBDEs) in U.S. mother's milk. Environ Health Perspect 111(14):1723–1729.

[104] Noren K, Meironyte D. 1998. Contaminants in Swedish human milk: decreasing levels of organochlorine and increasing levels of organobromine compounds. (Organohalogen Compounds) 38:1–4.

[105] Noren K, Meironyte D. 2000. Certain organochlorine and organobromine contaminants in Swedish human milk in perspective of past 20–30 years. Chemosphere 40:1111–1123.

[106] Schecter A, Pavuk M, Papke O, Ryan JJ, Birnbaum L, Rosen R. 2003. Polybrominated diphenyl ethers (PBDEs) in U.S. mother's milk. Environ Health Perspect 111(14):1723–1729.

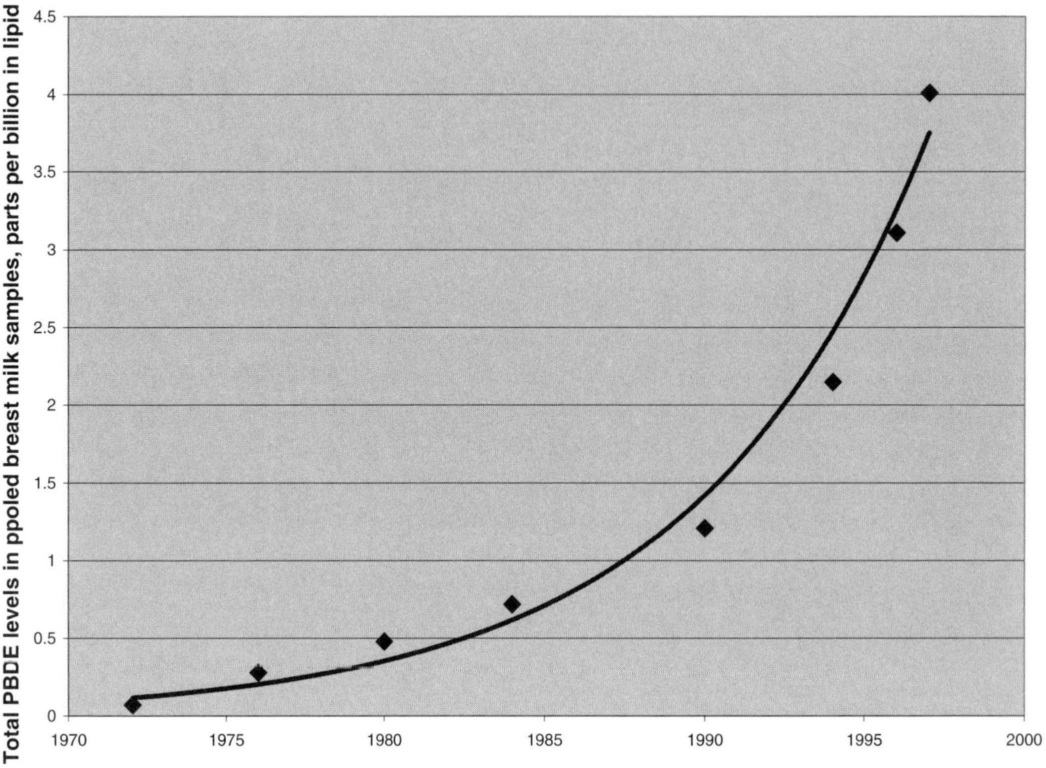

FIGURE 9-5 Dramatic increase in levels of PBDE fire retardants in Swedish women's breast milk, 1972–1997. From Meironyte D, Noren K, Bergman A. 1999. Analysis of polybrominated diphenyl ethers in Swedish human milk: A time related trend study 1972–1997. J Toxicol Environ Health Part A 58(6):329–341.

9.6.2 PCBs

PCBs, persistent and bioaccumulative compounds often found in breast milk, were used in industrial insulators and lubricants and have been banned since 1976. PCBs are known by the U.S. EPA to cause cancer and nervous system problems.[107] They can also cause infant death, birth defects, and brain damage. PCB exposure in the womb or during lactation is associated with decreased IQ, impaired psychomotor development, decreased immune function, and skin disease. They are probable human carcinogens, and concentrations of PCBs in breast milk can exceed those of PBDEs (Figure 9-6). The 2005 Region 9 PRG for PCBs is 3.4E-02 ppb in drinking water.

[107] United States Environmental Protection Agency. 2006. Health effects of PCBs. Available at www.epa.gov/pcb/pubs/effects.html [accessed March 24, 2006].

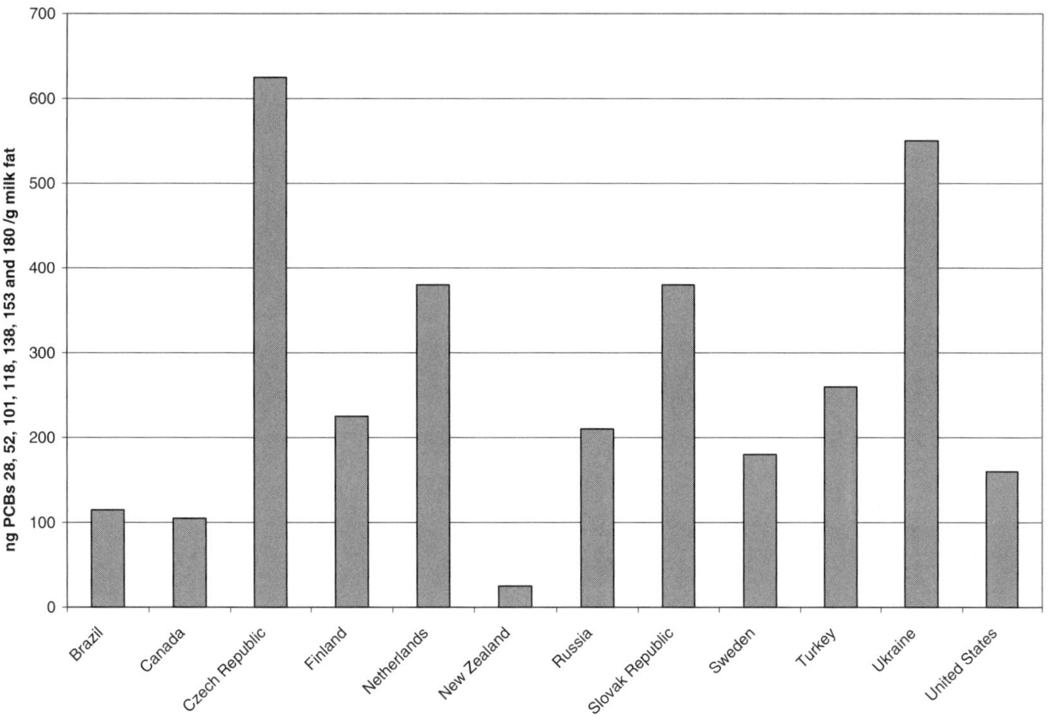

FIGURE 9-6 Breast milk concentrations of seven congeners of PCBs in breast milk around the world. From NRDC (Natural Resources Defense Council). 2005. Healthy Milk, Healthy Baby. Available at www.nrdc.org/breastmilk [accessed March 23, 2006].

9.6.3 Dioxins

These compounds can be found in flame retardants from plastics and are a by-product of incineration. They are very toxic to the developing endocrine system, cause birth defects in animals, disrupt reproductive development, and affect immune and hormonal systems.[108] They are also a known human carcinogen. Although the concentrations of dioxin found in breast milk may seem small compared to other chemicals drinking water (concentrations are less than 1 ppb), dioxin is potently more toxic. Dioxins' PRG is 4.5E-07 ppb. Figure 9-7 shows various concentrations of dioxins around the world. Countries all over the world have concentrations with risks that range in the tens of thousands.

[108] Agency for Toxic Substances and Disease Registry. 1998. Toxicological profile for chlorinated dibenzo-*p*-dioxins. U.S. Department of Health and Human Services.

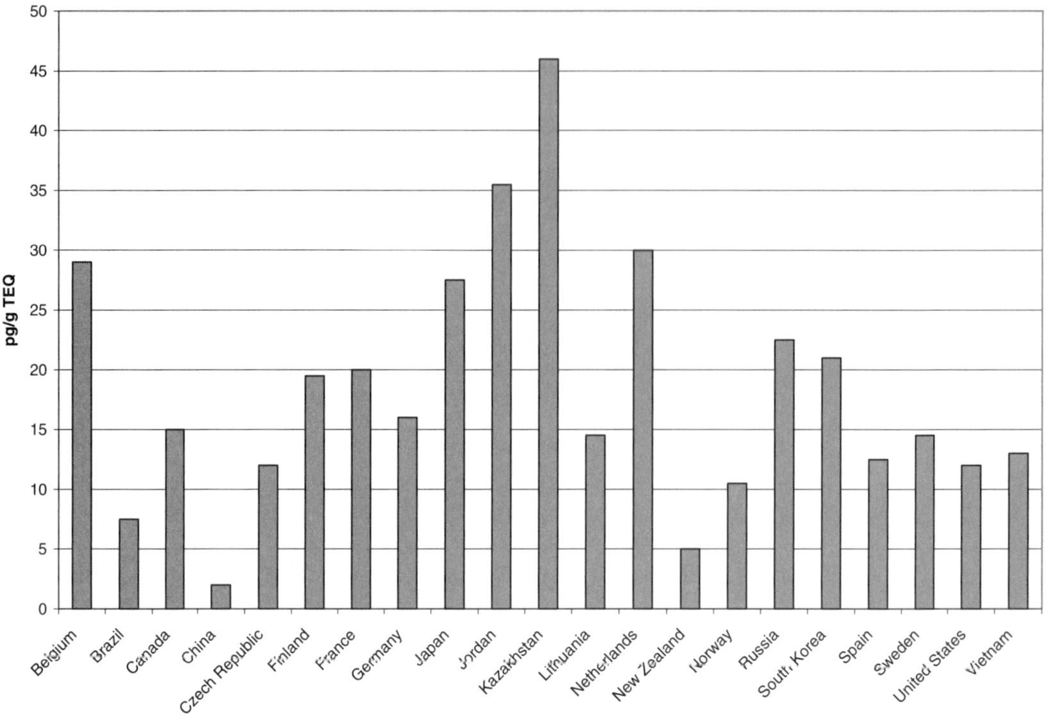

FIGURE 9-7 Dioxins and furans in breast milk around the world (late 1980s through the 1990s). From NRDC (Natural Resources Defense Council). 2005. Healthy Milk, Healthy Baby. Available at www.nrdc.org/breastmilk [accessed March 23, 2006].

9.6.4 DDT and DDE

DDT was used as a pesticide and enters the body through fruits, vegetables, fatty meat, fish, poultry, and contaminated drinking water. In the environment, DDT degrades into DDE and DDD. This pesticide is known as a possible human carcinogen and may cause decreased fertility, kidney and liver dysfunction, and immune system dysfunction.[109] DDT in breast milk is prevalent worldwide (Figure 9-8). Even after DDT was banned in 1972, it remains at a concentration of 264.45 ppb in breast milk. DDE is found at an even higher concentration, 790.3 ppb. The PRG for both DDE and DDT is 2.0E-01 ppb. The reason DDE concentrations are still high after being banned is related to the compound's high fat solubility and resistance to degradation. In developing

[109] Agency for Toxic Substances and Disease Registry. 2002. Toxicological profile for DDT, DDE, and DDD. U.S. Department of Health and Human Services.

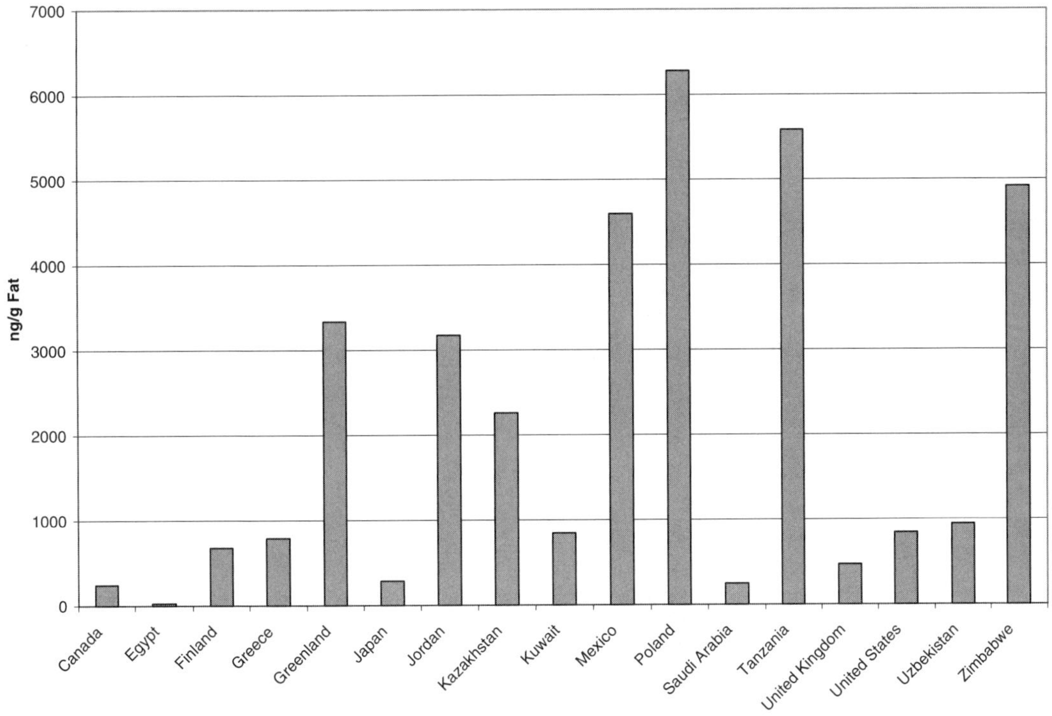

FIGURE 9-8 DDT and DDE in breast milk and adipose tissue around the world (1990s–2000s). From NRDC (Natural Resources Defense Council). 2005. Healthy Milk, Healthy Baby. Available at www.nrdc.org/breastmilk [accessed March 23, 2006].

countries, DDT use has increased in agriculture and for malaria control. It is still used today in Africa, Asia, and Latin America for vector control.

9.6.5 Exceeding Standards

Maternal milk is the main pathway through which fat-soluble toxic substances are eliminated from the body. As a result, concentrations are extremely high and can even surpass the FDA's standards for cow's milk set for adults (Figure 9-9). They can also exceed the federal drinking water standards (Table 9-5).

The increased use of synthetic compounds and their release to the environment, along with their persistence in the environment following their release, and the bioaccumulation of the compounds in biological systems (foodstuffs and humans), means that impacts from these contaminants will be present for generations. A study conducted by the National Academy of Sciences suggests that 28% of childhood disabilities can be attributed to environmental

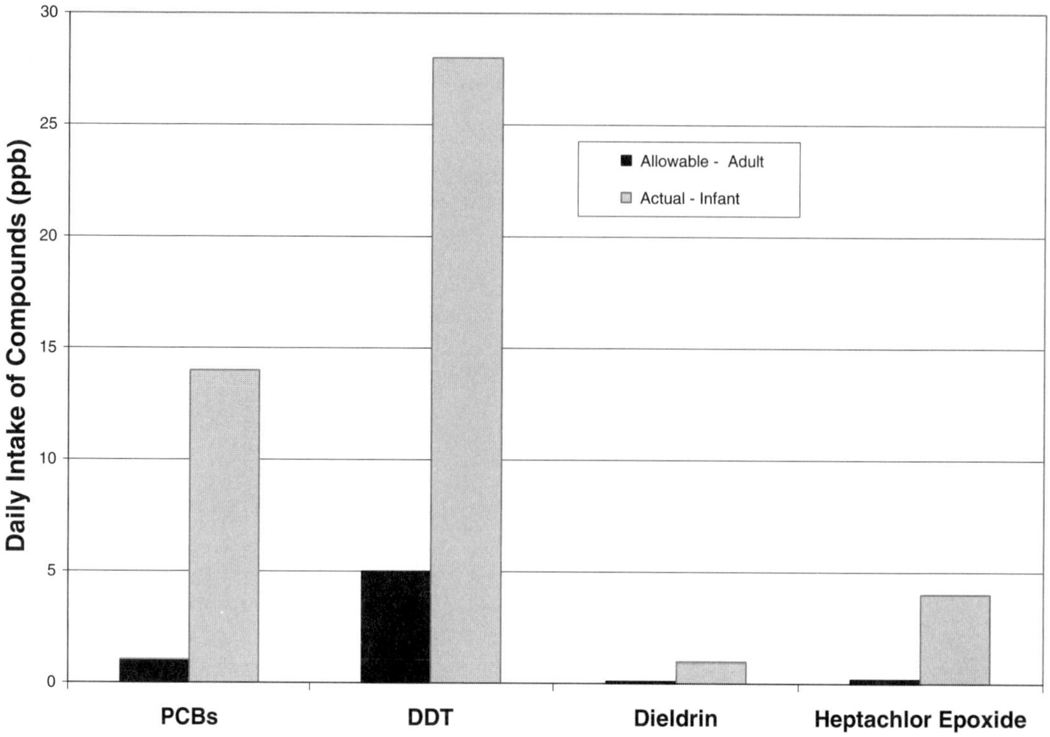

FIGURE 9-9 Comparison of the allowable daily intake of various contaminants and the actual intake seen in infants. From Rogan WJ, Bagniewska A, Damstra T. 1980. Pollutants in breast milk. New England J Med 302(26):1450–1453.

TABLE 9-5 Human breast milk exceeds federal drinking water standards

Contaminant	EPA Drinking Water Standard (ug/L) (EPA Region 9 PRGs)	Human Breast Milk Contaminant Levels (ug/L)
PCBs[110]	0.034	54
PBDEs[111]	0.07	73.9
Dioxin[112]	0.00000004	0.0156
DDT/DDE/DDD[113]	0.2	790.3
Mercury[114]	1.2	1.7

[110] Van Leeuween, FXR, Malish, R. 2002. Results of the third round of WHO-coordinated exposure study on the levels of PCBs, PCDDs, and PCDFs in human milk. Organohologen Compounds 56:311–316.

[111] Kalantzi O.I., Martin F.L., Thomas G.O., Alcock R.E., Tang H.R., Drury S.C., Carmichael P.L., Nicholson J.K., and Jones K.C., 2004. Different levels of polybrominated diphenyl ethers (PBDEs) and chlorinated compounds in breast milk from two U.K. regions. Environ. Health Perspectives, 112, 1085–1091

[112] LaKind, J., C.M. Berlin, D. Naiman. 2001. Infant Exposure to Chemicals in Breast Milk in the United States: What We Need to Learn From a Breast Milk Monitoring Program. Environmental Health Perspectives. 109:1

[113] Jaga, K. and C. Dharmani. 2003. Global Surveillance of DDT and DDE Levels in Human Tissues. International Journal of Occupational Medicine and Environmental Health. 16(1):7–20.

[114] Abadin, H.G., B.F. Hibbs, and H.R. Pohl. 1997. Breast-feeding Exposure of Infants to Cadmium, Lead, Mercury: A Public Health Viewpoint. Toxicol. Ind. Health. 13(4):495–517.

factors.[115] Since the 1980s there has been an increase in numerous childhood cancers.[116] Autism has increased by a factor of 10, male birth defects and childhood asthma by a factor of 2, acute lymphocytic leukemia by 62%, and childhood brain cancer by 40%. Although pollutants may not be the sole cause of these increases, they may have contributed to the growing problem.

As discussed previously, breast milk is a very good measure of the multiple contaminants to which humans are exposed. Thus, when evaluating the true toxic nature of synthetic chemicals in the environment, the degree to which chemical mixtures impact our health needs to be known.

9.7 MULTIPLE CONTAMINANTS AND INCREASING TOXIC RESPONSES

A basic simplifying assumption in toxicology has been to study one compound at a time to determine the effect of exposure. Exposures in the real world are not "neat," therefore the basis for the previous toxicological studies (and the predicted public health outcomes) are limited in their ability to predict disease outcomes in the general population. Contaminants are absorbed and utilized in the body via pathways in which a number of compounds can/are detoxified. If the specific detoxification pathway is overwhelmed (saturated with contaminants) the body pushes the remaining contaminant through other pathways, leading to the toxic outcomes. It is therefore logical that exposure to multiple contaminants could/will lead to the saturation of detoxification pathways, increasing toxic responses. Put another way, the exposure to an increasing number of toxins (no matter what the concentration) will lead to greater and greater toxicity (a lower dose to elicit a response).

In 2005, the Environmental Working Group (EWG) released the results of a biomonitoring program that confirmed that synthetic compounds were making their way into the bodies of newborns even before their birth. The cord blood from 10 infants born in August 2004 and September 2004 in the United States as analyzed for synthetic compounds and metals. A total of 287 different chemicals were detected in the cord blood, with an average of 200 different compounds being detected in the 10 newborns. Compounds detected included organochlorine pesticides, perfluorochemicals, PDBEs, PCBs, polychlorinated naphthalenes, polychlorinated and polybrominated dibenzodioxins and furans (PCDD/F and PBDD/F), and polyaromatic hydrocarbons (PAHs). According to EWG, 180 of the chemicals detected are carcinogens in animals or humans,

---

[115] National Academy of Sciences. 2000. Scientific frontiers in developmental toxicology and risk assessment. Committee on Developmental Toxicology. Washington, D.C.: National Academies Press.

[116] National Cancer Institute. 2002. National Cancer Institute Research on childhood cancers: fact sheet. Available at www.cancer.gov/cancertopics/factsheet/NCI-childhood-cancers-research [accessed March 23, 2006].

217 are known to be toxic to the brain and nervous system, and 208 cause birth defects or abnormal development in animal tests.

The EWG study noted that significant impacts (adverse health/diseases) likely to be expressed in the newborns tested were as follows:

- Because of a child's small body mass, chemical exposures are greater pound for pound than those of adults.
- The undeveloped blood-brain barrier in the infant allows greater chemical exposures to the developing brain.
- Children have fewer chemical-binding proteins allowing more of the chemical to reach "target organs."
- Because a baby's organs and systems are still developing, they are more susceptible to chemical damage.
- The systems that detoxify and excrete chemicals in adults are not fully developed in babies.

The Toxic Substances Control Act (TSCA) of 1976 mandated that manufacturers ensure that high production volume compounds be monitored from "cradle to grave." This term was meant to signify that the manufacturers and users could account for chemicals from the moment they were created to the moment that they were disposed of in a safe manner. The TSCA did not envision the accumulation of compounds in populations, nor the passage of those compounds from parent to child. Based on the results of the EWG study, this phrase should now be updated to "cradle to grave to cradle."

9.8 CHEMICAL MIXTURES AND TOXIC OUTCOMES

Exposure to a complete carcinogen or a procarcinogen will cause the toxic response from exposure to less "carcinogenic" contaminants to increase. According to the state of California, environmental tobacco smoke is a carcinogen that will cause adolescent females to have increased rates of breast cancer. Carcinogens in the smoke, including benzene, a known cause of breast cancer, damage the immune system of the adolescent, allowing other contaminants to express their toxicity at a later date.

A number of target organs affected by the chemicals of concern include the alimentary system, the development of a fetus, the endocrine system, the ocular system, the hematopoietic system, the immune system, the nervous system, the reproductive system, the respiratory system, and the dermal system. The impact to each system by synthetic chemicals must be accounted for when evaluating the toxicity of chemicals.

A number of toxic agents may affect similar organ systems or create the opportunity for adverse reactions. In the case of environments that may expose an individual to multiple synthetic compounds at the same time, the effects can be wide ranging. For example (Table 9-6), in an environment that may contain PCBs, polychlorinated dibenzofurans (PCDFs which are combustion products

TABLE 9-6 Target Organs Affected by Toxic Air Contaminants (Chronic Toxicity)*

	Contaminant				
	PCB	Dioxin-Like Compounds PCDF	Benzene	Formaldehyde	Potential Outcome
Organ system					
Alimentary	x	x			x
Cardiovascular					
Developmental	x	x	x		x
Endocrine	x	x			x
Ocular				x	x
Hematological	x	x	x		x
Nervous system			x		x
Reproductive	x	x			x
Respiratory	x	x		x	x

* Adapted from Table 10a of SCAQMD, 2005. Risk Assessment Procedures for Rules 1401 and 212, Version 7.0 (Latest Version), Procedures, Equations, and Assumptions Effective on or After July 1, 2005. See Attachment L. www.aqmd.gov/prdas/Risk%20Assessment/RiskAssessment.html.

of PCBs), benzene, and formaldehyde, the known affected target organs/systems include the alimentary system, the development of fetuses, the endocrine system, the ocular system, the nervous system, the reproductive system, and the respiratory system.

The time at which a chemical is delivered is extremely important in the outcome. For agents that cause developmental delays or dysfunction (e.g., PCBs and PCDFs or other dioxin-like compounds), exposure to the compounds prior to conception may lead to the dysfunction of necessary germ cells (abnormal development of sperm or eggs).

Silva et al. (2002)[117] demonstrated that when eight weak estrogenic compounds (including PCBs) were combined below their individual no-observable effect concentration (NOEC), significant effects were measured in the experimental system. In a practical sense, this means that the threshold for adverse health outcomes was lowered when exposures to multiple synthetic chemicals occurred.

The U.S. EPA has noted that the time in development at which an exposure occurs plays a significant role in the outcome of the exposure. For exposure to individuals who are less than 2 years of age, a multiplier of 10 should be used in the assessment; for individuals between the age of 2 and 16, a multiplier of 3 is used; and for individuals above the age of 16, no multiplier is used. The implications are that much lower environmental concentrations can lead to significant outcomes for the exposed populations. Children exposed to benzene in foods and air between the ages of 0 and 2 will receive a much greater impact than for exposures occurring after age 16.

[117] Silva E, Rajapakse N, Kortenkamp A. 2002. Something from "nothing" — eight weak estrogenic chemicals combined at concentrations below NOEC's produce significant mixture effects. Environ Sci Technol 36(9):179A–180A.

In summary, when dealing with chemical mixtures, a toxic outcome (e.g., cancer in a specific organ) becomes less specific with exposure to more toxins. In other words, a greater number of cancers will occur that are less specific to any one contaminant. Such outcomes are not amiable to the current regulatory reliance on single toxicity chemical standards.

9.9 WHAT DOSE IS SAFE?

For most chemicals in the environment, the greater the concentration, the more toxic the outcome will be. However, this is not necessarily true for chemical mixtures. For chemical mixtures in the environment, it is logical that, given the totality of exposure from compounds to which an individual is exposed, a lower dose is necessary to elicit a response. Thus, exposure assessments that take into account potential chemical interactions will be necessary for estimating the toxic dose of single and multiple chemical mixtures in the real world. This is a necessary step in biological systems that are already understood to be very dynamic.

As our understanding of the effects of chemicals on biological systems evolves, so does our understanding of what is "safe." As evidenced by Figure 9-10, it is clear that low doses of chemicals exert an effect on the system that

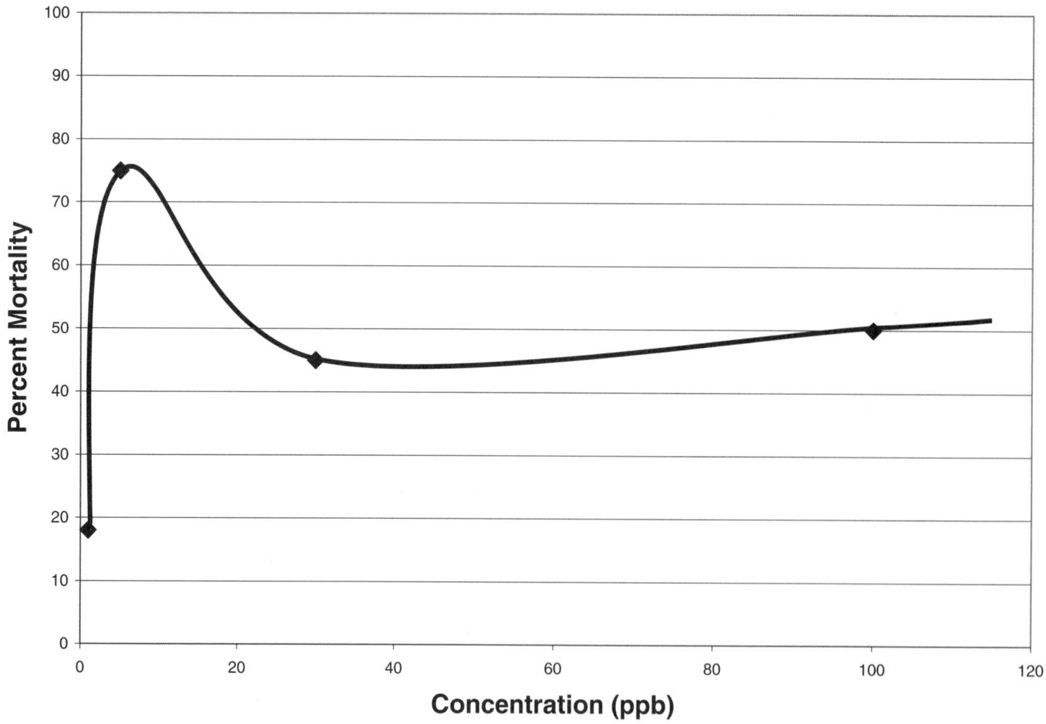

FIGURE 9-10 Example of toxic response for endocrine-disrupting compounds.

may not be expressed at higher doses. In the case of the compound above very low doses result in suppression of the endocrine system. Endocrine-disrupting compounds (EDC) express these effects in mammals by the inhibition of androgen receptors and the induction of aromatase, the enzyme that converts androgen to estrogen in mammals, amphibians, and potentially reptiles. This has been described as the demasculization of subject populations (the so-called "cross-dressing carp"). Many endocrine-disrupting chemicals do not have a definable toxicological threshold level below which effects do not exist. This is in contrast to classic toxins that are assumed to have a monotonic (linear) dose response. A study on the survivorship pattern of larval amphibians' exposure to low concentrations of atrazine (a known EDC) shows that atrazine exhibits a nonmonotonic dose-response curve (NMDRC) similar to an inverted U. The result is that at low doses, survival was lower than at high doses.

9.10 IT IS NOT JUST THE DOSE THAT MAKES THE POISON

Clearly the results of the water, air, and market basket studies cited earlier demonstrate that the environments of the cities evaluated are severely impacted by many synthetic chemicals. Individually, each chemical may not exceed a safe harbor level in one medium within a city (e.g, benzene may be below the MCL in drinking water), but when combined with the other media (e.g., benzene in food and air), the chemicals exceed the cumulative safe harbor level. The frequency with which the chemical insults occur, the variety of chemicals providing the insults, and the multiple pathways through which the insults occur make it evident that the cumulative dose of synthetic chemicals taken in daily by most Americans (the mixture of synthetic chemicals absorbed on a daily basis) will exceed any reasonable measure of what could be considered acceptable.

Biomonitoring of newborns and adults has clearly demonstrated that the cumulative body burden of synthetic compounds in residents of the United States is significant, yet the cumulative effects of exposure to all of these compounds simultaneously has yet to be comprehensively studied. Based on the existing information, which demonstrates that from birth, the synthetic chemical body burden is significant and the knowledge that the threshold for toxicity is lower for chemical mixtures than for single chemicals, the paradigm by which standards or acceptable levels are derived must be revised to include the totality of exposures and the interaction between chemicals.

10 THE BODY'S SYNTHETIC CHEMICAL BURDEN

In 1962, Rachel Carson wrote *Silent Spring* on the hazards of synthetic pesticides. In this book she contends that "We have subjected enormous numbers of people to contact with these poisons, without their consent and often without their knowledge . . . we have allowed these chemicals to be used with little or no advance investigation of their effect on soil, water, wildlife, and man himself."

There have been remarkable advances in understanding the toxicity and fate of individual pesticides in the environment, which continues to evolve since Carson's book was published. Yet most individuals are still unaware of the chemical mixtures to which they are exposed, and there is still "no advance investigation of their effect" as to how chemical mixtures impact human health.

Rachel Carson also surmised that "They [pesticides] occur in mother's milk and probably in the tissues of the unborn child. . . . This situation also means that today the average individual almost certainly starts life with the first deposit of the growing load of chemicals in his body. . . . There has been no such parallel situation in medical history. No one yet knows what the ultimate consequences may be." Today, we know that the blood of newborn babies is contaminated with more than 287 different chemicals,[1] and we still do not know the ultimate consequences of the growing chemical burden on our bodies. The increase in childhood disabilities, childhood cancers, and autism with the increased use of synthetic chemicals maybe more than a coincidence.

[1] The Environmental Working Group report (2005) titled "Body Burden, the Pollution in Newborns."

Carson also observed that "The common salad bowl may easily present a combination of organic phosphate insecticides. Residues well within the legally permissible limits may interact. The full scope of the dangerous interaction of chemicals is as yet little known." Once again, this surmised "dangerous interaction" is still unknown.

Since the potential for dangerous interactions of chemical mixtures was raised, there have been little or no advances in the medical sciences to address this issue. However, we do have more data on those chemicals to which we are exposed (as shown in Chapters 2 through 8) and have started to characterize the body's chemical burden.

10.1 THE BODY'S TOXIC BURDEN

The most recent study by the Centers for Disease Control and Prevention (CDC), published in July 2005, reported on the concentration of selected synthetic organic chemicals in blood and urine for the periods 1999–2000 and 2001–2002. The CDC evaluated the following basic types of chemicals (a list of the chemicals monitored is given in Appendix 10-1):

- Polycyclic aromatic hydrocarbons (PAHs) are distributed throughout the environment because they are manufactured, occur in petroleum fuels, and are also generated by the incomplete burning of organic materials (e.g., coal, wood, petroleum fuels, tobacco, and charcoal-broiled meat). PAHs can occur in air, water, and food. Vehicle exhaust and smoke (e.g., tobacco and fires) are the most common sources of air pollution. Water can be contaminated (1) by coming in contact with petroleum fuels or petroleum wastes and (2) by atmospheric deposition. Food is most commonly contaminated by PAHs from the method of preparation (i.e., smoking, charcoal broiling, or roasting, which are not monitored for by the Food and Drug Administration [FDA]) as well as possibly being contaminated by air or water that contains PAHs. It is important to note that PAH contamination occurs as a mixture of PAH chemicals.
- Dioxins, dibenzofurans, and polychlorinated biphenyls (PCBs) are all byproducts of industrial processes or chemical syntheses. Dioxins can also be created by incineration of waste and bleaching. PCBs (banned in 1979) were also extensively used in electrical/heat-exchange liquids. Individuals are primarily exposed to these chemicals in food as they accumulate in high-fat foods and fish with a high oil content.
- Phthalates are industrial chemicals that are added to plastics to enhance flexibility. Individuals are usually exposed through direct contact with plastic products, through food that has come into contact with plastics, and through water (e.g., phthalates are common in industrial wastewater discharges).

- Organochlorine and organophosphate pesticides are manufactured to be toxic. These compounds are distributed throughout the environment because they were directly applied to land and crops, occur in runoff from cropland, and are released from manufacturing and waste disposal facilities. These compounds primarily contaminate food and sometimes water.

When reviewing the list of chemicals monitored by the CDC in the blood and urine of individuals older than the age of 6, many of these chemicals are not looked for in our food and water. For example, although PAHs can occur in heated/charred foods, the FDA does not include these compounds in the Total Diet Study list of analyzed chemicals (see Appendix 2-1) and only benzo(a)pyrene is routinely monitored in drinking water (see Table 7-1). Furthermore, only the metabolites of PAHs are monitored because of the body's ability to break down these compounds. This also illustrates that the original synthetic chemical yields multiple chemical contaminants from the original compound.[2] The FDA also does not look for phthalates in foods and only looks for one phthalate in drinking water. Many of the chemicals monitored in food and water are not analyzed by the CDC in their blood and urine chemical investigations. This is particularly true for industrial chemicals.

A summary of the chemicals most frequently detected by the CDC is given in Table 10-1.[3] When reviewing this table, note that (1) the concentration of the indicated PAHs was determined from the levels of detected breakdown by-products and (2) the CDC could only characterize, for the most part, the metabolites of the organophosphate pesticides. Clearly, when trying to characterize the body's chemical burden, the CDC may have to look for not only the original chemical of interest but its associated by-products as well.

Because the chemical monitoring of blood and urine is both difficult (i.e., establishing analytical methods for very low detection limits) and very expensive, the CDC *cannot* monitor for all of the chemicals that are likely to be in the human body. This is clearly evident in its most recent study (2005) where only a limited number of chemical compounds were examined. Furthermore, most of the common industrial chemicals (i.e., single-ring aromatic hydrocarbons and chlorinated aromatic and aliphatic compounds) that occur in food, water, and air are not even monitored by the CDC. As a result, we will most likely never know the true nature of the chemical mixtures (i.e., the original chemical and its by-products) that really exist in the human body.

This means that the chemical contaminant data collected in our food, water, and air resources are even more important if we are going to try to forecast the likely chemical mixtures that occur in our bodies. Yet even these monitoring programs only analyze for a fraction of the 72,000 chemicals used in the United

---

[2] Just because a chemical is degraded does not mean the resulting compounds are harmless.

[3] The CDC was able to determine a mean concentration for the complete population sampled.

TABLE 10-1 Chemicals Detected in Humans (CDC, 2005)

Polycyclic Aromatic Hydrocarbons	1999–2000 ng/L	2001–2002 ng/L
Fluoranthene	13.4	
Fluorene	611	671
Phenanthrene	379.3	394.80
Pyrene	79.8	49.6
Naphthalene		2050
Dioxins, Furans, and PCBs		
OCDD		346
HpCDD		39
HxCDD		34.6
HpCDF		9.64
PCB 126		22.7
PCB 138 & 158		19.9
PCB 153		27.2
PCB 169		17.9
PCB 180		19.2
Phthalates		
Dimethyl phthalate		1.15
Diethyl phthalate	179	178
Butyl phthalates	24.6	21.6
Benzylbutyl phthalates	15.3	15.1
Diethylhexyl phthalate	3.43	37.8
Dioctyl phthalate		2.75
Pesticides		
Hexachlorocylcohexane (HCB)	9.68	
Trichlorophenol	2.85	
DDE	260	295
Chlordane		11.4
Organophoshphate metabolites		
Dimethylthiophosphate	1.82	
Diethylphosphate	1.03	
Diethylthiophosphate		0.46
Chlorpyrifos and chlorpyrifos-methyl	1.77	1.76
2,4-D	1.11	
Permethrin/cypermethrin		0.32
Phenylphenol	0.49	
Dichlorobenzene	6.01	

ng/L = nanograms per liter (in blood or urine).

States today.[4] Thus, it would appear that we will probably never be able to identify the true nature of the chemical mixtures in our bodies. In the meantime, the existing chemical monitoring data will have to serve as our only indicator of humankind's toxic legacy.

10.2 OUR TOXIC LEGACY

The synthetic chemical mixtures that have been identified in food, water, and air resources of the 36 cities do not represent an individual's actual exposure to specific chemicals. The amount of sampling and chemical analyses that would have to be completed to provide a statistically meaningful indication of our exposure to chemical mixtures in food, water, and air is beyond any cost a foundation, state agency, or the federal government would be willing to pay. The current data only provide an indication of what chemical mixtures individuals in the cities that were studied may have been exposed to during the time period over which the data were collected. This being said, the data clearly show that where you live will significantly influence the number of different synthetic chemicals that will be in your food, water, and air.

The data presented in the previous chapters show that each resource (food, water, or air) is contaminated with a unique set of different chemicals. Given this condition, Table 10-2 compares the most frequently encountered synthetic chemicals in each media. *This comparison shows that the most frequently detected compounds in all three media include 11 pesticides, 24 industrial chemicals, and 5 disinfection by-products. The synthetic chemicals that make up this mixture contribute to the total number of different compounds that may contaminate our bodies.*

When reviewing this list, note that two groups of chemicals occur in two or three of the media. The first group is made up of the single-ring aromatic compounds (benzene, ethylbenzene, toluene, xylenes, styrene, and trimethylbenzene). The second group is composed of chlorinated industrial chemicals (tetrachloroethylene, trichloroethylene, and dichloromethane). Exposure to these compounds is associated typically with depressed immune responses, neurological disorders, and cancers. This result is important for several reasons.

First of all, it is important to remember that there are other environmental sources of chemicals that contribute to the chemical body burden that are not routinely monitored by the federal government (i.e., as defined in the previous chapters). These sources include the following:

- The home environment provides multiple sources of exposure to synthetic chemicals. For example, volatile chemicals from building materials, carpets,

[4] National Research Council, *Identifying Future Drinking Water Contaminants*, National Academy Press, Washington, D.C., 1999.

TABLE 10-2 Comparison of Chemical Frequencies for All Resources

Most Frequent Synthetic Chemicals	Food	Water	Air
Pesticides (11)			
Atrazine		X	
Chlorpyrifos	X		
Chlorpyrifos-methyl	X		
DDE	X		
Dieldrin	X		
Diphenyl-2-ethylhexyl phosphate	X		
Endosulfan sulfate	X		
Iprodione	X		
Malathion	X		
Methoxychlor	X		
Permethrin	X		
Industrial Chemicals (24)			
Benzene	X		X
Butyl alcohol			X
Dichloroethylene		X	
Dichloromethane		X	X
Di(2-ethylhexyl)phthalate		X	
Diisocyanates			X
Dioxins*			X
Ethyl benzene	X	X	X
Ethylene glycol			X
Glycol ethers			X
Hexane			X
Methanol			X
Methyl ethyl ketone			X
Methyl isobutyl ketone			X
Methyl pyrrolidone			X
Methyl tertiary butyl ether (MTBE)		X	
Napthalene			X
PAHs*			X
Perchloroethylene	X	X	
Styrene	X		X
Trichloroethylene		X	X
Trimethylbenzene	X		X
Toluene	X	X	X
Xylenes	X	X	X
Disinfection By-Product (5)			
Chloroform	X		
Chloromethane		X	
Chlorodibromomethane		X	
Haloacetic acids*		X	
Totat trihalomethanes*		X	

* Mixture of chemicals.

drapes, and furniture; the use of pesticides, paints, solvents and cleaning agents; emissions/particulates from a fireplace or furnace; dry cleaned garments; and personal care products like cosmetics.

- A work environment provides similar sources of synthetic chemicals as a home environment with the addition of particulates from printers and copy machines and direct contact with chemicals or chemical emissions that result from daily operations.
- Both urban and rural settings have vehicular traffic and associated fueling facilities that release both particulates (containing PAHs) and aromatic hydrocarbons (e.g., benzene). The urban environment, however, would be expected to have more particulate matter and single-ring aromatics due to more vehicular traffic/emissions and the occurrence of more diesel trucks and buses.
- Individuals in agricultural areas (but not involved with agricultural operations) can also be exposed to pesticides during periods of application.

Because there are multiple sources of chemicals throughout the environment, there is the potential for unexpected interactions.

For example, as discussed in Chapter 9, if the detoxification pathway for benzene (a known human carcinogen) is saturated with other single benzene ring compounds (most probably toluene and xylene) benzene will not be destroyed and instead will be pushed through a secondary pathway that could lead to a toxic outcome. Furthermore, with the body potentially saturated with the single-ring aromatics (i.e., including benzene) from food (see Table 4-19), water, or industrial air sources, a small dose of benzene from automobiles while one is walking along a congested downtown street or filling his or her fuel tank could elicit a toxic response.

This same discussion could also be relevant to the occurrence of both tetrachloroethlyene and trichloroethylene (chemically, very similar compounds) in food, water, and air. Yet there is not another widespread source of these compounds in the environment like there is for the single-ring aromatics. However, individuals with a number of chlorinated industrial chemicals already in their bodies could possibly have a toxic response from receiving a small dose of tetrachloroethylene while picking up their dry cleaning.

Second, it is also important to remember that there are no governmental standards that limit the amount of single-ring aromatics that can occur in food or, for that matter, any industrial chemical. Furthermore, when considering the number and single-ring aromatics that occur in food along with the other sources of these chemicals in water, industrial air emissions, and vehicular sources, maybe the impact of single-ring aromatics on human health is more important than the occurrence of pesticide residues in food products. This may be particularly true in an urban environment.

When reviewing all the data summarized in Tables 4-16, 7-37, 8-2, and 10-2, it is certain that the synthetic chemical mixture found in our food, water,

TABLE 10-3 Basic Synthetic Chemical Mixture (Food, Water, and Air)

Pesticides
Chlorpyrifos
Chlorpyrifos-methyl
DDE
Dieldrin
Diphenyl-2-ethylhexyl phosphate
Endosulfan sulfate
Iprodione
Malathion
Methoxychlor
Permethrin
Single-Ring Aromatics
Benzene
Ethylbenzene
Styrene
Trimethylbenzene
Toluene
Xylenes
Disinfection By-Products
Chloroform
Chloromethane
Chlorodibromomethane
Haloacetic acids*
Total trihalomethanes*

* Several dozen to hundreds of compounds.

and industrial air emissions will include a combination of at least 10 pesticides (mostly banned organochlorine compounds and organophosphate compounds), dozens of disinfection by-products, and at least two single-ring aromatic hydrocarbons (xylenes and toluene). This basic chemical mixture (Table 10-3) should be the subject of future toxicity studies to evaluate the potential human health impact of our chronic exposure to these chemicals. This mixture is the toxic legacy of environmental policies that allow our environment to be polluted by low levels of individual compounds. As demonstrated in the previous chapters, the magnitude to which an individual is exposed to our chemical legacy depends on where you live.

10.3 A NATIONAL CONTAMINANT RANKING

When evaluating which factors influence the total number of different chemical contaminants that occur in a given city, toxic air emission from industrial facilities is very important. This is obvious because the absence or limited occurrence of industrial facilities and waste disposal sites in a city will determine the number of different toxic air pollutants that will be emitted. Thus, it would be expected that cities with numerous industrial facilities (e.g., Baton Rouge and Los Angeles) will have a significant number of toxic air contaminants as compared to a dense residential/commercial area (e.g., New York) or rural areas (e.g., Lafayette). This situation is well illustrated in Table 8-3.

Chemical contamination of water resources that are utilized for drinking water is highly dependent on surface watershed characteristics that are usually outside city limits (i.e., area of rural versus urban land use and industrial/municipal discharges) and groundwater aquifer characteristics near sources of chemical releases. Therefore, even areas that are generally agricultural with relatively little industrial development (e.g., Fresno) can have water resources that are as contaminated as dense urban/industrial areas such as Los Angeles. Yet Baton Rouge has a significant amount of industrial development, but drinking water quality is very good. In other words, it is not easy to predict drinking water quality based on land use.

This issue aside, one thing is certain: the occurrence of disinfection byproducts in tap water. Considering the fact that cities that monitored for bromodichloromethane and chloroform had average concentrations greater than California's safe harbor levels, it would behoove all water utilities to monitor for these chemicals. The chances are good that a significant portion of the population has been and continues to be consuming at least two carcinogens in their tap water above safe harbor levels.

The extent of chemical contamination in food is also difficult to predict based on city characteristics[5] and because food processing facilities distribute products nationally. However, dairy products and to a more limited degree protein products tend to be produced in regions closer to their points of distribution. Therefore, the chemical diversity found in dairy and protein products generally reflects regional pollution characteristics. *This is significant because dairy products contain the greatest number of toxic compounds that exceed the state of California safe harbor standards. It should also be noted that market basket 00-1 (Grand Rapids, Milwaukee, and Springfield) had the greatest number of contaminants in both dairy and protein products (see Figure 6-9).* As a result of the chemical diversity in both dairy and protein products, it is possible that dairy and protein products are indicators of regional pollution characteristics.

Given the complexity of the human environment, it is not easy to determine how regional land use and city characteristics may influence the chemical diversity found in each city. The contaminant profile of a city (i.e., extent of chemical mixtures in each media) is unique to that city and can only truly be

[5] For example, population, percentage of industrial, agricultural, commercial, or recreational land use areas, climate, and topography.

characterized by evaluating chemical data that have been collected for a specific city or region.

The final city rankings were based on the individual food, water, and air rankings given in Tables 6-2, 7-36, and 8-1. The final city ranking that incorporates all of the individual rankings is given in Table 10-4 and illustrated in Figure 10-1. In reviewing the city rankings, there is some indication that cities

TABLE 10-4 Total Chemical Contaminant Ranking

Cities	Final Ranking*	Total Points	Food Quality Contaminant Ranking	Water Quality Contaminant Ranking	Air Quality Contaminant Ranking
Lafayette, Louisiana	1	7	2	3	2
West Palm Beach, Florida	2	9	2	3	4
Sioux Falls, South Dakota	2	10	4	3	3
Long Beach, New Jersey	2	11	9	1	1
Peoria, Illinois	2	11	4	3	4
Ogden, Utah	2	12	5	2	5
Reno, Nevada	3	13	6	4	3
Baton Rouge, Louisiana	4	15	1	2	12
Charleston, South Carolina	4	15	2	4	9
Savannah, Georgia	4	15	1	6	8
Tulsa, Oklahoma	4	15	1	5	9
Binghamton, New York	5	17	10	5	2
Pueblo, Colorado	5	17	7	6	4
Worcester, Massachusetts	5	17	10	3	4
Fresno, California	6	18	3	11	4
Harrisburg, Pennsylvania	6	18	10	5	3
Providence, Rhode Island	6	18	9	2	7
Tacoma, Washington	6	18	5	4	9
Boise, Idaho	7	19	7	8	4
Phoenix, Arizona	7	19	3	7	9
Springfield, Illinois	7	19	8	4	7
Dayton, Ohio	8	20	4	7	9
Denver, Colorado	8	20	6	4	10
Jackson, Mississippi	8	20	12	2	6
Portland, Oregon	8	20	6	4	10
Washington, D.C.	9	21	11	8	2
Grand Rapids, Michigan	9	22	8	6	8
Oklahoma City, Oklahoma	9	22	11	4	7
New York, New York	9	22	9	11	2
San Antonio, Texas	9	22	3	12	7
Bakersfield, California	10	23	5	12	6
Columbus, Georgia	10	23	12	2	9
Orlando, Florida	11	25	11	8	6
Birmingham, Alabama	11	26	12	4	10
Milwaukee, Wisconsin	11	26	8	9	9
Los Angeles, California	12	28	7	10	11

* As the rank increases, there is an increasing risk of exposure to a more diverse group of chemicals.

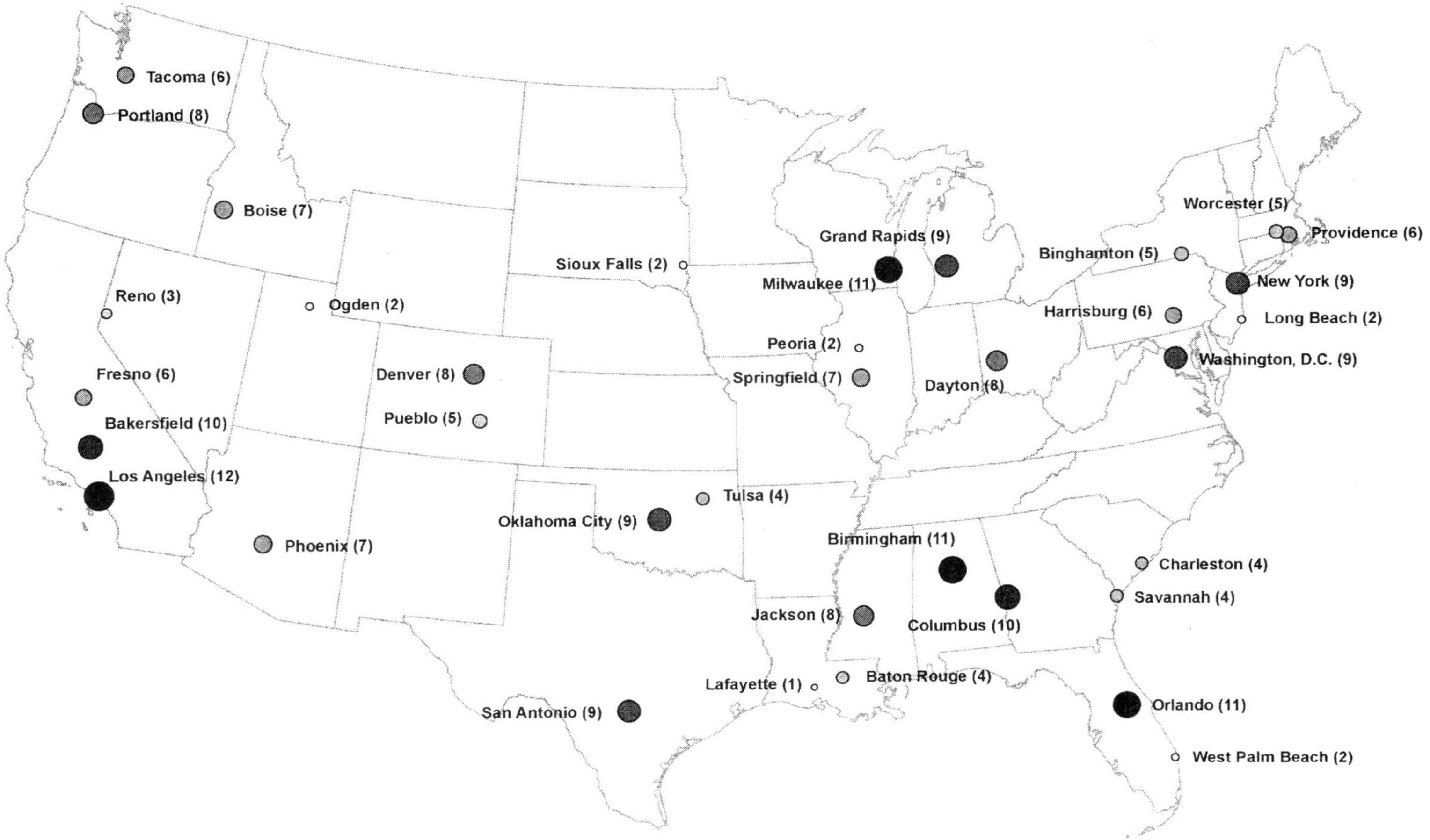

FIGURE 10-1 Total ranking.

with less industrial development (as measured by toxic industrial air emissions) do have less complex chemical mixtures contaminating their food, water, and air.

10.4 SUMMARY

Chemical mixtures in the environment are a reality and should not be ignored. The chemical mixtures characterized for the 36 cities are a reality, and residents of these areas should be aware of the degree to which synthetic chemical toxins permeate their environments. The relative magnitude of the chemical diversity found in other cities in the United States, however, can only be determined by using chemical data collected specifically for that city or region. For those individuals who live outside the 36 cities that have been characterized, it can be generally assumed that they are exposed, at a minimum, to the previously defined basic chemical mixture given in Table 10-3.

Fortunately, the freedom of personal choice allows individuals to limit their exposure to chemical mixtures by choosing safer foods and drinking bottled water (as well as treating their tap water). Minimizing our exposure to chemical mixtures in food and water, however, comes with an added economic cost. Thus, those unfortunate individuals or families who cannot afford to reduce their chemical exposure in food and water have few alternatives (i.e., they can avoid foods with a large number of individual chemical contaminants) but must continue to drink water contaminated with disinfection by-products and other potential chemical contaminants.

Limiting one's exposure to toxic air emissions from industrial sources is not easy. Also, the closer a home is to a toxic air source, the greater chance of exposure. Any individuals or families that want to limit their exposure to toxic industrial air emissions have only one effective choice — move to a city with little or no industrial development or move to another area of the same city that is farther away from industrial sources. Both choices may not be practical nor economically possible.

Since about the 1920s, federal and state laws have allowed industry and commercial businesses to pollute the environment with small amounts of individual chemical contaminants instead of bearing the cost of not polluting (i.e., to the degree technically possible). Because jobs are important, and given the competitive nature of a global economy, industries and commercial businesses will be allowed to continue to release low levels of known toxic compounds into the environment.

This means that the cost of controlling the contamination of our bodies is the responsibility of the individual. Given this ultimate responsibility, we believe that the information provided in this text will help individuals understand some of the chemical complexities associated with living in a man-made environment and assist individuals in making informed decisions.

APPENDIX

APPENDIX 2-1 MARKET BASKET CHEMICALS

Organochlorine Pesticides

1,2,3,5-tetrachlorobenzene
1,2,4,5-tetrachlorobenzene
2,3,5,6-tetrachloroaniline
2,3,5,6-tetrachloroanisidine
2,4-dichloro-6-nitrobenzenamine
2-chloroethyl caprate
2-chloroethyllaurate
2-chloroethyl linoleate
2-chloroethyl myristate
2-chloroethyl palmitate
2-chloroethyl stearate
alpha-cypermethrin
anilazine
atrazine
BHC, alph
BHC, beta
BHC, delta
bifenthrin
captan
chlordane, cis
chlordane, trans
chlorothalonil
chlorpropham
chlorpyrifos

chlorpyrifos-methyl
cyfluthrin
cypermethrin
DCPA
DDE, p,p′
DDT, o,p′
DDT, p,p′
deltamethrin
dicloran
dicofol, o,p′
dicofol, p,p′
dieldrin
endosulfan I
endosulfan II
endosulfan sulfate
endrin
endrin ketone
esfenvalerate
fenarimol
fenhexamid
fenvalerate
fluvalinate
folpet
heptachlor
heptachlor epoxide
hexachlorobenzene
iprodione
iprodione metabolite isomer
isopropyl(3-chloro-4-methoxyphenyl)carbamate
lambda-cyhalothrin
lindane
methoxychlor olefin
methoxychlor, o,p′
methoxychlor, p,p′
metolachlor
nonachlor, cis
nonachlor, trans
octachlor epoxide
oxyfluorfen
pentachloroaniline
pentachlorobenzene
pentachlorophenyl methyl ether
pentachlorophenyl methyl sulfide
permethrin, cis

permethrin, trans
polychlorinated biphenyls
procymidone
propiconazole
quintozene
simazine
TDE, o,p′
TDE, p,p′
TDE, p,p′-olefin
tecnazene
tetradifon
toxaphene
tridiphane
vinclozolin

Organophosphorus Pesticides

acephate
azinphos-ethyl
azinphos-methyl
bromophos
bromophos-ethyl
carbophenothion
carbophenothion oxygen analog sulfone
carbophenothion sulfone
carbophenothion sulfoxide
chlorfenvinphos, alpha
chlorfenvinphos, beta
chlorpyrifos
chlorpyrifos-methyl
coumaphos
crotoxyphos
cyanophos
demeton-S
demeton-S sulfone
diazinon
dichlofenthion
dichlorvos
dicrotophos
dimethoate
diphenyl 2-ethylhexyl phosphate
disulfoton sulfone
EPN

ethion
ethion oxygen analog
fenamiphos
fenitrothion
fenthion
fonofos
gardona
isofenphos
malathion
mecarbam
metasystox thiol
methidathion
mevinphos, (e)
mevinphos, (z)
monocrotophos
omethoate
parathion
parathion-methyl
phorate sulfone
phorate sulfoxide
phosalone
phosmet
phosphamidon
piperonyl butoxide
pirimiphos-ethyl
pirimiphos-methyl
propetamphos
ronnel
terbufos
tri(2-butoxyethyl) phosphate
tri(2-ethylhexyl) phosphate
tribufos
tributyl phosphate
trichlorfon
triphenyl phosphate
tris(2-butoxyethyl) phosphate
tris(beta-chloroethyl) phosphate
tris(chloropropyl) phosphate

N-Methylcarbamate Pesticides

3-hydroxycarbofuran
aldicarb

aldoxycarb
BPMC
bufencarb
butocarboxim
carbaryl
carbofuran
dioxacarb
isoprocarb
methiocarb
methomyl
metolcarb
oxamyl
promecarb

Chlorophenoxy Acid Pesticides

2,3,4,6-tetrachlorophenol
2,3,6-TBA
2,4,5-T
2,4,5-TB
2,4,5-trichlorophenol
2,4,6-trichlorophenol
2,4-D
2,4-DB
2,4-dichlorophenol
4-CPA
acifluorfen
bentazon
bromoxynil
chloramben
clopyralid
dicamba
dichlorooctadecanoic acids
dichlorprop
fenac
fluroxypyr
haloxyfop
MCPA
MCPB
mecoprop
pentachlorophenol
picloram
quinclorac

silvex
triclopyr

Pyrethroid Pesticides

alpha cypermethrin
bifenthrin
cypermethrin
cyfluthrin
deltamethrin
esfenvalerate
fenpropathrin
fenvalerate
fluvalinate
lambda-cyhalothrin
permethrin, cis- and trans-
tetramethrin
tralomethrin
zeta-cypermethrin

Substituted Urea Pesticides

chlorbromuron
chloroxuron
diuron
fenuron
linuron
metobromuron
metoxuron
monolinuron
monuron
neburon

Volatile Organic Compounds/Industrial Chemicals

1,1,1,2-tetrachloroethane
1,1,1-trichloroethane
1,1,2,2-tetrachloroethane
1,1,2-trichloroethane
1,1-dichloroethane

1,1-dichloropropene
1,2,3-trichloropropane
1,2,4-trimethylbenzene
1,2-dichloroethene, cis
1,2-dichloroethene, trans
1,2-dichloropropane
1,3,5-trimethylbenzene
1,3-dichlorobenzene
1,3-dichloropropene, cis
1,3-dichloropropene, trans
benzene
bromobenzene
bromochloromethane
bromodichloromethane
bromoform
butylbenzene, n
butylbenzene, sec
butylbenzene, tert
carbon tetrachloride
chlorobenzene
chloroform
chlorotoluene, o
chlorotoluene, p-(4-chlorotoluene)
cumene (isopropyl benzene)
dichlorobenzene, o
dichlorobenzene, p
ethyl benzene
ethylene dibromide
ethylene dichloride
methylene bromide (dibromomethane)
propylbenzene, n
polychlorinated biphenyls
styrene
tetrachloroethylene
toluene
trichloroethylene
xylene, m- and/or p
xylene, o

APPENDIX 2-2 Synthetic Chemicals Found in Market Basket Foods and Beverages

Pesticide	Abbreviation	Type	Class	Compound Formula
2,4-D	2,4-D	Herbicide	Organochlorine	2-(2,4-dichlorophenoxy)acetic acid
Acephate	Ace	Insecticide	Organophosphorus	O,S-dimethyl acetylphosphoramidothioate
Atrazine	Atz	Herbicide	Triazine	1-Chloro-3-ethylamino-5-isopropylamino-2,4,6-triazine
Azinphos-methyl	Azi	Insecticide	Organophosphorus	O,O-Dimethyl S-((4-oxo-1,2,3-benzotriazin-3(4H)-yl)methyl) phosphorodithioate
Benomyl	Ben	Fungicide	Benzimidazole	1-(N-Butylcarbamoyl)-2-(methoxy-carboxamido)benzimidazole
Benzene Hexachloride (Banned)*	BHC	Insecticide	Organochlorine	1,2,3,4,5,6-Hexachlorocyclohexane
Bifenthrin	Bif	Insecticide	Pyrethroid	3-(2-Chloro-3,3,3-trifluoro-1-propenyl)-2, 2-dimethylcyclopropanecarboxylic acid
Captan	Cap	Fungicide	Thiophthalimide	1H-Isoindole-1,2(2H)-dione, 3a,4,7,7a-tetrahydro-2-((trichloromethyl)thio)
Carbaryl	Car	Insecticide	N-methyl Carbamate	1-Naphthyl-N-methylcarbamate
Chlordane (Banned)*	Cd	Insecticide	Organochlorine	Octachloro-2,3,3a,4,7,7a-hexahydro-4,7-methano-1H-indene
Chloroethyl caprate*	Ccp	Pesticide	Organochlorine	2-chloroethyl decanoate
Chloroethyl laurate*	Cau	Pesticide	Organochlorine	2-chloroethyl dodecanoate
Chloroethyl palmitate*	Cpa	Pesticide	Organochlorine	2-chloroethyl hexadecanoate
Chloroethyl stearate*	Cte	Pesticide	Organochlorine	2-chloroethyl octadecanoate
Chloroethyl linoleate*	Cno	Pesticide	Organochlorine	2-chloroethyl octadecadienoate
Chloroethyl myristate*	Cmy	Pesticide	Organochlorine	2-chloroethyl tetradecanoate
Chlorothalonil*	Cil	Fungicide	Substituted benzene	2,4,5,6-Tetrachloro-1,3-dicyanobenzene
Chlorpropham*	Cam	Herbicide	Carbamate	isopropyl-N 93-chlorophenyl carbanamate
Chlorpyrifos*	Cpy	Insecticide	Organophosphorus	O,O-Diethyl-O-(3,5,6-trichloro-2-pyridinyl) phosphorothioate
Chlorpyrifos-methyl*	Cpm	Insecticide	Organophosphorus	O,O-Dimethyl O-(3,5,6-trichloro-2-pyridyl) phosphorothioate
Clopyralid*	Clo	Herbicide	Pyridinecarboxylic acid	3,6-dichloropyridine-2-carboxylic acid
Coumaphos*	Cou	Insecticide	Organophosphorus	O,O-Diethyl O-(3-chloro-4-methyl-2-oxo-2H-1-benzopyran-7-yl) phosphorothioate
Cumene (isopropyl benzene)	Cum	Pesticide	Petroleum derivative	Isopropylbenzene
Cyfluthrin*	Cyf	Insecticide	Pyrethroid	Cyano(4-fluoro-3-phenoxyphenyl)methyl 3-(2,2-dichloroethenyl)-2, 2-dimethylcyclopropanecarboxylate
Cypermethrin*	Cin	Insecticide	Pyrethroid	Cyclopropanecarboxylic acid, 3-(2,2-dichloroethenyl)-2, 2-dimethyl

Cyprodinil	Cil	Fungicide	Organoamine	2-Pyrimidinamine
DCPA*	DCP	Herbicide	Alkyl phthalate	Dimethyl 2,3,5,6-tetrachloro-1,4-benzenedicarboxylate
DDT (Banned)*	DDT	Insecticide	Organochlorine	1,1,1-Trichloro-2,2-bis(p-chlorophenyl)ethane
DDE*	DDE	Byproduct	Organochlorine	1,1-dichloro-2,2-bis(p-chlorophenyl)ethylene
Diazinon	Diz	Insecticide	Organophosphorus	Diethyl 2-isopropyl-4-methyl-6-pyrimidinyl phosphorothionate
Dicamba*	Dca	Herbicide	Benzoic acid	2,5-Dichloro-6-methoxybenzoic acid
Dicloran*	Dan	Fungicide	Substituted benzene	2,6-Dichloro-4-nitroaniline
Dieldrin (Banned)*	Die	Insecticide	Organochlorine	Hexachloroepoxyoctahydro-endo,exo-dimethanonaphthalene
Dicofol*	Dic	Insecticide	Organochlorine	1,1-Bis(chlorophenyl)-2,2,2-trichloroethanol
Dimethoate	Dat	Insecticide	Organophosphorus	O,O-Dimethyl phosphorodithioate
Diphenyl-2-ethylhexyl phosphate	Dip	Pesticide	Organophosphorus	Diphenyl-2-ethylhexyl phosphate
Endosulfan I*	EnI	Insecticide	Organochlorine	1,4,5,6,7,7-Hexachloro-5-norbornene-2,3-dimethanol, cyclic sulfite (alpha)
Endosulfan II*	EII	Insecticide	Organochlorine	1,4,5,6,7,7-Hexachloro-5-norbornene-2,3-dimethanol, cyclic sulfite (beta)
Endosulfan sulfate*	EnS	Byproduct	Organochlorine	1,4,5,6,7,7-Hexachloro-5-norbornene-2,3-dimethanol, cyclic sulfate
Endrin (Banned)*	Edr	Insecticide	Organochlorine	Hexachloro-1,4,4a,5,6,7,8,8a-octahydro-6,7-epoxy-1,4:5, 8-dimethanonaphthalene
Esfenvalerate*	Efv	Insecticide	Pyrethroid	(S)-alpha-Cyano-3-phenoxybenzyl (S)-2-(4-chlorophenyl)-3-methylbutyrate
Ethion	Etn	Insecticide	Organophosphorus	Bis(diethoxyphosphinothioylthio)methane
Ethylene dichloride (Banned)*	Edi	Insecticide	Organochlorine	1,2-Dichloroethane
Ethylene thiourea	Eth	Byproduct	Thiourea	1,3 ethylenethiourea
Fenarimol*	Far	Fungicide	Pyrimidine	2,4′-Dichloro-alpha-(5-pyrimidinyl)benzhydryl alcohol
Fenhexamid*	Fex	Fungicide	Anilide	N-(2,3-dichloro-4-hydroxyphenyl)-1-methyl-,Cyclohexanecarboxamide
Fenitrothion	Fit	Insecticide	Organophosphorus	O,O-Dimethyl O-(3-methyl-4-nitrophenyl) phosphorothioate
Fenvalerate*	Fen	Insecticide	Pyrethroid	alpha-Cyano-m-phenoxybenzyl 2-(p-chlorophenyl)-3-methylbutyrate
Folpet	Fol	Fungicide	Thiophthalimide	N-((Trichloromethyl)thio)phthalimide

APPENDIX 2-2 *Continued*

Pesticide	Abbreviation	Type	Class	Compound Formula
Heptachlor epoxide*	HE	Byproduct	Organochlorine	1,4,5,6,7,8,8-Heptachlorotetrahydro-4,7-methanoindene epoxide
Hexachlorobenzene (Banned)*	HCB	Insecticide	Organochlorine	Hexachlorobenzene
Iprodione*	Ipr	Fungicide	Dicarboximide	3-(3,5-Dichlorophenyl)-N-(1-methylethyl)-2,4-dioxo-1-imidazolidinecarboxamide
Iprodione metabolite*	Ipm	Byproduct	Organochlorine	1-(3,5-dichlorophenylcarbamoylmethyl)-3-isopropyl urea (one of several)
Isopropyl(3-chloro-4-methoxyphenyl)*	icm	Herbicide	Organochlorine	Isopropyl(3-chloro-4-methoxyphenyl)
lambda-cyhalothrin*	Lac	Insecticide	Pyrethroid	alpha-cyano-3-phenoxybenzyl-3-(2-chloro-3,3,3-trifluoropropy-1-enyl)-2,2-dimethylcyclopropanecarboxylate
Lindane (Restricted)*	Lin	Insecticide	Organochlorine	Gamma-1,2,3,4,5,6-hexachlorocyclohexane
Linuron*	Lrn	Herbicide	Urea	1-Methoxy-1-methyl-3-(3,4-dichlorophenyl)urea
Malathion	Mal	Insecticide	Organophosphorus	Dicarbethoxyethyl-O,O-dimethyldithiophosphate
Mecarbam	Mec	Insecticide	Organophosphorus	O,O-Diethyl S-(((ethoxycarbonyl)methylcarbamoyl)methyl) phosphorodithioate
Methomyl	Met	Insecticide	N-methyl Carbamate	N-((methylcarbamoyl)oxy)thio-,methyl ester
Methamidophos	Mos	Insecticide	Organophosphorus	O,S-Dimethyl phosphoramidothioate
Methidathion	Mat	Insecticide	Organophosphorus	S-((5-Methoxy-2-oxo-1,3,4-thiadiazol-3(2H)-yl)methyl) O,O-dimethyl phosphorodithioate
Methoxychlor*	Mxy	Insecticide	Organochlorine	1,1,1-Trichloro-2,2-bis(p-anisyl)ethane
Methoxychlor olefin*	Mxo	Byproduct	Organochlorine	1,1,1-Trichloro-2,2-bis(p-anisyl)ethane olefin
Neburon*	Neb	Herbicide	Urea	1-Butyl-3-(3,4-dichlorophenyl)-1-methylurea
Nonachlor-trans*	Nac	Insecticide	Organochlorine	4,7-Methano-1H-indene, 1,2,3,4,5,6,7,8,8-nonachloro-2,3,3a,4,7,7a-hexahydro
Octachlor epoxide*	Oct	Byproduct	Organochlorine	2,3,4,5,6,6a,7,7-octachloro-1a,1b,5,5a,6,6a-hexahydro-2,5-methano-2H-indeno[1,2-b]oxirene
Omethoate	Ome	Insecticide	Organophosphorus	O,O-Dimethyl S-((methylcarbamoyl)methyl) phosphorothioate
Oxamyl	Oxa	Insecticide	N-methyl Carbamate	2-Dimethylamino-1-(methylthio)glyoxal O-methylcarbamoylmonoxime

Parathion-methyl	Par	Insecticide	Organophosphorus	O,O-Dimethyl O-p-nitrophenyl thiophosphate
Pentachloroaniline*	Pal	Byproduct	Organochlorine	Pentachloroaniline
Pentachlorobenzene*	PBZ	Pesticide	Organochlorine	Pentachlorobenzene (used in production of pentachloronitrobenzene pesticide)
Pentachlorophenyl methyl ether*	Pme	Pesticide	Organochlorine	Pentachlorophenyl methyl ether
Pentachlorophenyl methyl sulfide*	Pms	Pesticide	Organochlorine	Pentachlorophenyl methyl sulfide
Permethrin*	P	Insecticide	Pyrethroid	3-Phenoxyphenyl)methyl cis,trans-(+-)-3-(2,2-dichloroethenyl)-2,2-dimethylcyclopropanecarboxylate
Phosalone	Pne	Insecticide	Organophosphorus	O,O-Diethyl S-(6-chloro-3-(mercaptomethyl)-2-benzoxazolinone) phosphorodithioate
Phosmet	Pho	Insecticide	Organophosphorus	N-(Mercaptomethyl)phthalimide S-(O,O-dimethyl phosphorodithioate)
Phosphamidon*	Pon	Insecticide	Organophosphorus	2-Chloro-2-diethylcarbamoyl-1-methylvinyl dimethyl phosphate
Pirimiphos-methyl	Pri	Insecticide	Organophosphorus	O,O-Dimethyl O-(2-(diethylamino)-6-methyl-4-pyrimidinyl)phosphorothioate
Propargite	Prg	Insecticide	Unclassified	2-(4-(1,1-Dimethylethyl)phenoxy)cyclohexyl 2-propynyl sulfite
Procymidone*	Pcy	Insecticide	Unclassified	1,2-Cyclopropanedicarboximide, N-(3,5-dichlorophenyl)-1, 2-dimethyl
Propiconazole*	Pzo	Fungicide	Azole	1-((2-(2,4-Dichlorophenyl)-4-propyl-1,3-dioxolan-2-yl)methyl)-1H-1,2,4-triazole
Quinclorac*	Qac	Herbicide	Unclassified	3,7-Dichloro-8-quinolinecarboxylic acid
Quintozene*	Qui	Fungicide	Substituted benzene	Pentachloronitrobenzene
TDE*	TDE	Byproduct	Organochlorine	1,1-Dichloro-2,2-bis(p-chlorophenyl)ethane
Tetrachloroaniline*	Tca	Byproduct	Organochlorine	Tetrachloroaniline
Tetradifon*	Ton	Insecticide	Unclassified	p-Chlorophenyl 2,4,5-trichlorophenyl sulfone
Thiabendazole	Thi	Fungicide	Benzimidazole	2-(4′-Thiazolyl)benzimidazole
Tribufos	Tri	Herbicide	Organophosphorus	S,S,S-tributyl phosphorotrithioate
Tributyl phosphate	Tbp	Pesticide	Unclassified	Tributyl phosphate
Triphenyl phosphate	Tpp	Pesticide	Unclassified	Triphenyl phosphate

APPENDIX 2-2 *Continued*

Pesticide	Abbreviation	Type	Class	Compound Formula
Toxaphene*	Tox	Insecticide	Organochlorine	Chlorocamphene
Vinclozolin*	Vin	Fungicide	Dicarboximide	2,4-Oxazolidinedione (chlorinated)
Industrial Chemical				
Benzene	BZ			
Bromobenzene	BRB			
Bromodichloromethane	BDM			
Butylbenzene, n-	BBZ			
Carbon Tetrachloride*	CTC			
Chlorobenzene*	CBZ			
Chloroform*	Cfo			
1,2-dichloroethene, trans-*	DCE			
Dichlorobenzene*	DBZ			
2,4-dichloro-6-nitrobenzenamine*	DNB			
Ethyl Benzene	EBZ			
Ethylene dichloride	EDC			
Perchloroethylene*	PCE			
Polychlorinated biphenyls (Banned)	PCB			
Propylbenzene	Pyl			
Styrene	Sty			
Tetrachlorobenzene*	RBZ			
Trichloroethylene	TCE			
1,1,1-trichloroethane*	TCA			
Trimethylbenzene	TBZ			
Toluene	Tlu			
Xylenes	Xyl			

* Compounds are chlorinated.

APPENDIX 4-1 Dairy Products with Very Low Contaminants

Product	Skim Milk			Chocolate Milk			Cottage Cheese (4%)				Chocolate Milk Shake			
Chemical Class			P			P			P	P			P	P
MB	Num	ppb	DDE	Num	ppb	DDE	Num	ppb	DDE	Die	Num	ppb	DDE	Die
96-1	1	0.2	0.2	1	0.4	0.4	1	1	0.6		1	0.7	0.7	
96-2	0	0		0	0		0	0			0	0		
97-1	0	0		1	1	1	1	5	5		0	0		
98-1	0	0		0	0		1	4	4		1	0.3	0.3	
98-2	0	0		0	0		0	0			0	0		
98-3	0	0		0	0		0	0			0	0		
98-4	0	0		1	0.1	0.1	0	0			1	0.1	0.1	
99-1	0	0		1	0.5	0.5	2	1	0.6	0.1	1	0.3	0.3	
99-3	0	0		0	0		0	0			1	0.2	0.2	
00-1	0	0		0	0		2	0	0.2	0.1	1	0.2	0.2	
00-3	0	0		0	0		0	0			0	0		
01-3	0	0		0	0		0	0			2	0.4	0.2	0.2
Totals	1	0.2		4	2		7	11			8	2.2		
Average													0.2	
Probably in product													x	

APPENDIX 4-2 Dairy Products — Yogurts

Product	Plain Yogurt (Low Fat)				Fruit-Flavored Yogurt				
Chemical Class			P	P			P	P	P
MB	Num	ppb	DDE	EnS	Num	ppb	DDE	Ipr	Vin
96-1	0	0			1	0.4	0.4		
96-2	0	0			0	0			
97-1	1	0.6	0.6		1	0.7	0.7		
98-1	1	0.3	0.3		2	2.2	0.2	2	
98-2	1	0.2	0.2		2	32.7		32	0.7
98-3	0	0			0	0			
98-4	0	0			2	1.4	0.4	1	
99-1	1	0.3	0.3		2	1.4	1	0.4	
99-3	0	0			0	0			
00-1	2	0.8	0.4	0.4	0	0			
00-3	0	0			1	2			2
01-3	0	0			0	0			
Totals	6	2.2			11	40.8			

APPENDIX 4-3 Dairy Products with Low Contaminants

Product	Evaporated Milk					Milk (2%)						
Chemical Class			P	P	IC			P	P	P	P
MB	Num	ppb	DDE	Die	Tol	Num	ppb	DDE	Die	EnS	Mxy
96-1	2	11	1		10	2	0.4	0.2	0.2		
96-2	1	1	1			0	0				
97-1	1	2	2			3	5.4	5	0.1	0.3	
98-1	1	0.4	0.4			1	0.5	0.5			
98-2	1	0.7	0.7			0	0				
98-3	1	4	4			0	0				
98-4	1	0.3	0.3			3	0.4		0.1	0.1	0.2
99-1	1	3	3			1	0.2	0.2			
99-3	1	0.2	0.2			1	0.1	0.1			
00-1	2	1	0.8	0.2		2	0.2	0.1	0.1		
00-3	0	0				0	0				
01-3	0	0				0	0				
Totals	12	24				13	7.2				
Average			1.1								
Probably in product			x								

APPENDIX 4-3 *Continued*

Product	Half-and-Half						Whole Milk										
Chemical Class			P	P	P	P			P	P	P	P	P	IC	IC	IC	IC
MB	Num	ppb	DDE	Die	HCB	HE	Num	ppb	DDE	Die	EnS	HE	Mxy	BZ	Cfo	Sty	Tol
96-1	1	3	3				1	0.7	0.7								
96-2	1	0.4	0.4				0	0									
97-1	1	13	13				4	31.5	9	0.1	0.4				22		
98-1	2	4.2	4	0.2			1	0.7	0.7								
98-2	1	0.9	0.9				1	22							22		
98-3	1	1	1				0	0									
98-4	3	1.8	0.3	1		0.5	5	12.7	0.3	0.1		0.1	0.2		12		
99-1	3	1	0.8	0.1	0.1		2	0.5	0.4	0.1							
99-3	1	0.4	0.4				2	0.2	0.1	0.1							
00-1	2	0.5		0.4		0.1	3	6.2	0.1	0.1					6		
00-3	2	0.4	0.2	0.2			1	9							9		
01-3	2	0.2	0.1	0.1			4	22						2	2	2	16
Totals	20	26.8					24	106.0									
Average			2						0.9								
Probably in product			x						x								

APPENDIX 4-4 Frozen Dairy Products

Product	Ice Cream (Light)							Sherbet									
Chemical Class			P	IC	IC	IC	IC			P	P	P	IC	IC	IC	IC	IC
MB	Num	ppb	DDE	BZ	Cfo	Tol	Xyl	Num	ppb	DDE	EnS	Mat	BZ	BDM	Cfo	Tol	Xyl
96-1	1	2	2					1	0.6	0.6							
96-2	2	41	3		38			3	97	1						31	65
97-1	2	52	2		50			3	12.3	2	0.3				10		
98-1	2	17	3		14			2	39.3	0.3						39	
98-2	1	14			14			2	25.8			0.8				25	
98-3	1	0.1	0.1					2	36				22			14	
98-4	2	21	1		20			1	0.6	0.6							
99-1	1	0.3	0.3					1	0.6	0.6							
99-3	2	11			4		7	2	19			9			10		
00-1	2	90			56		34	3	48						8	28	12
00-3	2	7			5		2	5	32			2		3	9	16	3
01-3	4	29		1	17	3	8	3	68						17	47	4
Totals	22	284						28	379								
Average			1		18.2											16.7	
Probably in product			x		x											x	

APPENDIX 4-4 *Continued*

Product	Vanilla Ice Cream													
Chemical Class			P	P	P	P	IC	IC	IC	IC	IC	IC	IC	IC
MB	Num	ppb	DDE	Die	EnS	HCB	Cfo	BDM	DBZ	PCE	Sty	TBZ	Tol	Xyl
96-1	1	2	2											
96-2	1	1	1											
97-1	3	34.3	10	0.3			24							
98-1	1	5	5											
98-2	2	29.2	0.2				29							
98-3	2	16.3	0.3				16							
98-4	4	54.3	1	0.3			23						30	
99-1	5	16.4	2	0.1	0.2	0.1	14							
99-3	5	21.8	0.6	0.2			12			6				3
00-1	5	32					11		2			7	4	8
00-3	7	35.5	0.3	0.2			25	3			2		3	2
01-3	5	32.1		0.1			16	3					7	6
Totals	41	280												
Average			1.9				14.2							
Probably in product			x				x							

APPENDIX 4-5 Dairy Products — Creams

Product Chemical Class	Sour Cream																
MB	Num	ppb	DDE	Die	EnS	HCB	HE	BZ	Cfo	BDM	CBZ	DBZ	PCE	Sty	TCA	Tol	Xyl
96-1	5	29	5	0.6		0.4			13						10		
96-2	2	15	1						14								
97-1	5	38.3	7	0.3	0.8	0.2			30								
98-1	4	22.5	3	0.2	0.3												19
98-2	2	59	2						57								
98-3	4	52.2	0.7	0.3		0.2			51								
98-4	4	1.2	0.2	0.8			0.2		t								
99-1	5	47.6	3	0.3		0.2	0.1		44								
99-3	6	45.2	1	0.2					30			2	7			5	
00-1	5	80.2	0.3	0.2				5	70		2						
00-3	6	76.3	0.2	0.1				3	64	4				5			
01-3	4	45							35	3						5	2
Totals	52	512															
Average			2	0.3					34								
Probably in product				x													
Occurs in product			x						x								

t = trace.

APPENDIX 4-5 *Continued*

Product Chemical Class	Cream Cheese																	
MB	Num	ppb	Cum	DDE	Die	EnS	HCB	HE	Mxy	Oct	P	BZ	Cfo	PCE	Sty	TCA	Tol	Xyl
96-1	6	28.5		6	0.9	0.7	0.5	0.4										20
96-2	2	7.7		7	0.7													
97-1	3	92.9		4	0.9								88					
98-1	4	69.7		7	0.7								51				11	
98-2	4	79.7		2	0.7								45				32	
98-3	5	30.4		2	0.4							10	t				18	
98-4	5	83.1		10	0.9			0.2					41				31	
99-1	7	40.9		13	0.4	2	0.3		1	0.2			24					
99-3	7	90.9		4	0.5				4		1.4		64	5			12	
00-1	10	96.3	2	0.9	1	0.3		0.3			1.8		64			6	17	3
00-3	6	93.3		7	0.3							17	56		2		11	
01-3	6	325		0.6	0.2							1	25				6	2
Totals	65	1038																
Average				5.3	0.63								38.2				11.5	
Probably in product													x				x	
Occurs in product				x	x													

APPENDIX 4-6 Dairy Products — Cheeses

Product	Swiss Cheese																		
Chemical Class			P	P	P	P	P	P	P	IC	IC	IC	IC	IC	IC	IC	IC	IC	IC
MB	Num	ppb	DDE	Die	HCB	HE	Mxy	Oct	P	BZ	BDM	Cfo	CBZ	PCE	Sty	TBZ	TCE	Tol	Xyl
96-1	2	2.5	2		0.5														
96-2	5	135	1	1	0.5							46						86	
97-1	5	24.5	2	0.5			1					10						11	
98-1	4	31	0.6	0.4												10		20	
98-2	3	14.6		0.5		0.1												14	
98-3	4	58.7	3	0.7														28	
98-4	4	39.9		0.7		0.2						29						10	
99-1	6	74.3	3	0.6	0.5			0.2				t						15	55
99-3	6	43.1	0.7	0.4								26		11	3			2	
00-1	8	54.3	1	1		0.5			1.8	6						20		21	3
00-3	8	91.6	0.3	0.3						35	3	30	2				4	17	
01-3	6	55.7	1	0.7								29			2			19	4
Totals	61	625																	
Average			1.2	0.6								14.2						20.3	
Probably in product			x									x							
Occurs in product				x														x	

t = trace.

APPENDIX 4-6 *Continued*

			American Cheese																		
Product			**P**	**P**	**P**	**P**	**P**	**P**	**P**	**IC**	**IC**	**IC**	**IC**	**IC**	**IC**	**IC**	**IC**	**IC**	**IC**	**IC**	**IC**
Chemical Class	**Num**	**ppb**	**DDE**	**Die**	**Dip**	**HCB**	**HE**	**EnS**	**Lin**	**BZ**	**BDM**	**Cfo**	**DBZ**	**EBZ**	**PCE**	**Sty**	**TBZ**	**TCA**	**TCE**	**Tol**	**Xyl**
MB96-1	5	52	6	0.6								16								13	16
MB96-2	3	81	2									54						25			
MB97-1	5	104	48	0.9				1				20								34	
MB98-1	4	157	0.9									22								60	74
MB98-2	3	52	1									11									40
MB98-3	6	52	12	0.8		0.3						16								29	23
MB98-4	4	61	4	0.2																32	25
MB99-1	5	82	3	0.5				0.5												49	39
MB99-3	9	87	20	1				1				14	3	3		11				27	7
MB00-1	10	81	10	0.5		0.2	0.2	0.4	0.2							3	4			23	39
MB00-3	10	165	2		80						3	12		4		2	6		2	21	33
MB01-3	8	82	10	0.3						3		11			2	4				27	5
Totals	72	1056																			
Average			9.9	0.4								14.7								26.3	25.1
Probably in product				x								x								x	x
Occurs in product			x																		

APPENDIX 4-6 *Continued*

Product Chemical Class MB	Cheddar Cheese Num	ppb	P Azi	P DDE	P Die	P EnS	P HE	P Mxy	P Oct	IC BZ	IC Cfo	IC CBZ	IC DBZ	IC DCE	IC EBZ	IC PCE	IC Sty	IC TCA	IC Tol	IC Xyl
96-1	5	43.4		6	0.8		0.6												13	23
96-2	3	56			1						27							28		
97-1	5	1790		10	1						13								1730	36
98-1	4	24.2		4	2														t	20
98-2	4	59.4		0.9	0.5						58								t	
98-3	5	1350	5		0.7						34			10					1300	
98-4	3	1		0.5	0.5						t									
99-1	10	51.8		4	1	0.7	0.2	0.6	0.3		29						t		t	16
99-3	6	109									46		2	24		8			18	11
00-1	8	119		0.4	0.5		0.3			9	44			42					8	14
00-3	6	105		0.7						47	40	3							9	5
01-3	5	67									42				2		3		6	14
Totals	64	3775																		
Average				2.2	0.7						28								257	11.6
Probably in product				x	x						x								x	x
Occurs in product																				

t = trace.

APPENDIX 4-7 Dairy Products — Butter

Product Chemical Class MB	Num	ppb	P BHC	P Cum	P DDE	P Die	P Dip	P EnS	P HCB	P HE	P Lin	P Mxy	P Nac	P Oct	P P	IC BZ	IC Cfc	IC DBZ	IC EBZ	IC PCB	IC PCE	IC Pyl	IC Sty	IC TBZ	IC TCA	IC TCE	IC Tol	IC Xyl
												Butter																
96-1	10	281			38	1		3	2								15		12					28	18		130	34
96-2	6	140			9	2			0.9								38								20		70	
97-1	8	172			38	4		3	2								56						11				35	23
98-1	11	280			28	0.9		2	0.8						5		48		16				19	33			48	79
98-2	10	360			52	2											70		14	120	11		11	10			t	70
98-3	9	189			3				7	0.9				5			70						20	30			30	23
98-4	10	197			19	2			0.9	0.9	0.3			0.4			67						17				63	26
99-1	13	177	0.4		26	1		2	1	0.4	0.5		0.4	0.4			51						10				44	40
99-3	18	334		3	10	2		0.6	1	0.7	0.6		0.4	0.4		9	42	18	11		20		14	42			114	45
00-1	20	252		6	3	2		0.4	0.5	0.6	2	1		0.2	4	2	67		9		4	3	18	14	15		62	40
00-3	14	394		8	10		90		0.4		0.2					13	51	13			25	4	21	22			128	28
01-3	15	221		2	10	2		1	0.7	0.5							77	13	4	20	3		19			2	54	13
Totals	144	1763																										
Average					20.5	1.6		1	1.4								54.3						13.3	14.9			64.8	35.1
Probably in product						x																	x	x				
Occurs in product				x	x			x									x										x	x

APPENDIX 4-8 Protein Products — Very Low Chemical Contaminants

Product	Chicken Breast					Veal Cutlet						Ham					Eggs (Fried)				Turkey Breast							Fish (Haddock)			
Chemical Class MB	Num	ppb	P DDE	P Mal	IC Tol	Num	ppb	P Cam	P DDE	P Lin	P P	Num	ppb	P DDE	P DDT	IC Tol	Num	ppb	P DDE	P Die	Num	ppb	P DDE	P Dip	IC BZ	IC Sty	IC Tol	Num	ppb	P Die	P HCB
96-1	1	20			20	0	0					1	12			12	1	0.6	0.6		3	68			34	13	21	2	0.5	0.2	0.3
96-2	0	0				0	0					1	0.7	0.7			0	0			0	0						0	0		
97-1	1	0.8	0.8			0	0					0	0				1	0.4	0.4		0	0						1	0.3		0.3
98-1	0	0				1	2		2			0	0				1	0.5	0.5		1	0.4	0.4					ND	ND		
98-2	0	0				1	4				4	0	0				0	0			0	0						ND	ND		
98-3	1	1		1		0	0					0	0				0	0			0	0						ND	ND		
98-4	0	0				0	0					0	0				0	0			1	60		60				ND	ND		
99-1	0	0				0	0					0	0				0	0			0	0						ND	ND		
99-3	0	0				1	0.4			0.4		2	0.4	0.2	0.2		0	0			0	0						ND	ND		
00-1	0	0				1	1	1				0	0				1	0.2		0.2	0	0						ND	ND		
00-3	0	0				0	0					0	0				1	0.2		0.2	0	0						ND	ND		
01-3	0	0				0	0					0	0				0	0			0	0						ND	ND		
Totals	3	21.8				4	7.4					4	13.1				5	1.9			5	128						3	0.8		

ND = No Data

APPENDIX 4-8 *Continued*

Product	Pork Roast							Shrimp					Pork Chop							Eggs (Hard boiled)				
Chemical Class			P	P	IC	IC	IC			P	P	P			P	P	P	P	IC			P	P	IC
MB	Num	ppb	Dan	Die	BZ	BBZ	Tol	Num	ppb	Atz	BHC	DDE	Num	ppb	DDE	Die	HE	Lin	Tol	Num	ppb	DDE	Die	Tol
96-1	4	235	2		30	169	34	2	1.6		1	0.6	2	12				1	11	2	10.5	0.5		10
96-2	0	0		1				0	0				0	0						0	0			
97-1	0	0						1	0.3			0.3	0	0						1	1			
98-1	0	0						0	0				0	0						2	0.4	0.2	0.2	
98-2	0	0						0	0				0	0						0	0			
98-3	0	0						0	0				0	0						0	0			
98-4	1	0.4		0.4				1	0.1		0.1		2	0.2		0.1	0.1			1	0.1		0.1	
99-1	0	0						0	0				0	0						1	0.5	0.5		
99-3	0	0						0	0				0	0						0	0			
00-1	0	0						0	0				0	0						1	0.1		0.1	
00-3	0	0						0	0				1	0.8	0.8					1	0.1	0.1		
01-3	0	0						2	1	0.9		0.1	1	0.2	0.2					0	0			
Totals	5	235						6	3				6	13.2						9	12.7			

APPENDIX 4-9 Protein Products — Low Chemical Contaminants

Product	Lamb Chop										Eggs (Scrambled)											
Chemical Class			P	P	P	P	P	P	P	IC			P	P	P	IC	IC	IC	IC	IC	IC	IC
MB	Num	ppb	Cpy	DDE	Die	Diz	EnS	HCB	Oct	Tol	Num	ppb	Cum	DDE	Die	BZ	DBZ	Pyl	Sty	TBZ	Tol	Xyl
96-1	3	15		2				2		11	1	0.4		0.4								
96-2	3	22		20	1			1			0	0										
97-1	1	0.3						0.3			1	2		2								
98-1	2	2.3		2				0.3			2	15.7		0.7							15	
98-2	2	0.8		0.7	0.1						1	43									43	
98-3	1	2		2							0	0										
98-4	2	1.1		1					0.1		3	44							10	13	21	
99-1	3	1.7	0.2	1				5			0	0										
99-3	3	3		0.7		2	0.3				4	25	4						5		14	2
00-1	0	0									5	42.1			0.1	13		7	16		6	
00-3	1	5		5							6	18.3		0.2	0.1	2	7		5			4
01-3	1	4		4							3	54							8		42	4
Totals	22	57.2									26	245										
Average				3.2																		
Probably in product				x																		

APPENDIX 4-9 *Continued*

Product	Tuna (Canned)											Pork Sausage											
Chemical Class			P	IC	IC	IC	IC	IC	IC	IC	IC			P	P	P	P	P	P	P	P	IC	IC
MB	Num	ppb	DDE	BZ	DBZ	Cfo	CTC	TBZ	TCE	Tol	Xyl	Num	ppb	BHC	Cno	Cpa	DDE	DDT	Etn	Lin	TDE	Tol	Xyl
96-1	2	100								83	17	2	64									51	13
96-2	2	12		12						t		1	2				2						
97-1	3	5.4	0.4	5						t		3	49				30	15			4		
98-1	1	42								42		0	0										
98-2	0	0										4	3.7	0.4			1		2	0.3			
98-3	1	t								t		0	0										
98-4	1	13		13								4	9.5		8	0.8	0.5	0.2					
99-1	1	21								21		5	6.9	0.2	4		0.5		2	0.2			
99-3	6	26		4	4		4	4		6	4	3	23.5		3	20	0.5						
00-1	4	24		6					9	6	3	1	10		10								
00-3	3	20		6					9	5		2	190		180	10							
01-3	4	24		12		3				7	2	3	70.9		60	10	0.9						
Totals	28	128										28	430										
Average				4.8						14.2							0.2						
Probably in product				x						x							x						

t = trace.

APPENDIX 4-10 Protein Products — Moderate Chemical Contaminants

Roast Beef

Product / Chemical Class / MB	Num	ppb	P Cam	P DDE	P Die	P HE	IC BZ	IC DBZ	IC EBZ	IC Sty	IC TCA	IC TCE	IC Tol	IC Xyl
96-1	3	105	7				51						47	
96-2	2	30					16				14			
97-1	6	215		1			66		14	11			36	87
98-1	3	65.1		0.5			31						34	
98-2	2	15.7		0.7									15	
98-3	2	31					17						14	
98-4	4	52.6		0.5		0.1	25						27	
99-1	4	184		0.3			99			31			54	
99-3	5	73		1			45			4			21	2
00-1	7	103					49		2	6	4	11	26	5
00-3	7	57.7		2	0.7		21	18		2		2	14	
01-3	3	94					57			3			34	
Totals	48	1026												
Average				0.5			39.8						26.8	
Probably in product				x										
Occurs in product							x						x	

Bacon

Product / Chemical Class / MB	Num	ppb	P DDE	P DDT	IC BZ	IC Cfo	IC DBZ	IC EBZ	IC PCE	IC Sty	IC TBZ	IC TCA	IC Tol	IC Xyl
96-1	3	113						16					62	35
96-2	3	35									11	24	t	
97-1	5	156	0.4						22	14			98	22
98-1	5	70	1	1						15	29		24	
98-2	2	35								17			18	
98-3	3	95	1								80		14	
98-4	2	36			11								25	
99-1	3	42.5	0.5										12	30
99-3	6	55					4	2		7	16		21	5
00-1	9	83.1	0.5	0.6	5			3		31	16	3	16	8
00-3	6	36			2	2	8			6			15	3
01-3	7	76.8	0.8		2			3	2	22			37	10
Totals	54	834												
Average										9.3			31.1	9.4
Probably in product										x				x
Occurs in product													x	

APPENDIX 4-10 *Continued*

Product Chemical Class MB	Salami Num	ppb	P Cmy	P Cno	P Cpa	P DDE	P Die	IC BZ	IC BDM	IC Cfo	IC CTC	IC EBZ	IC Pyl	IC PCE	IC Sty	IC Tol	IC Xyl
96-1	3	71											11			37	23
96-2	5	86	35		16	1		18								16	
97-1	5	110				0.5		19						10		52	28
98-1	2	20.4				0.4										20	
98-2	3	12.6				0.4	0.2									12	
98-3	2	32													19	13	
98-4	2	19.6				0.6										19	
99-1	5	60.8				0.4	0.4						10			30	20
99-3	8	208				2		24			5	8		8	43	82	36
00-1	6	30.4				0.4		3				2			3	10	12
00-3	6	44						3	4	7					6	17	7
01-3	8	93.4		40	10	0.2	0.2					2			3	33	5
Totals	55	788															
Average						0.5										28.4	10.9
Probably in product						x											x
Occurs in product																x	

APPENDIX 4-11 Protein Products — High Chemical Contaminants

Product											Hot Dogs											
Chemical Class			P	P	P	P	P	P	P	P	IC	IC	IC	IC	IC	IC	IC	IC	IC	IC	IC	IC
MB	Num	ppb	Cmy	Cno	Cpa	DDE	DDT	Die	HCB	HE	BZ	BBZ	BDM	Cfo	CTC	EBZ	PCE	Pyl	Sty	TCE	Tol	Xyl
96-1	3	36				8															12	16
96-2	7	1187	5	932	111			1		0.7							60				78	
97-1	3	27				1														10	16	
98-1	2	34				6															28	
98-2	2	29				1															28	
98-3	4	96		55	16	2															13	
98-4	7	65				0.4		0.2	0.2	0.2				14						18	32	
99-1	7	59				4		0.6	0.1	0.3	10							11			33	
99-3	9	122									2	3	5	12	4				4	5	61	26
00-1	11	55				0.4	0.4	0.3		0.1	3					3	2		5	4	23	14
00-3	11	184				0.7					4	4	11	7		4	8		6	105	15	19
01-3	13	121	5	30	8	0.6		0.2			4		4	7			16		3	15	17	14
Totals	79	2015																				
Average						2.1															29.7	
Probably in product						x																
Occurs in product																					x	

APPENDIX 4-11 *Continued*

Product	Ground Beef																			
Chemical Class			P	P	P	P	P	P	P	IC	IC	IC	IC	IC	IC	IC	IC	IC	IC	IC
MB	Num	ppb	Cpy	DDE	Die	Dip	HCB	HE	Oct	BZ	Cfo	DBZ	EBZ	PCE	Sty	TBZ	TCA	TCE	Tol	Xyl
96-1	3	27		4															12	11
96-2	2	37								18									19	
97-1	4	41.4		1	0.4					22									18	
98-1	2	34		2															32	
98-2	5	29.4		0.9	0.3			0.2		14									14	
98-3	4	18.9		2	0.8		0.1											16		
98-4	6	73.4		0.9	0.4	34		0.1		14									26	
99-1	8	89.5	0.5	2	0.4	71	0.2	0.3	0.1										15	
99-3	9	56.3		0.9	0.4					9		9		6	5			3	17	6
00-1	13	81.9		0.3	0.4		0.1	0.1		26		2	2	4	8	11	3		18	7
00-3	5	36								11	2				5				16	2
01-3	10	62.9		0.7	0.1			0.1		12		3	3		10			3	22	9
Totals	71	588																		
Average				1.2	0.3					10.5									17.4	
Probably in product				x	x					x										
Occurs in product																			x	

APPENDIX 4-12 Protein Products — Very High Chemical Contaminants

Product	Salmon (Fish)																	
Chemical Class			P	P	P	P	P	P	P	P	P	P	P	IC	IC	IC	IC	IC
MB	Num	ppb	BHC	Cam	Cd	DDE	DDT	Die	HCB	HE	Lin	Nac	Oct	TDE	PBZ	PCB	RBZ	TBZ
96-1	ND	ND																
96-2	ND	ND																
97-1	ND	ND																
98-1	7	70.9	0.5			15	5	2	0.4					6		42		
98-2	6	62.6	1			11		2	0.6					1		47		
98-3	10	22.2	0.6	1		5		2	0.9	0.1			0.2	0.3		9	0.4	
98-4	8	31.5	2			8		3	0.9	0.4	0.2			4		13		
99-1	9	76.3	2			15		4	1	0.5			0.4	5		48	0.4	
99-3	9	43.8	0.9			10			0.8	0.4	0.6	0.7	0.3		0.1	30		
00-1	10	50.5	1.3		1.1	10		2	1	0.3		0.5	0.3	4		30		
00-3	4	9.7				1		0.4	0.3									8
01-3	9	29.2	0.1		0.6	5		2	1	0.2			0.1		0.2	20		
Totals	72	397																
Average			0.9			8.9		1.9	0.8				0.1			26.6		
Probably in product			x					x					x			x		
Occurs in product						x			x									

ND = No Data

APPENDIX 4-13 Noncitrus Products with Very Low Contaminant Levels

Product Chemical Class MB	Pineapple (Canned)		Pineapple Juice			
					P	P
	Num	ppb	Num	ppb	Dip	EnS
96-1	0	0	1	0.5		0.5
96-2	0	0	1	0.5		0.5
97-1	0	0	0	0		
98-1	0	0	1	0.7		0.7
98-2	0	0	0	0		
98-3	0	0	0	0		
98-4	0	0	1	0.6		0.6
99-1	0	0	0	0		
99-3	0	0	0	0		
00-1	0	0	2	30.4	30	0.4
00-3	0	0	0	0		
01-3	0	0	1	0.2		0.2
Totals	0	0	7	32.9		

APPENDIX 4-14 Noncitrus Products with Very Low to Moderate Contaminant Levels

Product	Prune Juice							Prunes										
Chemical Class			P	P	P	P	IC			P	P	P	P	P	P	P	P	P
MB	Num	ppb	EnI	EnS	Ipr	Ipm	Xyl	Num	ppb	Azi	Cpy	DDE	Dan	Dic	EnI	Ell	EnS	Ipr
96-1	3	29			7	4	18	4	14		1			0.8	0			
96-2	0	0						7	123					3	0.3	0.4	2	27
97-1	2	7			4	3		6	163	7	1						0.8	64
98-1	0	0						2	27									21
98-2	0	0						1	4									4
98-3	0	0						1	3									3
98-4	1	1			1			1	2									2
99-1	1	1			1			8	69			0.1	0.8	2	0.2	0.6	1	57
99-3	0	0						4	86					14				10
00-1	0	0						0	0									
00-3	0	0						3	84					58				20
01-3	2	0.4	0.2	0.2				4	37					33		0.8	0.7	2
Totals	9	38.4						41	121									
Average																		18
Probably in product																		x
Occurs in product																		

APPENDIX 4-14 *Continued*

Product	Prunes					Plums												
Chemical Class	P	P	P	P	IC			P	P	P	P	P	P	P	P	P	P	P
MB	Ipm	P	Par	Prg	Tol	Num	ppb	Ben	Car	Cpy	Dan	Dic	DNB	EnS	Ipr	Ipm	Par	Prg
96-1			1		12	5	233			0.4	13				185	20		15
96-2	15			75		4	346	87			2				247	10		
97-1	37			53		6	452			4	225		1		79	137	6	
98-1	6					4	319			1	189				111	18		
98-2						6	825	302	3	2	60				454	4		
98-3						4	1067	111			345				219	392		
98-4						3	341				7				281	53		
99-1	7					4	68				2			0.4	52	14		
99-3	60	1.6				3	161				0.5				120	40		
00-1						5	540		10		60	190			200	80		
00-3	6					3	74.4					73			0.5	0.9		
01-3						4	332	70		4					250	8		
Totals						51	4759											
Average											75				183	64.7		
Probably in product											x							
Occurs in product															x	x		

APPENDIX 4-15 Citrus Products with Very Low to Low Contaminant Levels

Product	Lemonade				Grapefruit Juice					Grapefruit						
Chemical Class			P	IC			P	P	IC			P	P	P	P	P
MB	Num	ppb	Cpy	Tol	Num	ppb	Etn	Thi	BZ	Num	ppb	Dic	Diz	Etn	Mal	Thi
96-1	1	49		49	3	43	23	7	19	2	521		6			515
96-2	0	0			1	1	1			2	69				1	68
97-1	0	0			0	0				2	177	8				169
98-1	1	2	2		1	0.7	0.7			2	93.7			0.7		93
98-2	0	0			0	0				1	147					147
98-3	0	0			0	0				2	254			3		251
98-4	0	0			1	0.5	0.5			1	271					271
99-1	0	0			0	0				1	99					99
99-3	0	0			0	0				1	420					420
00-1	0	0			0	0				1	90					90
00-3	0	0			0	0				2	212	2				210
01-3	0	0			0	0				1	160					160
Totals	2	51			6	45.2				18	2514					
Average																208
Probably in product																x

APPENDIX 4-16 Noncitrus Products with Very Low to Moderate Contaminant Levels

Product	Pear (Canned)				Fruit Cocktail (Canned)									Pears															
Chemical Class			P	IC			P	P	P	P	P	IC	IC			P	P	P	P	P	P	P	P	P	P	P	P	P	IC
MB	Num	ppb	EnS	Tol	Num	ppb	Car	Dic	Ipr	Ipm	P	Tol	Xyl	Num	ppb	Azi	Ben	Cap	Car	Dan	Dic	EII	EnS	Ipr	P	Par	Pho	Thi	Xyl
96-1	1	13		13	3	66		8				38	20	6	329	227			4							1	38	59	29
96-2	0	0			4	19	7	8	2	2				4	338	37		3		2								295	
97-1	0	0			1	3				3				3	533	21										2		510	
98-1	0	0			2	19			11	8				3	442	2					326							114	
98-2	0	0			0	0								1	325													325	
98-3	0	0			0	0								4	258	15	124										23	96	
98-4	1	0.5	0.5		3	39		28	3	8				4	279		41									31	57	150	
99-1	0	0			2	10			6	4				2	93												31	62	
99-3	0	0			0	0								3	26			20					0.2		6				
00-1	0	0			1	2			2					5	260	60						10	10				10	170	
00-3	0	0			2	120		110	10					5	506	60			190					6			10	240	
01-3	0	0			1	2					2			3	421						50			20				350	
Totals	2	13.5			19	280								43	3810														
Average																35.2												197	
Probably in product																x												x	

APPENDIX 4-17 Melons

Product	Watermelon						Cantaloupe																	
Chemical Class			P	P	P	P			P	P	P	P	P	P	P	P	P	P	P	P	P	P	P
MB	Num	ppb	Dan	EII	EnS	Pal	Num	ppb	Cpy	Car	Dan	DCP	Die	Edr	EnI	EII	EnS	Mat	Mos	P	Pal	Thi	Tox
96-1	1	3	3				4	82.4	0.4				3					4					75
96-2	2	1.1		0.4	0.7		6	217			2				0.2	0.4	25		101			90	
97-1	1	0.9			0.9		4	30.3				1			0.3	1	28						
98-1	2	1.4			0.5	0.9	6	65	1				3	2			2		42				15
98-2	0	0					3	15.6							0.6	1	14						
98-3	0	0					2	51									16		35				
98-4	0	0					4	17			0.5				0.1		6						10
99-1	0	0					8	43				0.5	0.3		0.3	0.6	12			4	0.2		25
99-3	0	0					3	12		10			1				1						
00-1	0	0					0	0															
00-3	0	0					6	41.6	8						0.8	0.8	10					20	2
01-3	1	0.2			0.2		5	84.2								0.2	10		40			30	4
Totals	7	6.6					51	659															
Average																	10.3						
Probably in product																	x						

APPENDIX 4-18 Peaches

Product	Peach (Canned)						Peaches																					
Chemical Class			P	P	P	P			P	P	P	P	P	P	P	P	P	P	P	P	P	P	P	P	P	P	P	P
MB	Num	ppb	Car	Ipr	Ipm	Tbp	Num	ppb	Azi	Ben	Cap	Car	Cpy	DDE	Dan	Dic	Die	EnI	EII	EnS	Fen	Ipr	Ipm	Mos	P	Par	Pho	Prg
96-1	1	3		3			9	1,875	43			5			514			11	44	10		1,084					15	149
96-2	0	0					7	861	15	268		4	16								7	500	20					
97-1	1	17	17				8	1,056				14					3					868	34	5		20	30	82
98-1	2	3		2	1		8	1,112			4	60		5	191							710	15			2	125	
98-2	3	35		7	5	23	6	1,570	127	542			16									845	14				26	
98-3	0	0					4	1,826							987							756	80			3		
98-4	1	1		1			3	632		31																31	573	
99-1	1	14	14				6	1,343		588		13				3						710	20			9		
99-3	1	10	10				7	623		60					210							190	80		2	10	70	
00-1	1	6	6				5	5,810				10			650							4,950	150				50	
00-3	1	10	10				4	1,357							20							1,230				7	100	
01-3	1	8	8				4	3,604							4							3,460	80				60	
Totals	13	107					71	21,669																				
Average															215							1,275	41.1			6.8	87	
Probably in product															x							x	x			x	x	

APPENDIX 4-19 Grape Products

Product	Grape Juice							Grapes																				
Chemical Class			P	P	P	P	P			P	P	P	P	P	P	P	P	P	P	P	P	P	P	P	P	P	P	P
MB	Num	ppb	Car	Dat	Dic	Ipr	Ome	Num	ppb	Cap	Car	Cil	Cpy	DDE	Dat	Dic	Diz	EnI	EII	EnS	Far	Fex	Fol	Ipr	Ipm	Mos	Ome	Vin
96-1	2	9	8			1		4	93.7				0.7			6								30	57			
96-2	1	14	14					5	222	13											2			204	2			1
97-1	1	18	18					1	88							88												
98-1	0	0						2	87															68	19			
98-2	1	15	15					5	497	170	8				3									311	5			
98-3	1	8	8					9	631	18			41		107					16			23	330	5	35	107	
98-4	2	6	5			1		4	2.4				1	0.5				0.5	0.4									
99-1	4	17	5	2		7	3	1	0.5					0.5														
99-3	1	4	4					2	3						1												2	
00-1	2	11.6	4		7.6			3	20.5					0.5										10	10			
00-3	1	9	9					6	167	50	30				1		6							70			10	
01-3	1	4	4					5	450	40		120										110		160	20			
Totals	17	115.6						47	2262																			
Average			7.8																					98.6	9.8			
Probably in product			x																					x	x			

APPENDIX 4-19 *Continued*

Product	Raisins																								
Chemical Class			P	P	P	P	P	P	P	P	P	P	P	P	P	P	P	P	IC	IC	IC	IC	IC	IC	IC
MB	Num	ppb	Car	Cpy	DDE	DDT	Dic	Diz	EnI	EII	EnS	Far	Fen	Ipr	Ipm	Par	Pho	Prg	BZ	Cfo	PCE	Sty	TCE	Tol	Xyl
96-1	9	80	8	1	1		10	0.4	1	5	2							52							
96-2	5	185	9		0.7		53											108	14						
97-1	4	470		2	0.9													437						30	
98-1	2	192			1													191							
98-2	3	201	15									4						182							
98-3	7	1284	6		2	0.3						4				1	7	1260							
98-4	4	649			0.6		3											648	97						
99-1	4	319	28		0.5													280						14	
99-3	7	328			0.8		3							3	20		10	280			11				
00-1	5	503					190											280		15		2			18
00-3	9	196			0.6		10						0.9	3				160		7			3	2	9
01-3	5	796			1		510											270						1	14
Totals	64	5203																							
Average					0.8		64.9											346							
Probably in product					x		x																		
Occurs in product																		x							

APPENDIX 4-20 Orange Products

Orange Juice

Product / Chemical Class / MB	Num	ppb	P Car	P Dic	P Etn	IC BZ	IC Cfo	IC EBZ	IC Sty	IC Tol	IC Xyl
96-1	2	1.7		0.7	1						
96-2	0	0									
97-1	1	1			1						
98-1	1	2			2						
98-2	0	0									
98-3	0	0									
98-4	2	24.7			0.7					24	
99-1	1	0.7			0.7						
99-3	4	30				2	6		4	18	
00-1	3	16				7	4			5	
00-3	6	52	3			5		5	7	27	5
01-3	5	34	4		2		4			22	2
Totals	25	162.1									
Average											
Probably in product											

Oranges

Product / Chemical Class / MB	Num	ppb	P Car	P Cpy	P Dan	P Dic	P Dat	P EnS	P Mat	P Ome	P Thi	IC BZ	IC Cfo	IC Sty	IC TCE	IC Tol	IC Xyl
96-1	8	425	44	2	1	4			6		335					19	18
96-2	1	8	8														
97-1	3	146		0.9				0.5			132						
98-1	2	114		1							113						
98-2	3	126		14							58					54	
98-3	3	195									129					t	66
98-4	1	2		2													
99-1	2	112														86	26
99-3	10	85		1		5	2		3	2	10	11		2		28	21
00-1	4	325									210	1				86	28
00-3	6	282		1							270		2	3	2	2	
01-3	3	49									40		3			6	
Totals	46	1869															
Average				1.8							108					23	
Probably in product				x							x					x	

APPENDIX 4-21 Apple Products

Product	Apple Juice						Applesauce											
Chemical Class			P	P	P	P			P	P	P	P	P	P	P	P	P	P
MB	Num	ppb	Car	Dat	Eth	Thi	Num	ppb	Ben	Cap	Car	Cpy	Dat	EnS	Eth	Tbp	Thi	Xyl
96-1	2	243	22			221	4	157			4					26	115	12
96-2	2	16	4		12		3	58	44		11				3			
97-1	1	28	28				3	22					2	1		19		
98-1	2	207	5			202	2	248								26	222	
98-2	2	52	5			47	3	7		1	4		2					
98-3	1	3		3			3	25			6		1			18		
98-4	3	88	7	2		79	0	0										
99-1	2	96	4			92	3	149			4	2					143	
99-3	3	26	4	2		20	1	1				1						
00-1	2	10.6	10	0.6			1	10			10							
00-3	1	2	2				2	24			4					20		
01-3	3	183	20	3		160	2	18							8	10		
Totals	24	954.6					27	719										
Average			9.3			68.4					3.6							
Probably in product			x			x					x							
Occurs in product																		

APPENDIX 4-21 *Continued*

Product Chemical Class MB			Apples																					
			P	P	P	P	P	P	P	P	P	P	P	P	P	P	P	P	P	IC	IC	IC	IC	IC
	Num	ppb	Azi	Ben	Cap	Car	Cpy	Dan	Dic	EnI	EII	EnS	Far	Mxy	Pho	Pne	Pon	Prg	Thi	BZ	Cfo	DBZ	Tol	Xyl
96-1	9	314.4	88			20		0.9		1	2	1.9				27			153					22
96-2	8	769	120				9		14					101			4	56	417					48
97-1	5	429	32											2		51			332	12				
98-1	8	268	154			14	10			1	0.8			25					57					20
98-2	7	588	68				19			5	19	4		28					445					
98-3	5	735	6	49	4														651					25
98-4	6	881	117				8			0	0.3			55					660					
99-1	8	481	180				7			8	6	12		80					268					27
99-3	10	366		40	50		2			2	3	4							230	2			17	16
00-1	7	713	30		50									0.7	120				490		5			17
00-3	9	765	30										40	30	80				520	32		9	13	9
01-3	11	97	10	60						1	0.7	0.6			0.8				0.5	2	2		4	9
Totals	93	6406																						
Average			70							2	2.7			26.8					352					16
Probably in product										x	x			x										x
Occurs in product			x																x					

APPENDIX 4-22 Noncitrus Products with Low to Moderate Contaminant Levels

Product	Banana									
Chemical Class			P	IC	IC	IC	IC	IC	IC	IC
MB	Num	ppb	Thi	BZ	Cfo	EBZ	Sty	TCE	Tol	Xyl
96-1	1	43	43							
96-2	3	141	41	64					36	
97-1	2	35	19	16						
98-1	2	149	94	55						
98-2	0	0								
98-3	2	67	49	18						
98-4	3	146	80	54					12	
99-1	3	40	18	t						22
99-3	2	76	70						6	
00-1	5	120		104		2		6	6	2
00-3	4	54	30	11	8				5	
01-3	4	96	30	59			5			2
Totals	31	967								
Average			40	38						
Probably in product			x	x						

Product	Apricots																	
Chemical Class			P	P	P	P	P	P	P	P	P	P	P	P	P	P	P	P
MB	Num	ppb	Azi	Ben	Cap	Car	Cpy	Dan	Efv	EnI	EII	EnS	Fen	Ipr	Ipm	Par	Pho	Thi
96-1	ND	ND																
96-2	3	265							60					24			181	
97-1	8	1174	215		247		1			3	7	15		112			574	
98-1	4	989		170		350								119			350	
98-2	2	62												26	36			
98-3	2	114												112		2		
98-4	2	132												59			73	
99-1	4	83										0.4	36	38	9			
99-3	8	544	110	260	3			0.3			0.3		30				40	100
00-1	4	403	60			330								3			10	
00-3	3	66											10	6			50	
01-3	0	0																
Totals	40	3832																
Average														45			116	
Probably in product															x			x

ND = No Data

APPENDIX 4-23 Noncitrus Products with High to Very High Contaminant Levels

Product			Cherries																					
Chemical Class			P	P	P	P	P	P	P	P	P	P	P	P	P	P	P	P	P	P	P	P	P	P
MB	Num	ppb	Ace	Azi	Ben	Cap	Car	DDE	Dip	Dat	EnI	EII	EnS	Efv	Far	Fen	Ipr	Ipm	Mal	Mxy	Ome	P	Pho	Vin
96-1	ND	ND																						
96-2	5	770					7								19		710	31	3					
97-1	5	260		14	37												206		2	1				
98-1	5	169					51								3	110	3							2
98-2	4	769		79			321											355	14					
98-3	2	1650															1630	20						
98-4	9	144			76		0.6				0.5	2	6		25		30	3	1					
99-1	10	529		145			13	0.8		17	0.8	5	10				260	43			34			
99-3	8	149		40			50				2	7	10		9		30	1						
00-1	7	180.9		10			90					0.3	0.6					10				50	20	
00-3	8	412		10		8	50							4			230	4				6	10	
01-3	11	144	20	10			40		20			1	6			6	20	8	8			5		
Totals	74	5177																						
Average				28			57										284	43						
Probably in product				x			x										x	x						

APPENDIX 4-23 *Continued*

Product			Strawberries																															
Chemical Class			P	P	P	P	P	P	P	P	P	P	P	P	P	P	P	P	P	P	P	P	P	P	P	P	P	IC	IC	IC	IC	IC
MB	Num	ppb	Azi	Ben	Bif	Cap	Car	Cd	DDE	Dic	Die	Dip	Diz	EnI	EII	EnS	Fex	Fol	Ipr	Ipm	Mal	Mat	Met	Par	Prg	Tox	Vin	BZ	PCE	Sty	Tol	Xyl
96-1	5	288					34			36									16		9									190		
96-2	9	2390				81	5						3					7	1277	7			25			15	970					
97-1	12	1069	8	85		25								26	38	41			217		21		13	2			605			t		
98-1	7	1968	15						0.5		0.2								555	7	10									1380		
98-2	8	1073		110	84	774	44							12	32	7			10													
98-3	5	253				0.8	61												171								5			15		
98-4	7	682		242							0.6										10				379		27			17	16	
99-1	9	151		43	4	3	29			0.9		21							48	0.4							1					
99-3	11	212				10	160	0.5	0.3		0.5				0.5	2			10			10							5	13		
00-1	7	252.6							0.8		0.4			0.4		10			50											175		16
00-3	4	229				10	40												230											12		
01-3	5	875				40	470										350				4							11				
Totals	89	9443																														
Average						78.7	70												215											164		
Probably in product						x	x												x											x		

t = trace.

APPENDIX 4-24 Corn Products

Product	Corn Grits			Cornflakes					Creamed Corn					Corn Bread (HM)						
Chemical Class			P			P	P	IC			P	IC	IC			P	P	P	P	IC
MB	Num	ppb	Mal	Num	ppb	Mxy	Tbp	Xyl	Num	ppb	Dat	BZ	Tol	Num	ppb	Cpm	DDE	Mal	Pri	Xyl
96-1	0	0		1	11			11	2	57	2		55	2	11			1		10
96-2	0	0		0	0				0	0				0	0					
97-1	0	0		0	0				1	24			24	3	3		1	1	1	
98-1	0	0		0	0				0	0				2	4.3	29		12		
98-2	0	0		0	0				0	0				1	0.6	0.6				
98-3	0	0		0	0				0	0				0	0					
98-4	0	0		0	0				0	0				2	1.6	0.6		1		
99-1	1	3	3	0	0				0	0				2	2.3	0.3		2		
99-3	0	0		0	0				1	8		8		2	8	2		6		
00-1	0	0		0	0				1	1			1	2	10.2	0.2		10		
00-3	0	0		0	0				0	0				2	1.8	0.8		1		
01-3	0	0		2	42	2	40		0	0				3	13.9	0.9		3	10	
Totals	1	3		3	53				5	90				21	56.7					
Average																2.9		3.1		
Probably in product																x		x		

APPENDIX 4-24 *Continued*

Product			Popcorn (Oil)																	
Chemical Class			P	P	P	P	Pri	P	IC	IC	IC	IC	IC	IC	IC	IC	IC	IC	IC	IC
MB	Num	ppb	Die	Ipr	Mal	Mxy	Pri	Tox	BZ	BBZ	BDM	DBZ	EBZ	PCB	Sty	TBZ	TCA	TCE	Tol	Xyl
96-1	3	142				35													74	33
96-2	1	27															27			
97-1	3	227				1	210													16
98-1	2	52				2													48	
98-2	2	36					7												29	
98-3	3	252					163												65	24
98-4	3	210				0.8	163												46	
99-1	6	301			9		210	26	22							12			32	
99-3	5	469				300	120		22										20	7
00-1	12	328	0.2				240		1	6			3	20	5	6	16	3	29	15
00-3	12	509		130		5			5	6	5	292		9	2	5	4		36	10
01-3	8	79			20	2			6				2		5			4	30	10
Totals	60	2632																		
Average						28.8	92.8												34.1	9.6
Probably in product						x	x												x	x

APPENDIX 4-24 *Continued*

Product	Corn Chips																	
Chemical Class			P	P	P	P	Pri	IC	IC	IC	IC	IC	IC	IC	IC	IC	IC	IC
MB	Num	ppb	Cyp	Mal	Mxo	Mxy	Pri	BZ	BBZ	Cfo	DBZ	EBZ	Sty	TBZ	TCA	TCE	Tol	Xyl
96-1	5	104	3	4				13									66	18
96-2	1	24													24			
97-1	5	154		1	1	2	136										14	
98-1	2	66												16			51	
98-2	2	185					173										12	
98-3	0	0																
98-4	2	45					20										25	
99-1	1	37															37	
99-3	7	66					2	8			18		3	7			21	7
00-1	9	70					20	1				2	2	14	7	2	13	9
00-3	8	119						43	3	4			5	13		4	30	14
01-3	6	49						4		6		3	11				17	8
Totals	48	919																
Average																	23.8	
Probably in product																	x	

APPENDIX 4-25 Rice Products

Product	Crisped Rice Cereal			White Rice					
Chemical Class			P			P	P	P	P
MB	Num	ppb	Qac	Num	ppb	Dip	Ipm	Mxy	Qac
96-1	0	0		0	0				
96-2	0	0		1	0.5			0.5	
97-1	0	0		0	0				
98-1	0	0		1	22	22			
98-2	0	0		0	0				
98-3	1	4	4	2	5		2		3
98-4	0	0		1	3				3
99-1	0	0		1	6				6
99-3	1	7	7	1	8				8
00-1	0	0		0	0				
00-3	1	6	6	1	6				6
01-3	1	4	4	1	3				3
Totals	4	21		9	53.5				

APPENDIX 4-26 Oat Products

Product	Oatmeal							Oat Ring Cereal					Granola												
Chemical Class			P	P	P	P	IC			P	P	P			P	P	P	P	P	P	P	P	IC	IC	IC
MB	Num	ppb	EII	EnS	Mal	TBZ	Xyl	Num	ppb	2,4D	Clo	Dca	Num	ppb	Cpm	Cpy	Dic	Diz	Ipr	Mal	Myx	Pcy	Sty	Tol	Xyl
96-1	2	41				11	30	1	9			9	4	65						6			21	17	21
96-2	1	2			2			1	6			6	3	54.1	53	0.3		0.8							
97-1	0	0						1	11			11	2	16.7	16					0.7					
98-1	0	0						1	15			15	3	23.4	22	0.4				1					
98-2	0	0						1	4			4	1	27	27										
98-3	0	0						1	7			7	2	28	27	1									
98-4	0	0						1	10			10	2	160	148		12								
99-1	0	0						2	7	2		5	1	2						2					
99-3	2	5	2	3				2	18		8	10	2	30.9	30					0.9					
00-1	0	0						2	20		10	10	1	10	10										
00-3	0	0						2	20		10	10	1	10	10										
01-3	0	0						2	7		4	3	4	7.8			5		2		0.2	0.6			
Totals	5	48						17	134				26	435											
Average												8.3			28.6										
Probably in product												x			x										

APPENDIX 4-27 Pasta Products

Product	Macaroni						Noodles (Egg)							
Chemical Class			P	P	P	P			P	P	P	P	IC	IC
MB	Num	ppb	Cpm	Lin	Mal	Mxy	Num	ppb	Cpm	Diz	EnS	Mal	Tol	Xyl
96-1	2	1.8			0.8	1	2	28					13	15
96-2	0	0					1	3	3					
97-1	0	0					0	0						
98-1	2	21	18		3		1	0.5				0.5		
98-2	1	2	2				2	4	2			2		
98-3	0	0					2	10	7			3		
98-4	1	1	1				0	0						
99-1	1	2			2		2	5	3			2		
99-3	2	3	2		1		2	3	2			1		
00-1	1	0.6		0.6			0	0						
00-3	2	6	3		3		2	7	3			4		
01-3	2	8.9	8		0.9		4	22	8	8	2	4		
Totals	14	46.3					18	82.5						
Average									2.3			1.4		
Probably in product									x			x		

APPENDIX 4-28 Wheat-Based Cereals

Product	Wheat Cereal (instant)				Shredded Wheat										
Chemical Class			P	P			P	P	P	P	P	P	P	IC	IC
MB	Num	ppb	Cpm	Mal	Num	ppb	Cpm	Cpy	Diz	Mal	Myx	Tbp	TBZ	Tol	Xyl
96-1	0	0			5	53				2	2		12	13	17
96-2	0	0			0	0									
97-1	0	0			3	57	19			4	34				
98-1	0	0			3	49.3	32	0.3			17				
98-2	0	0			2	20	8			12					
98-3	0	0			2	90	56			34					
98-4	0	0			2	117	63			54					
99-1	2	8	5	3	3	104	48		3	53					
99-3	0	0			4	132	50		4	70		8			
00-1	0	0			2	9	80			10					
00-3	0	0			2	70	60			10					
01-3	0	0			2	170	100			70					
Totals	2	8			30	871									
Average							43			27					
Probably in product							x			x					

APPENDIX 4-28 *Continued*

Product Chemical Class MB	Num	ppb	Cpm	P Cpy	P Cum	P Dca	P EDC	P Edi	P Etn	P Mal	P Mat	P Mec	P Ton	IC BZ	IC Cfo	IC EBZ	IC Ply	IC Sty	IC TBZ	IC TCE	IC Tol	IC Xyl
	Fruit-Flavored Cereal																					
96-1	6	147	1							2				11	88						25	20
96-2	4	39						16		2				21							t	
97-1	1	36						36														
98-1	6	28.7	1	0.7		3		20		1	3											
98-2	0	0																				
98-3	6	79		2		3	66		2		3	4										
98-4	5	15		3		3			2		4	3										
99-1	6	71		2				23	1						12				20		13	
99-3	5	20			4	2												3			7	4
00-1	7	57						9			10			2			4	3			21	8
00-3	7	47				1		29						2	2					3	3	7
01-3	8	83.8				0.8							2	14	2	4		50			7	4
Totals	61	623.5																				
Average																					6.9	
Probably in product																					x	

t = trace.

APPENDIX 4-29 Breakfast Products

Product	Biscuit							English Muffin					Pancake Mix							
Chemical Class			P	P	P	P	IC			P	P	P			P	P	P	P	IC	IC
MB	Num	ppb	Cpm	Mal	Mxy	Pri	Cfo	Num	ppb	Cpm	Mal	Pri	Num	ppb	Cpm	DDE	Mal	Mxy	Tol	Xyl
96-1	2	14.5		0.5			14	2	18	9	9		5	39.5	3	0.5	3		19	14
96-2	0	0						2	15	8	7		1	5	5					
97-1	1	2	2					2	6	2	4		3	6	2	2	2			
98-1	2	3.9	0.9			3		3	15	3	2	10	3	2.9	0.5	0.5	2			
98-2	1	7	7					2	11	9	2		2	39	29		7			
98-3	2	12	11	1				0	0				2	38	20		18			
98-4	2	1	0.6		0.4			2	7	2	5		3	18.4	6		12	0.4		
99-1	2	1.7	1	0.7				3	6.5	2	4	0.5	2	3	2		1			
99-3	2	14	6	8				2	14	4	10		2	7	3		4			
00-1	0	0						2	13	7	5		0	0						
00-3	2	9	7	2				2	10	6	4		2	15	8		7			
01-3	1	10	10					2	30	20	10		2	8	2		6			
Totals	17	75.1						24	146				27	181.8						
Average			3.8							6	5.2				6.7		5.2			
Probably in product			x							x	x						x			
Occurs in product															x					

APPENDIX 4-29 *Continued*

Product	Bagel								Sweet Roll																	
Chemical Class			P	P	P	P	P	P			P	P	P	P	P	P	P	IC	IC	IC	IC	IC	IC	IC	IC	IC
MB	Num	ppb	Cpy	Cpm	Fit	Mal	Myx	Pri	Num	ppb	Cpm	Cpy	Cum	DDE	Mal	Mat	Mxy	BZ	DBZ	EBZ	PCE	Sty	TBZ	TCE	Tol	Xyl
96-1	2	7		3		4			4	69	3				3							41				22
96-2	2	14		6		8			3	7.7	1				6		0.7									
97-1	5	20	2	10	3	4		1	3	50					4							30			16	
98-1	4	14.7	0.7	6		6		2	5	74	11					2		11				41				14
98-2	1	1		1					5	96	3				13							51	19		t	
98-3	0	0							4	44	2				17							25			t	
98-4	4	10		2		4	2	2	6	167	4		25		6		0.9					53			78	
99-1	4	14.2	0.2	8		3		3	5	164	0.8				16							51	81		15	
99-3	3	6		3		2		1	8	79.4	4	0.3	8	0.1	2							46			15	4
00-1	2	11		10		1			8	115	1				2			2		3		53	26		19	9
00-3	2	12		10		2			9	241	7				7					2	8	46	33	2	121	15
01-3	2	4		2		2			9	220	9				5			6	19	5	9	41			121	8
Totals	31	113.9							69	1327																
Average				5.1		3					3.8				6.8							39.8			38.5	
Probably in product						x																			x	
Occurs in product				x							x				x							x				

APPENDIX 4-30 Pretzels and Crackers

Product	Pretzels							Saltine Crackers								
Chemical Class			P	P	P	P	P			P	P	P	P	Pri	P	P
MB	Num	ppb	Cpm	Diz	Mal	Mxy	Pri	Num	ppb	Cpm	Cpy	Mal	Myx	Pri	TBZ	Xyl
96-1	1	7	7					6	108	29	1	4	2		48	24
96-2	3	31	24		6	1		1	5			5				
97-1	2	26	17		9			2	10	2		8				
98-1	2	41	29		12			2	9	7		2				
98-2	3	62	26		24		12	2	10	6		4				
98-3	3	160	46	1	113			3	174	107	1	66				
98-4	3	28	17		10	1		3	36.5	25		11	0.5			
99-1	3	14.7	10		4	0.7		2	16	13		3				
99-3	3	80.9	70		10		0.9	2	9	2		7				
00-1	2	3.8	0.8		3			2	20	10		10				
00-3	3	31	20		10	1		3	16.4	6		4	0.4			
01-3	2	14	10		4			5	78.2	40	0.2	30	7	1		
Totals	30	499.4						33	492							
Average			23.1		17					20.6		12.8				
Occurs in product			x		x					x		x				

APPENDIX 4-30 *Continued*

Product	Graham Crackers																		
Chemical Class			P	P	P	P	P	P	P	IC	IC	IC	IC	IC	IC	IC	IC	IC	IC
MB	Num	ppb	Cpm	Cpy	Cum	Mal	Myx	Pri	Pyl	BZ	BBZ	Cfo	DBZ	EBZ	PCE	Sty	TBZ	Tol	Xyl
96-1	5	133	16	1		8											97		11
96-2	5	116	3			38	2							23					50
97-1	4	22.4	6			4		0.4										12	
98-1	5	78	17	2		4											20	35	
98-2	3	15	5	1		9													
98-3	3	83	5			11												67	
98-4	6	158	27			22	1									11	86	11	
99-1	4	129	41			23											12	53	
99-3	10	212	50			30				9		5	3	6		14	26	43	26
00-1	11	92.5	4		6	8	0.5		11	1				2		15	32	5	8
00-3	9	130	40			20			21		3		9		5		16	13	3
01-3	10	106	2			40	0.3			6	14		6	3		8		20	7
Totals	75	1275																	
Average			18			18											24.1	21.6	
Probably in product																	x	x	
Occurs in product			x			x													

APPENDIX 4-30 *Continued*

Product	Butter Crackers																			
Chemical Class			P	P	P	P	P	P	P	IC	IC	IC	IC	IC	IC	IC	IC	IC	IC	IC
MB	Num	ppb	Cpm	Cum	Dip	EnS	HCB	Mal	Myx	BZ	Cfo	DBZ	EBZ	PCE	Pyl	Sty	TBZ	TCE	Tol	Xyl
96-1	3	45	15					15											10	
96-2	2	14	6					8												
97-1	2	11	9					2												
98-1	3	36	28					4											28	
98-2	2	11	4					7												
98-3	3	28	5					5											18	
98-4	4	40	3					8											13	16
99-1	4	37	3					5									14		15	
99-3	7	35.5	4			0.2		10	0.3					5					14	2
00-1	10	89	20					20	1				3	5		2	14	2	10	12
00-3	11	99.5	4		20			10	0.5		4		6	2		2		2	22	27
01-3	13	123	3	3			0.8	20	0.2	3		2	7		7	29	24		11	13
Totals	64	569																		
Average			8.7					9.5											11.8	
Probably in product																			x	
Occurs in product			x					x												

APPENDIX 4-31 Wheat Products with Moderate Contaminant Levels

Product	Flour Tortilla									White Roll											
Chemical Class			P	P	P	P	P	IC	IC			P	P	P	P	P	P	P	P	P	IC
MB	Num	ppb	Cpm	Cpy	Dip	Mal	Mxy	Tol	Xyl	Num	ppb	Cam	Cpm	Cpy	Dip	Mal	Mxy	Par	Pri	Tpp	TBZ
96-1	4	34.8	0.8			4		10	20	3	24		5			7					12
96-2	3	26	1	1		18				3	48		12			34			2		
97-1	2	10	2			8				2	24		10			14					
98-1	3	101	9		90	2				5	62	6	7		39	3			3		
98-2	3	114	6		103	5				3	21	12				7		2			
98-3	3	275	5		257	13				4	52	5			39	6			2		
98-4	4	16.4	3	0.6		12	0.8			2	2		1			1					
99-1	2	42	13			29				3	30	12	12			6					
99-3	4	60.3	10		40	10	0.3			3	26	6	10			10					
00-1	1	190			190					3	260	10			220					30	
00-3	3	248	4		24	4				3	20.7		10	0.7		10					
01-3	4	81	10		60	10	1			5	65	6	10		40	8	1				
Totals	36	1199								39	635										
Average			5.3		64	9.6						4.8	6.4			8.8					
Probably in product					x							x	x								
Occurs in product			x			x										x					

APPENDIX 4-32 Breads

Product	Rye Bread										Whole-Wheat Bread									
Chemical Class			P	P	P	P	P	P	IC	IC			P	P	P	P	Par	Pri	IC	IC
MB	Num	ppb	Cpm	Dip	Lin	Mal	P	Pri	Tol	Xyl	Num	ppb	Cpm	Cpy	Dip	Mal	Par	Pri	TBZ	Xyl
96-1	5	55.5	11			11	1.5		16	16	5	127	40	2		40			23	22
96-2	2	17	6			11					3	232	25		193	13				
97-1	1	1	1								3	29	11	1		15				
98-1	3	14.9	9		0.9	5					5	177	31		84	59	2	1		
98-2	2	19	7			12					3	210	10		189	11				
98-3	2	32	23			9					4	188	76		89	19		4		
98-4	2	13	7			6					3	37	28			6		3		
99-1	4	59	14	34		10		1			3	260	29		204	27				
99-3	3	256	7	240		9					3	110	50		40	20				
00-1	2	20	10			10					3	160	30		120	10				
00-3	2	15	10			5					4	49	10	2	30	7				
01-3	3	309	9	280		20					4	202	20		170	8		4		
Totals	31	811.4									43	1781								
Average			9.5			9							30		93.3	20				
Probably in product			x			x									x					
Occurs in product													x			x				

APPENDIX 4-32 *Continued*

Product	Cracked-Wheat Bread											
Chemical Class			P	P	P	P	P	P	P	P	P	P
MB	Num	ppb	24D	Cpm	Cpy	Dca	Dip	Diz	Mal	Pri	P	Tpp
96-1	5	117.8		13	1		90		12		1.8	
96-2	2	55		34					21			
97-1	3	58		39		3			16			
98-1	4	22	5	9		2			6			
98-2	4	29	4	17		2			6			
98-3	4	41	7	16		3			15			
98-4	6	283	12	12		9	225		11			14
99-1	2	8	4					4				
99-3	4	36	4	10		2			20			
00-1	3	15		8					5	2		
00-3	3	21	3	9					9			
01-3	6	80	2	20	1		50		6	1		
Totals	46	648										
Average			3.4	15.6		1.8			11			
Probably in product			x			x						
Occurs in product				x					x			

APPENDIX 4-32 *Continued*

Product	White Bread																					
Chemical Class			P	P	P	P	P	P	P	P	P	P	P	IC	IC	IC	IC	IC	IC	IC	IC	IC
MB	Num	ppb	2,4-D	Cam	Cpm	Cpy	Cum	Dca	Dip	Eth	Ipr	Mal	Pri	BZ	DBZ	EBZ	Pyl	Sty	TBZ	TCE	Tol	Xyl
96-1	5	32		3	6	0.5						6										16
96-2	3	195			3				152			36										
97-1	2	307		9					289													
98-1	7	103	7		1			6	51			4							17		17	
98-2	3	83			8				62			13										
98-3	4	52			5				39			6	2									
98-4	5	42	4		9			4	19			6										
99-1	4	69	2		2			3	62													
99-3	9	291	2	20	10				210			20			14				8		4	3
00-1	7	174							150			10		1				2	6		2	3
00-3	10	542	3		10			2	470			20			21			2	9	2	3	
01-3	12	845	1		20		3		470	4	7	70			5	28	3	144				90
Totals	71	2735																				
Average					6.2				165			16										
Probably in product					x							x										
Occurs in product									x													

APPENDIX 4-33 Nut Products

Product	Mixed Nuts (No Peanuts)																						
Chemical Class			P	P	P	P	P	P	P	P	P	IC	IC	IC	IC	IC	IC	IC	IC	IC	IC	IC	IC
MB	Num	ppb	24D	BHC	Cpy	Cum	DDE	Die	Lin	Pal	PBZ	BZ	Cfo	DBZ	EBZ	PCE	Pyl	Sty	TBZ	TCA	TCE	Tol	Xyl
96-1	6	124	6				11		0.8									34				38	34
96-2	4	88		2			2											21				60	
97-1	7	187					9	0.5		0.7	0.3							80				50	46
98-1	7	185		0.6	1										12			48	19			52	52
98-2	7	257		1.1	0.9										20			69	24			62	80
98-3	6	152			1										21			42	16			t	72
98-4	6	141			0.9			0.6		0.5								66	32			41	
99-1	9	177		2.3	1				0.3					15	11			35	14			45	56
99-3	11	363		0.2	8	3								9	22	9		93	78	7		57	77
00-1	10	408				8						3			9	6	13	43	113		2	162	49
00-3	11	444			1	6						6	5	2	38			76	98		2	78	132
01-3	8	132			6	2									5			72	5		2	25	15
Totals	92	2657																					
Average					1.7										11.5			56.6	33.3			55.8	51.1
Probably in product					x										x				x				x
Occurs in product																		x				x	

t = trace.

APPENDIX 4-33 *Continued*

Product			Dry-Roasted Peanuts																	
Chemical Class			P	P	P	P	P	P	P	P	P	Pal	P	P	P	P	P	IC	IC	IC
MB	Num	ppb	BHC	Cno	Cpa	Cpy	Cte	DDE	Die	EnS	Mal	Pal	PBZ	Pme	Pms	Qui	Tox	PCB	Tol	Xyl
96-1	11	83.6				3			0.9		3	5	4	1	1	0.7	20		29	17
96-2	7	79.8	1	15		3		4	2				0.8				54			
97-1	7	50.4				5		3	2		3	0.5	0.9				36			
98-1	8	40				3		1	1	1		3	0.5	0.5			32			
98-2	8	28.6				3		3	0.3	1		2	0.8	0.5			18			
98-3	6	29.5						0.9	0.3			0.5	0.7	0.1			2			
98-4	6	8.6				0.5				0.7		0.4	0.8				6	0.2		
99-1	8	1077		1010	49		7			2		1	0.6	0.3			7			
99-3	7	32.8	0.2					1	0.5	0.4		0.4	0.3				30			
00-1	7	22.9	0.2					0.9		0.7		0.5	0.4	0.2			20			
00-3	7	166.9				1			0.6		3	1	1	0.3			160			
01-3	8	11.6				0.9		0.9	0.4	2		0.4	0.7	0.3			6			
Totals	90	1632																		
Average						1.6		1.2	0.7	0.7		1.2	1	0.3			32.6			
Probably in product						x		x	x	x				x						
Occurs in product												x	x				x			

APPENDIX 4-33 *Continued*

Product			Peanut Butter																											
Chemical Class			P	P	P	P	P	P	P	P	P	P	P	P	P	P	P	IC	IC	IC	IC	IC	IC	IC	IC	IC	IC	IC	IC	IC
MB	Num	ppb	2,4-D	BHC	Cpy	DDE	Die	Ell	EnS	Mal	Mxy	Pal	PBZ	Pme	Pms	Qui	Tox	BZ	BBZ	Cfo	DBZ	EBZ	PCB	Pyl	Sty	TBZ	TCA	TCE	Tol	Xyl
96-1	12	172			4	3	0.9					5	4	2	1	1	20								12				97	21
96-2	9	129		1	6	4	2			2		1					49										35		28	
97-1	11	143	9		3	4	1		0.5			0.9	0.8				58	3							19				53	
98-1	12	165			3	2	1		1	4		2	0.5				22								34	t	13		82	
98-2	10	92		0.3	5	3	0.4		0.5			1	0.6				39								18		24			
98-3	10	129			2	1	2				1	0.9	1	0.2			41								23				30	
98-4	12	185			0.9		0.3		1			0.4	0.3			0.2	14	19					0.1		25	62			62	
99-1	15	220			3		0.4		1	4		2	1	0.5		0.9	13					13		15	27.5	t			69	34
99-3	19	197		0.3	2	2	0.3	0.4	2	7		1	0.4				20	2		8	7	4			33	44	3		40	16
00-1	13	168		0.5			0.6		0.8			0.4	0.4	0.2			40	4				7		3			47		36	8
00-3	14	164			0.6		0.1					1	0.9	0.3			30	25	3			4			22	5		17	49	5
01-3	14	96			0.8	1						2	1	0.7	0.4	0.2	3	3		21		2			22				36	3
Totals	151	1860																												
Average					2.5	1.7	0.8		0.6			1.5	0.9				29								19.6				48.5	
Probably in product						x			x																x					
Occurs in product					x		x					x	x				x												x	

t = trace.

APPENDIX 4-34 Vegetable Products — Very Low Chemical Contaminants

Product	Onions				Cauliflower			Cabbage			Sauerkraut					Green Peas			
Chemical Class			P	P			P			P			P	P	P			P	P
MB	Num	ppb	Die	Ipm	Num	ppb	P	Num	ppb	EnS	Num	ppb	Dan	Die	P	Num	ppb	Dat	Mxy
96-1	1	4		4	0	0		0	0		1	8	8			1	3	3	
96-2	0	0			0	0		1	0.9	0.9	0	0				1	1	1	
97-1	0	0			0	0		0	0		0	0				1	2	2	
98-1	0	0			0	0		0	0		0	0				0	0		
98-2	0	0			0	0		1	1	1	0	0				1	27		27
98-3	0	0			0	0		0	0		0	0				1	7	7	
98-4	0	0			0	0		0	0		0	0				0	0		
99-1	1	0.3	0.3		1	0.6	0.6	1	0.4	0.4	1	0.4		0.4		1	4	4	
99-3	0	0			0	0		0	0		1	1.4			1.4	0	0		
00-1	0	0			0	0		1	0.5	0.5	0	0				0	0		
00-3	0	0			0	0		0	0		0	0				0	0		
01-3	0	0			1	4	4	0	0		1	0.2		0.2		0	0		
Totals	2	4.3			2	4.6		4	2.8		4	10				6	44		

APPENDIX 4-34 *Continued*

Product	Asparagus				Okra							Eggplant						Peas (Dried)						
Chemical Class			P	P			P	P	P	P	P			P	P	P	P			P	P	P	P	IC
MB	Num	ppb	Cpy	DDE	Num	ppb	Dat	Dip	Enl	Ell	EnS	Num	ppb	Ace	Ell	EnS	Mos	Num	ppb	Ace	Dip	Lin	Mos	Tol
96-1	1	1	1		0	0						0	0					1	10					10
96-2	0	0			0	0						1	7				7	0	0					
97-1	0	0			2	0.6				0.3	0.3	1	7				7	0	0					
98-1	0	0			0	0						0	0					3	39	6	30		3	
98-2	0	0			0	0						1	60				60	0	0					
98-3	0	0			2	2.6				0.6	2	1	7				7	0	0					
98-4	2	2	1	1	1	1	1					0	0					1	48		48			
99-1	1	0.4	0.4		0	0						0	0					1	0.4			0.4		
99-3	1	1	1		0	0						0	0					0	0					
00-1	0	0			0	0						1	7	7				1	70		70			
00-3	2	4.5	4	0.5	4	21		20	0.1	0.4	0.5	1	10				10	0	0					
01-3	0	0			0	0						2	0.7		0.2	0.5		2	300		300		0.2	
Totals	7	8.9			9	25.2						8	98.7					9	468					

APPENDIX 4-35 Dark-Yellow Vegetable Products

Carrots

MB	Num	ppb	P DDE	P Diz	P EnS	P Ipm	P Ipr	P Lrn
96-1	2	29				20	9	
96-2	1	14						14
97-1	1	0.9	0.9					
98-1	3	11	0.4			9	2	
98-2	2	10				7	3	
98-3	1	0.5	0.5					
98-4	3	31				14	3	14
99-1	1	7		7				
99-3	3	1.4	0.3		0.1	1		
00-1	0	0						
00-3	0	0						
01-3	0	0						
Totals	17	105						
Average								
Probably in product								

Sweet Potato

MB	Num	ppb	P Cam	P Cpy	P Dan	P Die	P DNB	Pal Pal	IC Tol
96-1	4	591			580	0.5		0.7	10
96-2	1	145			145				
97-1	0	0							
98-1	2	0.9			0.3	0.6			
98-2	1	20			20				
98-3	2	25.7			25		0.7		
98-4	3	24	2	1	21				
99-1	0	0							
99-3	2	171	1		170				
00-1	1	30			30				
00-3	1	1					1		
01-3	2	57	7		50				
Totals	19	1066							
Average					86.8				
Probably in product					x				

Winter Squash

MB	Num	P ppb	P Cam	P Cd	P Cpy	P DDE	P Dan	P Die	P EnS	P HE	P Nac	P Oct	Tox
96-1	2	64		58		6							
96-2	2	15.3						15		0.3			
97-1	3	3.3							2	1	0.3		
98-1	4	19.4		1.4		6				1			11
98-2	2	29					8		21				
98-3	2	8	2						6				
98-4	5	31.2		7				23	0.6	0.4		0.2	
99-1	5	32.7		10	4			16	2	0.7			
99-3	1	4						4					
00-1	3	12.4		1.5				10			0.9		
00-3	2	64							4				60
01-3	2	12						2	10				
Totals	33	295											
Average									3.8				
Probably in product									x				

APPENDIX 4-36 Vegetable Products with Low Contaminant Levels

Beets

MB	Num	ppb	P DCP	P DDE	P Die	P EnS	P Ipm	Pal Pal
96-1	1	2					2	
96-2	3	1.5	0.5		0.6	0.4		
97-1	3	6		2	1	3		
98-1	1	3		3				
98-2	2	2.4	2		0.4			
98-3	0	0						
98-4	2	1.1	0.4	0.7				
99-1	2	3			1			2
99-3	2	1.7			0.9	0.8		
00-1	2	0.5		0.3				0.2
00-3	0	0						
01-3	1	0.1			0.1			
Totals	19	21.3						
Average								
Probably in product								

Mixed Vegetable

MB	Num	ppb	P Ace	P DDE	P Die	P Mos	P P	P Vin
96-1	2	28	22			6		
96-2	2	7	5			2		
97-1	0	0						
98-1	4	8.3	6	0.9	0.4	1		
98-2	1	6	6					
98-3	3	18	13			4		1
98-4	3	26	20			4		2
99-1	2	5	3			2		
99-3	1	7					7	
00-1	0	0						
00-3	1	10	10					
01-3	2	60	10			50		
Totals	21	175						
Average			7.9			5.8		
Probably in product			x			x		

Mushrooms

MB	Num	ppb	P Ben	P Diz	P P	P Thi
96-1	2	164		28		136
96-2	1	19				19
97-1	3	1473	114		2.9	1356
98-1	2	576	296			280
98-2	1	121				121
98-3	2	217	151			66
98-4	1	4			4	
99-1	4	631	562	2	1.1	66
99-3	1	360				360
00-1	1	140	140			
00-3	2	200	60			140
01-3	1	50				50
Totals	21	3955				
Average						216
Probably in product						x

APPENDIX 4-36 *Continued*

Product	Broccoli											Brussels Sprouts														
Chemical Class			P	P	P	P	P	P	P	P	P				P	P	P	P	P	P	P	P	P	P	P	P
MB	Num	ppb	Cpy	DCP	DDE	Die	Diz	Enl	Ell	EnS	P	Num	ppb	Car	Cpy	DDE	Die	Dip	Enl	EII	EnS	Ipm	Ipr	Mos	P	Pal
96-1	0	0										3	6.3		5	0.5										0.8
96-2	3	12.3	11	0.9	0.4							1	15		15											
97-1	4	6.6			0.6			2	1	3		3	30				6		1		22					
98-1	1	3									3	3	13.2		5	0.2								8		
98-2	3	13.2		11	0.5						1.7	2	128								1				127	
98-3	4	28						1	4	11	12	1	3		3											
98-4	3	19.5		4	0.5						15	1	3		3											
99-1	8	18.2		2	0.5	0.1	8	0.2	0.2	0.2	7	2	6		2										4	
99-3	1	0.1			0.1							7	37.8		2				0.8	2	1	2	10		20	
00-1	2	0.6			0.4	0.2						3	11.3		7	0.3									4	
00-3	1	0.4			0.4							3	70.3		10		0.3	60								
01-3	1	0.3			0.3							3	16.2	9							0.2				7	
Totals	31	102										32	340													
Average					0.3										4.3											
Probably in product					x										x											

APPENDIX 4-36 *Continued*

Product	Radish												Lima Beans													
Chemical Class			P	P	P	P	P	P	P	P	P	Pal	P													
MB	Num	ppb	Cd	Cpy	Cyf	DCP	DDE	DDT	Die	Edr	EnS	Pal	TDE	Num	ppb	Ace	DCP	DDE	Dat	Dic	Die	Ipm	Ipr	Lin	Mos	Ome
96-1	5	20.1				9	4	2.5	4	0.6				2	20	14									6	
96-2	1	0.5					0.5							2	12	5									7	
97-1	4	5.7				4	1		0.4		0.3			6	36	18			2			0.5	9		6	0.5
98-1	3	19.9				17	2	0.9						4	44.3	26	0.3		12						8	
98-2	1	17				17								3	20	8			2	10						
98-3	3	1.6	0.8				0.3	0.5						2	3				2							1
98-4	6	22.2		2	13		5	1			0.7		0.5	3	1.1	t								0.2	0.9	
99-1	3	5.4		2			3		0.4					5	43	24			4		4				10	1
99-3	2	2.2					2		0.2					0	0											
00-1	4	30		20					1		4	5		4	51	40		0.8						0.2	10	
00-3	1	0.4					0.4							0	0											
01-3	0	0												3	230	200		0.4							30	
Totals	33	125												34	460											
Average							1.5									28									6.5	
Probably in product							x									x									x	

t = trace.

APPENDIX 4-36 *Continued*

Product	Avocado										Turnip										
Chemical Class			IC	IC	IC	IC	IC	IC	IC	IC			P	P	P	P	P	P	P	P	P
MB	Num	ppb	BZ	Cfo	DBZ	PCE	TCE	Sty	Tol	Xyl	Num	ppb	Cpy	DCP	DDE	DDT	Dat	Die	Ell	EnS	TDE
96-1	4	450	26	14			300	110			5	148	126	14	4			4			0.4
96-2	3	44		15			29	t			3	6		4	1					1	
97-1	1	34	34								5	47.6	46		0.4			0.3	0.5	0.4	
98-1	1	25				25					3	63	59		2			2			
98-2	3	181	10	13				158			0	0									
98-3	1	t						t			1	10		10							
98-4	2	30	30					t			5	17.3	14		0.8	0.9	1	0.8			
99-1	0	0									3	23.8	22		1			0.8			
99-3	4	28	14	5		3			3		1	3	3								
00-1	2	17	4			13					2	14	7	7							
00-3	4	29	7	10	10		2				2	1.3		0.9	0.4						
01-3	4	50	23	20					5	2	0	0									
Totals	29	154									30	334									
Average			12.3										23.1		0.8						
Probably in product			x										x		x						

t = trace.

APPENDIX 4-37 Tomato Products

Product	Tomato Juice							Stewed Tomato										
Chemical Class			P	P	P	P	P				P	P	P	P	P	P	P	P
MB	Num	ppb	Car	Enl	Ell	EnS	Mos	Num	ppb	Ace	Car	Cno	Cpa	Enl	EnS	Mos	P	BZ
96-1	1	4					4	1	10									10
96-2	1	4					4	0	0									
97-1	0	0						0	0									
98-1	2	6.9	6				0.9	2	6			5	1					
98-2	0	0						0	0									
98-3	0	0						0	0									
98-4	0	0						2	8			2	6					
99-1	0	0						2	6				2				4	
99-3	0	0						2	22			20	2					
00-1	3	1.1		0.2	0.4	0.5		0	0									
00-3	3	1.1		0.2	0.5	0.4		0	0									
01-3	3	6.7	6		0.5	0.2		7	62.8	6	1	40	8	0.2	0.6	7		
Totals	13	23.8						16	115									

APPENDIX 4-37 *Continued*

Tomato Sauce

Product / Chemical Class / MB	Num	ppb	P Ace	P Cin	P Cno	P Cpa	P Cpy	P Cte	P DDE	P Enl	P EII	P EnS	P Lac	P Mos	P P	IC Tol
96-1	5	53	4		6			16						1		26
96-2	0	0														
97-1	1	72												72		
98-1	2	8.8		8		0.8										
98-2	0	0														
98-3	2	45			42									3		
98-4	6	5.7					0.3		0.6	0.7	0.8	0.3	3			
99-1	0	0														
99-3	2	2.1							0.2						1.9	
00-1	2	10.7			10										0.7	
00-3	3	5.8								0.8	1			4		
01-3	0	0														
Totals	23	203														
Average																
Probably in product																

Tomato Catsup

Product / Chemical Class / MB	Num	ppb	P Ace	P Cam	P Cau	P Cd	P Cno	P Cpa	P Cpy	P Cte	P DDE	P Efv	P EII	P EnS	P Mos	P P
96-1	3	7	1									2			4	
96-2	2	3	1												2	
97-1	0	0														
98-1	4	96			1		73	10		12						
98-2	3	17					9	7	1							
98-3	3	24					17	2							2	
98-4	3	1.8				0.2			1		0.6					
99-1	1	1.6														1.6
99-3	0	0														
00-1	2	8					6									2
00-3	2	117		6			111									
01-3	3	2.8									0.3		2	0.5		
Totals	26	278														
Average																
Probably in product																

APPENDIX 4-37 *Continued*

Tomato

Product / Chemical Class / MB	Num	ppb	P Cam	P Cpy	P Dan	P Dic	P Efv	P Enl	P EII	P EnS	P Fen	P Mos	P Ome	P P	P Tpp	IC BZ	IC Cfo	IC EBZ	IC Sty	IC Tol	IC Xyl
96-1	8	30.6				1		0.6	1		4	5	5	4			10				
96-2	5	131		1					0.6	0.9		101		27							
97-1	0	0																			
98-1	5	63.7		20				0.7	2	1		40									
98-2	5	46		2				3	12	6				23							
98-3	7	108			4			2	11	41		24	2	24							
98-4	5	2.2		1	0.4			0.2	0.3	0.3											
99-1	7	27.4		0.4			5	4	6	4		3		5							
99-3	3	44.5				31										10				4	
00-1	7	252		20					10	10		30		20	150	3					
00-3	4	500				300								130		67	3				
01-3	5	22	6															2	7	2	5
Totals	61	1227																			
Average				3.7					3.6	5.3				19							
Probably in product				x					x	x				x							

APPENDIX 4-38 Vegetable Products with Moderate Contaminant Levels

Product	Green Beans																		
Chemical Class			P	P	P	P	P	P	P	P	P	P	P	P	P	P	P	P	P
MB	Num	ppb	Ace	Bif	Car	DCP	DDE	Dan	Dat	Enl	EII	EnS	Eth	Ipm	Ipr	Mos	Neb	Ome	Vin
96-1	0	0																	
96-2	3	18	10													5			3
97-1	3	463	357													96			10
98-1	6	181	92			2	1	34								48			4
98-2	0	0																	
98-3	4	75	40						15							15		5	
98-4	3	68	51													12			5
99-1	2	2.5					0.5												2
99-3	3	40.3	30									0.3				10			
00-1	2	100	80													20			
00-3	9	189		10	2			160		0.2	0.3	3	4	4					5
01-3	9	196	3		10					2	4	4	3	10	10		150		
Totals	44	1332																	
Average			55													17.2			
Probably in product			x													x			

APPENDIX 4-38 *Continued*

Product	Iceberg Lettuce															
Chemical Class			P	P	P	P	P	P	P	P	P	P	P	P	P	P	P
MB	Num	ppb	Ace	Cin	DCP	DDE	Dan	Dat	Die	Enl	EII	EnS	Met	Mos	Ome	P	Vin
96-1	4	7.3				2	0.5		0.7							4	
96-2	5	22.4	3							1	0.4	8	10				
97-1	5	131		13							0.9	0.6	46			70	
98-1	5	19	2					6		3	3	6					
98-2	8	35	2	24	0.9			1		0.3		0.8		2			4
98-3	5	30.4				0.4				1	1	10				18	
98-4	3	16.1	14			0.3										1.8	
99-1	5	40						6		6	3	11			14		
99-3	2	10	8											2			
00-1	0	0															
00-3	2	34	30											4			
01-3	3	1.7								0.4	0.3	1					
Totals	47	347															
Average												3.1					
Probably in product												x					

APPENDIX 4-39 Vegetable Products Associated with Oils

Product	Black Olives													Olive/Safflower Oil																		
Chemical Class			P	P	P	P	P	P	P	P	P	IC	IC				P	P	P	P	P	P	P	P	IC	IC	IC	IC	IC	IC	IC	IC
MB	Num	ppb	Cno	Cpa	Cpy	DCP	DDE	Dic	Enl	EII	EnS	Cfo	Tol	Num	ppb	Cpy	DDE	Diz	Enl	EII	EnS	Lin	Mat	Sim	BZ	Cfo	EBZ	PCE	Sty	TBZ	Tol	Xyl
96-1	6	69.4			0.6			1	0.8		3	11	53	1	11										11							
96-2	4	4.8			2		0.4		0.4		2			7	118		1				2		2				23		42		28	20
97-1	6	7.5			2	1	0.2	2	0.3		2			0	0																	
98-1	3	3.8			3		0.2				0.6			1	11																11	
98-2	5	6.7			1	0.7		4	0.2		0.8			4	60.5		2				10	0.5									48	
98-3	5	7.9			3	0.7		3	0.2		1			1	0.9						0.9											
98-4	4	5.1			2	0.4		2			0.7			9	101	1	2		2	2	6	2			12				32		18	
99-1	5	2.1			0.9	0.2	0.2	0.6			0.2			3	6		0.5			0.5	5											
99-3	2	2.9			2						0.9			10	116			4			10	0.6			12		4	7	19	8	25	26
00-1	3	1.2			0.5		0.4				0.3			10	124		0.4				3	0.3			10	2	5		44	13	18	28
00-3	2	0.9					0.2				0.7			3	15										8				5			2
01-3	3	24	20	2		2								10	55.7				0.3	0.4	4			20	4	2	2		6		4	13
Totals	48	136												59	618																	
Average					1.4						1										3.4										12.7	
Probably in product					x																x										x	
Occurs in product											x																					

APPENDIX 4-39 *Continued*

Product	Margarine																
Chemical Class			P	P	P	IC	IC	IC	IC	IC	IC	IC	IC	IC	IC	IC	IC
MB	Num	ppb	Cum	Dip	Tpp	BZ	BDM	BRB	Cfo	DBZ	EBZ	PCE	Pyl	Sty	TBZ	Tol	Xyl
96-1	2	32														11	21
96-2	4	2,634		2,540	65					11						18	
97-1	1	25														25	
98-1	1	15														15	
98-2	2	95													23	72	
98-3	7	1,444		1,190	84					25				11	84	28	22
98-4	4	2,304		1,960	250											62	32
99-1	2	46				30										16	
99-3	9	260	5							13	7	13	8	13	60	106	35
00-1	11	3,862	2	3,660	120	10	7	4			3	4		5		25	14
00-3	12	1,946	4	1,730	120	7			7	3	5	10		3	4	33	18
01-3	9	6,090	3	5,760	240				5		5	2		8		50	17
Totals	64	14,508															
Average																38.4	13
Probably in product																	x
Occurs in product																x	

APPENDIX 4-40 Vegetable Products with High Contaminant Levels

Product	Cucumbers																	
Chemical Class			P	P	P	P	P	P	P	P	P	P	P	P	P	P	P	P
MB	Num	ppb	BHC	Cd	Dan	Die	Diz	Edr	Enl	EII	EnS	HE	Met	Mos	Nac	Oct	Tox	Vin
96-1	4	59.3							0.9	0.4	2						56	
96-2	6	69			2	9			12	6	2							20
97-1	4	42							2	1	17			22				
98-1	4	18							3	1	7		7					
98-2	4	101							15	5	36			45				
98-3	4	38		26					2	1	9							
98-4	10	43.8	0.4	0.8		24	2		1	0.4	6	8			0.5	0.7		
99-1	8	35.9				3		1	5	2	6	0.9		3			15	
99-3	8	5.4	0.9	0.6		0.3			0.2	0.7	2	0.3			0.4			
00-1	4	60							10	10	30			10				
00-3	5	16.9		0.9					2	2	10				2			
01-3	3	15							2	3	10							
Totals	64	504.3																
Average									4.6	2.7	11.4							
Probably in product									x	x	x							
Occurs in product																		

APPENDIX 4-40 *Continued*

Product								Summer Squash									
Chemical Class		P	P	P	P	P	P	P	P	P	P	P	P	P	P	P	
MB	Num	ppb	Cd	DDE	DDT	Die	Edr	Enl	EII	EnS	HCB	HE	Nac	Oct	Qui	Pal	Tox
96-1	5	36.7		1		0.7		6	2	27							
96-2	6	87				8		13	5	26						4	31
97-1	3	29				6		1		22							
98-1	8	51.3		2				10	2	2	0.2				0.1	3	14
98-2	5	60.8				23		8		13						0.8	16
98-3	3	33				5				14							14
98-4	7	57.2	15			10		8	1	22		0.9		0.3			
99-1	8	65.6		8	3	8	2	6	0.6	17							21
99-3	7	5.6	1.4			0.8		0.5	0.4	1		0.6	0.9				
00-1	3	30.4						10		20						0.4	
00-3	5	83						9	3	20		1					50
01-3	5	23				2		2	2	10							7
Totals	65	563															
Average						7.1		6.1	1.3	16.2							13
Probably in product						x			x								x
Occurs in product								x		x							

APPENDIX 4-40 *Continued*

Product	Celery																						
Chemical Class			P	P	P	P	P	P	P	P	P	P	P	P	P	P	P	P	IC	IC	CI	IC	IC
MB	Num	ppb	Ace	Cam	Cil	DDE	DDT	Dan	Dip	Diz	Enl	EII	EnS	Mal	Met	Mos	Oxa	P	DBZ	DNB	PBZ	Tol	Xyl
96-1	7	218						10		7							106	52			15	14	14
96-2	1	15			15																		
97-1	6	156				4	2.4	137				1	1					11					
98-1	8	67	22			0.6			14		1	2	1		14			12					
98-2	5	325	165			0.7		135								22		2					
98-3	6	113	57			0.9		43								2		8		2			
98-4	8	48	12			3	1.5	9			4	5	6					7					
99-1	5	119	83		2	0.4										5		29					
99-3	6	106	20	5		0.4		80						2				1.7					
00-1	4	200	160					10								10		20					
00-3	4	150	50					90										13	2				
01-3	5	143	100		5			8								10		20					
Totals	65	1660																					
Average			55.8			0.8		43.5										14.6					
Probably in product		x				x		x															
Occurs in product																		x					

APPENDIX 4-40 *Continued*

Product			Green Peppers																				
Chemical Class			P	P	P	P	P	P	P	P	P	P	P	P	P	P	P	P	P	P	P	P	P
MB	Num	ppb	Ace	Bif	Car	Cam	Cpy	Cyf	DCP	DDE	DDT	Dat	Dic	Efv	Enl	EII	EnS	Fen	Lac	Met	Mos	Ome	P
96-1	10	68.9	8							0.9		5		9	13	8	4				2	19	64
96-2	9	240	26		84							8			23	45	29			20	10	4	
97-1	6	321	194						6	0.6			58								39		23
98-1	6	96	54												4	4	2			9	23		
98-2	6	447	296										83		5	12	11				70		
98-3	8	168				3	121					12		11	0.4	0.7	1					19	
98-4	4	41.4	13				2			0.4											26		
99-1	3	23	16								1										6		
99-3	3	90	50																		20		20
00-1	4	90	50																10		20		10
00-3	9	177			8		40	2							4	10	10	3		90	10		
01-3	6	76	30	2		4		2												30	8		
Totals	74	1838																					
Average			61.4																		19.5		
Probably in product			x																				
Occurs in product																					x		

APPENDIX 4-40 *Continued*

Product	Dill Pickles																		
Chemical Class			P	P	P	P	P	P	P	P	P	P	P	P	P	P	P	P	P
MB	Num	ppb	BHC	Car	Cd	DDE	DDT	Die	Enl	EII	EnS	Eth	HE	Lin	Nac	P	Tox	Tpp	Tol
96-1	6	65						13	5	3	13						18		13
96-2	4	69						16			1			3			49		
97-1	5	83						18	22	16	24						3		
98-1	6	37		8				6	4	2	16		2						
98-2	4	9.8						2	3	0.6	4								
98-3	9	96			10	6	4	3	4	10	25						24	10	
98-4	7	46.3	0.8		8.5	5				3	12			1			16		
99-1	6	35			15			3	2	1	8	0.6							
99-3	6	28.8						6	3	2	9			0.8			8		
00-1	7	35.1			0.5			4	10	10	10			0.4	0.2				
00-3	5	41.2		10					10	10	10					1.2			
01-3	4	32						8	3	1	20								
Totals	69	578.2																	
Average								6.6	5.5	4.9	12.7								
Probably in product								x	x										
Occurs in product										x	x								

APPENDIX 4-40 *Continued*

Product			Sweet Pickles															
Chemical Class			P	P	P	P	P	P	P	P	P	P	P	P	P	P	P	IC
MB	Num	ppb	BHC	Cd	Cin	Cpy	DDE	DDT	Die	Enl	EII	EnS	HE	Lin	Nac	Oct	Tox	BZ
96-1	8	29		0.2	0.3	2			8	2	1	4						11
96-2	5	39							3	0.8	0.6	2					34	
97-1	5	8							2	1	1	2		2				
98-1	5	15				3			1	3	2	6						
98-2	5	70							1	18	12	29					10	
98-3	10	14	0.9	1.2		1	0.9		0.4	0.6	1	7		0.4	0.3			
98-4	3	3.2							0.4		0.8	2						
99-1	5	14		3					2	2	2	5						
99-3	7	11		0.4					3	1	0.9	5			0.3	0.1		
00-1	3	4.3								0.3	1	3						
00-3	9	14	1	1.1			0.2	0.6				10	0.4	0.4	0.6	0.2		
01-3	8	27		0.4					4	2	2	10		0.5	0.6		7	
Totals	73	247																
Average									2.1	2.8	2	7.1						
Probably in product									x									
Occurs in product										x	x	x						

APPENDIX 4-41 Potato Products

Product	Potato Chips																						
Chemical Class MB	Num	ppb	P Cam	P DDE	P DDT	P Die	P EnS	P HE	P Icm	P Ipr	P Tri	P Tos	IC BZ	IC BBZ	IC Cfo	IC DBZ	IC EBZ	IC PCE	IC Sty	IC TBZ	IC TCA	IC TCE	IC Tol	IC Xyl
96-1	2	23					3																20	
96-2	5	2,290	2,245	0.5					11	1										23				
97-1	3	44	26								2												16	
98-1	6	1,952	19									6				10	9						1,870	38
98-2	4	485	450				2												16				17	
98-3	4	1,552	1,490	1	0.6																		60	
98-4	5	58	0.4			1	0.4	0.2															65	
99-1	4	4,514	0														19			16			4,440	38
99-3	9	63	0								4		2				2	7	6	4	7		24	7
00-1	12	74	0			0.5	1						2	4		6	4	3	6	12	8	3		24
00-3	10	332	290				2				2				12				2	4	3	6	11	2
01-3	10	244	70				0.6		6				9		8	6	26		86				14	18
Totals	74	11,631																						
Average			383																				545	
Probably in product			x																				x	
Occurs in product																								

APPENDIX 4-41 *Continued*

Product Chemical Class MB	Num	ppb	P Cam	P Cpy	P Cum	P DDE	P DDT	P Dan	P Die	P Diz	P EnS	P HE	P Icm	P Mal	IC BZ	IC BBZ	IC BDM	IC CBZ	IC Cfo	IC DBZ	IC EBZ	IC PBZ	IC PCE	IC Pyl	IC Sty	IC TBZ	IC TCE	IC Tol	IC Xyl
			French Fries (FF)																										
96-1	9	201.4		2		2	0.4			3	2										11				94			59	28
96-2	2	4.7		0.7							4																		
97-1	4	27	2								2														11			12	
98-1	3	63.2				0.2					3																	60	
98-2	6	77.9	37						0.1		0.7	0.2													18			22	
98-3	6	580	445								1							13		11					68			42	
98-4	6	598.7	450						0.7		13		22												27			95	
99-1	7	246	2					3			2														24	23		168	24
99-3	12	432	290		2						5				58	3		4			4		8		10	10		27	11
00-1	9	141			24						5			2							6			10	49		2	24	19
00-3	10	693.7	530						0.7						6		3				5	3			14	33		69	30
01-3	11	481	320			0.1			0.4		6		5						2	12	8				39			17	18
Totals	85	3547																											
Average			173								3.6														30			49.6	10.8
Probably in product			x																						x				x
Occurs in product											x																	x	

FF = Fast Food

APPENDIX 4-41 *Continued*

Product			French Fries (F)														
Chemical Class MB	Num	ppb	P Cam	P Cno	P Cpa	P DDE	P Dan	P Die	P Dip	P EnS	P Eth	P HE	P Icm	P Ipr	P Mal	IC RBZ	IC PBZ
96-1	3	659	658					0.4		0.6							
96-2	3	77	35							7				35			
97-1	4	430.3	427			0.3		1		2							
98-1	4	339	322							1			10		6		
98-2	4	111	2	96	7						6						
98-3	2	284	281							3							
98-4	3	82.6	81				0.9			0.7							
99-1	5	581.9	548							0.8	12	0.1	18				
99-3	6	636.6	590			0.2			30	3	13					0.4	
00-1	4	389.8	370							0.8	9		10				
00-3	4	107.1	90							8	9						0.1
01-3	4	43	30							2	5		6				
Totals	46	3741															
Average			286							2.4							
Probably in product			x							x							
Occurs in product																	

F = Frozen

APPENDIX 4-41 *Continued*

Product Chemical Class MB	Num	ppb	P Cam	P Cpy	P DDE	P DDT	P EnI	P EII	P EnS	P Eth	P Icm	P Ipr	P Ipm	P Nac	P Mos	Pal Pal	P Thi
			White Potato (with Skin)														
96-1	6	3856	3850		1	0.6			0.4		3		0.8				
96-2	7	1090	580		2	2.3			5	5		6					490
97-1	4	154.5	31		0.7				0.8								122
98-1	5	179.3	45				2	0.1	2								132
98-2	7	484.1	253		1	0.7			0.4	8		2					219
98-3	5	333	173						1		1				20		138
98-4	7	68.2	41		6	1.2	1		2	7				10			
99-1	4	96.5	10		0.3											0.2	86
99-3	4	32.5	20			0.3			0.2	12							
00-1	3	95	20						5								70
00-3	6	184.2	50	9					0.2	25	10						90
01-3	6	378	290	2	2	0.8			3		80						
Totals	64	6951															
Average			447		1.1				1.7								112
Probably in product			x		x												x
Occurs in product									x								

APPENDIX 4-41 *Continued*

Product	Mashed Potato									White Potato (No Skin)										
Chemical Class MB	Num	ppb	P Cam	P DDE	P Dip	P EnS	P Eth	P Icm	P Tpp	Num	ppb	P Cam	P DDE	P Ell	P EnS	P Eth	P Icm	P Mos	P Nac	P Thi
96-1	2	33	31	2						3	705	704			0.2		0.8			
96-2	2	117	33		84					5	217.5	94	0.5		2		3			118
97-1	3	164.9	114	0.9	50					1	0.9				0.9					
98-1	2	226	202					24		2	4	2			2					
98-2	2	4.9	4			0.9				1	48	48								
98-3	3	243	106		135	2				3	49	24			1			24		
98-4	5	285.7	51		221	0.7		1	12	5	33.1	32	0.3	0.4	0.2				0.2	
99-1	2	51.8	51			0.8				1	2	2								
99-3	2	22	20			2				2	14	2				12				
00-1	3	27.4	20			0.4		7		3	13	4			1	8				
00-3	3	175.9	5		170	0.9				3	67	50				13	4			
01-3	5	354	20	0.1	330	0.6	3			3	73.9	70			0.9		3			
Totals	34	1706								32	1227									
Average			54.8			0.7						86			0.7					
Probably in product						x									x					
Occurs in product			x									x								

APPENDIX 4-42 Dark-Green Leafy Vegetables

Product Chemical Class MB	Collard Greens																						
	Num	ppb	P Cin	P Cd	P Cpy	P DCP	P DDE	P DDT	P Dat	P Die	P Efv	P Enl	P EII	P EnS	P Eth	P Fen	P Nac	P Ome	P P	P Pal	P Qui	P TDE	P Tox
96-1	8	63.6			7	18	29	2		0.7				0.9					4			2	
96-2	5	183				27	2									26			112				16
97-1	4	950.5				64	5							0.5					881				
98-1	8	670	610		4	8	23	2											8			7	8
98-2	5	1,767	1,247			2	5							2					411				
98-3	5	346	98			34	1									24			189				
98-4	11	2,073	52	0.4	19	152	5	3.7	4	2				0.6				7	1,827				
99-1	14	2,094	578	0.9	7	1	2	0.8		0.3		0.2		0.4		44	0.4		1,456	0.3	0.2		
99-3	6	1,212	40			0.8	0.3			1	50								1,120				
00-1	3	50.4	10				0.4												40				
00-3	3	1,866													26	130			1,710				
01-3	9	1,867				0.4	0.3			2		0.2	0.1	3		10			1,850				0.5
Totals	81	13,143																					
Average			220			25.6	6.1												801				
Probably in product			x			x																	
Occurs in product							x												x				

APPENDIX 4-42 *Continued*

Product	Spinach																							
Chemical Class MB	Num	ppb	P Ace	P Cd	P Cpy	P DCP	P DDE	P DDT	P Dan	P Dat	P Die	P Diz	P Enl	P EII	P EnS	P Eth	P Lin	P Nac	P Ome	P P	P Pal	P TDE	P Tox	IC Cfo
96-1	8	110	6		0.9		26	2.5							0.6					64		0.5		10
96-2	4	1,748			1		11	1												1,771				
97-1	13	350			1	6	37	8	0.3	1	4	1		1	9	15			5	252				
98-1	10	1,591		5	4	7	20	10							0.7			2		1,531		4	8	
98-2	5	27				2	12	2							6					5				
98-3	7	2,107					0.9			4			1	2	25				0.7	2,073				
98-4	7	1,884		1.8			8	4												1,861	1	4	4	
99-1	9	1,229			2	1	10	3			2				1					1,204		3	3	
99-3	5	67					5	0.8			0.9									60		0.5		
00-1	6	2,640					30	10			10								10	2,450			130	
00-3	6	2,406			0.8		10	1.5			1								3	2,390				
01-3	7	23					9	0.7			0.3			0.1	10		0.6			2.5				
Totals	87	14,182																						
Average					0.8		14.9	3.6			1.5				4.4					1,139				
Probably in product					x						x				x									
Occurs in product							x	x												x				

APPENDIX 4-43 Mixed Food Products with Very Low Contaminant Levels

Product	Bean and Bacon Soup			Tomato Soup				Spaghetti and Tomato					Chocolate Pudding				Fried Chicken (HM)					
Chemical		P			P	P			P	P	IC			P	IC			P	P	IC	IC	
Class MB	Num	ppb	Die	Num	ppb	Ell	EnS	Num	ppb	Cpm	Ell	Xyl	Num	ppb	DDE	Tol	Num	ppb	DDE	Lin	Tol	Xyl
96-1	0	0		0	0			1	12			12	1	12		12	3	22.5	0.5		12	10
96-2	0	0		0	0			0	0				0	0			0	0				
97-1	0	0		0	0			0	0				1	4	4		1	0.5	0.5			
98-1	0	0		0	0			0	0				1	0.7	0.7		0	0				
98-2	0	0		0	0			2	1.2	0.8	0.4		0	0			0	0				
98-3	0	0		0	0			0	0				0	0			0	0				
98-4	0	0		1	0.2		0.2	0	0				0	0			0	0				
99-1	1	0.1	0.1	0	0			0	0				1	0.3	0.3		2	0.7	0.5	0.2		
99-3	0	0		0	0			0	0				0	0			0	0				
00-1	0	0		0	0			0	0				0	0			0	0				
00-3	0	0		1	0.5	0.5		0	0				0	0			0	0				
01-3	0	0		0	0			1	0.4		0.4		0	0			0	0				
Totals	1	0.1		2	0.7			4	13.6				4	17.0			6	23.7				

HM = Home Made

APPENDIX 4-43 *Continued*

Product	Mushroom Soup							Suckers							Pork and Beans						Vegetable Beef Soup					
Chemical Class MB	Num	ppb	P Cpm	P Cpy	P DDE	P Diz	P Thi	Num	ppb	P Cpy	P Etn	P Dip	P Mat	IC Tol	Num	ppb	P Cpa	P Cno	P Cpy	IC Xyl	Num	ppb	P Cam	P DDE	P Tpp	P Vin
96-1	0	0						1	33					33	1	13				13	0	0				
96-2	0	0						1	0.8		0.8				0	0					1	13	13			
97-1	1	2			2			2	23			22	1		0	0					0	0				
98-1	2	3		2		1		1	0.4		0.4				0	0					1	38			38	
98-2	0	0						0	0						0	0					1	8	8			
98-3	1	4	4					0	0						0	0					1	15	15			
98-4	1	18					18	0	0						2	50	t	50			3	17.7	17	0.2		0.5
99-1	0	0						0	0						1	23		23			0	0				
99-3	0	0						0	0						0	0					1	5	5			
00-1	1	20					20	0	0						1	10		10			1	10	10			
00-3	0	0						1	1	1					0	0					1	8	8			
01-3	0	0						0	0						1	0.2			0.2		1	20	20			
Totals	6	47						6	58.2						6	96.2					11	135				
Average																							8			
Probably in product																							x			

t = trace.

APPENDIX 4-44 Mixed Food Products with Low Contaminant Levels

Product	Brown Gravy (HM)									Spaghetti and Meatballs											Mayonnaise					
Chemical			P	P	P	P	P	P	IC			P	P	P	P	P	P	IC	IC	IC			P	P	P	P
Class MB	Num	ppb	BHC	Cpm	Cpy	DDE	Die	Mal	Tol	Num	ppb	Cpm	Cpy	DDE	Ell	Lin	Mal	BZ	Tol	Xyl	Num	ppb	Die	Dip	EnS	Mxy
96-1	2	29		16					13	4	48.6			0.6				22	14	12	0	0				
96-2	0	0								0	0										2	361	1	360		
97-1	0	0								1	1			1							1	1	1			
98-1	1	0.3			0.3					0	0										1	0.3	0.3			
98-2	2	0.9		0.6		0.3				1	0.7		0.7								0	0				
98-3	0	0								0	0										2	0.8	0.5		0.3	
98-4	2	2.7		0.7		2				0	0										0	0				
99-1	1	0.1	0.1							5	1.1	0.1	0.3	0.1		0.1	0.5				3	2.2	0.8		0.4	1
99-3	2	3		1				2	1	0.3				0.3							3	2.1	0.6		0.5	1
00-1	0	0								1	1.3			0.3							1	0.2	0.2			
00-3	2	0.8				0.5	0.3			0	0										1	0.3	0.3			
01-3	1	0.2		0.2						3	0.8	0.4		0.2	0.2						2	0.9	0.6		0.3	
Totals	13	37								16	53.8										16	369				
Average																							0.4			
Probably in product																							x			

HM = Home Made

APPENDIX 4-44 *Continued*

Product	Yellow Cake						Lasagna and Meat (HM)												Beef Stew								
Chemical Class MB	Num	ppb	P Cpm	P Die	P Mal	IC Xyl	Num	ppb	P Cno	P Cpa	P Cpm	P Cpy	P DDE	P Die	P Ell	P EnS	IC BZ	IC Xyl	Num	ppb	P Cam	P Cpm	P DDE	P Die	P EnS	P Icm	P Ipr
96-1	3	24	3		1	25	3	31					1				19	11	2	65		51					14
96-2	1	2			2		1	1							1				1	51		51					
97-1	1	1		1			1	7					7						1	0.4					0.4		
98-1	2	7	3		4		1	0.8					0.8						2	0.4			0.3		0.1		
98-2	2	12	4		8		0	0											1	3	3						
98-3	2	6	2		4		2	1.9			1		0.9						1	0.9		0.9					
98-4	2	7	3		4		1	0.2					0.2						2	0.4			0.2		0.2		
99-1	1	1	1				1	0.5			0.5								1	1	1						
99-3	2	3	1		2		1	0.5					0.5						1	0.3			0.3				
00-1	2	3	1		2		3	4.5	2	2			0.5						2	2.1				0.1			2
00-3	2	5	3		2		2	1.4			0.9		0.5						4	80.6	70		0.6	0.3	0.7		
01-3	0	0					5	1.9			0.4	0.6	0.6	0.1		0.2			3	44.5	40				0.5	4	
Totals	20	71					21	50.7											21	250.0							
Average			1.8		2.4								1														
Probably in product			x		x								x														

HM = Home Made

APPENDIX 4-44 *Continued*

Product Chemical Class MB	Italian Salad Dressing (Low Calorie)									White Sauce (HM)							
	Num	ppb	P Cau	P Cpa	P Cmy	P Cno	P Cum	P DCP	P Etn	Num	ppb	P Cpm	P DDE	P Die	P Dip	P Mal	P Tpp
96-1	1	9				9				1	0.6		0.6				
96-2	0	0								1	97				97		
97-1	1	1						1		2	110		2		108		
98-1	2	3.2				3	0.2			1	0.5		0.5				
98-2	0	0								1	0.7	0.7					
98-3	0	0								3	62.4	0.4			60	2	
98-4	3	6		1	2	3				3	675	0.3		0.1	675		
99-1	3	22		1	1	20				2	0.6	0.4	0.2				
99-3	3	80.2		10		70			0.2	2	1.8	0.8				1	
00-1	1	7				7				2	391				390	0.9	
00-3	4	62		9	2	50			1	1	220				220		
01-3	4	36.9	0.9	10	6	20				3	2000	0.4			1950		50
Totals	22	227.3								22	3560						
Average						15									292		
Probably in product						x									x		

HM = Home Made

APPENDIX 4-44 *Continued*

Product Chemical Class MB	Beef Chow Mein (FF)													
	Num	ppb	P Cpm	P Cpy	P Cyf	P DDE	P Dan	P Diz	P Ell	P EnS	P P	P Mal	IC Sty	IC Tol
96-1	2	59											41	17
96-2	0	0												
97-1	3	8			4		1				2			
98-1	4	17.6	0.9	0.7			7	9						
98-2	1	3					3							
98-3	3	7.5							0.5	2	5			
98-4	1	0.4					0.4							
99-1	4	3.5	0.6	1		1						0.9		
99-3	2	2.8		2						0.8				
00-1	0	0												
00-3	2	10.4	0.4				10							
01-3	2	1.8								0.3	1.5			
Totals	24	114												
Average														
Probably in product														

FF = Fast Food

APPENDIX 4-44 *Continued*

Product	Macaroni and Cheese								Turkey Dinner (Frozen)												
Chemical Class MB	Num	ppb	P Cpm	P Cpy	P DDE	P Dip	P Mal	IC Xyl	Num	ppb	P Cam	P Cpm	P Cpy	P DDE	P Die	P Enl	P EnS	P Mal	P Vin	IC Cfo	IC Tol
96-1	5	22.4	0.7	0.2	2		0.7	17	3	50	34									14	12
96-2	2	147.5			0.5	147			0	0											
97-1	3	4.6	3		1		0.6		0	0											
98-1	1	0.9	0.9						1	33	33										
98-2	1	1	1						2	35	33								2		
98-3	1	10	10						1	9	9										
98-4	2	1.6	0.6				1		2	25.3	25		0.3								
99-1	3	1.8	0.5	0.6	0.7				4	58.4	56			0.2			0.2	2			
99-3	2	6		5			1		4	8.6	8			0.1		2	0.1				
00-1	2	1.7	1				0.7		5	40.9	40	0.2	0.3		0.1		0.3				
00-3	3	8.2	4		0.2		4		3	3	2						0.4		0.6		
01-3	1	3	3						4	12.2	10	0.7					0.6	0.9			
Totals	26	208.7							29	275											
Average			2.1								20.8										
Probably in product			x								x										

APPENDIX 4-44 *Continued*

Product Chemical Class MB	Beef Stroganoff (HM) Num	ppb	P Cam	P Cmy	P Cpm	P DDE	P DDT	P Die	P Dip	P Diz	P EnS	P HE	P Ipr	P Mal
96-1	3	8.9				2	0.9			6				
96-2	1	0.9			0.9									
97-1	3	4.9				0.9				1			3	
98-1	1	0.2				0.2								
98-2	2	1.8		1										0.8
98-3	1	2			2									
98-4	3	1			0.7			0.3				1		
99-1	3	3			1	1								1
99-3	3	5.8			0.8								3	2
00-1	3	5.8			2	0.8								3
00-3	3	22			1				10					1
01-3	5	15.4	10		3	0.2					0.2			2
Totals	31	71.7												
Average					1									
Probably in product					x									

Product Chemical Class MB	Scalloped Potatoes (HM) Num	ppb	P Cam	P DDE	P Dan	P Die	P Dip	P EnS	P Icm	P Tpp	IC Tol
96-1	3	79.6	67	0.5							12
96-2	2	105	103					2			
97-1	2	23				1	22				
98-1	3	5.7	5	0.1				0.6			
98-2	3	24	3		20			1			
98-3	1	51	51								
98-4	3	192.4					183	0.4		9	
99-1	2	3.1	3	0.1							
99-3	1	2	2								
00-1	3	82.2	2				80	0.2			
00-3	3	180	70				100		10		
01-3	5	471.2	150	0.2			310	1	10		
Totals	31	1219									
Average			38								
Probably in product			x								

HM = Home Made

APPENDIX 4-44 *Continued*

Product Chemical Class MB	Caramel Candy													
	Num	ppb	P DDE	P Dip	P Tpp	P BZ	P CTC	P Cfo	P EBZ	PCE PCE	P Sty	P TCE	IC Tol	IC Xyl
96-1	2	3,418		3,360	58									
96-2	2	1,565		1,520	45									
97-1	2	1,845		1,820	25									
98-1	3	2,544	0.4	2,440	104									
98-2	3	179.3		4.3	161					14				
98-3	2	1,179		1,140	39									
98-4	2	1,880		1,840	40									
99-1	0	0												
99-3	3	1,486		1,470			9					7		
00-1	3	10					4			4				2
00-3	4	34.6		0.6	30	2		2						
01-3	8	827		770	40	5		3	2		2	2	3	
Totals	34	14,968												
Average				1,197	45.2									
Probably in product				x	x									

APPENDIX 4-44 *Continued*

Product	Popsicle								Jelly											Clam Chowder								
Chemical Class MB	Num	ppb	P Cum	P Dic	P BZ	P BDM	IC Cfo	IC Tol	Num	ppb	P Car	P Dat	P Dip	P EnI	P EII	P EnS	P Ipr	P Ipm	IC BZ	Num	ppb	P Cam	P Cpm	P DDE	P Dan	P Die	P EnS	IC BZ
96-1	2	112	12					100	1	25	25									2	37	24						13
96-2	1	36						36	2	80	71						9			1	85	85						
97-1	2	18	17	1					5	40		1	20	6	9	4				1	2			2				
98-1	2	75	63					12	2	8	7						1			1	0.2			0.2				
98-2	1	18			18				1	14									14	1	0.3						0.3	
98-3	0	0							0	0										1	1	1						
98-4	1	16					16		1	1							1			1	51	51						
99-1	0	0							2	38	25						13			2	5.3	5		0.3				
99-3	0	0							0	0										1	2	2						
00-1	3	7			2		3	2	0	0										1	10	10						
00-3	2	4			1	3			1	10	10									2	71	70			1			
01-3	3	23				6	14	3	3	12	7						2	3		4	34	30	0.1			0.1	0.2	
Totals	17	309							18	228										18	299							
Average																						23.2						
Probably in product																						x						

APPENDIX 4-45 Mixed Food Products with Moderate Contaminant Levels

Product	Chili and Beans (HM)												
Chemical			P	P	P	P	P	P	P	P	P	IC	IC
Class MB	Num	ppb	Cau	Cmy	Cno	Cpa	Cte	DDE	Die	Dip	Ipr	Tol	Xyl
96-1	3	26.8						0.8				16	10
96-2	0	0											
97-1	3	106.3			97	9		0.3					
98-1	3	128.4			117	11		0.4					
98-2	0	0											
98-3	4	999	4	28	779	188							
98-4	5	1366	7	16	1045	165	133						
99-1	6	2749	5	30	2400	280		0.8		33			
99-3	5	864.4	4	10	740	110		0.4					
00-1	2	2.1							0.1		2		
00-3	5	1333	2	20	1310	0.4		0.2					
01-3	3	87.2			80	7		0.2					
Totals	39	7662											
Average					547	64		0.3					
Probably in product					x	x		x					

HM = Home Made

APPENDIX 4-45 *Continued*

Product	Peppers, Rice, and Beef (HM)																	
Chemical			P	P	P	P	P	P	P	P	P	P	P	P	P	P	P	P
Class MB	Num	ppb	Azi	Cpm	Cpy	Cyf	DCP	DDE	DDT	Dic	Die	Efv	Enl	Ell	EnS	Fen	Lac	P
96-1	3	2.7						0.7						1	1			
96-2	3	51											11	25	15			
97-1	4	82.7					2	0.7		74								6
98-1	4	15.5						0.5					4	8	3			
98-2	6	192								170			1	2	2	4		13
98-3	4	27	1	1	22							3						
98-4	5	2.5		1	0.4			0.5	0.4		0.2							
99-1	1	3						3										
99-3	2	8				4												4
00-1	2	11															9	2
00-3	5	19			8								1	2	2	6		
01-3	1	3		3														
Totals	40	417																
Average																		
Probably in product																		

HM = Home Made

APPENDIX 4-45 *Continued*

Product	Salisbury Steak and Gravy														
Chemical Class MB	Num	ppb	P Azi	P Cam	P Cno	P Cpm	P DDE	P Dan	P Die	P EnS	P Mal	P Vin	IC Cfo	IC Tol	IC Xyl
96-1	5	91.5		54					0.5				11	14	12
96-2	3	28.6		27		0.9					0.7				
97-1	4	8.2		2	5		0.7			0.5					
98-1	1	33		33											
98-2	5	36.1	0.3	33			0.7		0.1			2			
98-3	3	14.7		14		5				0.2					
98-4	2	25.5		25		0.5									
99-1	3	57.6		56			0.6					1			
99-3	6	14.2		10	1		0.3			0.1	2	0.8			
00-1	4	32.3		30					0.1	0.1		2			
00-3	3	10.5		10			0.3			0.2					
01-3	4	11.9		10			0.1	0.8		1					
Totals	43	364.1													
Average				25.3											
Occurs in product				x											

APPENDIX 4-45 *Continued*

Product Chemical Class MB	Raisin Bran													
	Num	ppb	P 24-D	P Cap	P Cpy	P Cpm	P DDE	P Dan	P Dca	P Dic	P EnS	P Ipr	P Mal	Prg Prg	IC TBZ
96-1	5	33.4			2	8					0.4		3		20
96-2	6	141		1		6				2	0.6	7	124		
97-1	3	21		7		2								12	
98-1	4	85	4			26							7	48	
98-2	3	86	3			10								73	
98-3	4	26	7			10		6					3		
98-4	5	99.7			0.3	34	0.4						15	5	
99-1	4	92	3			60							6	23	
99-3	2	15				8							7		
00-1	3	103				10			10					80	
00-3	2	28				8								20	
01-3	3	40				10				10				20	
Totals	44	770													
Average						16							14	23	
Probably in product													x	x	
Occurs in product						x									

APPENDIX 4-45 *Continued*

Product	French Salad Dressing																
Chemical			P	P	P	P	P	P	P	P	P	P	P	P	P	P	P	IC
Class MB	Num	ppb	Azi	BHC	Cau	Cin	Cmy	Cno	Cpa	Cpy	Cte	DDE	DDT	Dic	Die	Etn	Lin	Xyl
96-1	4	632					14	496	102									31
96-2	1	0.4													0.4			
97-1	4	38.5					2	30	6						0.5			
98-1	5	208.3	0.3				2	176	28							2		
98-2	0	0																
98-3	1	57						57										
98-4	5	7,403					43	4,320	1,370		1,670					2		
99-1	3	403					12	326	65									
99-3	5	576			3		20	430	120							3		
00-1	2	90						80	10									
00-3	11	347.2		1		10	4	260	40	3		0.4	0.5	20		8	0.2	
01-3	4	291			1		10	260	20									
Totals	45	10,046																
Average							8.9	536	147									
Probably in product							x	x	x									

APPENDIX 4-45 *Continued*

Product	Chocolate Snack Cake											
Chemical Class MB	**Num**	**ppb**	**P Cpm**	**P Lin**	**P Mal**	**IC BZ**	**IC Cfo**	**IC PCE**	**IC Sty**	**IC TBZ**	**IC Tol**	**IC Xyl**
96-1	2	27									12	15
96-2	2	65								17	48	
97-1	4	71.9	6	0.9						49	16	
98-1	1	45									45	
98-2	3	40	4		5						31	
98-3	1	t										
98-4	4	45.8	0.8		1					22	22	
99-1	6	95	3		11	10			11		22	48
99-3	5	45	2					7	11		22	3
00-1	6	62	1		1				5	32	18	5
00-3	6	36.7	0.7		1	2			3		28	2
01-3	6	31	7		4		2		3		12	3
Totals	46	564										
Average			2								23	
Probably in product			x									
Occurs in product											x	

APPENDIX 4-45 *Continued*

Product	Fried Chicken (Fast Food)																	
Chemical			P	P	P	P	P	P	P	IC	IC	IC	IC	IC	IC	IC	IC	IC
Class MB	Num	ppb	Cpa	Cno	Cpm	Cum	DDE	Die	Mal	BZ	Cfo	EBZ	Pyl	Sty	TCA	TCE	Tol	Xyl
96-1	2	32															20	12
96-2	5	85.1	8	60			0.4	0.7									16	
97-1	3	76								14					10		52	
98-1	3	9		6					3								t	
98-2	1	29															29	
98-3	2	92												53			39	
98-4	3	24.5			0.5				1								23	
99-1	3	44.5			0.5				1								43	
99-3	6	53				12				5		2		10			17	7
00-1	10	65.2			0.2	14				8		2	3	6	6	2	16	8
00-3	6	61								11	4			5		2	32	7
01-3	5	81								6		7		20			39	10
Totals	49	652.3																
Average																	27.2	
Probably in product																	x	
Occurs in product																		

t = trace.

APPENDIX 4-45 *Continued*

Product Chemical Class MB	Coleslaw with Dressing (HM)																			
	Num	ppb	P BHC	P Cam	P Cpa	P Cpy	P Cno	P DDE	P DDT	P Die	P EnS	P Lin	P P	IC BZ	IC Cfo	IC CTC	IC DBZ	IC Sty	IC Tol	IC Xyl
96-1	2	83.3	0.3											83						
96-2	10	82.2	0.5	22	0.9	0.4	10	0.9		0.3	2	0.2		45						
97-1	6	70.7	0.3		3		19	0.4						29					19	
98-1	2	28.3	0.3											28						
98-2	3	35.3	0.3								1			34						
98-3	1	11												11						
98-4	3	30.1								0.1				14					16	
99-1	7	24.8	0.1					0.1	0.1	0.4	1	0.1		23						
99-3	4	15.3	0.1							0.3	0.9			14						
00-1	5	21.5									0.5			12		5			2	2
00-3	3	75.4									0.4			44			31			
01-3	7	54.9						0.9					13	16	6			6	11	2
Totals	53	533																		
Average			0.2											29.4						
Probably in product			x																	
Occurs in product														x						

HM = Home Made

APPENDIX 4-45 *Continued*

Product	Mustard														
Chemical			P	P	P	P	P	P	P	P	P	P	P	P	P	IC
Class MB	Num	ppb	Azi	BHC	Cpa	Cno	Cpm	Cmy	Cpy	DDE	DDT	EnI	EII	EnS	Lin	Tol
96-1	6	945		2.3	88	805			28						0.7	21
96-2	2	3.8													0.9	2.9
97-1	6	817		1.1	68	745		2			0.2				0.7	
98-1	6	42.1		1.3	3	23		14			0.3				0.5	
98-2	8	408.6	0.6		30	371	1	5	0.4	0.3					0.3	
98-3	4	2.6		1.2								0.4	0.5		0.5	
98-4	3	11.7		1.8		9									0.9	
99-1	5	30.5		0.3	4	13		13							0.2	
99-3	2	1		0.6											0.4	
00-1	5	1651		3	80	1570					0.4				0.3	
00-3	3	78.4			8	70									0.4	
01-3	3	34.5			4	30								0.5		
Totals	53	4026														
Average				1	23.8	303									0.5	
Probably in product				x	x	x										
Occurs in product															x	

Product	Fish and Bun (Fast Food)													
Chemical			P	P	P	P	P	P	P	P	P	P	P	IC
Class MB	Num	ppb	BHC	Cam	Cpm	Cpy	DDE	DDT	Die	Diz	Enl	EnS	Mal	Xyl
96-1	5	44.4			4		2		0.4				2	36
96-2	6	10.1	0.3		3	1	2		0.8				3	
97-1	3	8.8			0.8		2						6	
98-1	3	4			1		2						1	
98-2	4	10.5			3		1		0.5				6	
98-3	5	8.3			2				0.2		0.2	0.9	5	
98-4	7	10.2		2	2	2	0.4		0.5	0.3			3	
99-1	4	7.2			3		2		0.2				2	
99-3	5	8.5			1		0.2	0.1	0.1				7	
00-1	3	2.6			0.4		0.2						2	
00-3	3	7.4			5		0.4						2	
01-3	6	12.8		2	3		0.2	0.2	0.4				7	
Totals	54	135												
Average					2.4		1		0.3				3.8	
Probably in product									x					
Occurs in product					x		x						x	

APPENDIX 4-45 *Continued*

Product Chemical Class MB	Sandwich Cookies with Crème Filling Num	ppb	P Cpm	P Mal	IC BZ	IC CBZ	IC Cfo	IC PCE	IC Sty	IC TBZ	IC Tol	IC Xyl
96-1	4	170	16	7					110			37
96-2	3	36	1	19					16			
97-1	3	85	2						67			16
98-1	5	57	5	3					13	11	25	
98-2	4	29	14	3					t		12	
98-3	2	48	2	46								
98-4	4	178	3	86					77		12	
99-1	4	64	2	10					41	11		
99-3	4	64		7					47		5	5
00-1	6	62	2	10	3				33		11	3
00-3	8	143	1	20	39	2	4		64		10	3
01-3	8	84	3	8	6		3	3	50		8	3
Totals	55	1020										
Average			4.3	18					43.2		6.9	
Probably in product											x	
Occurs in product			x	x					x			

Product Chemical Class MB	Tuna Noodle Casserole (Home Made) Num	ppb	P Cno	P Ccp	P Cpm	P Cpy	P DDE	P DDT	P Dan	P Die	P Enl	P Ell	P EnS	P Mal	P P	P Vin	IC Tol
96-1	3	26					1		4								21
96-2	4	65	54	7			2						2				
97-1	2	23					3		20								
98-1	3	3.5					2						0.2		1.3		
98-2	8	43.7			1		0.4		40	0.1	0.2	0.6	0.4	1			
98-3	6	18.1			2		0.8		13				0.3	1		1	
98-4	8	7.7			0.5		0.9	0.3	2		0.2	0.2	0.6		3		
99-1	4	14.5			0.5		1		10						3		
99-3	3	3.6			0.6		1		2								
00-1	6	6.3			0.7		2		0.4				0.2	1	4		
00-3	4	7.6			1		0.3		6				0.3				
01-3	6	5.1			2	0.2	0.9		0.9	0.1					1		
Totals	57	224															
Average					0.7		1.3		8.2				0.3				
Probably in product					x				x				x				
Occurs in product							x										

APPENDIX 4-45 *Continued*

Product	Hot Dog and Bun (Home Made)														
Chemical			P	P	P	P	P	P	P	P	P	P	P	P	IC	IC
Class MB	Num	ppb	BHC	Cno	Cpm	Cpy	DDE	DDT	Die	EnS	HCB	Lin	Mal	Pri	Tol	Xyl
96-1	4	92			3								2		62	25
96-2	5	16		4	3	0.6	0.4						8			
97-1	6	7	0.1			2	0.5	0.3			0.1		4			
98-1	4	17.8		10	4		0.8						3			
98-2	4	15.6			4	1	0.6						10			
98-3	4	11.3			6	2			0.3				3			
98-4	5	5.7			3	0.3	0.3		0.1				2			
99-1	6	14.5			8	0.5	0.8				0.2		3	2		
99-3	8	11.9	0.1	3	5		0.5	0.1	0.1			0.2	3			
00-1	3	8			3		2						3			
00-3	3	8.9			6		0.9						2			
01-3	7	9.8			2		2	0.2	0.3	0.2	0.1		5			
Totals	59	218.5														
Average					3.9		0.7						4			
Probably in product							x									
Occurs in product					x								x			

APPENDIX 4-45 *Continued*

Product	Muffin, Egg, Ham, and Cheese (Fast Food)											Chicken Potpie (Frozen)														
Chemical			P	P	P	P	P	P	P	Pri	IC			P	P	P	P	P	P	P	P	P	P	P	Pri	IC
Class MB	Num	ppb	Cpm	Cpy	DDE	Die	Diz	Mal	Mxy	Pri	Xyl	Num	ppb	Cam	Cno	Cpa	Cpm	Cpy	Die	Diz	EnS	Lin	Mal	Mxy	Pri	Xyl
96-1	4	19	4		4			1			10	3	33				6						4			23
96-2	3	10	2		4			4				3	11.6	9				0.6					2			
97-1	1	6			6							3	10.8	8			0.8						2			
98-1	2	3.6	0.6		3							3	16	3			5						8			
98-2	4	18.2	2		4	0.2		12				2	21	7									14			
98-3	3	5.1	1		0.1			4				2	4.7				0.7						4			
98-4	7	10.4	1	0.8	5	0.3	0.7	2		0.6		3	23	9			4						10			
99-1	2	6			4			2				4	42	11			15						15	1		
99-3	3	5.3	2		0.3			3				3	5				1						3		1	
00-1	4	7.1	3		3	0.2		0.9				9	40.5	10	10	2	6	0.9		1	0.3		10	0.3		
00-3	4	10.2	1		2	0.2		7				2	60				40						20			
01-3	4	11.3	5		3			3	0.3			5	20.7	0.6			10		0.1			1	9			
Totals	41	112.2										42	288													
Average			1.8		3.2			3.2						4.8			7.4						101			
Probably in product			x					x						x			x									
Occurs in product					x																		x			

APPENDIX 4-46 Mixed Food Products with High Contaminant Levels

Product								Apple Pie							
Chemical Class			P	P	P	IC	IC	IC	IC	IC	IC	IC	IC	IC	IC
MB	Num	ppb	Cpm	Cum	Mal	BZ	BBZ	Cfo	DBZ	EBZ	PCE	Sty	TBZ	Tol	Xyl
96-1	4	91	8		1								60		22
96-2	2	25	12		13										
97-1	7	132	15		29		11		11			11	34	21	
98-1	4	51	11		4							23	13		
98-2	2	39	20		19										
98-3	2	20			3								17		
98-4	3	78	52		10							16			
99-1	4	53	9		16							17	11		
99-3	10	209	4	22	3	2	8			10		14	34	23	89
00-1	8	57		2	1	2				3		20	13	5	11
00-3	8	68	3		4	11					5	10	24	4	7
01-3	8	36	6		2	3		5		2		7		5	6
Totals	62	859													
Average			11.7		8.8							9.8	17.2		
Probably in product			x									x	x		
Occurs in product					x										

APPENDIX 4-46 *Continued*

Product	Chicken Nuggets (Fast Food)																					
Chemical Class			P	P	P	P	P	P	P	P	IC	IC	IC	IC	IC	IC	IC	IC	IC	IC	IC	IC
MB	Num	ppb	Cpa	Cno	Cpm	Cum	DDE	DDT	Etn	Mal	BZ	BBZ	BDM	Cfo	EBZ	PCE	Pyl	Sty	TCA	TCE	Tol	Xyl
96-1	2	111									100										11	
96-2	4	46				16				1								16			13	
97-1	3	59																10			27	22
98-1	4	129.4							0.4		16							16			97	
98-2	2	60																21			39	
98-3	3	79.4			0.4													10			69	
98-4	2	14																14			t	
99-1	5	92.4			0.4					2								11			57	22
99-3	14	92.3	1	5	3	8	0.4	0.9		1	3				3	7		12	4		36	10
00-1	11	126			1					10	2	5			4		4	34	6	3	55	16
00-3	14	163.9	2	6	1	17				0.9	5		3	4	5		4	18		5	71	22
01-3	9	230			1					2	2		4	5	9			39			151	17
Totals	73	1203																				
Average																		16.8			52.2	
Probably in product																		x			x	
Occurs in product																						

t = trace.

APPENDIX 4-46 *Continued*

Product			Chocolate Cake																	
Chemical Class			P	P	P	P	P	P	IC	IC	IC	IC	IC	IC	IC	IC	IC	IC	IC	IC
MB	Num	ppb	Cpm	Cum	DDE	Die	Lin	Mal	BZ	BBZ	CBZ	Cfo	DBZ	EBZ	PCE	Sty	TBZ	TCE	Tol	Xyl
96-1	5	234	2													41	103		34	54
96-2	5	39.4	1				0.4	2											14	22
97-1	8	172	2		0.5					14			20		21	27			33	33
98-1	4	115														14	29		54	18
98-2	4	51	3					5								18			25	
98-3	4	44	3					3									18		20	
98-4	5	92.5	0.5			0.2		0.8								14			77	
99-1	4	25.3	2		0.3											12			11	
99-3	12	164	2					2	23		2	3	10	2	13	6	15		73	10
00-1	6	77						1						5		14	10		36	11
00-3	10	197		4								4	3	4	29	7	11	57	57	21
01-3	6	110							3					5	5	12			65	20
Totals	73	1321																		
Average			1.3													13.8			41.6	15.8
Probably in product			x													x				x
Occurs in product																			x	

APPENDIX 4-46 *Continued*

Product			Doughnut — Cake and Icing																		
Chemical Class			P	P	P	P	P	P	P	P	IC	IC	IC	IC	IC	IC	IC	IC	IC	IC	IC
MB	Num	ppb	Cmy	Cpm	Cpy	Cum	Diz	Mal	Mxy	Pri	BZ	BBZ	Cfo	EBZ	PCE	Pyl	Sty	TBZ	TCE	Tol	Xyl
96-1	3	91		3				8												80	
96-2	3	95		7				7												81	
97-1	7	184		10	3		3								40		34			61	33
98-1	5	81		1				4									12			44	20
98-2	5	82.7		2	0.7			3										19		58	
98-3	6	97		8				31									16	18		t	24
98-4	5	88	8	33				32									45			t	
99-1	6	131		5				5								11	29			27	54
99-3	7	45		6				9		2		5					6			9	8
00-1	8	103.6						0.6			2			6		7	21		4	33	30
00-3	10	200		2		7		3			3	5		3			37	7		125	8
01-3	11	81.4	5	10				10	0.4		2		3	2			18	5		17	12
Totals	76	1280																			
Average				7.3				9.4									18.2			44.6	15.8
Probably in product																	x				x
Occurs in product				x				x												x	

APPENDIX 4-46 *Continued*

Product			Meatloaf (Home Made)																									
Chemical Class			P	P	P	P	P	P	P	P	P	P	P	P	P	P	P	P	P	IC	IC	IC	IC	IC	IC	IC	IC	IC
MB	Num	ppb	BHC	Cam	Cmy	Cno	Cpa	Cpy	Cum	DCP	DDE	Die	Dip	Diz	HCB	HE	Ipr	Oct	P	BZ	Cfo	EBZ	PCE	Sty	TCA	TCE	Tol	Xyl
96-1	3	71									2									56								13
96-2	4	39			2	28	8				1																	
97-1	4	3.5									2	0.4			0.1				1									
98-1	2	34									2																32	
98-2	5	23.2						2			1	0.3		0.9													19	
98-3	7	64.1						0.8			2	0.5		0.8						26				19			15	
98-4	5	37.5									1	0.4				0.1				10							26	
99-1	10	92.3	0.1					0.9			2	0.4	69		0.3	0.2		0.1	1.3								18	
99-3	9	59.1							23		0.9				0.1		2			5			6	6			12	4
00-1	13	53.9		6							0.5	0.5				0.1	0.8			5		2	3	7	4	3	13	9
00-3	6	38								2	1									5	4			9			17	
01-3	9	47.9									0.6	0.2				0.1				4	3			9		6	18	7
Totals	77	563.5																										
Average											1.3	0.2								9.3							14.2	
Probably in product												x								x							x	
Occurs in product											x																	

APPENDIX 4-46 *Continued*

Product			Brownies																
Chemical Class			P	P	P	P	P	IC	IC	IC	IC	IC	IC	IC	IC	IC	IC	IC	IC
MB	Num	ppb	Cpm	Cum	Diz	Lin	Mal	BZ	BBZ	CBZ	Cfo	DBZ	EBZ	PCE	Sty	TBZ	TCE	Tol	Xyl
96-1	7	130	13				2						11		25	12		22	55
96-2	4	62	3			2	10											47	
97-1	5	36	2				2		6					13		13			
98-1	5	137	33				3									27		33	57
98-2	3	52	4				18											30	
98-3	8	198	3		2		19						10		21	20		77	46
98-4	4	51	3				2	25										21	
99-1	4	89	1				7									48		33	
99-3	8	59.9	0.9	3			1						2		7	17		19	9
00-1	9	56	2				1	1					2		6	21	2	13	8
00-3	11	119	9				9	40		3	4		2	3	4	10		23	12
01-3	9	91	7				20	2			6	7	3		8			30	6
Totals	77	1081																	
Average			6.7				7.8									14		29	16.1
Probably in product																x			x
Occurs in product			x				x											x	

APPENDIX 4-46 *Continued*

Product	Blueberry Muffin																				
Chemical Class			P	P	P	P	P	Pri	Pho	IC	IC	IC	IC	IC	IC	IC	IC	IC	IC	IC	IC
MB	Num	ppb	Azi	Cpm	Cum	Mal	Mxy	Pri	Pho	BZ	Cfo	DBZ	EBZ	PCE	Pyl	Sty	TBZ	TCA	TCE	Tol	Xyl
96-1	6	107		6		2							11	38						17	33
96-2	5	157				3							14				24			31	85
97-1	6	59		7		1		1								18				13	19
98-1	6	159	29	5												16	20		67	22	
98-2	4	96		5		2										18				71	
98-3	3	66		5		3														58	
98-4	5	34		0.2		1	0.3		6											26	
99-1	4	98		7		2										32				56	
99-3	8	112		2	2	3				3		10				20				64	8
00-1	9	53.6			4			0.6		1			2	4		17		4		10	11
00-3	9	426		3		1				8			7	3		46	19			298	41
01-3	12	1311		20	5	5				14	10	11	224	10	5	510				130	367
Totals	77	2679																			
Average				5		1.9										56.4				66.3	47
Probably in product				x		x										x					x
Occurs in product																				x	

APPENDIX 4-46 *Continued*

Product			Sugar Cookies																
Chemical Class			P	P	P	P	P	P	IC	IC	IC	IC	IC	IC	IC	IC	IC	IC	IC
MB	Num	ppb	Cau	Cmy	Cpm	DDE	Lin	Mal	BBZ	Cfo	DBZ	EBZ	PCE	Pyl	Sty	TBZ	TCE	Tol	Xyl
96-1	4	84			8			7							32			37	
96-2	4	43			3			6							25			9	
97-1	4	150			22										59			24	45
98-1	6	186			5					10		19			11			46	95
98-2	3	104				0.6		8							95				
98-3	5	136			48			23					23		24			18	
98-4	6	84			17			16		10					t	20		21	
99-1	6	137	8	102	20			7							t			t	
99-3	6	109			6			10			2				82			6	3
00-1	9	123			3	0.3		7	4			3		3	73			18	12
00-3	11	184			0.6		3	20		3		2	2		35	9	4	88	17
01-3	8	126			9			3		19		3	3		21			60	8
Totals	72	1466																	
Average					11.8			8.9							38.1			27.3	
Probably in product								x											
Occurs in product					x										x			x	

t = trace.

APPENDIX 4-47 Mixed Food Products with Very High Contaminant Levels

Product	Pumpkin Pie																
Chemical Class			P	P	P	P	P	P	P	P	P	P	P	P	IC	IC	IC
MB	Num	ppb	BHC	Cam	Cau	Cmy	Cno	Cpa	Cpm	Cte	DDE	Die	Mal	P	Sty	Tol	Xyl
96-1	10	145.8			4	56	11	8	5			0.8	1		37	13	10
96-2	5	29				11	2		3			4	9				
97-1	6	168				2	12	2	2			2	16				
98-1	6	32			1	21		4	3			1	2				
98-2	6	51		19		4	12	3				2	11				
98-3	9	42.9	0.1		2	22	9	9			0.8	2	1	3			
98-4	8	116		13		39	24	7		12	3		7	11			
99-1	8	129.6			1	73	41	8	2			1	2	1.6			
99-3	8	51.7			10	30	10	4	2			1	3	1.7			
00-1	6	214			8	100	50	20		30		6					
00-3	8	160.6			2	60	80	10	0.8			1	0.8	6			
01-3	6	21.7				6	6		4			0.7	3	2			
Totals	86	1162															
Average					2.3	35.3	21.4	6.3	1.8			1.8	4.7				
Probably in product					x				x								
Occurs in product						x	x	x				x	x				

APPENDIX 4-47 *Continued*

Product	Chocolate Chip Cookies																		
Chemical Class			P	P	P	P	P	P	P	P	IC	IC	IC	IC	IC	IC	IC	IC	IC
MB	Num	ppb	BHC	Cpm	Cpy	Cum	Dip	Lin	Mal	Myx	Cfo	DBZ	EBZ	PCE	Sty	TBZ	TCE	Tol	Xyl
96-1	5	127		5					3						37			71	11
96-2	8	70	0.6	1	1			0.7	2						15	24		25	
97-1	6	154		4					4			27			67	22		31	
98-1	6	179		48				0.7	4						87	34		48	
98-2	6	132	0.1					0.4	19					18	28			66	
98-3	5	78		11				0.3	14						25			27	
98-4	6	153		7				0.1	4						98	12		32	
99-1	6	120	0.3	3				1	2						94			20	
99-3	10	183		6			30	0.3	9	4		8			29	68		19	10
00-1	9	151		0.7		2		0.3	10			4			87	8		21	18
00-3	11	184		0.6				3	20		3		2	2	35	9	4	88	17
01-3	10	523		30				0.7	30		4	176	33	2	198			27	22
Totals	88	2054																	
Average				9.7				0.6	10.1						66.7	14.8		39.6	
Probably in product								x								x			
Occurs in product				x					x						x			x	

APPENDIX 4-47 *Continued*

Product / Chemical Class / MB	Num	ppb	P Cmy	P Cno	P Cpa	P Cpm	P Cpy	P Cum	P DDE	P Die	P Diz	P EnI	P EII	P EnS	P Mal	P P	IC BZ	IC Cfo	IC EBZ	IC PCE	IC Sty	IC TBZ	IC TCA	IC TCE	IC Tol	IC Xyl
			Cheese Pizza (Fast Food)																							
96-1	8	97.9				10	0.9		3						6				15		25				17	21
96-2	11	77.8				7	1		2		1	0.5	0.7	0.6	7			11			11				36	
97-1	6	39.7		11		3	0.7		7						5										13	
98-1	4	68				2			2											12					52	
98-2	5	59				4			4						2										22	27
98-3	7	55.5		32	0.9	2	0.3								2	0.6									18	
98-4	8	111	6	11	4	3			1						7						13				66	
99-1	9	166.8		60	2	2			2	0.9					4	1.9									72	22
99-3	9	55.2				4		9	0.2						1		1			16	3				17	4
00-1	12	99		6		8		7	2						2				2	4	8		3	2	42	13
00-3	10	160		80	30	4			4						3		2	3			8				18	8
01-3	14	137.5				2	0.2		1	0.1				0.2	2		4	9	5	2	23	4			74	11
Totals	103	1127																								
Average						4.3			2.4						3.4						7.6				37.3	8.8
Probably in product																					x					x
Occurs in product						x			x						x										x	

APPENDIX 4-47 *Continued*

Product / Chemical Class / MB	Num	ppb	P BHC	P Cpy	P DDE	P DDT	P EnI	P EnS	P Lin	P Pir	P Pme	P TDE	IC BZ	IC BBZ	IC BDM	IC CBZ	IC Cfo	IC DBZ	IC EBZ	IC PCE	IC Sty	IC TBZ	IC TCA	IC TCE	IC Tol	IC Xyl
			Milk Chocolate Candy Bar																							
96-1	8	104.9	0.3		1	2			0.9			0.7							12						36	50
96-2	8	45.3	0.8		1	1			4			1						12		10		16				
97-1	6	108			1				1											16	21				24	45
98-1	7	183.9			0.5	0.4			1												75	41			47	19
98-2	3	68							2												26				40	
98-3	9	153.9		0.4	0.3	0.2			1	1										43	26				33	59
98-4	8	258			3				2										13	38	31	25			78	68
99-1	11	113.5	0.1	0.8	2	0.5		0.5	2		0.2	0.4									42	26			39	
99-3	11	98.1		0.6	0.3			0.5	0.7									3	2	11	13	31			27	9
00-1	9	94.8							0.8										3	22	11	13	4	2	27	3
00-3	11	177.8			0.3				0.5				26			4			3	14	16	22		2	76	14
01-3	12	75.8					0.4	0.9	0.5				1	3	4		3		3	4	14				27	15
Totals	103	1483																								
Average					0.8				1.4											13.2	22.9	14.5			37.8	23.5
Probably in product					x															x	x	x				x
Occurs in product									x																	x

APPENDIX 4-47 *Continued*

Product Chemical Class MB	Num	ppb	P BHC	P Cam	P Cpm	P Cpy	P Cum	P DDE	P DDT	P Die	P EnI	P EII	P EnS	P HCB	P HE	P Lin	P Mal	P Pri	IC BZ	IC BBZ	IC Cfo	IC EBZ	IC PCE	IC Pyl	IC Sty	IC TBZ	IC TCA	IC TCE	IC Tol	IC Xyl
			1/4-lb Hamburger and Bun (Fast Food)																											
96-1	7	53.3			4			2		0.3							2								13				14	18
96-2	7	86.5			2			2								0.5	2		47		14						19			
97-1	8	62.3	0.9		0.4			5									7	4	18						12				15	
98-1	6	56.3			2			5		0.3							1		19										32	
98-2	5	61.2			2			3		0.2							5												51	
98-3	6	20.3			0.9			0.7		0.4			0.3				4												14	
98-4	7	73.2			0.5	0.2		0.3		0.2							2								18				52	
99-1	9	53.8		2	4			2		0.4			0.2	0.2			2												23	20
99-3	14	100.4				2		0.6	0.2	0.4	0.3	0.3	0.5		0.1		7		13			2			12				54	8
00-1	17	277.2			0.7		12	0.5	0.4				0.5		0.1		1		8	11		21	6	12	6	23		2	53	116
00-3	8	81			9			1									7		4		14				4				40	2
01-3	13	105.3			2		2	0.6	0.1	0.2			0.3		0.1		10				5	3			28				50	4
Totals	107	1031																												
Average					2.3			22.7		0.2							4.2								7.8				33.2	
Probably in product										x															x					
Occurs in product					x			x									x												x	

APPENDIX 4-47 *Continued*

Product Chemical Class MB	Num	ppb	P BHC	P Cau	P Ccp	P Cmy	P Cno	P Cpa	P Cpm	P Cpy	P Cte	P Cum	P DDE	P Die	P EnI	P EII	P EnS	P Mal	P P	P Pri	IC BZ	IC BBZ	IC BDM	IC CBZ	IC Cfo	IC EBZ	IC PCE	IC Pyl	IC Sty	IC TCE	IC Tol	IC Xyl
			Taco Tostada (Fast Food)																													
96-1	4	68					7						2																		32	23
96-2	4	65			7		54						2				2															
97-1	10	200					150	19					2		0.3	0.7	2	4	2										11		9	
98-1	9	339.9					251	4		1	81		1		0.4	0.5	1														t	
98-2	8	54.8					20			1			1		0.4	0.4	2	6													24	
98-3	8	181.2				0.8	58	6				15	0.4																44		34	23
98-4	8	1789	0.2			7	1390	135			136		0.4																24		96	
99-1	4	256					234	17					3					2														
99-3	11	103					20	1	6			5	2												11	4	6		8		26	14
00-1	14	465.9				4	270	50	0.3		70	9	0.9					0.7		1						3			20	2	18	17
00-3	14	125.1										27	1	0.2					0.9		23	3	4	2	2	2		3	26		19	12
01-3	14	148.5		3	0.5	0.9	30	3					0.6	0.1			0.4				22				8	8			20		33	19
Totals	108	3796																														
Average							207	19.6					1.4																12.8		24.3	9
Probably in product								x																					x		x	x
Occurs in product							x						x																			

t = trace.

APPENDIX 4-47 *Continued*

Product			1/4-lb Cheeseburger and Bun (Fast Food)																											
Chemical Class			P	P	P	P	P	P	P	P	P	P	P	P	P	P	P	P	IC	IC	IC	IC	IC	IC	IC	IC	IC	IC	IC	IC
MB	Num	ppb	BHC	Cpm	Cpy	Cum	DDE	DDT	Die	EnI	EII	EnS	HCB	HE	Lin	Mal	Oct	Pri	BZ	BBZ	CBZ	Cfo	DBZ	EBZ	PCE	Pyl	Sty	TCE	Tol	Xyl
96-1	8	103.9		54			7		0.5			0.4				2			54										22	15
96-2	8	12.6		2	0.6		6		0.5			0.7			0.8	2													t	
97-1	8	78.6	0.2				8		0.4							4		2	37								11		16	
98-1	4	43.4		2			12		0.4																				29	
98-2	8	47.7		2			0.4		0.2	0.2	0.3	0.6				6													38	
98-3	5	23.5		2			1		0.8							5													15	
98-4	6	28		0.5	0.3		5		0.2							2													20	
99-1	10	33.5		3			6		0.5		0.1	0.4	0.2	0.2		2	0.1												21	
99-3	14	125.6		1			0.6		0.2	0.2	0.3	0.2				5			34	4		2		3			13		49	13
00-1	17	191.4		0.5		10	2		0.9			1				2			26	6	5		4	11	8	8	4	2	37	63
00-3	9	75.5		4			6		0.5							3			10			15					5		30	2
01-3	15	106		4			3	0.2	0.4			0.1	0.1	0.1	0.1	8			21			6		3			16		36	8
Totals	112	869.7																												
Average				6.3			4.8		0.5			0.3				3.4													26.1	
Probably in product												x																	x	
Occurs in product				x			x		x							x														

t = trace.

APPENDIX 4-47 *Continued*

Product			Cheese & Pepperoni Pizza (Fast Food)																											
Chemical Class			P	P	P	P	P	P	P	P	P	P	P	P	P	P	P	P	P	P	P	IC	IC	IC	IC	IC	IC	IC	IC
MB	Num	ppb	BHC	Ccp	Cmy	Cno	Cpa	Cpm	Cpy	Cte	Cum	DDE	Dan	Die	Diz	EnI	EII	EnS	Etn	Mal	P	BZ	Cfo	EBZ	PCE	Sty	TCE	Tol	Xyl
96-1	5	37		3				3				2								5									18
96-2	9	35						5	2			2			0.8	0.5	0.7			6								t	18
97-1	9	129.8		4		47		3	0.8			5								5						11		22	32
98-1	8	230.7				126	14	4	0.7			1								5					20			60	
98-2	6	65.6						3				2							0.6	3								29	28
98-3	13	157.4	0.2			53	2	2	0.5			0.4		0.2		0.2	0.2			1	0.7							40	56
98-4	10	370			3	179	23	3		43			0.4						2	6						19		92	
99-1	9	220.5				82	4	4				2								5	1.5					11		64	27
99-3	11	107.2				30	2	4			8	0.2								2				2	19	9		23	8
00-1	11	89				10		7		14		1								2				2	2	7	2	30	11
00-3	12	193				90	30	4				4								3		8	3	2		13	2	23	11
01-3	14	193				7		2	0.3			0.9		0.1		0.1	0.4	0.2		1		16		5		28		116	11
Totals	117	1828																											
Average						52		3.7				1.7								3.7						8.2		41.6	18.3
Probably in product						x																				x			x
Occurs in product								x				x								x								x	

t = trace.

APPENDIX 4-47 *Continued*

Product	Fish Sticks																				
Chemical Class			P	P	P	P	P	IC	IC	IC	IC	IC	IC	IC	IC	IC	IC	IC	IC	IC	IC
MB	Num	ppb	Cpm	Cum	DDE	DDT	Mal	BZ	BBZ	CBZ	Cfo	CTC	DBZ	EBZ	PCE	Pyl	Sty	TBZ	TCA	Tol	Xyl
96-1	7	132	0.8				2		34	18								37		15	25
96-2	5	65.8	0.8				4			25								24	12		
97-1	5	41.5	2		0.2	0.3	5													34	
98-1	7	132	2				0.8				15		11					55		24	24
98-2	5	87	2				3				47							13		22	
98-3	5	58	2				1				19							17		19	
98-4	6	150	6				6				63							24		16	37
99-1	4	44	2				1											17		24	
99-3	11	67	2				4		6		6	5	3		9		3	14		3	12
00-1	10	56		4			10	2					3	2	3		3	13		5	11
00-3	12	165	6				6	8	9		33			2	9	4	10	28		40	10
01-3	10	134	4				4	2			34		7	3	2		7		56	15	
Totals	87	1132																			
Average			2.5				3.9				18.1							22		18.1	
Probably in product											x							x			
Occurs in product			x				x													x	

APPENDIX 4-48 Child/Infant Products — No Chemical Contaminants

Product Chemical Class MB	Bananas and Tapioca		Creamed Corn (Strained)		Custard Pudding (Strained)	
	Num	ppb	Num	ppb	Num	ppb
96-1	0	0	0	0	0	0
96-2	0	0	0	0	0	0
97-1	0	0	0	0	0	0
98-1	0	0	0	0	0	0
98-2	0	0	0	0	0	0
98-3	0	0	0	0	0	0
98-4	0	0	0	0	0	0
99-1	0	0	0	0	0	0
99-3	0	0	0	0	0	0
00-1	0	0	0	0	0	0
00-3	0	0	0	0	0	0
01-3	0	0	ND	ND	0	0
Totals	0	0	0	0	0	0

ND = No Data

APPENDIX 4-49 Child/Infant Products — Strained Vegetables with Very Low Contaminants

Product Chemical Class MB	Peas (Strained)			Mixed Vegetables (Strained)						Sweet Potatoes (Strained)			
	Num	ppb	IC Xyl	Num	ppb	P Cam	P Mos	P Tca	IC Xyl	Num	ppb	P Dan	P DDE
96-1	1	17	17	2	11	1			10	1	7	7	
96-2	0	0		1	10	10				1	28	28	
97-1	0	0		1	9	9				1	2		2
98-1	0	0		1	47	47				1	5	5	
98-2	0	0		1	11	11				1	48	48	
98-3	0	0		1	3		3			1	22	22	
98-4	0	0		1	2	2				1	10	10	
99-1	0	0		0	0					1	11	11	
99-3	0	0		1	3			3		1	7	7	
00-1	0	0		0	0					1	7	7	
00-3	0	0		0	0					0	0		
01-3	0	0		0	0					1	10	10	
Totals	1	17		9	96					11	157		
Average												12.9	
Probably in product												x	

APPENDIX 4-50 Child/Infant Products — Fruit Juices

Product	Orange Juice (Strained)				Apple Juice (Strained)										
Chemical Class			P	IC			P	P	P	P	IC	IC	IC	IC	IC
MB	Num	ppb	Eth	Xyl	Num	ppb	Car	Dat	Ome	Thi	BZ	Cfo	CTC	PCE	Sty
96-1	2	18	1	17	3	58		4	2	53					
96-2	0	0			3	175	3	3		169					
97-1	0	0			2	28		6		22					
98-1	0	0			2	415	8			407					
98-2	0	0			2	24		3		21					
98-3	0	0			2	13		3				10			
98-4	0	0			1	3		3							
99-1	0	0			2	31		2		29					
99-3	0	0			4	19					2	8	4	5	
00-1	0	0			3	31				20	9				2
00-3	0	0			2	4					2	2			
01-3	0	0			1	10				10					
Totals	2	18			27	811									
Average								2		60.9					
Probably in product								x		x					

APPENDIX 4-51 Child/Infant Products — Formula

Product	Milk-Based Formula					Soy-Based Formula						
Chemical Class			P	P	IC			P	IC	IC	IC	IC
MB	Num	ppb	BZ	Cfo	Tol	Num	ppb	Cpm	BZ	Cfo	Sty	Tol
96-1	0	0				0	0					
96-2	0	0				1	13			13		
97-1	0	0				0	0					
98-1	0	0				0	0					
98-2	0	0				0	0					
98-3	0	0				0	0					
98-4	0	0				0	0					
99-1	0	0				0	0					
99-3	1	3		3		0	0					
00-1	0	0				2	8		6		2	
00-3	1	1	1			1	3		3			
01-3	1	2			2	4	3.4	0.4	1	2		1
Totals	3	6				8	27.4					

APPENDIX 4-52 Child/Infant Products — Protein

Product	Chicken (Strained)					Beef (Strained)									
Chemical Class			P	P	IC			P	IC	IC	IC	IC	IC	IC	IC
MB	Num	ppb	DDE	P	Tol	Num	ppb	DDE	BZ	BDM	Cfo	Ply	TBZ	Tol	Xyl
96-1	1	11			11	1	13							13	
96-2	0	0				3	21.6	0.6			10			11	
97-1	1	0.2	0.2			1	0.3	0.3							
98-1	0	0				4	61	2	16				8	35	
98-2	0	0				2	24.7	0.7						24	
98-3	0	0				1	45							45	
98-4	1	1.8		1.8		2	43.4	0.4						43	
99-1	0	0				2	13	1						12	
99-3	0	0				4	28.2	0.2	3		14			11	
00-1	0	0				0	0								
00-3	0	0				5	46.6	0.6	1				7	36	2
01-3	0	0				7	80	1		3	20	3	8	38	7
Totals	3	13				32	377								
Average								0.6						22.3	
Probably in product								x						x	

APPENDIX 4-53 Child/Infant Products — Dinners

Product	Macaroni/Tomato/ Beef			Chicken Noodle Dinner					Turkey and Rice								
Chemical Class			P			P	P	P			P	P	P	P	IC	IC	IC
MB	Num	ppb	DDE	Num	ppb	Cam	DDE	Die	Num	ppb	Cam	Die	EnI	EnII	BZ	Cfo	Xyl
96-1	0	0		1	6	6			4	37.3		0.3			12	10	15
96-2	0	0		1	4	4			0	0							
97-1	1	0.7	0.7	0	0				1	4	4						
98-1	0	0		0	0				0	0							
98-2	0	0		1	0.4			0.4	0	0							
98-3	0	0		0	0				1	4	4						
98-4	0	0		0	0				3	1.7		0.5	1	0.2			
99-1	0	0		0	0				1	3	3						
99-3	0	0		1	0.6			0.6	0	0							
00-1	1	0.2	0.2	0	0				0	0							
00-3	1	0.1	0.1	1	0.1		0.1		0	0							
01-3	1	0.2	0.2	0	0				1	10	10						
Totals	4	1.2		5	11.1				11	60							

APPENDIX 4-54 Child/Infant Products — Infant Products

Product	Rice Cereal (Infant)				Teething Biscuits								
Chemical Class			P	P			P	P	P	P	IC	IC	IC
MB	Num	ppb	DDE	Qac	Num	ppb	Cpm	Dip	Mal	Mxy	BBZ	TBZ	Xyl
96-1	0	0			3	38	9		2				27
96-2	0	0			2	6	3		3				
97-1	1	3	3		4	48	6		10		19	13	
98-1	1	0.5	0.5		3	91	3	86	2				
98-2	0	0			2	3	1		2				
98-3	1	5		5	2	7	3		4				
98-4	0	0			2	6	2		4				
99-1	1	2		2	2	6	3		3				
99-3	0	0			2	5	3		2				
00-1	0	0			2	5	3		2				
00-3	0	0			3	9	3		3	3			
01-3	1	4		4	2	316	310		6				
Totals	5	14.5			29	540							
Average							29.1		3.6				
Probably in product							x		x				

APPENDIX 4-55 Child/Infant Products — Vegetable Dinners

Product	Vegetables and Beef					Vegetables and Ham							Vegetables and Chicken					
Chemical Class			P	P	IC			P	P	P	P	P			P	P	P	P
MB	Num	ppb	Cam	DDE	Xyl	Num	ppb	Cam	DDE	Die	Ipr	Ipm	Num	ppb	Cam	DDE	Die	Ipr
96-1	1	16			16	1	8	8					2	7.5	7		0.5	
96-2	1	14	14			1	10	10					1	9	9			
97-1	1	59	59			1	10	10					1	30	30			
98-1	1	0.5		0.5		0	0						0	0				
98-2	0	0				1	2	2					0	0				
98-3	1	0.1		0.1		1	8	8					2	1.2	1		0.2	
98-4	2	1.1	0.9	0.2		3	7.6	7	0.4	0.2			3	3.5	2	1	0.5	
99-1	0	0				1	5	5					0	0				
99-3	0	0				2	10.5	10	0.5				1	1		1		
00-1	2	2.3	2	0.3		1	0.1		0.1				1	4	4			
00-3	1	0.3		0.3		3	2.1	1			0.5	0.6	2	1.4	1	0.4		
01-3	1	0.2		0.2		1	7	7					3	18.2	9		0.2	9
Totals	11	93.5				16	70.3						16	75.8				
Average								5.7							5.3			
Probably in product								x							x			

APPENDIX 4-56 Child/Infant Products — Strained Vegetables with Low Contaminants

Product	Carrots (Strained)								Spinach (Strained)					Green Beans (Strained)							Squash (Strained)								
Chemical Class			P	IC	IC	IC	CI	CI			P	P	IC			P	P	P	P	P			P	P	P	P	P	P	P
MB	Num	ppb	Cum	BZ	Cfo	Sty	Tol	Xyl	Num	ppb	DDE	P	Xyl	Num	ppb	Ace	Bif	Die	Mos	Vin	Num	ppb	DDE	Die	EnS	HCB	Pal	P	TDE
96-1	0	0							1	15			15	1	4				4		1	10		10					
96-2	0	0							0	0				1	8				8		0	0							
97-1	0	0							2	2.2	0.9	1.3		2	26	9			17		0	0							
98-1	0	0							2	93	3	90		0	0						1	1		1					
98-2	0	0							2	260	2	258		0	0						2	1.2		0.7	0.5				
98-3	0	0							1	0.4	0.4			2	26	3			23		1	2			2				
98-4	0	0							2	1.6	0.4	1.2		0	0						0	0							
99-1	0	0							1	0.7	0.7			0	0						4	8.9			2	0.2	5	1.7	
99-3	5	20		5	2	7	3	3	1	0.3	0.3			1	3					3	2	1.6		0.6	1				
00-1	3	12	3		8		1		1	0.1	0.1			3	63.5	60	3			0.5	1	10		10					
00-3	2	3		1			2		1	0.7	0.7			3	41	10	1		30		1	0.8			0.8				
01-3	2	8			6		2		0	0				2	20.2	20		0.2			3	1.6	0.4			1			0.2
Totals	12	43							14	374				15	192						16	37.1							
Average											0.7																		
Probably in product											x																		

APPENDIX 4-57 Child/Infant Products — Strained Pudding and Cereal

Product Chemical Class	Fruit Pudding (Strained)												Rice w/Apple Cereal (Strained)									
MB	Num	ppb	Cpy	Dan	Dat	Efv	EnI	EnII	EnS	Ipr	Ipm	P	Num	ppb	Ben	Cap	Car	Cpy	Dat	EnI	EnII	EnS
96-1	3	15				2				4		9	0	0								
96-2	6	10		4	2			0.2	0.6	2		1.2	4	4.5				1		0.6	0.9	2
97-1	4	22	8							6	4	4	3	3.4						0.6	2	0.8
98-1	1	8								8			ND	ND								
98-2	0	0											3	8.6				7			0.6	1
98-3	1	7										7	3	3.7			3	0.3				0.4
98-4	5	7.1					0.6	0.9	2	2	1.6		4	7				1		2	3	1
99-1	1	2										2	5	7.9				1	6	0.2	0.3	0.4
99-3	2	2.1								1		2.1	4	2.2				1		0.2	0.5	0.5
00-1	1	13										13	2	1.5		0.5						1
00-3	6	25.3		10				0.1	0.2	10	1	4	6	36.5	30		4	0.8		0.3	0.6	0.8
01-3	0	0											2	0.8							0.5	0.3
Totals	30	112											36	76.1								
Average										2.8		3.5						1			0.7	0.7
Probably in product										x		x						x			x	x

ND = No Data

APPENDIX 4-58 Child/Infant Products — Strained Fruits with Low and Moderate Contaminants

Product	Pears (Strained)															
Chemical Class			P	P	P	P	P	P	P	P	P	P	P	P	CI	CI
MB	Num	ppb	Ace	Azi	Dat	Dic	EnI	EnII	EnS	Eth	P	Par	Pho	Thi	Tol	Xyl
96-1	3	37								13					12	12
96-2	9	63.6		5	30	6	0.6	2	5	7		3	5			
97-1	3	18	5							11		2				
98-1	1	19								19						
98-2	2	28		16						12						
98-3	0	0														
98-4	6	94.4						0.2	0.4	6	1.8		9	77		
99-1	3	17					1	4	12							
99-3	4	13.5					0.5	2	3	8						
00-1	2	23									3		20			
00-3	1	10								10						
01-3	1	11								11						
Totals	35	335														
Average										8						
Probably in product										x						

APPENDIX 4-58 *Continued*

Product	Applesauce (Strained)												
Chemical Class			P	P	P	P	P	P	P	P	P	IC	IC
MB	Num	ppb	Ben	Car	Cpy	Dat	EnI	EnII	EnS	Eth	Ome	Tol	Xyl
96-1	4	40.4				7					4	13	20
96-2	4	4.4			2	1		0.4	1				
97-1	5	21.2			1	12			0.2	6	2		
98-1	2	6				2					4		
98-2	4	11.5			2			0.5	1	9			
98-3	3	57	48	6	3								
98-4	3	3.4			4			0.4	1				
99-1	4	28			12		4	7	5				
99-3	1	10								10			
00-1	2	23		4						19			
00-3	4	39.8	30					0.2	0.6	9			
01-3	2	7		1						6			
Totals	38	252											

APPENDIX 4-58 *Continued*

Product	Peaches (Strained)																			
Chemical Class			P	P	P	P	P	P	P	P	P	P	P	P	P	P	IC	IC	IC	IC
MB	Num	ppb	Car	Cpm	Cpy	Efv	EnI	EnII	EnS	Fen	Ipr	Ipm	P	Par	Prg	Pzo	Cfo	TCA	Tol	Xyl
96-1	4	77															16	17	19	25
96-2	4	20.3									2	2	1.3		15					
97-1	2	68									45	23								
98-1	3	20									1		15	4						
98-2	1	11											11							
98-3	4	107			2						14	4	87							
98-4	4	6.1						0.5	1		0.6					4				
99-1	4	28	5						2				11			4				
99-3	6	63.9		0.8			0.6	2	0.5	10			50							
00-1	3	12	2						1				9							
00-3	2	20	6										14							
01-3	3	22.4	0.4			10							12							
Totals	40	456																		
Average													17.5							
Probably in product													x							

APPENDIX 7-1 Current Federal Drinking Water Criteria to Control Microbial Hazards

Constituent	MCL	Characteristics
Cryptosporidium	TT	Gastrointestinal illness (e.g., diarrhea, vomiting, cramps).
Giardia lamblia	TT	Gastrointestinal illness (e.g., diarrhea, vomiting, cramps).
Heterotrophic plate count	TT	HPC has no health effects; it is an analytic method used to measure the variety of bacteria that is common in water. The lower the concentration of bacteria in drinking water, the better maintained the water system is. HPC measures a range of bacteria that are naturally present in the environment.
Legionella	TT	Legionnaire's disease, a type of pneumonia, found naturally in water; multiplies in heating systems.
Total Coliforms	5.0%*	Not a health threat in itself; it is used to indicate whether other potentially harmful bacteria may be present. Coliforms are naturally present in the environment, as well as feces. Fecal coliforms and *E. coli* only come from human and animal fecal waste.
Turbidity	TT	Turbidity is a measure of the cloudiness of water. It is used to indicate water quality and filtration effectiveness (e.g., whether disease-causing organisms are present). Higher turbidity levels are often associated with higher levels of disease-causing microorganisms such as viruses, parasites, and some bacteria. These organisms can cause symptoms such as nausea, cramps, diarrhea, and associated headaches.
Viruses (enteric)	TT	Gastrointestinal illness (e.g., diarrhea, vomiting, cramps).
TT Definition		

* No more than 5% samples total coliform-positive in a month. (For water systems that collect fewer than 40 routine samples per month, no more than one sample can be total coliform-positive per month.) Every sample that has total coliform must be analyzed for either fecal coliforms or *E. coli* if two consecutive TC-positive samples, and one is also positive for *E. coli* fecal coliforms. System has an acute MCL violation.

EPA's surface water treatment rules require systems using surface water or groundwater under the direct influence of surface water to (1) disinfect their water and (2) filter their water or meet criteria for avoiding filtration so that the following contaminants are controlled at the following levels:

- Cryptosporidium: (as of 1/1/02 for systems serving >10,000 and 1/14/05 for systems serving <10,000) 99% removal.
- *Giardia lamblia*: 99.9% removal/inactivation.
- Viruses: 99.99% removal/inactivation.
- Legionella: No limit, but EPA believes that if *Giardia* and viruses are removed/inactivated, *Legionella* will also be controlled.
- Turbidity: At no time can turbidity (cloudiness of water) go above 5 nephelolometric turbidity units (NTU); systems that filter must ensure that the turbidity go no higher than 1 NTU (0.5 NTU for conventional or direct filtration) in at least 95% of the daily samples in any month. As of January 1, 2002, turbidity may never exceed 1 NTU, and must not exceed 0.3 NTU in 95% of daily samples in any month.
- HPC: No more than 500 bacterial colonies per milliliter.

APPENDIX 7-2 Bakersfield Groundwater Data

State	PWS_ID	Location	POE	Constituent	Level	Unit	COLL_DATE	PWSType	Pop
CA	CA1502471	MESA MARIN RACEWAY	WELL 01	2,4-D	65	UG/L	3/5/2003	TNCWS	3000
CA	CA1503304	WM. BOLTHOUSE FARMS, INC. — WEST PLANT	NORTH WELL	ACENAPHTHENE	65	UG/L	4/1/2003	NTNCWS	300
CA	CA1503304	WM. BOLTHOUSE FARMS, INC. — WEST PLANT	NORTH WELL	ACENAPHTHENE	104	UG/L	7/1/2003	NTNCWS	300
CA	CA1502615	Frito-Lay No.America, Inc. Water System	SOUTH WELL (WELL 01)	ACENAPHTHENE	103	UG/L	7/16/2003	NTNCWS	400
CA	CA1502615	Frito-Lay No.America, Inc. Water System	NORTH WELL (WELL 02)	ACENAPHTHENE	103	UG/L	7/16/2003	NTNCWS	400
CA	CA1503304	WM. BOLTHOUSE FARMS, INC. — WEST PLANT	SOUTH WELL	ACENAPHTHENE	104	UG/L	7/1/2003	NTNCWS	300
CA	CA1503304	WM. BOLTHOUSE FARMS, INC. — WEST PLANT	SOUTH WELL	ACENAPHTHENE	100	UG/L	4/1/2003	NTNCWS	300
CA	CA1502094	GENERAL SHAFTER SCHOOL	WELL 01	ACENAPHTHENE	98	UG/L	4/13/2003	NTNCWS	240
CA	CA1503349	WM. BOLTHOUSE FARMS, INC.	LIMI WELL	ACENAPHTHENE	99	UG/L	4/1/2003	NTNCWS	1100
CA	CA1503349	WM. BOLTHOUSE FARMS, INC.	PORTER WELL	ACENAPHTHENE	103	UG/L	4/1/2003	NTNCWS	1100
CA	CA1500535	MANON MANOR MUTUAL WATER COMPANY	WELL 01	ACENAPHTHENE	101	UG/L	5/7/2003	CWS	200
CA	CA1502471	MESA MARIN RACEWAY	WELL 01	ACENAPHTHENE	99	UG/L	3/5/2003	TNCWS	3000
CA	CA1503264	POOLWELL SERVICES — JAMES ROAD	WELL 01	ACENAPHTHENE	103	UG/L	4/16/2003	NTNCWS	60
CA	CA1500096	OLD RIVER MUTUAL WATER COMPANY	WELL 01	ACENAPHTHENE	102	UG/L	5/1/2003	CWS	66
CA	CA1500344	SOUTH KERN MUTUAL WATER COMPANY	WELL 01	ACENAPHTHENE	103	UG/L	3/13/2003	CWS	35
					99.1				
CA	CA1503304	WM. BOLTHOUSE FARMS, INC. — WEST PLANT	NORTH WELL	ACENAPHTHYLENE	97	UG/L	7/1/2003	NTNCWS	300
CA	CA1502615	Frito-Lay No.America, Inc. Water System	SOUTH WELL (WELL 01)	ACENAPHTHYLENE	100	UG/L	7/16/2003	NTNCWS	400
CA	CA1502615	Frito-Lay No.America, Inc. Water System	NORTH WELL (WELL 02)	ACENAPHTHYLENE	100	UG/L	7/16/2003	NTNCWS	400
CA	CA1503304	WM. BOLTHOUSE FARMS, INC. — WEST PLANT	NORTH WELL	ACENAPHTHYLENE	97	UG/L	4/1/2003	NTNCWS	300
CA	CA1503304	WM. BOLTHOUSE FARMS, INC. — WEST PLANT	SOUTH WELL	ACENAPHTHYLENE	91	UG/L	7/1/2003	NTNCWS	300
CA	CA1503304	WM. BOLTHOUSE FARMS, INC. — WEST PLANT	SOUTH WELL	ACENAPHTHYLENE	100	UG/L	4/1/2003	NTNCWS	300
CA	CA1502094	GENERAL SHAFTER SCHOOL	WELL 01	ACENAPHTHYLENE	99	UG/L	4/13/2003	NTNCWS	240
CA	CA1503349	WM. BOLTHOUSE FARMS, INC.	LIMI WELL	ACENAPHTHYLENE	100	UG/L	4/1/2003	NTNCWS	1100
CA	CA1503349	WM. BOLTHOUSE FARMS, INC.	PORTER WELL	ACENAPHTHYLENE	102	UG/L	4/1/2003	NTNCWS	1100
CA	CA1500535	MANON MANOR MUTUAL WATER COMPANY	WELL 01	ACENAPHTHYLENE	100	UG/L	5/7/2003	CWS	200
CA	CA1502471	MESA MARIN RACEWAY	WELL 01	ACENAPHTHYLENE	101	UG/L	3/5/2003	TNCWS	3000
CA	CA1500096	OLD RIVER MUTUAL WATER COMPANY	WELL 01	ACENAPHTHYLENE	96	UG/L	5/1/2003	CWS	66
CA	CA1503264	POOLWELL SERVICES — JAMES ROAD	WELL 01	ACENAPHTHYLENE	101	UG/L	4/16/2003	NTNCWS	60
					98.8				
CA	CA1503182	SUN PACIFIC SHIPPERS-KERN DIVISION	WELL 01	ACETONE	2	UG/L	11/12/2002	NTNCWS	250
CA	CA1502482	BONANZA FARMS WATER SYSTEM	WELL 01	ALDICARB SULFONE	83	UG/L	5/16/2003	CWS	60
CA	CA1503093	GRIMMWAY ENTERPRISES-MALAGA WATER SYSTEM	WELL 01	ALDICARB SULFONE	86	UG/L	4/10/2003	NTNCWS	1500
CA	CA1503093	GRIMMWAY ENTERPRISES-MALAGA WATER SYSTEM	WELL 02	ALDICARB SULFONE	86	UG/L	4/10/2003	NTNCWS	1500
CA	CA1503093	GRIMMWAY ENTERPRISES-MALAGA WATER SYSTEM	WELL 03	ALDICARB SULFONE	85	UG/L	4/10/2003	NTNCWS	1500
CA	CA1500584	GOOSELAKE WATER COMPANY	WELL 01 STEPHANIE ST.	ALDICARB SULFONE	84	UG/L	5/21/2003	CWS	102
CA	CA1500546	HARVEST MOON MUTUAL WATER CO	WELL 01 (HEATH/BLACKHAWK)	ALDICARB SULFONE	88	UG/L	6/30/2003	CWS	180
CA	CA1500491	INTERSTATE 5 PROPERTIES	WELL 01	ALDICARB SULFONE	83	UG/L	5/6/2003	NTNCWS	800
CA	CA1500491	INTERSTATE 5 PROPERTIES	WELL 02	ALDICARB SULFONE	82	UG/L	5/6/2003	NTNCWS	800
CA	CA1500600	J.G. BOSWELL COMPANY WATER SYSTEM	WELL 01	ALDICARB SULFONE	86	UG/L	5/22/2003	NTNCWS	50
CA	CA1502471	MESA MARIN RACEWAY	WELL 01	ALDICARB SULFONE	97	UG/L	3/5/2003	TNCWS	3000
					86				
CA	CA1503304	WM. BOLTHOUSE FARMS, INC. — WEST PLANT	NORTH WELL	ALDRIN	102	UG/L	4/1/2003	NTNCWS	300
CA	CA1503304	WM. BOLTHOUSE FARMS, INC. — WEST PLANT	NORTH WELL	ALDRIN	99	UG/L	7/1/2003	NTNCWS	300
CA	CA1502615	Frito-Lay No.America, Inc. Water System	SOUTH WELL (WELL 01)	ALDRIN	98	UG/L	7/16/2003	NTNCWS	400
CA	CA1502615	Frito-Lay No.America, Inc. Water System	NORTH WELL (WELL 02)	ALDRIN	99	UG/L	7/16/2003	NTNCWS	400

CA	CA1502094	GENERAL SHAFTER SCHOOL	WELL 01	ALDRIN	102	UG/L	4/13/2003	NTNCWS	240
CA	CA1503349	WM. BOLTHOUSE FARMS, INC.	LIMI WELL	ALDRIN	100	UG/L	4/1/2003	NTNCWS	1100
CA	CA1503349	WM. BOLTHOUSE FARMS, INC.	PORTER WELL	ALDRIN	99	UG/L	4/1/2003	NTNCWS	1100
CA	CA1503304	WM. BOLTHOUSE FARMS, INC. — WEST PLANT	SOUTH WELL	ALDRIN	100	UG/L	7/1/2003	NTNCWS	300
CA	CA1503304	WM. BOLTHOUSE FARMS, INC. — WEST PLANT	SOUTH WELL	ALDRIN	102	UG/L	4/1/2003	NTNCWS	300
CA	CA1500535	MANON MANOR MUTUAL WATER COMPANY	WELL 01	ALDRIN	100	UG/L	5/7/2003	CWS	200
CA	CA1502471	MESA MARIN RACEWAY	WELL 01	ALDRIN	107	UG/L	3/5/2003	TNCWS	3000
CA	CA1500096	OLD RIVER MUTUAL WATER COMPANY	WELL 01	ALDRIN	100	UG/L	5/1/2003	CWS	66
CA	CA1503264	POOLWELL SERVICES — JAMES ROAD	WELL 01	ALDRIN	100	UG/L	4/16/2003	NTNCWS	60
					100.6				
CA	CA1503100	Chevron-TEXACO G & W WATER SYSTEM	DEL REY #100 — INACTIVE	BENZENE	0.67	UG/L	8/7/2001	NTNCWS	146
CA	CA1502471	MESA MARIN RACEWAY	WELL 01	BROMOBENZENE	83	UG/L	3/5/2003	TNCWS	3000
CA	CA1510004	Casa Loma Water Co, Inc.	WELL 03 — RAW (DEL MAR DR)	BROMODICHLOROMETHANE (THM)	3.6	UG/L	3/20/2003	CWS	600
CA	CA1502482	BONANZA FARMS WATER SYSTEM	WELL 01	BROMODICHLOROMETHANE (THM)	0.54	UG/L	5/1/2000	CWS	60
CA	CA1500546	HARVEST MOON MUTUAL WATER CO	WELL 01 (HEATH/BLACKHAWK)	BROMODICHLOROMETHANE (THM)	0.56	UG/L	6/8/2000	CWS	180
CA	CA1503330	GCG WATER SYSTEM	WELL 01	BROMODICHLOROMETHANE (THM)	0.89	UG/L	8/26/2004	NTNCWS	40
CA	CA1502320	LERDO SHERIFF'S FACILITY	WELL 02 — CL2 TREATMENT	BROMODICHLOROMETHANE (THM)	0.67	UG/L	7/7/2004	CWS	3000
CA	CA1500578	LONG CANYON WATER COMPANY CORP.	WELL 02	BROMODICHLOROMETHANE (THM)	6.1	UG/L	1/1/2001	CWS	197
CA	CA1500378	MAHER MUTUAL WATER COMPANY	WELL 01	BROMODICHLOROMETHANE (THM)	0.61	UG/L	8/17/2004	CWS	150
CA	CA1503182	SUN PACIFIC SHIPPERS-KERN DIVISION	WELL 01	BROMODICHLOROMETHANE (THM)	1.72	UG/L	8/3/2004	NTNCWS	250
CA	CA1503182	SUN PACIFIC SHIPPERS-KERN DIVISION	WELL 01	BROMODICHLOROMETHANE (THM)	2.4	UG/L	11/12/2002	NTNCWS	250
CA	CA1502465	PANAMA ROAD PROPERTY OWNERS ASSOC	WELL 01	BROMODICHLOROMETHANE (THM)	3.31	UG/L	6/13/2000	CWS	50
CA	CA1502663	TOWN & COUNTRY WATER COMPANY	WELL 02 — BOREL — STANDBY	BROMODICHLOROMETHANE (THM)	0.6	UG/L	8/27/2004	CWS	80
					1.91				
CA	CA1500483	ALLEN ROAD MUTUAL WATER SYSTEM	WELL 01	BROMOFORM (THM)	1.05	UG/L	7/14/1999	CWS	200
CA	CA1500544	ENOS LANE PUBLIC UTILITY DISTRICT	WELL 02 — WEST	BROMOFORM (THM)	3.4	UG/L	8/27/2004	CWS	240
CA	CA1510004	Casa Loma Water Co, Inc.	WELL 03 — RAW (DEL MAR DR)	BROMOFORM (THM)	2.2	UG/L	3/20/2003	CWS	600
CA	CA1503349	WM. BOLTHOUSE FARMS, INC.	LIMI WELL	BROMOFORM (THM)	1.1	UG/L	5/5/1998	NTNCWS	1100
CA	CA1500289	ATHAL MUTUAL WATER SYSTEM	WELL 01	BROMOFORM (THM)	0.52	UG/L	8/17/2004	CWS	85
CA	CA1500409	BROCK MUTUAL WATER COMPANY	WELL 02 — NORTH	BROMOFORM (THM)	1.8	UG/L	1/4/2002	CWS	300
CA	CA1502615	Frito-Lay No.America, Inc. Water System	SOUTH WELL (WELL 01)	BROMOFORM (THM)	0.7	UG/L	2/28/2005	NTNCWS	400
CA	CA1503336	GOLDEN EMPIRE CONCRETE COMPANY	WELL 01	BROMOFORM (THM)	0.53	UG/L	8/26/2004	NTNCWS	44
CA	CA1502615	Frito-Lay No.America, Inc. Water System	SOUTH WELL (WELL 01) — CL2 TREATMENT	BROMOFORM (THM)	0.8	UG/L	5/28/2003	NTNCWS	400
CA	CA1502320	LERDO SHERIFF'S FACILITY	WELL 02 — CL2 TREATMENT	BROMOFORM (THM)	2.5	UG/L	7/7/2004	CWS	3000
CA	CA1502615	Frito-Lay No.America, Inc. Water System	SOUTH WELL (WELL 01) — CL2 TREATMENT	BROMOFORM (THM)	0.7	UG/L	11/18/2002	NTNCWS	400
CA	CA1502615	Frito-Lay No.America, Inc. Water System	NORTH WELL (WELL 02)	BROMOFORM (THM)	0.8	UG/L	2/28/2005	NTNCWS	400
CA	CA1502615	Frito-Lay No.America, Inc. Water System	NORTH WELL (WELL 02) — CL2 TREATMENT	BROMOFORM (THM)	0.6	UG/L	4/1/2003	NTNCWS	400
CA	CA1502615	Frito-Lay No.America, Inc. Water System	NORTH WELL (WELL 02) — CL2 TREATMENT	BROMOFORM (THM)	0.6	UG/L	5/28/2003	NTNCWS	400
CA	CA1502615	Frito-Lay No.America, Inc. Water System	NORTH WELL (WELL 02) — CL2 TREATMENT	BROMOFORM (THM)	1	UG/L	11/18/2002	NTNCWS	400
CA	CA1500296	FULLER ACRES MUTUAL WATER COMPANY	WELL 01	BROMOFORM (THM)	0.82	UG/L	8/16/2004	CWS	194
CA	CA1500546	HARVEST MOON MUTUAL WATER CO	WELL 01 (HEATH/BLACKHAWK)	BROMOFORM (THM)	2.78	UG/L	6/8/2000	CWS	180
CA	CA1500578	LONG CANYON WATER COMPANY CORP.	WELL 02	BROMOFORM (THM)	0.98	UG/L	1/1/2001	CWS	197
CA	CA1502465	PANAMA ROAD PROPERTY OWNERS ASSOC	WELL 01	BROMOFORM (THM)	4.4	UG/L	6/13/2000	CWS	50
CA	CA1503182	SUN PACIFIC SHIPPERS-KERN DIVISION	WELL 01	BROMOFORM (THM)	7	UG/L	11/12/2002	NTNCWS	250
CA	CA1503182	SUN PACIFIC SHIPPERS-KERN DIVISION	WELL 01	BROMOFORM (THM)	13.73	UG/L	8/3/2004	NTNCWS	250
CA	CA1502663	TOWN & COUNTRY WATER COMPANY	WELL 02 — BOREL — STANDBY	BROMOFORM (THM)	6.7	UG/L	8/27/2004	CWS	80
CA	CA1510029	Vaughn WC INC	TORREY WELL — BEFORE GAC	BROMOFORM (THM)	0.7	UG/L	4/11/2000	CWS	19130
CA	CA1500494	WILSON ROAD WATER COMMUNITY	WELL 01	BROMOFORM (THM)	1	UG/L	8/17/2004	CWS	60
					2.35				

APPENDIX 7-2 *Continued*

State	PWS_ID	Location	POE	Constituent	Level	Unit	COLL_DATE	PWSType	Pop
CA	CA1510004	Casa Loma Water Co, Inc.	WELL 03 — RAW (DEL MAR DR)	CHLOROFORM (THM)	5	UG/L	3/20/2003	CWS	600
CA	CA1502398	FARMER JOHN EGG RANCH #2	WELL 01	CHLOROFORM (THM)	0.93	UG/L	6/17/1999	NTNCWS	30
CA	CA1503330	GCG WATER SYSTEM	WELL 01	CHLOROFORM (THM)	0.9	UG/L	8/26/2004	NTNCWS	40
CA	CA1500597	GRIMMWAY FARMS FROZEN FOODS	WELL 01	CHLOROFORM (THM)	0.66	UG/L	5/10/2004	NTNCWS	350
CA	CA1500578	LONG CANYON WATER COMPANY CORP.	WELL 02	CHLOROFORM (THM)	3.5	UG/L	1/1/2001	CWS	197
CA	CA1500378	MAHER MUTUAL WATER COMPANY	WELL 01	CHLOROFORM (THM)	0.73	UG/L	8/17/2004	CWS	150
CA	CA1500452	RIVERDALE VILLAGE	WELL 01	CHLOROFORM (THM)	1.1	UG/L	4/4/2001	CWS	760
CA	CA1500452	RIVERDALE VILLAGE	WELL 01	CHLOROFORM (THM)	1.9	UG/L	4/7/2004	CWS	760
CA	CA1502465	PANAMA ROAD PROPERTY OWNERS ASSOC	WELL 01	CHLOROFORM (THM)	1.33	UG/L	6/13/2000	CWS	50
CA	CA1503182	SUN PACIFIC SHIPPERS-KERN DIVISION	WELL 01	CHLOROFORM (THM)	1.7	UG/L	11/12/2002	NTNCWS	250
CA	CA1503182	SUN PACIFIC SHIPPERS-KERN DIVISION	WELL 01	CHLOROFORM (THM)	0.67	UG/L	8/3/2004	NTNCWS	250
CA	CA1502663	TOWN & COUNTRY WATER COMPANY	WELL 02 — BOREL — STANDBY	CHLOROFORM (THM)	0.58	UG/L	8/27/2004	CWS	80
CA	CA0104012	Vulcan Materials, CalMat Division	WELL 02 — PENDING	CHLOROFORM (THM)	0.94	UG/L	3/25/2003	NTNCWS	45
CA	CA1500494	WILSON ROAD WATER COMMUNITY	WELL 01	CHLOROFORM (THM)	0.85	UG/L	9/24/1997	CWS	60
					1.49				
CA	CA1502615	Frito-Lay No.America, Inc. Water System	NORTH WELL (WELL 02) — CL2 TREATMENT	CHLOROMETHANE	0.6	UG/L	11/18/2002	NTNCWS	400
CA	CA1500544	ENOS LANE PUBLIC UTILITY DISTRICT	WELL 02 — WEST	DIBROMOACETIC ACID (DBAA)	1.2	UG/L	8/27/2004	CWS	240
CA	CA1503182	SUN PACIFIC SHIPPERS-KERN DIVISION	WELL 01	DIBROMOACETIC ACID (DBAA)	3.2	UG/L	8/3/2004	NTNCWS	250
CA	CA1502663	TOWN & COUNTRY WATER COMPANY	WELL 02 — BOREL — STANDBY	DIBROMOACETIC ACID (DBAA)	1.4	UG/L	8/27/2004	CWS	80
					1.9				
CA	CA1510004	Casa Loma Water Co, Inc.	WELL 03 — RAW (DEL MAR DR)	DIBROMOCHLOROMETHANE (THM)	3.1	UG/L	3/20/2003	CWS	600
CA	CA1500544	ENOS LANE PUBLIC UTILITY DISTRICT	WELL 02 — WEST	DIBROMOCHLOROMETHANE (THM)	0.84	UG/L	8/27/2004	CWS	240
CA	CA1502615	Frito-Lay No.America, Inc. Water System	NORTH WELL (WELL 02)	DIBROMOCHLOROMETHANE (THM)	0.5	UG/L	2/28/2005	NTNCWS	400
CA	CA1502615	Frito-Lay No.America, Inc. Water System	NORTH WELL (WELL 02) — CL2 TREATMENT	DIBROMOCHLOROMETHANE (THM)	0.6	UG/L	11/18/2002	NTNCWS	400
CA	CA1500296	FULLER ACRES MUTUAL WATER COMPANY	WELL 01	DIBROMOCHLOROMETHANE (THM)	0.65	UG/L	8/16/2004	CWS	194
CA	CA1503330	GCG WATER SYSTEM	WELL 01	DIBROMOCHLOROMETHANE (THM)	1.1	UG/L	8/26/2004	NTNCWS	40
CA	CA1503336	GOLDEN EMPIRE CONCRETE COMPANY	WELL 01	BROMODICHLOROMETHANE (THM)	0.63	UG/L	8/26/2004	NTNCWS	44
CA	CA1503336	GOLDEN EMPIRE CONCRETE COMPANY	WELL 01	DIBROMOCHLOROMETHANE (THM)	1.3	UG/L	8/26/2004	NTNCWS	44
CA	CA1500546	HARVEST MOON MUTUAL WATER CO	WELL 01 (HEATH/BLACKHAWK)	DIBROMOCHLOROMETHANE (THM)	1.72	UG/L	6/8/2000	CWS	180
CA	CA1502320	LERDO SHERIFF'S FACILITY	WELL 02 — CL2 TREATMENT	DIBROMOCHLOROMETHANE (THM)	1.5	UG/L	7/7/2004	CWS	3000
CA	CA1500578	LONG CANYON WATER COMPANY CORP.	WELL 02	DIBROMOCHLOROMETHANE (THM)	5.9	UG/L	1/1/2001	CWS	197
CA	CA1500378	MAHER MUTUAL WATER COMPANY	WELL 01	DIBROMOCHLOROMETHANE (THM)	0.62	UG/L	8/17/2004	CWS	150
CA	CA1502465	PANAMA ROAD PROPERTY OWNERS ASSOC	WELL 01	DIBROMOCHLOROMETHANE (THM)	5.99	UG/L	6/13/2000	CWS	50
CA	CA1503182	SUN PACIFIC SHIPPERS-KERN DIVISION	WELL 01	DIBROMOCHLOROMETHANE (THM)	4.5	UG/L	11/12/2002	NTNCWS	250
CA	CA1503182	SUN PACIFIC SHIPPERS-KERN DIVISION	WELL 01	DIBROMOCHLOROMETHANE (THM)	4.1	UG/L	8/3/2004	NTNCWS	250
CA	CA1502663	TOWN & COUNTRY WATER COMPANY	WELL 02 — BOREL — STANDBY	DIBROMOCHLOROMETHANE (THM)	1.2	UG/L	8/27/2004	CWS	80
					2.14				
CA	CA1502615	Frito-Lay No.America, Inc. Water System	NORTH WELL (WELL 02) — CL2 TREATMENT	DICHLOROACETIC ACID (DCAA)	1.4	UG/L	4/1/2003	NTNCWS	400
CA	CA1500544	ENOS LANE PUBLIC UTILITY DISTRICT	WELL 01 EAST — STANDBY	DICHLOROMETHANE	0.51	UG/L	9/7/1999	CWS	240
CA	CA1500491	INTERSTATE 5 PROPERTIES	WELL 01	DICHLOROMETHANE	1.1	UG/L	10/7/1998	NTNCWS	800
					0.81				
CA	CA1500584	GOOSELAKE WATER COMPANY	WELL 01 STEPHANIE ST.	DIBROMOCHLOROPROPANE (DBCP)	0.03	UG/L	5/11/2000	CWS	102
CA	CA1503093	GRIMMWAY ENTERPRISES-MALAGA WATER SYSTEM	WELL 01	DIBROMOCHLOROPROPANE (DBCP)	0.01	UG/L	9/5/2003	NTNCWS	1500
CA	CA1503093	GRIMMWAY ENTERPRISES-MALAGA WATER SYSTEM	WELL 01	DIBROMOCHLOROPROPANE (DBCP)	0.01	UG/L	6/6/2003	NTNCWS	1500
CA	CA1503093	GRIMMWAY ENTERPRISES-MALAGA WATER SYSTEM	WELL 01	DIBROMOCHLOROPROPANE (DBCP)	0.01	UG/L	8/8/2003	NTNCWS	1500
CA	CA1503093	GRIMMWAY ENTERPRISES-MALAGA WATER SYSTEM	WELL 01	DIBROMOCHLOROPROPANE (DBCP)	0.01	UG/L	5/2/2003	NTNCWS	1500

CA	CA1503093	GRIMMWAY ENTERPRISES-MALAGA WATER SYSTEM	WELL 01	DIBROMOCHLOROPROPANE (DBCP)	0.01	UG/L	8/13/2004	NTNCWS	1500
CA	CA1503093	GRIMMWAY ENTERPRISES-MALAGA WATER SYSTEM	WELL 01	DIBROMOCHLOROPROPANE (DBCP)	0.04	UG/L	11/5/2004	NTNCWS	1500
CA	CA1503093	GRIMMWAY ENTERPRISES-MALAGA WATER SYSTEM	WELL 01	DIBROMOCHLOROPROPANE (DBCP)	0.01	UG/L	11/7/2003	NTNCWS	1500
CA	CA1503093	GRIMMWAY ENTERPRISES-MALAGA WATER SYSTEM	WELL 01	DIBROMOCHLOROPROPANE (DBCP)	0.02	UG/L	5/7/2004	NTNCWS	1500
CA	CA1503093	GRIMMWAY ENTERPRISES-MALAGA WATER SYSTEM	WELL 01	DIBROMOCHLOROPROPANE (DBCP)	0.01	UG/L	2/12/2004	NTNCWS	1500
CA	CA1503093	GRIMMWAY ENTERPRISES-MALAGA WATER SYSTEM	WELL 01	DIBROMOCHLOROPROPANE (DBCP)	0.01	UG/L	4/2/2004	NTNCWS	1500
CA	CA1503093	GRIMMWAY ENTERPRISES-MALAGA WATER SYSTEM	WELL 01	DIBROMOCHLOROPROPANE (DBCP)	0.01	UG/L	10/24/2003	NTNCWS	1500
CA	CA1503093	GRIMMWAY ENTERPRISES-MALAGA WATER SYSTEM	WELL 01	DIBROMOCHLOROPROPANE (DBCP)	0.02	UG/L	2/1/2002	NTNCWS	1500
CA	CA1503093	GRIMMWAY ENTERPRISES-MALAGA WATER SYSTEM	WELL 01	DIBROMOCHLOROPROPANE (DBCP)	0.23	UG/L	3/8/2002	NTNCWS	1500
CA	CA1503093	GRIMMWAY ENTERPRISES-MALAGA WATER SYSTEM	WELL 01	DIBROMOCHLOROPROPANE (DBCP)	0.02	UG/L	7/6/2001	NTNCWS	1500
CA	CA1503093	GRIMMWAY ENTERPRISES-MALAGA WATER SYSTEM	WELL 01	DIBROMOCHLOROPROPANE (DBCP)	0.01	UG/L	4/8/2002	NTNCWS	1500
CA	CA1503093	GRIMMWAY ENTERPRISES-MALAGA WATER SYSTEM	WELL 01	DIBROMOCHLOROPROPANE (DBCP)	0.01	UG/L	5/3/2002	NTNCWS	1500
CA	CA1503093	GRIMMWAY ENTERPRISES-MALAGA WATER SYSTEM	WELL 01	DIBROMOCHLOROPROPANE (DBCP)	0.01	UG/L	1/7/2002	NTNCWS	1500
CA	CA1503093	GRIMMWAY ENTERPRISES-MALAGA WATER SYSTEM	WELL 01	DIBROMOCHLOROPROPANE (DBCP)	0.02	UG/L	11/10/2000	NTNCWS	1500
CA	CA1503093	GRIMMWAY ENTERPRISES-MALAGA WATER SYSTEM	WELL 01	DIBROMOCHLOROPROPANE (DBCP)	0.01	UG/L	1/3/2003	NTNCWS	1500
CA	CA1503093	GRIMMWAY ENTERPRISES-MALAGA WATER SYSTEM	WELL 01	DIBROMOCHLOROPROPANE (DBCP)	0.04	UG/L	8/9/2002	NTNCWS	1500
CA	CA1503093	GRIMMWAY ENTERPRISES-MALAGA WATER SYSTEM	WELL 02	DIBROMOCHLOROPROPANE (DBCP)	0.02	UG/L	4/10/2003	NTNCWS	1500
CA	CA1503093	GRIMMWAY ENTERPRISES-MALAGA WATER SYSTEM	WELL 02	DIBROMOCHLOROPROPANE (DBCP)	0.02	UG/L	6/6/2003	NTNCWS	1500
CA	CA1503093	GRIMMWAY ENTERPRISES-MALAGA WATER SYSTEM	WELL 02	DIBROMOCHLOROPROPANE (DBCP)	0.05	UG/L	8/8/2003	NTNCWS	1500
CA	CA1503093	GRIMMWAY ENTERPRISES-MALAGA WATER SYSTEM	WELL 02	DIBROMOCHLOROPROPANE (DBCP)	0.02	UG/L	3/7/2003	NTNCWS	1500
CA	CA1503093	GRIMMWAY ENTERPRISES-MALAGA WATER SYSTEM	WELL 02	DIBROMOCHLOROPROPANE (DBCP)	0.05	UG/L	11/5/2004	NTNCWS	1500
CA	CA1503093	GRIMMWAY ENTERPRISES-MALAGA WATER SYSTEM	WELL 02	DIBROMOCHLOROPROPANE (DBCP)	0.02	UG/L	9/5/2003	NTNCWS	1500
CA	CA1503093	GRIMMWAY ENTERPRISES-MALAGA WATER SYSTEM	WELL 02	DIBROMOCHLOROPROPANE (DBCP)	0.04	UG/L	4/4/2003	NTNCWS	1500
CA	CA1503093	GRIMMWAY ENTERPRISES-MALAGA WATER SYSTEM	WELL 02	DIBROMOCHLOROPROPANE (DBCP)	0.02	UG/L	8/13/2004	NTNCWS	1500
CA	CA1503093	GRIMMWAY ENTERPRISES-MALAGA WATER SYSTEM	WELL 02	DIBROMOCHLOROPROPANE (DBCP)	0.04	UG/L	5/7/2004	NTNCWS	1500
CA	CA1503093	GRIMMWAY ENTERPRISES-MALAGA WATER SYSTEM	WELL 02	DIBROMOCHLOROPROPANE (DBCP)	0.03	UG/L	4/2/2004	NTNCWS	1500
CA	CA1503093	GRIMMWAY ENTERPRISES-MALAGA WATER SYSTEM	WELL 02	DIBROMOCHLOROPROPANE (DBCP)	0.04	UG/L	12/5/2003	NTNCWS	1500
CA	CA1503093	GRIMMWAY ENTERPRISES-MALAGA WATER SYSTEM	WELL 02	DIBROMOCHLOROPROPANE (DBCP)	0.03	UG/L	10/24/2003	NTNCWS	1500
CA	CA1503093	GRIMMWAY ENTERPRISES-MALAGA WATER SYSTEM	WELL 02	DIBROMOCHLOROPROPANE (DBCP)	0.01	UG/L	11/7/2003	NTNCWS	1500
CA	CA1503093	GRIMMWAY ENTERPRISES-MALAGA WATER SYSTEM	WELL 02	DIBROMOCHLOROPROPANE (DBCP)	0.02	UG/L	2/12/2004	NTNCWS	1500
CA	CA1503093	GRIMMWAY ENTERPRISES-MALAGA WATER SYSTEM	WELL 02	DIBROMOCHLOROPROPANE (DBCP)	0.03	UG/L	3/5/2004	NTNCWS	1500
CA	CA1503093	GRIMMWAY ENTERPRISES-MALAGA WATER SYSTEM	WELL 02	DIBROMOCHLOROPROPANE (DBCP)	0.04	UG/L	4/8/2002	NTNCWS	1500
CA	CA1503093	GRIMMWAY ENTERPRISES-MALAGA WATER SYSTEM	WELL 02	DIBROMOCHLOROPROPANE (DBCP)	0.03	UG/L	2/1/2002	NTNCWS	1500
CA	CA1503093	GRIMMWAY ENTERPRISES-MALAGA WATER SYSTEM	WELL 02	DIBROMOCHLOROPROPANE (DBCP)	0.02	UG/L	3/8/2002	NTNCWS	1500
CA	CA1503093	GRIMMWAY ENTERPRISES-MALAGA WATER SYSTEM	WELL 02	DIBROMOCHLOROPROPANE (DBCP)	0.01	UG/L	5/3/2002	NTNCWS	1500
CA	CA1503093	GRIMMWAY ENTERPRISES-MALAGA WATER SYSTEM	WELL 02	DIBROMOCHLOROPROPANE (DBCP)	0.01	UG/L	6/10/2002	NTNCWS	1500
CA	CA1503093	GRIMMWAY ENTERPRISES-MALAGA WATER SYSTEM	WELL 02	DIBROMOCHLOROPROPANE (DBCP)	0.03	UG/L	11/10/2000	NTNCWS	1500
CA	CA1503093	GRIMMWAY ENTERPRISES-MALAGA WATER SYSTEM	WELL 02	DIBROMOCHLOROPROPANE (DBCP)	0.03	UG/L	7/6/2001	NTNCWS	1500
CA	CA1503093	GRIMMWAY ENTERPRISES-MALAGA WATER SYSTEM	WELL 02	DIBROMOCHLOROPROPANE (DBCP)	0.03	UG/L	1/7/2002	NTNCWS	1500
CA	CA1503093	GRIMMWAY ENTERPRISES-MALAGA WATER SYSTEM	WELL 02	DIBROMOCHLOROPROPANE (DBCP)	0.01	UG/L	1/3/2003	NTNCWS	1500
CA	CA1503093	GRIMMWAY ENTERPRISES-MALAGA WATER SYSTEM	WELL 02	DIBROMOCHLOROPROPANE (DBCP)	0.01	UG/L	6/10/2002	NTNCWS	1500
CA	CA1503093	GRIMMWAY ENTERPRISES-MALAGA WATER SYSTEM	WELL 02	DIBROMOCHLOROPROPANE (DBCP)	0.01	UG/L	11/4/2002	NTNCWS	1500
CA	CA1503093	GRIMMWAY ENTERPRISES-MALAGA WATER SYSTEM	WELL 02	DIBROMOCHLOROPROPANE (DBCP)	0.02	UG/L	8/9/2002	NTNCWS	1500
CA	CA1503093	GRIMMWAY ENTERPRISES-MALAGA WATER SYSTEM	WELL 03	DIBROMOCHLOROPROPANE (DBCP)	0.02	UG/L	4/10/2003	NTNCWS	1500
CA	CA1503093	GRIMMWAY ENTERPRISES-MALAGA WATER SYSTEM	WELL 03	DIBROMOCHLOROPROPANE (DBCP)	0.02	UG/L	9/5/2003	NTNCWS	1500
CA	CA1503093	GRIMMWAY ENTERPRISES-MALAGA WATER SYSTEM	WELL 03	DIBROMOCHLOROPROPANE (DBCP)	0.02	UG/L	10/24/2003	NTNCWS	1500
CA	CA1503093	GRIMMWAY ENTERPRISES-MALAGA WATER SYSTEM	WELL 03	DIBROMOCHLOROPROPANE (DBCP)	0.03	UG/L	8/8/2003	NTNCWS	1500
CA	CA1503093	GRIMMWAY ENTERPRISES-MALAGA WATER SYSTEM	WELL 03	DIBROMOCHLOROPROPANE (DBCP)	0.01	UG/L	5/2/2003	NTNCWS	1500
CA	CA1503093	GRIMMWAY ENTERPRISES-MALAGA WATER SYSTEM	WELL 03	DIBROMOCHLOROPROPANE (DBCP)	0.04	UG/L	4/8/2002	NTNCWS	1500
CA	CA1503093	GRIMMWAY ENTERPRISES-MALAGA WATER SYSTEM	WELL 03	DIBROMOCHLOROPROPANE (DBCP)	0.01	UG/L	8/9/2002	NTNCWS	1500

APPENDIX 7-2 *Continued*

State	PWS_ID	Location	POE	Constituent	Level	Unit	COLL_DATE	PWSType	Pop
CA	CA1503093	GRIMMWAY ENTERPRISES-MALAGA WATER SYSTEM	WELL 03	DIBROMOCHLOROPROPANE (DBCP)	0.02	UG/L	3/8/2002	NTNCWS	1500
CA	CA1503093	GRIMMWAY ENTERPRISES-MALAGA WATER SYSTEM	WELL 03	DIBROMOCHLOROPROPANE (DBCP)	0.07	UG/L	2/1/2002	NTNCWS	1500
CA	CA1503093	GRIMMWAY ENTERPRISES-MALAGA WATER SYSTEM	WELL 03	DIBROMOCHLOROPROPANE (DBCP)	0.01	UG/L	11/7/2003	NTNCWS	1500
CA	CA1503093	GRIMMWAY ENTERPRISES-MALAGA WATER SYSTEM	WELL 03	DIBROMOCHLOROPROPANE (DBCP)	0.01	UG/L	1/7/2003	NTNCWS	1500
CA	CA1503093	GRIMMWAY ENTERPRISES-MALAGA WATER SYSTEM	WELL 03	DIBROMOCHLOROPROPANE (DBCP)	0.01	UG/L	3/7/2003	NTNCWS	1500
CA	CA1503093	GRIMMWAY ENTERPRISES-MALAGA WATER SYSTEM	WELL 03	DIBROMOCHLOROPROPANE (DBCP)	0.02	UG/L	12/6/2002	NTNCWS	1500
CA	CA1503093	GRIMMWAY ENTERPRISES-MALAGA WATER SYSTEM	WELL 03	DIBROMOCHLOROPROPANE (DBCP)	0.03	UG/L	11/4/2002	NTNCWS	1500
CA	CA1503093	GRIMMWAY ENTERPRISES-MALAGA WATER SYSTEM	WELL 03	DIBROMOCHLOROPROPANE (DBCP)	0.03	UG/L	11/5/2004	NTNCWS	1500
CA	CA1503093	GRIMMWAY ENTERPRISES-MALAGA WATER SYSTEM	WELL 03	DIBROMOCHLOROPROPANE (DBCP)	0.01	UG/L	8/13/2004	NTNCWS	1500
CA	CA1503093	GRIMMWAY ENTERPRISES-MALAGA WATER SYSTEM	WELL 03	DIBROMOCHLOROPROPANE (DBCP)	0.02	UG/L	5/7/2004	NTNCWS	1500
CA	CA1503093	GRIMMWAY ENTERPRISES-MALAGA WATER SYSTEM	WELL 03	DIBROMOCHLOROPROPANE (DBCP)	0.02	UG/L	2/12/2004	NTNCWS	1500
CA	CA1503093	GRIMMWAY ENTERPRISES-MALAGA WATER SYSTEM	WELL 03	DIBROMOCHLOROPROPANE (DBCP)	0.02	UG/L	12/5/2003	NTNCWS	1500
CA	CA1503093	GRIMMWAY ENTERPRISES-MALAGA WATER SYSTEM	WELL 03	DIBROMOCHLOROPROPANE (DBCP)	0.03	UG/L	3/5/2004	NTNCWS	1500
CA	CA1503093	GRIMMWAY ENTERPRISES-MALAGA WATER SYSTEM	WELL 03	DIBROMOCHLOROPROPANE (DBCP)	0.01	UG/L	4/2/2004	NTNCWS	1500
CA	CA1500597	GRIMMWAY FARMS FROZEN FOODS	WELL 01	DIBROMOCHLOROPROPANE (DBCP)	0.03	UG/L	3/12/2004	NTNCWS	350
CA	CA1500597	GRIMMWAY FARMS FROZEN FOODS	WELL 01	DIBROMOCHLOROPROPANE (DBCP)	0.04	UG/L	3/22/2001	NTNCWS	350
CA	CA1500597	GRIMMWAY FARMS FROZEN FOODS	WELL 01	DIBROMOCHLOROPROPANE (DBCP)	0.03	UG/L	6/7/1999	NTNCWS	350
CA	CA1500597	GRIMMWAY FARMS FROZEN FOODS	WELL 01	DIBROMOCHLOROPROPANE (DBCP)	0.02	UG/L	9/6/2000	NTNCWS	350
CA	CA1500597	GRIMMWAY FARMS FROZEN FOODS	WELL 01	DIBROMOCHLOROPROPANE (DBCP)	0.1	UG/L	12/8/2000	NTNCWS	350
CA	CA1500597	GRIMMWAY FARMS FROZEN FOODS	WELL 02	DIBROMOCHLOROPROPANE (DBCP)	0.02	UG/L	6/7/1999	NTNCWS	350
CA	CA1500597	GRIMMWAY FARMS FROZEN FOODS	WELL 02	DIBROMOCHLOROPROPANE (DBCP)	0.03	UG/L	12/8/2000	NTNCWS	350
CA	CA1500597	GRIMMWAY FARMS FROZEN FOODS	WELL 02	DIBROMOCHLOROPROPANE (DBCP)	0.02	UG/L	3/22/2001	NTNCWS	350
CA	CA1500597	GRIMMWAY FARMS FROZEN FOODS	WELL 02	DIBROMOCHLOROPROPANE (DBCP)	0.01	UG/L	9/6/2000	NTNCWS	350
CA	CA1500546	HARVEST MOON MUTUAL WATER CO	WELL 01 (HEATH/BLACKHAWK)	DIBROMOCHLOROPROPANE (DBCP)	0.04	UG/L	6/30/2003	CWS	180
CA	CA1500575	SAN JOAQUIN ESTATES MUTUAL	WELL 01	DIBROMOCHLOROPROPANE (DBCP)	0.01	UG/L	7/24/2001	CWS	220
CA	CA1502545	SCHWEIKART WATER SYSTEM	WELL 01	DIBROMOCHLOROPROPANE (DBCP)	0.03	UG/L	12/30/1997	CWS	27
CA	CA1502545	SCHWEIKART WATER SYSTEM	WELL 01	DIBROMOCHLOROPROPANE (DBCP)	0.04	UG/L	9/26/1997	CWS	27
CA	CA1502545	SCHWEIKART WATER SYSTEM	WELL 01	DIBROMOCHLOROPROPANE (DBCP)	0.03	UG/L	6/24/1997	CWS	27
CA	CA1502273	SUN WORLD INTERNATIONAL, INC.-COM CENTER	WELL 99-2	DIBROMOCHLOROPROPANE (DBCP)	0.47	UG/L	8/26/2003	NTNCWS	260
CA	CA1502273	SUN WORLD INTERNATIONAL, INC.-COM CENTER	WELL 99-2	DIBROMOCHLOROPROPANE (DBCP)	0.82	UG/L	7/7/2003	NTNCWS	260
CA	CA1502273	SUN WORLD INTERNATIONAL, INC.-COM CENTER	WELL 99-2	DIBROMOCHLOROPROPANE (DBCP)	0.33	UG/L	11/26/2003	NTNCWS	260
CA	CA1502273	SUN WORLD INTERNATIONAL, INC.-COM CENTER	WELL 99-2	DIBROMOCHLOROPROPANE (DBCP)	0.21	UG/L	2/2/2004	NTNCWS	260
CA	CA1502273	SUN WORLD INTERNATIONAL, INC.-COM CENTER	WELL 99-2	DIBROMOCHLOROPROPANE (DBCP)	0.35	UG/L	5/3/2004	NTNCWS	260
CA	CA1502273	SUN WORLD INTERNATIONAL, INC.-COM CENTER	WELL 99-2	DIBROMOCHLOROPROPANE (DBCP)	0.5	UG/L	8/2/2004	NTNCWS	260
CA	CA1502273	SUN WORLD INTERNATIONAL, INC.-COM CENTER	WELL 99-2	DIBROMOCHLOROPROPANE (DBCP)	0.46	UG/L	11/1/2004	NTNCWS	260
CA	CA1502273	SUN WORLD INTERNATIONAL, INC.-COM CENTER	WELL 99-2	DIBROMOCHLOROPROPANE (DBCP)	0.79	UG/L	2/4/2005	NTNCWS	260
CA	CA1502273	SUN WORLD INTERNATIONAL, INC.-COM CENTER	WELL 99-2	DIBROMOCHLOROPROPANE (DBCP)	0.54	UG/L	5/3/2005	NTNCWS	260
CA	CA1502273	SUN WORLD INTERNATIONAL, INC.-COM CENTER	WELL 99-2	DIBROMOCHLOROPROPANE (DBCP)	0.1	UG/L	3/12/1997	NTNCWS	260
CA	CA1502273	SUN WORLD INTERNATIONAL, INC.-COM CENTER	WELL 99-2	DIBROMOCHLOROPROPANE (DBCP)	0.28	UG/L	6/11/1997	NTNCWS	260
CA	CA1502273	SUN WORLD INTERNATIONAL, INC.-COM CENTER	WELL 99-2	DIBROMOCHLOROPROPANE (DBCP)	0.28	UG/L	9/10/1997	NTNCWS	260
CA	CA1502273	SUN WORLD INTERNATIONAL, INC.-COM CENTER	WELL 99-2	DIBROMOCHLOROPROPANE (DBCP)	0.13	UG/L	12/10/1997	NTNCWS	260
CA	CA1502273	SUN WORLD INTERNATIONAL, INC.-COM CENTER	WELL 99-2	DIBROMOCHLOROPROPANE (DBCP)	0.06	UG/L	3/11/1998	NTNCWS	260
CA	CA1502273	SUN WORLD INTERNATIONAL, INC.-COM CENTER	WELL 99-2	DIBROMOCHLOROPROPANE (DBCP)	0.07	UG/L	7/11/2000	NTNCWS	260
CA	CA1502273	SUN WORLD INTERNATIONAL, INC.-COM CENTER	WELL 99-2	DIBROMOCHLOROPROPANE (DBCP)	0.18	UG/L	6/9/1998	NTNCWS	260
CA	CA1502273	SUN WORLD INTERNATIONAL, INC.-COM CENTER	WELL 99-2	DIBROMOCHLOROPROPANE (DBCP)	0.06	UG/L	7/19/1999	NTNCWS	260
					0.08				

CA	CA1510029	Vaughn WC INC	TORREY WELL — BEFORE GAC	DIBROMOCHLOROPROPANE (DBCP)	0.87	UG/L	4/29/2003	CWS	19130
CA	CA1510029	Vaughn WC INC	TORREY WELL — BEFORE GAC	DIBROMOCHLOROPROPANE (DBCP)	0.56	UG/L	3/25/2003	CWS	19130
CA	CA1510029	Vaughn WC INC	TORREY WELL — BEFORE GAC	DIBROMOCHLOROPROPANE (DBCP)	0.94	UG/L	2/27/2003	CWS	19130
CA	CA1510029	Vaughn WC INC	TORREY WELL — BEFORE GAC	DIBROMOCHLOROPROPANE (DBCP)	1.13	UG/L	1/29/2003	CWS	19130
CA	CA1510029	Vaughn WC INC	TORREY WELL — BEFORE GAC	DIBROMOCHLOROPROPANE (DBCP)	0.81	UG/L	9/29/2003	CWS	19130
CA	CA1510029	Vaughn WC INC	TORREY WELL — BEFORE GAC	DIBROMOCHLOROPROPANE (DBCP)	0.78	UG/L	12/30/2002	CWS	19130
CA	CA1510029	Vaughn WC INC	TORREY WELL — BEFORE GAC	DIBROMOCHLOROPROPANE (DBCP)	1.02	UG/L	8/26/2003	CWS	19130
CA	CA1510029	Vaughn WC INC	TORREY WELL — BEFORE GAC	DIBROMOCHLOROPROPANE (DBCP)	1.53	UG/L	7/29/2003	CWS	19130
CA	CA1510029	Vaughn WC INC	TORREY WELL — BEFORE GAC	DIBROMOCHLOROPROPANE (DBCP)	0.63	UG/L	6/27/2003	CWS	19130
CA	CA1510029	Vaughn WC INC	TORREY WELL — BEFORE GAC	DIBROMOCHLOROPROPANE (DBCP)	0.89	UG/L	5/27/2003	CWS	19130
CA	CA1510029	Vaughn WC INC	TORREY WELL — BEFORE GAC	DIBROMOCHLOROPROPANE (DBCP)	0.09	UG/L	5/23/2003	CWS	19130
CA	CA1510029	Vaughn WC INC	TORREY WELL — BEFORE GAC	DIBROMOCHLOROPROPANE (DBCP)	1.42	UG/L	8/29/2002	CWS	19130
CA	CA1510029	Vaughn WC INC	TORREY WELL — BEFORE GAC	DIBROMOCHLOROPROPANE (DBCP)	1.2	UG/L	7/23/2002	CWS	19130
CA	CA1510029	Vaughn WC INC	TORREY WELL — BEFORE GAC	DIBROMOCHLOROPROPANE (DBCP)	1.47	UG/L	9/17/2002	CWS	19130
CA	CA1510029	Vaughn WC INC	TORREY WELL — BEFORE GAC	DIBROMOCHLOROPROPANE (DBCP)	0.73	UG/L	10/30/2003	CWS	19130
CA	CA1510029	Vaughn WC INC	TORREY WELL — BEFORE GAC	DIBROMOCHLOROPROPANE (DBCP)	1.11	UG/L	6/24/2002	CWS	19130
CA	CA1510029	Vaughn WC INC	TORREY WELL — BEFORE GAC	DIBROMOCHLOROPROPANE (DBCP)	0.85	UG/L	11/19/2002	CWS	19130
CA	CA1510029	Vaughn WC INC	TORREY WELL — BEFORE GAC	DIBROMOCHLOROPROPANE (DBCP)	0.95	UG/L	10/31/2002	CWS	19130
CA	CA1510029	Vaughn WC INC	TORREY WELL — BEFORE GAC	DIBROMOCHLOROPROPANE (DBCP)	1.2	UG/L	8/19/2004	CWS	19130
CA	CA1510029	Vaughn WC INC	TORREY WELL — BEFORE GAC	DIBROMOCHLOROPROPANE (DBCP)	0.97	UG/L	7/28/2004	CWS	19130
CA	CA1510029	Vaughn WC INC	TORREY WELL — BEFORE GAC	DIBROMOCHLOROPROPANE (DBCP)	0.94	UG/L	10/5/2000	CWS	19130
CA	CA1510029	Vaughn WC INC	TORREY WELL — BEFORE GAC	DIBROMOCHLOROPROPANE (DBCP)	0.71	UG/L	1/26/2005	CWS	19130
CA	CA1510029	Vaughn WC INC	TORREY WELL — BEFORE GAC	DIBROMOCHLOROPROPANE (DBCP)	0.67	UG/L	9/30/2004	CWS	19130
CA	CA1510029	Vaughn WC INC	TORREY WELL — BEFORE GAC	DIBROMOCHLOROPROPANE (DBCP)	1.28	UG/L	12/23/2004	CWS	19130
CA	CA1510029	Vaughn WC INC	TORREY WELL — BEFORE GAC	DIBROMOCHLOROPROPANE (DBCP)	0.58	UG/L	11/23/2004	CWS	19130
CA	CA1510029	Vaughn WC INC	TORREY WELL — BEFORE GAC	DIBROMOCHLOROPROPANE (DBCP)	0.78	UG/L	10/27/2004	CWS	19130
CA	CA1510029	Vaughn WC INC	TORREY WELL — BEFORE GAC	DIBROMOCHLOROPROPANE (DBCP)	0.86	UG/L	12/29/2003	CWS	19130
CA	CA1510029	Vaughn WC INC	TORREY WELL — BEFORE GAC	DIBROMOCHLOROPROPANE (DBCP)	0.86	UG/L	3/31/2004	CWS	19130
CA	CA1510029	Vaughn WC INC	TORREY WELL — BEFORE GAC	DIBROMOCHLOROPROPANE (DBCP)	0.85	UG/L	2/25/2004	CWS	19130
CA	CA1510029	Vaughn WC INC	TORREY WELL — BEFORE GAC	DIBROMOCHLOROPROPANE (DBCP)	0.78	UG/L	11/24/2003	CWS	19130
CA	CA1510029	Vaughn WC INC	TORREY WELL — BEFORE GAC	DIBROMOCHLOROPROPANE (DBCP)	0.83	UG/L	1/29/2004	CWS	19130
CA	CA1510029	Vaughn WC INC	TORREY WELL — BEFORE GAC	DIBROMOCHLOROPROPANE (DBCP)	0.73	UG/L	4/30/2004	CWS	19130
CA	CA1510029	Vaughn WC INC	TORREY WELL — BEFORE GAC	DIBROMOCHLOROPROPANE (DBCP)	1.29	UG/L	6/30/2004	CWS	19130
CA	CA1510029	Vaughn WC INC	TORREY WELL — BEFORE GAC	DIBROMOCHLOROPROPANE (DBCP)	0.83	UG/L	5/28/2004	CWS	19130
CA	CA1510029	Vaughn WC INC	TORREY WELL — BEFORE GAC	DIBROMOCHLOROPROPANE (DBCP)	1	UG/L	10/1/1998	CWS	19130
CA	CA1510029	Vaughn WC INC	TORREY WELL — BEFORE GAC	DIBROMOCHLOROPROPANE (DBCP)	0.86	UG/L	9/2/1998	CWS	19130
CA	CA1510029	Vaughn WC INC	TORREY WELL — BEFORE GAC	DIBROMOCHLOROPROPANE (DBCP)	0.92	UG/L	8/4/1998	CWS	19130
CA	CA1510029	Vaughn WC INC	TORREY WELL — BEFORE GAC	DIBROMOCHLOROPROPANE (DBCP)	0.96	UG/L	11/5/1998	CWS	19130
CA	CA1510029	Vaughn WC INC	TORREY WELL — BEFORE GAC	DIBROMOCHLOROPROPANE (DBCP)	0.98	UG/L	12/1/1998	CWS	19130
CA	CA1510029	Vaughn WC INC	TORREY WELL — BEFORE GAC	DIBROMOCHLOROPROPANE (DBCP)	1.09	UG/L	6/1/1999	CWS	19130
CA	CA1510029	Vaughn WC INC	TORREY WELL — BEFORE GAC	DIBROMOCHLOROPROPANE (DBCP)	2.68	UG/L	11/2/1999	CWS	19130
CA	CA1510029	Vaughn WC INC	TORREY WELL — BEFORE GAC	DIBROMOCHLOROPROPANE (DBCP)	1.29	UG/L	10/6/1999	CWS	19130
CA	CA1510029	Vaughn WC INC	TORREY WELL — BEFORE GAC	DIBROMOCHLOROPROPANE (DBCP)	1.06	UG/L	9/2/1999	CWS	19130
CA	CA1510029	Vaughn WC INC	TORREY WELL — BEFORE GAC	DIBROMOCHLOROPROPANE (DBCP)	1.18	UG/L	8/3/1999	CWS	19130
CA	CA1510029	Vaughn WC INC	TORREY WELL — BEFORE GAC	DIBROMOCHLOROPROPANE (DBCP)	0.53	UG/L	7/27/1999	CWS	19130
CA	CA1510029	Vaughn WC INC	TORREY WELL — BEFORE GAC	DIBROMOCHLOROPROPANE (DBCP)	1.11	UG/L	7/1/1999	CWS	19130
CA	CA1510029	Vaughn WC INC	TORREY WELL — BEFORE GAC	DIBROMOCHLOROPROPANE (DBCP)	1.1	UG/L	5/4/1999	CWS	19130
CA	CA1510029	Vaughn WC INC	TORREY WELL — BEFORE GAC	DIBROMOCHLOROPROPANE (DBCP)	1.6	UG/L	4/6/1999	CWS	19130
CA	CA1510029	Vaughn WC INC	TORREY WELL — BEFORE GAC	DIBROMOCHLOROPROPANE (DBCP)	1.5	UG/L	2/2/1999	CWS	19130

APPENDIX 7-2 *Continued*

State	PWS_ID	Location	POE	Constituent	Level	Unit	COLL_DATE	PWSType	Pop
CA	CA1510029	Vaughn WC INC	TORREY WELL — BEFORE GAC	DIBROMOCHLOROPROPANE (DBCP)	1.8	UG/L	1/5/1999	CWS	19130
CA	CA1510029	Vaughn WC INC	TORREY WELL — BEFORE GAC	DIBROMOCHLOROPROPANE (DBCP)	0.8	UG/L	4/3/1997	CWS	19130
CA	CA1510029	Vaughn WC INC	TORREY WELL — BEFORE GAC	DIBROMOCHLOROPROPANE (DBCP)	0.77	UG/L	6/3/1997	CWS	19130
CA	CA1510029	Vaughn WC INC	TORREY WELL — BEFORE GAC	DIBROMOCHLOROPROPANE (DBCP)	0.78	UG/L	2/5/1997	CWS	19130
CA	CA1510029	Vaughn WC INC	TORREY WELL — BEFORE GAC	DIBROMOCHLOROPROPANE (DBCP)	0.68	UG/L	6/19/1997	CWS	19130
CA	CA1510029	Vaughn WC INC	TORREY WELL — BEFORE GAC	DIBROMOCHLOROPROPANE (DBCP)	0.92	UG/L	1/6/1997	CWS	19130
CA	CA1510029	Vaughn WC INC	TORREY WELL — BEFORE GAC	DIBROMOCHLOROPROPANE (DBCP)	0.94	UG/L	5/30/2002	CWS	19130
CA	CA1510029	Vaughn WC INC	TORREY WELL — BEFORE GAC	DIBROMOCHLOROPROPANE (DBCP)	0.93	UG/L	11/3/2000	CWS	19130
CA	CA1510029	Vaughn WC INC	TORREY WELL — BEFORE GAC	DIBROMOCHLOROPROPANE (DBCP)	0.69	UG/L	3/5/1997	CWS	19130
CA	CA1510029	Vaughn WC INC	TORREY WELL — BEFORE GAC	DIBROMOCHLOROPROPANE (DBCP)	1.1	UG/L	4/7/1998	CWS	19130
CA	CA1510029	Vaughn WC INC	TORREY WELL — BEFORE GAC	DIBROMOCHLOROPROPANE (DBCP)	0.84	UG/L	7/2/1998	CWS	19130
CA	CA1510029	Vaughn WC INC	TORREY WELL — BEFORE GAC	DIBROMOCHLOROPROPANE (DBCP)	0.84	UG/L	5/6/1997	CWS	19130
CA	CA1510029	Vaughn WC INC	TORREY WELL — BEFORE GAC	DIBROMOCHLOROPROPANE (DBCP)	0.94	UG/L	5/15/1998	CWS	19130
CA	CA1510029	Vaughn WC INC	TORREY WELL — BEFORE GAC	DIBROMOCHLOROPROPANE (DBCP)	0.88	UG/L	12/4/1997	CWS	19130
CA	CA1510029	Vaughn WC INC	TORREY WELL — BEFORE GAC	DIBROMOCHLOROPROPANE (DBCP)	0.94	UG/L	11/4/1997	CWS	19130
CA	CA1510029	Vaughn WC INC	TORREY WELL — BEFORE GAC	DIBROMOCHLOROPROPANE (DBCP)	0.69	UG/L	10/1/1997	CWS	19130
CA	CA1510029	Vaughn WC INC	TORREY WELL — BEFORE GAC	DIBROMOCHLOROPROPANE (DBCP)	0.9	UG/L	9/2/1997	CWS	19130
CA	CA1510029	Vaughn WC INC	TORREY WELL — BEFORE GAC	DIBROMOCHLOROPROPANE (DBCP)	0.69	UG/L	8/5/1997	CWS	19130
CA	CA1510029	Vaughn WC INC	TORREY WELL — BEFORE GAC	DIBROMOCHLOROPROPANE (DBCP)	0.72	UG/L	7/1/1997	CWS	19130
CA	CA1510029	Vaughn WC INC	TORREY WELL — BEFORE GAC	DIBROMOCHLOROPROPANE (DBCP)	0.82	UG/L	6/10/1998	CWS	19130
CA	CA1510029	Vaughn WC INC	TORREY WELL — BEFORE GAC	DIBROMOCHLOROPROPANE (DBCP)	1.27	UG/L	7/12/2001	CWS	19130
CA	CA1510029	Vaughn WC INC	TORREY WELL — BEFORE GAC	DIBROMOCHLOROPROPANE (DBCP)	0.88	UG/L	4/30/2002	CWS	19130
CA	CA1510029	Vaughn WC INC	TORREY WELL — BEFORE GAC	DIBROMOCHLOROPROPANE (DBCP)	0.87	UG/L	3/28/2002	CWS	19130
CA	CA1510029	Vaughn WC INC	TORREY WELL — BEFORE GAC	DIBROMOCHLOROPROPANE (DBCP)	0.95	UG/L	2/25/2002	CWS	19130
CA	CA1510029	Vaughn WC INC	TORREY WELL — BEFORE GAC	DIBROMOCHLOROPROPANE (DBCP)	0.97	UG/L	1/29/2002	CWS	19130
CA	CA1510029	Vaughn WC INC	TORREY WELL — BEFORE GAC	DIBROMOCHLOROPROPANE (DBCP)	0.95	UG/L	9/26/2001	CWS	19130
CA	CA1510029	Vaughn WC INC	TORREY WELL — BEFORE GAC	DIBROMOCHLOROPROPANE (DBCP)	1.15	UG/L	11/15/1999	CWS	19130
CA	CA1510029	Vaughn WC INC	TORREY WELL — BEFORE GAC	DIBROMOCHLOROPROPANE (DBCP)	1.34	UG/L	2/1/2000	CWS	19130
CA	CA1510029	Vaughn WC INC	TORREY WELL — BEFORE GAC	DIBROMOCHLOROPROPANE (DBCP)	1.12	UG/L	7/5/2000	CWS	19130
CA	CA1510029	Vaughn WC INC	TORREY WELL — BEFORE GAC	DIBROMOCHLOROPROPANE (DBCP)	1.27	UG/L	6/5/2000	CWS	19130
CA	CA1510029	Vaughn WC INC	TORREY WELL — BEFORE GAC	DIBROMOCHLOROPROPANE (DBCP)	1.11	UG/L	5/3/2000	CWS	19130
CA	CA1510029	Vaughn WC INC	TORREY WELL — BEFORE GAC	DIBROMOCHLOROPROPANE (DBCP)	1.14	UG/L	8/3/2000	CWS	19130
CA	CA1510029	Vaughn WC INC	TORREY WELL — BEFORE GAC	DIBROMOCHLOROPROPANE (DBCP)	2.82	UG/L	3/1/2000	CWS	19130
CA	CA1510029	Vaughn WC INC	TORREY WELL — BEFORE GAC	DIBROMOCHLOROPROPANE (DBCP)	1.89	UG/L	2/15/2000	CWS	19130
CA	CA1510029	Vaughn WC INC	TORREY WELL — BEFORE GAC	DIBROMOCHLOROPROPANE (DBCP)	1.14	UG/L	1/4/2000	CWS	19130
CA	CA1510029	Vaughn WC INC	TORREY WELL — BEFORE GAC	DIBROMOCHLOROPROPANE (DBCP)	0.34	UG/L	12/15/1999	CWS	19130
CA	CA1510029	Vaughn WC INC	TORREY WELL — BEFORE GAC	DIBROMOCHLOROPROPANE (DBCP)	1.4	UG/L	12/15/1999	CWS	19130
CA	CA1510029	Vaughn WC INC	TORREY WELL — BEFORE GAC	DIBROMOCHLOROPROPANE (DBCP)	1.12	UG/L	4/4/2000	CWS	19130
CA	CA1510029	Vaughn WC INC	TORREY WELL — BEFORE GAC	DIBROMOCHLOROPROPANE (DBCP)	1.06	UG/L	6/13/2001	CWS	19130
CA	CA1510029	Vaughn WC INC	TORREY WELL — BEFORE GAC	DIBROMOCHLOROPROPANE (DBCP)	1.06	UG/L	5/14/2001	CWS	19130
CA	CA1510029	Vaughn WC INC	TORREY WELL — BEFORE GAC	DIBROMOCHLOROPROPANE (DBCP)	1.15	UG/L	1/17/2000	CWS	19130
CA	CA1510029	Vaughn WC INC	TORREY WELL — BEFORE GAC	DIBROMOCHLOROPROPANE (DBCP)	1.12	UG/L	9/7/2000	CWS	19130
CA	CA1510029	Vaughn WC INC	TORREY WELL — BEFORE GAC	DIBROMOCHLOROPROPANE (DBCP)	1.07	UG/L	4/9/2001	CWS	19130
CA	CA1510029	Vaughn WC INC	TORREY WELL — BEFORE GAC	DIBROMOCHLOROPROPANE (DBCP)	1.08	UG/L	2/13/2001	CWS	19130
CA	CA1510029	Vaughn WC INC	TORREY WELL — BEFORE GAC	DIBROMOCHLOROPROPANE (DBCP)	1.07	UG/L	12/12/2000	CWS	19130
CA	CA1510029	Vaughn WC INC	TORREY WELL — BEFORE GAC	DIBROMOCHLOROPROPANE (DBCP)	1.11	UG/L	1/4/2001	CWS	19130

CA	CA1510029	Vaughn WC INC	TORREY WELL — BEFORE GAC	DIBROMOCHLOROPROPANE (DBCP)	2.2	UG/L	12/1/1999	CWS	19130
					1.03				
CA	CA1500585	OASIS PROPERTY OWNERS ASSOCIATION	WELL 03	1,4-DICHLOROBENZENE	1.1	UG/L	9/26/2001	CWS	60
CA	CA1500373	SEVENTH STANDARD MUTUAL	WELL 02 — STANDBY	1,2-DICHLOROPROPANE	0.56	UG/L	3/2/1998	CWS	25
CA	CA1510029	Vaughn WC INC	TORREY WELL — AFTER COMBINED GAC 1 & 2	1,2-DICHLOROPROPANE	0.6	UG/L	4/3/1997	CWS	19130
					0.58				
CA	CA1503290	THE GARLIC COMPANY	WELL 02	ETHYLBENZENE	0.83	UG/L	7/27/1999	NTNCWS	200
CA	CA1500535	MANON MANOR MUTUAL WATER COMPANY	WELL 01	ETHYLENE DIBROMIDE (EDB)	0.02	UG/L	10/20/1998	CWS	200
CA	CA1500535	MANON MANOR MUTUAL WATER COMPANY	WELL 01	ETHYLENE DIBROMIDE (EDB)	0.02	UG/L	7/24/2001	CWS	200
CA	CA1510029	Vaughn WC INC	CLARISSE NO. 1-AFTER GAC 1 & 2 COMBINED	ETHYLENE DIBROMIDE (EDB)	0.02	UG/L	7/5/2000	CWS	19130
CA	CA1510029	Vaughn WC INC	CLARISSE NO. 1-AFTER GAC 1 & 2 COMBINED	ETHYLENE DIBROMIDE (EDB)	0.02	UG/L	8/3/2000	CWS	19130
CA	CA1510029	Vaughn WC INC	CLARISSE NO. 1-AFTER GAC 1 & 2 COMBINED	ETHYLENE DIBROMIDE (EDB)	0.03	UG/L	9/7/2000	CWS	19130
CA	CA1510029	Vaughn WC INC	CLARISSE WELL 2-AFTER GAC 1(EAST) VESSEL	ETHYLENE DIBROMIDE (EDB)	0.05	UG/L	8/18/1997	CWS	19130
CA	CA1510029	Vaughn WC INC	CLARISSE WELL 2-AFTER GAC 1(EAST) VESSEL	ETHYLENE DIBROMIDE (EDB)	0.02	UG/L	10/6/1999	CWS	19130
CA	CA1510029	Vaughn WC INC	CLARISSE WELL 2-AFTER GAC 1(EAST) VESSEL	ETHYLENE DIBROMIDE (EDB)	0.03	UG/L	7/1/1999	CWS	19130
CA	CA1510029	Vaughn WC INC	CLARISSE WELL 2-AFTER GAC 1(EAST) VESSEL	ETHYLENE DIBROMIDE (EDB)	0.01	UG/L	7/27/1999	CWS	19130
CA	CA1510029	Vaughn WC INC	CLARISSE WELL 2-AFTER GAC 1(EAST) VESSEL	ETHYLENE DIBROMIDE (EDB)	0.09	UG/L	9/2/1997	CWS	19130
CA	CA1510029	Vaughn WC INC	CLARISSE WELL 2-AFTER GAC 1(EAST) VESSEL	ETHYLENE DIBROMIDE (EDB)	0.04	UG/L	8/25/1997	CWS	19130
CA	CA1510029	Vaughn WC INC	CLARISSE WELL 2-AFTER GAC 1(EAST) VESSEL	ETHYLENE DIBROMIDE (EDB)	0.04	UG/L	7/21/1997	CWS	19130
CA	CA1510029	Vaughn WC INC	CLARISSE WELL 2-AFTER GAC 1(EAST) VESSEL	ETHYLENE DIBROMIDE (EDB)	0.05	UG/L	7/1/1997	CWS	19130
CA	CA1510029	Vaughn WC INC	CLARISSE WELL 2-AFTER GAC 1(EAST) VESSEL	ETHYLENE DIBROMIDE (EDB)	0.04	UG/L	6/19/1997	CWS	19130
CA	CA1510029	Vaughn WC INC	CLARISSE WELL 2-AFTER GAC 1(EAST) VESSEL	ETHYLENE DIBROMIDE (EDB)	0.04	UG/L	6/3/1997	CWS	19130
CA	CA1510029	Vaughn WC INC	CLARISSE WELL 2-AFTER GAC 1(EAST) VESSEL	ETHYLENE DIBROMIDE (EDB)	0.03	UG/L	3/5/1997	CWS	19130
CA	CA1510029	Vaughn WC INC	CLARISSE WELL 2-AFTER GAC 1(EAST) VESSEL	ETHYLENE DIBROMIDE (EDB)	0.06	UG/L	8/5/1997	CWS	19130
CA	CA1510029	Vaughn WC INC	CLARISSE WELL 2-AFTER GAC 1(EAST) VESSEL	ETHYLENE DIBROMIDE (EDB)	0.02	UG/L	8/3/1999	CWS	19130
CA	CA1510029	Vaughn WC INC	CLARISSE WELL 2-AFTER GAC 2(WEST) VESSEL	ETHYLENE DIBROMIDE (EDB)	0.12	UG/L	9/2/1997	CWS	19130
CA	CA1510029	Vaughn WC INC	CLARISSE WELL 2-AFTER GAC 2(WEST) VESSEL	ETHYLENE DIBROMIDE (EDB)	0.05	UG/L	8/3/1999	CWS	19130
CA	CA1510029	Vaughn WC INC	CLARISSE WELL 2-AFTER GAC 2(WEST) VESSEL	ETHYLENE DIBROMIDE (EDB)	0.03	UG/L	8/29/2002	CWS	19130
CA	CA1510029	Vaughn WC INC	CLARISSE WELL 2-AFTER GAC 2(WEST) VESSEL	ETHYLENE DIBROMIDE (EDB)	0.02	UG/L	1/4/2000	CWS	19130
CA	CA1510029	Vaughn WC INC	CLARISSE WELL 2-AFTER GAC 2(WEST) VESSEL	ETHYLENE DIBROMIDE (EDB)	0.02	UG/L	8/3/1999	CWS	19130
CA	CA1510029	Vaughn WC INC	CLARISSE WELL 2-AFTER GAC 2(WEST) VESSEL	ETHYLENE DIBROMIDE (EDB)	0.01	UG/L	7/27/1999	CWS	19130
CA	CA1510029	Vaughn WC INC	CLARISSE WELL 2-AFTER GAC 2(WEST) VESSEL	ETHYLENE DIBROMIDE (EDB)	0.05	UG/L	7/1/1999	CWS	19130
CA	CA1510029	Vaughn WC INC	CLARISSE WELL 2-AFTER GAC 2(WEST) VESSEL	ETHYLENE DIBROMIDE (EDB)	0.02	UG/L	11/5/1998	CWS	19130
CA	CA1510029	Vaughn WC INC	CLARISSE WELL 2-AFTER GAC 2(WEST) VESSEL	ETHYLENE DIBROMIDE (EDB)	0.02	UG/L	9/17/2002	CWS	19130
CA	CA1510029	Vaughn WC INC	CLARISSE WELL 2-AFTER GAC 2(WEST) VESSEL	ETHYLENE DIBROMIDE (EDB)	0.06	UG/L	8/18/1997	CWS	19130
CA	CA1510029	Vaughn WC INC	CLARISSE WELL 2-AFTER GAC 2(WEST) VESSEL	ETHYLENE DIBROMIDE (EDB)	0.08	UG/L	8/5/1997	CWS	19130
CA	CA1510029	Vaughn WC INC	CLARISSE WELL 2-AFTER GAC 2(WEST) VESSEL	ETHYLENE DIBROMIDE (EDB)	0.06	UG/L	7/21/1997	CWS	19130
CA	CA1510029	Vaughn WC INC	CLARISSE WELL 2-AFTER GAC 2(WEST) VESSEL	ETHYLENE DIBROMIDE (EDB)	0.02	UG/L	9/2/1999	CWS	19130
CA	CA1510029	Vaughn WC INC	CLARISSE WELL 2-AFTER GAC 2(WEST) VESSEL	ETHYLENE DIBROMIDE (EDB)	0.07	UG/L	7/1/1997	CWS	19130
CA	CA1510029	Vaughn WC INC	CLARISSE WELL 2-AFTER GAC 2(WEST) VESSEL	ETHYLENE DIBROMIDE (EDB)	0.05	UG/L	8/25/1997	CWS	19130
CA	CA1510029	Vaughn WC INC	CLARISSE WELL 2-AFTER GAC 2(WEST) VESSEL	ETHYLENE DIBROMIDE (EDB)	0.04	UG/L	6/19/1997	CWS	19130
CA	CA1510029	Vaughn WC INC	CLARISSE WELL 2-AFTER GAC 2(WEST) VESSEL	ETHYLENE DIBROMIDE (EDB)	0.06	UG/L	6/3/1997	CWS	19130
CA	CA1510029	Vaughn WC INC	CLARISSE WELL 2-AFTER GAC 2(WEST) VESSEL	ETHYLENE DIBROMIDE (EDB)	0.02	UG/L	2/5/1997	CWS	19130
CA	CA1510029	Vaughn WC INC	CLARISSE WELL 2-AFTER GAC 2(WEST) VESSEL	ETHYLENE DIBROMIDE (EDB)	0.03	UG/L	1/6/1997	CWS	19130
CA	CA1510029	Vaughn WC INC	CLARISSE WELL 2-AFTER GAC 1 & 2 COMBINED	ETHYLENE DIBROMIDE (EDB)	0.04	UG/L	8/5/1997	CWS	19130
CA	CA1510029	Vaughn WC INC	CLARISSE WELL 2-AFTER GAC 1 & 2 COMBINED	ETHYLENE DIBROMIDE (EDB)	0.05	UG/L	8/25/1997	CWS	19130
CA	CA1510029	Vaughn WC INC	CLARISSE WELL 2-AFTER GAC 1 & 2 COMBINED	ETHYLENE DIBROMIDE (EDB)	0.02	UG/L	7/21/1997	CWS	19130
CA	CA1510029	Vaughn WC INC	CLARISSE WELL 2-AFTER GAC 1 & 2 COMBINED	ETHYLENE DIBROMIDE (EDB)	0.03	UG/L	7/1/1997	CWS	19130
CA	CA1510029	Vaughn WC INC	CLARISSE WELL 2-AFTER GAC 1 & 2 COMBINED	ETHYLENE DIBROMIDE (EDB)	0.08	UG/L	9/2/1997	CWS	19130
CA	CA1510029	Vaughn WC INC	CLARISSE NO. 1-AFTER GAC1 (NORTH) VESSEL	ETHYLENE DIBROMIDE (EDB)	0.05	UG/L	10/5/2000	CWS	19130

APPENDIX 7-2 *Continued*

State	PWS_ID	Location	POE	Constituent	Level	Unit	COLL_DATE	PWSType	Pop
CA	CA1510029	Vaughn WC INC	CLARISSE NO. 1-AFTER GAC1 (NORTH) VESSEL	ETHYLENE DIBROMIDE (EDB)	0.02	UG/L	5/3/2000	CWS	19130
CA	CA1510029	Vaughn WC INC	CLARISSE NO. 1-AFTER GAC1 (NORTH) VESSEL	ETHYLENE DIBROMIDE (EDB)	0.03	UG/L	9/7/2000	CWS	19130
CA	CA1510029	Vaughn WC INC	CLARISSE NO. 1-AFTER GAC2 (SOUTH) VESSEL	ETHYLENE DIBROMIDE (EDB)	0.05	UG/L	6/5/2000	CWS	19130
CA	CA1510029	Vaughn WC INC	CLARISSE NO. 1-AFTER GAC2 (SOUTH) VESSEL	ETHYLENE DIBROMIDE (EDB)	0.05	UG/L	10/5/2000	CWS	19130
CA	CA1510029	Vaughn WC INC	TORREY WELL — AFTER GAC 2 (WEST) VESSEL	ETHYLENE DIBROMIDE (EDB)	0.03	UG/L	7/29/2003	CWS	19130
					0.04				
CA	CA1510029	Vaughn WC INC	CLARISSE WELL 01 — BEFORE GAC	ETHYLENE DIBROMIDE (EDB)	0.04	UG/L	5/30/2002	CWS	19130
CA	CA1510029	Vaughn WC INC	CLARISSE WELL 01 — BEFORE GAC	ETHYLENE DIBROMIDE (EDB)	0.05	UG/L	7/23/2002	CWS	19130
CA	CA1510029	Vaughn WC INC	CLARISSE WELL 01 — BEFORE GAC	ETHYLENE DIBROMIDE (EDB)	0.03	UG/L	9/17/2002	CWS	19130
CA	CA1510029	Vaughn WC INC	CLARISSE WELL 01 — BEFORE GAC	ETHYLENE DIBROMIDE (EDB)	0.05	UG/L	3/28/2002	CWS	19130
CA	CA1510029	Vaughn WC INC	CLARISSE WELL 01 — BEFORE GAC	ETHYLENE DIBROMIDE (EDB)	0.03	UG/L	11/19/2002	CWS	19130
CA	CA1510029	Vaughn WC INC	CLARISSE WELL 01 — BEFORE GAC	ETHYLENE DIBROMIDE (EDB)	0.04	UG/L	12/30/2002	CWS	19130
CA	CA1510029	Vaughn WC INC	CLARISSE WELL 01 — BEFORE GAC	ETHYLENE DIBROMIDE (EDB)	0.04	UG/L	1/29/2003	CWS	19130
CA	CA1510029	Vaughn WC INC	CLARISSE WELL 01 — BEFORE GAC	ETHYLENE DIBROMIDE (EDB)	0.05	UG/L	8/29/2002	CWS	19130
CA	CA1510029	Vaughn WC INC	CLARISSE WELL 01 — BEFORE GAC	ETHYLENE DIBROMIDE (EDB)	0.04	UG/L	4/30/2002	CWS	19130
CA	CA1510029	Vaughn WC INC	CLARISSE WELL 01 — BEFORE GAC	ETHYLENE DIBROMIDE (EDB)	0.03	UG/L	1/29/2002	CWS	19130
CA	CA1510029	Vaughn WC INC	CLARISSE WELL 01 — BEFORE GAC	ETHYLENE DIBROMIDE (EDB)	0.05	UG/L	2/25/2002	CWS	19130
CA	CA1510029	Vaughn WC INC	CLARISSE WELL 01 — BEFORE GAC	ETHYLENE DIBROMIDE (EDB)	0.02	UG/L	6/24/2002	CWS	19130
CA	CA1510029	Vaughn WC INC	CLARISSE WELL 01 — BEFORE GAC	ETHYLENE DIBROMIDE (EDB)	0.03	UG/L	2/27/2003	CWS	19130
CA	CA1510029	Vaughn WC INC	CLARISSE WELL 01 — BEFORE GAC	ETHYLENE DIBROMIDE (EDB)	0.03	UG/L	5/27/2003	CWS	19130
CA	CA1510029	Vaughn WC INC	CLARISSE WELL 01 — BEFORE GAC	ETHYLENE DIBROMIDE (EDB)	0.02	UG/L	6/27/2003	CWS	19130
CA	CA1510029	Vaughn WC INC	CLARISSE WELL 01 — BEFORE GAC	ETHYLENE DIBROMIDE (EDB)	0.03	UG/L	6/4/2004	CWS	19130
CA	CA1510029	Vaughn WC INC	CLARISSE WELL 01 — BEFORE GAC	ETHYLENE DIBROMIDE (EDB)	0.02	UG/L	6/30/2004	CWS	19130
CA	CA1510029	Vaughn WC INC	CLARISSE WELL 01 — BEFORE GAC	ETHYLENE DIBROMIDE (EDB)	0.09	UG/L	4/6/1999	CWS	19130
CA	CA1510029	Vaughn WC INC	CLARISSE WELL 01 — BEFORE GAC	ETHYLENE DIBROMIDE (EDB)	0.05	UG/L	9/23/1998	CWS	19130
CA	CA1510029	Vaughn WC INC	CLARISSE WELL 01 — BEFORE GAC	ETHYLENE DIBROMIDE (EDB)	0.17	UG/L	5/4/1999	CWS	19130
CA	CA1510029	Vaughn WC INC	CLARISSE WELL 01 — BEFORE GAC	ETHYLENE DIBROMIDE (EDB)	0.15	UG/L	6/1/1999	CWS	19130
CA	CA1510029	Vaughn WC INC	CLARISSE WELL 01 — BEFORE GAC	ETHYLENE DIBROMIDE (EDB)	0.15	UG/L	7/1/1999	CWS	19130
CA	CA1510029	Vaughn WC INC	CLARISSE WELL 01 — BEFORE GAC	ETHYLENE DIBROMIDE (EDB)	0.12	UG/L	8/3/1999	CWS	19130
CA	CA1510029	Vaughn WC INC	CLARISSE WELL 01 — BEFORE GAC	ETHYLENE DIBROMIDE (EDB)	0.11	UG/L	9/2/1999	CWS	19130
CA	CA1510029	Vaughn WC INC	CLARISSE WELL 01 — BEFORE GAC	ETHYLENE DIBROMIDE (EDB)	0.06	UG/L	2/13/2001	CWS	19130
CA	CA1510029	Vaughn WC INC	CLARISSE WELL 01 — BEFORE GAC	ETHYLENE DIBROMIDE (EDB)	0.06	UG/L	4/5/2001	CWS	19130
CA	CA1510029	Vaughn WC INC	CLARISSE WELL 01 — BEFORE GAC	ETHYLENE DIBROMIDE (EDB)	0.03	UG/L	12/11/2000	CWS	19130
CA	CA1510029	Vaughn WC INC	CLARISSE WELL 01 — BEFORE GAC	ETHYLENE DIBROMIDE (EDB)	0.05	UG/L	7/12/2001	CWS	19130
CA	CA1510029	Vaughn WC INC	CLARISSE WELL 01 — BEFORE GAC	ETHYLENE DIBROMIDE (EDB)	0.11	UG/L	10/6/1999	CWS	19130
CA	CA1510029	Vaughn WC INC	CLARISSE WELL 01 — BEFORE GAC	ETHYLENE DIBROMIDE (EDB)	0.03	UG/L	6/13/2001	CWS	19130
CA	CA1510029	Vaughn WC INC	CLARISSE WELL 01 — BEFORE GAC	ETHYLENE DIBROMIDE (EDB)	0.09	UG/L	5/3/2000	CWS	19130
CA	CA1510029	Vaughn WC INC	CLARISSE WELL 01 — BEFORE GAC	ETHYLENE DIBROMIDE (EDB)	0.11	UG/L	12/1/1999	CWS	19130
CA	CA1510029	Vaughn WC INC	CLARISSE WELL 01 — BEFORE GAC	ETHYLENE DIBROMIDE (EDB)	0.06	UG/L	5/14/2001	CWS	19130
CA	CA1510029	Vaughn WC INC	CLARISSE WELL 01 — BEFORE GAC	ETHYLENE DIBROMIDE (EDB)	0.03	UG/L	11/3/2000	CWS	19130
CA	CA1510029	Vaughn WC INC	CLARISSE WELL 01 — BEFORE GAC	ETHYLENE DIBROMIDE (EDB)	0.12	UG/L	2/1/2000	CWS	19130
CA	CA1510029	Vaughn WC INC	CLARISSE WELL 01 — BEFORE GAC	ETHYLENE DIBROMIDE (EDB)	0.11	UG/L	3/1/2000	CWS	19130
CA	CA1510029	Vaughn WC INC	CLARISSE WELL 01 — BEFORE GAC	ETHYLENE DIBROMIDE (EDB)	0.11	UG/L	11/2/1999	CWS	19130
CA	CA1510029	Vaughn WC INC	CLARISSE WELL 01 — BEFORE GAC	ETHYLENE DIBROMIDE (EDB)	0.09	UG/L	4/4/2000	CWS	19130
CA	CA1510029	Vaughn WC INC	CLARISSE WELL 01 — BEFORE GAC	ETHYLENE DIBROMIDE (EDB)	0.12	UG/L	1/4/2000	CWS	19130
CA	CA1510029	Vaughn WC INC	CLARISSE WELL 01 — BEFORE GAC	ETHYLENE DIBROMIDE (EDB)	0.1	UG/L	6/5/2000	CWS	19130

CA	CA1510029	Vaughn WC INC	CLARISSE WELL 01 — BEFORE GAC	ETHYLENE DIBROMIDE (EDB)	0.08	UG/L	7/5/2000	CWS	19130
CA	CA1510029	Vaughn WC INC	CLARISSE WELL 01 — BEFORE GAC	ETHYLENE DIBROMIDE (EDB)	0.07	UG/L	8/3/2000	CWS	19130
CA	CA1510029	Vaughn WC INC	CLARISSE WELL 01 — BEFORE GAC	ETHYLENE DIBROMIDE (EDB)	0.07	UG/L	9/7/2000	CWS	19130
CA	CA1510029	Vaughn WC INC	CLARISSE WELL 01 — BEFORE GAC	ETHYLENE DIBROMIDE (EDB)	0.07	UG/L	10/5/2000	CWS	19130
CA	CA1510029	Vaughn WC INC	CLARISSE WELL 02 — BEFORE GAC	ETHYLENE DIBROMIDE (EDB)	0.03	UG/L	5/30/2002	CWS	19130
CA	CA1510029	Vaughn WC INC	CLARISSE WELL 02 — BEFORE GAC	ETHYLENE DIBROMIDE (EDB)	0.03	UG/L	8/29/2002	CWS	19130
CA	CA1510029	Vaughn WC INC	CLARISSE WELL 02 — BEFORE GAC	ETHYLENE DIBROMIDE (EDB)	0.02	UG/L	8/26/2003	CWS	19130
CA	CA1510029	Vaughn WC INC	CLARISSE WELL 02 — BEFORE GAC	ETHYLENE DIBROMIDE (EDB)	0.02	UG/L	5/27/2003	CWS	19130
CA	CA1510029	Vaughn WC INC	CLARISSE WELL 02 — BEFORE GAC	ETHYLENE DIBROMIDE (EDB)	0.02	UG/L	6/27/2003	CWS	19130
CA	CA1510029	Vaughn WC INC	CLARISSE WELL 02 — BEFORE GAC	ETHYLENE DIBROMIDE (EDB)	0.02	UG/L	6/28/2005	CWS	19130
CA	CA1510029	Vaughn WC INC	CLARISSE WELL 02 — BEFORE GAC	ETHYLENE DIBROMIDE (EDB)	0.02	UG/L	4/30/2004	CWS	19130
CA	CA1510029	Vaughn WC INC	CLARISSE WELL 02 — BEFORE GAC	ETHYLENE DIBROMIDE (EDB)	0.08	UG/L	8/4/1998	CWS	19130
CA	CA1510029	Vaughn WC INC	CLARISSE WELL 02 — BEFORE GAC	ETHYLENE DIBROMIDE (EDB)	0.02	UG/L	5/5/1998	CWS	19130
CA	CA1510029	Vaughn WC INC	CLARISSE WELL 02 — BEFORE GAC	ETHYLENE DIBROMIDE (EDB)	0.09	UG/L	12/4/1997	CWS	19130
CA	CA1510029	Vaughn WC INC	CLARISSE WELL 02 — BEFORE GAC	ETHYLENE DIBROMIDE (EDB)	0.09	UG/L	6/3/1997	CWS	19130
CA	CA1510029	Vaughn WC INC	CLARISSE WELL 02 — BEFORE GAC	ETHYLENE DIBROMIDE (EDB)	0.04	UG/L	1/6/1997	CWS	19130
CA	CA1510029	Vaughn WC INC	CLARISSE WELL 02 — BEFORE GAC	ETHYLENE DIBROMIDE (EDB)	0.05	UG/L	2/5/1997	CWS	19130
CA	CA1510029	Vaughn WC INC	CLARISSE WELL 02 — BEFORE GAC	ETHYLENE DIBROMIDE (EDB)	0.14	UG/L	3/5/1997	CWS	19130
CA	CA1510029	Vaughn WC INC	CLARISSE WELL 02 — BEFORE GAC	ETHYLENE DIBROMIDE (EDB)	0.03	UG/L	7/2/1998	CWS	19130
CA	CA1510029	Vaughn WC INC	CLARISSE WELL 02 — BEFORE GAC	ETHYLENE DIBROMIDE (EDB)	0.05	UG/L	6/10/1998	CWS	19130
CA	CA1510029	Vaughn WC INC	CLARISSE WELL 02 — BEFORE GAC	ETHYLENE DIBROMIDE (EDB)	0.06	UG/L	6/19/1997	CWS	19130
CA	CA1510029	Vaughn WC INC	CLARISSE WELL 02 — BEFORE GAC	ETHYLENE DIBROMIDE (EDB)	0.11	UG/L	7/1/1997	CWS	19130
CA	CA1510029	Vaughn WC INC	CLARISSE WELL 02 — BEFORE GAC	ETHYLENE DIBROMIDE (EDB)	0.13	UG/L	8/5/1997	CWS	19130
CA	CA1510029	Vaughn WC INC	CLARISSE WELL 02 — BEFORE GAC	ETHYLENE DIBROMIDE (EDB)	0.09	UG/L	8/25/1997	CWS	19130
CA	CA1510029	Vaughn WC INC	CLARISSE WELL 02 — BEFORE GAC	ETHYLENE DIBROMIDE (EDB)	0.17	UG/L	9/2/1997	CWS	19130
CA	CA1510029	Vaughn WC INC	CLARISSE WELL 02 — BEFORE GAC	ETHYLENE DIBROMIDE (EDB)	0.08	UG/L	9/16/1997	CWS	19130
CA	CA1510029	Vaughn WC INC	CLARISSE WELL 02 — BEFORE GAC	ETHYLENE DIBROMIDE (EDB)	0.11	UG/L	10/1/1997	CWS	19130
CA	CA1510029	Vaughn WC INC	CLARISSE WELL 02 — BEFORE GAC	ETHYLENE DIBROMIDE (EDB)	0.07	UG/L	7/5/2000	CWS	19130
CA	CA1510029	Vaughn WC INC	CLARISSE WELL 02 — BEFORE GAC	ETHYLENE DIBROMIDE (EDB)	0.06	UG/L	8/3/2000	CWS	19130
CA	CA1510029	Vaughn WC INC	CLARISSE WELL 02 — BEFORE GAC	ETHYLENE DIBROMIDE (EDB)	0.06	UG/L	10/5/2000	CWS	19130
CA	CA1510029	Vaughn WC INC	CLARISSE WELL 02 — BEFORE GAC	ETHYLENE DIBROMIDE (EDB)	0.06	UG/L	11/3/2000	CWS	19130
CA	CA1510029	Vaughn WC INC	CLARISSE WELL 02 — BEFORE GAC	ETHYLENE DIBROMIDE (EDB)	0.02	UG/L	12/11/2000	CWS	19130
CA	CA1510029	Vaughn WC INC	CLARISSE WELL 02 — BEFORE GAC	ETHYLENE DIBROMIDE (EDB)	0.07	UG/L	6/5/2000	CWS	19130
CA	CA1510029	Vaughn WC INC	CLARISSE WELL 02 — BEFORE GAC	ETHYLENE DIBROMIDE (EDB)	0.04	UG/L	7/12/2001	CWS	19130
CA	CA1510029	Vaughn WC INC	CLARISSE WELL 02 — BEFORE GAC	ETHYLENE DIBROMIDE (EDB)	0.03	UG/L	4/9/2001	CWS	19130
CA	CA1510029	Vaughn WC INC	CLARISSE WELL 02 — BEFORE GAC	ETHYLENE DIBROMIDE (EDB)	0.06	UG/L	5/14/2001	CWS	19130
CA	CA1510029	Vaughn WC INC	CLARISSE WELL 02 — BEFORE GAC	ETHYLENE DIBROMIDE (EDB)	0.03	UG/L	6/13/2001	CWS	19130
CA	CA1510029	Vaughn WC INC	CLARISSE WELL 02 — BEFORE GAC	ETHYLENE DIBROMIDE (EDB)	0.08	UG/L	9/2/1998	CWS	19130
CA	CA1510029	Vaughn WC INC	CLARISSE WELL 02 — BEFORE GAC	ETHYLENE DIBROMIDE (EDB)	0.06	UG/L	2/5/1998	CWS	19130
CA	CA1510029	Vaughn WC INC	CLARISSE WELL 02 — BEFORE GAC	ETHYLENE DIBROMIDE (EDB)	0.02	UG/L	2/13/2001	CWS	19130
CA	CA1510029	Vaughn WC INC	CLARISSE WELL 02 — BEFORE GAC	ETHYLENE DIBROMIDE (EDB)	0.03	UG/L	12/1/1998	CWS	19130
CA	CA1510029	Vaughn WC INC	CLARISSE WELL 02 — BEFORE GAC	ETHYLENE DIBROMIDE (EDB)	0.03	UG/L	11/5/1998	CWS	19130
CA	CA1510029	Vaughn WC INC	CLARISSE WELL 02 — BEFORE GAC	ETHYLENE DIBROMIDE (EDB)	0.08	UG/L	1/5/1999	CWS	19130
CA	CA1510029	Vaughn WC INC	CLARISSE WELL 02 — BEFORE GAC	ETHYLENE DIBROMIDE (EDB)	0.03	UG/L	2/2/1999	CWS	19130
CA	CA1510029	Vaughn WC INC	CLARISSE WELL 02 — BEFORE GAC	ETHYLENE DIBROMIDE (EDB)	0.09	UG/L	5/4/1999	CWS	19130
CA	CA1510029	Vaughn WC INC	CLARISSE WELL 02 — BEFORE GAC	ETHYLENE DIBROMIDE (EDB)	0.06	UG/L	6/1/1999	CWS	19130
CA	CA1510029	Vaughn WC INC	CLARISSE WELL 02 — BEFORE GAC	ETHYLENE DIBROMIDE (EDB)	0.08	UG/L	7/1/1999	CWS	19130
CA	CA1510029	Vaughn WC INC	CLARISSE WELL 02 — BEFORE GAC	ETHYLENE DIBROMIDE (EDB)	0.02	UG/L	7/27/1999	CWS	19130
CA	CA1510029	Vaughn WC INC	CLARISSE WELL 02 — BEFORE GAC	ETHYLENE DIBROMIDE (EDB)	0.08	UG/L	8/3/1999	CWS	19130
CA	CA1510029	Vaughn WC INC	CLARISSE WELL 02 — BEFORE GAC	ETHYLENE DIBROMIDE (EDB)	0.07	UG/L	5/3/2000	CWS	19130

APPENDIX 7-2 *Continued*

State	PWS_ID	Location	POE	Constituent	Level	Unit	COLL_DATE	PWSType	Pop
CA	CA1510029	Vaughn WC INC	CLARISSE WELL 02 — BEFORE GAC	ETHYLENE DIBROMIDE (EDB)	0.08	UG/L	10/6/1999	CWS	19130
CA	CA1510029	Vaughn WC INC	CLARISSE WELL 02 — BEFORE GAC	ETHYLENE DIBROMIDE (EDB)	0.03	UG/L	12/1/1999	CWS	19130
CA	CA1510029	Vaughn WC INC	CLARISSE WELL 02 — BEFORE GAC	ETHYLENE DIBROMIDE (EDB)	0.03	UG/L	1/4/2000	CWS	19130
CA	CA1510029	Vaughn WC INC	CLARISSE WELL 02 — BEFORE GAC	ETHYLENE DIBROMIDE (EDB)	0.03	UG/L	2/1/2000	CWS	19130
CA	CA1510029	Vaughn WC INC	CLARISSE WELL 02 — BEFORE GAC	ETHYLENE DIBROMIDE (EDB)	0.03	UG/L	3/1/2000	CWS	19130
CA	CA1510029	Vaughn WC INC	CLARISSE WELL 02 — BEFORE GAC	ETHYLENE DIBROMIDE (EDB)	0.07	UG/L	4/4/2000	CWS	19130
CA	CA1510029	Vaughn WC INC	CLARISSE WELL 02 — BEFORE GAC	ETHYLENE DIBROMIDE (EDB)	0.07	UG/L	10/1/1998	CWS	19130
CA	CA1510029	Vaughn WC INC	CLARISSE WELL 02 — BEFORE GAC	ETHYLENE DIBROMIDE (EDB)	0.03	UG/L	9/2/1999	CWS	19130
CA	CA1510029	Vaughn WC INC	HEATH WELL — BEFORE GAC	ETHYLENE DIBROMIDE (EDB)	0.14	UG/L	4/29/2003	CWS	19130
CA	CA1510029	Vaughn WC INC	HEATH WELL — BEFORE GAC	ETHYLENE DIBROMIDE (EDB)	0.11	UG/L	3/25/2003	CWS	19130
CA	CA1510029	Vaughn WC INC	HEATH WELL — BEFORE GAC	ETHYLENE DIBROMIDE (EDB)	0.15	UG/L	2/27/2003	CWS	19130
CA	CA1510029	Vaughn WC INC	HEATH WELL — BEFORE GAC	ETHYLENE DIBROMIDE (EDB)	0.03	UG/L	5/27/2003	CWS	19130
CA	CA1510029	Vaughn WC INC	HEATH WELL — BEFORE GAC	ETHYLENE DIBROMIDE (EDB)	0.12	UG/L	9/29/2003	CWS	19130
CA	CA1510029	Vaughn WC INC	HEATH WELL — BEFORE GAC	ETHYLENE DIBROMIDE (EDB)	0.02	UG/L	8/26/2003	CWS	19130
CA	CA1510029	Vaughn WC INC	HEATH WELL — BEFORE GAC	ETHYLENE DIBROMIDE (EDB)	0.1	UG/L	10/30/2003	CWS	19130
CA	CA1510029	Vaughn WC INC	HEATH WELL — BEFORE GAC	ETHYLENE DIBROMIDE (EDB)	0.18	UG/L	1/29/2003	CWS	19130
CA	CA1510029	Vaughn WC INC	HEATH WELL — BEFORE GAC	ETHYLENE DIBROMIDE (EDB)	0.17	UG/L	7/23/2002	CWS	19130
CA	CA1510029	Vaughn WC INC	HEATH WELL — BEFORE GAC	ETHYLENE DIBROMIDE (EDB)	0.14	UG/L	4/30/2002	CWS	19130
CA	CA1510029	Vaughn WC INC	HEATH WELL — BEFORE GAC	ETHYLENE DIBROMIDE (EDB)	0.03	UG/L	6/24/2002	CWS	19130
CA	CA1510029	Vaughn WC INC	HEATH WELL — BEFORE GAC	ETHYLENE DIBROMIDE (EDB)	0.11	UG/L	10/31/2002	CWS	19130
CA	CA1510029	Vaughn WC INC	HEATH WELL — BEFORE GAC	ETHYLENE DIBROMIDE (EDB)	0.08	UG/L	11/24/2003	CWS	19130
CA	CA1510029	Vaughn WC INC	HEATH WELL — BEFORE GAC	ETHYLENE DIBROMIDE (EDB)	0.16	UG/L	12/30/2002	CWS	19130
CA	CA1510029	Vaughn WC INC	HEATH WELL — BEFORE GAC	ETHYLENE DIBROMIDE (EDB)	0.16	UG/L	5/30/2002	CWS	19130
CA	CA1510029	Vaughn WC INC	HEATH WELL — BEFORE GAC	ETHYLENE DIBROMIDE (EDB)	0.16	UG/L	11/19/2002	CWS	19130
CA	CA1510029	Vaughn WC INC	HEATH WELL — BEFORE GAC	ETHYLENE DIBROMIDE (EDB)	0.08	UG/L	5/26/2005	CWS	19130
CA	CA1510029	Vaughn WC INC	HEATH WELL — BEFORE GAC	ETHYLENE DIBROMIDE (EDB)	0.06	UG/L	9/30/2004	CWS	19130
CA	CA1510029	Vaughn WC INC	HEATH WELL — BEFORE GAC	ETHYLENE DIBROMIDE (EDB)	0.05	UG/L	4/27/2005	CWS	19130
CA	CA1510029	Vaughn WC INC	HEATH WELL — BEFORE GAC	ETHYLENE DIBROMIDE (EDB)	0.09	UG/L	10/27/2004	CWS	19130
CA	CA1510029	Vaughn WC INC	HEATH WELL — BEFORE GAC	ETHYLENE DIBROMIDE (EDB)	0.09	UG/L	11/22/2004	CWS	19130
CA	CA1510029	Vaughn WC INC	HEATH WELL — BEFORE GAC	ETHYLENE DIBROMIDE (EDB)	0.08	UG/L	12/23/2004	CWS	19130
CA	CA1510029	Vaughn WC INC	HEATH WELL — BEFORE GAC	ETHYLENE DIBROMIDE (EDB)	0.08	UG/L	1/26/2005	CWS	19130
CA	CA1510029	Vaughn WC INC	HEATH WELL — BEFORE GAC	ETHYLENE DIBROMIDE (EDB)	0.09	UG/L	2/25/2005	CWS	19130
CA	CA1510029	Vaughn WC INC	HEATH WELL — BEFORE GAC	ETHYLENE DIBROMIDE (EDB)	0.09	UG/L	3/31/2005	CWS	19130
CA	CA1510029	Vaughn WC INC	HEATH WELL — BEFORE GAC	ETHYLENE DIBROMIDE (EDB)	0.1	UG/L	1/29/2004	CWS	19130
CA	CA1510029	Vaughn WC INC	HEATH WELL — BEFORE GAC	ETHYLENE DIBROMIDE (EDB)	0.11	UG/L	2/25/2004	CWS	19130
CA	CA1510029	Vaughn WC INC	HEATH WELL — BEFORE GAC	ETHYLENE DIBROMIDE (EDB)	0.13	UG/L	3/31/2004	CWS	19130
CA	CA1510029	Vaughn WC INC	HEATH WELL — BEFORE GAC	ETHYLENE DIBROMIDE (EDB)	0.02	UG/L	4/30/2004	CWS	19130
CA	CA1510029	Vaughn WC INC	HEATH WELL — BEFORE GAC	ETHYLENE DIBROMIDE (EDB)	0.12	UG/L	5/28/2004	CWS	19130
CA	CA1510029	Vaughn WC INC	HEATH WELL — BEFORE GAC	ETHYLENE DIBROMIDE (EDB)	0.02	UG/L	7/28/2004	CWS	19130
CA	CA1510029	Vaughn WC INC	HEATH WELL — BEFORE GAC	ETHYLENE DIBROMIDE (EDB)	0.12	UG/L	12/29/2003	CWS	19130
CA	CA1510029	Vaughn WC INC	HEATH WELL — BEFORE GAC	ETHYLENE DIBROMIDE (EDB)	0.13	UG/L	6/30/2004	CWS	19130
CA	CA1510029	Vaughn WC INC	HEATH WELL — BEFORE GAC	ETHYLENE DIBROMIDE (EDB)	0.11	UG/L	4/6/1999	CWS	19130
CA	CA1510029	Vaughn WC INC	HEATH WELL — BEFORE GAC	ETHYLENE DIBROMIDE (EDB)	0.03	UG/L	7/21/1998	CWS	19130
CA	CA1510029	Vaughn WC INC	HEATH WELL — BEFORE GAC	ETHYLENE DIBROMIDE (EDB)	0.06	UG/L	8/21/1998	CWS	19130
CA	CA1510029	Vaughn WC INC	HEATH WELL — BEFORE GAC	ETHYLENE DIBROMIDE (EDB)	0.02	UG/L	9/23/1998	CWS	19130
CA	CA1510029	Vaughn WC INC	HEATH WELL — BEFORE GAC	ETHYLENE DIBROMIDE (EDB)	0.03	UG/L	11/5/1998	CWS	19130

CA	CA1510029	Vaughn WC INC	HEATH WELL — BEFORE GAC	ETHYLENE DIBROMIDE (EDB)	0.17	UG/L	1/5/1999	CWS	19130
CA	CA1510029	Vaughn WC INC	HEATH WELL — BEFORE GAC	ETHYLENE DIBROMIDE (EDB)	0.18	UG/L	4/4/2000	CWS	19130
CA	CA1510029	Vaughn WC INC	HEATH WELL — BEFORE GAC	ETHYLENE DIBROMIDE (EDB)	0.09	UG/L	1/22/1999	CWS	19130
CA	CA1510029	Vaughn WC INC	HEATH WELL — BEFORE GAC	ETHYLENE DIBROMIDE (EDB)	0.15	UG/L	5/4/1999	CWS	19130
CA	CA1510029	Vaughn WC INC	HEATH WELL — BEFORE GAC	ETHYLENE DIBROMIDE (EDB)	0.16	UG/L	6/1/1999	CWS	19130
CA	CA1510029	Vaughn WC INC	HEATH WELL — BEFORE GAC	ETHYLENE DIBROMIDE (EDB)	0.19	UG/L	7/1/1999	CWS	19130
CA	CA1510029	Vaughn WC INC	HEATH WELL — BEFORE GAC	ETHYLENE DIBROMIDE (EDB)	0.21	UG/L	10/6/1999	CWS	19130
CA	CA1510029	Vaughn WC INC	HEATH WELL — BEFORE GAC	ETHYLENE DIBROMIDE (EDB)	0.18	UG/L	12/1/1999	CWS	19130
CA	CA1510029	Vaughn WC INC	HEATH WELL — BEFORE GAC	ETHYLENE DIBROMIDE (EDB)	0.19	UG/L	1/4/2000	CWS	19130
CA	CA1510029	Vaughn WC INC	HEATH WELL — BEFORE GAC	ETHYLENE DIBROMIDE (EDB)	0.19	UG/L	2/1/2000	CWS	19130
CA	CA1510029	Vaughn WC INC	HEATH WELL — BEFORE GAC	ETHYLENE DIBROMIDE (EDB)	0.06	UG/L	1/21/1999	CWS	19130
CA	CA1510029	Vaughn WC INC	HEATH WELL — BEFORE GAC	ETHYLENE DIBROMIDE (EDB)	0.04	UG/L	8/29/2002	CWS	19130
CA	CA1510029	Vaughn WC INC	HEATH WELL — BEFORE GAC	ETHYLENE DIBROMIDE (EDB)	0.19	UG/L	1/29/2002	CWS	19130
CA	CA1510029	Vaughn WC INC	HEATH WELL — BEFORE GAC	ETHYLENE DIBROMIDE (EDB)	0.17	UG/L	2/25/2002	CWS	19130
CA	CA1510029	Vaughn WC INC	HEATH WELL — BEFORE GAC	ETHYLENE DIBROMIDE (EDB)	0.19	UG/L	3/28/2002	CWS	19130
CA	CA1510029	Vaughn WC INC	HEATH WELL — BEFORE GAC	ETHYLENE DIBROMIDE (EDB)	0.17	UG/L	2/13/2001	CWS	19130
CA	CA1510029	Vaughn WC INC	HEATH WELL — BEFORE GAC	ETHYLENE DIBROMIDE (EDB)	0.13	UG/L	1/18/2001	CWS	19130
CA	CA1510029	Vaughn WC INC	HEATH WELL — BEFORE GAC	ETHYLENE DIBROMIDE (EDB)	0.14	UG/L	4/5/2001	CWS	19130
CA	CA1510029	Vaughn WC INC	HEATH WELL — BEFORE GAC	ETHYLENE DIBROMIDE (EDB)	0.07	UG/L	5/14/2001	CWS	19130
CA	CA1510029	Vaughn WC INC	HEATH WELL — BEFORE GAC	ETHYLENE DIBROMIDE (EDB)	0.05	UG/L	6/13/2001	CWS	19130
CA	CA1510029	Vaughn WC INC	HEATH WELL — BEFORE GAC	ETHYLENE DIBROMIDE (EDB)	0.05	UG/L	7/12/2001	CWS	19130
CA	CA1510029	Vaughn WC INC	TORREY WELL — BEFORE GAC	ETHYLENE DIBROMIDE (EDB)	0.04	UG/L	9/17/2002	CWS	19130
CA	CA1510029	Vaughn WC INC	TORREY WELL — BEFORE GAC	ETHYLENE DIBROMIDE (EDB)	0.02	UG/L	5/23/2003	CWS	19130
CA	CA1510029	Vaughn WC INC	TORREY WELL — BEFORE GAC	ETHYLENE DIBROMIDE (EDB)	0.02	UG/L	8/29/2002	CWS	19130
CA	CA1510029	Vaughn WC INC	TORREY WELL — BEFORE GAC	ETHYLENE DIBROMIDE (EDB)	0.02	UG/L	7/28/2004	CWS	19130
CA	CA1510029	Vaughn WC INC	TORREY WELL — BEFORE GAC	ETHYLENE DIBROMIDE (EDB)	0.03	UG/L	8/19/2004	CWS	19130
CA	CA1510029	Vaughn WC INC	TORREY WELL — BEFORE GAC	ETHYLENE DIBROMIDE (EDB)	0.03	UG/L	12/23/2004	CWS	19130
CA	CA1510029	Vaughn WC INC	TORREY WELL — BEFORE GAC	ETHYLENE DIBROMIDE (EDB)	0.02	UG/L	1/26/2005	CWS	19130
CA	CA1510029	Vaughn WC INC	TORREY WELL — BEFORE GAC	ETHYLENE DIBROMIDE (EDB)	0.02	UG/L	11/23/2004	CWS	19130
CA	CA1510029	Vaughn WC INC	TORREY WELL — BEFORE GAC	ETHYLENE DIBROMIDE (EDB)	0.02	UG/L	3/31/2004	CWS	19130
CA	CA1510029	Vaughn WC INC	TORREY WELL — BEFORE GAC	ETHYLENE DIBROMIDE (EDB)	0.02	UG/L	2/25/2004	CWS	19130
CA	CA1510029	Vaughn WC INC	TORREY WELL — BEFORE GAC	ETHYLENE DIBROMIDE (EDB)	0.02	UG/L	4/30/2004	CWS	19130
CA	CA1510029	Vaughn WC INC	TORREY WELL — BEFORE GAC	ETHYLENE DIBROMIDE (EDB)	0.03	UG/L	6/30/2004	CWS	19130
CA	CA1510029	Vaughn WC INC	TORREY WELL — BEFORE GAC	ETHYLENE DIBROMIDE (EDB)	0.02	UG/L	5/28/2004	CWS	19130
CA	CA1510029	Vaughn WC INC	TORREY WELL — BEFORE GAC	ETHYLENE DIBROMIDE (EDB)	0.02	UG/L	11/2/1999	CWS	19130
CA	CA1510029	Vaughn WC INC	TORREY WELL — BEFORE GAC	ETHYLENE DIBROMIDE (EDB)	0.03	UG/L	4/9/2001	CWS	19130
CA	CA1510029	Vaughn WC INC	TORREY WELL — BEFORE GAC	ETHYLENE DIBROMIDE (EDB)	0.02	UG/L	2/13/2001	CWS	19130
					0.08				
CA	CA1503182	SUN PACIFIC SHIPPERS-KERN DIVISION	WELL 01	HALOACETIC ACIDS (5) (HAA5)	3.2	UG/L	8/3/2004	NTNCWS	250
CA	CA1500296	FULLER ACRES MUTUAL WATER COMPANY	WELL 01	MONOBROMOACETIC ACID (MBAA)	1.1	UG/L	8/16/2004	CWS	194
CA	CA1510004	Casa Loma Water Co, Inc.	WELL 03 — RAW (DEL MAR DR)	TETRACHLOROETHYLENE	2.8	UG/L	3/20/2003	CWS	600
CA	CA1510004	Casa Loma Water Co, Inc.	WELL 03 — RAW (DEL MAR DR)	TETRACHLOROETHYLENE	0.86	UG/L	9/12/2003	CWS	600
CA	CA1510004	Casa Loma Water Co, Inc.	WELL 03 — RAW (DEL MAR DR)	TETRACHLOROETHYLENE	2.9	UG/L	4/11/2005	CWS	600
CA	CA1510004	Casa Loma Water Co, Inc.	WELL 03 — RAW (DEL MAR DR)	TETRACHLOROETHYLENE	2.4	UG/L	1/21/2005	CWS	600
CA	CA1510004	Casa Loma Water Co, Inc.	WELL 03 — RAW (DEL MAR DR)	TETRACHLOROETHYLENE	1.1	UG/L	9/27/2004	CWS	600
CA	CA1510004	Casa Loma Water Co, Inc.	WELL 03 — RAW (DEL MAR DR)	TETRACHLOROETHYLENE	1.3	UG/L	6/21/2004	CWS	600
CA	CA1510004	Casa Loma Water Co, Inc.	WELL 03 — RAW (DEL MAR DR)	TETRACHLOROETHYLENE	2.7	UG/L	12/18/2003	CWS	600
CA	CA1510004	Casa Loma Water Co, Inc.	WELL 03 — RAW (DEL MAR DR)	TETRACHLOROETHYLENE	2.1	UG/L	5/23/2003	CWS	600
CA	CA1510004	Casa Loma Water Co, Inc.	WELL 03 — RAW (DEL MAR DR)	TETRACHLOROETHYLENE	2.3	UG/L	3/9/2004	CWS	600
CA	CA1510004	Casa Loma Water Co, Inc.	WELL 03 — RAW (DEL MAR DR)	TETRACHLOROETHYLENE	4.5	UG/L	2/25/2000	CWS	600

APPENDIX 7-2 *Continued*

State	PWS_ID	Location	POE	Constituent	Level	Unit	COLL_DATE	PWSType	Pop
CA	CA1510004	Casa Loma Water Co, Inc.	WELL 03 — RAW (DEL MAR DR)	TETRACHLOROETHYLENE	3.6	UG/L	12/13/2002	CWS	600
CA	CA1510004	Casa Loma Water Co, Inc.	WELL 03 — RAW (DEL MAR DR)	TETRACHLOROETHYLENE	5	UG/L	7/24/2002	CWS	600
CA	CA1510004	Casa Loma Water Co, Inc.	WELL 03 — RAW (DEL MAR DR)	TETRACHLOROETHYLENE	3.1	UG/L	5/16/2002	CWS	600
CA	CA1510004	Casa Loma Water Co, Inc.	WELL 03 — RAW (DEL MAR DR)	TETRACHLOROETHYLENE	9.1	UG/L	3/11/2002	CWS	600
CA	CA1510004	Casa Loma Water Co, Inc.	WELL 03 — RAW (DEL MAR DR)	TETRACHLOROETHYLENE	7	UG/L	1/2/2002	CWS	600
CA	CA1510004	Casa Loma Water Co, Inc.	WELL 03 — RAW (DEL MAR DR)	TETRACHLOROETHYLENE	9.2	UG/L	8/24/2001	CWS	600
CA	CA1510004	Casa Loma Water Co, Inc.	WELL 03 — RAW (DEL MAR DR)	TETRACHLOROETHYLENE	11	UG/L	5/24/2001	CWS	600
CA	CA1510004	Casa Loma Water Co, Inc.	WELL 03 — RAW (DEL MAR DR)	TETRACHLOROETHYLENE	9.1	UG/L	12/11/2000	CWS	600
CA	CA1510004	Casa Loma Water Co, Inc.	WELL 03 — RAW (DEL MAR DR)	TETRACHLOROETHYLENE	8.9	UG/L	9/19/2000	CWS	600
CA	CA1510004	Casa Loma Water Co, Inc.	WELL 03 — RAW (DEL MAR DR)	TETRACHLOROETHYLENE	7.1	UG/L	12/10/2001	CWS	600
CA	CA1510004	Casa Loma Water Co, Inc.	WELL 03 — RAW (DEL MAR DR)	TETRACHLOROETHYLENE	2.5	UG/L	6/9/2000	CWS	600
CA	CA1510004	Casa Loma Water Co, Inc.	WELL 04 — RAW (LOMITA DR)	TETRACHLOROETHYLENE	1.5	UG/L	9/26/2002	CWS	600
CA	CA1510004	Casa Loma Water Co, Inc.	WELL 04 — RAW (LOMITA DR)	TETRACHLOROETHYLENE	0.96	UG/L	9/12/2003	CWS	600
CA	CA1510004	Casa Loma Water Co, Inc.	WELL 04 — RAW (LOMITA DR)	TETRACHLOROETHYLENE	0.84	UG/L	9/27/2004	CWS	600
CA	CA1510004	Casa Loma Water Co, Inc.	WELL 04 — RAW (LOMITA DR)	TETRACHLOROETHYLENE	1.8	UG/L	9/27/2001	CWS	600
CA	CA1510004	Casa Loma Water Co, Inc.	WELL 04 — RAW (LOMITA DR)	TETRACHLOROETHYLENE	1.5	UG/L	3/25/2002	CWS	600
CA	CA1510004	Casa Loma Water Co, Inc.	WELL 04 — RAW (LOMITA DR)	TETRACHLOROETHYLENE	2.1	UG/L	8/24/2001	CWS	600
CA	CA1510004	Casa Loma Water Co, Inc.	WELL 04 — RAW (LOMITA DR)	TETRACHLOROETHYLENE	1.8	UG/L	5/16/2002	CWS	600
					3.9				
CA	CA1502615	Frito-Lay No.America, Inc. Water System	NORTH WELL (WELL 02) — CL2 TREATMENT	TRICHLOROACETIC ACID (TCAA)	1	UG/L	4/1/2003	NTNCWS	400
CA	CA1502017	WHEELER FARMS HEADQUARTERS	WELL 01	TRICHLOROETHYLENE	0.6	UG/L	9/19/2000	CWS	25
CA	CA1503304	WM. BOLTHOUSE FARMS, INC. — WEST PLANT	NORTH WELL	1,2,3-TRICHLOROPROPANE	0.03	UG/L	5/23/2002	NTNCWS	300
CA	CA1503304	WM. BOLTHOUSE FARMS, INC. — WEST PLANT	NORTH WELL	1,2,3-TRICHLOROPROPANE	0.21	UG/L	11/5/2002	NTNCWS	300
CA	CA1503304	WM. BOLTHOUSE FARMS, INC. — WEST PLANT	SOUTH WELL	1,2,3-TRICHLOROPROPANE	0.06	UG/L	5/23/2002	NTNCWS	300
CA	CA1503304	WM. BOLTHOUSE FARMS, INC. — WEST PLANT	SOUTH WELL	1,2,3-TRICHLOROPROPANE	0.01	UG/L	11/5/2002	NTNCWS	300
CA	CA1500597	GRIMMWAY FARMS FROZEN FOODS	WELL 01	1,2,3-TRICHLOROPROPANE	0.18	UG/L	8/9/2001	NTNCWS	350
CA	CA1500597	GRIMMWAY FARMS FROZEN FOODS	WELL 02	1,2,3-TRICHLOROPROPANE	0.17	UG/L	7/19/2001	NTNCWS	350
CA	CA1500560	KRANENBURG WATER SYSTEM	WELL 01 NORD RD	1,2,3-TRICHLOROPROPANE	0.01	UG/L	4/28/2005	CWS	39
CA	CA1500546	HARVEST MOON MUTUAL WATER CO	WELL 01 (HEATH/BLACKHAWK)	1,2,3-TRICHLOROPROPANE	8.00E-03	UG/L	10/5/2004	CWS	180
CA	CA1510029	Vaughn WC INC	CLARISSE NO. 1-AFTER GAC 1 & 2 COMBINED	1,2,3-TRICHLOROPROPANE	0.01	UG/L	8/13/2004	CWS	19130
CA	CA1510029	Vaughn WC INC	ATAKAPA WELL — BEFORE OZONATION	1,2,3-TRICHLOROPROPANE	0.03	UG/L	11/26/2002	CWS	19130
CA	CA1510029	Vaughn WC INC	ATAKAPA WELL — BEFORE OZONATION	1,2,3-TRICHLOROPROPANE	0.02	UG/L	7/23/2002	CWS	19130
CA	CA1510029	Vaughn WC INC	ATAKAPA WELL — BEFORE OZONATION	1,2,3-TRICHLOROPROPANE	0.01	UG/L	8/13/2004	CWS	19130
CA	CA1510029	Vaughn WC INC	CLARISSE WELL 01 — BEFORE GAC	1,2,3-TRICHLOROPROPANE	0.03	UG/L	7/23/2002	CWS	19130
CA	CA1510029	Vaughn WC INC	CLARISSE WELL 01 — BEFORE GAC	1,2,3-TRICHLOROPROPANE	0.02	UG/L	8/13/2004	CWS	19130
CA	CA1510029	Vaughn WC INC	CLARISSE WELL 02 — BEFORE GAC	1,2,3-TRICHLOROPROPANE	0.01	UG/L	7/23/2002	CWS	19130
CA	CA1510029	Vaughn WC INC	CLARISSE WELL 02 — BEFORE GAC	1,2,3-TRICHLOROPROPANE	7.00E-03	UG/L	8/13/2004	CWS	19130
CA	CA1510029	Vaughn WC INC	HEATH WELL — BEFORE GAC	1,2,3-TRICHLOROPROPANE	0.01	UG/L	7/23/2002	CWS	19130
CA	CA1510029	Vaughn WC INC	HEATH WELL — BEFORE GAC	1,2,3-TRICHLOROPROPANE	0.01	UG/L	8/13/2004	CWS	19130
CA	CA1510029	Vaughn WC INC	TORREY WELL — BEFORE GAC	1,2,3-TRICHLOROPROPANE	0.14	UG/L	7/23/2002	CWS	19130
CA	CA1510029	Vaughn WC INC	TORREY WELL — BEFORE GAC	1,2,3-TRICHLOROPROPANE	0.14	UG/L	11/19/2002	CWS	19130
CA	CA1510029	Vaughn WC INC	TORREY WELL — BEFORE GAC	1,2,3-TRICHLOROPROPANE	0.09	UG/L	8/13/2004	CWS	19130
CA	CA1500535	MANON MANOR MUTUAL WATER COMPANY	WELL 01	1,2,3-TRICHLOROPROPANE	0.13	UG/L	7/24/2001	CWS	200
CA	CA1500535	MANON MANOR MUTUAL WATER COMPANY	WELL 01	1,2,3-TRICHLOROPROPANE	0.13	UG/L	7/24/2001	CWS	200
CA	CA1500575	SAN JOAQUIN ESTATES MUTUAL	WELL 01	1,2,3-TRICHLOROPROPANE	0.29	UG/L	7/24/2001	CWS	220

CA	CA1500575	SAN JOAQUIN ESTATES MUTUAL	WELL 01	1,2,3-TRICHLOROPROPANE	0.31	UG/L	7/24/2001	CWS	220
CA	CA1500373	SEVENTH STANDARD MUTUAL	WELL 02 — STANDBY	1,2,3-TRICHLOROPROPANE	0.09	UG/L	10/11/1999	CWS	25
CA	CA1500373	SEVENTH STANDARD MUTUAL	WELL 02 — STANDBY	1,2,3-TRICHLOROPROPANE	0.08	UG/L	10/11/1999	CWS	25
CA	CA1503349	WM. BOLTHOUSE FARMS, INC.	PORTER WELL	1,2,3-TRICHLOROPROPANE	9.00E-03	UG/L	5/23/2002	NTNCWS	1100
CA	CA1502017	WHEELER FARMS HEADQUARTERS	WELL 01	1,2,3-TRICHLOROPROPANE	0.13	UG/L	7/23/2001	CWS	25
CA	CA1502017	WHEELER FARMS HEADQUARTERS	WELL 01	1,2,3-TRICHLOROPROPANE	0.13	UG/L	7/23/2001	CWS	25
CA	CA1502017	WHEELER FARMS HEADQUARTERS	WELL 01	1,2,3-TRICHLOROPROPANE	0.13	UG/L	7/23/2001	CWS	25
CA	CA1500231	VICTORY MUTUAL WATER COMPANY	WELL 02 — SHIRLEY LANE	1,2,3-TRICHLOROPROPANE	0.11	UG/L	11/15/2001	CWS	700
					0.09				
CA	CA0104012	Vulcan Materials, CalMat Division	WELL 02 — PENDING	TOTAL TRIHALOMETHANES	0.94	UG/L	3/25/2003	NTNCWS	45
CA	CA1500483	ALLEN ROAD MUTUAL WATER SYSTEM	WELL 01	TOTAL TRIHALOMETHANES	1.05	UG/L	7/14/1999	CWS	200
CA	CA1500289	ATHAL MUTUAL WATER SYSTEM	WELL 01	TOTAL TRIHALOMETHANES	0.52	UG/L	8/17/2004	CWS	85
CA	CA1500409	BROCK MUTUAL WATER COMPANY	WELL 02 — NORTH	TOTAL TRIHALOMETHANES	2	UG/L	1/4/2002	CWS	300
CA	CA1503349	WM. BOLTHOUSE FARMS, INC.	LIMI WELL	TOTAL TRIHALOMETHANES	1.1	UG/L	5/5/1998	NTNCWS	1100
CA	CA1510004	Casa Loma Water Co, Inc.	WELL 03 — RAW (DEL MAR DR)	TOTAL TRIHALOMETHANES	14	UG/L	3/20/2003	CWS	600
CA	CA1500544	ENOS LANE PUBLIC UTILITY DISTRICT	WELL 02 — WEST	TOTAL TRIHALOMETHANES	4.4	UG/L	8/27/2004	CWS	240
CA	CA1503330	GCG WATER SYSTEM	WELL 01	TOTAL TRIHALOMETHANES	2.89	UG/L	8/26/2004	NTNCWS	40
CA	CA1503336	GOLDEN EMPIRE CONCRETE COMPANY	WELL 01	TOTAL TRIHALOMETHANES	2.46	UG/L	8/26/2004	NTNCWS	44
CA	CA1500597	GRIMMWAY FARMS FROZEN FOODS	WELL 01	TOTAL TRIHALOMETHANES	0.66	UG/L	5/10/2004	NTNCWS	350
CA	CA1500546	HARVEST MOON MUTUAL WATER CO	WELL 01 (HEATH/BLACKHAWK)	TOTAL TRIHALOMETHANES	5.06	UG/L	6/8/2000	CWS	180
CA	CA1503182	SUN PACIFIC SHIPPERS-KERN DIVISION	WELL 01	TOTAL TRIHALOMETHANES	15.6	UG/L	11/12/2002	NTNCWS	250
CA	CA1503182	SUN PACIFIC SHIPPERS-KERN DIVISION	WELL 01	TOTAL TRIHALOMETHANES	20.22	UG/L	8/3/2004	NTNCWS	250
CA	CA1502663	TOWN & COUNTRY WATER COMPANY	WELL 02 — BOREL — STANDBY	TOTAL TRIHALOMETHANES	9	UG/L	8/27/2004	CWS	80
CA	CA1502320	LERDO SHERIFF'S FACILITY	WELL 02 — CL2 TREATMENT	TOTAL TRIHALOMETHANES	4.7	UG/L	7/7/2004	CWS	3000
CA	CA1500578	LONG CANYON WATER COMPANY CORP.	WELL 02	TOTAL TRIHALOMETHANES	16.48	UG/L	1/1/2001	CWS	197
CA	CA1500378	MAHER MUTUAL WATER COMPANY	WELL 01	TOTAL TRIHALOMETHANES	2.2	UG/L	8/17/2004	CWS	150
CA	CA1502465	PANAMA ROAD PROPERTY OWNERS ASSOC	WELL 01	TOTAL TRIHALOMETHANES	15.03	UG/L	6/13/2000	CWS	50
CA	CA1500452	RIVERDALE VILLAGE	WELL 01	TOTAL TRIHALOMETHANES	1.1	UG/L	4/4/2001	CWS	760
CA	CA1500452	RIVERDALE VILLAGE	WELL 01	TOTAL TRIHALOMETHANES	1.9	UG/L	4/7/2004	CWS	760
CA	CA1500494	WILSON ROAD WATER COMMUNITY	WELL 01	TOTAL TRIHALOMETHANES	1	UG/L	8/17/2004	CWS	60
CA	CA1500494	WILSON ROAD WATER COMMUNITY	WELL 01	TOTAL TRIHALOMETHANES	0.85	UG/L	9/24/1997	CWS	60
CA	CA1510029	Vaughn WC INC	TORREY WELL — BEFORE GAC	TOTAL TRIHALOMETHANES	0.7	UG/L	4/11/2000	CWS	19130
					5.4				
CA	CA1503290	THE GARLIC COMPANY	WELL 02	M,P-XYLENE	14.3	UG/L	7/27/1999	NTNCWS	200
CA	CA1503290	THE GARLIC COMPANY	WELL 02	O-XYLENE	5.3	UG/L	7/27/1999	NTNCWS	200
CA	CA1503290	THE GARLIC COMPANY	WELL 02	XYLENES (TOTAL)	19.6	UG/L	7/27/1999	NTNCWS	200
					13.1				

APPENDIX 7-3 Bakersfield Surface Water Data

State	PWS_ID	Utility	POE	Constituent	LEVEL_NO	Unit	COLL_DATE	PWS	Pop
CA	CA1510040	Kern County Water Agency	ID4 RAW FK	1,2,3-TRICHLOROPROPANE	0.01	UG/L	9/11/2001	CWS	0
CA	CA1510040	Kern County Water Agency	ID4 RAW KWB	1,2,3-TRICHLOROPROPANE	6.00E-03	UG/L	9/11/2001	CWS	0
CA	CA1510006	East Niles CSD	WELL 20-RAW	ATRAZINE	1.4	UG/L	8/4/1998	CWS	24,316
CA	CA1510006	East Niles CSD	THM SAMP SITE-PACIFIC BELL (E9)-ENS	BROMOCHLOROACETIC ACID (BCAA)	13	UG/L	6/18/2002	CWS	24,316
CA	CA1510006	East Niles CSD	THM SAMP SITE-PACIFIC BELL (E9)-ENS	BROMOCHLOROACETIC ACID (BCAA)	1	UG/L	12/18/2001	CWS	24,316
CA	CA1510006	East Niles CSD	THM SAMP SITE-8717 BUTTERNUT (E10)-ENS	BROMOCHLOROACETIC ACID (BCAA)	8	UG/L	6/18/2002	CWS	24,316
CA	CA1510006	East Niles CSD	THM SAMP SITE-8717 BUTTERNUT (E10)-ENS	BROMOCHLOROACETIC ACID (BCAA)	2	UG/L	12/18/2001	CWS	24,316
CA	CA1510006	East Niles CSD	THM SAMP SITE-11845 SCHOOL (E11)-ENS	BROMOCHLOROACETIC ACID (BCAA)	10	UG/L	6/18/2002	CWS	24,316
CA	CA1510006	East Niles CSD	THM SAMP SITE-11845 SCHOOL (E11)-ENS	BROMOCHLOROACETIC ACID (BCAA)	2	UG/L	12/18/2001	CWS	24,316
CA	CA1510006	East Niles CSD	THM SAMP SITE-11845 SCHOOL (E11)-ENS	BROMOCHLOROACETIC ACID (BCAA)	2	UG/L	9/24/2001	CWS	24,316
CA	CA1510006	East Niles CSD	THM SAMP SITE-1417 VALE (E7)-KCWAS	BROMOCHLOROACETIC ACID (BCAA)	16	UG/L	6/18/2002	CWS	24,316
CA	CA1510006	East Niles CSD	THM SAMP SITE-1417 VALE (E7)-KCWAS	BROMOCHLOROACETIC ACID (BCAA)	6	UG/L	12/18/2001	CWS	24,316
CA	CA1510006	East Niles CSD	THM SAMP SITE-5310 WINGFOOT (E6)-KCWAS	BROMOCHLOROACETIC ACID (BCAA)	14	UG/L	6/18/2002	CWS	24,316
CA	CA1510006	East Niles CSD	THM SAMP SITE-5310 WINGFOOT (E6)-KCWAS	BROMOCHLOROACETIC ACID (BCAA)	6	UG/L	12/18/2001	CWS	24,316
CA	CA1510006	East Niles CSD	THM SAMP SITE-3209 FLINTRIDGE (E4)-KCWAS	BROMOCHLOROACETIC ACID (BCAA)	17	UG/L	6/18/2002	CWS	24,316
CA	CA1510006	East Niles CSD	THM SAMP SITE-3209 FLINTRIDGE (E4)-KCWAS	BROMOCHLOROACETIC ACID (BCAA)	7	UG/L	12/18/2001	CWS	24,316
					8				
CA	CA1510006	East Niles CSD	THM SAMP SITE-DEACON (E8)-ENG	BROMODICHLOROMETHANE (THM)	7.5	UG/L	12/16/2003	CWS	24,316
CA	CA1510006	East Niles CSD	THM SAMP SITE-DEACON (E8)-ENG	BROMODICHLOROMETHANE (THM)	4.58	UG/L	2/26/1998	CWS	24,316
CA	CA1510006	East Niles CSD	THM SAMP SITE-PACIFIC BELL (E9)-ENS	BROMODICHLOROMETHANE (THM)	0.9	UG/L	12/18/2002	CWS	24,316
CA	CA1510006	East Niles CSD	THM SAMP SITE-PACIFIC BELL (E9)-ENS	BROMODICHLOROMETHANE (THM)	34.1	UG/L	6/18/2002	CWS	24,316
CA	CA1510006	East Niles CSD	THM SAMP SITE-PACIFIC BELL (E9)-ENS	BROMODICHLOROMETHANE (THM)	9.5	UG/L	5/8/2002	CWS	24,316
CA	CA1510006	East Niles CSD	THM SAMP SITE-PACIFIC BELL (E9)-ENS	BROMODICHLOROMETHANE (THM)	18	UG/L	6/11/2003	CWS	24,316
CA	CA1510006	East Niles CSD	THM SAMP SITE-PACIFIC BELL (E9)-ENS	BROMODICHLOROMETHANE (THM)	2.8	UG/L	3/18/2003	CWS	24,316
CA	CA1510006	East Niles CSD	THM SAMP SITE-PACIFIC BELL (E9)-ENS	BROMODICHLOROMETHANE (THM)	6.66	UG/L	8/26/1998	CWS	24,316
CA	CA1510006	East Niles CSD	THM SAMP SITE-PACIFIC BELL (E9)-ENS	BROMODICHLOROMETHANE (THM)	6.69	UG/L	6/23/1998	CWS	24,316
CA	CA1510006	East Niles CSD	THM SAMP SITE-PACIFIC BELL (E9)-ENS	BROMODICHLOROMETHANE (THM)	6.76	UG/L	2/26/1998	CWS	24,316
CA	CA1510006	East Niles CSD	THM SAMP SITE-PACIFIC BELL (E9)-ENS	BROMODICHLOROMETHANE (THM)	3.1	UG/L	9/11/1997	CWS	24,316
CA	CA1510006	East Niles CSD	THM SAMP SITE-PACIFIC BELL (E9)-ENS	BROMODICHLOROMETHANE (THM)	6.66	UG/L	3/9/1999	CWS	24,316
CA	CA1510006	East Niles CSD	THM SAMP SITE-PACIFIC BELL (E9)-ENS	BROMODICHLOROMETHANE (THM)	1.1	UG/L	12/18/2001	CWS	24,316
CA	CA1510006	East Niles CSD	THM SAMP SITE-PACIFIC BELL (E9)-ENS	BROMODICHLOROMETHANE (THM)	0.5	UG/L	9/24/2001	CWS	24,316
CA	CA1510006	East Niles CSD	THM SAMP SITE-PACIFIC BELL (E9)-ENS	BROMODICHLOROMETHANE (THM)	8.7	UG/L	11/3/1998	CWS	24,316
CA	CA1510006	East Niles CSD	THM SAMP SITE-8717 BUTTERNUT (E10)-ENS	BROMODICHLOROMETHANE (THM)	25.3	UG/L	6/18/2002	CWS	24,316
CA	CA1510006	East Niles CSD	THM SAMP SITE-8717 BUTTERNUT (E10)-ENS	BROMODICHLOROMETHANE (THM)	4.4	UG/L	5/8/2002	CWS	24,316
CA	CA1510006	East Niles CSD	THM SAMP SITE-8717 BUTTERNUT (E10)-ENS	BROMODICHLOROMETHANE (THM)	0.7	UG/L	12/18/2002	CWS	24,316
CA	CA1510006	East Niles CSD	THM SAMP SITE-8717 BUTTERNUT (E10)-ENS	BROMODICHLOROMETHANE (THM)	4.3	UG/L	3/18/2003	CWS	24,316
CA	CA1510006	East Niles CSD	THM SAMP SITE-8717 BUTTERNUT (E10)-ENS	BROMODICHLOROMETHANE (THM)	3.01	UG/L	8/26/1998	CWS	24,316
CA	CA1510006	East Niles CSD	THM SAMP SITE-8717 BUTTERNUT (E10)-ENS	BROMODICHLOROMETHANE (THM)	4.5	UG/L	9/11/1997	CWS	24,316
CA	CA1510006	East Niles CSD	THM SAMP SITE-8717 BUTTERNUT (E10)-ENS	BROMODICHLOROMETHANE (THM)	4.42	UG/L	2/26/1998	CWS	24,316
CA	CA1510006	East Niles CSD	THM SAMP SITE-8717 BUTTERNUT (E10)-ENS	BROMODICHLOROMETHANE (THM)	3.75	UG/L	6/23/1998	CWS	24,316
CA	CA1510006	East Niles CSD	THM SAMP SITE-8717 BUTTERNUT (E10)-ENS	BROMODICHLOROMETHANE (THM)	1.9	UG/L	12/18/2001	CWS	24,316

CA	CA1510006	East Niles CSD	THM SAMP SITE-8717 BUTTERNUT (E10)-ENS	BROMODICHLOROMETHANE (THM)	6.3	UG/L	9/24/2001	CWS	24,316
CA	CA1510006	East Niles CSD	THM SAMP SITE-8717 BUTTERNUT (E10)-ENS	BROMODICHLOROMETHANE (THM)	8.43	UG/L	3/9/1999	CWS	24,316
CA	CA1510006	East Niles CSD	THM SAMP SITE-8717 BUTTERNUT (E10)-ENS	BROMODICHLOROMETHANE (THM)	5.1	UG/L	11/3/1998	CWS	24,316
CA	CA1510006	East Niles CSD	THM SAMP SITE-11845 SCHOOL (E11)-ENS	BROMODICHLOROMETHANE (THM)	54	UG/L	6/18/2002	CWS	24,316
CA	CA1510006	East Niles CSD	THM SAMP SITE-11845 SCHOOL (E11)-ENS	BROMODICHLOROMETHANE (THM)	2.3	UG/L	12/18/2002	CWS	24,316
CA	CA1510006	East Niles CSD	THM SAMP SITE-11845 SCHOOL (E11)-ENS	BROMODICHLOROMETHANE (THM)	19.2	UG/L	5/8/2002	CWS	24,316
CA	CA1510006	East Niles CSD	THM SAMP SITE-11845 SCHOOL (E11)-ENS	BROMODICHLOROMETHANE (THM)	2.2	UG/L	3/18/2003	CWS	24,316
CA	CA1510006	East Niles CSD	THM SAMP SITE-11845 SCHOOL (E11)-ENS	BROMODICHLOROMETHANE (THM)	4.1	UG/L	8/26/1998	CWS	24,316
CA	CA1510006	East Niles CSD	THM SAMP SITE-11845 SCHOOL (E11)-ENS	BROMODICHLOROMETHANE (THM)	7	UG/L	11/3/1998	CWS	24,316
CA	CA1510006	East Niles CSD	THM SAMP SITE-11845 SCHOOL (E11)-ENS	BROMODICHLOROMETHANE (THM)	2.2	UG/L	12/18/2001	CWS	24,316
CA	CA1510006	East Niles CSD	THM SAMP SITE-11845 SCHOOL (E11)-ENS	BROMODICHLOROMETHANE (THM)	3.5	UG/L	9/11/1997	CWS	24,316
CA	CA1510006	East Niles CSD	THM SAMP SITE-11845 SCHOOL (E11)-ENS	BROMODICHLOROMETHANE (THM)	5.32	UG/L	6/23/1998	CWS	24,316
CA	CA1510006	East Niles CSD	THM SAMP SITE-11845 SCHOOL (E11)-ENS	BROMODICHLOROMETHANE (THM)	8.79	UG/L	2/26/1998	CWS	24,316
CA	CA1510006	East Niles CSD	THM SAMP SITE-11845 SCHOOL (E11)-ENS	BROMODICHLOROMETHANE (THM)	10.87	UG/L	3/9/1999	CWS	24,316
CA	CA1510006	East Niles CSD	THM SAMP SITE-11845 SCHOOL (E11)-ENS	BROMODICHLOROMETHANE (THM)	1.1	UG/L	9/24/2001	CWS	24,316
CA	CA1510006	East Niles CSD	THM SAMP SITE-FAIRFAX SCHOOL (E12)-ENG	BROMODICHLOROMETHANE (THM)	0.5	UG/L	6/18/2002	CWS	24,316
CA	CA1510006	East Niles CSD	THM SAMP SITE-FAIRFAX SCHOOL (E12)-ENG	BROMODICHLOROMETHANE (THM)	0.54	UG/L	2/26/1998	CWS	24,316
CA	CA1510006	East Niles CSD	THM SAMP SITE-FAIRFAX SCHOOL (E12)-ENG	BROMODICHLOROMETHANE (THM)	0.65	UG/L	3/10/1999	CWS	24,316
CA	CA1510006	East Niles CSD	THM SAMP SITE-1417 VALE (E7)-KCWAS	BROMODICHLOROMETHANE (THM)	1.1	UG/L	12/18/2002	CWS	24,316
CA	CA1510006	East Niles CSD	THM SAMP SITE-1417 VALE (E7)-KCWAS	BROMODICHLOROMETHANE (THM)	7.5	UG/L	5/8/2002	CWS	24,316
CA	CA1510006	East Niles CSD	THM SAMP SITE-1417 VALE (E7)-KCWAS	BROMODICHLOROMETHANE (THM)	34.8	UG/L	6/18/2002	CWS	24,316
CA	CA1510006	East Niles CSD	THM SAMP SITE-1417 VALE (E7)-KCWAS	BROMODICHLOROMETHANE (THM)	3.47	UG/L	6/23/1998	CWS	24,316
CA	CA1510006	East Niles CSD	THM SAMP SITE-1417 VALE (E7)-KCWAS	BROMODICHLOROMETHANE (THM)	0.6	UG/L	3/18/2003	CWS	24,316
CA	CA1510006	East Niles CSD	THM SAMP SITE-1417 VALE (E7)-KCWAS	BROMODICHLOROMETHANE (THM)	8.99	UG/L	3/9/1999	CWS	24,316
CA	CA1510006	East Niles CSD	THM SAMP SITE-1417 VALE (E7)-KCWAS	BROMODICHLOROMETHANE (THM)	6.5	UG/L	11/3/1998	CWS	24,316
CA	CA1510006	East Niles CSD	THM SAMP SITE-1417 VALE (E7)-KCWAS	BROMODICHLOROMETHANE (THM)	3.65	UG/L	2/26/1998	CWS	24,316
CA	CA1510006	East Niles CSD	THM SAMP SITE-1417 VALE (E7)-KCWAS	BROMODICHLOROMETHANE (THM)	4.19	UG/L	8/26/1998	CWS	24,316
CA	CA1510006	East Niles CSD	THM SAMP SITE-1417 VALE (E7)-KCWAS	BROMODICHLOROMETHANE (THM)	3.6	UG/L	9/11/1997	CWS	24,316
CA	CA1510006	East Niles CSD	THM SAMP SITE-1417 VALE (E7)-KCWAS	BROMODICHLOROMETHANE (THM)	7.6	UG/L	12/18/2001	CWS	24,316
CA	CA1510006	East Niles CSD	THM SAMP SITE-1417 VALE (E7)-KCWAS	BROMODICHLOROMETHANE (THM)	38.6	UG/L	9/24/2001	CWS	24,316
CA	CA1510006	East Niles CSD	BRECKENRIDGE TP-RAW-OFFLINE	BROMODICHLOROMETHANE (THM)	4.8	UG/L	3/13/2001	CWS	24,316
CA	CA1510006	East Niles CSD	BRECKENRIDGE TP-RAW-OFFLINE	BROMODICHLOROMETHANE (THM)	7.6	UG/L	10/1/2003	CWS	24,316
CA	CA1510006	East Niles CSD	BRECKENRIDGE TP-RAW-OFFLINE	BROMODICHLOROMETHANE (THM)	12	UG/L	3/23/1998	CWS	24,316
CA	CA1510006	East Niles CSD	BRECKENRIDGE TP-RAW-OFFLINE	BROMODICHLOROMETHANE (THM)	1.8	UG/L	3/19/1997	CWS	24,316
CA	CA1510006	East Niles CSD	BRECKENRIDGE TP-RAW-OFFLINE	BROMODICHLOROMETHANE (THM)	5.4	UG/L	3/15/1999	CWS	24,316
CA	CA1510006	East Niles CSD	THM SAMP SITE-5310 WINGFOOT (E6)-KCWAS	BROMODICHLOROMETHANE (THM)	8.6	UG/L	12/18/2002	CWS	24,316
CA	CA1510006	East Niles CSD	THM SAMP SITE-5310 WINGFOOT (E6)-KCWAS	BROMODICHLOROMETHANE (THM)	37.1	UG/L	6/18/2002	CWS	24,316
CA	CA1510006	East Niles CSD	THM SAMP SITE-5310 WINGFOOT (E6)-KCWAS	BROMODICHLOROMETHANE (THM)	6.8	UG/L	5/8/2002	CWS	24,316
CA	CA1510006	East Niles CSD	THM SAMP SITE-5310 WINGFOOT (E6)-KCWAS	BROMODICHLOROMETHANE (THM)	4.8	UG/L	12/16/2003	CWS	24,316
CA	CA1510006	East Niles CSD	THM SAMP SITE-5310 WINGFOOT (E6)-KCWAS	BROMODICHLOROMETHANE (THM)	9.8	UG/L	3/18/2003	CWS	24,316
CA	CA1510006	East Niles CSD	THM SAMP SITE-5310 WINGFOOT (E6)-KCWAS	BROMODICHLOROMETHANE (THM)	3.48	UG/L	6/23/1998	CWS	24,316
CA	CA1510006	East Niles CSD	THM SAMP SITE-5310 WINGFOOT (E6)-KCWAS	BROMODICHLOROMETHANE (THM)	5.9	UG/L	11/3/1998	CWS	24,316
CA	CA1510006	East Niles CSD	THM SAMP SITE-5310 WINGFOOT (E6)-KCWAS	BROMODICHLOROMETHANE (THM)	5.9	UG/L	11/3/1998	CWS	24,316
CA	CA1510006	East Niles CSD	THM SAMP SITE-5310 WINGFOOT (E6)-KCWAS	BROMODICHLOROMETHANE (THM)	4.2	UG/L	8/26/1998	CWS	24,316

APPENDIX 7-3 *Continued*

State	PWS_ID	Utility	POE	Constituent	LEVEL_NO	Unit	COLL_DATE	PWS	Pop
CA	CA1510006	East Niles CSD	THM SAMP SITE-5310 WINGFOOT (E6)-KCWAS	BROMODICHLOROMETHANE (THM)	5.7	UG/L	2/26/1998	CWS	24,316
CA	CA1510006	East Niles CSD	THM SAMP SITE-5310 WINGFOOT (E6)-KCWAS	BROMODICHLOROMETHANE (THM)	3.7	UG/L	9/11/1997	CWS	24,316
CA	CA1510006	East Niles CSD	THM SAMP SITE-5310 WINGFOOT (E6)-KCWAS	BROMODICHLOROMETHANE (THM)	8.8	UG/L	12/18/2001	CWS	24,316
CA	CA1510006	East Niles CSD	THM SAMP SITE-5310 WINGFOOT (E6)-KCWAS	BROMODICHLOROMETHANE (THM)	29.4	UG/L	9/24/2001	CWS	24,316
CA	CA1510006	East Niles CSD	THM SAMP SITE-5310 WINGFOOT (E6)-KCWAS	BROMODICHLOROMETHANE (THM)	5.59	UG/L	3/10/1999	CWS	24,316
CA	CA1510006	East Niles CSD	WELL 13-RAW	BROMODICHLOROMETHANE (THM)	2.7	UG/L	6/18/2002	CWS	24,316
CA	CA1510006	East Niles CSD	THM SAMP SITE-3209 FLINTRIDGE (E4)-KCWAS	BROMODICHLOROMETHANE (THM)	9.1	UG/L	12/18/2002	CWS	24,316
CA	CA1510006	East Niles CSD	THM SAMP SITE-3209 FLINTRIDGE (E4)-KCWAS	BROMODICHLOROMETHANE (THM)	37.5	UG/L	6/18/2002	CWS	24,316
CA	CA1510006	East Niles CSD	THM SAMP SITE-3209 FLINTRIDGE (E4)-KCWAS	BROMODICHLOROMETHANE (THM)	5.2	UG/L	12/16/2003	CWS	24,316
CA	CA1510006	East Niles CSD	THM SAMP SITE-3209 FLINTRIDGE (E4)-KCWAS	BROMODICHLOROMETHANE (THM)	11.3	UG/L	3/18/2003	CWS	24,316
CA	CA1510006	East Niles CSD	THM SAMP SITE-3209 FLINTRIDGE (E4)-KCWAS	BROMODICHLOROMETHANE (THM)	7.91	UG/L	3/10/1999	CWS	24,316
CA	CA1510006	East Niles CSD	THM SAMP SITE-3209 FLINTRIDGE (E4)-KCWAS	BROMODICHLOROMETHANE (THM)	4.8	UG/L	11/3/1998	CWS	24,316
CA	CA1510006	East Niles CSD	THM SAMP SITE-3209 FLINTRIDGE (E4)-KCWAS	BROMODICHLOROMETHANE (THM)	2.7	UG/L	9/11/1997	CWS	24,316
CA	CA1510006	East Niles CSD	THM SAMP SITE-3209 FLINTRIDGE (E4)-KCWAS	BROMODICHLOROMETHANE (THM)	4.16	UG/L	8/26/1998	CWS	24,316
CA	CA1510006	East Niles CSD	THM SAMP SITE-3209 FLINTRIDGE (E4)-KCWAS	BROMODICHLOROMETHANE (THM)	6.01	UG/L	2/26/1998	CWS	24,316
CA	CA1510006	East Niles CSD	THM SAMP SITE-3209 FLINTRIDGE (E4)-KCWAS	BROMODICHLOROMETHANE (THM)	3.93	UG/L	6/23/1998	CWS	24,316
CA	CA1510006	East Niles CSD	THM SAMP SITE-3209 FLINTRIDGE (E4)-KCWAS	BROMODICHLOROMETHANE (THM)	10.1	UG/L	12/18/2001	CWS	24,316
CA	CA1510006	East Niles CSD	THM SAMP SITE-3209 FLINTRIDGE (E4)-KCWAS	BROMODICHLOROMETHANE (THM)	30.5	UG/L	9/24/2001	CWS	24,316
CA	CA1510006	East Niles CSD	BRECKENRIDGE TP-TREATED	BROMODICHLOROMETHANE (THM)	7.1	UG/L	3/28/2002	CWS	24,316
CA	CA1510040	Kern County Water Agency	ID4 WEN. & COLUMBAS (T7)-DIST. SYS.	BROMODICHLOROMETHANE (THM)	4.15	UG/L	5/29/1998	CWS	0
CA	CA1510040	Kern County Water Agency	ID4 WEN. & COLUMBAS (T7)-DIST. SYS.	BROMODICHLOROMETHANE (THM)	2.47	UG/L	6/22/1998	CWS	0
CA	CA1510040	Kern County Water Agency	ID4 WEN. & COLUMBAS (T7)-DIST. SYS.	BROMODICHLOROMETHANE (THM)	3.02	UG/L	7/20/1998	CWS	0
CA	CA1510040	Kern County Water Agency	ID4 WEN. & COLUMBAS (T7)-DIST. SYS.	BROMODICHLOROMETHANE (THM)	2.76	UG/L	8/17/1998	CWS	0
CA	CA1510040	Kern County Water Agency	ID4 WEN. & COLUMBAS (T7)-DIST. SYS.	BROMODICHLOROMETHANE (THM)	4.2	UG/L	9/28/1998	CWS	0
CA	CA1510040	Kern County Water Agency	ID4 WEN. & COLUMBAS (T7)-DIST. SYS.	BROMODICHLOROMETHANE (THM)	5.1	UG/L	3/8/1999	CWS	0
CA	CA1510040	Kern County Water Agency	ID4 WEN. & COLUMBAS (T7)-DIST. SYS.	BROMODICHLOROMETHANE (THM)	2.41	UG/L	3/2/1998	CWS	0
CA	CA1510040	Kern County Water Agency	ID4 WEN. & COLUMBAS (T7)-DIST. SYS.	BROMODICHLOROMETHANE (THM)	2.2	UG/L	6/2/1997	CWS	0
CA	CA1510040	Kern County Water Agency	ID4 WEN. & COLUMBAS (T7)-DIST. SYS.	BROMODICHLOROMETHANE (THM)	1.8	UG/L	7/7/1997	CWS	0
CA	CA1510040	Kern County Water Agency	ID4 WEN. & COLUMBAS (T7)-DIST. SYS.	BROMODICHLOROMETHANE (THM)	4	UG/L	9/23/1997	CWS	0
CA	CA1510040	Kern County Water Agency	ID4 WEN. & COLUMBAS (T7)-DIST. SYS.	BROMODICHLOROMETHANE (THM)	5.34	UG/L	4/6/1998	CWS	0
CA	CA1510040	Kern County Water Agency	ID4 WEN. & COLUMBAS (T7)-DIST. SYS.	BROMODICHLOROMETHANE (THM)	24.77	UG/L	1/5/1998	CWS	0
CA	CA1510040	Kern County Water Agency	ID4 WEN. & COLUMBAS (T7)-DIST. SYS.	BROMODICHLOROMETHANE (THM)	4.43	UG/L	2/2/1998	CWS	0
CA	CA1510040	Kern County Water Agency	ID4 CORNER TANK-T6-TREATED	BROMODICHLOROMETHANE (THM)	3.09	UG/L	3/2/1998	CWS	0
CA	CA1510040	Kern County Water Agency	ID4 CORNER TANK-T6-TREATED	BROMODICHLOROMETHANE (THM)	3.65	UG/L	6/22/1998	CWS	0
CA	CA1510040	Kern County Water Agency	ID4 CORNER TANK-T6-TREATED	BROMODICHLOROMETHANE (THM)	3.98	UG/L	5/29/1998	CWS	0
CA	CA1510040	Kern County Water Agency	ID4 CORNER TANK-T6-TREATED	BROMODICHLOROMETHANE (THM)	6.72	UG/L	4/6/1998	CWS	0
CA	CA1510040	Kern County Water Agency	ID4 CORNER TANK-T6-TREATED	BROMODICHLOROMETHANE (THM)	3.31	UG/L	7/20/1998	CWS	0
CA	CA1510040	Kern County Water Agency	ID4 CORNER TANK-T6-TREATED	BROMODICHLOROMETHANE (THM)	2.82	UG/L	8/17/1998	CWS	0
CA	CA1510040	Kern County Water Agency	ID4 CORNER TANK-T6-TREATED	BROMODICHLOROMETHANE (THM)	4.4	UG/L	9/28/1998	CWS	0
CA	CA1510040	Kern County Water Agency	ID4 CORNER TANK-T6-TREATED	BROMODICHLOROMETHANE (THM)	9.8	UG/L	3/8/1999	CWS	0
CA	CA1510040	Kern County Water Agency	ID4 CORNER TANK-T6-TREATED	BROMODICHLOROMETHANE (THM)	6.18	UG/L	2/2/1998	CWS	0
CA	CA1510040	Kern County Water Agency	ID4 CORNER TANK-T6-TREATED	BROMODICHLOROMETHANE (THM)	6.34	UG/L	1/31/1997	CWS	0

CA	CA1510040	Kern County Water Agency	ID4 CORNER TANK-T6-TREATED	BROMODICHLOROMETHANE (THM)	6.8	UG/L	2/24/1997	CWS	0
CA	CA1510040	Kern County Water Agency	ID4 CORNER TANK-T6-TREATED	BROMODICHLOROMETHANE (THM)	2.46	UG/L	3/24/1997	CWS	0
CA	CA1510040	Kern County Water Agency	ID4 CORNER TANK-T6-TREATED	BROMODICHLOROMETHANE (THM)	3.2	UG/L	4/14/1997	CWS	0
CA	CA1510040	Kern County Water Agency	ID4 CORNER TANK-T6-TREATED	BROMODICHLOROMETHANE (THM)	3.2	UG/L	5/12/1997	CWS	0
CA	CA1510040	Kern County Water Agency	ID4 CORNER TANK-T6-TREATED	BROMODICHLOROMETHANE (THM)	33.82	UG/L	1/5/1998	CWS	0
CA	CA1510040	Kern County Water Agency	ID4 CORNER TANK-T6-TREATED	BROMODICHLOROMETHANE (THM)	4	UG/L	9/8/1997	CWS	0
CA	CA1510040	Kern County Water Agency	ID4 CORNER TANK-T6-TREATED	BROMODICHLOROMETHANE (THM)	3.7	UG/L	7/7/1997	CWS	0
CA	CA1510040	Kern County Water Agency	ID4 CORNER TANK-T6-TREATED	BROMODICHLOROMETHANE (THM)	3.1	UG/L	6/2/1997	CWS	0
CA	CA1510040	Kern County Water Agency	ID4 BELLE-SEQUOIA TANK-T2-TREATED	BROMODICHLOROMETHANE (THM)	5.66	UG/L	5/29/1998	CWS	0
CA	CA1510040	Kern County Water Agency	ID4 BELLE-SEQUOIA TANK-T2-TREATED	BROMODICHLOROMETHANE (THM)	3.31	UG/L	2/2/1998	CWS	0
CA	CA1510040	Kern County Water Agency	ID4 BELLE-SEQUOIA TANK-T2-TREATED	BROMODICHLOROMETHANE (THM)	8.45	UG/L	4/6/1998	CWS	0
CA	CA1510040	Kern County Water Agency	ID4 BELLE-SEQUOIA TANK-T2-TREATED	BROMODICHLOROMETHANE (THM)	2.19	UG/L	3/2/1998	CWS	0
CA	CA1510040	Kern County Water Agency	ID4 BELLE-SEQUOIA TANK-T2-TREATED	BROMODICHLOROMETHANE (THM)	2.27	UG/L	6/22/1998	CWS	0
CA	CA1510040	Kern County Water Agency	ID4 BELLE-SEQUOIA TANK-T2-TREATED	BROMODICHLOROMETHANE (THM)	3.2	UG/L	4/14/1997	CWS	0
CA	CA1510040	Kern County Water Agency	ID4 BELLE-SEQUOIA TANK-T2-TREATED	BROMODICHLOROMETHANE (THM)	2.62	UG/L	8/17/1998	CWS	0
CA	CA1510040	Kern County Water Agency	ID4 BELLE-SEQUOIA TANK-T2-TREATED	BROMODICHLOROMETHANE (THM)	4	UG/L	9/28/1998	CWS	0
CA	CA1510040	Kern County Water Agency	ID4 BELLE-SEQUOIA TANK-T2-TREATED	BROMODICHLOROMETHANE (THM)	2.52	UG/L	7/20/1998	CWS	0
CA	CA1510040	Kern County Water Agency	ID4 BELLE-SEQUOIA TANK-T2-TREATED	BROMODICHLOROMETHANE (THM)	4.7	UG/L	3/8/1999	CWS	0
CA	CA1510040	Kern County Water Agency	ID4 BELLE-SEQUOIA TANK-T2-TREATED	BROMODICHLOROMETHANE (THM)	7.5	UG/L	2/24/1997	CWS	0
CA	CA1510040	Kern County Water Agency	ID4 BELLE-SEQUOIA TANK-T2-TREATED	BROMODICHLOROMETHANE (THM)	2.73	UG/L	3/24/1997	CWS	0
CA	CA1510040	Kern County Water Agency	ID4 BELLE-SEQUOIA TANK-T2-TREATED	BROMODICHLOROMETHANE (THM)	4.82	UG/L	1/31/1997	CWS	0
CA	CA1510040	Kern County Water Agency	ID4 BELLE-SEQUOIA TANK-T2-TREATED	BROMODICHLOROMETHANE (THM)	2.3	UG/L	5/12/1997	CWS	0
CA	CA1510040	Kern County Water Agency	ID4 BELLE-SEQUOIA TANK-T2-TREATED	BROMODICHLOROMETHANE (THM)	2.3	UG/L	6/2/1997	CWS	0
CA	CA1510040	Kern County Water Agency	ID4 BELLE-SEQUOIA TANK-T2-TREATED	BROMODICHLOROMETHANE (THM)	20.18	UG/L	1/5/1998	CWS	0
CA	CA1510040	Kern County Water Agency	ID4 BELLE-SEQUOIA TANK-T2-TREATED	BROMODICHLOROMETHANE (THM)	4.5	UG/L	9/8/1997	CWS	0
CA	CA1510040	Kern County Water Agency	ID4 BELLE-SEQUOIA TANK-T2-TREATED	BROMODICHLOROMETHANE (THM)	5	UG/L	9/8/1997	CWS	0
CA	CA1510040	Kern County Water Agency	ID4 BELLE-SEQUOIA TANK-T2-TREATED	BROMODICHLOROMETHANE (THM)	3.2	UG/L	7/7/1997	CWS	0
CA	CA1510040	Kern County Water Agency	ID4 OSWELL TANK-T5-TREATED	BROMODICHLOROMETHANE (THM)	4.49	UG/L	5/29/1998	CWS	0
CA	CA1510040	Kern County Water Agency	ID4 OSWELL TANK-T5-TREATED	BROMODICHLOROMETHANE (THM)	5.33	UG/L	4/6/1998	CWS	0
CA	CA1510040	Kern County Water Agency	ID4 OSWELL TANK-T5-TREATED	BROMODICHLOROMETHANE (THM)	2.87	UG/L	3/2/1998	CWS	0
CA	CA1510040	Kern County Water Agency	ID4 OSWELL TANK-T5-TREATED	BROMODICHLOROMETHANE (THM)	6.05	UG/L	2/2/1998	CWS	0
CA	CA1510040	Kern County Water Agency	ID4 OSWELL TANK-T5-TREATED	BROMODICHLOROMETHANE (THM)	3.49	UG/L	6/22/1998	CWS	0
CA	CA1510040	Kern County Water Agency	ID4 OSWELL TANK-T5-TREATED	BROMODICHLOROMETHANE (THM)	3.09	UG/L	7/20/1998	CWS	0
CA	CA1510040	Kern County Water Agency	ID4 OSWELL TANK-T5-TREATED	BROMODICHLOROMETHANE (THM)	6	UG/L	9/28/1998	CWS	0
CA	CA1510040	Kern County Water Agency	ID4 OSWELL TANK-T5-TREATED	BROMODICHLOROMETHANE (THM)	7.6	UG/L	3/8/1999	CWS	0
CA	CA1510040	Kern County Water Agency	ID4 OSWELL TANK-T5-TREATED	BROMODICHLOROMETHANE (THM)	2.72	UG/L	8/17/1998	CWS	0
CA	CA1510040	Kern County Water Agency	ID4 OSWELL TANK-T5-TREATED	BROMODICHLOROMETHANE (THM)	7	UG/L	2/24/1997	CWS	0
CA	CA1510040	Kern County Water Agency	ID4 OSWELL TANK-T5-TREATED	BROMODICHLOROMETHANE (THM)	6.99	UG/L	1/31/1997	CWS	0
CA	CA1510040	Kern County Water Agency	ID4 OSWELL TANK-T5-TREATED	BROMODICHLOROMETHANE (THM)	3.2	UG/L	4/14/1997	CWS	0
CA	CA1510040	Kern County Water Agency	ID4 OSWELL TANK-T5-TREATED	BROMODICHLOROMETHANE (THM)	2.5	UG/L	5/12/1997	CWS	0
CA	CA1510040	Kern County Water Agency	ID4 OSWELL TANK-T5-TREATED	BROMODICHLOROMETHANE (THM)	2.54	UG/L	3/24/1997	CWS	0
CA	CA1510040	Kern County Water Agency	ID4 OSWELL TANK-T5-TREATED	BROMODICHLOROMETHANE (THM)	36.47	UG/L	1/5/1998	CWS	0
CA	CA1510040	Kern County Water Agency	ID4 OSWELL TANK-T5-TREATED	BROMODICHLOROMETHANE (THM)	3.3	UG/L	7/7/1997	CWS	0
CA	CA1510040	Kern County Water Agency	ID4 OSWELL TANK-T5-TREATED	BROMODICHLOROMETHANE (THM)	2.7	UG/L	6/2/1997	CWS	0
CA	CA1510040	Kern County Water Agency	ID4 OSWELL TANK-T5-TREATED	BROMODICHLOROMETHANE (THM)	5.1	UG/L	9/8/1997	CWS	0

APPENDIX 7-3 *Continued*

State	PWS_ID	Utility	POE	Constituent	LEVEL_NO	Unit	COLL_DATE	PWS	Pop
CA	CA1510040	Kern County Water Agency	ID4 NOR TANK (600 SITE) T3-TREATED	BROMODICHLOROMETHANE (THM)	3.15	UG/L	6/22/1998	CWS	0
CA	CA1510040	Kern County Water Agency	ID4 NOR TANK (600 SITE) T3-TREATED	BROMODICHLOROMETHANE (THM)	5.61	UG/L	5/29/1998	CWS	0
CA	CA1510040	Kern County Water Agency	ID4 NOR TANK (600 SITE) T3-TREATED	BROMODICHLOROMETHANE (THM)	8.08	UG/L	4/6/1998	CWS	0
CA	CA1510040	Kern County Water Agency	ID4 NOR TANK (600 SITE) T3-TREATED	BROMODICHLOROMETHANE (THM)	2.49	UG/L	3/2/1998	CWS	0
CA	CA1510040	Kern County Water Agency	ID4 NOR TANK (600 SITE) T3-TREATED	BROMODICHLOROMETHANE (THM)	2.86	UG/L	7/20/1998	CWS	0
CA	CA1510040	Kern County Water Agency	ID4 NOR TANK (600 SITE) T3-TREATED	BROMODICHLOROMETHANE (THM)	4.26	UG/L	8/17/1998	CWS	0
CA	CA1510040	Kern County Water Agency	ID4 NOR TANK (600 SITE) T3-TREATED	BROMODICHLOROMETHANE (THM)	5.2	UG/L	9/28/1998	CWS	0
CA	CA1510040	Kern County Water Agency	ID4 NOR TANK (600 SITE) T3-TREATED	BROMODICHLOROMETHANE (THM)	7.5	UG/L	3/8/1999	CWS	0
CA	CA1510040	Kern County Water Agency	ID4 NOR TANK (600 SITE) T3-TREATED	BROMODICHLOROMETHANE (THM)	7.9	UG/L	2/24/1997	CWS	0
CA	CA1510040	Kern County Water Agency	ID4 NOR TANK (600 SITE) T3-TREATED	BROMODICHLOROMETHANE (THM)	6.31	UG/L	1/31/1997	CWS	0
CA	CA1510040	Kern County Water Agency	ID4 NOR TANK (600 SITE) T3-TREATED	BROMODICHLOROMETHANE (THM)	3.28	UG/L	3/24/1997	CWS	0
CA	CA1510040	Kern County Water Agency	ID4 NOR TANK (600 SITE) T3-TREATED	BROMODICHLOROMETHANE (THM)	2.6	UG/L	5/12/1997	CWS	0
CA	CA1510040	Kern County Water Agency	ID4 NOR TANK (600 SITE) T3-TREATED	BROMODICHLOROMETHANE (THM)	3.3	UG/L	4/14/1997	CWS	0
CA	CA1510040	Kern County Water Agency	ID4 NOR TANK (600 SITE) T3-TREATED	BROMODICHLOROMETHANE (THM)	39.78	UG/L	1/5/1998	CWS	0
CA	CA1510040	Kern County Water Agency	ID4 NOR TANK (600 SITE) T3-TREATED	BROMODICHLOROMETHANE (THM)	5.87	UG/L	2/2/1998	CWS	0
CA	CA1510040	Kern County Water Agency	ID4 NOR TANK (600 SITE) T3-TREATED	BROMODICHLOROMETHANE (THM)	3.3	UG/L	7/7/1997	CWS	0
CA	CA1510040	Kern County Water Agency	ID4 NOR TANK (600 SITE) T3-TREATED	BROMODICHLOROMETHANE (THM)	2.7	UG/L	6/2/1997	CWS	0
CA	CA1510040	Kern County Water Agency	ID4 TREATED-T1	BROMODICHLOROMETHANE (THM)	2.25	UG/L	3/2/1998	CWS	0
CA	CA1510040	Kern County Water Agency	ID4 TREATED-T1	BROMODICHLOROMETHANE (THM)	20.28	UG/L	1/5/1998	CWS	0
CA	CA1510040	Kern County Water Agency	ID4 TREATED-T1	BROMODICHLOROMETHANE (THM)	5.7	UG/L	9/8/1997	CWS	0
CA	CA1510040	Kern County Water Agency	ID4 TREATED-T1	BROMODICHLOROMETHANE (THM)	1.65	UG/L	2/2/1998	CWS	0
CA	CA1510040	Kern County Water Agency	ID4 TREATED-T1	BROMODICHLOROMETHANE (THM)	2.36	UG/L	6/22/1998	CWS	0
CA	CA1510040	Kern County Water Agency	ID4 TREATED-T1	BROMODICHLOROMETHANE (THM)	3.74	UG/L	5/29/1998	CWS	0
CA	CA1510040	Kern County Water Agency	ID4 TREATED-T1	BROMODICHLOROMETHANE (THM)	5.37	UG/L	4/6/1998	CWS	0
CA	CA1510040	Kern County Water Agency	ID4 TREATED-T1	BROMODICHLOROMETHANE (THM)	3.9	UG/L	2/24/1997	CWS	0
CA	CA1510040	Kern County Water Agency	ID4 TREATED-T1	BROMODICHLOROMETHANE (THM)	2.76	UG/L	3/24/1997	CWS	0
CA	CA1510040	Kern County Water Agency	ID4 TREATED-T1	BROMODICHLOROMETHANE (THM)	4.23	UG/L	1/31/1997	CWS	0
CA	CA1510040	Kern County Water Agency	ID4 TREATED-T1	BROMODICHLOROMETHANE (THM)	4.5	UG/L	7/7/1997	CWS	0
CA	CA1510040	Kern County Water Agency	ID4 TREATED-T1	BROMODICHLOROMETHANE (THM)	3.1	UG/L	4/14/1997	CWS	0
CA	CA1510040	Kern County Water Agency	ID4 TREATED-T1	BROMODICHLOROMETHANE (THM)	2.7	UG/L	6/2/1997	CWS	0
CA	CA1510040	Kern County Water Agency	ID4 TREATED-T1	BROMODICHLOROMETHANE (THM)	3.2	UG/L	5/12/1997	CWS	0
CA	CA1510040	Kern County Water Agency	ID4 TREATED-T1	BROMODICHLOROMETHANE (THM)	11.4	UG/L	8/30/2002	CWS	0
CA	CA1510040	Kern County Water Agency	ID4 TREATED-T1	BROMODICHLOROMETHANE (THM)	6.2	UG/L	8/30/1999	CWS	0
CA	CA1510040	Kern County Water Agency	ID4 TREATED-T1	BROMODICHLOROMETHANE (THM)	2.62	UG/L	7/20/1998	CWS	0
CA	CA1510040	Kern County Water Agency	ID4 TREATED-T1	BROMODICHLOROMETHANE (THM)	5	UG/L	9/28/1998	CWS	0
CA	CA1510040	Kern County Water Agency	ID4 TREATED-T1	BROMODICHLOROMETHANE (THM)	2.76	UG/L	8/17/1998	CWS	0
CA	CA1510040	Kern County Water Agency	ID4 TREATED-T1	BROMODICHLOROMETHANE (THM)	4.8	UG/L	3/8/1999	CWS	0
CA	CA1510040	Kern County Water Agency	ID4 TREATED-T1	BROMODICHLOROMETHANE (THM)	6.6	UG/L	11/15/2000	CWS	0
CA	CA1510041	North of the River MWD	THM SAMP SITE-2841 PEGASUS (N1)-TRTD	BROMODICHLOROMETHANE (THM)	6.5	UG/L	11/26/2001	CWS	7,000
CA	CA1510041	North of the River MWD	THM SAMP SITE-2841 PEGASUS (N1)-TRTD	BROMODICHLOROMETHANE (THM)	7.3	UG/L	8/22/2001	CWS	7,000
CA	CA1510041	North of the River MWD	THM SAMP SITE-2841 PEGASUS (N1)-TRTD	BROMODICHLOROMETHANE (THM)	9.2	UG/L	5/17/2002	CWS	7,000
CA	CA1510041	North of the River MWD	THM SAMP SITE-2841 PEGASUS (N1)-TRTD	BROMODICHLOROMETHANE (THM)	9.6	UG/L	8/27/2002	CWS	7,000

CA	CA1510041	North of the River MWD	THM SAMP SITE-2841 PEGASUS (N1)-TRTD	BROMODICHLOROMETHANE (THM)	11.3	UG/L	11/7/2000	CWS	7,000
CA	CA1510041	North of the River MWD	THM SAMP SITE-2841 PEGASUS (N1)-TRTD	BROMODICHLOROMETHANE (THM)	9.5	UG/L	5/8/2001	CWS	7,000
CA	CA1510041	North of the River MWD	THM SAMP SITE-2841 PEGASUS (N1)-TRTD	BROMODICHLOROMETHANE (THM)	4.54	UG/L	2/25/1998	CWS	7,000
CA	CA1510041	North of the River MWD	THM SAMP SITE-2841 PEGASUS (N1)-TRTD	BROMODICHLOROMETHANE (THM)	4.21	UG/L	6/23/1998	CWS	7,000
CA	CA1510041	North of the River MWD	THM SAMP SITE-2841 PEGASUS (N1)-TRTD	BROMODICHLOROMETHANE (THM)	6	UG/L	11/3/1998	CWS	7,000
CA	CA1510041	North of the River MWD	THM SAMP SITE-2841 PEGASUS (N1)-TRTD	BROMODICHLOROMETHANE (THM)	7.67	UG/L	3/10/1999	CWS	7,000
CA	CA1510041	North of the River MWD	THM SAMP SITE-2841 PEGASUS (N1)-TRTD	BROMODICHLOROMETHANE (THM)	10.2	UG/L	6/7/1999	CWS	7,000
CA	CA1510041	North of the River MWD	THM SAMP SITE-900 TANKSITE (N4)-TRTD	BROMODICHLOROMETHANE (THM)	15.4	UG/L	11/26/2001	CWS	7,000
CA	CA1510041	North of the River MWD	THM SAMP SITE-900 TANKSITE (N4)-TRTD	BROMODICHLOROMETHANE (THM)	20.5	UG/L	8/22/2001	CWS	7,000
CA	CA1510041	North of the River MWD	THM SAMP SITE-900 TANKSITE (N4)-TRTD	BROMODICHLOROMETHANE (THM)	9	UG/L	5/17/2002	CWS	7,000
CA	CA1510041	North of the River MWD	THM SAMP SITE-900 TANKSITE (N4)-TRTD	BROMODICHLOROMETHANE (THM)	14.1	UG/L	8/27/2002	CWS	7,000
CA	CA1510041	North of the River MWD	THM SAMP SITE-900 TANKSITE (N4)-TRTD	BROMODICHLOROMETHANE (THM)	10.87	UG/L	11/7/2000	CWS	7,000
CA	CA1510041	North of the River MWD	THM SAMP SITE-900 TANKSITE (N4)-TRTD	BROMODICHLOROMETHANE (THM)	10	UG/L	5/8/2001	CWS	7,000
CA	CA1510041	North of the River MWD	THM SAMP SITE-900 TANKSITE (N4)-TRTD	BROMODICHLOROMETHANE (THM)	7.1	UG/L	9/23/1997	CWS	7,000
CA	CA1510041	North of the River MWD	THM SAMP SITE-900 TANKSITE (N4)-TRTD	BROMODICHLOROMETHANE (THM)	11.3	UG/L	2/25/1998	CWS	7,000
CA	CA1510041	North of the River MWD	THM SAMP SITE-900 TANKSITE (N4)-TRTD	BROMODICHLOROMETHANE (THM)	6.93	UG/L	6/23/1998	CWS	7,000
CA	CA1510041	North of the River MWD	THM SAMP SITE-900 TANKSITE (N4)-TRTD	BROMODICHLOROMETHANE (THM)	6.3	UG/L	11/3/1998	CWS	7,000
CA	CA1510041	North of the River MWD	THM SAMP SITE-900 TANKSITE (N4)-TRTD	BROMODICHLOROMETHANE (THM)	9.5	UG/L	6/7/1999	CWS	7,000
CA	CA1510041	North of the River MWD	THM SAMP SITE-900 TANKSITE (N4)-TRTD	BROMODICHLOROMETHANE (THM)	8.72	UG/L	3/10/1999	CWS	7,000
CA	CA1510041	North of the River MWD	THM SAMP SITE-3418 N CHESTER (N3)-TRTD	BROMODICHLOROMETHANE (THM)	5.5	UG/L	11/26/2001	CWS	7,000
CA	CA1510041	North of the River MWD	THM SAMP SITE-3418 N CHESTER (N3)-TRTD	BROMODICHLOROMETHANE (THM)	7.2	UG/L	8/22/2001	CWS	7,000
CA	CA1510041	North of the River MWD	THM SAMP SITE-3418 N CHESTER (N3)-TRTD	BROMODICHLOROMETHANE (THM)	9.7	UG/L	8/27/2002	CWS	7,000
CA	CA1510041	North of the River MWD	THM SAMP SITE-3418 N CHESTER (N3)-TRTD	BROMODICHLOROMETHANE (THM)	8.3	UG/L	5/17/2002	CWS	7,000
CA	CA1510041	North of the River MWD	THM SAMP SITE-3418 N CHESTER (N3)-TRTD	BROMODICHLOROMETHANE (THM)	8.4	UG/L	5/8/2001	CWS	7,000
CA	CA1510041	North of the River MWD	THM SAMP SITE-3418 N CHESTER (N3)-TRTD	BROMODICHLOROMETHANE (THM)	3.8	UG/L	9/11/1997	CWS	7,000
CA	CA1510041	North of the River MWD	THM SAMP SITE-3418 N CHESTER (N3)-TRTD	BROMODICHLOROMETHANE (THM)	3.25	UG/L	2/25/1998	CWS	7,000
CA	CA1510041	North of the River MWD	THM SAMP SITE-3418 N CHESTER (N3)-TRTD	BROMODICHLOROMETHANE (THM)	3.55	UG/L	6/23/1998	CWS	7,000
CA	CA1510041	North of the River MWD	THM SAMP SITE-3418 N CHESTER (N3)-TRTD	BROMODICHLOROMETHANE (THM)	5.3	UG/L	11/3/1998	CWS	7,000
CA	CA1510041	North of the River MWD	THM SAMP SITE-3418 N CHESTER (N3)-TRTD	BROMODICHLOROMETHANE (THM)	9.1	UG/L	6/7/1999	CWS	7,000
CA	CA1510041	North of the River MWD	THM SAMP SITE-3418 N CHESTER (N3)-TRTD	BROMODICHLOROMETHANE (THM)	7.47	UG/L	3/10/1999	CWS	7,000
CA	CA1510041	North of the River MWD	THM SAMP SITE-3418 N CHESTER (N3)-TRTD	BROMODICHLOROMETHANE (THM)	10.52	UG/L	11/7/2000	CWS	7,000
CA	CA1510041	North of the River MWD	THM SAMP SITE-528 CHARLANA (N2)-TRTD	BROMODICHLOROMETHANE (THM)	7.3	UG/L	11/26/2001	CWS	7,000
CA	CA1510041	North of the River MWD	THM SAMP SITE-528 CHARLANA (N2)-TRTD	BROMODICHLOROMETHANE (THM)	6.7	UG/L	8/22/2001	CWS	7,000
CA	CA1510041	North of the River MWD	THM SAMP SITE-528 CHARLANA (N2)-TRTD	BROMODICHLOROMETHANE (THM)	8.4	UG/L	5/17/2002	CWS	7,000
CA	CA1510041	North of the River MWD	THM SAMP SITE-528 CHARLANA (N2)-TRTD	BROMODICHLOROMETHANE (THM)	9.8	UG/L	8/27/2002	CWS	7,000
CA	CA1510041	North of the River MWD	THM SAMP SITE-528 CHARLANA (N2)-TRTD	BROMODICHLOROMETHANE (THM)	9.1	UG/L	5/8/2001	CWS	7,000
CA	CA1510041	North of the River MWD	THM SAMP SITE-528 CHARLANA (N2)-TRTD	BROMODICHLOROMETHANE (THM)	4.97	UG/L	2/25/1998	CWS	7,000
CA	CA1510041	North of the River MWD	THM SAMP SITE-528 CHARLANA (N2)-TRTD	BROMODICHLOROMETHANE (THM)	2.89	UG/L	6/23/1998	CWS	7,000
CA	CA1510041	North of the River MWD	THM SAMP SITE-528 CHARLANA (N2)-TRTD	BROMODICHLOROMETHANE (THM)	4.3	UG/L	11/3/1998	CWS	7,000
CA	CA1510041	North of the River MWD	THM SAMP SITE-528 CHARLANA (N2)-TRTD	BROMODICHLOROMETHANE (THM)	5.87	UG/L	3/10/1999	CWS	7,000
CA	CA1510041	North of the River MWD	THM SAMP SITE-528 CHARLANA (N2)-TRTD	BROMODICHLOROMETHANE (THM)	10.79	UG/L	11/7/2000	CWS	7,000
CA	CA1510041	North of the River MWD	THM SAMP SITE-528 CHARLANA (N2)-TRTD	BROMODICHLOROMETHANE (THM)	5.1	UG/L	6/7/1999	CWS	7,000
CA	CA1510041	North of the River MWD	THM SAMP SITE-528 CHARLANA (N2)-TRTD	BROMODICHLOROMETHANE (THM)	2.5	UG/L	9/11/1997	CWS	7,000
					7.17				
CA	CA1510006	East Niles CSD	THM SAMP SITE-DEACON (E8)-ENG	BROMOFORM (THM)	2.2	UG/L	12/18/2002	CWS	24,316
CA	CA1510006	East Niles CSD	THM SAMP SITE-DEACON (E8)-ENG	BROMOFORM (THM)	1.8	UG/L	3/18/2003	CWS	24,316

APPENDIX 7-3 *Continued*

State	PWS_ID	Utility	POE	Constituent	LEVEL_NO	Unit	COLL_DATE	PWS	Pop
CA	CA1510006	East Niles CSD	THM SAMP SITE-DEACON (E8)-ENG	BROMOFORM (THM)	1.6	UG/L	6/18/2002	CWS	24,316
CA	CA1510006	East Niles CSD	THM SAMP SITE-DEACON (E8)-ENG	BROMOFORM (THM)	0.8	UG/L	12/16/2003	CWS	24,316
CA	CA1510006	East Niles CSD	THM SAMP SITE-DEACON (E8)-ENG	BROMOFORM (THM)	1.2	UG/L	5/8/2002	CWS	24,316
CA	CA1510006	East Niles CSD	THM SAMP SITE-DEACON (E8)-ENG	BROMOFORM (THM)	0.62	UG/L	2/26/1998	CWS	24,316
CA	CA1510006	East Niles CSD	THM SAMP SITE-DEACON (E8)-ENG	BROMOFORM (THM)	1.8	UG/L	11/3/1998	CWS	24,316
CA	CA1510006	East Niles CSD	THM SAMP SITE-DEACON (E8)-ENG	BROMOFORM (THM)	1.2	UG/L	9/24/2001	CWS	24,316
CA	CA1510006	East Niles CSD	THM SAMP SITE-DEACON (E8)-ENG	BROMOFORM (THM)	2.4	UG/L	12/18/2001	CWS	24,316
CA	CA1510006	East Niles CSD	THM SAMP SITE-PACIFIC BELL (E9)-ENS	BROMOFORM (THM)	18.3	UG/L	6/18/2002	CWS	24,316
CA	CA1510006	East Niles CSD	THM SAMP SITE-PACIFIC BELL (E9)-ENS	BROMOFORM (THM)	8.8	UG/L	12/18/2002	CWS	24,316
CA	CA1510006	East Niles CSD	THM SAMP SITE-PACIFIC BELL (E9)-ENS	BROMOFORM (THM)	13.8	UG/L	5/8/2002	CWS	24,316
CA	CA1510006	East Niles CSD	THM SAMP SITE-PACIFIC BELL (E9)-ENS	BROMOFORM (THM)	18.8	UG/L	6/11/2003	CWS	24,316
CA	CA1510006	East Niles CSD	THM SAMP SITE-PACIFIC BELL (E9)-ENS	BROMOFORM (THM)	19.7	UG/L	3/18/2003	CWS	24,316
CA	CA1510006	East Niles CSD	THM SAMP SITE-PACIFIC BELL (E9)-ENS	BROMOFORM (THM)	10.09	UG/L	2/26/1998	CWS	24,316
CA	CA1510006	East Niles CSD	THM SAMP SITE-PACIFIC BELL (E9)-ENS	BROMOFORM (THM)	12.76	UG/L	8/26/1998	CWS	24,316
CA	CA1510006	East Niles CSD	THM SAMP SITE-PACIFIC BELL (E9)-ENS	BROMOFORM (THM)	3.6	UG/L	9/11/1997	CWS	24,316
CA	CA1510006	East Niles CSD	THM SAMP SITE-PACIFIC BELL (E9)-ENS	BROMOFORM (THM)	9.4	UG/L	12/18/2001	CWS	24,316
CA	CA1510006	East Niles CSD	THM SAMP SITE-PACIFIC BELL (E9)-ENS	BROMOFORM (THM)	12.76	UG/L	3/9/1999	CWS	24,316
CA	CA1510006	East Niles CSD	THM SAMP SITE-PACIFIC BELL (E9)-ENS	BROMOFORM (THM)	6.2	UG/L	9/24/2001	CWS	24,316
CA	CA1510006	East Niles CSD	THM SAMP SITE-8717 BUTTERNUT (E10)-ENS	BROMOFORM (THM)	28.6	UG/L	6/18/2002	CWS	24,316
CA	CA1510006	East Niles CSD	THM SAMP SITE-8717 BUTTERNUT (E10)-ENS	BROMOFORM (THM)	2.8	UG/L	5/8/2002	CWS	24,316
CA	CA1510006	East Niles CSD	THM SAMP SITE-8717 BUTTERNUT (E10)-ENS	BROMOFORM (THM)	7.2	UG/L	3/18/2003	CWS	24,316
CA	CA1510006	East Niles CSD	THM SAMP SITE-8717 BUTTERNUT (E10)-ENS	BROMOFORM (THM)	4.4	UG/L	12/18/2002	CWS	24,316
CA	CA1510006	East Niles CSD	THM SAMP SITE-8717 BUTTERNUT (E10)-ENS	BROMOFORM (THM)	4.89	UG/L	8/26/1998	CWS	24,316
CA	CA1510006	East Niles CSD	THM SAMP SITE-8717 BUTTERNUT (E10)-ENS	BROMOFORM (THM)	3.48	UG/L	6/23/1998	CWS	24,316
CA	CA1510006	East Niles CSD	THM SAMP SITE-8717 BUTTERNUT (E10)-ENS	BROMOFORM (THM)	4.6	UG/L	12/18/2001	CWS	24,316
CA	CA1510006	East Niles CSD	THM SAMP SITE-8717 BUTTERNUT (E10)-ENS	BROMOFORM (THM)	7.2	UG/L	9/24/2001	CWS	24,316
CA	CA1510006	East Niles CSD	THM SAMP SITE-11845 SCHOOL (E11)-ENS	BROMOFORM (THM)	13.9	UG/L	6/18/2002	CWS	24,316
CA	CA1510006	East Niles CSD	THM SAMP SITE-11845 SCHOOL (E11)-ENS	BROMOFORM (THM)	11.1	UG/L	5/8/2002	CWS	24,316
CA	CA1510006	East Niles CSD	THM SAMP SITE-11845 SCHOOL (E11)-ENS	BROMOFORM (THM)	16.4	UG/L	3/18/2003	CWS	24,316
CA	CA1510006	East Niles CSD	THM SAMP SITE-11845 SCHOOL (E11)-ENS	BROMOFORM (THM)	11.1	UG/L	12/18/2002	CWS	24,316
CA	CA1510006	East Niles CSD	THM SAMP SITE-11845 SCHOOL (E11)-ENS	BROMOFORM (THM)	12.18	UG/L	2/26/1998	CWS	24,316
CA	CA1510006	East Niles CSD	THM SAMP SITE-11845 SCHOOL (E11)-ENS	BROMOFORM (THM)	10.6	UG/L	12/18/2001	CWS	24,316
CA	CA1510006	East Niles CSD	THM SAMP SITE-11845 SCHOOL (E11)-ENS	BROMOFORM (THM)	7.9	UG/L	9/24/2001	CWS	24,316
CA	CA1510006	East Niles CSD	THM SAMP SITE-FAIRFAX SCHOOL (E12)-ENG	BROMOFORM (THM)	4.9	UG/L	12/18/2002	CWS	24,316
CA	CA1510006	East Niles CSD	THM SAMP SITE-FAIRFAX SCHOOL (E12)-ENG	BROMOFORM (THM)	3.8	UG/L	12/16/2003	CWS	24,316
CA	CA1510006	East Niles CSD	THM SAMP SITE-FAIRFAX SCHOOL (E12)-ENG	BROMOFORM (THM)	3.7	UG/L	6/18/2002	CWS	24,316
CA	CA1510006	East Niles CSD	THM SAMP SITE-FAIRFAX SCHOOL (E12)-ENG	BROMOFORM (THM)	6.9	UG/L	3/18/2003	CWS	24,316
CA	CA1510006	East Niles CSD	THM SAMP SITE-FAIRFAX SCHOOL (E12)-ENG	BROMOFORM (THM)	3.3	UG/L	5/8/2002	CWS	24,316
CA	CA1510006	East Niles CSD	THM SAMP SITE-FAIRFAX SCHOOL (E12)-ENG	BROMOFORM (THM)	3.7	UG/L	2/26/1998	CWS	24,316
CA	CA1510006	East Niles CSD	THM SAMP SITE-FAIRFAX SCHOOL (E12)-ENG	BROMOFORM (THM)	1	UG/L	6/23/1998	CWS	24,316
CA	CA1510006	East Niles CSD	THM SAMP SITE-FAIRFAX SCHOOL (E12)-ENG	BROMOFORM (THM)	1.4	UG/L	8/26/1998	CWS	24,316
CA	CA1510006	East Niles CSD	THM SAMP SITE-FAIRFAX SCHOOL (E12)-ENG	BROMOFORM (THM)	1.8	UG/L	9/24/2001	CWS	24,316

CA	CA1510006	East Niles CSD	THM SAMP SITE-FAIRFAX SCHOOL (E12)-ENG	BROMOFORM (THM)	3.2	UG/L	11/3/1998	CWS	24,316
CA	CA1510006	East Niles CSD	THM SAMP SITE-FAIRFAX SCHOOL (E12)-ENG	BROMOFORM (THM)	2.2	UG/L	12/18/2001	CWS	24,316
CA	CA1510006	East Niles CSD	THM SAMP SITE-FAIRFAX SCHOOL (E12)-ENG	BROMOFORM (THM)	1.87	UG/L	3/10/1999	CWS	24,316
CA	CA1510006	East Niles CSD	THM SAMP SITE-1417 VALE (E7)-KCWAS	BROMOFORM (THM)	4.3	UG/L	6/18/2002	CWS	24,316
CA	CA1510006	East Niles CSD	THM SAMP SITE-1417 VALE (E7)-KCWAS	BROMOFORM (THM)	4.3	UG/L	12/18/2002	CWS	24,316
CA	CA1510006	East Niles CSD	THM SAMP SITE-1417 VALE (E7)-KCWAS	BROMOFORM (THM)	2.1	UG/L	12/16/2003	CWS	24,316
CA	CA1510006	East Niles CSD	THM SAMP SITE-1417 VALE (E7)-KCWAS	BROMOFORM (THM)	4	UG/L	3/18/2003	CWS	24,316
CA	CA1510006	East Niles CSD	THM SAMP SITE-1417 VALE (E7)-KCWAS	BROMOFORM (THM)	9.8	UG/L	12/18/2001	CWS	24,316
CA	CA1510006	East Niles CSD	THM SAMP SITE-1417 VALE (E7)-KCWAS	BROMOFORM (THM)	28.4	UG/L	9/24/2001	CWS	24,316
CA	CA1510006	East Niles CSD	BRECKENRIDGE TP-RAW-OFFLINE	BROMOFORM (THM)	0.5	UG/L	3/23/1998	CWS	24,316
CA	CA1510006	East Niles CSD	THM SAMP SITE-5310 WINGFOOT (E6)-KCWAS	BROMOFORM (THM)	4.2	UG/L	6/18/2002	CWS	24,316
CA	CA1510006	East Niles CSD	THM SAMP SITE-5310 WINGFOOT (E6)-KCWAS	BROMOFORM (THM)	3.7	UG/L	12/18/2002	CWS	24,316
CA	CA1510006	East Niles CSD	THM SAMP SITE-5310 WINGFOOT (E6)-KCWAS	BROMOFORM (THM)	3.2	UG/L	12/16/2003	CWS	24,316
CA	CA1510006	East Niles CSD	THM SAMP SITE-5310 WINGFOOT (E6)-KCWAS	BROMOFORM (THM)	8.3	UG/L	12/18/2001	CWS	24,316
CA	CA1510006	East Niles CSD	THM SAMP SITE-5310 WINGFOOT (E6)-KCWAS	BROMOFORM (THM)	24.1	UG/L	9/24/2001	CWS	24,316
CA	CA1510006	East Niles CSD	WELL 13-RAW	BROMOFORM (THM)	4.1	UG/L	6/18/2002	CWS	24,316
CA	CA1510006	East Niles CSD	THM SAMP SITE-3209 FLINTRIDGE (E4)-KCWAS	BROMOFORM (THM)	4.2	UG/L	6/18/2002	CWS	24,316
CA	CA1510006	East Niles CSD	THM SAMP SITE-3209 FLINTRIDGE (E4)-KCWAS	BROMOFORM (THM)	2.2	UG/L	5/8/2002	CWS	24,316
CA	CA1510006	East Niles CSD	THM SAMP SITE-3209 FLINTRIDGE (E4)-KCWAS	BROMOFORM (THM)	3.6	UG/L	12/18/2002	CWS	24,316
CA	CA1510006	East Niles CSD	THM SAMP SITE-3209 FLINTRIDGE (E4)-KCWAS	BROMOFORM (THM)	10.8	UG/L	12/18/2001	CWS	24,316
CA	CA1510006	East Niles CSD	THM SAMP SITE-3209 FLINTRIDGE (E4)-KCWAS	BROMOFORM (THM)	22.6	UG/L	9/24/2001	CWS	24,316
CA	CA1510006	East Niles CSD	BRECKENRIDGE TP-TREATED	BROMOFORM (THM)	1.6	UG/L	3/28/2002	CWS	24,316
CA	CA1510040	Kern County Water Agency	ID4 WEN. & COLUMBAS (T7)-DIST. SYS.	BROMOFORM (THM)	5.54	UG/L	1/5/1998	CWS	0
CA	CA1510040	Kern County Water Agency	ID4 CORNER TANK-T6-TREATED	BROMOFORM (THM)	7.21	UG/L	1/5/1998	CWS	0
CA	CA1510040	Kern County Water Agency	ID4 BELLE-SEQUOIA TANK-T2-TREATED	BROMOFORM (THM)	5.18	UG/L	1/5/1998	CWS	0
CA	CA1510040	Kern County Water Agency	ID4 OSWELL TANK-T5-TREATED	BROMOFORM (THM)	8.01	UG/L	1/5/1998	CWS	0
CA	CA1510040	Kern County Water Agency	ID4 NOR TANK (600 SITE) T3-TREATED	BROMOFORM (THM)	7.4	UG/L	9/28/1998	CWS	0
CA	CA1510040	Kern County Water Agency	ID4 NOR TANK (600 SITE) T3-TREATED	BROMOFORM (THM)	8.8	UG/L	1/5/1998	CWS	0
CA	CA1510040	Kern County Water Agency	ID4 TREATED-T1	BROMOFORM (THM)	5.99	UG/L	1/5/1998	CWS	0
CA	CA1510041	North of the River MWD	THM SAMP SITE-2841 PEGASUS (N1)-TRTD	BROMOFORM (THM)	2.6	UG/L	8/22/2001	CWS	7,000
CA	CA1510041	North of the River MWD	THM SAMP SITE-2841 PEGASUS (N1)-TRTD	BROMOFORM (THM)	4.6	UG/L	11/26/2001	CWS	7,000
CA	CA1510041	North of the River MWD	THM SAMP SITE-900 TANKSITE (N4)-TRTD	BROMOFORM (THM)	5.4	UG/L	8/22/2001	CWS	7,000
CA	CA1510041	North of the River MWD	THM SAMP SITE-900 TANKSITE (N4)-TRTD	BROMOFORM (THM)	6.5	UG/L	11/26/2001	CWS	7,000
CA	CA1510041	North of the River MWD	THM SAMP SITE-3418 N CHESTER (N3)-TRTD	BROMOFORM (THM)	3.7	UG/L	8/22/2001	CWS	7,000
CA	CA1510041	North of the River MWD	THM SAMP SITE-3418 N CHESTER (N3)-TRTD	BROMOFORM (THM)	3.6	UG/L	11/26/2001	CWS	7,000
CA	CA1510041	North of the River MWD	THM SAMP SITE-528 CHARLANA (N2)-TRTD	BROMOFORM (THM)	4.3	UG/L	11/26/2001	CWS	7,000
CA	CA1510041	North of the River MWD	THM SAMP SITE-528 CHARLANA (N2)-TRTD	BROMOFORM (THM)	3.4	UG/L	8/22/2001	CWS	7,000
					6.92				
CA	CA1510006	East Niles CSD	BRECKENRIDGE TP-RAW-OFFLINE	BROMOMETHANE	11.1	UG/L	3/13/2001	CWS	24,316
CA	CA1510006	East Niles CSD	THM SAMP SITE-DEACON (E8)-ENG	CHLOROFORM (THM)	29	UG/L	12/16/2003	CWS	24,316
CA	CA1510006	East Niles CSD	THM SAMP SITE-DEACON (E8)-ENG	CHLOROFORM (THM)	0.9	UG/L	6/18/2002	CWS	24,316
CA	CA1510006	East Niles CSD	THM SAMP SITE-DEACON (E8)-ENG	CHLOROFORM (THM)	36.46	UG/L	2/26/1998	CWS	24,316
CA	CA1510006	East Niles CSD	THM SAMP SITE-DEACON (E8)-ENG	CHLOROFORM (THM)	0.9	UG/L	9/11/1997	CWS	24,316
CA	CA1510006	East Niles CSD	THM SAMP SITE-PACIFIC BELL (E9)-ENS	CHLOROFORM (THM)	37.6	UG/L	6/18/2002	CWS	24,316
CA	CA1510006	East Niles CSD	THM SAMP SITE-PACIFIC BELL (E9)-ENS	CHLOROFORM (THM)	6.3	UG/L	3/18/2003	CWS	24,316

APPENDIX 7-3 *Continued*

State	PWS_ID	Utility	POE	Constituent	LEVEL_NO	Unit	COLL_DATE	PWS	Pop
CA	CA1510006	East Niles CSD	THM SAMP SITE-PACIFIC BELL (E9)-ENS	CHLOROFORM (THM)	25.3	UG/L	5/8/2002	CWS	24,316
CA	CA1510006	East Niles CSD	THM SAMP SITE-PACIFIC BELL (E9)-ENS	CHLOROFORM (THM)	66.5	UG/L	11/3/1998	CWS	24,316
CA	CA1510006	East Niles CSD	THM SAMP SITE-PACIFIC BELL (E9)-ENS	CHLOROFORM (THM)	51	UG/L	6/11/2003	CWS	24,316
CA	CA1510006	East Niles CSD	THM SAMP SITE-PACIFIC BELL (E9)-ENS	CHLOROFORM (THM)	0.7	UG/L	12/18/2002	CWS	24,316
CA	CA1510006	East Niles CSD	THM SAMP SITE-PACIFIC BELL (E9)-ENS	CHLOROFORM (THM)	16.38	UG/L	6/23/1998	CWS	24,316
CA	CA1510006	East Niles CSD	THM SAMP SITE-PACIFIC BELL (E9)-ENS	CHLOROFORM (THM)	20.14	UG/L	8/26/1998	CWS	24,316
CA	CA1510006	East Niles CSD	THM SAMP SITE-PACIFIC BELL (E9)-ENS	CHLOROFORM (THM)	25	UG/L	9/11/1997	CWS	24,316
CA	CA1510006	East Niles CSD	THM SAMP SITE-PACIFIC BELL (E9)-ENS	CHLOROFORM (THM)	27.64	UG/L	2/26/1998	CWS	24,316
CA	CA1510006	East Niles CSD	THM SAMP SITE-PACIFIC BELL (E9)-ENS	CHLOROFORM (THM)	0.8	UG/L	12/18/2001	CWS	24,316
CA	CA1510006	East Niles CSD	THM SAMP SITE-PACIFIC BELL (E9)-ENS	CHLOROFORM (THM)	20.14	UG/L	3/9/1999	CWS	24,316
CA	CA1510006	East Niles CSD	THM SAMP SITE-8717 BUTTERNUT (E10)-ENS	CHLOROFORM (THM)	21.2	UG/L	5/8/2002	CWS	24,316
CA	CA1510006	East Niles CSD	THM SAMP SITE-8717 BUTTERNUT (E10)-ENS	CHLOROFORM (THM)	19.2	UG/L	6/18/2002	CWS	24,316
CA	CA1510006	East Niles CSD	THM SAMP SITE-8717 BUTTERNUT (E10)-ENS	CHLOROFORM (THM)	1.2	UG/L	12/18/2001	CWS	24,316
CA	CA1510006	East Niles CSD	THM SAMP SITE-8717 BUTTERNUT (E10)-ENS	CHLOROFORM (THM)	17.5	UG/L	3/18/2003	CWS	24,316
CA	CA1510006	East Niles CSD	THM SAMP SITE-8717 BUTTERNUT (E10)-ENS	CHLOROFORM (THM)	0.6	UG/L	12/18/2002	CWS	24,316
CA	CA1510006	East Niles CSD	THM SAMP SITE-8717 BUTTERNUT (E10)-ENS	CHLOROFORM (THM)	38.67	UG/L	2/26/1998	CWS	24,316
CA	CA1510006	East Niles CSD	THM SAMP SITE-8717 BUTTERNUT (E10)-ENS	CHLOROFORM (THM)	37.8	UG/L	6/23/1998	CWS	24,316
CA	CA1510006	East Niles CSD	THM SAMP SITE-8717 BUTTERNUT (E10)-ENS	CHLOROFORM (THM)	44.4	UG/L	9/11/1997	CWS	24,316
CA	CA1510006	East Niles CSD	THM SAMP SITE-8717 BUTTERNUT (E10)-ENS	CHLOROFORM (THM)	15.15	UG/L	8/26/1998	CWS	24,316
CA	CA1510006	East Niles CSD	THM SAMP SITE-8717 BUTTERNUT (E10)-ENS	CHLOROFORM (THM)	2.6	UG/L	9/24/2001	CWS	24,316
CA	CA1510006	East Niles CSD	THM SAMP SITE-8717 BUTTERNUT (E10)-ENS	CHLOROFORM (THM)	21.41	UG/L	3/9/1999	CWS	24,316
CA	CA1510006	East Niles CSD	THM SAMP SITE-8717 BUTTERNUT (E10)-ENS	CHLOROFORM (THM)	39.3	UG/L	11/3/1998	CWS	24,316
CA	CA1510006	East Niles CSD	THM SAMP SITE-11845 SCHOOL (E11)-ENS	CHLOROFORM (THM)	71	UG/L	6/18/2002	CWS	24,316
CA	CA1510006	East Niles CSD	THM SAMP SITE-11845 SCHOOL (E11)-ENS	CHLOROFORM (THM)	25.5	UG/L	5/8/2002	CWS	24,316
CA	CA1510006	East Niles CSD	THM SAMP SITE-11845 SCHOOL (E11)-ENS	CHLOROFORM (THM)	2.3	UG/L	12/18/2002	CWS	24,316
CA	CA1510006	East Niles CSD	THM SAMP SITE-11845 SCHOOL (E11)-ENS	CHLOROFORM (THM)	4.9	UG/L	3/18/2003	CWS	24,316
CA	CA1510006	East Niles CSD	THM SAMP SITE-11845 SCHOOL (E11)-ENS	CHLOROFORM (THM)	32.24	UG/L	2/26/1998	CWS	24,316
CA	CA1510006	East Niles CSD	THM SAMP SITE-11845 SCHOOL (E11)-ENS	CHLOROFORM (THM)	68.94	UG/L	8/26/1998	CWS	24,316
CA	CA1510006	East Niles CSD	THM SAMP SITE-11845 SCHOOL (E11)-ENS	CHLOROFORM (THM)	77.57	UG/L	6/23/1998	CWS	24,316
CA	CA1510006	East Niles CSD	THM SAMP SITE-11845 SCHOOL (E11)-ENS	CHLOROFORM (THM)	59.6	UG/L	9/11/1997	CWS	24,316
CA	CA1510006	East Niles CSD	THM SAMP SITE-11845 SCHOOL (E11)-ENS	CHLOROFORM (THM)	67.1	UG/L	11/3/1998	CWS	24,316
CA	CA1510006	East Niles CSD	THM SAMP SITE-11845 SCHOOL (E11)-ENS	CHLOROFORM (THM)	1.2	UG/L	12/18/2001	CWS	24,316
CA	CA1510006	East Niles CSD	THM SAMP SITE-11845 SCHOOL (E11)-ENS	CHLOROFORM (THM)	0.8	UG/L	9/24/2001	CWS	24,316
CA	CA1510006	East Niles CSD	THM SAMP SITE-11845 SCHOOL (E11)-ENS	CHLOROFORM (THM)	38.43	UG/L	3/9/1999	CWS	24,316
CA	CA1510006	East Niles CSD	THM SAMP SITE-FAIRFAX SCHOOL (E12)-ENG	CHLOROFORM (THM)	2	UG/L	6/18/2002	CWS	24,316
CA	CA1510006	East Niles CSD	THM SAMP SITE-FAIRFAX SCHOOL (E12)-ENG	CHLOROFORM (THM)	0.9	UG/L	9/11/1997	CWS	24,316
CA	CA1510006	East Niles CSD	THM SAMP SITE-FAIRFAX SCHOOL (E12)-ENG	CHLOROFORM (THM)	1.85	UG/L	2/26/1998	CWS	24,316
CA	CA1510006	East Niles CSD	THM SAMP SITE-1417 VALE (E7)-KCWAS	CHLOROFORM (THM)	28.6	UG/L	6/18/2002	CWS	24,316
CA	CA1510006	East Niles CSD	THM SAMP SITE-1417 VALE (E7)-KCWAS	CHLOROFORM (THM)	52	UG/L	5/8/2002	CWS	24,316
CA	CA1510006	East Niles CSD	THM SAMP SITE-1417 VALE (E7)-KCWAS	CHLOROFORM (THM)	2.4	UG/L	3/18/2003	CWS	24,316
CA	CA1510006	East Niles CSD	THM SAMP SITE-1417 VALE (E7)-KCWAS	CHLOROFORM (THM)	0.8	UG/L	12/18/2002	CWS	24,316
CA	CA1510006	East Niles CSD	THM SAMP SITE-1417 VALE (E7)-KCWAS	CHLOROFORM (THM)	46.8	UG/L	11/3/1998	CWS	24,316

CA	CA1510006	East Niles CSD	THM SAMP SITE-1417 VALE (E7)-KCWAS	CHLOROFORM (THM)	48.42	UG/L	6/23/1998	CWS	24,316
CA	CA1510006	East Niles CSD	THM SAMP SITE-1417 VALE (E7)-KCWAS	CHLOROFORM (THM)	53.49	UG/L	8/26/1998	CWS	24,316
CA	CA1510006	East Niles CSD	THM SAMP SITE-1417 VALE (E7)-KCWAS	CHLOROFORM (THM)	41.19	UG/L	2/26/1998	CWS	24,316
CA	CA1510006	East Niles CSD	THM SAMP SITE-1417 VALE (E7)-KCWAS	CHLOROFORM (THM)	62	UG/L	9/11/1997	CWS	24,316
CA	CA1510006	East Niles CSD	THM SAMP SITE-1417 VALE (E7)-KCWAS	CHLOROFORM (THM)	2.9	UG/L	12/18/2001	CWS	24,316
CA	CA1510006	East Niles CSD	THM SAMP SITE-1417 VALE (E7)-KCWAS	CHLOROFORM (THM)	35.19	UG/L	3/9/1999	CWS	24,316
CA	CA1510006	East Niles CSD	THM SAMP SITE-1417 VALE (E7)-KCWAS	CHLOROFORM (THM)	14.4	UG/L	9/24/2001	CWS	24,316
CA	CA1510006	East Niles CSD	BRECKENRIDGE TP-RAW-OFFLINE	CHLOROFORM (THM)	24.6	UG/L	3/13/2001	CWS	24,316
CA	CA1510006	East Niles CSD	BRECKENRIDGE TP-RAW-OFFLINE	CHLOROFORM (THM)	27.7	UG/L	10/1/2003	CWS	24,316
CA	CA1510006	East Niles CSD	BRECKENRIDGE TP-RAW-OFFLINE	CHLOROFORM (THM)	53	UG/L	3/23/1998	CWS	24,316
CA	CA1510006	East Niles CSD	BRECKENRIDGE TP-RAW-OFFLINE	CHLOROFORM (THM)	27	UG/L	3/19/1997	CWS	24,316
CA	CA1510006	East Niles CSD	BRECKENRIDGE TP-RAW-OFFLINE	CHLOROFORM (THM)	15	UG/L	3/15/1999	CWS	24,316
CA	CA1510006	East Niles CSD	THM SAMP SITE-5310 WINGFOOT (E6)-KCWAS	CHLOROFORM (THM)	32.2	UG/L	6/18/2002	CWS	24,316
CA	CA1510006	East Niles CSD	THM SAMP SITE-5310 WINGFOOT (E6)-KCWAS	CHLOROFORM (THM)	48	UG/L	5/8/2002	CWS	24,316
CA	CA1510006	East Niles CSD	THM SAMP SITE-5310 WINGFOOT (E6)-KCWAS	CHLOROFORM (THM)	116	UG/L	3/18/2003	CWS	24,316
CA	CA1510006	East Niles CSD	THM SAMP SITE-5310 WINGFOOT (E6)-KCWAS	CHLOROFORM (THM)	32.8	UG/L	12/16/2003	CWS	24,316
CA	CA1510006	East Niles CSD	THM SAMP SITE-5310 WINGFOOT (E6)-KCWAS	CHLOROFORM (THM)	5.5	UG/L	12/18/2002	CWS	24,316
CA	CA1510006	East Niles CSD	THM SAMP SITE-5310 WINGFOOT (E6)-KCWAS	CHLOROFORM (THM)	44.8	UG/L	11/3/1998	CWS	24,316
CA	CA1510006	East Niles CSD	THM SAMP SITE-5310 WINGFOOT (E6)-KCWAS	CHLOROFORM (THM)	54.23	UG/L	8/26/1998	CWS	24,316
CA	CA1510006	East Niles CSD	THM SAMP SITE-5310 WINGFOOT (E6)-KCWAS	CHLOROFORM (THM)	44.8	UG/L	11/3/1998	CWS	24,316
CA	CA1510006	East Niles CSD	THM SAMP SITE-5310 WINGFOOT (E6)-KCWAS	CHLOROFORM (THM)	49.76	UG/L	6/23/1998	CWS	24,316
CA	CA1510006	East Niles CSD	THM SAMP SITE-5310 WINGFOOT (E6)-KCWAS	CHLOROFORM (THM)	58.1	UG/L	9/11/1997	CWS	24,316
CA	CA1510006	East Niles CSD	THM SAMP SITE-5310 WINGFOOT (E6)-KCWAS	CHLOROFORM (THM)	55.37	UG/L	2/26/1998	CWS	24,316
CA	CA1510006	East Niles CSD	THM SAMP SITE-5310 WINGFOOT (E6)-KCWAS	CHLOROFORM (THM)	3.5	UG/L	12/18/2001	CWS	24,316
CA	CA1510006	East Niles CSD	THM SAMP SITE-5310 WINGFOOT (E6)-KCWAS	CHLOROFORM (THM)	10.7	UG/L	9/24/2001	CWS	24,316
CA	CA1510006	East Niles CSD	THM SAMP SITE-5310 WINGFOOT (E6)-KCWAS	CHLOROFORM (THM)	16.06	UG/L	3/10/1999	CWS	24,316
CA	CA1510006	East Niles CSD	WELL 13-RAW	CHLOROFORM (THM)	5.8	UG/L	6/18/2002	CWS	24,316
CA	CA1510006	East Niles CSD	THM SAMP SITE-3209 FLINTRIDGE (E4)-KCWAS	CHLOROFORM (THM)	31.4	UG/L	6/18/2002	CWS	24,316
CA	CA1510006	East Niles CSD	THM SAMP SITE-3209 FLINTRIDGE (E4)-KCWAS	CHLOROFORM (THM)	33	UG/L	12/16/2003	CWS	24,316
CA	CA1510006	East Niles CSD	THM SAMP SITE-3209 FLINTRIDGE (E4)-KCWAS	CHLOROFORM (THM)	129	UG/L	3/18/2003	CWS	24,316
CA	CA1510006	East Niles CSD	THM SAMP SITE-3209 FLINTRIDGE (E4)-KCWAS	CHLOROFORM (THM)	6	UG/L	12/18/2002	CWS	24,316
CA	CA1510006	East Niles CSD	THM SAMP SITE-3209 FLINTRIDGE (E4)-KCWAS	CHLOROFORM (THM)	29.4	UG/L	11/3/1998	CWS	24,316
CA	CA1510006	East Niles CSD	THM SAMP SITE-3209 FLINTRIDGE (E4)-KCWAS	CHLOROFORM (THM)	48.15	UG/L	8/26/1998	CWS	24,316
CA	CA1510006	East Niles CSD	THM SAMP SITE-3209 FLINTRIDGE (E4)-KCWAS	CHLOROFORM (THM)	51.67	UG/L	2/26/1998	CWS	24,316
CA	CA1510006	East Niles CSD	THM SAMP SITE-3209 FLINTRIDGE (E4)-KCWAS	CHLOROFORM (THM)	37.4	UG/L	9/11/1997	CWS	24,316
CA	CA1510006	East Niles CSD	THM SAMP SITE-3209 FLINTRIDGE (E4)-KCWAS	CHLOROFORM (THM)	53	UG/L	6/23/1998	CWS	24,316
CA	CA1510006	East Niles CSD	THM SAMP SITE-3209 FLINTRIDGE (E4)-KCWAS	CHLOROFORM (THM)	44.4	UG/L	3/10/1999	CWS	24,316
CA	CA1510006	East Niles CSD	THM SAMP SITE-3209 FLINTRIDGE (E4)-KCWAS	CHLOROFORM (THM)	4.1	UG/L	12/18/2001	CWS	24,316
CA	CA1510006	East Niles CSD	THM SAMP SITE-3209 FLINTRIDGE (E4)-KCWAS	CHLOROFORM (THM)	11.7	UG/L	9/24/2001	CWS	24,316
CA	CA1510006	East Niles CSD	BRECKENRIDGE TP-TREATED	CHLOROFORM (THM)	4.8	UG/L	3/28/2002	CWS	24,316
CA	CA1510040	Kern County Water Agency	ID4 WEN. & COLUMBAS (T7)-DIST. SYS.	CHLOROFORM (THM)	36.58	UG/L	7/20/1998	CWS	0
CA	CA1510040	Kern County Water Agency	ID4 WEN. & COLUMBAS (T7)-DIST. SYS.	CHLOROFORM (THM)	41.52	UG/L	5/29/1998	CWS	0
CA	CA1510040	Kern County Water Agency	ID4 WEN. & COLUMBAS (T7)-DIST. SYS.	CHLOROFORM (THM)	31.49	UG/L	6/22/1998	CWS	0
CA	CA1510040	Kern County Water Agency	ID4 WEN. & COLUMBAS (T7)-DIST. SYS.	CHLOROFORM (THM)	34.79	UG/L	8/17/1998	CWS	0
CA	CA1510040	Kern County Water Agency	ID4 WEN. & COLUMBAS (T7)-DIST. SYS.	CHLOROFORM (THM)	29.2	UG/L	9/28/1998	CWS	0

APPENDIX 7-3 *Continued*

State	PWS_ID	Utility	POE	Constituent	LEVEL_NO	Unit	COLL_DATE	PWS	Pop
CA	CA1510040	Kern County Water Agency	ID4 WEN. & COLUMBAS (T7)-DIST. SYS.	CHLOROFORM (THM)	20.1	UG/L	3/8/1999	CWS	0
CA	CA1510040	Kern County Water Agency	ID4 WEN. & COLUMBAS (T7)-DIST. SYS.	CHLOROFORM (THM)	27.55	UG/L	2/2/1998	CWS	0
CA	CA1510040	Kern County Water Agency	ID4 WEN. & COLUMBAS (T7)-DIST. SYS.	CHLOROFORM (THM)	36	UG/L	9/23/1997	CWS	0
CA	CA1510040	Kern County Water Agency	ID4 WEN. & COLUMBAS (T7)-DIST. SYS.	CHLOROFORM (THM)	40.4	UG/L	6/2/1997	CWS	0
CA	CA1510040	Kern County Water Agency	ID4 WEN. & COLUMBAS (T7)-DIST. SYS.	CHLOROFORM (THM)	37.3	UG/L	7/7/1997	CWS	0
CA	CA1510040	Kern County Water Agency	ID4 WEN. & COLUMBAS (T7)-DIST. SYS.	CHLOROFORM (THM)	39.58	UG/L	4/6/1998	CWS	0
CA	CA1510040	Kern County Water Agency	ID4 WEN. & COLUMBAS (T7)-DIST. SYS.	CHLOROFORM (THM)	26.69	UG/L	3/2/1998	CWS	0
CA	CA1510040	Kern County Water Agency	ID4 WEN. & COLUMBAS (T7)-DIST. SYS.	CHLOROFORM (THM)	13.23	UG/L	1/5/1998	CWS	0
CA	CA1510040	Kern County Water Agency	ID4 CORNER TANK-T6-TREATED	CHLOROFORM (THM)	48.87	UG/L	6/22/1998	CWS	0
CA	CA1510040	Kern County Water Agency	ID4 CORNER TANK-T6-TREATED	CHLOROFORM (THM)	41.62	UG/L	5/29/1998	CWS	0
CA	CA1510040	Kern County Water Agency	ID4 CORNER TANK-T6-TREATED	CHLOROFORM (THM)	48.04	UG/L	4/6/1998	CWS	0
CA	CA1510040	Kern County Water Agency	ID4 CORNER TANK-T6-TREATED	CHLOROFORM (THM)	38.76	UG/L	2/2/1998	CWS	0
CA	CA1510040	Kern County Water Agency	ID4 CORNER TANK-T6-TREATED	CHLOROFORM (THM)	33.6	UG/L	3/2/1998	CWS	0
CA	CA1510040	Kern County Water Agency	ID4 CORNER TANK-T6-TREATED	CHLOROFORM (THM)	41.6	UG/L	7/20/1998	CWS	0
CA	CA1510040	Kern County Water Agency	ID4 CORNER TANK-T6-TREATED	CHLOROFORM (THM)	34.73	UG/L	8/17/1998	CWS	0
CA	CA1510040	Kern County Water Agency	ID4 CORNER TANK-T6-TREATED	CHLOROFORM (THM)	34.5	UG/L	9/28/1998	CWS	0
CA	CA1510040	Kern County Water Agency	ID4 CORNER TANK-T6-TREATED	CHLOROFORM (THM)	36.8	UG/L	3/8/1999	CWS	0
CA	CA1510040	Kern County Water Agency	ID4 CORNER TANK-T6-TREATED	CHLOROFORM (THM)	22.5	UG/L	2/24/1997	CWS	0
CA	CA1510040	Kern County Water Agency	ID4 CORNER TANK-T6-TREATED	CHLOROFORM (THM)	93.3	UG/L	1/31/1997	CWS	0
CA	CA1510040	Kern County Water Agency	ID4 CORNER TANK-T6-TREATED	CHLOROFORM (THM)	38.64	UG/L	3/24/1997	CWS	0
CA	CA1510040	Kern County Water Agency	ID4 CORNER TANK-T6-TREATED	CHLOROFORM (THM)	58.4	UG/L	4/14/1997	CWS	0
CA	CA1510040	Kern County Water Agency	ID4 CORNER TANK-T6-TREATED	CHLOROFORM (THM)	45.9	UG/L	5/12/1997	CWS	0
CA	CA1510040	Kern County Water Agency	ID4 CORNER TANK-T6-TREATED	CHLOROFORM (THM)	17.84	UG/L	1/5/1998	CWS	0
CA	CA1510040	Kern County Water Agency	ID4 CORNER TANK-T6-TREATED	CHLOROFORM (THM)	60	UG/L	9/8/1997	CWS	0
CA	CA1510040	Kern County Water Agency	ID4 CORNER TANK-T6-TREATED	CHLOROFORM (THM)	47.9	UG/L	7/7/1997	CWS	0
CA	CA1510040	Kern County Water Agency	ID4 CORNER TANK-T6-TREATED	CHLOROFORM (THM)	46.1	UG/L	6/2/1997	CWS	0
CA	CA1510040	Kern County Water Agency	ID4 BELLE-SEQUOIA TANK-T2-TREATED	CHLOROFORM (THM)	62.29	UG/L	5/29/1998	CWS	0
CA	CA1510040	Kern County Water Agency	ID4 BELLE-SEQUOIA TANK-T2-TREATED	CHLOROFORM (THM)	19.14	UG/L	2/2/1998	CWS	0
CA	CA1510040	Kern County Water Agency	ID4 BELLE-SEQUOIA TANK-T2-TREATED	CHLOROFORM (THM)	22.38	UG/L	3/2/1998	CWS	0
CA	CA1510040	Kern County Water Agency	ID4 BELLE-SEQUOIA TANK-T2-TREATED	CHLOROFORM (THM)	60.47	UG/L	4/6/1998	CWS	0
CA	CA1510040	Kern County Water Agency	ID4 BELLE-SEQUOIA TANK-T2-TREATED	CHLOROFORM (THM)	26.4	UG/L	9/28/1998	CWS	0
CA	CA1510040	Kern County Water Agency	ID4 BELLE-SEQUOIA TANK-T2-TREATED	CHLOROFORM (THM)	10.43	UG/L	1/5/1998	CWS	0
CA	CA1510040	Kern County Water Agency	ID4 BELLE-SEQUOIA TANK-T2-TREATED	CHLOROFORM (THM)	18.3	UG/L	3/8/1999	CWS	0
CA	CA1510040	Kern County Water Agency	ID4 BELLE-SEQUOIA TANK-T2-TREATED	CHLOROFORM (THM)	27.52	UG/L	6/22/1998	CWS	0
CA	CA1510040	Kern County Water Agency	ID4 BELLE-SEQUOIA TANK-T2-TREATED	CHLOROFORM (THM)	33.39	UG/L	8/17/1998	CWS	0
CA	CA1510040	Kern County Water Agency	ID4 BELLE-SEQUOIA TANK-T2-TREATED	CHLOROFORM (THM)	31.38	UG/L	7/20/1998	CWS	0
CA	CA1510040	Kern County Water Agency	ID4 BELLE-SEQUOIA TANK-T2-TREATED	CHLOROFORM (THM)	47.95	UG/L	3/24/1997	CWS	0
CA	CA1510040	Kern County Water Agency	ID4 BELLE-SEQUOIA TANK-T2-TREATED	CHLOROFORM (THM)	87.1	UG/L	1/31/1997	CWS	0
CA	CA1510040	Kern County Water Agency	ID4 BELLE-SEQUOIA TANK-T2-TREATED	CHLOROFORM (THM)	20.3	UG/L	2/24/1997	CWS	0
CA	CA1510040	Kern County Water Agency	ID4 BELLE-SEQUOIA TANK-T2-TREATED	CHLOROFORM (THM)	54.6	UG/L	4/14/1997	CWS	0
CA	CA1510040	Kern County Water Agency	ID4 BELLE-SEQUOIA TANK-T2-TREATED	CHLOROFORM (THM)	31.9	UG/L	5/12/1997	CWS	0
CA	CA1510040	Kern County Water Agency	ID4 BELLE-SEQUOIA TANK-T2-TREATED	CHLOROFORM (THM)	38	UG/L	6/2/1997	CWS	0

CA	CA1510040	Kern County Water Agency	ID4 BELLE-SEQUOIA TANK-T2-TREATED	CHLOROFORM (THM)	54.3	UG/L	9/8/1997	CWS	0
CA	CA1510040	Kern County Water Agency	ID4 BELLE-SEQUOIA TANK-T2-TREATED	CHLOROFORM (THM)	45.6	UG/L	9/8/1997	CWS	0
CA	CA1510040	Kern County Water Agency	ID4 BELLE-SEQUOIA TANK-T2-TREATED	CHLOROFORM (THM)	35.1	UG/L	7/7/1997	CWS	0
CA	CA1510040	Kern County Water Agency	ID4 OSWELL TANK-T5-TREATED	CHLOROFORM (THM)	44.89	UG/L	5/29/1998	CWS	0
CA	CA1510040	Kern County Water Agency	ID4 OSWELL TANK-T5-TREATED	CHLOROFORM (THM)	38.3	UG/L	4/6/1998	CWS	0
CA	CA1510040	Kern County Water Agency	ID4 OSWELL TANK-T5-TREATED	CHLOROFORM (THM)	32.21	UG/L	3/2/1998	CWS	0
CA	CA1510040	Kern County Water Agency	ID4 OSWELL TANK-T5-TREATED	CHLOROFORM (THM)	37.43	UG/L	2/2/1998	CWS	0
CA	CA1510040	Kern County Water Agency	ID4 OSWELL TANK-T5-TREATED	CHLOROFORM (THM)	75.3	UG/L	9/28/1998	CWS	0
CA	CA1510040	Kern County Water Agency	ID4 OSWELL TANK-T5-TREATED	CHLOROFORM (THM)	48.9	UG/L	6/22/1998	CWS	0
CA	CA1510040	Kern County Water Agency	ID4 OSWELL TANK-T5-TREATED	CHLOROFORM (THM)	35.97	UG/L	8/17/1998	CWS	0
CA	CA1510040	Kern County Water Agency	ID4 OSWELL TANK-T5-TREATED	CHLOROFORM (THM)	45.1	UG/L	3/8/1999	CWS	0
CA	CA1510040	Kern County Water Agency	ID4 OSWELL TANK-T5-TREATED	CHLOROFORM (THM)	19.54	UG/L	1/5/1998	CWS	0
CA	CA1510040	Kern County Water Agency	ID4 OSWELL TANK-T5-TREATED	CHLOROFORM (THM)	38.12	UG/L	7/20/1998	CWS	0
CA	CA1510040	Kern County Water Agency	ID4 OSWELL TANK-T5-TREATED	CHLOROFORM (THM)	103	UG/L	1/31/1997	CWS	0
CA	CA1510040	Kern County Water Agency	ID4 OSWELL TANK-T5-TREATED	CHLOROFORM (THM)	23.1	UG/L	2/24/1997	CWS	0
CA	CA1510040	Kern County Water Agency	ID4 OSWELL TANK-T5-TREATED	CHLOROFORM (THM)	38.06	UG/L	3/24/1997	CWS	0
CA	CA1510040	Kern County Water Agency	ID4 OSWELL TANK-T5-TREATED	CHLOROFORM (THM)	58.7	UG/L	4/14/1997	CWS	0
CA	CA1510040	Kern County Water Agency	ID4 OSWELL TANK-T5-TREATED	CHLOROFORM (THM)	39	UG/L	7/7/1997	CWS	0
CA	CA1510040	Kern County Water Agency	ID4 OSWELL TANK-T5-TREATED	CHLOROFORM (THM)	40.4	UG/L	5/12/1997	CWS	0
CA	CA1510040	Kern County Water Agency	ID4 OSWELL TANK-T5-TREATED	CHLOROFORM (THM)	61.2	UG/L	9/8/1997	CWS	0
CA	CA1510040	Kern County Water Agency	ID4 OSWELL TANK-T5-TREATED	CHLOROFORM (THM)	41.6	UG/L	6/2/1997	CWS	0
CA	CA1510040	Kern County Water Agency	ID4 NOR TANK (600 SITE) T3-TREATED	CHLOROFORM (THM)	42.66	UG/L	6/22/1998	CWS	0
CA	CA1510040	Kern County Water Agency	ID4 NOR TANK (600 SITE) T3-TREATED	CHLOROFORM (THM)	59.2	UG/L	5/29/1998	CWS	0
CA	CA1510040	Kern County Water Agency	ID4 NOR TANK (600 SITE) T3-TREATED	CHLOROFORM (THM)	26.15	UG/L	3/2/1998	CWS	0
CA	CA1510040	Kern County Water Agency	ID4 NOR TANK (600 SITE) T3-TREATED	CHLOROFORM (THM)	57.69	UG/L	4/6/1998	CWS	0
CA	CA1510040	Kern County Water Agency	ID4 NOR TANK (600 SITE) T3-TREATED	CHLOROFORM (THM)	35.76	UG/L	7/20/1998	CWS	0
CA	CA1510040	Kern County Water Agency	ID4 NOR TANK (600 SITE) T3-TREATED	CHLOROFORM (THM)	67.21	UG/L	8/17/1998	CWS	0
CA	CA1510040	Kern County Water Agency	ID4 NOR TANK (600 SITE) T3-TREATED	CHLOROFORM (THM)	46.4	UG/L	9/28/1998	CWS	0
CA	CA1510040	Kern County Water Agency	ID4 NOR TANK (600 SITE) T3-TREATED	CHLOROFORM (THM)	18.7	UG/L	3/8/1999	CWS	0
CA	CA1510040	Kern County Water Agency	ID4 NOR TANK (600 SITE) T3-TREATED	CHLOROFORM (THM)	37.17	UG/L	2/2/1998	CWS	0
CA	CA1510040	Kern County Water Agency	ID4 NOR TANK (600 SITE) T3-TREATED	CHLOROFORM (THM)	94.1	UG/L	1/31/1997	CWS	0
CA	CA1510040	Kern County Water Agency	ID4 NOR TANK (600 SITE) T3-TREATED	CHLOROFORM (THM)	55.52	UG/L	3/24/1997	CWS	0
CA	CA1510040	Kern County Water Agency	ID4 NOR TANK (600 SITE) T3-TREATED	CHLOROFORM (THM)	50.5	UG/L	4/14/1997	CWS	0
CA	CA1510040	Kern County Water Agency	ID4 NOR TANK (600 SITE) T3-TREATED	CHLOROFORM (THM)	24	UG/L	2/24/1997	CWS	0
CA	CA1510040	Kern County Water Agency	ID4 NOR TANK (600 SITE) T3-TREATED	CHLOROFORM (THM)	35.7	UG/L	5/12/1997	CWS	0
CA	CA1510040	Kern County Water Agency	ID4 NOR TANK (600 SITE) T3-TREATED	CHLOROFORM (THM)	20.78	UG/L	1/5/1998	CWS	0
CA	CA1510040	Kern County Water Agency	ID4 NOR TANK (600 SITE) T3-TREATED	CHLOROFORM (THM)	38.9	UG/L	7/7/1997	CWS	0
CA	CA1510040	Kern County Water Agency	ID4 NOR TANK (600 SITE) T3-TREATED	CHLOROFORM (THM)	44.1	UG/L	6/2/1997	CWS	0
CA	CA1510040	Kern County Water Agency	ID4 TREATED-T1	CHLOROFORM (THM)	21.15	UG/L	2/2/1998	CWS	0
CA	CA1510040	Kern County Water Agency	ID4 TREATED-T1	CHLOROFORM (THM)	23	UG/L	3/2/1998	CWS	0
CA	CA1510040	Kern County Water Agency	ID4 TREATED-T1	CHLOROFORM (THM)	7.97	UG/L	1/5/1998	CWS	0
CA	CA1510040	Kern County Water Agency	ID4 TREATED-T1	CHLOROFORM (THM)	54.2	UG/L	9/8/1997	CWS	0
CA	CA1510040	Kern County Water Agency	ID4 TREATED-T1	CHLOROFORM (THM)	35.73	UG/L	5/29/1998	CWS	0
CA	CA1510040	Kern County Water Agency	ID4 TREATED-T1	CHLOROFORM (THM)	31.32	UG/L	6/22/1998	CWS	0
CA	CA1510040	Kern County Water Agency	ID4 TREATED-T1	CHLOROFORM (THM)	35.76	UG/L	4/6/1998	CWS	0
CA	CA1510040	Kern County Water Agency	ID4 TREATED-T1	CHLOROFORM (THM)	18.2	UG/L	2/24/1997	CWS	0

APPENDIX 7-3 *Continued*

State	PWS_ID	Utility	POE	Constituent	LEVEL_NO	Unit	COLL_DATE	PWS	Pop
CA	CA1510040	Kern County Water Agency	ID4 TREATED-T1	CHLOROFORM (THM)	47.95	UG/L	3/24/1997	CWS	0
CA	CA1510040	Kern County Water Agency	ID4 TREATED-T1	CHLOROFORM (THM)	79.1	UG/L	1/31/1997	CWS	0
CA	CA1510040	Kern County Water Agency	ID4 TREATED-T1	CHLOROFORM (THM)	55.4	UG/L	7/7/1997	CWS	0
CA	CA1510040	Kern County Water Agency	ID4 TREATED-T1	CHLOROFORM (THM)	57	UG/L	4/14/1997	CWS	0
CA	CA1510040	Kern County Water Agency	ID4 TREATED-T1	CHLOROFORM (THM)	44.4	UG/L	6/2/1997	CWS	0
CA	CA1510040	Kern County Water Agency	ID4 TREATED-T1	CHLOROFORM (THM)	57.3	UG/L	5/12/1997	CWS	0
CA	CA1510040	Kern County Water Agency	ID4 TREATED-T1	CHLOROFORM (THM)	50.1	UG/L	8/30/2002	CWS	0
CA	CA1510040	Kern County Water Agency	ID4 TREATED-T1	CHLOROFORM (THM)	42.5	UG/L	8/30/1999	CWS	0
CA	CA1510040	Kern County Water Agency	ID4 TREATED-T1	CHLOROFORM (THM)	35.1	UG/L	9/28/1998	CWS	0
CA	CA1510040	Kern County Water Agency	ID4 TREATED-T1	CHLOROFORM (THM)	35.51	UG/L	7/20/1998	CWS	0
CA	CA1510040	Kern County Water Agency	ID4 TREATED-T1	CHLOROFORM (THM)	38.01	UG/L	8/17/1998	CWS	0
CA	CA1510040	Kern County Water Agency	ID4 TREATED-T1	CHLOROFORM (THM)	18.5	UG/L	3/8/1999	CWS	0
CA	CA1510040	Kern County Water Agency	ID4 TREATED-T1	CHLOROFORM (THM)	15	UG/L	11/15/2000	CWS	0
CA	CA1510041	North of the River MWD	THM SAMP SITE-2841 PEGASUS (N1)-TRTD	CHLOROFORM (THM)	4.4	UG/L	8/22/2001	CWS	7,000
CA	CA1510041	North of the River MWD	THM SAMP SITE-2841 PEGASUS (N1)-TRTD	CHLOROFORM (THM)	54.7	UG/L	5/8/2001	CWS	7,000
CA	CA1510041	North of the River MWD	THM SAMP SITE-2841 PEGASUS (N1)-TRTD	CHLOROFORM (THM)	36.1	UG/L	6/7/1999	CWS	7,000
CA	CA1510041	North of the River MWD	THM SAMP SITE-2841 PEGASUS (N1)-TRTD	CHLOROFORM (THM)	2.9	UG/L	11/26/2001	CWS	7,000
CA	CA1510041	North of the River MWD	THM SAMP SITE-2841 PEGASUS (N1)-TRTD	CHLOROFORM (THM)	63.4	UG/L	5/17/2002	CWS	7,000
CA	CA1510041	North of the River MWD	THM SAMP SITE-2841 PEGASUS (N1)-TRTD	CHLOROFORM (THM)	59.5	UG/L	8/27/2002	CWS	7,000
CA	CA1510041	North of the River MWD	THM SAMP SITE-2841 PEGASUS (N1)-TRTD	CHLOROFORM (THM)	59.86	UG/L	6/23/1998	CWS	7,000
CA	CA1510041	North of the River MWD	THM SAMP SITE-2841 PEGASUS (N1)-TRTD	CHLOROFORM (THM)	56.87	UG/L	2/25/1998	CWS	7,000
CA	CA1510041	North of the River MWD	THM SAMP SITE-2841 PEGASUS (N1)-TRTD	CHLOROFORM (THM)	43.9	UG/L	11/3/1998	CWS	7,000
CA	CA1510041	North of the River MWD	THM SAMP SITE-2841 PEGASUS (N1)-TRTD	CHLOROFORM (THM)	49.42	UG/L	11/7/2000	CWS	7,000
CA	CA1510041	North of the River MWD	THM SAMP SITE-2841 PEGASUS (N1)-TRTD	CHLOROFORM (THM)	44.25	UG/L	3/10/1999	CWS	7,000
CA	CA1510041	North of the River MWD	THM SAMP SITE-900 TANKSITE (N4)-TRTD	CHLOROFORM (THM)	23	UG/L	8/22/2001	CWS	7,000
CA	CA1510041	North of the River MWD	THM SAMP SITE-900 TANKSITE (N4)-TRTD	CHLOROFORM (THM)	51.3	UG/L	5/8/2001	CWS	7,000
CA	CA1510041	North of the River MWD	THM SAMP SITE-900 TANKSITE (N4)-TRTD	CHLOROFORM (THM)	10.8	UG/L	11/26/2001	CWS	7,000
CA	CA1510041	North of the River MWD	THM SAMP SITE-900 TANKSITE (N4)-TRTD	CHLOROFORM (THM)	89.3	UG/L	5/17/2002	CWS	7,000
CA	CA1510041	North of the River MWD	THM SAMP SITE-900 TANKSITE (N4)-TRTD	CHLOROFORM (THM)	84.2	UG/L	8/27/2002	CWS	7,000
CA	CA1510041	North of the River MWD	THM SAMP SITE-900 TANKSITE (N4)-TRTD	CHLOROFORM (THM)	42.44	UG/L	11/7/2000	CWS	7,000
CA	CA1510041	North of the River MWD	THM SAMP SITE-900 TANKSITE (N4)-TRTD	CHLOROFORM (THM)	96.9	UG/L	9/23/1997	CWS	7,000
CA	CA1510041	North of the River MWD	THM SAMP SITE-900 TANKSITE (N4)-TRTD	CHLOROFORM (THM)	65.07	UG/L	2/25/1998	CWS	7,000
CA	CA1510041	North of the River MWD	THM SAMP SITE-900 TANKSITE (N4)-TRTD	CHLOROFORM (THM)	77.77	UG/L	6/23/1998	CWS	7,000
CA	CA1510041	North of the River MWD	THM SAMP SITE-900 TANKSITE (N4)-TRTD	CHLOROFORM (THM)	108.2	UG/L	6/7/1999	CWS	7,000
CA	CA1510041	North of the River MWD	THM SAMP SITE-900 TANKSITE (N4)-TRTD	CHLOROFORM (THM)	78.86	UG/L	3/10/1999	CWS	7,000
CA	CA1510041	North of the River MWD	THM SAMP SITE-900 TANKSITE (N4)-TRTD	CHLOROFORM (THM)	60.6	UG/L	11/3/1998	CWS	7,000
CA	CA1510041	North of the River MWD	THM SAMP SITE-3418 N CHESTER (N3)-TRTD	CHLOROFORM (THM)	4	UG/L	8/22/2001	CWS	7,000
CA	CA1510041	North of the River MWD	THM SAMP SITE-3418 N CHESTER (N3)-TRTD	CHLOROFORM (THM)	46.8	UG/L	5/8/2001	CWS	7,000
CA	CA1510041	North of the River MWD	THM SAMP SITE-3418 N CHESTER (N3)-TRTD	CHLOROFORM (THM)	2.5	UG/L	11/26/2001	CWS	7,000
CA	CA1510041	North of the River MWD	THM SAMP SITE-3418 N CHESTER (N3)-TRTD	CHLOROFORM (THM)	59.1	UG/L	8/27/2002	CWS	7,000
CA	CA1510041	North of the River MWD	THM SAMP SITE-3418 N CHESTER (N3)-TRTD	CHLOROFORM (THM)	39.38	UG/L	11/7/2000	CWS	7,000
CA	CA1510041	North of the River MWD	THM SAMP SITE-3418 N CHESTER (N3)-TRTD	CHLOROFORM (THM)	55.7	UG/L	5/17/2002	CWS	7,000

CA	CA1510041	North of the River MWD	THM SAMP SITE-3418 N CHESTER (N3)-TRTD	CHLOROFORM (THM)	74.8	UG/L	9/11/1997	CWS	7,000
CA	CA1510041	North of the River MWD	THM SAMP SITE-3418 N CHESTER (N3)-TRTD	CHLOROFORM (THM)	32.49	UG/L	2/25/1998	CWS	7,000
CA	CA1510041	North of the River MWD	THM SAMP SITE-3418 N CHESTER (N3)-TRTD	CHLOROFORM (THM)	49.23	UG/L	6/23/1998	CWS	7,000
CA	CA1510041	North of the River MWD	THM SAMP SITE-3418 N CHESTER (N3)-TRTD	CHLOROFORM (THM)	37.7	UG/L	11/3/1998	CWS	7,000
CA	CA1510041	North of the River MWD	THM SAMP SITE-3418 N CHESTER (N3)-TRTD	CHLOROFORM (THM)	27.9	UG/L	6/7/1999	CWS	7,000
CA	CA1510041	North of the River MWD	THM SAMP SITE-3418 N CHESTER (N3)-TRTD	CHLOROFORM (THM)	38.33	UG/L	3/10/1999	CWS	7,000
CA	CA1510041	North of the River MWD	THM SAMP SITE-528 CHARLANA (N2)-TRTD	CHLOROFORM (THM)	3.6	UG/L	8/22/2001	CWS	7,000
CA	CA1510041	North of the River MWD	THM SAMP SITE-528 CHARLANA (N2)-TRTD	CHLOROFORM (THM)	3.6	UG/L	11/26/2001	CWS	7,000
CA	CA1510041	North of the River MWD	THM SAMP SITE-528 CHARLANA (N2)-TRTD	CHLOROFORM (THM)	55.7	UG/L	5/17/2002	CWS	7,000
CA	CA1510041	North of the River MWD	THM SAMP SITE-528 CHARLANA (N2)-TRTD	CHLOROFORM (THM)	63.4	UG/L	8/27/2002	CWS	7,000
CA	CA1510041	North of the River MWD	THM SAMP SITE-528 CHARLANA (N2)-TRTD	CHLOROFORM (THM)	36.9	UG/L	9/11/1997	CWS	7,000
CA	CA1510041	North of the River MWD	THM SAMP SITE-528 CHARLANA (N2)-TRTD	CHLOROFORM (THM)	53.12	UG/L	2/25/1998	CWS	7,000
CA	CA1510041	North of the River MWD	THM SAMP SITE-528 CHARLANA (N2)-TRTD	CHLOROFORM (THM)	39.19	UG/L	6/23/1998	CWS	7,000
CA	CA1510041	North of the River MWD	THM SAMP SITE-528 CHARLANA (N2)-TRTD	CHLOROFORM (THM)	28.2	UG/L	11/3/1998	CWS	7,000
CA	CA1510041	North of the River MWD	THM SAMP SITE-528 CHARLANA (N2)-TRTD	CHLOROFORM (THM)	37.56	UG/L	11/7/2000	CWS	7,000
CA	CA1510041	North of the River MWD	THM SAMP SITE-528 CHARLANA (N2)-TRTD	CHLOROFORM (THM)	47.5	UG/L	5/8/2001	CWS	7,000
CA	CA1510041	North of the River MWD	THM SAMP SITE-528 CHARLANA (N2)-TRTD	CHLOROFORM (THM)	15	UG/L	6/7/1999	CWS	7,000
CA	CA1510041	North of the River MWD	THM SAMP SITE-528 CHARLANA (N2)-TRTD	CHLOROFORM (THM)	25.02	UG/L	3/10/1999	CWS	7,000
					37.9				
CA	CA1510006	East Niles CSD	THM SAMP SITE-DEACON (E8)-ENG	DIBROMOACETIC ACID (DBAA)	1	UG/L	6/18/2002	CWS	24,316
CA	CA1510006	East Niles CSD	THM SAMP SITE-DEACON (E8)-ENG	DIBROMOACETIC ACID (DBAA)	2	UG/L	12/18/2002	CWS	24,316
CA	CA1510006	East Niles CSD	THM SAMP SITE-PACIFIC BELL (E9)-ENS	DIBROMOACETIC ACID (DBAA)	9	UG/L	6/18/2002	CWS	24,316
CA	CA1510006	East Niles CSD	THM SAMP SITE-PACIFIC BELL (E9)-ENS	DIBROMOACETIC ACID (DBAA)	4	UG/L	3/18/2003	CWS	24,316
CA	CA1510006	East Niles CSD	THM SAMP SITE-PACIFIC BELL (E9)-ENS	DIBROMOACETIC ACID (DBAA)	3	UG/L	12/18/2002	CWS	24,316
CA	CA1510006	East Niles CSD	THM SAMP SITE-PACIFIC BELL (E9)-ENS	DIBROMOACETIC ACID (DBAA)	3	UG/L	12/18/2001	CWS	24,316
CA	CA1510006	East Niles CSD	THM SAMP SITE-PACIFIC BELL (E9)-ENS	DIBROMOACETIC ACID (DBAA)	2	UG/L	9/24/2001	CWS	24,316
CA	CA1510006	East Niles CSD	THM SAMP SITE-8717 BUTTERNUT (E10)-ENS	DIBROMOACETIC ACID (DBAA)	2	UG/L	12/18/2001	CWS	24,316
CA	CA1510006	East Niles CSD	THM SAMP SITE-8717 BUTTERNUT (E10)-ENS	DIBROMOACETIC ACID (DBAA)	3	UG/L	3/18/2003	CWS	24,316
CA	CA1510006	East Niles CSD	THM SAMP SITE-8717 BUTTERNUT (E10)-ENS	DIBROMOACETIC ACID (DBAA)	3	UG/L	12/18/2002	CWS	24,316
CA	CA1510006	East Niles CSD	THM SAMP SITE-8717 BUTTERNUT (E10)-ENS	DIBROMOACETIC ACID (DBAA)	8	UG/L	6/18/2002	CWS	24,316
CA	CA1510006	East Niles CSD	THM SAMP SITE-8717 BUTTERNUT (E10)-ENS	DIBROMOACETIC ACID (DBAA)	5	UG/L	9/24/2001	CWS	24,316
CA	CA1510006	East Niles CSD	THM SAMP SITE-11845 SCHOOL (E11)-ENS	DIBROMOACETIC ACID (DBAA)	5	UG/L	6/18/2002	CWS	24,316
CA	CA1510006	East Niles CSD	THM SAMP SITE-11845 SCHOOL (E11)-ENS	DIBROMOACETIC ACID (DBAA)	3	UG/L	9/24/2001	CWS	24,316
CA	CA1510006	East Niles CSD	THM SAMP SITE-11845 SCHOOL (E11)-ENS	DIBROMOACETIC ACID (DBAA)	4	UG/L	12/18/2001	CWS	24,316
CA	CA1510006	East Niles CSD	THM SAMP SITE-11845 SCHOOL (E11)-ENS	DIBROMOACETIC ACID (DBAA)	4	UG/L	3/18/2003	CWS	24,316
CA	CA1510006	East Niles CSD	THM SAMP SITE-11845 SCHOOL (E11)-ENS	DIBROMOACETIC ACID (DBAA)	4	UG/L	12/18/2002	CWS	24,316
CA	CA1510006	East Niles CSD	THM SAMP SITE-FAIRFAX SCHOOL (E12)-ENG	DIBROMOACETIC ACID (DBAA)	1	UG/L	6/18/2002	CWS	24,316
CA	CA1510006	East Niles CSD	THM SAMP SITE-FAIRFAX SCHOOL (E12)-ENG	DIBROMOACETIC ACID (DBAA)	2	UG/L	12/18/2002	CWS	24,316
CA	CA1510006	East Niles CSD	THM SAMP SITE-FAIRFAX SCHOOL (E12)-ENG	DIBROMOACETIC ACID (DBAA)	2	UG/L	3/18/2003	CWS	24,316
CA	CA1510006	East Niles CSD	THM SAMP SITE-FAIRFAX SCHOOL (E12)-ENG	DIBROMOACETIC ACID (DBAA)	1	UG/L	12/18/2001	CWS	24,316
CA	CA1510006	East Niles CSD	THM SAMP SITE-1417 VALE (E7)-KCWAS	DIBROMOACETIC ACID (DBAA)	8	UG/L	6/18/2002	CWS	24,316
CA	CA1510006	East Niles CSD	THM SAMP SITE-1417 VALE (E7)-KCWAS	DIBROMOACETIC ACID (DBAA)	8	UG/L	12/18/2001	CWS	24,316
CA	CA1510006	East Niles CSD	THM SAMP SITE-1417 VALE (E7)-KCWAS	DIBROMOACETIC ACID (DBAA)	3	UG/L	12/18/2002	CWS	24,316
CA	CA1510006	East Niles CSD	THM SAMP SITE-1417 VALE (E7)-KCWAS	DIBROMOACETIC ACID (DBAA)	2	UG/L	3/18/2003	CWS	24,316
CA	CA1510006	East Niles CSD	THM SAMP SITE-1417 VALE (E7)-KCWAS	DIBROMOACETIC ACID (DBAA)	29	UG/L	9/24/2001	CWS	24,316

APPENDIX 7-3 *Continued*

State	PWS_ID	Utility	POE	Constituent	LEVEL_NO	Unit	COLL_DATE	PWS	Pop
CA	CA1510006	East Niles CSD	THM SAMP SITE-5310 WINGFOOT (E6)-KCWAS	DIBROMOACETIC ACID (DBAA)	7	UG/L	6/18/2002	CWS	24,316
CA	CA1510006	East Niles CSD	THM SAMP SITE-5310 WINGFOOT (E6)-KCWAS	DIBROMOACETIC ACID (DBAA)	4	UG/L	12/18/2002	CWS	24,316
CA	CA1510006	East Niles CSD	THM SAMP SITE-5310 WINGFOOT (E6)-KCWAS	DIBROMOACETIC ACID (DBAA)	7	UG/L	12/18/2001	CWS	24,316
CA	CA1510006	East Niles CSD	THM SAMP SITE-5310 WINGFOOT (E6)-KCWAS	DIBROMOACETIC ACID (DBAA)	23	UG/L	9/24/2001	CWS	24,316
CA	CA1510006	East Niles CSD	THM SAMP SITE-3209 FLINTRIDGE (E4)-KCWAS	DIBROMOACETIC ACID (DBAA)	8	UG/L	6/18/2002	CWS	24,316
CA	CA1510006	East Niles CSD	THM SAMP SITE-3209 FLINTRIDGE (E4)-KCWAS	DIBROMOACETIC ACID (DBAA)	7	UG/L	12/18/2001	CWS	24,316
CA	CA1510006	East Niles CSD	THM SAMP SITE-3209 FLINTRIDGE (E4)-KCWAS	DIBROMOACETIC ACID (DBAA)	2	UG/L	12/18/2002	CWS	24,316
CA	CA1510006	East Niles CSD	THM SAMP SITE-3209 FLINTRIDGE (E4)-KCWAS	DIBROMOACETIC ACID (DBAA)	19	UG/L	9/24/2001	CWS	24,316
CA	CA1600034	PARAMOUNT FARMING CO	TREATMENT PLANT	DIBROMOACETIC ACID (DBAA)	17	UG/L	11/4/2004	CWS	100
CA	CA1600034	PARAMOUNT FARMING CO	TREATMENT PLANT	DIBROMOACETIC ACID (DBAA)	11	UG/L	7/14/2004	CWS	100
					6.28				
CA	CA1510006	East Niles CSD	THM SAMP SITE-DEACON (E8)-ENG	DIBROMOCHLOROMETHANE (THM)	2.6	UG/L	12/16/2003	CWS	24,316
CA	CA1510006	East Niles CSD	THM SAMP SITE-DEACON (E8)-ENG	DIBROMOCHLOROMETHANE (THM)	0.6	UG/L	11/3/1998	CWS	24,316
CA	CA1510006	East Niles CSD	THM SAMP SITE-DEACON (E8)-ENG	DIBROMOCHLOROMETHANE (THM)	0.6	UG/L	12/18/2001	CWS	24,316
CA	CA1510006	East Niles CSD	THM SAMP SITE-DEACON (E8)-ENG	DIBROMOCHLOROMETHANE (THM)	1.12	UG/L	2/26/1998	CWS	24,316
CA	CA1510006	East Niles CSD	THM SAMP SITE-PACIFIC BELL (E9)-ENS	DIBROMOCHLOROMETHANE (THM)	32	UG/L	6/18/2002	CWS	24,316
CA	CA1510006	East Niles CSD	THM SAMP SITE-PACIFIC BELL (E9)-ENS	DIBROMOCHLOROMETHANE (THM)	11.6	UG/L	5/8/2002	CWS	24,316
CA	CA1510006	East Niles CSD	THM SAMP SITE-PACIFIC BELL (E9)-ENS	DIBROMOCHLOROMETHANE (THM)	23.6	UG/L	6/11/2003	CWS	24,316
CA	CA1510006	East Niles CSD	THM SAMP SITE-PACIFIC BELL (E9)-ENS	DIBROMOCHLOROMETHANE (THM)	3.3	UG/L	12/18/2002	CWS	24,316
CA	CA1510006	East Niles CSD	THM SAMP SITE-PACIFIC BELL (E9)-ENS	DIBROMOCHLOROMETHANE (THM)	7.7	UG/L	3/18/2003	CWS	24,316
CA	CA1510006	East Niles CSD	THM SAMP SITE-PACIFIC BELL (E9)-ENS	DIBROMOCHLOROMETHANE (THM)	7.26	UG/L	8/26/1998	CWS	24,316
CA	CA1510006	East Niles CSD	THM SAMP SITE-PACIFIC BELL (E9)-ENS	DIBROMOCHLOROMETHANE (THM)	6.24	UG/L	2/26/1998	CWS	24,316
CA	CA1510006	East Niles CSD	THM SAMP SITE-PACIFIC BELL (E9)-ENS	DIBROMOCHLOROMETHANE (THM)	2.2	UG/L	9/11/1997	CWS	24,316
CA	CA1510006	East Niles CSD	THM SAMP SITE-PACIFIC BELL (E9)-ENS	DIBROMOCHLOROMETHANE (THM)	4.6	UG/L	12/18/2001	CWS	24,316
CA	CA1510006	East Niles CSD	THM SAMP SITE-PACIFIC BELL (E9)-ENS	DIBROMOCHLOROMETHANE (THM)	2.2	UG/L	9/24/2001	CWS	24,316
CA	CA1510006	East Niles CSD	THM SAMP SITE-PACIFIC BELL (E9)-ENS	DIBROMOCHLOROMETHANE (THM)	0.9	UG/L	11/3/1998	CWS	24,316
CA	CA1510006	East Niles CSD	THM SAMP SITE-PACIFIC BELL (E9)-ENS	DIBROMOCHLOROMETHANE (THM)	7.26	UG/L	3/9/1999	CWS	24,316
CA	CA1510006	East Niles CSD	THM SAMP SITE-8717 BUTTERNUT (E10)-ENS	DIBROMOCHLOROMETHANE (THM)	26.1	UG/L	6/18/2002	CWS	24,316
CA	CA1510006	East Niles CSD	THM SAMP SITE-8717 BUTTERNUT (E10)-ENS	DIBROMOCHLOROMETHANE (THM)	2.1	UG/L	5/8/2002	CWS	24,316
CA	CA1510006	East Niles CSD	THM SAMP SITE-8717 BUTTERNUT (E10)-ENS	DIBROMOCHLOROMETHANE (THM)	4.3	UG/L	12/18/2001	CWS	24,316
CA	CA1510006	East Niles CSD	THM SAMP SITE-8717 BUTTERNUT (E10)-ENS	DIBROMOCHLOROMETHANE (THM)	4.3	UG/L	3/18/2003	CWS	24,316
CA	CA1510006	East Niles CSD	THM SAMP SITE-8717 BUTTERNUT (E10)-ENS	DIBROMOCHLOROMETHANE (THM)	1.9	UG/L	12/18/2002	CWS	24,316
CA	CA1510006	East Niles CSD	THM SAMP SITE-8717 BUTTERNUT (E10)-ENS	DIBROMOCHLOROMETHANE (THM)	1.92	UG/L	6/23/1998	CWS	24,316
CA	CA1510006	East Niles CSD	THM SAMP SITE-8717 BUTTERNUT (E10)-ENS	DIBROMOCHLOROMETHANE (THM)	2.5	UG/L	8/26/1998	CWS	24,316
CA	CA1510006	East Niles CSD	THM SAMP SITE-8717 BUTTERNUT (E10)-ENS	DIBROMOCHLOROMETHANE (THM)	0.58	UG/L	2/26/1998	CWS	24,316
CA	CA1510006	East Niles CSD	THM SAMP SITE-8717 BUTTERNUT (E10)-ENS	DIBROMOCHLOROMETHANE (THM)	1.2	UG/L	9/11/1997	CWS	24,316
CA	CA1510006	East Niles CSD	THM SAMP SITE-8717 BUTTERNUT (E10)-ENS	DIBROMOCHLOROMETHANE (THM)	11.7	UG/L	9/24/2001	CWS	24,316
CA	CA1510006	East Niles CSD	THM SAMP SITE-8717 BUTTERNUT (E10)-ENS	DIBROMOCHLOROMETHANE (THM)	1.73	UG/L	3/9/1999	CWS	24,316
CA	CA1510006	East Niles CSD	THM SAMP SITE-11845 SCHOOL (E11)-ENS	DIBROMOCHLOROMETHANE (THM)	37.2	UG/L	6/18/2002	CWS	24,316
CA	CA1510006	East Niles CSD	THM SAMP SITE-11845 SCHOOL (E11)-ENS	DIBROMOCHLOROMETHANE (THM)	20.3	UG/L	5/8/2002	CWS	24,316
CA	CA1510006	East Niles CSD	THM SAMP SITE-11845 SCHOOL (E11)-ENS	DIBROMOCHLOROMETHANE (THM)	6.3	UG/L	3/18/2003	CWS	24,316
CA	CA1510006	East Niles CSD	THM SAMP SITE-11845 SCHOOL (E11)-ENS	DIBROMOCHLOROMETHANE (THM)	6.2	UG/L	12/18/2002	CWS	24,316

CA	CA1510006	East Niles CSD	THM SAMP SITE-11845 SCHOOL (E11)-ENS	DIBROMOCHLOROMETHANE (THM)	0.93	UG/L	6/23/1998	CWS	24,316
CA	CA1510006	East Niles CSD	THM SAMP SITE-11845 SCHOOL (E11)-ENS	DIBROMOCHLOROMETHANE (THM)	9.47	UG/L	2/26/1998	CWS	24,316
CA	CA1510006	East Niles CSD	THM SAMP SITE-11845 SCHOOL (E11)-ENS	DIBROMOCHLOROMETHANE (THM)	0.5	UG/L	9/11/1997	CWS	24,316
CA	CA1510006	East Niles CSD	THM SAMP SITE-11845 SCHOOL (E11)-ENS	DIBROMOCHLOROMETHANE (THM)	7	UG/L	12/18/2001	CWS	24,316
CA	CA1510006	East Niles CSD	THM SAMP SITE-11845 SCHOOL (E11)-ENS	DIBROMOCHLOROMETHANE (THM)	1.71	UG/L	3/9/1999	CWS	24,316
CA	CA1510006	East Niles CSD	THM SAMP SITE-11845 SCHOOL (E11)-ENS	DIBROMOCHLOROMETHANE (THM)	3.5	UG/L	9/24/2001	CWS	24,316
CA	CA1510006	East Niles CSD	THM SAMP SITE-FAIRFAX SCHOOL (E12)-ENG	DIBROMOCHLOROMETHANE (THM)	1.1	UG/L	6/18/2002	CWS	24,316
CA	CA1510006	East Niles CSD	THM SAMP SITE-FAIRFAX SCHOOL (E12)-ENG	DIBROMOCHLOROMETHANE (THM)	0.7	UG/L	5/8/2002	CWS	24,316
CA	CA1510006	East Niles CSD	THM SAMP SITE-FAIRFAX SCHOOL (E12)-ENG	DIBROMOCHLOROMETHANE (THM)	1.4	UG/L	12/18/2002	CWS	24,316
CA	CA1510006	East Niles CSD	THM SAMP SITE-FAIRFAX SCHOOL (E12)-ENG	DIBROMOCHLOROMETHANE (THM)	2.3	UG/L	3/18/2003	CWS	24,316
CA	CA1510006	East Niles CSD	THM SAMP SITE-FAIRFAX SCHOOL (E12)-ENG	DIBROMOCHLOROMETHANE (THM)	1.5	UG/L	12/16/2003	CWS	24,316
CA	CA1510006	East Niles CSD	THM SAMP SITE-FAIRFAX SCHOOL (E12)-ENG	DIBROMOCHLOROMETHANE (THM)	0.62	UG/L	8/26/1998	CWS	24,316
CA	CA1510006	East Niles CSD	THM SAMP SITE-FAIRFAX SCHOOL (E12)-ENG	DIBROMOCHLOROMETHANE (THM)	1.06	UG/L	2/26/1998	CWS	24,316
CA	CA1510006	East Niles CSD	THM SAMP SITE-FAIRFAX SCHOOL (E12)-ENG	DIBROMOCHLOROMETHANE (THM)	0.6	UG/L	12/18/2001	CWS	24,316
CA	CA1510006	East Niles CSD	THM SAMP SITE-FAIRFAX SCHOOL (E12)-ENG	DIBROMOCHLOROMETHANE (THM)	1.22	UG/L	3/10/1999	CWS	24,316
CA	CA1510006	East Niles CSD	THM SAMP SITE-FAIRFAX SCHOOL (E12)-ENG	DIBROMOCHLOROMETHANE (THM)	0.8	UG/L	11/3/1998	CWS	24,316
CA	CA1510006	East Niles CSD	THM SAMP SITE-1417 VALE (E7)-KCWAS	DIBROMOCHLOROMETHANE (THM)	29.4	UG/L	6/18/2002	CWS	24,316
CA	CA1510006	East Niles CSD	THM SAMP SITE-1417 VALE (E7)-KCWAS	DIBROMOCHLOROMETHANE (THM)	0.8	UG/L	12/16/2003	CWS	24,316
CA	CA1510006	East Niles CSD	THM SAMP SITE-1417 VALE (E7)-KCWAS	DIBROMOCHLOROMETHANE (THM)	1.5	UG/L	3/18/2003	CWS	24,316
CA	CA1510006	East Niles CSD	THM SAMP SITE-1417 VALE (E7)-KCWAS	DIBROMOCHLOROMETHANE (THM)	2.6	UG/L	12/18/2002	CWS	24,316
CA	CA1510006	East Niles CSD	THM SAMP SITE-1417 VALE (E7)-KCWAS	DIBROMOCHLOROMETHANE (THM)	2.34	UG/L	3/9/1999	CWS	24,316
CA	CA1510006	East Niles CSD	THM SAMP SITE-1417 VALE (E7)-KCWAS	DIBROMOCHLOROMETHANE (THM)	16.5	UG/L	12/18/2001	CWS	24,316
CA	CA1510006	East Niles CSD	THM SAMP SITE-1417 VALE (E7)-KCWAS	DIBROMOCHLOROMETHANE (THM)	48	UG/L	9/24/2001	CWS	24,316
CA	CA1510006	East Niles CSD	BRECKENRIDGE TP-RAW-OFFLINE	DIBROMOCHLOROMETHANE (THM)	1.5	UG/L	10/1/2003	CWS	24,316
CA	CA1510006	East Niles CSD	BRECKENRIDGE TP-RAW-OFFLINE	DIBROMOCHLOROMETHANE (THM)	1.7	UG/L	3/23/1998	CWS	24,316
CA	CA1510006	East Niles CSD	BRECKENRIDGE TP-RAW-OFFLINE	DIBROMOCHLOROMETHANE (THM)	1.3	UG/L	3/15/1999	CWS	24,316
CA	CA1510006	East Niles CSD	THM SAMP SITE-5310 WINGFOOT (E6)-KCWAS	DIBROMOCHLOROMETHANE (THM)	29.9	UG/L	6/18/2002	CWS	24,316
CA	CA1510006	East Niles CSD	THM SAMP SITE-5310 WINGFOOT (E6)-KCWAS	DIBROMOCHLOROMETHANE (THM)	1	UG/L	12/16/2003	CWS	24,316
CA	CA1510006	East Niles CSD	THM SAMP SITE-5310 WINGFOOT (E6)-KCWAS	DIBROMOCHLOROMETHANE (THM)	10.7	UG/L	12/18/2002	CWS	24,316
CA	CA1510006	East Niles CSD	THM SAMP SITE-5310 WINGFOOT (E6)-KCWAS	DIBROMOCHLOROMETHANE (THM)	16.5	UG/L	12/18/2001	CWS	24.316
CA	CA1510006	East Niles CSD	THM SAMP SITE-5310 WINGFOOT (E6)-KCWAS	DIBROMOCHLOROMETHANE (THM)	0.68	UG/L	3/10/1999	CWS	24,316
CA	CA1510006	East Niles CSD	THM SAMP SITE-5310 WINGFOOT (E6)-KCWAS	DIBROMOCHLOROMETHANE (THM)	48	UG/L	9/24/2001	CWS	24,316
CA	CA1510006	East Niles CSD	THM SAMP SITE-3209 FLINTRIDGE (E4)-KCWAS	DIBROMOCHLOROMETHANE (THM)	29.7	UG/L	6/18/2002	CWS	24,316
CA	CA1510006	East Niles CSD	THM SAMP SITE-3209 FLINTRIDGE (E4)-KCWAS	DIBROMOCHLOROMETHANE (THM)	0.6	UG/L	5/8/2002	CWS	24,316
CA	CA1510006	East Niles CSD	THM SAMP SITE-3209 FLINTRIDGE (E4)-KCWAS	DIBROMOCHLOROMETHANE (THM)	0.5	UG/L	3/18/2003	CWS	24,316
CA	CA1510006	East Niles CSD	THM SAMP SITE-3209 FLINTRIDGE (E4)-KCWAS	DIBROMOCHLOROMETHANE (THM)	11	UG/L	12/18/2002	CWS	24,316
CA	CA1510006	East Niles CSD	THM SAMP SITE-3209 FLINTRIDGE (E4)-KCWAS	DIBROMOCHLOROMETHANE (THM)	0.81	UG/L	3/10/1999	CWS	24,316
CA	CA1510006	East Niles CSD	THM SAMP SITE-3209 FLINTRIDGE (E4)-KCWAS	DIBROMOCHLOROMETHANE (THM)	1.1	UG/L	12/16/2003	CWS	24,316
CA	CA1510006	East Niles CSD	THM SAMP SITE-3209 FLINTRIDGE (E4)-KCWAS	DIBROMOCHLOROMETHANE (THM)	18.3	UG/L	12/18/2001	CWS	24,316
CA	CA1510006	East Niles CSD	THM SAMP SITE-3209 FLINTRIDGE (E4)-KCWAS	DIBROMOCHLOROMETHANE (THM)	46	UG/L	9/24/2001	CWS	24,316
CA	CA1510006	East Niles CSD	BRECKENRIDGE TP-TREATED	DIBROMOCHLOROMETHANE (THM)	6.4	UG/L	3/28/2002	CWS	24,316
CA	CA1510040	Kern County Water Agency	ID4 WEN. & COLUMBAS (T7)-DIST. SYS.	DIBROMOCHLOROMETHANE (THM)	0.6	UG/L	3/8/1999	CWS	0
CA	CA1510040	Kern County Water Agency	ID4 WEN. & COLUMBAS (T7)-DIST. SYS.	DIBROMOCHLOROMETHANE (THM)	26.86	UG/L	1/5/1998	CWS	0
CA	CA1510040	Kern County Water Agency	ID4 CORNER TANK-T6-TREATED	DIBROMOCHLOROMETHANE (THM)	1.3	UG/L	3/8/1999	CWS	0
CA	CA1510040	Kern County Water Agency	ID4 CORNER TANK-T6-TREATED	DIBROMOCHLOROMETHANE (THM)	0.5	UG/L	2/24/1997	CWS	0

APPENDIX 7-3 *Continued*

State	PWS_ID	Utility	POE	Constituent	LEVEL_NO	Unit	COLL_DATE	PWS	Pop
CA	CA1510040	Kern County Water Agency	ID4 CORNER TANK-T6-TREATED	DIBROMOCHLOROMETHANE (THM)	36.45	UG/L	1/5/1998	CWS	0
CA	CA1510040	Kern County Water Agency	ID4 BELLE-SEQUOIA TANK-T2-TREATED	DIBROMOCHLOROMETHANE (THM)	0.5	UG/L	3/8/1999	CWS	0
CA	CA1510040	Kern County Water Agency	ID4 BELLE-SEQUOIA TANK-T2-TREATED	DIBROMOCHLOROMETHANE (THM)	22.89	UG/L	1/5/1998	CWS	0
CA	CA1510040	Kern County Water Agency	ID4 BELLE-SEQUOIA TANK-T2-TREATED	DIBROMOCHLOROMETHANE (THM)	0.8	UG/L	9/8/1997	CWS	0
CA	CA1510040	Kern County Water Agency	ID4 OSWELL TANK-T5-TREATED	DIBROMOCHLOROMETHANE (THM)	0.7	UG/L	3/8/1999	CWS	0
CA	CA1510040	Kern County Water Agency	ID4 OSWELL TANK-T5-TREATED	DIBROMOCHLOROMETHANE (THM)	39.38	UG/L	1/5/1998	CWS	0
CA	CA1510040	Kern County Water Agency	ID4 OSWELL TANK-T5-TREATED	DIBROMOCHLOROMETHANE (THM)	0.8	UG/L	2/24/1997	CWS	0
CA	CA1510040	Kern County Water Agency	ID4 OSWELL TANK-T5-TREATED	DIBROMOCHLOROMETHANE (THM)	0.8	UG/L	9/8/1997	CWS	0
CA	CA1510040	Kern County Water Agency	ID4 NOR TANK (600 SITE) T3-TREATED	DIBROMOCHLOROMETHANE (THM)	43.96	UG/L	1/5/1998	CWS	0
CA	CA1510040	Kern County Water Agency	ID4 TREATED-T1	DIBROMOCHLOROMETHANE (THM)	22.48	UG/L	1/5/1998	CWS	0
CA	CA1510040	Kern County Water Agency	ID4 TREATED-T1	DIBROMOCHLOROMETHANE (THM)	0.7	UG/L	9/8/1997	CWS	0
CA	CA1510040	Kern County Water Agency	ID4 TREATED-T1	DIBROMOCHLOROMETHANE (THM)	1.7	UG/L	8/30/2002	CWS	0
CA	CA1510040	Kern County Water Agency	ID4 TREATED-T1	DIBROMOCHLOROMETHANE (THM)	0.6	UG/L	8/30/1999	CWS	0
CA	CA1510040	Kern County Water Agency	ID4 TREATED-T1	DIBROMOCHLOROMETHANE (THM)	0.6	UG/L	3/8/1999	CWS	0
CA	CA1510040	Kern County Water Agency	ID4 TREATED-T1	DIBROMOCHLOROMETHANE (THM)	0.64	UG/L	11/15/2000	CWS	0
CA	CA1510041	North of the River MWD	THM SAMP SITE-2841 PEGASUS (N1)-TRTD	DIBROMOCHLOROMETHANE (THM)	8.3	UG/L	8/22/2001	CWS	7,000
CA	CA1510041	North of the River MWD	THM SAMP SITE-2841 PEGASUS (N1)-TRTD	DIBROMOCHLOROMETHANE (THM)	1.1	UG/L	5/17/2002	CWS	7,000
CA	CA1510041	North of the River MWD	THM SAMP SITE-2841 PEGASUS (N1)-TRTD	DIBROMOCHLOROMETHANE (THM)	1.1	UG/L	8/27/2002	CWS	7,000
CA	CA1510041	North of the River MWD	THM SAMP SITE-2841 PEGASUS (N1)-TRTD	DIBROMOCHLOROMETHANE (THM)	1	UG/L	5/8/2001	CWS	7,000
CA	CA1510041	North of the River MWD	THM SAMP SITE-2841 PEGASUS (N1)-TRTD	DIBROMOCHLOROMETHANE (THM)	10	UG/L	11/26/2001	CWS	7,000
CA	CA1510041	North of the River MWD	THM SAMP SITE-2841 PEGASUS (N1)-TRTD	DIBROMOCHLOROMETHANE (THM)	1.31	UG/L	11/7/2000	CWS	7,000
CA	CA1510041	North of the River MWD	THM SAMP SITE-2841 PEGASUS (N1)-TRTD	DIBROMOCHLOROMETHANE (THM)	0.91	UG/L	3/10/1999	CWS	7,000
CA	CA1510041	North of the River MWD	THM SAMP SITE-2841 PEGASUS (N1)-TRTD	DIBROMOCHLOROMETHANE (THM)	3.5	UG/L	6/7/1999	CWS	7,000
CA	CA1510041	North of the River MWD	THM SAMP SITE-900 TANKSITE (N4)-TRTD	DIBROMOCHLOROMETHANE (THM)	16.6	UG/L	8/22/2001	CWS	7,000
CA	CA1510041	North of the River MWD	THM SAMP SITE-900 TANKSITE (N4)-TRTD	DIBROMOCHLOROMETHANE (THM)	1.1	UG/L	5/8/2001	CWS	7,000
CA	CA1510041	North of the River MWD	THM SAMP SITE-900 TANKSITE (N4)-TRTD	DIBROMOCHLOROMETHANE (THM)	18.7	UG/L	11/26/2001	CWS	7,000
CA	CA1510041	North of the River MWD	THM SAMP SITE-900 TANKSITE (N4)-TRTD	DIBROMOCHLOROMETHANE (THM)	2.3	UG/L	8/27/2002	CWS	7,000
CA	CA1510041	North of the River MWD	THM SAMP SITE-900 TANKSITE (N4)-TRTD	DIBROMOCHLOROMETHANE (THM)	1.32	UG/L	11/7/2000	CWS	7,000
CA	CA1510041	North of the River MWD	THM SAMP SITE-900 TANKSITE (N4)-TRTD	DIBROMOCHLOROMETHANE (THM)	0.9	UG/L	5/17/2002	CWS	7,000
CA	CA1510041	North of the River MWD	THM SAMP SITE-900 TANKSITE (N4)-TRTD	DIBROMOCHLOROMETHANE (THM)	2.5	UG/L	2/25/1998	CWS	7,000
CA	CA1510041	North of the River MWD	THM SAMP SITE-900 TANKSITE (N4)-TRTD	DIBROMOCHLOROMETHANE (THM)	0.5	UG/L	9/23/1997	CWS	7,000
CA	CA1510041	North of the River MWD	THM SAMP SITE-900 TANKSITE (N4)-TRTD	DIBROMOCHLOROMETHANE (THM)	0.71	UG/L	6/23/1998	CWS	7,000
CA	CA1510041	North of the River MWD	THM SAMP SITE-900 TANKSITE (N4)-TRTD	DIBROMOCHLOROMETHANE (THM)	1.9	UG/L	6/7/1999	CWS	7,000
CA	CA1510041	North of the River MWD	THM SAMP SITE-900 TANKSITE (N4)-TRTD	DIBROMOCHLOROMETHANE (THM)	0.95	UG/L	3/10/1999	CWS	7,000
CA	CA1510041	North of the River MWD	THM SAMP SITE-3418 N CHESTER (N3)-TRTD	DIBROMOCHLOROMETHANE (THM)	9.6	UG/L	8/22/2001	CWS	7,000
CA	CA1510041	North of the River MWD	THM SAMP SITE-3418 N CHESTER (N3)-TRTD	DIBROMOCHLOROMETHANE (THM)	0.8	UG/L	5/8/2001	CWS	7,000
CA	CA1510041	North of the River MWD	THM SAMP SITE-3418 N CHESTER (N3)-TRTD	DIBROMOCHLOROMETHANE (THM)	8.7	UG/L	11/26/2001	CWS	7,000
CA	CA1510041	North of the River MWD	THM SAMP SITE-3418 N CHESTER (N3)-TRTD	DIBROMOCHLOROMETHANE (THM)	1.1	UG/L	5/17/2002	CWS	7,000
CA	CA1510041	North of the River MWD	THM SAMP SITE-3418 N CHESTER (N3)-TRTD	DIBROMOCHLOROMETHANE (THM)	1.1	UG/L	8/27/2002	CWS	7,000
CA	CA1510041	North of the River MWD	THM SAMP SITE-3418 N CHESTER (N3)-TRTD	DIBROMOCHLOROMETHANE (THM)	1.13	UG/L	11/7/2000	CWS	7,000
CA	CA1510041	North of the River MWD	THM SAMP SITE-3418 N CHESTER (N3)-TRTD	DIBROMOCHLOROMETHANE (THM)	3.3	UG/L	6/7/1999	CWS	7,000
CA	CA1510041	North of the River MWD	THM SAMP SITE-3418 N CHESTER (N3)-TRTD	DIBROMOCHLOROMETHANE (THM)	0.78	UG/L	3/10/1999	CWS	7,000

CA	CA1510041	North of the River MWD	THM SAMP SITE-528 CHARLANA (N2)-TRTD	DIBROMOCHLOROMETHANE (THM)	9.4	UG/L	8/22/2001	CWS	7,000
CA	CA1510041	North of the River MWD	THM SAMP SITE-528 CHARLANA (N2)-TRTD	DIBROMOCHLOROMETHANE (THM)	0.9	UG/L	5/8/2001	CWS	7,000
CA	CA1510041	North of the River MWD	THM SAMP SITE-528 CHARLANA (N2)-TRTD	DIBROMOCHLOROMETHANE (THM)	10.5	UG/L	11/26/2001	CWS	7,000
CA	CA1510041	North of the River MWD	THM SAMP SITE-528 CHARLANA (N2)-TRTD	DIBROMOCHLOROMETHANE (THM)	1.1	UG/L	5/17/2002	CWS	7,000
CA	CA1510041	North of the River MWD	THM SAMP SITE-528 CHARLANA (N2)-TRTD	DIBROMOCHLOROMETHANE (THM)	1.1	UG/L	8/27/2002	CWS	7,000
CA	CA1510041	North of the River MWD	THM SAMP SITE-528 CHARLANA (N2)-TRTD	DIBROMOCHLOROMETHANE (THM)	0.8	UG/L	3/10/1999	CWS	7,000
CA	CA1510041	North of the River MWD	THM SAMP SITE-528 CHARLANA (N2)-TRTD	DIBROMOCHLOROMETHANE (THM)	1.21	UG/L	11/7/2000	CWS	7,000
CA	CA1510041	North of the River MWD	THM SAMP SITE-528 CHARLANA (N2)-TRTD	DIBROMOCHLOROMETHANE (THM)	1.6	UG/L	6/7/1999	CWS	7,000
					7.35				
CA	CA1510006	East Niles CSD	BRECKENRIDGE TP-TREATED	DIBROMOCHLOROPROPANE (DBCP)	0.02	UG/L	3/28/2002	CWS	24,316
CA	CA1510006	East Niles CSD	THM SAMP SITE-DEACON (E8)-ENG	DICHLOROACETIC ACID (DCAA)	1	UG/L	6/18/2002	CWS	24,316
CA	CA1510006	East Niles CSD	THM SAMP SITE-PACIFIC BELL (E9)-ENS	DICHLOROACETIC ACID (DCAA)	19	UG/L	6/18/2002	CWS	24,316
CA	CA1510006	East Niles CSD	THM SAMP SITE-PACIFIC BELL (E9)-ENS	DICHLOROACETIC ACID (DCAA)	4	UG/L	3/18/2003	CWS	24,316
CA	CA1510006	East Niles CSD	THM SAMP SITE-8717 BUTTERNUT (E10)-ENS	DICHLOROACETIC ACID (DCAA)	7	UG/L	6/18/2002	CWS	24,316
CA	CA1510006	East Niles CSD	THM SAMP SITE-8717 BUTTERNUT (E10)-ENS	DICHLOROACETIC ACID (DCAA)	8	UG/L	3/18/2003	CWS	24,316
CA	CA1510006	East Niles CSD	THM SAMP SITE-8717 BUTTERNUT (E10)-ENS	DICHLOROACETIC ACID (DCAA)	1	UG/L	9/24/2001	CWS	24,316
CA	CA1510006	East Niles CSD	THM SAMP SITE-11845 SCHOOL (E11)-ENS	DICHLOROACETIC ACID (DCAA)	18	UG/L	6/18/2002	CWS	24,316
CA	CA1510006	East Niles CSD	THM SAMP SITE-11845 SCHOOL (E11)-ENS	DICHLOROACETIC ACID (DCAA)	2	UG/L	3/18/2003	CWS	24,316
CA	CA1510006	East Niles CSD	THM SAMP SITE-11845 SCHOOL (E11)-ENS	DICHLOROACETIC ACID (DCAA)	1	UG/L	12/18/2002	CWS	24,316
CA	CA1510006	East Niles CSD	THM SAMP SITE-FAIRFAX SCHOOL (E12)-ENG	DICHLOROACETIC ACID (DCAA)	2	UG/L	6/18/2002	CWS	24,316
CA	CA1510006	East Niles CSD	THM SAMP SITE-FAIRFAX SCHOOL (E12)-ENG	DICHLOROACETIC ACID (DCAA)	3	UG/L	12/18/2002	CWS	24,316
CA	CA1510006	East Niles CSD	THM SAMP SITE-1417 VALE (E7)-KCWAS	DICHLOROACETIC ACID (DCAA)	18	UG/L	6/18/2002	CWS	24,316
CA	CA1510006	East Niles CSD	THM SAMP SITE-1417 VALE (E7)-KCWAS	DICHLOROACETIC ACID (DCAA)	9	UG/L	12/16/2003	CWS	24,316
CA	CA1510006	East Niles CSD	THM SAMP SITE-1417 VALE (E7)-KCWAS	DICHLOROACETIC ACID (DCAA)	12	UG/L	9/24/2001	CWS	24,316
CA	CA1510006	East Niles CSD	THM SAMP SITE-1417 VALE (E7)-KCWAS	DICHLOROACETIC ACID (DCAA)	3	UG/L	12/18/2001	CWS	24,316
CA	CA1510006	East Niles CSD	THM SAMP SITE-5310 WINGFOOT (E6)-KCWAS	DICHLOROACETIC ACID (DCAA)	17	UG/L	6/18/2002	CWS	24,316
CA	CA1510006	East Niles CSD	THM SAMP SITE-5310 WINGFOOT (E6)-KCWAS	DICHLOROACETIC ACID (DCAA)	4	UG/L	12/18/2001	CWS	24,316
CA	CA1510006	East Niles CSD	THM SAMP SITE-5310 WINGFOOT (E6)-KCWAS	DICHLOROACETIC ACID (DCAA)	13	UG/L	12/16/2003	CWS	24,316
CA	CA1510006	East Niles CSD	THM SAMP SITE-5310 WINGFOOT (E6)-KCWAS	DICHLOROACETIC ACID (DCAA)	25	UG/L	3/18/2003	CWS	24,316
CA	CA1510006	East Niles CSD	THM SAMP SITE-5310 WINGFOOT (E6)-KCWAS	DICHLOROACETIC ACID (DCAA)	3	UG/L	12/18/2002	CWS	24,316
CA	CA1510006	East Niles CSD	THM SAMP SITE-5310 WINGFOOT (E6)-KCWAS	DICHLOROACETIC ACID (DCAA)	10	UG/L	9/24/2001	CWS	24,316
CA	CA1510006	East Niles CSD	THM SAMP SITE-3209 FLINTRIDGE (E4)-KCWAS	DICHLOROACETIC ACID (DCAA)	19	UG/L	6/18/2002	CWS	24,316
CA	CA1510006	East Niles CSD	THM SAMP SITE-3209 FLINTRIDGE (E4)-KCWAS	DICHLOROACETIC ACID (DCAA)	22	UG/L	12/16/2003	CWS	24,316
CA	CA1510006	East Niles CSD	THM SAMP SITE-3209 FLINTRIDGE (E4)-KCWAS	DICHLOROACETIC ACID (DCAA)	44	UG/L	3/18/2003	CWS	24,316
CA	CA1510006	East Niles CSD	THM SAMP SITE-3209 FLINTRIDGE (E4)-KCWAS	DICHLOROACETIC ACID (DCAA)	4	UG/L	12/18/2001	CWS	24,316
CA	CA1510006	East Niles CSD	THM SAMP SITE-3209 FLINTRIDGE (E4)-KCWAS	DICHLOROACETIC ACID (DCAA)	9	UG/L	9/24/2001	CWS	24,316
CA	CA1600034	PARAMOUNT FARMING CO	TREATMENT PLANT	DICHLOROACETIC ACID (DCAA)	6.4	UG/L	11/4/2004	CWS	100
CA	CA1600034	PARAMOUNT FARMING CO	TREATMENT PLANT	DICHLOROACETIC ACID (DCAA)	9.7	UG/L	7/14/2004	CWS	100
					10.5				
CA	CA1503212	DWR-EDMONSTON	CA AQUEDUCT-RAW (EDMONSTON)	DIURON	0.29	UG/L	3/14/2000	NTNCWS	35
CA	CA1510040	Kern County Water Agency	ID4 RAW FK	DIURON	7.7	UG/L	3/31/2003	CWS	0
CA	CA1510040	Kern County Water Agency	ID4 RAW FK	DIURON	1	UG/L	2/20/2002	CWS	0
					3				
CA	CA1600034	PARAMOUNT FARMING CO	TREATMENT PLANT	HALOACETIC ACIDS (5) (HAA5)	27	UG/L	11/4/2004	CWS	100
CA	CA1600034	PARAMOUNT FARMING CO	TREATMENT PLANT	HALOACETIC ACIDS (5) (HAA5)	25	UG/L	7/14/2004	CWS	100

APPENDIX 7-3 *Continued*

State	PWS_ID	Utility	POE	Constituent	LEVEL_NO	Unit	COLL_DATE	PWS	Pop
CA	CA1510006	East Niles CSD	THM SAMP SITE-PACIFIC BELL (E9)-ENS	MONOBROMOACETIC ACID (MBAA)	3	UG/L	3/18/2003	CWS	24,316
CA	CA1510006	East Niles CSD	THM SAMP SITE-PACIFIC BELL (E9)-ENS	MONOBROMOACETIC ACID (MBAA)	3	UG/L	12/18/2002	CWS	24,316
CA	CA1510006	East Niles CSD	THM SAMP SITE-8717 BUTTERNUT (E10)-ENS	MONOBROMOACETIC ACID (MBAA)	1	UG/L	3/18/2003	CWS	24,316
CA	CA1510006	East Niles CSD	THM SAMP SITE-11845 SCHOOL (E11)-ENS	MONOBROMOACETIC ACID (MBAA)	4	UG/L	12/18/2002	CWS	24,316
CA	CA1510006	East Niles CSD	THM SAMP SITE-11845 SCHOOL (E11)-ENS	MONOBROMOACETIC ACID (MBAA)	3	UG/L	3/18/2003	CWS	24,316
CA	CA1510006	East Niles CSD	THM SAMP SITE-FAIRFAX SCHOOL (E12)-ENG	MONOBROMOACETIC ACID (MBAA)	2	UG/L	12/18/2002	CWS	24,316
CA	CA1510006	East Niles CSD	THM SAMP SITE-1417 VALE (E7)-KCWAS	MONOBROMOACETIC ACID (MBAA)	2	UG/L	9/24/2001	CWS	24,316
CA	CA1510006	East Niles CSD	THM SAMP SITE-5310 WINGFOOT (E6)-KCWAS	MONOBROMOACETIC ACID (MBAA)	2	UG/L	12/18/2002	CWS	24,316
CA	CA1510006	East Niles CSD	THM SAMP SITE-5310 WINGFOOT (E6)-KCWAS	MONOBROMOACETIC ACID (MBAA)	2	UG/L	9/24/2001	CWS	24,316
CA	CA1510006	East Niles CSD	THM SAMP SITE-3209 FLINTRIDGE (E4)-KCWAS	MONOBROMOACETIC ACID (MBAA)	2	UG/L	9/24/2001	CWS	24,316
					2.4				
CA	CA1510006	East Niles CSD	THM SAMP SITE-3209 FLINTRIDGE (E4)-KCWAS	MONOCHLOROACETIC ACID (MCAA)	5	UG/L	12/18/2002	CWS	24,316
CA	CA1510006	East Niles CSD	THM SAMP SITE-DEACON (E8)-ENG	TOTAL TRIHALOMETHANES	2.5	UG/L	6/18/2002	CWS	24,316
CA	CA1510006	East Niles CSD	THM SAMP SITE-DEACON (E8)-ENG	TOTAL TRIHALOMETHANES	2.2	UG/L	12/18/2002	CWS	24,316
CA	CA1510006	East Niles CSD	THM SAMP SITE-DEACON (E8)-ENG	TOTAL TRIHALOMETHANES	1.8	UG/L	3/18/2003	CWS	24,316
CA	CA1510006	East Niles CSD	THM SAMP SITE-DEACON (E8)-ENG	TOTAL TRIHALOMETHANES	1.2	UG/L	5/8/2002	CWS	24,316
CA	CA1510006	East Niles CSD	THM SAMP SITE-DEACON (E8)-ENG	TOTAL TRIHALOMETHANES	39.9	UG/L	12/16/2003	CWS	24,316
CA	CA1510006	East Niles CSD	THM SAMP SITE-DEACON (E8)-ENG	TOTAL TRIHALOMETHANES	42.78	UG/L	2/26/1998	CWS	24,316
CA	CA1510006	East Niles CSD	THM SAMP SITE-DEACON (E8)-ENG	TOTAL TRIHALOMETHANES	0.9	UG/L	9/11/1997	CWS	24,316
CA	CA1510006	East Niles CSD	THM SAMP SITE-DEACON (E8)-ENG	TOTAL TRIHALOMETHANES	2.4	UG/L	11/3/1998	CWS	24,316
CA	CA1510006	East Niles CSD	THM SAMP SITE-DEACON (E8)-ENG	TOTAL TRIHALOMETHANES	1.2	UG/L	9/24/2001	CWS	24,316
CA	CA1510006	East Niles CSD	THM SAMP SITE-DEACON (E8)-ENG	TOTAL TRIHALOMETHANES	3	UG/L	12/18/2001	CWS	24,316
CA	CA1510006	East Niles CSD	THM SAMP SITE-PACIFIC BELL (E9)-ENS	TOTAL TRIHALOMETHANES	111	UG/L	6/11/2003	CWS	24,316
CA	CA1510006	East Niles CSD	THM SAMP SITE-PACIFIC BELL (E9)-ENS	TOTAL TRIHALOMETHANES	122	UG/L	6/18/2002	CWS	24,316
CA	CA1510006	East Niles CSD	THM SAMP SITE-PACIFIC BELL (E9)-ENS	TOTAL TRIHALOMETHANES	60.2	UG/L	5/8/2002	CWS	24,316
CA	CA1510006	East Niles CSD	THM SAMP SITE-PACIFIC BELL (E9)-ENS	TOTAL TRIHALOMETHANES	36.5	UG/L	3/18/2003	CWS	24,316
CA	CA1510006	East Niles CSD	THM SAMP SITE-PACIFIC BELL (E9)-ENS	TOTAL TRIHALOMETHANES	13.7	UG/L	12/18/2002	CWS	24,316
CA	CA1510006	East Niles CSD	THM SAMP SITE-PACIFIC BELL (E9)-ENS	TOTAL TRIHALOMETHANES	23.07	UG/L	6/23/1998	CWS	24,316
CA	CA1510006	East Niles CSD	THM SAMP SITE-PACIFIC BELL (E9)-ENS	TOTAL TRIHALOMETHANES	50.73	UG/L	2/26/1998	CWS	24,316
CA	CA1510006	East Niles CSD	THM SAMP SITE-PACIFIC BELL (E9)-ENS	TOTAL TRIHALOMETHANES	33.9	UG/L	9/11/1997	CWS	24,316
CA	CA1510006	East Niles CSD	THM SAMP SITE-PACIFIC BELL (E9)-ENS	TOTAL TRIHALOMETHANES	46.82	UG/L	8/26/1998	CWS	24,316
CA	CA1510006	East Niles CSD	THM SAMP SITE-PACIFIC BELL (E9)-ENS	TOTAL TRIHALOMETHANES	15.9	UG/L	12/18/2001	CWS	24,316
CA	CA1510006	East Niles CSD	THM SAMP SITE-PACIFIC BELL (E9)-ENS	TOTAL TRIHALOMETHANES	8.9	UG/L	9/24/2001	CWS	24,316
CA	CA1510006	East Niles CSD	THM SAMP SITE-PACIFIC BELL (E9)-ENS	TOTAL TRIHALOMETHANES	46.82	UG/L	3/9/1999	CWS	24,316
CA	CA1510006	East Niles CSD	THM SAMP SITE-PACIFIC BELL (E9)-ENS	TOTAL TRIHALOMETHANES	76.1	UG/L	11/3/1998	CWS	24,316
CA	CA1510006	East Niles CSD	THM SAMP SITE-8717 BUTTERNUT (E10)-ENS	TOTAL TRIHALOMETHANES	30.5	UG/L	5/8/2002	CWS	24,316
CA	CA1510006	East Niles CSD	THM SAMP SITE-8717 BUTTERNUT (E10)-ENS	TOTAL TRIHALOMETHANES	99.2	UG/L	6/18/2002	CWS	24,316
CA	CA1510006	East Niles CSD	THM SAMP SITE-8717 BUTTERNUT (E10)-ENS	TOTAL TRIHALOMETHANES	7.6	UG/L	12/18/2002	CWS	24,316
CA	CA1510006	East Niles CSD	THM SAMP SITE-8717 BUTTERNUT (E10)-ENS	TOTAL TRIHALOMETHANES	33.3	UG/L	3/18/2003	CWS	24,316
CA	CA1510006	East Niles CSD	THM SAMP SITE-8717 BUTTERNUT (E10)-ENS	TOTAL TRIHALOMETHANES	12	UG/L	12/18/2001	CWS	24,316
CA	CA1510006	East Niles CSD	THM SAMP SITE-8717 BUTTERNUT (E10)-ENS	TOTAL TRIHALOMETHANES	46.95	UG/L	6/23/1998	CWS	24,316
CA	CA1510006	East Niles CSD	THM SAMP SITE-8717 BUTTERNUT (E10)-ENS	TOTAL TRIHALOMETHANES	43.67	UG/L	2/26/1998	CWS	24,316

CA	CA1510006	East Niles CSD	THM SAMP SITE-8717 BUTTERNUT (E10)-ENS	TOTAL TRIHALOMETHANES	50.1	UG/L	9/11/1997	CWS	24,316
CA	CA1510006	East Niles CSD	THM SAMP SITE-8717 BUTTERNUT (E10)-ENS	TOTAL TRIHALOMETHANES	27.8	UG/L	9/24/2001	CWS	24,316
CA	CA1510006	East Niles CSD	THM SAMP SITE-8717 BUTTERNUT (E10)-ENS	TOTAL TRIHALOMETHANES	44.4	UG/L	11/3/1998	CWS	24,316
CA	CA1510006	East Niles CSD	THM SAMP SITE-8717 BUTTERNUT (E10)-ENS	TOTAL TRIHALOMETHANES	31.57	UG/L	3/9/1999	CWS	24,316
CA	CA1510006	East Niles CSD	THM SAMP SITE-8717 BUTTERNUT (E10)-ENS	TOTAL TRIHALOMETHANES	25.55	UG/L	8/26/1998	CWS	24,316
CA	CA1510006	East Niles CSD	THM SAMP SITE-11845 SCHOOL (E11)-ENS	TOTAL TRIHALOMETHANES	176	UG/L	6/18/2002	CWS	24,316
CA	CA1510006	East Niles CSD	THM SAMP SITE-11845 SCHOOL (E11)-ENS	TOTAL TRIHALOMETHANES	76.1	UG/L	5/8/2002	CWS	24,316
CA	CA1510006	East Niles CSD	THM SAMP SITE-11845 SCHOOL (E11)-ENS	TOTAL TRIHALOMETHANES	29.8	UG/L	3/18/2003	CWS	24,316
CA	CA1510006	East Niles CSD	THM SAMP SITE-11845 SCHOOL (E11)-ENS	TOTAL TRIHALOMETHANES	21.9	UG/L	12/18/2002	CWS	24,316
CA	CA1510006	East Niles CSD	THM SAMP SITE-11845 SCHOOL (E11)-ENS	TOTAL TRIHALOMETHANES	73.04	UG/L	8/26/1998	CWS	24,316
CA	CA1510006	East Niles CSD	THM SAMP SITE-11845 SCHOOL (E11)-ENS	TOTAL TRIHALOMETHANES	83.82	UG/L	6/23/1998	CWS	24,316
CA	CA1510006	East Niles CSD	THM SAMP SITE-11845 SCHOOL (E11)-ENS	TOTAL TRIHALOMETHANES	62.68	UG/L	2/26/1998	CWS	24,316
CA	CA1510006	East Niles CSD	THM SAMP SITE-11845 SCHOOL (E11)-ENS	TOTAL TRIHALOMETHANES	21	UG/L	12/18/2001	CWS	24,316
CA	CA1510006	East Niles CSD	THM SAMP SITE-11845 SCHOOL (E11)-ENS	TOTAL TRIHALOMETHANES	63.6	UG/L	9/11/1997	CWS	24,316
CA	CA1510006	East Niles CSD	THM SAMP SITE-11845 SCHOOL (E11)-ENS	TOTAL TRIHALOMETHANES	13.3	UG/L	9/24/2001	CWS	24,316
CA	CA1510006	East Niles CSD	THM SAMP SITE-11845 SCHOOL (E11)-ENS	TOTAL TRIHALOMETHANES	51.01	UG/L	3/9/1999	CWS	24,316
CA	CA1510006	East Niles CSD	THM SAMP SITE-11845 SCHOOL (E11)-ENS	TOTAL TRIHALOMETHANES	74.1	UG/L	11/3/1998	CWS	24,316
CA	CA1510006	East Niles CSD	THM SAMP SITE-FAIRFAX SCHOOL (E12)-ENG	TOTAL TRIHALOMETHANES	7.3	UG/L	6/18/2002	CWS	24,316
CA	CA1510006	East Niles CSD	THM SAMP SITE-FAIRFAX SCHOOL (E12)-ENG	TOTAL TRIHALOMETHANES	4	UG/L	5/8/2002	CWS	24,316
CA	CA1510006	East Niles CSD	THM SAMP SITE-FAIRFAX SCHOOL (E12)-ENG	TOTAL TRIHALOMETHANES	6.3	UG/L	12/18/2002	CWS	24,316
CA	CA1510006	East Niles CSD	THM SAMP SITE-FAIRFAX SCHOOL (E12)-ENG	TOTAL TRIHALOMETHANES	5.3	UG/L	12/16/2003	CWS	24,316
CA	CA1510006	East Niles CSD	THM SAMP SITE-FAIRFAX SCHOOL (E12)-ENG	TOTAL TRIHALOMETHANES	9.2	UG/L	3/18/2003	CWS	24,316
CA	CA1510006	East Niles CSD	THM SAMP SITE-FAIRFAX SCHOOL (E12)-ENG	TOTAL TRIHALOMETHANES	7.15	UG/L	2/26/1998	CWS	24,316
CA	CA1510006	East Niles CSD	THM SAMP SITE-FAIRFAX SCHOOL (E12)-ENG	TOTAL TRIHALOMETHANES	0.9	UG/L	9/11/1997	CWS	24,316
CA	CA1510006	East Niles CSD	THM SAMP SITE-FAIRFAX SCHOOL (E12)-ENG	TOTAL TRIHALOMETHANES	1	UG/L	6/23/1998	CWS	24,316
CA	CA1510006	East Niles CSD	THM SAMP SITE-FAIRFAX SCHOOL (E12)-ENG	TOTAL TRIHALOMETHANES	2.02	UG/L	8/26/1998	CWS	24,316
CA	CA1510006	East Niles CSD	THM SAMP SITE-FAIRFAX SCHOOL (E12)-ENG	TOTAL TRIHALOMETHANES	1.8	UG/L	9/24/2001	CWS	24,316
CA	CA1510006	East Niles CSD	THM SAMP SITE-FAIRFAX SCHOOL (E12)-ENG	TOTAL TRIHALOMETHANES	3.74	UG/L	3/10/1999	CWS	24,316
CA	CA1510006	East Niles CSD	THM SAMP SITE-FAIRFAX SCHOOL (E12)-ENG	TOTAL TRIHALOMETHANES	2.8	UG/L	12/18/2001	CWS	24,316
CA	CA1510006	East Niles CSD	THM SAMP SITE-FAIRFAX SCHOOL (E12)-ENG	TOTAL TRIHALOMETHANES	4	UG/L	11/3/1998	CWS	24,316
CA	CA1510006	East Niles CSD	THM SAMP SITE-1417 VALE (E7)-KCWAS	TOTAL TRIHALOMETHANES	97.1	UG/L	6/18/2002	CWS	24,316
CA	CA1510006	East Niles CSD	THM SAMP SITE-1417 VALE (E7)-KCWAS	TOTAL TRIHALOMETHANES	59.5	UG/L	5/8/2002	CWS	24,316
CA	CA1510006	East Niles CSD	THM SAMP SITE-1417 VALE (E7)-KCWAS	TOTAL TRIHALOMETHANES	8.5	UG/L	3/18/2003	CWS	24,316
CA	CA1510006	East Niles CSD	THM SAMP SITE-1417 VALE (E7)-KCWAS	TOTAL TRIHALOMETHANES	2.9	UG/L	12/16/2003	CWS	24,316
CA	CA1510006	East Niles CSD	THM SAMP SITE-1417 VALE (E7)-KCWAS	TOTAL TRIHALOMETHANES	8.8	UG/L	12/18/2002	CWS	24,316
CA	CA1510006	East Niles CSD	THM SAMP SITE-1417 VALE (E7)-KCWAS	TOTAL TRIHALOMETHANES	51.89	UG/L	6/23/1998	CWS	24,316
CA	CA1510006	East Niles CSD	THM SAMP SITE-1417 VALE (E7)-KCWAS	TOTAL TRIHALOMETHANES	53.3	UG/L	11/3/1998	CWS	24,316
CA	CA1510006	East Niles CSD	THM SAMP SITE-1417 VALE (E7)-KCWAS	TOTAL TRIHALOMETHANES	57.69	UG/L	8/26/1998	CWS	24,316
CA	CA1510006	East Niles CSD	THM SAMP SITE-1417 VALE (E7)-KCWAS	TOTAL TRIHALOMETHANES	44.84	UG/L	2/26/1998	CWS	24,316
CA	CA1510006	East Niles CSD	THM SAMP SITE-1417 VALE (E7)-KCWAS	TOTAL TRIHALOMETHANES	65.6	UG/L	9/11/1997	CWS	24,316
CA	CA1510006	East Niles CSD	THM SAMP SITE-1417 VALE (E7)-KCWAS	TOTAL TRIHALOMETHANES	36.8	UG/L	12/18/2001	CWS	24,316
CA	CA1510006	East Niles CSD	THM SAMP SITE-1417 VALE (E7)-KCWAS	TOTAL TRIHALOMETHANES	129	UG/L	9/24/2001	CWS	24,316
CA	CA1510006	East Niles CSD	THM SAMP SITE-1417 VALE (E7)-KCWAS	TOTAL TRIHALOMETHANES	46.52	UG/L	3/9/1999	CWS	24,316
CA	CA1510006	East Niles CSD	BRECKENRIDGE TP-RAW-OFFLINE	TOTAL TRIHALOMETHANES	29.4	UG/L	3/13/2001	CWS	24,316
CA	CA1510006	East Niles CSD	BRECKENRIDGE TP-RAW-OFFLINE	TOTAL TRIHALOMETHANES	36.8	UG/L	10/1/2003	CWS	24,316
CA	CA1510006	East Niles CSD	BRECKENRIDGE TP-RAW-OFFLINE	TOTAL TRIHALOMETHANES	67	UG/L	3/23/1998	CWS	24,316

APPENDIX 7-3 *Continued*

State	PWS_ID	Utility	POE	Constituent	LEVEL_NO	Unit	COLL_DATE	PWS	Pop
CA	CA1510006	East Niles CSD	BRECKENRIDGE TP-RAW-OFFLINE	TOTAL TRIHALOMETHANES	29	UG/L	3/19/1997	CWS	24,316
CA	CA1510006	East Niles CSD	BRECKENRIDGE TP-RAW-OFFLINE	TOTAL TRIHALOMETHANES	21.7	UG/L	3/15/1999	CWS	24,316
CA	CA1510006	East Niles CSD	THM SAMP SITE-5310 WINGFOOT (E6)-KCWAS	TOTAL TRIHALOMETHANES	103	UG/L	6/18/2002	CWS	24,316
CA	CA1510006	East Niles CSD	THM SAMP SITE-5310 WINGFOOT (E6)-KCWAS	TOTAL TRIHALOMETHANES	54.8	UG/L	5/8/2002	CWS	24,316
CA	CA1510006	East Niles CSD	THM SAMP SITE-5310 WINGFOOT (E6)-KCWAS	TOTAL TRIHALOMETHANES	41.8	UG/L	12/16/2003	CWS	24,316
CA	CA1510006	East Niles CSD	THM SAMP SITE-5310 WINGFOOT (E6)-KCWAS	TOTAL TRIHALOMETHANES	28.5	UG/L	12/18/2002	CWS	24,316
CA	CA1510006	East Niles CSD	THM SAMP SITE-5310 WINGFOOT (E6)-KCWAS	TOTAL TRIHALOMETHANES	126	UG/L	3/18/2003	CWS	24,316
CA	CA1510006	East Niles CSD	THM SAMP SITE-5310 WINGFOOT (E6)-KCWAS	TOTAL TRIHALOMETHANES	50.7	UG/L	11/3/1998	CWS	24,316
CA	CA1510006	East Niles CSD	THM SAMP SITE-5310 WINGFOOT (E6)-KCWAS	TOTAL TRIHALOMETHANES	58.43	UG/L	8/26/1998	CWS	24,316
CA	CA1510006	East Niles CSD	THM SAMP SITE-5310 WINGFOOT (E6)-KCWAS	TOTAL TRIHALOMETHANES	61.07	UG/L	2/26/1998	CWS	24,316
CA	CA1510006	East Niles CSD	THM SAMP SITE-5310 WINGFOOT (E6)-KCWAS	TOTAL TRIHALOMETHANES	61.8	UG/L	9/11/1997	CWS	24,316
CA	CA1510006	East Niles CSD	THM SAMP SITE-5310 WINGFOOT (E6)-KCWAS	TOTAL TRIHALOMETHANES	37.1	UG/L	12/18/2001	CWS	24,316
CA	CA1510006	East Niles CSD	THM SAMP SITE-5310 WINGFOOT (E6)-KCWAS	TOTAL TRIHALOMETHANES	53.24	UG/L	6/23/1998	CWS	24,316
CA	CA1510006	East Niles CSD	THM SAMP SITE-5310 WINGFOOT (E6)-KCWAS	TOTAL TRIHALOMETHANES	50.7	UG/L	11/3/1998	CWS	24,316
CA	CA1510006	East Niles CSD	THM SAMP SITE-5310 WINGFOOT (E6)-KCWAS	TOTAL TRIHALOMETHANES	112	UG/L	9/24/2001	CWS	24,316
CA	CA1510006	East Niles CSD	THM SAMP SITE-5310 WINGFOOT (E6)-KCWAS	TOTAL TRIHALOMETHANES	22.33	UG/L	3/10/1999	CWS	24,316
CA	CA1510006	East Niles CSD	THM SAMP SITE-3209 FLINTRIDGE (E4)-KCWAS	TOTAL TRIHALOMETHANES	103	UG/L	6/18/2002	CWS	24,316
CA	CA1510006	East Niles CSD	THM SAMP SITE-3209 FLINTRIDGE (E4)-KCWAS	TOTAL TRIHALOMETHANES	2.8	UG/L	5/8/2002	CWS	24,316
CA	CA1510006	East Niles CSD	THM SAMP SITE-3209 FLINTRIDGE (E4)-KCWAS	TOTAL TRIHALOMETHANES	39.3	UG/L	12/16/2003	CWS	24,316
CA	CA1510006	East Niles CSD	THM SAMP SITE-3209 FLINTRIDGE (E4)-KCWAS	TOTAL TRIHALOMETHANES	141	UG/L	3/18/2003	CWS	24,316
CA	CA1510006	East Niles CSD	THM SAMP SITE-3209 FLINTRIDGE (E4)-KCWAS	TOTAL TRIHALOMETHANES	29.7	UG/L	12/18/2002	CWS	24,316
CA	CA1510006	East Niles CSD	THM SAMP SITE-3209 FLINTRIDGE (E4)-KCWAS	TOTAL TRIHALOMETHANES	57.68	UG/L	2/26/1998	CWS	24,316
CA	CA1510006	East Niles CSD	THM SAMP SITE-3209 FLINTRIDGE (E4)-KCWAS	TOTAL TRIHALOMETHANES	34.2	UG/L	11/3/1998	CWS	24,316
CA	CA1510006	East Niles CSD	THM SAMP SITE-3209 FLINTRIDGE (E4)-KCWAS	TOTAL TRIHALOMETHANES	52.31	UG/L	8/26/1998	CWS	24,316
CA	CA1510006	East Niles CSD	THM SAMP SITE-3209 FLINTRIDGE (E4)-KCWAS	TOTAL TRIHALOMETHANES	40.1	UG/L	9/11/1997	CWS	24,316
CA	CA1510006	East Niles CSD	THM SAMP SITE-3209 FLINTRIDGE (E4)-KCWAS	TOTAL TRIHALOMETHANES	43.3	UG/L	12/18/2001	CWS	24,316
CA	CA1510006	East Niles CSD	THM SAMP SITE-3209 FLINTRIDGE (E4)-KCWAS	TOTAL TRIHALOMETHANES	53.12	UG/L	3/10/1999	CWS	24,316
CA	CA1510006	East Niles CSD	THM SAMP SITE-3209 FLINTRIDGE (E4)-KCWAS	TOTAL TRIHALOMETHANES	111	UG/L	9/24/2001	CWS	24,316
CA	CA1510006	East Niles CSD	BRECKENRIDGE TP-TREATED	TOTAL TRIHALOMETHANES	19.9	UG/L	3/28/2002	CWS	24,316
CA	CA1510040	Kern County Water Agency	ID4 WEN. & COLUMBAS (T7)-DIST. SYS.	TOTAL TRIHALOMETHANES	45.67	UG/L	5/29/1998	CWS	0
CA	CA1510040	Kern County Water Agency	ID4 WEN. & COLUMBAS (T7)-DIST. SYS.	TOTAL TRIHALOMETHANES	33.96	UG/L	6/22/1998	CWS	0
CA	CA1510040	Kern County Water Agency	ID4 WEN. & COLUMBAS (T7)-DIST. SYS.	TOTAL TRIHALOMETHANES	25.8	UG/L	3/8/1999	CWS	0
CA	CA1510040	Kern County Water Agency	ID4 WEN. & COLUMBAS (T7)-DIST. SYS.	TOTAL TRIHALOMETHANES	39.6	UG/L	7/20/1998	CWS	0
CA	CA1510040	Kern County Water Agency	ID4 WEN. & COLUMBAS (T7)-DIST. SYS.	TOTAL TRIHALOMETHANES	37.55	UG/L	8/17/1998	CWS	0
CA	CA1510040	Kern County Water Agency	ID4 WEN. & COLUMBAS (T7)-DIST. SYS.	TOTAL TRIHALOMETHANES	33.4	UG/L	9/28/1998	CWS	0
CA	CA1510040	Kern County Water Agency	ID4 WEN. & COLUMBAS (T7)-DIST. SYS.	TOTAL TRIHALOMETHANES	44.92	UG/L	4/6/1998	CWS	0
CA	CA1510040	Kern County Water Agency	ID4 WEN. & COLUMBAS (T7)-DIST. SYS.	TOTAL TRIHALOMETHANES	42.6	UG/L	6/2/1997	CWS	0
CA	CA1510040	Kern County Water Agency	ID4 WEN. & COLUMBAS (T7)-DIST. SYS.	TOTAL TRIHALOMETHANES	40	UG/L	9/23/1997	CWS	0
CA	CA1510040	Kern County Water Agency	ID4 WEN. & COLUMBAS (T7)-DIST. SYS.	TOTAL TRIHALOMETHANES	31.98	UG/L	2/2/1998	CWS	0
CA	CA1510040	Kern County Water Agency	ID4 WEN. & COLUMBAS (T7)-DIST. SYS.	TOTAL TRIHALOMETHANES	39.1	UG/L	7/7/1997	CWS	0
CA	CA1510040	Kern County Water Agency	ID4 WEN. & COLUMBAS (T7)-DIST. SYS.	TOTAL TRIHALOMETHANES	70.4	UG/L	1/5/1998	CWS	0
CA	CA1510040	Kern County Water Agency	ID4 WEN. & COLUMBAS (T7)-DIST. SYS.	TOTAL TRIHALOMETHANES	29.1	UG/L	3/2/1998	CWS	0

CA	CA1510040	Kern County Water Agency	ID4 CORNER TANK-T6-TREATED	TOTAL TRIHALOMETHANES	45.6	UG/L	5/29/1998	CWS	0
CA	CA1510040	Kern County Water Agency	ID4 CORNER TANK-T6-TREATED	TOTAL TRIHALOMETHANES	54.76	UG/L	4/6/1998	CWS	0
CA	CA1510040	Kern County Water Agency	ID4 CORNER TANK-T6-TREATED	TOTAL TRIHALOMETHANES	52.52	UG/L	6/22/1998	CWS	0
CA	CA1510040	Kern County Water Agency	ID4 CORNER TANK-T6-TREATED	TOTAL TRIHALOMETHANES	44.94	UG/L	2/2/1998	CWS	0
CA	CA1510040	Kern County Water Agency	ID4 CORNER TANK-T6-TREATED	TOTAL TRIHALOMETHANES	36.69	UG/L	3/2/1998	CWS	0
CA	CA1510040	Kern County Water Agency	ID4 CORNER TANK-T6-TREATED	TOTAL TRIHALOMETHANES	44.91	UG/L	7/20/1998	CWS	0
CA	CA1510040	Kern County Water Agency	ID4 CORNER TANK-T6-TREATED	TOTAL TRIHALOMETHANES	37.55	UG/L	8/17/1998	CWS	0
CA	CA1510040	Kern County Water Agency	ID4 CORNER TANK-T6-TREATED	TOTAL TRIHALOMETHANES	38.9	UG/L	9/28/1998	CWS	0
CA	CA1510040	Kern County Water Agency	ID4 CORNER TANK-T6-TREATED	TOTAL TRIHALOMETHANES	48.4	UG/L	3/8/1999	CWS	0
CA	CA1510040	Kern County Water Agency	ID4 CORNER TANK-T6-TREATED	TOTAL TRIHALOMETHANES	99.64	UG/L	1/31/1997	CWS	0
CA	CA1510040	Kern County Water Agency	ID4 CORNER TANK-T6-TREATED	TOTAL TRIHALOMETHANES	29.8	UG/L	2/24/1997	CWS	0
CA	CA1510040	Kern County Water Agency	ID4 CORNER TANK-T6-TREATED	TOTAL TRIHALOMETHANES	95.32	UG/L	1/5/1998	CWS	0
CA	CA1510040	Kern County Water Agency	ID4 CORNER TANK-T6-TREATED	TOTAL TRIHALOMETHANES	41.1	UG/L	3/24/1997	CWS	0
CA	CA1510040	Kern County Water Agency	ID4 CORNER TANK-T6-TREATED	TOTAL TRIHALOMETHANES	61.6	UG/L	4/14/1997	CWS	0
CA	CA1510040	Kern County Water Agency	ID4 CORNER TANK-T6-TREATED	TOTAL TRIHALOMETHANES	64	UG/L	9/8/1997	CWS	0
CA	CA1510040	Kern County Water Agency	ID4 CORNER TANK-T6-TREATED	TOTAL TRIHALOMETHANES	49.1	UG/L	5/12/1997	CWS	0
CA	CA1510040	Kern County Water Agency	ID4 CORNER TANK-T6-TREATED	TOTAL TRIHALOMETHANES	51.6	UG/L	7/7/1997	CWS	0
CA	CA1510040	Kern County Water Agency	ID4 CORNER TANK-T6-TREATED	TOTAL TRIHALOMETHANES	49.2	UG/L	6/2/1997	CWS	0
CA	CA1510040	Kern County Water Agency	ID4 BELLE-SEQUOIA TANK-T2-TREATED	TOTAL TRIHALOMETHANES	24.57	UG/L	3/2/1998	CWS	0
CA	CA1510040	Kern County Water Agency	ID4 BELLE-SEQUOIA TANK-T2-TREATED	TOTAL TRIHALOMETHANES	67.95	UG/L	5/29/1998	CWS	0
CA	CA1510040	Kern County Water Agency	ID4 BELLE-SEQUOIA TANK-T2-TREATED	TOTAL TRIHALOMETHANES	58.68	UG/L	1/5/1998	CWS	0
CA	CA1510040	Kern County Water Agency	ID4 BELLE-SEQUOIA TANK-T2-TREATED	TOTAL TRIHALOMETHANES	68.92	UG/L	4/6/1998	CWS	0
CA	CA1510040	Kern County Water Agency	ID4 BELLE-SEQUOIA TANK-T2-TREATED	TOTAL TRIHALOMETHANES	22.45	UG/L	2/2/1998	CWS	0
CA	CA1510040	Kern County Water Agency	ID4 BELLE-SEQUOIA TANK-T2-TREATED	TOTAL TRIHALOMETHANES	36.01	UG/L	8/17/1998	CWS	0
CA	CA1510040	Kern County Water Agency	ID4 BELLE-SEQUOIA TANK-T2-TREATED	TOTAL TRIHALOMETHANES	23	UG/L	3/8/1999	CWS	0
CA	CA1510040	Kern County Water Agency	ID4 BELLE-SEQUOIA TANK-T2-TREATED	TOTAL TRIHALOMETHANES	30.4	UG/L	9/28/1998	CWS	0
CA	CA1510040	Kern County Water Agency	ID4 BELLE-SEQUOIA TANK-T2-TREATED	TOTAL TRIHALOMETHANES	33.9	UG/L	7/20/1998	CWS	0
CA	CA1510040	Kern County Water Agency	ID4 BELLE-SEQUOIA TANK-T2-TREATED	TOTAL TRIHALOMETHANES	29.79	UG/L	6/22/1998	CWS	0
CA	CA1510040	Kern County Water Agency	ID4 BELLE-SEQUOIA TANK-T2-TREATED	TOTAL TRIHALOMETHANES	57.8	UG/L	4/14/1997	CWS	0
CA	CA1510040	Kern County Water Agency	ID4 BELLE-SEQUOIA TANK-T2-TREATED	TOTAL TRIHALOMETHANES	91.92	UG/L	1/31/1997	CWS	0
CA	CA1510040	Kern County Water Agency	ID4 BELLE-SEQUOIA TANK-T2-TREATED	TOTAL TRIHALOMETHANES	50.68	UG/L	3/24/1997	CWS	0
CA	CA1510040	Kern County Water Agency	ID4 BELLE-SEQUOIA TANK-T2-TREATED	TOTAL TRIHALOMETHANES	34.2	UG/L	5/12/1997	CWS	0
CA	CA1510040	Kern County Water Agency	ID4 BELLE-SEQUOIA TANK-T2-TREATED	TOTAL TRIHALOMETHANES	27.8	UG/L	2/24/1997	CWS	0
CA	CA1510040	Kern County Water Agency	ID4 BELLE-SEQUOIA TANK-T2-TREATED	TOTAL TRIHALOMETHANES	58.8	UG/L	9/8/1997	CWS	0
CA	CA1510040	Kern County Water Agency	ID4 BELLE-SEQUOIA TANK-T2-TREATED	TOTAL TRIHALOMETHANES	51.4	UG/L	9/8/1997	CWS	0
CA	CA1510040	Kern County Water Agency	ID4 BELLE-SEQUOIA TANK-T2-TREATED	TOTAL TRIHALOMETHANES	38.3	UG/L	7/7/1997	CWS	0
CA	CA1510040	Kern County Water Agency	ID4 BELLE-SEQUOIA TANK-T2-TREATED	TOTAL TRIHALOMETHANES	40.3	UG/L	6/2/1997	CWS	0
CA	CA1510040	Kern County Water Agency	ID4 OSWELL TANK-T5-TREATED	TOTAL TRIHALOMETHANES	43.63	UG/L	4/6/1998	CWS	0
CA	CA1510040	Kern County Water Agency	ID4 OSWELL TANK-T5-TREATED	TOTAL TRIHALOMETHANES	35.08	UG/L	3/2/1998	CWS	0
CA	CA1510040	Kern County Water Agency	ID4 OSWELL TANK-T5-TREATED	TOTAL TRIHALOMETHANES	43.48	UG/L	2/2/1998	CWS	0
CA	CA1510040	Kern County Water Agency	ID4 OSWELL TANK-T5-TREATED	TOTAL TRIHALOMETHANES	49.38	UG/L	5/29/1998	CWS	0
CA	CA1510040	Kern County Water Agency	ID4 OSWELL TANK-T5-TREATED	TOTAL TRIHALOMETHANES	103.4	UG/L	1/5/1998	CWS	0
CA	CA1510040	Kern County Water Agency	ID4 OSWELL TANK-T5-TREATED	TOTAL TRIHALOMETHANES	52.39	UG/L	6/22/1998	CWS	0
CA	CA1510040	Kern County Water Agency	ID4 OSWELL TANK-T5-TREATED	TOTAL TRIHALOMETHANES	41.21	UG/L	7/20/1998	CWS	0
CA	CA1510040	Kern County Water Agency	ID4 OSWELL TANK-T5-TREATED	TOTAL TRIHALOMETHANES	53.4	UG/L	3/8/1999	CWS	0

APPENDIX 7-3 *Continued*

State	PWS_ID	Utility	POE	Constituent	LEVEL_NO	Unit	COLL_DATE	PWS	Pop
CA	CA1510040	Kern County Water Agency	ID4 OSWELL TANK-T5-TREATED	TOTAL TRIHALOMETHANES	81.3	UG/L	9/28/1998	CWS	0
CA	CA1510040	Kern County Water Agency	ID4 OSWELL TANK-T5-TREATED	TOTAL TRIHALOMETHANES	38.7	UG/L	8/17/1998	CWS	0
CA	CA1510040	Kern County Water Agency	ID4 OSWELL TANK-T5-TREATED	TOTAL TRIHALOMETHANES	40.6	UG/L	3/24/1997	CWS	0
CA	CA1510040	Kern County Water Agency	ID4 OSWELL TANK-T5-TREATED	TOTAL TRIHALOMETHANES	109.99	UG/L	1/31/1997	CWS	0
CA	CA1510040	Kern County Water Agency	ID4 OSWELL TANK-T5-TREATED	TOTAL TRIHALOMETHANES	30.9	UG/L	2/24/1997	CWS	0
CA	CA1510040	Kern County Water Agency	ID4 OSWELL TANK-T5-TREATED	TOTAL TRIHALOMETHANES	61.9	UG/L	4/14/1997	CWS	0
CA	CA1510040	Kern County Water Agency	ID4 OSWELL TANK-T5-TREATED	TOTAL TRIHALOMETHANES	42.3	UG/L	7/7/1997	CWS	0
CA	CA1510040	Kern County Water Agency	ID4 OSWELL TANK-T5-TREATED	TOTAL TRIHALOMETHANES	67.1	UG/L	9/8/1997	CWS	0
CA	CA1510040	Kern County Water Agency	ID4 OSWELL TANK-T5-TREATED	TOTAL TRIHALOMETHANES	44.3	UG/L	6/2/1997	CWS	0
CA	CA1510040	Kern County Water Agency	ID4 OSWELL TANK-T5-TREATED	TOTAL TRIHALOMETHANES	42.9	UG/L	5/12/1997	CWS	0
CA	CA1510040	Kern County Water Agency	ID4 NOR TANK (600 SITE) T3-TREATED	TOTAL TRIHALOMETHANES	28.64	UG/L	3/2/1998	CWS	0
CA	CA1510040	Kern County Water Agency	ID4 NOR TANK (600 SITE) T3-TREATED	TOTAL TRIHALOMETHANES	45.81	UG/L	6/22/1998	CWS	0
CA	CA1510040	Kern County Water Agency	ID4 NOR TANK (600 SITE) T3-TREATED	TOTAL TRIHALOMETHANES	64.81	UG/L	5/29/1998	CWS	0
CA	CA1510040	Kern County Water Agency	ID4 NOR TANK (600 SITE) T3-TREATED	TOTAL TRIHALOMETHANES	65.77	UG/L	4/6/1998	CWS	0
CA	CA1510040	Kern County Water Agency	ID4 NOR TANK (600 SITE) T3-TREATED	TOTAL TRIHALOMETHANES	38.62	UG/L	7/20/1998	CWS	0
CA	CA1510040	Kern County Water Agency	ID4 NOR TANK (600 SITE) T3-TREATED	TOTAL TRIHALOMETHANES	71.47	UG/L	8/17/1998	CWS	0
CA	CA1510040	Kern County Water Agency	ID4 NOR TANK (600 SITE) T3-TREATED	TOTAL TRIHALOMETHANES	59	UG/L	9/28/1998	CWS	0
CA	CA1510040	Kern County Water Agency	ID4 NOR TANK (600 SITE) T3-TREATED	TOTAL TRIHALOMETHANES	51.8	UG/L	3/8/1999	CWS	0
CA	CA1510040	Kern County Water Agency	ID4 NOR TANK (600 SITE) T3-TREATED	TOTAL TRIHALOMETHANES	43.04	UG/L	2/2/1998	CWS	0
CA	CA1510040	Kern County Water Agency	ID4 NOR TANK (600 SITE) T3-TREATED	TOTAL TRIHALOMETHANES	53.8	UG/L	4/14/1997	CWS	0
CA	CA1510040	Kern County Water Agency	ID4 NOR TANK (600 SITE) T3-TREATED	TOTAL TRIHALOMETHANES	31.9	UG/L	2/24/1997	CWS	0
CA	CA1510040	Kern County Water Agency	ID4 NOR TANK (600 SITE) T3-TREATED	TOTAL TRIHALOMETHANES	100.41	UG/L	1/31/1997	CWS	0
CA	CA1510040	Kern County Water Agency	ID4 NOR TANK (600 SITE) T3-TREATED	TOTAL TRIHALOMETHANES	58.8	UG/L	3/24/1997	CWS	0
CA	CA1510040	Kern County Water Agency	ID4 NOR TANK (600 SITE) T3-TREATED	TOTAL TRIHALOMETHANES	42.2	UG/L	7/7/1997	CWS	0
CA	CA1510040	Kern County Water Agency	ID4 NOR TANK (600 SITE) T3-TREATED	TOTAL TRIHALOMETHANES	113.32	UG/L	1/5/1998	CWS	0
CA	CA1510040	Kern County Water Agency	ID4 NOR TANK (600 SITE) T3-TREATED	TOTAL TRIHALOMETHANES	38.3	UG/L	5/12/1997	CWS	0
CA	CA1510040	Kern County Water Agency	ID4 NOR TANK (600 SITE) T3-TREATED	TOTAL TRIHALOMETHANES	46.8	UG/L	6/2/1997	CWS	0
CA	CA1510040	Kern County Water Agency	ID4 TREATED-T1	TOTAL TRIHALOMETHANES	22.8	UG/L	2/2/1998	CWS	0
CA	CA1510040	Kern County Water Agency	ID4 TREATED-T1	TOTAL TRIHALOMETHANES	25.25	UG/L	3/2/1998	CWS	0
CA	CA1510040	Kern County Water Agency	ID4 TREATED-T1	TOTAL TRIHALOMETHANES	60.6	UG/L	9/8/1997	CWS	0
CA	CA1510040	Kern County Water Agency	ID4 TREATED-T1	TOTAL TRIHALOMETHANES	56.72	UG/L	1/5/1998	CWS	0
CA	CA1510040	Kern County Water Agency	ID4 TREATED-T1	TOTAL TRIHALOMETHANES	39.47	UG/L	5/29/1998	CWS	0
CA	CA1510040	Kern County Water Agency	ID4 TREATED-T1	TOTAL TRIHALOMETHANES	33.68	UG/L	6/22/1998	CWS	0
CA	CA1510040	Kern County Water Agency	ID4 TREATED-T1	TOTAL TRIHALOMETHANES	41.13	UG/L	4/6/1998	CWS	0
CA	CA1510040	Kern County Water Agency	ID4 TREATED-T1	TOTAL TRIHALOMETHANES	22.1	UG/L	2/24/1997	CWS	0
CA	CA1510040	Kern County Water Agency	ID4 TREATED-T1	TOTAL TRIHALOMETHANES	50.71	UG/L	3/24/1997	CWS	0
CA	CA1510040	Kern County Water Agency	ID4 TREATED-T1	TOTAL TRIHALOMETHANES	83.33	UG/L	1/31/1997	CWS	0
CA	CA1510040	Kern County Water Agency	ID4 TREATED-T1	TOTAL TRIHALOMETHANES	59.9	UG/L	7/7/1997	CWS	0
CA	CA1510040	Kern County Water Agency	ID4 TREATED-T1	TOTAL TRIHALOMETHANES	60.1	UG/L	4/14/1997	CWS	0
CA	CA1510040	Kern County Water Agency	ID4 TREATED-T1	TOTAL TRIHALOMETHANES	60.5	UG/L	5/12/1997	CWS	0
CA	CA1510040	Kern County Water Agency	ID4 TREATED-T1	TOTAL TRIHALOMETHANES	47.1	UG/L	6/2/1997	CWS	0
CA	CA1510040	Kern County Water Agency	ID4 TREATED-T1	TOTAL TRIHALOMETHANES	63.2	UG/L	8/30/2002	CWS	0

CA	CA1510040	Kern County Water Agency	ID4 TREATED-T1	TOTAL TRIHALOMETHANES	49.3	UG/L	8/30/1999	CWS	0
CA	CA1510040	Kern County Water Agency	ID4 TREATED-T1	TOTAL TRIHALOMETHANES	40.1	UG/L	9/28/1998	CWS	0
CA	CA1510040	Kern County Water Agency	ID4 TREATED-T1	TOTAL TRIHALOMETHANES	38.13	UG/L	7/20/1998	CWS	0
CA	CA1510040	Kern County Water Agency	ID4 TREATED-T1	TOTAL TRIHALOMETHANES	40.77	UG/L	8/17/1998	CWS	0
CA	CA1510040	Kern County Water Agency	ID4 TREATED-T1	TOTAL TRIHALOMETHANES	23.9	UG/L	3/8/1999	CWS	0
CA	CA1510040	Kern County Water Agency	ID4 TREATED-T1	TOTAL TRIHALOMETHANES	22.24	UG/L	11/15/2000	CWS	0
CA	CA1510041	North of the River MWD	HOLSON WELL-RAW	TOTAL TRIHALOMETHANES	1.95	UG/L	3/18/1998	CWS	7,000
CA	CA1510041	North of the River MWD	HOLSON WELL-RAW	TOTAL TRIHALOMETHANES	5.84	UG/L	8/15/2000	CWS	7,000
CA	CA1510041	North of the River MWD	HOLSON WELL-RAW	TOTAL TRIHALOMETHANES	2.42	UG/L	6/12/2000	CWS	7,000
CA	CA1510041	North of the River MWD	HOLSON WELL-RAW	TOTAL TRIHALOMETHANES	0.63	UG/L	2/8/2000	CWS	7,000
CA	CA1510041	North of the River MWD	THM SAMP SITE-2841 PEGASUS (N1)-TRTD	TOTAL TRIHALOMETHANES	70.3	UG/L	8/27/2002	CWS	7,000
CA	CA1510041	North of the River MWD	THM SAMP SITE-2841 PEGASUS (N1)-TRTD	TOTAL TRIHALOMETHANES	22.6	UG/L	8/22/2001	CWS	7,000
CA	CA1510041	North of the River MWD	THM SAMP SITE-2841 PEGASUS (N1)-TRTD	TOTAL TRIHALOMETHANES	24.1	UG/L	11/26/2001	CWS	7,000
CA	CA1510041	North of the River MWD	THM SAMP SITE-2841 PEGASUS (N1)-TRTD	TOTAL TRIHALOMETHANES	65.2	UG/L	5/8/2001	CWS	7,000
CA	CA1510041	North of the River MWD	THM SAMP SITE-2841 PEGASUS (N1)-TRTD	TOTAL TRIHALOMETHANES	73.7	UG/L	5/17/2002	CWS	7,000
CA	CA1510041	North of the River MWD	THM SAMP SITE-2841 PEGASUS (N1)-TRTD	TOTAL TRIHALOMETHANES	64.07	UG/L	6/23/1998	CWS	7,000
CA	CA1510041	North of the River MWD	THM SAMP SITE-2841 PEGASUS (N1)-TRTD	TOTAL TRIHALOMETHANES	49.9	UG/L	11/3/1998	CWS	7,000
CA	CA1510041	North of the River MWD	THM SAMP SITE-2841 PEGASUS (N1)-TRTD	TOTAL TRIHALOMETHANES	61.41	UG/L	2/25/1998	CWS	7,000
CA	CA1510041	North of the River MWD	THM SAMP SITE-2841 PEGASUS (N1)-TRTD	TOTAL TRIHALOMETHANES	62.03	UG/L	11/7/2000	CWS	7,000
CA	CA1510041	North of the River MWD	THM SAMP SITE-2841 PEGASUS (N1)-TRTD	TOTAL TRIHALOMETHANES	49.8	UG/L	6/7/1999	CWS	7,000
CA	CA1510041	North of the River MWD	THM SAMP SITE-2841 PEGASUS (N1)-TRTD	TOTAL TRIHALOMETHANES	52.83	UG/L	3/10/1999	CWS	7,000
CA	CA1510041	North of the River MWD	THM SAMP SITE-900 TANKSITE (N4)-TRTD	TOTAL TRIHALOMETHANES	65.4	UG/L	8/22/2001	CWS	7,000
CA	CA1510041	North of the River MWD	THM SAMP SITE-900 TANKSITE (N4)-TRTD	TOTAL TRIHALOMETHANES	62.3	UG/L	5/8/2001	CWS	7,000
CA	CA1510041	North of the River MWD	THM SAMP SITE-900 TANKSITE (N4)-TRTD	TOTAL TRIHALOMETHANES	54.63	UG/L	11/7/2000	CWS	7,000
CA	CA1510041	North of the River MWD	THM SAMP SITE-900 TANKSITE (N4)-TRTD	TOTAL TRIHALOMETHANES	99.1	UG/L	5/17/2002	CWS	7,000
CA	CA1510041	North of the River MWD	THM SAMP SITE-900 TANKSITE (N4)-TRTD	TOTAL TRIHALOMETHANES	51.3	UG/L	11/26/2001	CWS	7,000
CA	CA1510041	North of the River MWD	THM SAMP SITE-900 TANKSITE (N4)-TRTD	TOTAL TRIHALOMETHANES	100.6	UG/L	8/27/2002	CWS	7,000
CA	CA1510041	North of the River MWD	THM SAMP SITE-900 TANKSITE (N4)-TRTD	TOTAL TRIHALOMETHANES	78.87	UG/L	2/25/1998	CWS	7,000
CA	CA1510041	North of the River MWD	THM SAMP SITE-900 TANKSITE (N4)-TRTD	TOTAL TRIHALOMETHANES	85.41	UG/L	6/23/1998	CWS	7,000
CA	CA1510041	North of the River MWD	THM SAMP SITE-900 TANKSITE (N4)-TRTD	TOTAL TRIHALOMETHANES	104.5	UG/L	9/23/1997	CWS	7,000
CA	CA1510041	North of the River MWD	THM SAMP SITE-900 TANKSITE (N4)-TRTD	TOTAL TRIHALOMETHANES	119.6	UG/L	6/7/1999	CWS	7,000
CA	CA1510041	North of the River MWD	THM SAMP SITE-900 TANKSITE (N4)-TRTD	TOTAL TRIHALOMETHANES	88.53	UG/L	3/10/1999	CWS	7,000
CA	CA1510041	North of the River MWD	THM SAMP SITE-900 TANKSITE (N4)-TRTD	TOTAL TRIHALOMETHANES	66.9	UG/L	11/3/1998	CWS	7,000
CA	CA1510041	North of the River MWD	THM SAMP SITE-3418 N CHESTER (N3)-TRTD	TOTAL TRIHALOMETHANES	24.4	UG/L	8/22/2001	CWS	7,000
CA	CA1510041	North of the River MWD	THM SAMP SITE-3418 N CHESTER (N3)-TRTD	TOTAL TRIHALOMETHANES	56.1	UG/L	5/8/2001	CWS	7,000
CA	CA1510041	North of the River MWD	THM SAMP SITE-3418 N CHESTER (N3)-TRTD	TOTAL TRIHALOMETHANES	20.4	UG/L	11/26/2001	CWS	7,000
CA	CA1510041	North of the River MWD	THM SAMP SITE-3418 N CHESTER (N3)-TRTD	TOTAL TRIHALOMETHANES	65.1	UG/L	5/17/2002	CWS	7,000
CA	CA1510041	North of the River MWD	THM SAMP SITE-3418 N CHESTER (N3)-TRTD	TOTAL TRIHALOMETHANES	69.8	UG/L	8/27/2002	CWS	7,000
CA	CA1510041	North of the River MWD	THM SAMP SITE-3418 N CHESTER (N3)-TRTD	TOTAL TRIHALOMETHANES	51.02	UG/L	11/7/2000	CWS	7,000
CA	CA1510041	North of the River MWD	THM SAMP SITE-3418 N CHESTER (N3)-TRTD	TOTAL TRIHALOMETHANES	78.6	UG/L	9/11/1997	CWS	7,000
CA	CA1510041	North of the River MWD	THM SAMP SITE-3418 N CHESTER (N3)-TRTD	TOTAL TRIHALOMETHANES	35.74	UG/L	2/25/1998	CWS	7,000
CA	CA1510041	North of the River MWD	THM SAMP SITE-3418 N CHESTER (N3)-TRTD	TOTAL TRIHALOMETHANES	52.78	UG/L	6/23/1998	CWS	7,000
CA	CA1510041	North of the River MWD	THM SAMP SITE-3418 N CHESTER (N3)-TRTD	TOTAL TRIHALOMETHANES	43	UG/L	11/3/1998	CWS	7,000
CA	CA1510041	North of the River MWD	THM SAMP SITE-3418 N CHESTER (N3)-TRTD	TOTAL TRIHALOMETHANES	33.3	UG/L	6/7/1999	CWS	7,000
CA	CA1510041	North of the River MWD	THM SAMP SITE-3418 N CHESTER (N3)-TRTD	TOTAL TRIHALOMETHANES	46.58	UG/L	3/10/1999	CWS	7,000

APPENDIX 7-3 *Continued*

State	PWS_ID	Utility	POE	Constituent	LEVEL_NO	Unit	COLL_DATE	PWS	Pop
CA	CA1510041	North of the River MWD	THM SAMP SITE-528 CHARLANA (N2)-TRTD	TOTAL TRIHALOMETHANES	57.5	UG/L	5/8/2001	CWS	7,000
CA	CA1510041	North of the River MWD	THM SAMP SITE-528 CHARLANA (N2)-TRTD	TOTAL TRIHALOMETHANES	23.1	UG/L	8/22/2001	CWS	7,000
CA	CA1510041	North of the River MWD	THM SAMP SITE-528 CHARLANA (N2)-TRTD	TOTAL TRIHALOMETHANES	49.55	UG/L	11/7/2000	CWS	7,000
CA	CA1510041	North of the River MWD	THM SAMP SITE-528 CHARLANA (N2)-TRTD	TOTAL TRIHALOMETHANES	25.7	UG/L	11/26/2001	CWS	7,000
CA	CA1510041	North of the River MWD	THM SAMP SITE-528 CHARLANA (N2)-TRTD	TOTAL TRIHALOMETHANES	65.2	UG/L	5/17/2002	CWS	7,000
CA	CA1510041	North of the River MWD	THM SAMP SITE-528 CHARLANA (N2)-TRTD	TOTAL TRIHALOMETHANES	74.3	UG/L	8/27/2002	CWS	7,000
CA	CA1510041	North of the River MWD	THM SAMP SITE-528 CHARLANA (N2)-TRTD	TOTAL TRIHALOMETHANES	39.4	UG/L	9/11/1997	CWS	7,000
CA	CA1510041	North of the River MWD	THM SAMP SITE-528 CHARLANA (N2)-TRTD	TOTAL TRIHALOMETHANES	58.09	UG/L	2/25/1998	CWS	7,000
CA	CA1510041	North of the River MWD	THM SAMP SITE-528 CHARLANA (N2)-TRTD	TOTAL TRIHALOMETHANES	42.08	UG/L	6/23/1998	CWS	7,000
CA	CA1510041	North of the River MWD	THM SAMP SITE-528 CHARLANA (N2)-TRTD	TOTAL TRIHALOMETHANES	32.5	UG/L	11/3/1998	CWS	7,000
CA	CA1510041	North of the River MWD	THM SAMP SITE-528 CHARLANA (N2)-TRTD	TOTAL TRIHALOMETHANES	21.7	UG/L	6/7/1999	CWS	7,000
CA	CA1510041	North of the River MWD	THM SAMP SITE-528 CHARLANA (N2)-TRTD	TOTAL TRIHALOMETHANES	31.69	UG/L	3/10/1999	CWS	7,000
CA	CA1600034	PARAMOUNT FARMING CO	TREATMENT PLANT	TOTAL TRIHALOMETHANES	152.9	UG/L	11/4/2004	CWS	100
					47.3				
CA	CA1510006	East Niles CSD	THM SAMP SITE-PACIFIC BELL (E9)-ENS	TRICHLOROACETIC ACID (TCAA)	11	UG/L	6/18/2002	CWS	24,316
CA	CA1510006	East Niles CSD	THM SAMP SITE-PACIFIC BELL (E9)-ENS	TRICHLOROACETIC ACID (TCAA)	4	UG/L	3/18/2003	CWS	24,316
CA	CA1510006	East Niles CSD	THM SAMP SITE-8717 BUTTERNUT (E10)-ENS	TRICHLOROACETIC ACID (TCAA)	10	UG/L	3/18/2003	CWS	24,316
CA	CA1510006	East Niles CSD	THM SAMP SITE-8717 BUTTERNUT (E10)-ENS	TRICHLOROACETIC ACID (TCAA)	8	UG/L	6/18/2002	CWS	24,316
CA	CA1510006	East Niles CSD	THM SAMP SITE-11845 SCHOOL (E11)-ENS	TRICHLOROACETIC ACID (TCAA)	10	UG/L	6/18/2002	CWS	24,316
CA	CA1510006	East Niles CSD	THM SAMP SITE-11845 SCHOOL (E11)-ENS	TRICHLOROACETIC ACID (TCAA)	3	UG/L	3/18/2003	CWS	24,316
CA	CA1510006	East Niles CSD	THM SAMP SITE-11845 SCHOOL (E11)-ENS	TRICHLOROACETIC ACID (TCAA)	2	UG/L	12/18/2002	CWS	24,316
CA	CA1510006	East Niles CSD	THM SAMP SITE-FAIRFAX SCHOOL (E12)-ENG	TRICHLOROACETIC ACID (TCAA)	1	UG/L	6/18/2002	CWS	24,316
CA	CA1510006	East Niles CSD	THM SAMP SITE-1417 VALE (E7)-KCWAS	TRICHLOROACETIC ACID (TCAA)	15	UG/L	6/18/2002	CWS	24,316
CA	CA1510006	East Niles CSD	THM SAMP SITE-1417 VALE (E7)-KCWAS	TRICHLOROACETIC ACID (TCAA)	2	UG/L	12/18/2001	CWS	24,316
CA	CA1510006	East Niles CSD	THM SAMP SITE-1417 VALE (E7)-KCWAS	TRICHLOROACETIC ACID (TCAA)	10	UG/L	12/16/2003	CWS	24,316
CA	CA1510006	East Niles CSD	THM SAMP SITE-1417 VALE (E7)-KCWAS	TRICHLOROACETIC ACID (TCAA)	1	UG/L	12/18/2002	CWS	24,316
CA	CA1510006	East Niles CSD	THM SAMP SITE-1417 VALE (E7)-KCWAS	TRICHLOROACETIC ACID (TCAA)	7	UG/L	9/24/2001	CWS	24,316
CA	CA1510006	East Niles CSD	THM SAMP SITE-5310 WINGFOOT (E6)-KCWAS	TRICHLOROACETIC ACID (TCAA)	14	UG/L	6/18/2002	CWS	24,316
CA	CA1510006	East Niles CSD	THM SAMP SITE-5310 WINGFOOT (E6)-KCWAS	TRICHLOROACETIC ACID (TCAA)	2	UG/L	12/18/2001	CWS	24,316
CA	CA1510006	East Niles CSD	THM SAMP SITE-5310 WINGFOOT (E6)-KCWAS	TRICHLOROACETIC ACID (TCAA)	14	UG/L	12/16/2003	CWS	24,316
CA	CA1510006	East Niles CSD	THM SAMP SITE-5310 WINGFOOT (E6)-KCWAS	TRICHLOROACETIC ACID (TCAA)	75	UG/L	3/18/2003	CWS	24,316
CA	CA1510006	East Niles CSD	THM SAMP SITE-5310 WINGFOOT (E6)-KCWAS	TRICHLOROACETIC ACID (TCAA)	3	UG/L	12/18/2002	CWS	24,316
CA	CA1510006	East Niles CSD	THM SAMP SITE-5310 WINGFOOT (E6)-KCWAS	TRICHLOROACETIC ACID (TCAA)	5	UG/L	9/24/2001	CWS	24,316
CA	CA1510006	East Niles CSD	THM SAMP SITE-3209 FLINTRIDGE (E4)-KCWAS	TRICHLOROACETIC ACID (TCAA)	17	UG/L	6/18/2002	CWS	24,316
CA	CA1510006	East Niles CSD	THM SAMP SITE-3209 FLINTRIDGE (E4)-KCWAS	TRICHLOROACETIC ACID (TCAA)	2	UG/L	12/18/2001	CWS	24,316
CA	CA1510006	East Niles CSD	THM SAMP SITE-3209 FLINTRIDGE (E4)-KCWAS	TRICHLOROACETIC ACID (TCAA)	84	UG/L	3/18/2003	CWS	24,316
CA	CA1510006	East Niles CSD	THM SAMP SITE-3209 FLINTRIDGE (E4)-KCWAS	TRICHLOROACETIC ACID (TCAA)	18	UG/L	12/16/2003	CWS	24,316
CA	CA1510006	East Niles CSD	THM SAMP SITE-3209 FLINTRIDGE (E4)-KCWAS	TRICHLOROACETIC ACID (TCAA)	5	UG/L	9/24/2001	CWS	24,316
CA	CA1600034	PARAMOUNT FARMING CO	TREATMENT PLANT	TRICHLOROACETIC ACID (TCAA)	3.6	UG/L	11/4/2004	CWS	100
CA	CA1600034	PARAMOUNT FARMING CO	TREATMENT PLANT	TRICHLOROACETIC ACID (TCAA)	3.9	UG/L	7/14/2004	CWS	100
					12.7				

APPENDIX 7-4 Binghamton Groundwater Data

State	PWSID	System	Analyte	CONC	UOM	Date	PWSType	Pop
NY	NY2630054	DIELECTRIC LABORATORIES	BROMOMETHANE	2.46	UG/L	22-Jan-01	NTNCWS	350
NY	NY0310336	HEATHS TRAILER PARK	DICHLOROMETHANE	1.5	UG/L	30-Jun-03	CWS	92
NY	NY0310336	HEATHS TRAILER PARK	DICHLOROMETHANE	1.6	UG/L	30-Jun-03	CWS	92
NY	NY0310351	SHADY MAPLE TRAILER PARK	METHYL TERT BUTYL ETHER (MTBE)	1	UG/L	13-Feb-04	CWS	72
NY	NY0310336	HEATHS TRAILER PARK	TOLUENE	0.6	UG/L	30-Jun-03	CWS	92
NY	NY2630054	DIELECTRIC LABORATORIES	TOLUENE	2.65	UG/L	26-Jan-01	NTNCWS	350

APPENDIX 7-5 Binghamton Surface Water Data

State	PWSID	System	Analyte	CONC	UOM	Date	PWSType	Pop
NY	NY0301651	BINGHAMTON, CITY OF	1,1-DICHLOROETHANE	0.6	UG/L	03-Aug-04	CWS	47,600
NY	NY0301651	BINGHAMTON, CITY OF	1,1-DICHLOROETHANE	0.6	UG/L	14-Nov-02	CWS	47,600
NY	NY0301651	BINGHAMTON, CITY OF	METHYL TERT BUTYL ETHER (MTBE)	1.4	UG/L	25-Oct-99	CWS	47,600
NY	NY0301651	BINGHAMTON, CITY OF	TRICHLOROETHYLENE	0.7	UG/L	11-Apr-00	CWS	47,600
NY	NY0301651	BINGHAMTON, CITY OF	TRICHLOROETHYLENE	0.7	UG/L	14-Nov-02	CWS	47,600
NY	NY0301651	BINGHAMTON, CITY OF	TRICHLOROETHYLENE	0.7	UG/L	15-Aug-02	CWS	47,600

APPENDIX 7-6 Birmingham Groundwater Data

State	PWSID	Water System Name	Source	Analyte	Concentration	Unit	Sample Date	PWSType	Pop
AL	AL0001491	FARLEY NUCLEAR PLANT	CONSTRUCTION WELL 1 (EAST) + TREATMENT P	BROMODICHLOROMETHANE	2.26	ug/L	06-Jun-05	NTNCWS	1000
AL	AL0001491	FARLEY NUCLEAR PLANT	PRODUCTION WELL 2 + TREATMENT PLANT	BROMODICHLOROMETHANE	0.64 1.45	ug/L	06-Jun-05	NTNCWS	1000
AL	AL0001491	FARLEY NUCLEAR PLANT	CONSTRUCTION WELL 1 (EAST) + TREATMENT P	BROMOFORM	1.63	ug/L	06-Jun-05	NTNCWS	1000
AL	AL0001491	FARLEY NUCLEAR PLANT	PRODUCTION WELL 2 + TREATMENT PLANT	BROMOFORM	0.72 1.18	ug/L	06-Jun-05	NTNCWS	1000
AL	AL0001491	FARLEY NUCLEAR PLANT	CONSTRUCTION WELL 1 (EAST) + TREATMENT P	CHLOROFORM	2.99	ug/L	06-Jun-05	NTNCWS	1000
AL	AL0001491	FARLEY NUCLEAR PLANT	PRODUCTION WELL 2 + TREATMENT PLANT	CHLOROFORM	0.58 1.79	ug/L	06-Jun-05	NTNCWS	1000
AL	AL0001491	FARLEY NUCLEAR PLANT	DISTRIBUTION SYSTEM	TOTAL HALOACETIC ACIDS (HAA5)	0.8	ug/L	08-Sep-04	NTNCWS	1000
AL	AL0001491	FARLEY NUCLEAR PLANT	DISTRIBUTION SYSTEM	TOTAL HALOACETIC ACIDS (HAA5)	0.6 0.7	ug/L	08-Sep-04	NTNCWS	1000
AL	AL0001491	FARLEY NUCLEAR PLANT	DISTRIBUTION SYSTEM	TOTAL TRIHALOMETHANES (TTHM)	5.3	ug/L	08-Sep-04	NTNCWS	1000
AL	AL0001491	FARLEY NUCLEAR PLANT	DISTRIBUTION SYSTEM	TOTAL TRIHALOMETHANES (TTHM)	1.6 3.45	ug/L	08-Sep-04	NTNCWS	1000

APPENDIX 7-7 Birmingham Surface Water Data

State	PWSID	Water System Name	Source	Analyte	Level	Unit	Sample Date	PWSType	Pop
AL	AL0000738	BIRMINGHAM WATER BOARD	CARSON (INLAND LAKE)	BROMODICHLOROMETHANE	4.45	ug/L	25-Jan-00	CWS	585,000
AL	AL0000738	BIRMINGHAM WATER BOARD	PUTNAM WATER TREATMENT PLANT	BROMODICHLOROMETHANE	4.65	ug/L	25-Jan-00	CWS	585,000
AL	AL0000738	BIRMINGHAM WATER BOARD	SHADES MTN (CAHABA RIVER & LAKE PURDY)	BROMODICHLOROMETHANE	4.78	ug/L	25-Jan-00	CWS	585,000
AL	AL0000738	BIRMINGHAM WATER BOARD	WESTERN (MULBERRY & SIPSEY FORKS)	BROMODICHLOROMETHANE	8.59	ug/L	25-Jan-00	CWS	585,000
AL	AL0000738	BIRMINGHAM WATER BOARD	CARSON (INLAND LAKE)	BROMODICHLOROMETHANE	5.26	ug/L	17-May-00	CWS	585,000
AL	AL0000738	BIRMINGHAM WATER BOARD	PUTNAM WATER TREATMENT PLANT	BROMODICHLOROMETHANE	4.3	ug/L	17-May-00	CWS	585,000
AL	AL0000738	BIRMINGHAM WATER BOARD	SHADES MTN (CAHABA RIVER & LAKE PURDY)	BROMODICHLOROMETHANE	9.54	ug/L	17-May-00	CWS	585,000
AL	AL0000738	BIRMINGHAM WATER BOARD	WESTERN (MULBERRY & SIPSEY FORKS)	BROMODICHLOROMETHANE	7.56	ug/L	17-May-00	CWS	585,000
AL	AL0000738	BIRMINGHAM WATER BOARD	CARSON (INLAND LAKE)	BROMODICHLOROMETHANE	2.72	ug/L	16-Jan-01	CWS	585,000
AL	AL0000738	BIRMINGHAM WATER BOARD	PUTNAM WATER TREATMENT PLANT	BROMODICHLOROMETHANE	2.81	ug/L	16-Jan-01	CWS	585,000
AL	AL0000738	BIRMINGHAM WATER BOARD	SHADES MTN (CAHABA RIVER & LAKE PURDY)	BROMODICHLOROMETHANE	4.39	ug/L	16-Jan-01	CWS	585,000
AL	AL0000738	BIRMINGHAM WATER BOARD	WESTERN (MULBERRY & SIPSEY FORKS)	BROMODICHLOROMETHANE	2.59	ug/L	16-Jan-01	CWS	585,000
AL	AL0000738	BIRMINGHAM WATER BOARD	CARSON (INLAND LAKE)	BROMODICHLOROMETHANE	3.01	ug/L	21-Aug-01	CWS	585,000
AL	AL0000738	BIRMINGHAM WATER BOARD	PUTNAM WATER TREATMENT PLANT	BROMODICHLOROMETHANE	3.15	ug/L	21-Aug-01	CWS	585,000
AL	AL0000738	BIRMINGHAM WATER BOARD	SHADES MTN (CAHABA RIVER & LAKE PURDY)	BROMODICHLOROMETHANE	8.22	ug/L	21-Aug-01	CWS	585,000
AL	AL0000738	BIRMINGHAM WATER BOARD	WESTERN (MULBERRY & SIPSEY FORKS)	BROMODICHLOROMETHANE	4.06	ug/L	21-Aug-01	CWS	585,000
AL	AL0000738	BIRMINGHAM WATER BOARD	CARSON (INLAND LAKE)	BROMODICHLOROMETHANE	3.01	ug/L	19-Mar-02	CWS	585,000
AL	AL0000738	BIRMINGHAM WATER BOARD	PUTNAM WATER TREATMENT PLANT	BROMODICHLOROMETHANE	3.15	ug/L	19-Mar-02	CWS	585,000
AL	AL0000738	BIRMINGHAM WATER BOARD	SHADES MTN (CAHABA RIVER & LAKE PURDY)	BROMODICHLOROMETHANE	8.22	ug/L	19-Mar-02	CWS	585,000
AL	AL0000738	BIRMINGHAM WATER BOARD	WESTERN (MULBERRY & SIPSEY FORKS)	BROMODICHLOROMETHANE	4.06	ug/L	19-Mar-02	CWS	585,000
AL	AL0000738	BIRMINGHAM WATER BOARD	CARSON (INLAND LAKE)	BROMODICHLOROMETHANE	2.91	ug/L	27-Sep-02	CWS	585,000
AL	AL0000738	BIRMINGHAM WATER BOARD	PUTNAM WATER TREATMENT PLANT	BROMODICHLOROMETHANE	1.66	ug/L	27-Sep-02	CWS	585,000
AL	AL0000738	BIRMINGHAM WATER BOARD	SHADES MTN (CAHABA RIVER & LAKE PURDY)	BROMODICHLOROMETHANE	8.14	ug/L	27-Sep-02	CWS	585,000
AL	AL0000738	BIRMINGHAM WATER BOARD	WESTERN (MULBERRY & SIPSEY FORKS)	BROMODICHLOROMETHANE	3.62	ug/L	27-Sep-02	CWS	585,000
AL	AL0000738	BIRMINGHAM WATER BOARD	CARSON (INLAND LAKE)	BROMODICHLOROMETHANE	1.6	ug/L	19-Nov-02	CWS	585,000
AL	AL0000738	BIRMINGHAM WATER BOARD	PUTNAM WATER TREATMENT PLANT	BROMODICHLOROMETHANE	1.6	ug/L	19-Nov-02	CWS	585,000
AL	AL0000738	BIRMINGHAM WATER BOARD	SHADES MTN (CAHABA RIVER & LAKE PURDY)	BROMODICHLOROMETHANE	4.2	ug/L	19-Nov-02	CWS	585,000
AL	AL0000738	BIRMINGHAM WATER BOARD	WESTERN (MULBERRY & SIPSEY FORKS)	BROMODICHLOROMETHANE	2.3	ug/L	19-Nov-02	CWS	585,000
AL	AL0000738	BIRMINGHAM WATER BOARD	CARSON (INLAND LAKE)	BROMODICHLOROMETHANE	1.6	ug/L	22-Jan-03	CWS	585,000
AL	AL0000738	BIRMINGHAM WATER BOARD	PUTNAM WATER TREATMENT PLANT	BROMODICHLOROMETHANE	1.6	ug/L	22-Jan-03	CWS	585,000
AL	AL0000738	BIRMINGHAM WATER BOARD	SHADES MTN (CAHABA RIVER & LAKE PURDY)	BROMODICHLOROMETHANE	4.2	ug/L	22-Jan-03	CWS	585,000
AL	AL0000738	BIRMINGHAM WATER BOARD	WESTERN (MULBERRY & SIPSEY FORKS)	BROMODICHLOROMETHANE	2.3	ug/L	22-Jan-03	CWS	585,000
AL	AL0000738	BIRMINGHAM WATER BOARD	CARSON (INLAND LAKE)	BROMODICHLOROMETHANE	2	ug/L	13-May-03	CWS	585,000
AL	AL0000738	BIRMINGHAM WATER BOARD	PUTNAM WATER TREATMENT PLANT	BROMODICHLOROMETHANE	1.4	ug/L	13-May-03	CWS	585,000
AL	AL0000738	BIRMINGHAM WATER BOARD	SHADES MTN (CAHABA RIVER & LAKE PURDY)	BROMODICHLOROMETHANE	5.1	ug/L	13-May-03	CWS	585,000
AL	AL0000738	BIRMINGHAM WATER BOARD	WESTERN (MULBERRY & SIPSEY FORKS)	BROMODICHLOROMETHANE	5.4	ug/L	13-May-03	CWS	585,000
AL	AL0000738	BIRMINGHAM WATER BOARD	CARSON (INLAND LAKE)	BROMODICHLOROMETHANE	1.6	ug/L	19-Aug-03	CWS	585,000
AL	AL0000738	BIRMINGHAM WATER BOARD	PUTNAM WATER TREATMENT PLANT	BROMODICHLOROMETHANE	1.1	ug/L	19-Aug-03	CWS	585,000
AL	AL0000738	BIRMINGHAM WATER BOARD	SHADES MTN (CAHABA RIVER & LAKE PURDY)	BROMODICHLOROMETHANE	7	ug/L	19-Aug-03	CWS	585,000
AL	AL0000738	BIRMINGHAM WATER BOARD	WESTERN (MULBERRY & SIPSEY FORKS)	BROMODICHLOROMETHANE	3	ug/L	19-Aug-03	CWS	585,000
AL	AL0000738	BIRMINGHAM WATER BOARD	CARSON (INLAND LAKE)	BROMODICHLOROMETHANE	2.1	ug/L	09-Dec-03	CWS	585,000

APPENDIX 7-7 *Continued*

State	PWSID	Water System Name	Source	Analyte	Level	Unit	Sample Date	PWSType	Pop
AL	AL0000738	BIRMINGHAM WATER BOARD	PUTNAM WATER TREATMENT PLANT	BROMODICHLOROMETHANE	1.4	ug/L	09-Dec-03	CWS	585,000
AL	AL0000738	BIRMINGHAM WATER BOARD	SHADES MTN (CAHABA RIVER & LAKE PURDY)	BROMODICHLOROMETHANE	5.4	ug/L	09-Dec-03	CWS	585,000
AL	AL0000738	BIRMINGHAM WATER BOARD	WESTERN (MULBERRY & SIPSEY FORKS)	BROMODICHLOROMETHANE	2.8	ug/L	09-Dec-03	CWS	585,000
AL	AL0000738	BIRMINGHAM WATER BOARD	CARSON (INLAND LAKE)	BROMODICHLOROMETHANE	1.87	ug/L	10-Feb-04	CWS	585,000
AL	AL0000738	BIRMINGHAM WATER BOARD	PUTNAM WATER TREATMENT PLANT	BROMODICHLOROMETHANE	1.81	ug/L	10-Feb-04	CWS	585,000
AL	AL0000738	BIRMINGHAM WATER BOARD	SHADES MTN (CAHABA RIVER & LAKE PURDY)	BROMODICHLOROMETHANE	3.77	ug/L	10-Feb-04	CWS	585,000
AL	AL0000738	BIRMINGHAM WATER BOARD	WESTERN (MULBERRY & SIPSEY FORKS)	BROMODICHLOROMETHANE	3.5	ug/L	10-Feb-04	CWS	585,000
AL	AL0000738	BIRMINGHAM WATER BOARD	CARSON WATER TREATMENT PLANT	BROMODICHLOROMETHANE	1.84	ug/L	11-May-04	CWS	585,000
AL	AL0000738	BIRMINGHAM WATER BOARD	PUTNAM WATER TREATMENT PLANT	BROMODICHLOROMETHANE	3.12	ug/L	11-May-04	CWS	585,000
AL	AL0000738	BIRMINGHAM WATER BOARD	SHADES MTN WATER TREATMENT PLANT	BROMODICHLOROMETHANE	5.21	ug/L	11-May-04	CWS	585,000
AL	AL0000738	BIRMINGHAM WATER BOARD	WESTERN WATER TREATMENT PLANT	BROMODICHLOROMETHANE	4.6	ug/L	11-May-04	CWS	585,000
AL	AL0000738	BIRMINGHAM WATER BOARD	CARSON WATER TREATMENT PLANT	BROMODICHLOROMETHANE	2.83	ug/L	19-Aug-04	CWS	585,000
AL	AL0000738	BIRMINGHAM WATER BOARD	PUTNAM WATER TREATMENT PLANT	BROMODICHLOROMETHANE	1.79	ug/L	19-Aug-04	CWS	585,000
AL	AL0000738	BIRMINGHAM WATER BOARD	SHADES MTN WATER TREATMENT PLANT	BROMODICHLOROMETHANE	10.02	ug/L	19-Aug-04	CWS	585,000
AL	AL0000738	BIRMINGHAM WATER BOARD	WESTERN WATER TREATMENT PLANT	BROMODICHLOROMETHANE	4.76	ug/L	19-Aug-04	CWS	585,000
AL	AL0000738	BIRMINGHAM WATER BOARD	CARSON WATER TREATMENT PLANT	BROMODICHLOROMETHANE	2.62	ug/L	09-Nov-04	CWS	585,000
AL	AL0000738	BIRMINGHAM WATER BOARD	PUTNAM WATER TREATMENT PLANT	BROMODICHLOROMETHANE	3.98	ug/L	09-Nov-04	CWS	585,000
AL	AL0000738	BIRMINGHAM WATER BOARD	SHADES MTN WATER TREATMENT PLANT	BROMODICHLOROMETHANE	6.1	ug/L	09-Nov-04	CWS	585,000
AL	AL0000738	BIRMINGHAM WATER BOARD	WESTERN WATER TREATMENT PLANT	BROMODICHLOROMETHANE	3.74	ug/L	09-Nov-04	CWS	585,000
AL	AL0000738	BIRMINGHAM WATER BOARD	CARSON WATER TREATMENT PLANT	BROMODICHLOROMETHANE	2.3	ug/L	11-Jan-05	CWS	585,000
AL	AL0000738	BIRMINGHAM WATER BOARD	PUTNAM WATER TREATMENT PLANT	BROMODICHLOROMETHANE	4.4	ug/L	11-Jan-05	CWS	585,000
AL	AL0000738	BIRMINGHAM WATER BOARD	SHADES MTN WATER TREATMENT PLANT	BROMODICHLOROMETHANE	4.85	ug/L	11-Jan-05	CWS	585,000
AL	AL0000738	BIRMINGHAM WATER BOARD	WESTERN WATER TREATMENT PLANT	BROMODICHLOROMETHANE	5.1	ug/L	11-Jan-05	CWS	585,000
AL	AL0000738	BIRMINGHAM WATER BOARD	CARSON WATER TREATMENT PLANT	BROMODICHLOROMETHANE	2.62	ug/L	25-May-05	CWS	585,000
AL	AL0000738	BIRMINGHAM WATER BOARD	PUTNAM WATER TREATMENT PLANT	BROMODICHLOROMETHANE	3.98	ug/L	25-May-05	CWS	585,000
AL	AL0000738	BIRMINGHAM WATER BOARD	SHADES MTN WATER TREATMENT PLANT	BROMODICHLOROMETHANE	6.1	ug/L	25-May-05	CWS	585,000
AL	AL0000738	BIRMINGHAM WATER BOARD	WESTERN WATER TREATMENT PLANT	BROMODICHLOROMETHANE	3.74	ug/L	25-May-05	CWS	585,000
					3.93				
AL	AL0000738	BIRMINGHAM WATER BOARD	CARSON (INLAND LAKE)	CHLORODIBROMOMETHANE	1.51	ug/L	25-Jan-00	CWS	585,000
AL	AL0000738	BIRMINGHAM WATER BOARD	PUTNAM WATER TREATMENT PLANT	CHLORODIBROMOMETHANE	0.93	ug/L	25-Jan-00	CWS	585,000
AL	AL0000738	BIRMINGHAM WATER BOARD	SHADES MTN (CAHABA RIVER & LAKE PURDY)	CHLORODIBROMOMETHANE	0.95	ug/L	25-Jan-00	CWS	585,000
AL	AL0000738	BIRMINGHAM WATER BOARD	WESTERN (MULBERRY & SIPSEY FORKS)	CHLORODIBROMOMETHANE	4.42	ug/L	25-Jan-00	CWS	585,000
AL	AL0000738	BIRMINGHAM WATER BOARD	CARSON (INLAND LAKE)	CHLORODIBROMOMETHANE	1.84	ug/L	17-May-00	CWS	585,000
AL	AL0000738	BIRMINGHAM WATER BOARD	PUTNAM WATER TREATMENT PLANT	CHLORODIBROMOMETHANE	0.8	ug/L	17-May-00	CWS	585,000
AL	AL0000738	BIRMINGHAM WATER BOARD	SHADES MTN (CAHABA RIVER & LAKE PURDY)	CHLORODIBROMOMETHANE	1.89	ug/L	17-May-00	CWS	585,000
AL	AL0000738	BIRMINGHAM WATER BOARD	WESTERN (MULBERRY & SIPSEY FORKS)	CHLORODIBROMOMETHANE	1.57	ug/L	17-May-00	CWS	585,000
AL	AL0000738	BIRMINGHAM WATER BOARD	CARSON (INLAND LAKE)	CHLORODIBROMOMETHANE	0.78	ug/L	16-Jan-01	CWS	585,000
AL	AL0000738	BIRMINGHAM WATER BOARD	PUTNAM WATER TREATMENT PLANT	CHLORODIBROMOMETHANE	0.8	ug/L	16-Jan-01	CWS	585,000
AL	AL0000738	BIRMINGHAM WATER BOARD	SHADES MTN (CAHABA RIVER & LAKE PURDY)	CHLORODIBROMOMETHANE	0.83	ug/L	16-Jan-01	CWS	585,000
AL	AL0000738	BIRMINGHAM WATER BOARD	CARSON (INLAND LAKE)	CHLORODIBROMOMETHANE	0.91	ug/L	27-Sep-02	CWS	585,000
AL	AL0000738	BIRMINGHAM WATER BOARD	SHADES MTN (CAHABA RIVER & LAKE PURDY)	CHLORODIBROMOMETHANE	0.88	ug/L	27-Sep-02	CWS	585,000
AL	AL0000738	BIRMINGHAM WATER BOARD	SHADES MTN (CAHABA RIVER & LAKE PURDY)	CHLORODIBROMOMETHANE	0.9	ug/L	19-Nov-02	CWS	585,000

AL	AL0000738	BIRMINGHAM WATER BOARD	SHADES MTN (CAHABA RIVER & LAKE PURDY)	CHLORODIBROMOMETHANE	0.9	ug/L	22-Jan-03	CWS	585,000
AL	AL0000738	BIRMINGHAM WATER BOARD	CARSON (INLAND LAKE)	CHLORODIBROMOMETHANE	0.9	ug/L	13-May-03	CWS	585,000
AL	AL0000738	BIRMINGHAM WATER BOARD	SHADES MTN (CAHABA RIVER & LAKE PURDY)	CHLORODIBROMOMETHANE	0.6	ug/L	13-May-03	CWS	585,000
AL	AL0000738	BIRMINGHAM WATER BOARD	WESTERN (MULBERRY & SIPSEY FORKS)	CHLORODIBROMOMETHANE	0.7	ug/L	13-May-03	CWS	585,000
AL	AL0000738	BIRMINGHAM WATER BOARD	SHADES MTN (CAHABA RIVER & LAKE PURDY)	CHLORODIBROMOMETHANE	1.4	ug/L	19-Aug-03	CWS	585,000
AL	AL0000738	BIRMINGHAM WATER BOARD	WESTERN (MULBERRY & SIPSEY FORKS)	CHLORODIBROMOMETHANE	0.5	ug/L	19-Aug-03	CWS	585,000
AL	AL0000738	BIRMINGHAM WATER BOARD	CARSON (INLAND LAKE)	CHLORODIBROMOMETHANE	0.5	ug/L	09-Dec-03	CWS	585,000
AL	AL0000738	BIRMINGHAM WATER BOARD	SHADES MTN (CAHABA RIVER & LAKE PURDY)	CHLORODIBROMOMETHANE	0.9	ug/L	09-Dec-03	CWS	585,000
AL	AL0000738	BIRMINGHAM WATER BOARD	CARSON (INLAND LAKE)	CHLORODIBROMOMETHANE	0.55	ug/L	10-Feb-04	CWS	585,000
AL	AL0000738	BIRMINGHAM WATER BOARD	SHADES MTN (CAHABA RIVER & LAKE PURDY)	CHLORODIBROMOMETHANE	0.52	ug/L	10-Feb-04	CWS	585,000
AL	AL0000738	BIRMINGHAM WATER BOARD	WESTERN (MULBERRY & SIPSEY FORKS)	CHLORODIBROMOMETHANE	0.6	ug/L	10-Feb-04	CWS	585,000
AL	AL0000738	BIRMINGHAM WATER BOARD	CARSON WATER TREATMENT PLANT	CHLORODIBROMOMETHANE	0.5	ug/L	11-May-04	CWS	585,000
AL	AL0000738	BIRMINGHAM WATER BOARD	PUTNAM WATER TREATMENT PLANT	CHLORODIBROMOMETHANE	1.64	ug/L	11-May-04	CWS	585,000
AL	AL0000738	BIRMINGHAM WATER BOARD	SHADES MTN WATER TREATMENT PLANT	CHLORODIBROMOMETHANE	0.94	ug/L	11-May-04	CWS	585,000
AL	AL0000738	BIRMINGHAM WATER BOARD	WESTERN WATER TREATMENT PLANT	CHLORODIBROMOMETHANE	1.32	ug/L	11-May-04	CWS	585,000
AL	AL0000738	BIRMINGHAM WATER BOARD	SHADES MTN WATER TREATMENT PLANT	CHLORODIBROMOMETHANE	2.16	ug/L	19-Aug-04	CWS	585,000
AL	AL0000738	BIRMINGHAM WATER BOARD	WESTERN WATER TREATMENT PLANT	CHLORODIBROMOMETHANE	0.97	ug/L	19-Aug-04	CWS	585,000
AL	AL0000738	BIRMINGHAM WATER BOARD	PUTNAM WATER TREATMENT PLANT	CHLORODIBROMOMETHANE	1.1	ug/L	09-Nov-04	CWS	585,000
AL	AL0000738	BIRMINGHAM WATER BOARD	SHADES MTN WATER TREATMENT PLANT	CHLORODIBROMOMETHANE	0.75	ug/L	09-Nov-04	CWS	585,000
AL	AL0000738	BIRMINGHAM WATER BOARD	WESTERN WATER TREATMENT PLANT	CHLORODIBROMOMETHANE	0.54	ug/L	09-Nov-04	CWS	585,000
AL	AL0000738	BIRMINGHAM WATER BOARD	PUTNAM WATER TREATMENT PLANT	CHLORODIBROMOMETHANE	1.2	ug/L	11-Jan-05	CWS	585,000
AL	AL0000738	BIRMINGHAM WATER BOARD	SHADES MTN WATER TREATMENT PLANT	CHLORODIBROMOMETHANE	0.98	ug/L	11-Jan-05	CWS	585,000
AL	AL0000738	BIRMINGHAM WATER BOARD	WESTERN WATER TREATMENT PLANT	CHLORODIBROMOMETHANE	0.51	ug/L	11-Jan-05	CWS	585,000
AL	AL0000738	BIRMINGHAM WATER BOARD	PUTNAM WATER TREATMENT PLANT	CHLORODIBROMOMETHANE	1.1	ug/L	25-May-05	CWS	585,000
AL	AL0000738	BIRMINGHAM WATER BOARD	SHADES MTN WATER TREATMENT PLANT	CHLORODIBROMOMETHANE	0.75	ug/L	25-May-05	CWS	585,000
AL	AL0000738	BIRMINGHAM WATER BOARD	WESTERN WATER TREATMENT PLANT	CHLORODIBROMOMETHANE	0.54	ug/L	25-May-05	CWS	585,000
					1.05				
AL	AL0000738	BIRMINGHAM WATER BOARD	CARSON (INLAND LAKE)	CHLOROFORM	12.5	ug/L	25-Jan-00	CWS	585,000
AL	AL0000738	BIRMINGHAM WATER BOARD	PUTNAM WATER TREATMENT PLANT	CHLOROFORM	14.2	ug/L	25-Jan-00	CWS	585,000
AL	AL0000738	BIRMINGHAM WATER BOARD	SHADES MTN (CAHABA RIVER & LAKE PURDY)	CHLOROFORM	12.9	ug/L	25-Jan-00	CWS	585,000
AL	AL0000738	BIRMINGHAM WATER BOARD	WESTERN (MULBERRY & SIPSEY FORKS)	CHLOROFORM	9.85	ug/L	25-Jan-00	CWS	585,000
AL	AL0000738	BIRMINGHAM WATER BOARD	CARSON (INLAND LAKE)	CHLOROFORM	17.7	ug/L	17-May-00	CWS	585,000
AL	AL0000738	BIRMINGHAM WATER BOARD	PUTNAM WATER TREATMENT PLANT	CHLOROFORM	17.2	ug/L	17-May-00	CWS	585,000
AL	AL0000738	BIRMINGHAM WATER BOARD	SHADES MTN (CAHABA RIVER & LAKE PURDY)	CHLOROFORM	26.8	ug/L	17-May-00	CWS	585,000
AL	AL0000738	BIRMINGHAM WATER BOARD	WESTERN (MULBERRY & SIPSEY FORKS)	CHLOROFORM	24	ug/L	17-May-00	CWS	585,000
AL	AL0000738	BIRMINGHAM WATER BOARD	CARSON (INLAND LAKE)	CHLOROFORM	6.51	ug/L	16-Jan-01	CWS	585,000
AL	AL0000738	BIRMINGHAM WATER BOARD	PUTNAM WATER TREATMENT PLANT	CHLOROFORM	8.82	ug/L	16-Jan-01	CWS	585,000
AL	AL0000738	BIRMINGHAM WATER BOARD	SHADES MTN (CAHABA RIVER & LAKE PURDY)	CHLOROFORM	12	ug/L	16-Jan-01	CWS	585,000
AL	AL0000738	BIRMINGHAM WATER BOARD	WESTERN (MULBERRY & SIPSEY FORKS)	CHLOROFORM	11	ug/L	16-Jan-01	CWS	585,000
AL	AL0000738	BIRMINGHAM WATER BOARD	CARSON (INLAND LAKE)	CHLOROFORM	13.8	ug/L	21-Aug-01	CWS	585,000
AL	AL0000738	BIRMINGHAM WATER BOARD	PUTNAM WATER TREATMENT PLANT	CHLOROFORM	19.6	ug/L	21-Aug-01	CWS	585,000
AL	AL0000738	BIRMINGHAM WATER BOARD	SHADES MTN (CAHABA RIVER & LAKE PURDY)	CHLOROFORM	47.2	ug/L	21-Aug-01	CWS	585,000
AL	AL0000738	BIRMINGHAM WATER BOARD	WESTERN (MULBERRY & SIPSEY FORKS)	CHLOROFORM	18.1	ug/L	21-Aug-01	CWS	585,000
AL	AL0000738	BIRMINGHAM WATER BOARD	CARSON (INLAND LAKE)	CHLOROFORM	13.8	ug/L	19-Mar-02	CWS	585,000
AL	AL0000738	BIRMINGHAM WATER BOARD	PUTNAM WATER TREATMENT PLANT	CHLOROFORM	19.6	ug/L	19-Mar-02	CWS	585,000

APPENDIX 7-7 *Continued*

State	PWSID	Water System Name	Source	Analyte	Level	Unit	Sample Date	PWSType	Pop
AL	AL0000738	BIRMINGHAM WATER BOARD	SHADES MTN (CAHABA RIVER & LAKE PURDY)	CHLOROFORM	47.2	ug/L	19-Mar-02	CWS	585,000
AL	AL0000738	BIRMINGHAM WATER BOARD	WESTERN (MULBERRY & SIPSEY FORKS)	CHLOROFORM	18.1	ug/L	19-Mar-02	CWS	585,000
AL	AL0000738	BIRMINGHAM WATER BOARD	CARSON (INLAND LAKE)	CHLOROFORM	9.64	ug/L	27-Sep-02	CWS	585,000
AL	AL0000738	BIRMINGHAM WATER BOARD	PUTNAM WATER TREATMENT PLANT	CHLOROFORM	6.98	ug/L	27-Sep-02	CWS	585,000
AL	AL0000738	BIRMINGHAM WATER BOARD	SHADES MTN (CAHABA RIVER & LAKE PURDY)	CHLOROFORM	44.9	ug/L	27-Sep-02	CWS	585,000
AL	AL0000738	BIRMINGHAM WATER BOARD	WESTERN (MULBERRY & SIPSEY FORKS)	CHLOROFORM	16.1	ug/L	27-Sep-02	CWS	585,000
AL	AL0000738	BIRMINGHAM WATER BOARD	CARSON (INLAND LAKE)	CHLOROFORM	6.5	ug/L	19-Nov-02	CWS	585,000
AL	AL0000738	BIRMINGHAM WATER BOARD	PUTNAM WATER TREATMENT PLANT	CHLOROFORM	6.5	ug/L	19-Nov-02	CWS	585,000
AL	AL0000738	BIRMINGHAM WATER BOARD	SHADES MTN (CAHABA RIVER & LAKE PURDY)	CHLOROFORM	11.6	ug/L	19-Nov-02	CWS	585,000
AL	AL0000738	BIRMINGHAM WATER BOARD	WESTERN (MULBERRY & SIPSEY FORKS)	CHLOROFORM	7.2	ug/L	19-Nov-02	CWS	585,000
AL	AL0000738	BIRMINGHAM WATER BOARD	CARSON (INLAND LAKE)	CHLOROFORM	6.5	ug/L	22-Jan-03	CWS	585,000
AL	AL0000738	BIRMINGHAM WATER BOARD	PUTNAM WATER TREATMENT PLANT	CHLOROFORM	6.5	ug/L	22-Jan-03	CWS	585,000
AL	AL0000738	BIRMINGHAM WATER BOARD	SHADES MTN (CAHABA RIVER & LAKE PURDY)	CHLOROFORM	11.6	ug/L	22-Jan-03	CWS	585,000
AL	AL0000738	BIRMINGHAM WATER BOARD	WESTERN (MULBERRY & SIPSEY FORKS)	CHLOROFORM	7.2	ug/L	22-Jan-03	CWS	585,000
AL	AL0000738	BIRMINGHAM WATER BOARD	CARSON (INLAND LAKE)	CHLOROFORM	6.1	ug/L	13-May-03	CWS	585,000
AL	AL0000738	BIRMINGHAM WATER BOARD	PUTNAM WATER TREATMENT PLANT	CHLOROFORM	4.6	ug/L	13-May-03	CWS	585,000
AL	AL0000738	BIRMINGHAM WATER BOARD	SHADES MTN (CAHABA RIVER & LAKE PURDY)	CHLOROFORM	26	ug/L	13-May-03	CWS	585,000
AL	AL0000738	BIRMINGHAM WATER BOARD	WESTERN (MULBERRY & SIPSEY FORKS)	CHLOROFORM	21	ug/L	13-May-03	CWS	585,000
AL	AL0000738	BIRMINGHAM WATER BOARD	CARSON (INLAND LAKE)	CHLOROFORM	9.4	ug/L	19-Aug-03	CWS	585,000
AL	AL0000738	BIRMINGHAM WATER BOARD	PUTNAM WATER TREATMENT PLANT	CHLOROFORM	6.8	ug/L	19-Aug-03	CWS	585,000
AL	AL0000738	BIRMINGHAM WATER BOARD	SHADES MTN (CAHABA RIVER & LAKE PURDY)	CHLOROFORM	329	ug/L	19-Aug-03	CWS	585,000
AL	AL0000738	BIRMINGHAM WATER BOARD	WESTERN (MULBERRY & SIPSEY FORKS)	CHLOROFORM	13.7	ug/L	19-Aug-03	CWS	585,000
AL	AL0000738	BIRMINGHAM WATER BOARD	CARSON (INLAND LAKE)	CHLOROFORM	12.4	ug/L	09-Dec-03	CWS	585,000
AL	AL0000738	BIRMINGHAM WATER BOARD	PUTNAM WATER TREATMENT PLANT	CHLOROFORM	7.1	ug/L	09-Dec-03	CWS	585,000
AL	AL0000738	BIRMINGHAM WATER BOARD	SHADES MTN (CAHABA RIVER & LAKE PURDY)	CHLOROFORM	17.5	ug/L	09-Dec-03	CWS	585,000
AL	AL0000738	BIRMINGHAM WATER BOARD	WESTERN (MULBERRY & SIPSEY FORKS)	CHLOROFORM	9	ug/L	09-Dec-03	CWS	585,000
AL	AL0000738	BIRMINGHAM WATER BOARD	CARSON (INLAND LAKE)	CHLOROFORM	7.57	ug/L	10-Feb-04	CWS	585,000
AL	AL0000738	BIRMINGHAM WATER BOARD	PUTNAM WATER TREATMENT PLANT	CHLOROFORM	6.97	ug/L	10-Feb-04	CWS	585,000
AL	AL0000738	BIRMINGHAM WATER BOARD	SHADES MTN (CAHABA RIVER & LAKE PURDY)	CHLOROFORM	15.7	ug/L	10-Feb-04	CWS	585,000
AL	AL0000738	BIRMINGHAM WATER BOARD	WESTERN (MULBERRY & SIPSEY FORKS)	CHLOROFORM	14	ug/L	10-Feb-04	CWS	585,000
AL	AL0000738	BIRMINGHAM WATER BOARD	CARSON WATER TREATMENT PLANT	CHLOROFORM	6.75	ug/L	11-May-04	CWS	585,000
AL	AL0000738	BIRMINGHAM WATER BOARD	PUTNAM WATER TREATMENT PLANT	CHLOROFORM	7.49	ug/L	11-May-04	CWS	585,000
AL	AL0000738	BIRMINGHAM WATER BOARD	SHADES MTN WATER TREATMENT PLANT	CHLOROFORM	20.5	ug/L	11-May-04	CWS	585,000
AL	AL0000738	BIRMINGHAM WATER BOARD	WESTERN WATER TREATMENT PLANT	CHLOROFORM	10.7	ug/L	11-May-04	CWS	585,000
AL	AL0000738	BIRMINGHAM WATER BOARD	CARSON WATER TREATMENT PLANT	CHLOROFORM	11.41	ug/L	19-Aug-04	CWS	585,000
AL	AL0000738	BIRMINGHAM WATER BOARD	PUTNAM WATER TREATMENT PLANT	CHLOROFORM	7.37	ug/L	19-Aug-04	CWS	585,000
AL	AL0000738	BIRMINGHAM WATER BOARD	SHADES MTN WATER TREATMENT PLANT	CHLOROFORM	34.2	ug/L	19-Aug-04	CWS	585,000
AL	AL0000738	BIRMINGHAM WATER BOARD	WESTERN WATER TREATMENT PLANT	CHLOROFORM	15.6	ug/L	19-Aug-04	CWS	585,000
AL	AL0000738	BIRMINGHAM WATER BOARD	CARSON WATER TREATMENT PLANT	CHLOROFORM	16.72	ug/L	09-Nov-04	CWS	585,000
AL	AL0000738	BIRMINGHAM WATER BOARD	PUTNAM WATER TREATMENT PLANT	CHLOROFORM	12.87	ug/L	09-Nov-04	CWS	585,000
AL	AL0000738	BIRMINGHAM WATER BOARD	SHADES MTN WATER TREATMENT PLANT	CHLOROFORM	30.7	ug/L	09-Nov-04	CWS	585,000
AL	AL0000738	BIRMINGHAM WATER BOARD	WESTERN WATER TREATMENT PLANT	CHLOROFORM	13.3	ug/L	09-Nov-04	CWS	585,000

AL	AL0000738	BIRMINGHAM WATER BOARD	CARSON WATER TREATMENT PLANT	CHLOROFORM	10.4	ug/L	11-Jan-05	CWS	585,000
AL	AL0000738	BIRMINGHAM WATER BOARD	PUTNAM WATER TREATMENT PLANT	CHLOROFORM	11.1	ug/L	11-Jan-05	CWS	585,000
AL	AL0000738	BIRMINGHAM WATER BOARD	SHADES MTN WATER TREATMENT PLANT	CHLOROFORM	15.3	ug/L	11-Jan-05	CWS	585,000
AL	AL0000738	BIRMINGHAM WATER BOARD	WESTERN WATER TREATMENT PLANT	CHLOROFORM	11.3	ug/L	11-Jan-05	CWS	585,000
AL	AL0000738	BIRMINGHAM WATER BOARD	CARSON WATER TREATMENT PLANT	CHLOROFORM	16.72	ug/L	25-May-05	CWS	585,000
AL	AL0000738	BIRMINGHAM WATER BOARD	PUTNAM WATER TREATMENT PLANT	CHLOROFORM	12.87	ug/L	25-May-05	CWS	585,000
AL	AL0000738	BIRMINGHAM WATER BOARD	SHADES MTN WATER TREATMENT PLANT	CHLOROFORM	30.7	ug/L	25-May-05	CWS	585,000
AL	AL0000738	BIRMINGHAM WATER BOARD	WESTERN WATER TREATMENT PLANT	CHLOROFORM	13.3	ug/L	25-May-05	CWS	585,000
					19.5				
AL	AL0000738	BIRMINGHAM WATER BOARD	CARSON WATER TREATMENT PLANT	DICHLORODIFLUOROMETHANE	0.86	ug/L	19-Aug-04	CWS	585,000
AL	AL0000738	BIRMINGHAM WATER BOARD	CARSON WATER TREATMENT PLANT	DICHLORODIFLUOROMETHANE	0.9	ug/L	09-Nov-04	CWS	585,000
AL	AL0000738	BIRMINGHAM WATER BOARD	CARSON WATER TREATMENT PLANT	DICHLORODIFLUOROMETHANE	0.9	ug/L	25-May-05	CWS	585,000
					0.89				
AL	AL0000738	BIRMINGHAM WATER BOARD	CARSON (INLAND LAKE)	DICHLOROMETHANE	0.57	ug/L	16-Jan-01	CWS	585,000
AL	AL0000738	BIRMINGHAM WATER BOARD	PUTNAM WATER TREATMENT PLANT	DICHLOROMETHANE	0.71	ug/L	16-Jan-01	CWS	585,000
AL	AL0000738	BIRMINGHAM WATER BOARD	SHADES MTN WATER TREATMENT PLANT	DICHLOROMETHANE	3.5	ug/L	25-May-05	CWS	585,000
					1.59				
AL	AL0000738	BIRMINGHAM WATER BOARD	CARSON WATER TREATMENT PLANT	TOTAL HALOACETIC ACIDS (HAA5)	11	ug/L	10-Feb-04	CWS	585,000
AL	AL0000738	BIRMINGHAM WATER BOARD	DISTRIBUTION SYSTEM	TOTAL HALOACETIC ACIDS (HAA5)	40.1	ug/L	10-Feb-04	CWS	585,000
AL	AL0000738	BIRMINGHAM WATER BOARD	DISTRIBUTION SYSTEM	TOTAL HALOACETIC ACIDS (HAA5)	37	ug/L	10-Feb-04	CWS	585,000
AL	AL0000738	BIRMINGHAM WATER BOARD	DISTRIBUTION SYSTEM	TOTAL HALOACETIC ACIDS (HAA5)	36	ug/L	10-Feb-04	CWS	585,000
AL	AL0000738	BIRMINGHAM WATER BOARD	DISTRIBUTION SYSTEM	TOTAL HALOACETIC ACIDS (HAA5)	35	ug/L	10-Feb-04	CWS	585,000
AL	AL0000738	BIRMINGHAM WATER BOARD	DISTRIBUTION SYSTEM	TOTAL HALOACETIC ACIDS (HAA5)	32	ug/L	10-Feb-04	CWS	585,000
AL	AL0000738	BIRMINGHAM WATER BOARD	DISTRIBUTION SYSTEM	TOTAL HALOACETIC ACIDS (HAA5)	22	ug/L	10-Feb-04	CWS	585,000
AL	AL0000738	BIRMINGHAM WATER BOARD	DISTRIBUTION SYSTEM	TOTAL HALOACETIC ACIDS (HAA5)	22	ug/L	10-Feb-04	CWS	585,000
AL	AL0000738	BIRMINGHAM WATER BOARD	DISTRIBUTION SYSTEM	TOTAL HALOACETIC ACIDS (HAA5)	19	ug/L	10-Feb-04	CWS	585,000
AL	AL0000738	BIRMINGHAM WATER BOARD	DISTRIBUTION SYSTEM	TOTAL HALOACETIC ACIDS (HAA5)	18.1	ug/L	10-Feb-04	CWS	585,000
AL	AL0000738	BIRMINGHAM WATER BOARD	DISTRIBUTION SYSTEM	TOTAL HALOACETIC ACIDS (HAA5)	18	ug/L	10-Feb-04	CWS	585,000
AL	AL0000738	BIRMINGHAM WATER BOARD	DISTRIBUTION SYSTEM	TOTAL HALOACETIC ACIDS (HAA5)	14.7	ug/L	10-Feb-04	CWS	585,000
AL	AL0000738	BIRMINGHAM WATER BOARD	DISTRIBUTION SYSTEM	TOTAL HALOACETIC ACIDS (HAA5)	14	ug/L	10-Feb-04	CWS	585,000
AL	AL0000738	BIRMINGHAM WATER BOARD	DISTRIBUTION SYSTEM	TOTAL HALOACETIC ACIDS (HAA5)	13.8	ug/L	10-Feb-04	CWS	585,000
AL	AL0000738	BIRMINGHAM WATER BOARD	DISTRIBUTION SYSTEM	TOTAL HALOACETIC ACIDS (HAA5)	13	ug/L	10-Feb-04	CWS	585,000
AL	AL0000738	BIRMINGHAM WATER BOARD	DISTRIBUTION SYSTEM	TOTAL HALOACETIC ACIDS (HAA5)	13	ug/L	10-Feb-04	CWS	585,000
AL	AL0000738	BIRMINGHAM WATER BOARD	DISTRIBUTION SYSTEM	TOTAL HALOACETIC ACIDS (HAA5)	13	ug/L	10-Feb-04	CWS	585,000
AL	AL0000738	BIRMINGHAM WATER BOARD	PUTNAM WATER TREATMENT PLANT	TOTAL HALOACETIC ACIDS (HAA5)	11	ug/L	10-Feb-04	CWS	585,000
AL	AL0000738	BIRMINGHAM WATER BOARD	SHADES MTN WATER TREATMENT PLANT	TOTAL HALOACETIC ACIDS (HAA5)	26	ug/L	10-Feb-04	CWS	585,000
AL	AL0000738	BIRMINGHAM WATER BOARD	WESTERN WATER TREATMENT PLANT	TOTAL HALOACETIC ACIDS (HAA5)	13	ug/L	10-Feb-04	CWS	585,000
AL	AL0000738	BIRMINGHAM WATER BOARD	CARSON WATER TREATMENT PLANT	TOTAL HALOACETIC ACIDS (HAA5)	6.98	ug/L	11-May-04	CWS	585,000
AL	AL0000738	BIRMINGHAM WATER BOARD	DISTRIBUTION SYSTEM	TOTAL HALOACETIC ACIDS (HAA5)	52.3	ug/L	11-May-04	CWS	585,000
AL	AL0000738	BIRMINGHAM WATER BOARD	DISTRIBUTION SYSTEM	TOTAL HALOACETIC ACIDS (HAA5)	47	ug/L	11-May-04	CWS	585,000
AL	AL0000738	BIRMINGHAM WATER BOARD	DISTRIBUTION SYSTEM	TOTAL HALOACETIC ACIDS (HAA5)	45.3	ug/L	11-May-04	CWS	585,000
AL	AL0000738	BIRMINGHAM WATER BOARD	DISTRIBUTION SYSTEM	TOTAL HALOACETIC ACIDS (HAA5)	34.7	ug/L	11-May-04	CWS	585,000
AL	AL0000738	BIRMINGHAM WATER BOARD	DISTRIBUTION SYSTEM	TOTAL HALOACETIC ACIDS (HAA5)	29.4	ug/L	11-May-04	CWS	585,000
AL	AL0000738	BIRMINGHAM WATER BOARD	DISTRIBUTION SYSTEM	TOTAL HALOACETIC ACIDS (HAA5)	25.3	ug/L	11-May-04	CWS	585,000

APPENDIX 7-7 *Continued*

State	PWSID	Water System Name	Source	Analyte	Level	Unit	Sample Date	PWSType	Pop
AL	AL0000738	BIRMINGHAM WATER BOARD	DISTRIBUTION SYSTEM	TOTAL HALOACETIC ACIDS (HAA5)	24.3	ug/L	11-May-04	CWS	585,000
AL	AL0000738	BIRMINGHAM WATER BOARD	DISTRIBUTION SYSTEM	TOTAL HALOACETIC ACIDS (HAA5)	23.6	ug/L	11-May-04	CWS	585,000
AL	AL0000738	BIRMINGHAM WATER BOARD	DISTRIBUTION SYSTEM	TOTAL HALOACETIC ACIDS (HAA5)	23	ug/L	11-May-04	CWS	585,000
AL	AL0000738	BIRMINGHAM WATER BOARD	DISTRIBUTION SYSTEM	TOTAL HALOACETIC ACIDS (HAA5)	21	ug/L	11-May-04	CWS	585,000
AL	AL0000738	BIRMINGHAM WATER BOARD	DISTRIBUTION SYSTEM	TOTAL HALOACETIC ACIDS (HAA5)	20.4	ug/L	11-May-04	CWS	585,000
AL	AL0000738	BIRMINGHAM WATER BOARD	DISTRIBUTION SYSTEM	TOTAL HALOACETIC ACIDS (HAA5)	19.7	ug/L	11-May-04	CWS	585,000
AL	AL0000738	BIRMINGHAM WATER BOARD	DISTRIBUTION SYSTEM	TOTAL HALOACETIC ACIDS (HAA5)	18.9	ug/L	11-May-04	CWS	585,000
AL	AL0000738	BIRMINGHAM WATER BOARD	DISTRIBUTION SYSTEM	TOTAL HALOACETIC ACIDS (HAA5)	17.3	ug/L	11-May-04	CWS	585,000
AL	AL0000738	BIRMINGHAM WATER BOARD	DISTRIBUTION SYSTEM	TOTAL HALOACETIC ACIDS (HAA5)	16	ug/L	11-May-04	CWS	585,000
AL	AL0000738	BIRMINGHAM WATER BOARD	DISTRIBUTION SYSTEM	TOTAL HALOACETIC ACIDS (HAA5)	10.3	ug/L	11-May-04	CWS	585,000
AL	AL0000738	BIRMINGHAM WATER BOARD	PUTNAM WATER TREATMENT PLANT	TOTAL HALOACETIC ACIDS (HAA5)	12.9	ug/L	11-May-04	CWS	585,000
AL	AL0000738	BIRMINGHAM WATER BOARD	SHADES MTN WATER TREATMENT PLANT	TOTAL HALOACETIC ACIDS (HAA5)	44.4	ug/L	11-May-04	CWS	585,000
AL	AL0000738	BIRMINGHAM WATER BOARD	WESTERN WATER TREATMENT PLANT	TOTAL HALOACETIC ACIDS (HAA5)	16.9	ug/L	11-May-04	CWS	585,000
AL	AL0000738	BIRMINGHAM WATER BOARD	CARSON WATER TREATMENT PLANT	TOTAL HALOACETIC ACIDS (HAA5)	13.6	ug/L	19-Aug-04	CWS	585,000
AL	AL0000738	BIRMINGHAM WATER BOARD	DISTRIBUTION SYSTEM	TOTAL HALOACETIC ACIDS (HAA5)	39.5	ug/L	19-Aug-04	CWS	585,000
AL	AL0000738	BIRMINGHAM WATER BOARD	DISTRIBUTION SYSTEM	TOTAL HALOACETIC ACIDS (HAA5)	33.8	ug/L	19-Aug-04	CWS	585,000
AL	AL0000738	BIRMINGHAM WATER BOARD	DISTRIBUTION SYSTEM	TOTAL HALOACETIC ACIDS (HAA5)	30.4	ug/L	19-Aug-04	CWS	585,000
AL	AL0000738	BIRMINGHAM WATER BOARD	DISTRIBUTION SYSTEM	TOTAL HALOACETIC ACIDS (HAA5)	28.2	ug/L	19-Aug-04	CWS	585,000
AL	AL0000738	BIRMINGHAM WATER BOARD	DISTRIBUTION SYSTEM	TOTAL HALOACETIC ACIDS (HAA5)	27.4	ug/L	19-Aug-04	CWS	585,000
AL	AL0000738	BIRMINGHAM WATER BOARD	DISTRIBUTION SYSTEM	TOTAL HALOACETIC ACIDS (HAA5)	24.8	ug/L	19-Aug-04	CWS	585,000
AL	AL0000738	BIRMINGHAM WATER BOARD	DISTRIBUTION SYSTEM	TOTAL HALOACETIC ACIDS (HAA5)	21.6	ug/L	19-Aug-04	CWS	585,000
AL	AL0000738	BIRMINGHAM WATER BOARD	DISTRIBUTION SYSTEM	TOTAL HALOACETIC ACIDS (HAA5)	20.9	ug/L	19-Aug-04	CWS	585,000
AL	AL0000738	BIRMINGHAM WATER BOARD	DISTRIBUTION SYSTEM	TOTAL HALOACETIC ACIDS (HAA5)	20.8	ug/L	19-Aug-04	CWS	585,000
AL	AL0000738	BIRMINGHAM WATER BOARD	DISTRIBUTION SYSTEM	TOTAL HALOACETIC ACIDS (HAA5)	19	ug/L	19-Aug-04	CWS	585,000
AL	AL0000738	BIRMINGHAM WATER BOARD	DISTRIBUTION SYSTEM	TOTAL HALOACETIC ACIDS (HAA5)	16.4	ug/L	19-Aug-04	CWS	585,000
AL	AL0000738	BIRMINGHAM WATER BOARD	DISTRIBUTION SYSTEM	TOTAL HALOACETIC ACIDS (HAA5)	15.8	ug/L	19-Aug-04	CWS	585,000
AL	AL0000738	BIRMINGHAM WATER BOARD	DISTRIBUTION SYSTEM	TOTAL HALOACETIC ACIDS (HAA5)	14	ug/L	19-Aug-04	CWS	585,000
AL	AL0000738	BIRMINGHAM WATER BOARD	DISTRIBUTION SYSTEM	TOTAL HALOACETIC ACIDS (HAA5)	12.7	ug/L	19-Aug-04	CWS	585,000
AL	AL0000738	BIRMINGHAM WATER BOARD	DISTRIBUTION SYSTEM	TOTAL HALOACETIC ACIDS (HAA5)	10.5	ug/L	19-Aug-04	CWS	585,000
AL	AL0000738	BIRMINGHAM WATER BOARD	PUTNAM WATER TREATMENT PLANT	TOTAL HALOACETIC ACIDS (HAA5)	11.41	ug/L	19-Aug-04	CWS	585,000
AL	AL0000738	BIRMINGHAM WATER BOARD	SHADES MTN WATER TREATMENT PLANT	TOTAL HALOACETIC ACIDS (HAA5)	63.7	ug/L	19-Aug-04	CWS	585,000
AL	AL0000738	BIRMINGHAM WATER BOARD	WESTERN WATER TREATMENT PLANT	TOTAL HALOACETIC ACIDS (HAA5)	12.4	ug/L	19-Aug-04	CWS	585,000
AL	AL0000738	BIRMINGHAM WATER BOARD	DISTRIBUTION SYSTEM	TOTAL HALOACETIC ACIDS (HAA5)	13.7	ug/L	31-Aug-04	CWS	585,000
AL	AL0000738	BIRMINGHAM WATER BOARD	CARSON WATER TREATMENT PLANT	TOTAL HALOACETIC ACIDS (HAA5)	11	ug/L	08-Dec-04	CWS	585,000
AL	AL0000738	BIRMINGHAM WATER BOARD	DISTRIBUTION SYSTEM	TOTAL HALOACETIC ACIDS (HAA5)	25	ug/L	08-Dec-04	CWS	585,000
AL	AL0000738	BIRMINGHAM WATER BOARD	DISTRIBUTION SYSTEM	TOTAL HALOACETIC ACIDS (HAA5)	25	ug/L	08-Dec-04	CWS	585,000
AL	AL0000738	BIRMINGHAM WATER BOARD	DISTRIBUTION SYSTEM	TOTAL HALOACETIC ACIDS (HAA5)	24.8	ug/L	08-Dec-04	CWS	585,000
AL	AL0000738	BIRMINGHAM WATER BOARD	DISTRIBUTION SYSTEM	TOTAL HALOACETIC ACIDS (HAA5)	23	ug/L	08-Dec-04	CWS	585,000
AL	AL0000738	BIRMINGHAM WATER BOARD	DISTRIBUTION SYSTEM	TOTAL HALOACETIC ACIDS (HAA5)	21	ug/L	08-Dec-04	CWS	585,000
AL	AL0000738	BIRMINGHAM WATER BOARD	DISTRIBUTION SYSTEM	TOTAL HALOACETIC ACIDS (HAA5)	20	ug/L	08-Dec-04	CWS	585,000
AL	AL0000738	BIRMINGHAM WATER BOARD	DISTRIBUTION SYSTEM	TOTAL HALOACETIC ACIDS (HAA5)	20	ug/L	08-Dec-04	CWS	585,000

AL	AL0000738	BIRMINGHAM WATER BOARD	DISTRIBUTION SYSTEM	TOTAL HALOACETIC ACIDS (HAA5)	18	ug/L	08-Dec-04	CWS	585,000
AL	AL0000738	BIRMINGHAM WATER BOARD	DISTRIBUTION SYSTEM	TOTAL HALOACETIC ACIDS (HAA5)	18	ug/L	08-Dec-04	CWS	585,000
AL	AL0000738	BIRMINGHAM WATER BOARD	DISTRIBUTION SYSTEM	TOTAL HALOACETIC ACIDS (HAA5)	18	ug/L	08-Dec-04	CWS	585,000
AL	AL0000738	BIRMINGHAM WATER BOARD	DISTRIBUTION SYSTEM	TOTAL HALOACETIC ACIDS (HAA5)	17	ug/L	08-Dec-04	CWS	585,000
AL	AL0000738	BIRMINGHAM WATER BOARD	DISTRIBUTION SYSTEM	TOTAL HALOACETIC ACIDS (HAA5)	16	ug/L	08-Dec-04	CWS	585,000
AL	AL0000738	BIRMINGHAM WATER BOARD	DISTRIBUTION SYSTEM	TOTAL HALOACETIC ACIDS (HAA5)	14	ug/L	08-Dec-04	CWS	585,000
AL	AL0000738	BIRMINGHAM WATER BOARD	DISTRIBUTION SYSTEM	TOTAL HALOACETIC ACIDS (HAA5)	13	ug/L	08-Dec-04	CWS	585,000
AL	AL0000738	BIRMINGHAM WATER BOARD	DISTRIBUTION SYSTEM	TOTAL HALOACETIC ACIDS (HAA5)	13	ug/L	08-Dec-04	CWS	585,000
AL	AL0000738	BIRMINGHAM WATER BOARD	DISTRIBUTION SYSTEM	TOTAL HALOACETIC ACIDS (HAA5)	12	ug/L	08-Dec-04	CWS	585,000
AL	AL0000738	BIRMINGHAM WATER BOARD	PUTNAM WATER TREATMENT PLANT	TOTAL HALOACETIC ACIDS (HAA5)	9	ug/L	08-Dec-04	CWS	585,000
AL	AL0000738	BIRMINGHAM WATER BOARD	SHADES MTN WATER TREATMENT PLANT	TOTAL HALOACETIC ACIDS (HAA5)	26	ug/L	08-Dec-04	CWS	585,000
AL	AL0000738	BIRMINGHAM WATER BOARD	WESTERN WATER TREATMENT PLANT	TOTAL HALOACETIC ACIDS (HAA5)	9	ug/L	08-Dec-04	CWS	585,000
AL	AL0000738	BIRMINGHAM WATER BOARD	CARSON WATER TREATMENT PLANT	TOTAL HALOACETIC ACIDS (HAA5)	17.1	ug/L	15-Jan-05	CWS	585,000
AL	AL0000738	BIRMINGHAM WATER BOARD	DISTRIBUTION SYSTEM	TOTAL HALOACETIC ACIDS (HAA5)	31.3	ug/L	15-Jan-05	CWS	585,000
AL	AL0000738	BIRMINGHAM WATER BOARD	DISTRIBUTION SYSTEM	TOTAL HALOACETIC ACIDS (HAA5)	28	ug/L	15-Jan-05	CWS	585,000
AL	AL0000738	BIRMINGHAM WATER BOARD	DISTRIBUTION SYSTEM	TOTAL HALOACETIC ACIDS (HAA5)	26.7	ug/L	15-Jan-05	CWS	585,000
AL	AL0000738	BIRMINGHAM WATER BOARD	DISTRIBUTION SYSTEM	TOTAL HALOACETIC ACIDS (HAA5)	26.5	ug/L	15-Jan-05	CWS	585,000
AL	AL0000738	BIRMINGHAM WATER BOARD	DISTRIBUTION SYSTEM	TOTAL HALOACETIC ACIDS (HAA5)	26	ug/L	15-Jan-05	CWS	585,000
AL	AL0000738	BIRMINGHAM WATER BOARD	DISTRIBUTION SYSTEM	TOTAL HALOACETIC ACIDS (HAA5)	25.5	ug/L	15-Jan-05	CWS	585,000
AL	AL0000738	BIRMINGHAM WATER BOARD	DISTRIBUTION SYSTEM	TOTAL HALOACETIC ACIDS (HAA5)	25	ug/L	15-Jan-05	CWS	585,000
AL	AL0000738	BIRMINGHAM WATER BOARD	DISTRIBUTION SYSTEM	TOTAL HALOACETIC ACIDS (HAA5)	24.1	ug/L	15-Jan-05	CWS	585,000
AL	AL0000738	BIRMINGHAM WATER BOARD	DISTRIBUTION SYSTEM	TOTAL HALOACETIC ACIDS (HAA5)	22	ug/L	15-Jan-05	CWS	585,000
AL	AL0000738	BIRMINGHAM WATER BOARD	DISTRIBUTION SYSTEM	TOTAL HALOACETIC ACIDS (HAA5)	20.5	ug/L	15-Jan-05	CWS	585,000
AL	AL0000738	BIRMINGHAM WATER BOARD	DISTRIBUTION SYSTEM	TOTAL HALOACETIC ACIDS (HAA5)	20.4	ug/L	15-Jan-05	CWS	585,000
AL	AL0000738	BIRMINGHAM WATER BOARD	DISTRIBUTION SYSTEM	TOTAL HALOACETIC ACIDS (HAA5)	20.2	ug/L	15-Jan-05	CWS	585,000
AL	AL0000738	BIRMINGHAM WATER BOARD	DISTRIBUTION SYSTEM	TOTAL HALOACETIC ACIDS (HAA5)	18.9	ug/L	15-Jan-05	CWS	585,000
AL	AL0000738	BIRMINGHAM WATER BOARD	DISTRIBUTION SYSTEM	TOTAL HALOACETIC ACIDS (HAA5)	18	ug/L	15-Jan-05	CWS	585,000
AL	AL0000738	BIRMINGHAM WATER BOARD	DISTRIBUTION SYSTEM	TOTAL HALOACETIC ACIDS (HAA5)	17.5	ug/L	15-Jan-05	CWS	585,000
AL	AL0000738	BIRMINGHAM WATER BOARD	DISTRIBUTION SYSTEM	TOTAL HALOACETIC ACIDS (HAA5)	17.1	ug/L	15-Jan-05	CWS	585,000
AL	AL0000738	BIRMINGHAM WATER BOARD	PUTNAM WATER TREATMENT PLANT	TOTAL HALOACETIC ACIDS (HAA5)	14.9	ug/L	15-Jan-05	CWS	585,000
AL	AL0000738	BIRMINGHAM WATER BOARD	SHADES MTN WATER TREATMENT PLANT	TOTAL HALOACETIC ACIDS (HAA5)	22.3	ug/L	15-Jan-05	CWS	585,000
AL	AL0000738	BIRMINGHAM WATER BOARD	WESTERN WATER TREATMENT PLANT	TOTAL HALOACETIC ACIDS (HAA5)	165	ug/L	15-Jan-05	CWS	585,000
AL	AL0000738	BIRMINGHAM WATER BOARD	CARSON WATER TREATMENT PLANT	TOTAL HALOACETIC ACIDS (HAA5)	19.3	ug/L	10-May-05	CWS	585,000
AL	AL0000738	BIRMINGHAM WATER BOARD	DISTRIBUTION SYSTEM	TOTAL HALOACETIC ACIDS (HAA5)	45.1	ug/L	10-May-05	CWS	585,000
AL	AL0000738	BIRMINGHAM WATER BOARD	DISTRIBUTION SYSTEM	TOTAL HALOACETIC ACIDS (HAA5)	39.3	ug/L	10-May-05	CWS	585,000
AL	AL0000738	BIRMINGHAM WATER BOARD	DISTRIBUTION SYSTEM	TOTAL HALOACETIC ACIDS (HAA5)	34.9	ug/L	10-May-05	CWS	585,000
AL	AL0000738	BIRMINGHAM WATER BOARD	DISTRIBUTION SYSTEM	TOTAL HALOACETIC ACIDS (HAA5)	33.8	ug/L	10-May-05	CWS	585,000
AL	AL0000738	BIRMINGHAM WATER BOARD	DISTRIBUTION SYSTEM	TOTAL HALOACETIC ACIDS (HAA5)	32.9	ug/L	10-May-05	CWS	585,000
AL	AL0000738	BIRMINGHAM WATER BOARD	DISTRIBUTION SYSTEM	TOTAL HALOACETIC ACIDS (HAA5)	32.1	ug/L	10-May-05	CWS	585,000
AL	AL0000738	BIRMINGHAM WATER BOARD	DISTRIBUTION SYSTEM	TOTAL HALOACETIC ACIDS (HAA5)	31.5	ug/L	10-May-05	CWS	585,000
AL	AL0000738	BIRMINGHAM WATER BOARD	DISTRIBUTION SYSTEM	TOTAL HALOACETIC ACIDS (HAA5)	30.6	ug/L	10-May-05	CWS	585,000
AL	AL0000738	BIRMINGHAM WATER BOARD	DISTRIBUTION SYSTEM	TOTAL HALOACETIC ACIDS (HAA5)	29	ug/L	10-May-05	CWS	585,000
AL	AL0000738	BIRMINGHAM WATER BOARD	DISTRIBUTION SYSTEM	TOTAL HALOACETIC ACIDS (HAA5)	28.9	ug/L	10-May-05	CWS	585,000
AL	AL0000738	BIRMINGHAM WATER BOARD	DISTRIBUTION SYSTEM	TOTAL HALOACETIC ACIDS (HAA5)	28.2	ug/L	10-May-05	CWS	585,000
AL	AL0000738	BIRMINGHAM WATER BOARD	DISTRIBUTION SYSTEM	TOTAL HALOACETIC ACIDS (HAA5)	26.7	ug/L	10-May-05	CWS	585,000

APPENDIX 7-7 *Continued*

State	PWSID	Water System Name	Source	Analyte	Level	Unit	Sample Date	PWSType	Pop
AL	AL0000738	BIRMINGHAM WATER BOARD	DISTRIBUTION SYSTEM	TOTAL HALOACETIC ACIDS (HAA5)	26.5	ug/L	10-May-05	CWS	585,000
AL	AL0000738	BIRMINGHAM WATER BOARD	DISTRIBUTION SYSTEM	TOTAL HALOACETIC ACIDS (HAA5)	25.7	ug/L	10-May-05	CWS	585,000
AL	AL0000738	BIRMINGHAM WATER BOARD	DISTRIBUTION SYSTEM	TOTAL HALOACETIC ACIDS (HAA5)	24.9	ug/L	10-May-05	CWS	585,000
AL	AL0000738	BIRMINGHAM WATER BOARD	DISTRIBUTION SYSTEM	TOTAL HALOACETIC ACIDS (HAA5)	22.1	ug/L	10-May-05	CWS	585,000
AL	AL0000738	BIRMINGHAM WATER BOARD	PUTNAM WATER TREATMENT PLANT	TOTAL HALOACETIC ACIDS (HAA5)	19.7	ug/L	10-May-05	CWS	585,000
AL	AL0000738	BIRMINGHAM WATER BOARD	SHADES MTN WATER TREATMENT PLANT	TOTAL HALOACETIC ACIDS (HAA5)	15.9	ug/L	10-May-05	CWS	585,000
AL	AL0000738	BIRMINGHAM WATER BOARD	WESTERN WATER TREATMENT PLANT	TOTAL HALOACETIC ACIDS (HAA5)	29	ug/L	10-May-05	CWS	585,000
					24.1				
AL	AL0000738	BIRMINGHAM WATER BOARD	CARSON WATER TREATMENT PLANT	TOTAL TRIHALOMETHANES (TTHM)	10.6	ug/L	10-Feb-04	CWS	585,000
AL	AL0000738	BIRMINGHAM WATER BOARD	DISTRIBUTION SYSTEM	TOTAL TRIHALOMETHANES (TTHM)	42.9	ug/L	10-Feb-04	CWS	585,000
AL	AL0000738	BIRMINGHAM WATER BOARD	DISTRIBUTION SYSTEM	TOTAL TRIHALOMETHANES (TTHM)	39.9	ug/L	10-Feb-04	CWS	585,000
AL	AL0000738	BIRMINGHAM WATER BOARD	DISTRIBUTION SYSTEM	TOTAL TRIHALOMETHANES (TTHM)	35.9	ug/L	10-Feb-04	CWS	585,000
AL	AL0000738	BIRMINGHAM WATER BOARD	DISTRIBUTION SYSTEM	TOTAL TRIHALOMETHANES (TTHM)	35.6	ug/L	10-Feb-04	CWS	585,000
AL	AL0000738	BIRMINGHAM WATER BOARD	DISTRIBUTION SYSTEM	TOTAL TRIHALOMETHANES (TTHM)	33.1	ug/L	10-Feb-04	CWS	585,000
AL	AL0000738	BIRMINGHAM WATER BOARD	DISTRIBUTION SYSTEM	TOTAL TRIHALOMETHANES (TTHM)	32.2	ug/L	10-Feb-04	CWS	585,000
AL	AL0000738	BIRMINGHAM WATER BOARD	DISTRIBUTION SYSTEM	TOTAL TRIHALOMETHANES (TTHM)	30	ug/L	10-Feb-04	CWS	585,000
AL	AL0000738	BIRMINGHAM WATER BOARD	DISTRIBUTION SYSTEM	TOTAL TRIHALOMETHANES (TTHM)	29.8	ug/L	10-Feb-04	CWS	585,000
AL	AL0000738	BIRMINGHAM WATER BOARD	DISTRIBUTION SYSTEM	TOTAL TRIHALOMETHANES (TTHM)	26.8	ug/L	10-Feb-04	CWS	585,000
AL	AL0000738	BIRMINGHAM WATER BOARD	DISTRIBUTION SYSTEM	TOTAL TRIHALOMETHANES (TTHM)	26	ug/L	10-Feb-04	CWS	585,000
AL	AL0000738	BIRMINGHAM WATER BOARD	DISTRIBUTION SYSTEM	TOTAL TRIHALOMETHANES (TTHM)	26	ug/L	10-Feb-04	CWS	585,000
AL	AL0000738	BIRMINGHAM WATER BOARD	DISTRIBUTION SYSTEM	TOTAL TRIHALOMETHANES (TTHM)	26	ug/L	10-Feb-04	CWS	585,000
AL	AL0000738	BIRMINGHAM WATER BOARD	DISTRIBUTION SYSTEM	TOTAL TRIHALOMETHANES (TTHM)	19.6	ug/L	10-Feb-04	CWS	585,000
AL	AL0000738	BIRMINGHAM WATER BOARD	DISTRIBUTION SYSTEM	TOTAL TRIHALOMETHANES (TTHM)	18.3	ug/L	10-Feb-04	CWS	585,000
AL	AL0000738	BIRMINGHAM WATER BOARD	DISTRIBUTION SYSTEM	TOTAL TRIHALOMETHANES (TTHM)	17.6	ug/L	10-Feb-04	CWS	585,000
AL	AL0000738	BIRMINGHAM WATER BOARD	DISTRIBUTION SYSTEM	TOTAL TRIHALOMETHANES (TTHM)	17.2	ug/L	10-Feb-04	CWS	585,000
AL	AL0000738	BIRMINGHAM WATER BOARD	PUTNAM WATER TREATMENT PLANT	TOTAL TRIHALOMETHANES (TTHM)	10.6	ug/L	10-Feb-04	CWS	585,000
AL	AL0000738	BIRMINGHAM WATER BOARD	SHADES MTN WATER TREATMENT PLANT	TOTAL TRIHALOMETHANES (TTHM)	23.3	ug/L	10-Feb-04	CWS	585,000
AL	AL0000738	BIRMINGHAM WATER BOARD	WESTERN WATER TREATMENT PLANT	TOTAL TRIHALOMETHANES (TTHM)	18.1	ug/L	10-Feb-04	CWS	585,000
AL	AL0000738	BIRMINGHAM WATER BOARD	CARSON WATER TREATMENT PLANT	TOTAL TRIHALOMETHANES (TTHM)	9.72	ug/L	11-May-04	CWS	585,000
AL	AL0000738	BIRMINGHAM WATER BOARD	DISTRIBUTION SYSTEM	TOTAL TRIHALOMETHANES (TTHM)	53.5	ug/L	11-May-04	CWS	585,000
AL	AL0000738	BIRMINGHAM WATER BOARD	DISTRIBUTION SYSTEM	TOTAL TRIHALOMETHANES (TTHM)	45.5	ug/L	11-May-04	CWS	585,000
AL	AL0000738	BIRMINGHAM WATER BOARD	DISTRIBUTION SYSTEM	TOTAL TRIHALOMETHANES (TTHM)	40.3	ug/L	11-May-04	CWS	585,000
AL	AL0000738	BIRMINGHAM WATER BOARD	DISTRIBUTION SYSTEM	TOTAL TRIHALOMETHANES (TTHM)	40.1	ug/L	11-May-04	CWS	585,000
AL	AL0000738	BIRMINGHAM WATER BOARD	DISTRIBUTION SYSTEM	TOTAL TRIHALOMETHANES (TTHM)	38.1	ug/L	11-May-04	CWS	585,000
AL	AL0000738	BIRMINGHAM WATER BOARD	DISTRIBUTION SYSTEM	TOTAL TRIHALOMETHANES (TTHM)	37	ug/L	11-May-04	CWS	585,000
AL	AL0000738	BIRMINGHAM WATER BOARD	DISTRIBUTION SYSTEM	TOTAL TRIHALOMETHANES (TTHM)	36.9	ug/L	11-May-04	CWS	585,000
AL	AL0000738	BIRMINGHAM WATER BOARD	DISTRIBUTION SYSTEM	TOTAL TRIHALOMETHANES (TTHM)	36.8	ug/L	11-May-04	CWS	585,000
AL	AL0000738	BIRMINGHAM WATER BOARD	DISTRIBUTION SYSTEM	TOTAL TRIHALOMETHANES (TTHM)	35.8	ug/L	11-May-04	CWS	585,000
AL	AL0000738	BIRMINGHAM WATER BOARD	DISTRIBUTION SYSTEM	TOTAL TRIHALOMETHANES (TTHM)	35.5	ug/L	11-May-04	CWS	585,000
AL	AL0000738	BIRMINGHAM WATER BOARD	DISTRIBUTION SYSTEM	TOTAL TRIHALOMETHANES (TTHM)	34.4	ug/L	11-May-04	CWS	585,000
AL	AL0000738	BIRMINGHAM WATER BOARD	DISTRIBUTION SYSTEM	TOTAL TRIHALOMETHANES (TTHM)	30.5	ug/L	11-May-04	CWS	585,000
AL	AL0000738	BIRMINGHAM WATER BOARD	DISTRIBUTION SYSTEM	TOTAL TRIHALOMETHANES (TTHM)	30.2	ug/L	11-May-04	CWS	585,000

AL	AL0000738	BIRMINGHAM WATER BOARD	DISTRIBUTION SYSTEM	TOTAL TRIHALOMETHANES (TTHM)	29.4	ug/L	11-May-04	CWS	585,000
AL	AL0000738	BIRMINGHAM WATER BOARD	DISTRIBUTION SYSTEM	TOTAL TRIHALOMETHANES (TTHM)	26.7	ug/L	11-May-04	CWS	585,000
AL	AL0000738	BIRMINGHAM WATER BOARD	DISTRIBUTION SYSTEM	TOTAL TRIHALOMETHANES (TTHM)	16	ug/L	11-May-04	CWS	585,000
AL	AL0000738	BIRMINGHAM WATER BOARD	PUTNAM WATER TREATMENT PLANT	TOTAL TRIHALOMETHANES (TTHM)	12.5	ug/L	11-May-04	CWS	585,000
AL	AL0000738	BIRMINGHAM WATER BOARD	SHADES MTN WATER TREATMENT PLANT	TOTAL TRIHALOMETHANES (TTHM)	31.2	ug/L	11-May-04	CWS	585,000
AL	AL0000738	BIRMINGHAM WATER BOARD	WESTERN WATER TREATMENT PLANT	TOTAL TRIHALOMETHANES (TTHM)	18.9	ug/L	11-May-04	CWS	585,000
AL	AL0000738	BIRMINGHAM WATER BOARD	CARSON WATER TREATMENT PLANT	TOTAL TRIHALOMETHANES (TTHM)	13.8	ug/L	19-Aug-04	CWS	585,000
AL	AL0000738	BIRMINGHAM WATER BOARD	DISTRIBUTION SYSTEM	TOTAL TRIHALOMETHANES (TTHM)	63.5	ug/L	19-Aug-04	CWS	585,000
AL	AL0000738	BIRMINGHAM WATER BOARD	DISTRIBUTION SYSTEM	TOTAL TRIHALOMETHANES (TTHM)	61.5	ug/L	19-Aug-04	CWS	585,000
AL	AL0000738	BIRMINGHAM WATER BOARD	DISTRIBUTION SYSTEM	TOTAL TRIHALOMETHANES (TTHM)	43.7	ug/L	19-Aug-04	CWS	585,000
AL	AL0000738	BIRMINGHAM WATER BOARD	DISTRIBUTION SYSTEM	TOTAL TRIHALOMETHANES (TTHM)	38.7	ug/L	19-Aug-04	CWS	585,000
AL	AL0000738	BIRMINGHAM WATER BOARD	DISTRIBUTION SYSTEM	TOTAL TRIHALOMETHANES (TTHM)	36.4	ug/L	19-Aug-04	CWS	585,000
AL	AL0000738	BIRMINGHAM WATER BOARD	DISTRIBUTION SYSTEM	TOTAL TRIHALOMETHANES (TTHM)	36.2	ug/L	19-Aug-04	CWS	585,000
AL	AL0000738	BIRMINGHAM WATER BOARD	DISTRIBUTION SYSTEM	TOTAL TRIHALOMETHANES (TTHM)	30.6	ug/L	19-Aug-04	CWS	585,000
AL	AL0000738	BIRMINGHAM WATER BOARD	DISTRIBUTION SYSTEM	TOTAL TRIHALOMETHANES (TTHM)	30.6	ug/L	19-Aug-04	CWS	585,000
AL	AL0000738	BIRMINGHAM WATER BOARD	DISTRIBUTION SYSTEM	TOTAL TRIHALOMETHANES (TTHM)	30.6	ug/L	19-Aug-04	CWS	585,000
AL	AL0000738	BIRMINGHAM WATER BOARD	DISTRIBUTION SYSTEM	TOTAL TRIHALOMETHANES (TTHM)	27.8	ug/L	19-Aug-04	CWS	585,000
AL	AL0000738	BIRMINGHAM WATER BOARD	DISTRIBUTION SYSTEM	TOTAL TRIHALOMETHANES (TTHM)	26.2	ug/L	19-Aug-04	CWS	585,000
AL	AL0000738	BIRMINGHAM WATER BOARD	DISTRIBUTION SYSTEM	TOTAL TRIHALOMETHANES (TTHM)	24.6	ug/L	19-Aug-04	CWS	585,000
AL	AL0000738	BIRMINGHAM WATER BOARD	DISTRIBUTION SYSTEM	TOTAL TRIHALOMETHANES (TTHM)	21.4	ug/L	19-Aug-04	CWS	585,000
AL	AL0000738	BIRMINGHAM WATER BOARD	DISTRIBUTION SYSTEM	TOTAL TRIHALOMETHANES (TTHM)	20.5	ug/L	19-Aug-04	CWS	585,000
AL	AL0000738	BIRMINGHAM WATER BOARD	DISTRIBUTION SYSTEM	TOTAL TRIHALOMETHANES (TTHM)	8.2	ug/L	19-Aug-04	CWS	585,000
AL	AL0000738	BIRMINGHAM WATER BOARD	PUTNAM WATER TREATMENT PLANT	TOTAL TRIHALOMETHANES (TTHM)	9.33	ug/L	19-Aug-04	CWS	585,000
AL	AL0000738	BIRMINGHAM WATER BOARD	SHADES MTN WATER TREATMENT PLANT	TOTAL TRIHALOMETHANES (TTHM)	42.9	ug/L	19-Aug-04	CWS	585,000
AL	AL0000738	BIRMINGHAM WATER BOARD	WESTERN WATER TREATMENT PLANT	TOTAL TRIHALOMETHANES (TTHM)	19.1	ug/L	19-Aug-04	CWS	585,000
AL	AL0000738	BIRMINGHAM WATER BOARD	DISTRIBUTION SYSTEM	TOTAL TRIHALOMETHANES (TTHM)	57.3	ug/L	31-Aug-04	CWS	585,000
AL	AL0000738	BIRMINGHAM WATER BOARD	CARSON WATER TREATMENT PLANT	TOTAL TRIHALOMETHANES (TTHM)	22.8	ug/L	09-Nov-04	CWS	585,000
AL	AL0000738	BIRMINGHAM WATER BOARD	DISTRIBUTION SYSTEM	TOTAL TRIHALOMETHANES (TTHM)	73.2	ug/L	09-Nov-04	CWS	585,000
AL	AL0000738	BIRMINGHAM WATER BOARD	DISTRIBUTION SYSTEM	TOTAL TRIHALOMETHANES (TTHM)	69.6	ug/L	09-Nov-04	CWS	585,000
AL	AL0000738	BIRMINGHAM WATER BOARD	DISTRIBUTION SYSTEM	TOTAL TRIHALOMETHANES (TTHM)	68.6	ug/L	09-Nov-04	CWS	585,000
AL	AL0000738	BIRMINGHAM WATER BOARD	DISTRIBUTION SYSTEM	TOTAL TRIHALOMETHANES (TTHM)	60.9	ug/L	09-Nov-04	CWS	585,000
AL	AL0000738	BIRMINGHAM WATER BOARD	DISTRIBUTION SYSTEM	TOTAL TRIHALOMETHANES (TTHM)	56.9	ug/L	09-Nov-04	CWS	585,000
AL	AL0000738	BIRMINGHAM WATER BOARD	DISTRIBUTION SYSTEM	TOTAL TRIHALOMETHANES (TTHM)	52.4	ug/L	09-Nov-04	CWS	585,000
AL	AL0000738	BIRMINGHAM WATER BOARD	DISTRIBUTION SYSTEM	TOTAL TRIHALOMETHANES (TTHM)	48.7	ug/L	09-Nov-04	CWS	585,000
AL	AL0000738	BIRMINGHAM WATER BOARD	DISTRIBUTION SYSTEM	TOTAL TRIHALOMETHANES (TTHM)	40.1	ug/L	09-Nov-04	CWS	585,000
AL	AL0000738	BIRMINGHAM WATER BOARD	DISTRIBUTION SYSTEM	TOTAL TRIHALOMETHANES (TTHM)	38.4	ug/L	09-Nov-04	CWS	585,000
AL	AL0000738	BIRMINGHAM WATER BOARD	DISTRIBUTION SYSTEM	TOTAL TRIHALOMETHANES (TTHM)	38.2	ug/L	09-Nov-04	CWS	585,000
AL	AL0000738	BIRMINGHAM WATER BOARD	DISTRIBUTION SYSTEM	TOTAL TRIHALOMETHANES (TTHM)	35.8	ug/L	09-Nov-04	CWS	585,000
AL	AL0000738	BIRMINGHAM WATER BOARD	DISTRIBUTION SYSTEM	TOTAL TRIHALOMETHANES (TTHM)	32.1	ug/L	09-Nov-04	CWS	585,000
AL	AL0000738	BIRMINGHAM WATER BOARD	DISTRIBUTION SYSTEM	TOTAL TRIHALOMETHANES (TTHM)	30.3	ug/L	09-Nov-04	CWS	585,000
AL	AL0000738	BIRMINGHAM WATER BOARD	DISTRIBUTION SYSTEM	TOTAL TRIHALOMETHANES (TTHM)	29.4	ug/L	09-Nov-04	CWS	585,000
AL	AL0000738	BIRMINGHAM WATER BOARD	DISTRIBUTION SYSTEM	TOTAL TRIHALOMETHANES (TTHM)	28.7	ug/L	09-Nov-04	CWS	585,000
AL	AL0000738	BIRMINGHAM WATER BOARD	DISTRIBUTION SYSTEM	TOTAL TRIHALOMETHANES (TTHM)	24	ug/L	09-Nov-04	CWS	585,000
AL	AL0000738	BIRMINGHAM WATER BOARD	PUTNAM WATER TREATMENT PLANT	TOTAL TRIHALOMETHANES (TTHM)	20.1	ug/L	09-Nov-04	CWS	585,000
AL	AL0000738	BIRMINGHAM WATER BOARD	SHADES MTN WATER TREATMENT PLANT	TOTAL TRIHALOMETHANES (TTHM)	42.6	ug/L	09-Nov-04	CWS	585,000

APPENDIX 7-7 *Continued*

State	PWSID	Water System Name	Source	Analyte	Level	Unit	Sample Date	PWSType	Pop
AL	AL0000738	BIRMINGHAM WATER BOARD	WESTERN WATER TREATMENT PLANT	TOTAL TRIHALOMETHANES (TTHM)	19.8	ug/L	09-Nov-04	CWS	585,000
AL	AL0000738	BIRMINGHAM WATER BOARD	CARSON WATER TREATMENT PLANT	TOTAL TRIHALOMETHANES (TTHM)	13.9	ug/L	15-Jan-05	CWS	585,000
AL	AL0000738	BIRMINGHAM WATER BOARD	DISTRIBUTION SYSTEM	TOTAL TRIHALOMETHANES (TTHM)	70.1	ug/L	15-Jan-05	CWS	585,000
AL	AL0000738	BIRMINGHAM WATER BOARD	DISTRIBUTION SYSTEM	TOTAL TRIHALOMETHANES (TTHM)	53.7	ug/L	15-Jan-05	CWS	585,000
AL	AL0000738	BIRMINGHAM WATER BOARD	DISTRIBUTION SYSTEM	TOTAL TRIHALOMETHANES (TTHM)	44.3	ug/L	15-Jan-05	CWS	585,000
AL	AL0000738	BIRMINGHAM WATER BOARD	DISTRIBUTION SYSTEM	TOTAL TRIHALOMETHANES (TTHM)	44.3	ug/L	15-Jan-05	CWS	585,000
AL	AL0000738	BIRMINGHAM WATER BOARD	DISTRIBUTION SYSTEM	TOTAL TRIHALOMETHANES (TTHM)	39.5	ug/L	15-Jan-05	CWS	585,000
AL	AL0000738	BIRMINGHAM WATER BOARD	DISTRIBUTION SYSTEM	TOTAL TRIHALOMETHANES (TTHM)	37.9	ug/L	15-Jan-05	CWS	585,000
AL	AL0000738	BIRMINGHAM WATER BOARD	DISTRIBUTION SYSTEM	TOTAL TRIHALOMETHANES (TTHM)	35.1	ug/L	15-Jan-05	CWS	585,000
AL	AL0000738	BIRMINGHAM WATER BOARD	DISTRIBUTION SYSTEM	TOTAL TRIHALOMETHANES (TTHM)	33.7	ug/L	15-Jan-05	CWS	585,000
AL	AL0000738	BIRMINGHAM WATER BOARD	DISTRIBUTION SYSTEM	TOTAL TRIHALOMETHANES (TTHM)	30.1	ug/L	15-Jan-05	CWS	585,000
AL	AL0000738	BIRMINGHAM WATER BOARD	DISTRIBUTION SYSTEM	TOTAL TRIHALOMETHANES (TTHM)	28.3	ug/L	15-Jan-05	CWS	585,000
AL	AL0000738	BIRMINGHAM WATER BOARD	DISTRIBUTION SYSTEM	TOTAL TRIHALOMETHANES (TTHM)	27.6	ug/L	15-Jan-05	CWS	585,000
AL	AL0000738	BIRMINGHAM WATER BOARD	DISTRIBUTION SYSTEM	TOTAL TRIHALOMETHANES (TTHM)	27.1	ug/L	15-Jan-05	CWS	585,000
AL	AL0000738	BIRMINGHAM WATER BOARD	DISTRIBUTION SYSTEM	TOTAL TRIHALOMETHANES (TTHM)	25.7	ug/L	15-Jan-05	CWS	585,000
AL	AL0000738	BIRMINGHAM WATER BOARD	DISTRIBUTION SYSTEM	TOTAL TRIHALOMETHANES (TTHM)	23.2	ug/L	15-Jan-05	CWS	585,000
AL	AL0000738	BIRMINGHAM WATER BOARD	DISTRIBUTION SYSTEM	TOTAL TRIHALOMETHANES (TTHM)	22.5	ug/L	15-Jan-05	CWS	585,000
AL	AL0000738	BIRMINGHAM WATER BOARD	DISTRIBUTION SYSTEM	TOTAL TRIHALOMETHANES (TTHM)	22.1	ug/L	15-Jan-05	CWS	585,000
AL	AL0000738	BIRMINGHAM WATER BOARD	PUTNAM WATER TREATMENT PLANT	TOTAL TRIHALOMETHANES (TTHM)	18.7	ug/L	15-Jan-05	CWS	585,000
AL	AL0000738	BIRMINGHAM WATER BOARD	SHADES MTN WATER TREATMENT PLANT	TOTAL TRIHALOMETHANES (TTHM)	22.5	ug/L	15-Jan-05	CWS	585,000
AL	AL0000738	BIRMINGHAM WATER BOARD	WESTERN WATER TREATMENT PLANT	TOTAL TRIHALOMETHANES (TTHM)	19.8	ug/L	15-Jan-05	CWS	585,000
AL	AL0000738	BIRMINGHAM WATER BOARD	CARSON WATER TREATMENT PLANT	TOTAL TRIHALOMETHANES (TTHM)	22.7	ug/L	10-May-05	CWS	585,000
AL	AL0000738	BIRMINGHAM WATER BOARD	DISTRIBUTION SYSTEM	TOTAL TRIHALOMETHANES (TTHM)	77.3	ug/L	10-May-05	CWS	585,000
AL	AL0000738	BIRMINGHAM WATER BOARD	DISTRIBUTION SYSTEM	TOTAL TRIHALOMETHANES (TTHM)	63.1	ug/L	10-May-05	CWS	585,000
AL	AL0000738	BIRMINGHAM WATER BOARD	DISTRIBUTION SYSTEM	TOTAL TRIHALOMETHANES (TTHM)	62.6	ug/L	10-May-05	CWS	585,000
AL	AL0000738	BIRMINGHAM WATER BOARD	DISTRIBUTION SYSTEM	TOTAL TRIHALOMETHANES (TTHM)	62.1	ug/L	10-May-05	CWS	585,000
AL	AL0000738	BIRMINGHAM WATER BOARD	DISTRIBUTION SYSTEM	TOTAL TRIHALOMETHANES (TTHM)	57.8	ug/L	10-May-05	CWS	585,000
AL	AL0000738	BIRMINGHAM WATER BOARD	DISTRIBUTION SYSTEM	TOTAL TRIHALOMETHANES (TTHM)	54.6	ug/L	10-May-05	CWS	585,000
AL	AL0000738	BIRMINGHAM WATER BOARD	DISTRIBUTION SYSTEM	TOTAL TRIHALOMETHANES (TTHM)	42.8	ug/L	10-May-05	CWS	585,000
AL	AL0000738	BIRMINGHAM WATER BOARD	DISTRIBUTION SYSTEM	TOTAL TRIHALOMETHANES (TTHM)	41.1	ug/L	10-May-05	CWS	585,000
AL	AL0000738	BIRMINGHAM WATER BOARD	DISTRIBUTION SYSTEM	TOTAL TRIHALOMETHANES (TTHM)	41.1	ug/L	10-May-05	CWS	585,000
AL	AL0000738	BIRMINGHAM WATER BOARD	DISTRIBUTION SYSTEM	TOTAL TRIHALOMETHANES (TTHM)	34.8	ug/L	10-May-05	CWS	585,000
AL	AL0000738	BIRMINGHAM WATER BOARD	DISTRIBUTION SYSTEM	TOTAL TRIHALOMETHANES (TTHM)	32.3	ug/L	10-May-05	CWS	585,000
AL	AL0000738	BIRMINGHAM WATER BOARD	DISTRIBUTION SYSTEM	TOTAL TRIHALOMETHANES (TTHM)	28	ug/L	10-May-05	CWS	585,000
AL	AL0000738	BIRMINGHAM WATER BOARD	DISTRIBUTION SYSTEM	TOTAL TRIHALOMETHANES (TTHM)	27	ug/L	10-May-05	CWS	585,000
AL	AL0000738	BIRMINGHAM WATER BOARD	DISTRIBUTION SYSTEM	TOTAL TRIHALOMETHANES (TTHM)	25.9	ug/L	10-May-05	CWS	585,000
AL	AL0000738	BIRMINGHAM WATER BOARD	DISTRIBUTION SYSTEM	TOTAL TRIHALOMETHANES (TTHM)	25.8	ug/L	10-May-05	CWS	585,000
AL	AL0000738	BIRMINGHAM WATER BOARD	DISTRIBUTION SYSTEM	TOTAL TRIHALOMETHANES (TTHM)	25	ug/L	10-May-05	CWS	585,000
AL	AL0000738	BIRMINGHAM WATER BOARD	PUTNAM WATER TREATMENT PLANT	TOTAL TRIHALOMETHANES (TTHM)	31.4	ug/L	10-May-05	CWS	585,000
AL	AL0000738	BIRMINGHAM WATER BOARD	SHADES MTN WATER TREATMENT PLANT	TOTAL TRIHALOMETHANES (TTHM)	21.4	ug/L	10-May-05	CWS	585,000
AL	AL0000738	BIRMINGHAM WATER BOARD	WESTERN WATER TREATMENT PLANT	TOTAL TRIHALOMETHANES (TTHM)	44.2	ug/L	10-May-05	CWS	585,000
					34.1				

APPENDIX 7-8 Boise Groundwater Data

State	PWSNUMBER	PWSNAME	SAMPLING_LOCATION	ANALYTE_NAME	RESULT	RESULT_UNIT	COLLECTION_DATE	PWSType	Pop
ID	ID4010147	TERRA GRANDE WATER COMPANY	WELL #1 HATCH	CARBON TETRACHLORIDE	1.6	UG/L	16-Jun-04	CWS	410
ID	ID4010147	TERRA GRANDE WATER COMPANY	WELL #3 PENNIGER	1,1-DICHLOROETHYLENE	0.7	UG/L	16-Nov-00	CWS	410
ID	ID4010147	TERRA GRANDE WATER COMPANY	WELL #1 HATCH	1,1-DICHLOROETHYLENE	1.3	UG/L	16-Nov-00	CWS	410
ID	ID4010147	TERRA GRANDE WATER COMPANY	WELL #1 HATCH	1,1-DICHLOROETHYLENE	10.9	UG/L	08-Dec-04	CWS	410
ID	ID4010147	TERRA GRANDE WATER COMPANY	WELL #3 PENNIGER	1,1-DICHLOROETHYLENE	9.3	UG/L	09-Dec-04	CWS	410
ID	ID4010147	TERRA GRANDE WATER COMPANY	WELL #3 PENNIGER	1,1-DICHLOROETHYLENE	7.6	UG/L	20-Jan-05	CWS	410
ID	ID4010147	TERRA GRANDE WATER COMPANY	WELL #1 HATCH	1,1-DICHLOROETHYLENE	13.4	UG/L	26-Jan-05	CWS	410
ID	ID4010147	TERRA GRANDE WATER COMPANY	WELL #2 ASH PARK	1,1-DICHLOROETHYLENE	9.8	UG/L	08-Dec-04	CWS	410
ID	ID4010147	TERRA GRANDE WATER COMPANY	WELL #3 PENNIGER	1,1-DICHLOROETHYLENE	1.7	UG/L	20-Jan-05	CWS	410
ID	ID4010147	TERRA GRANDE WATER COMPANY	WELL #2 ASH PARK	1,1-DICHLOROETHYLENE	11.4	UG/L	20-Jan-05	CWS	410
ID	ID4010147	TERRA GRANDE WATER COMPANY	WELL #1 HATCH	1,1-DICHLOROETHYLENE	8.9	UG/L	20-Jan-05	CWS	410
ID	ID4010147	TERRA GRANDE WATER COMPANY	WELL #2 ASH PARK	1,1-DICHLOROETHYLENE	9.5	UG/L	20-Jan-05	CWS	410
ID	ID4010147	TERRA GRANDE WATER COMPANY	WELL #1 HATCH	1,1-DICHLOROETHYLENE	5.4	UG/L	16-Jun-04	CWS	410
					7.49				
ID	ID4010107	NORTHWEST BLDG SYSTEMS	WELL #1	CIS-1,2-DICHLOROETHYLENE	0.9	UG/L	21-Jul-05	NTNCWS	65
ID	ID4010107	NORTHWEST BLDG SYSTEMS	WELL #1	CIS-1,2-DICHLOROETHYLENE	1.1	UG/L	28-Jun-05	NTNCWS	65
ID	ID4010107	NORTHWEST BLDG SYSTEMS	WELL #1	CIS-1,2-DICHLOROETHYLENE	1	UG/L	09-Mar-05	NTNCWS	65
ID	ID4010107	NORTHWEST BLDG SYSTEMS	WELL #1	CIS-1,2-DICHLOROETHYLENE	1	MG/L	01-Dec-04	NTNCWS	65
ID	ID4010107	NORTHWEST BLDG SYSTEMS	WELL #1	CIS-1,2-DICHLOROETHYLENE	0.8	UG/L	25-Aug-04	NTNCWS	65
ID	ID4010107	NORTHWEST BLDG SYSTEMS	WELL #1	CIS-1,2-DICHLOROETHYLENE	0.8	UG/L	23-Jun-04	NTNCWS	65
ID	ID4010107	NORTHWEST BLDG SYSTEMS	WELL #1	CIS-1,2-DICHLOROETHYLENE	0.8	MG/L	28-May-04	NTNCWS	65
ID	ID4010107	NORTHWEST BLDG SYSTEMS	WELL #1	CIS-1,2-DICHLOROETHYLENE	1.1	MG/L	12-Nov-03	NTNCWS	65
ID	ID4010107	NORTHWEST BLDG SYSTEMS	WELL #1	CIS-1,2-DICHLOROETHYLENE	1.2	MG/L	26-Sep-03	NTNCWS	65
ID	ID4010107	NORTHWEST BLDG SYSTEMS	WELL #1	CIS-1,2-DICHLOROETHYLENE	1.2	UG/L	11-Jun-03	NTNCWS	65
ID	ID4010107	NORTHWEST BLDG SYSTEMS	DISTRIBUTION SYSTEM	CIS-1,2-DICHLOROETHYLENE	0.8	UG/L	24-Mar-03	NTNCWS	65
ID	ID4010107	NORTHWEST BLDG SYSTEMS	DISTRIBUTION SYSTEM	CIS-1,2-DICHLOROETHYLENE	1.2	UG/L	22-Oct-02	NTNCWS	65
ID	ID4010107	NORTHWEST BLDG SYSTEMS	WELL #1	CIS-1,2-DICHLOROETHYLENE	0.9	UG/L	28-Mar-02	NTNCWS	65
ID	ID4010147	TERRA GRANDE WATER COMPANY	WELL #1 HATCH	CIS-1,2-DICHLOROETHYLENE	0.6	UG/L	16-Nov-00	CWS	410
ID	ID4010147	TERRA GRANDE WATER COMPANY	WELL #1 HATCH	CIS-1,2-DICHLOROETHYLENE	3.8	UG/L	16-Jun-04	CWS	410
ID	ID4010147	TERRA GRANDE WATER COMPANY	WELL #3 PENNIGER	CIS-1,2-DICHLOROETHYLENE	6.7	UG/L	20-Jan-05	CWS	410
ID	ID4010147	TERRA GRANDE WATER COMPANY	WELL #1 HATCH	CIS-1,2-DICHLOROETHYLENE	6.6	UG/L	20-Jan-05	CWS	410
ID	ID4010147	TERRA GRANDE WATER COMPANY	WELL #2 ASH PARK	CIS-1,2-DICHLOROETHYLENE	6.6	UG/L	20-Jan-05	CWS	410
ID	ID4010147	TERRA GRANDE WATER COMPANY	WELL #1 HATCH	CIS-1,2-DICHLOROETHYLENE	6	UG/L	20-Jan-05	CWS	410
ID	ID4010147	TERRA GRANDE WATER COMPANY	WELL #2 ASH PARK	CIS-1,2-DICHLOROETHYLENE	6.2	UG/L	20-Jan-05	CWS	410
ID	ID4010147	TERRA GRANDE WATER COMPANY	WELL #3 PENNIGER	CIS-1,2-DICHLOROETHYLENE	6	UG/L	20-Jan-05	CWS	410
ID	ID4010147	TERRA GRANDE WATER COMPANY	WELL #1 HATCH	CIS-1,2-DICHLOROETHYLENE	8.6	UG/L	26-Jan-05	CWS	410
ID	ID4010147	TERRA GRANDE WATER COMPANY	WELL #2 ASH PARK	CIS-1,2-DICHLOROETHYLENE	7.3	UG/L	08-Dec-04	CWS	410
ID	ID4010147	TERRA GRANDE WATER COMPANY	WELL #1 HATCH	CIS-1,2-DICHLOROETHYLENE	8.7	UG/L	08-Dec-04	CWS	410
ID	ID4010147	TERRA GRANDE WATER COMPANY	WELL #3 PENNIGER	CIS-1,2-DICHLOROETHYLENE	7	UG/L	09-Dec-04	CWS	410
					3.48				

APPENDIX 7-8 *Continued*

State	PWSNUMBER	PWSNAME	SAMPLING_LOCATION	ANALYTE_NAME	RESULT	RESULT_UNIT	COLLECTION_DATE	PWSType	Pop
ID	ID4010017	BRIAN WATER CORP	WELL #1	DI(2-ETHYLHEXYL) — PHTHALATE	1.2	UG/L	16-Dec-04	CWS	140
ID	ID4010062	FLYING H TRAILER RANCH INC	WELL #4	DI(2-ETHYLHEXYL) — PHTHALATE	3	UG/L	15-Mar-01	CWS	500
ID	ID4010098	ALI MERIDIAN HEIGHTS WSA	WELL #2	DI(2-ETHYLHEXYL) — PHTHALATE	0.92	UG/L	08-Sep-04	CWS	790
					1.71				
ID	ID4010003	BOISE STAGE STOP	WELL #2	TETRACHLOROETHYLENE	0.9	UG/L	18-Sep-01	NTNCWS	100
ID	ID4010022	CAPITOL WATER CORP	WELL #5	TETRACHLOROETHYLENE	1.7	UG/L	30-Aug-05	CWS	7400
ID	ID4010022	CAPITOL WATER CORP	WELL #5	TETRACHLOROETHYLENE	2.3	UG/L	21-Sep-04	CWS	7400
ID	ID4010022	CAPITOL WATER CORP	WELL #7	TETRACHLOROETHYLENE	0.7	UG/L	08-May-02	CWS	7400
ID	ID4010022	CAPITOL WATER CORP	WELL #7	TETRACHLOROETHYLENE	0.7	UG/L	04-Dec-01	CWS	7400
ID	ID4010022	CAPITOL WATER CORP	WELL #4	TETRACHLOROETHYLENE	0.6	UG/L	12-Dec-00	CWS	7400
ID	ID4010059	FIESTA RV PARK	WELL #1-FRONT (BACK UP)	TETRACHLOROETHYLENE	0.8	UG/L	14-Dec-04	CWS	150
ID	ID4010059	FIESTA RV PARK	WELL #1-FRONT (BACK UP)	TETRACHLOROETHYLENE	1.2	UG/L	11-Apr-05	CWS	150
ID	ID4010059	FIESTA RV PARK	WELL #1-FRONT (BACK UP)	TETRACHLOROETHYLENE	1.1	UG/L	18-Mar-05	CWS	150
ID	ID4010107	NORTHWEST BLDG SYSTEMS	DISTRIBUTION SYSTEM	TETRACHLOROETHYLENE	6.2	UG/L	24-Mar-03	NTNCWS	65
					1.62				
ID	ID4010147	TERRA GRANDE WATER COMPANY	WELL #1 HATCH	1,1,1-TRICHLOROETHANE	1.4	UG/L	21-Dec-01	CWS	410
ID	ID4010147	TERRA GRANDE WATER COMPANY	WELL #3 PENNIGER	1,1,1-TRICHLOROETHANE	1.6	UG/L	16-Nov-00	CWS	410
ID	ID4010147	TERRA GRANDE WATER COMPANY	WELL #1 HATCH	1,1,1-TRICHLOROETHANE	2.9	UG/L	16-Nov-00	CWS	410
ID	ID4010147	TERRA GRANDE WATER COMPANY	WELL #1 HATCH	1,1,1-TRICHLOROETHANE	32.8	UG/L	08-Dec-04	CWS	410
ID	ID4010147	TERRA GRANDE WATER COMPANY	WELL #3 PENNIGER	1,1,1-TRICHLOROETHANE	24.6	UG/L	09-Dec-04	CWS	410
ID	ID4010147	TERRA GRANDE WATER COMPANY	WELL #2 ASH PARK	1,1,1-TRICHLOROETHANE	26.2	UG/L	08-Dec-04	CWS	410
ID	ID4010147	TERRA GRANDE WATER COMPANY	WELL #2 ASH PARK	1,1,1-TRICHLOROETHANE	0.8	UG/L	20-Jul-04	CWS	410
ID	ID4010147	TERRA GRANDE WATER COMPANY	WELL #1 HATCH	1,1,1-TRICHLOROETHANE	0.8	UG/L	20-Jul-04	CWS	410
ID	ID4010147	TERRA GRANDE WATER COMPANY	WELL #3 PENNIGER	1,1,1-TRICHLOROETHANE	21.1	UG/L	20-Jan-05	CWS	410
ID	ID4010147	TERRA GRANDE WATER COMPANY	WELL #1 HATCH	1,1,1-TRICHLOROETHANE	34.3	UG/L	26-Jan-05	CWS	410
ID	ID4010147	TERRA GRANDE WATER COMPANY	WELL #2 ASH PARK	1,1,1-TRICHLOROETHANE	29.6	UG/L	20-Jan-05	CWS	410
ID	ID4010147	TERRA GRANDE WATER COMPANY	WELL #1 HATCH	1,1,1-TRICHLOROETHANE	23.2	UG/L	20-Jan-05	CWS	410
ID	ID4010147	TERRA GRANDE WATER COMPANY	WELL #3 PENNIGER	1,1,1-TRICHLOROETHANE	27.5	UG/L	20-Jan-05	CWS	410
ID	ID4010147	TERRA GRANDE WATER COMPANY	WELL #1 HATCH	1,1,1-TRICHLOROETHANE	27.9	UG/L	20-Jan-05	CWS	410
ID	ID4010147	TERRA GRANDE WATER COMPANY	WELL #2 ASH PARK	1,1,1-TRICHLOROETHANE	23.4	UG/L	20-Jan-05	CWS	410
ID	ID4010147	TERRA GRANDE WATER COMPANY	WELL #3 PENNIGER	1,1,1-TRICHLOROETHANE	0.6	UG/L	20-Jul-04	CWS	410
ID	ID4010147	TERRA GRANDE WATER COMPANY	WELL #3 PENNIGER	1,1,1-TRICHLOROETHANE	0.5	UG/L	16-Jun-04	CWS	410
ID	ID4010147	TERRA GRANDE WATER COMPANY	WELL #3 PENNIGER	1,1,1-TRICHLOROETHANE	0.6	UG/L	20-Jul-04	CWS	410
ID	ID4010147	TERRA GRANDE WATER COMPANY	WELL #2 ASH PARK	1,1,1-TRICHLOROETHANE	1	UG/L	16-Jun-04	CWS	410
ID	ID4010147	TERRA GRANDE WATER COMPANY	WELL #1 HATCH	1,1,1-TRICHLOROETHANE	16.7	UG/L	16-Jun-04	CWS	410
					14.9				
ID	ID4010107	NORTHWEST BLDG SYSTEMS	WELL #1	TRICHLOROETHYLENE	6.9	UG/L	10-Jan-01	NTNCWS	65
ID	ID4010107	NORTHWEST BLDG SYSTEMS	WELL #1	TRICHLOROETHYLENE	6.8	UG/L	04-May-00	NTNCWS	65

ID	ID4010107	NORTHWEST BLDG SYSTEMS	WELL #1	TRICHLOROETHYLENE	13.8	UG/L	04-Jan-00	NTNCWS	65
ID	ID4010107	NORTHWEST BLDG SYSTEMS	WELL #1	TRICHLOROETHYLENE	7.8	UG/L	02-Oct-00	NTNCWS	65
ID	ID4010107	NORTHWEST BLDG SYSTEMS	WELL #1	TRICHLOROETHYLENE	10.6	UG/L	06-Jul-00	NTNCWS	65
ID	ID4010107	NORTHWEST BLDG SYSTEMS	WELL #1	TRICHLOROETHYLENE	6.8	UG/L	21-Jul-05	NTNCWS	65
ID	ID4010107	NORTHWEST BLDG SYSTEMS	WELL #1	TRICHLOROETHYLENE	8.9	UG/L	28-Jun-05	NTNCWS	65
ID	ID4010107	NORTHWEST BLDG SYSTEMS	WELL #1	TRICHLOROETHYLENE	7.3	UG/L	09-Mar-05	NTNCWS	65
ID	ID4010107	NORTHWEST BLDG SYSTEMS	WELL #1	TRICHLOROETHYLENE	7.1	MG/L	01-Dec-04	NTNCWS	65
ID	ID4010107	NORTHWEST BLDG SYSTEMS	WELL #1	TRICHLOROETHYLENE	6.6	UG/L	25-Aug-04	NTNCWS	65
ID	ID4010107	NORTHWEST BLDG SYSTEMS	WELL #1	TRICHLOROETHYLENE	6.4	UG/L	23-Jun-04	NTNCWS	65
ID	ID4010107	NORTHWEST BLDG SYSTEMS	WELL #1	TRICHLOROETHYLENE	6.2	MG/L	28-May-04	NTNCWS	65
ID	ID4010107	NORTHWEST BLDG SYSTEMS	WELL #1	TRICHLOROETHYLENE	7.1	MG/L	12-Nov-03	NTNCWS	65
ID	ID4010107	NORTHWEST BLDG SYSTEMS	WELL #1	TRICHLOROETHYLENE	8.3	MG/L	26-Sep-03	NTNCWS	65
ID	ID4010107	NORTHWEST BLDG SYSTEMS	WELL #1	TRICHLOROETHYLENE	7.8	UG/L	11-Jun-03	NTNCWS	65
ID	ID4010107	NORTHWEST BLDG SYSTEMS	DISTRIBUTION SYSTEM	TRICHLOROETHYLENE	5.1	UG/L	22-Oct-02	NTNCWS	65
ID	ID4010107	NORTHWEST BLDG SYSTEMS	WELL #1	TRICHLOROETHYLENE	8	UG/L	28-Mar-02	NTNCWS	65
ID	ID4010147	TERRA GRANDE WATER COMPANY	WELL #1 HATCH	TRICHLOROETHYLENE	5.5	UG/L	16-Nov-00	CWS	410
ID	ID4010147	TERRA GRANDE WATER COMPANY	WELL #1 HATCH	TRICHLOROETHYLENE	62	UG/L	20-Jan-05	CWS	410
ID	ID4010147	TERRA GRANDE WATER COMPANY	WELL #2 ASH PARK	TRICHLOROETHYLENE	57.4	UG/L	20-Jan-05	CWS	410
ID	ID4010147	TERRA GRANDE WATER COMPANY	WELL #1 HATCH	TRICHLOROETHYLENE	0.0006	UG/L	21-Jul-04	CWS	410
ID	ID4010147	TERRA GRANDE WATER COMPANY	WELL #1 HATCH	TRICHLOROETHYLENE	66.1	UG/L	20-Jan-05	CWS	410
ID	ID4010147	TERRA GRANDE WATER COMPANY	WELL #2 ASH PARK	TRICHLOROETHYLENE	65.1	UG/L	20-Jan-05	CWS	410
ID	ID4010147	TERRA GRANDE WATER COMPANY	WELL #3 PENNIGER	TRICHLOROETHYLENE	53.5	UG/L	20-Jan-05	CWS	410
ID	ID4010147	TERRA GRANDE WATER COMPANY	WELL #1 HATCH	TRICHLOROETHYLENE	70.9	UG/L	26-Jan-05	CWS	410
ID	ID4010147	TERRA GRANDE WATER COMPANY	WELL #2 ASH PARK	TRICHLOROETHYLENE	68.9	UG/L	08-Dec-04	CWS	410
ID	ID4010147	TERRA GRANDE WATER COMPANY	WELL #1 HATCH	TRICHLOROETHYLENE	96.1	UG/L	08-Dec-04	CWS	410
ID	ID4010147	TERRA GRANDE WATER COMPANY	WELL #3 PENNIGER	TRICHLOROETHYLENE	68.7	UG/L	09-Dec-04	CWS	410
ID	ID4010147	TERRA GRANDE WATER COMPANY	WELL #2 ASH PARK	TRICHLOROETHYLENE	0.6	UG/L	20-Jul-04	CWS	410
ID	ID4010147	TERRA GRANDE WATER COMPANY	WELL #1 HATCH	TRICHLOROETHYLENE	0.5	UG/L	20-Jul-04	CWS	410
ID	ID4010147	TERRA GRANDE WATER COMPANY	WELL #3 PENNIGER	TRICHLOROETHYLENE	56	UG/L	20-Jan-05	CWS	410
ID	ID4010147	TERRA GRANDE WATER COMPANY	WELL #2 ASH PARK	TRICHLOROETHYLENE	2.6	UG/L	16-Jun-04	CWS	410
ID	ID4010147	TERRA GRANDE WATER COMPANY	WELL #1 HATCH	TRICHLOROETHYLENE	94.5	UG/L	16-Jun-04	CWS	410
					27.3				
ID	ID4010124	RIVERLAND TERRACE NONPROFIT WATER CORP	WELL #3	XYLENES	5.79	UG/L	11-Jul-05	CWS	99
ID	ID4010141	IDOC IDAHO ST CORRECTIONAL INST IMSI	WELL #3	XYLENES	1.1	UG/L	29-Sep-04	CWS	2000
					3.45				

APPENDIX 7-9 Boise Surface Water Data

State	PWSNUMBER	PWSNAME	SAMPLING_LOCATION	ANALYTE_NAME	RESULT	RESULT_UNIT	COLLECTION_DATE	PWSType	Pop
ID	ID4010016	UNITED WATER IDAHO INC	HIDDEN VALLEY 1	ATRAZINE	0.54	UG/L	27-May-04	CWS	186,000
ID	ID4010016	UNITED WATER IDAHO INC	TERTELING WELL	ATRAZINE	0.55	UG/L	09-Aug-04	CWS	186,000
ID	ID4010016	UNITED WATER IDAHO INC	HIDDEN VALLEY 1	ATRAZINE	0.54	UG/L	22-Oct-03	CWS	186,000
					0.54				
ID	ID4010016	UNITED WATER IDAHO INC	DANSKIN RIDGE	DI(2-ETHYLHEXYL) — PHTHALATE	9.8	UG/L	23-May-00	CWS	186,000
ID	ID4010016	UNITED WATER IDAHO INC	COLUMBIA WTP	HEXACHLOROCYCLOPENTADIENE	0.1	UG/L	15-Jun-05	CWS	186,000
ID	ID4010016	UNITED WATER IDAHO INC	COLUMBIA WTP	HEXACHLOROCYCLOPENTADIENE	0.07	UG/L	16-Mar-05	CWS	186,000
					0.09				
ID	ID4010016	UNITED WATER IDAHO INC	CHAMBERLAIN #2	TETRACHLOROETHYLENE	0.6	UG/L	19-Aug-02	CWS	186,000
ID	ID4010016	UNITED WATER IDAHO INC	SWIFT #2 — ASR SUMMER WELL	TETRACHLOROETHYLENE	2.7	UG/L	26-Jul-04	CWS	186,000
ID	ID4010016	UNITED WATER IDAHO INC	SWIFT #2 — ASR SUMMER WELL	TETRACHLOROETHYLENE	4	UG/L	07-Aug-03	CWS	186,000
ID	ID4010016	UNITED WATER IDAHO INC	SWIFT #2 — ASR SUMMER WELL	TETRACHLOROETHYLENE	6.3	UG/L	26-Sep-02	CWS	186,000
ID	ID4010016	UNITED WATER IDAHO INC	FRANKLIN PARK	TETRACHLOROETHYLENE	0.5	UG/L	15-Aug-02	CWS	186,000
ID	ID4010016	UNITED WATER IDAHO INC	SWIFT #2 — ASR SUMMER WELL	TETRACHLOROETHYLENE	3.4	UG/L	17-Oct-02	CWS	186,000
ID	ID4010016	UNITED WATER IDAHO INC	SWIFT #2 — ASR SUMMER WELL	TETRACHLOROETHYLENE	2.4	UG/L	28-Oct-02	CWS	186,000
ID	ID4010016	UNITED WATER IDAHO INC	COUNTRY CLUB	TETRACHLOROETHYLENE	0.5	UG/L	03-Aug-04	CWS	186,000
ID	ID4010016	UNITED WATER IDAHO INC	SWIFT #2 — ASR SUMMER WELL	TETRACHLOROETHYLENE	4	UG/L	02-Aug-04	CWS	186,000
ID	ID4010016	UNITED WATER IDAHO INC	WESTMORELAND	TETRACHLOROETHYLENE	4	UG/L	05-Aug-04	CWS	186,000
ID	ID4010016	UNITED WATER IDAHO INC	CLIFFSIDE BACK UP WELL	TETRACHLOROETHYLENE	4.9	UG/L	13-Aug-04	CWS	186,000
ID	ID4010016	UNITED WATER IDAHO INC	COUNTRY CLUB	TETRACHLOROETHYLENE	0.9	UG/L	01-Nov-05	CWS	186000
ID	ID4010016	UNITED WATER IDAHO INC	WESTMORELAND	TETRACHLOROETHYLENE	6.5	UG/L	08-Sep-05	CWS	186,000
ID	ID4010016	UNITED WATER IDAHO INC	SWIFT #2 — ASR SUMMER WELL	TETRACHLOROETHYLENE	8.5	UG/L	19-Aug-05	CWS	186,000
ID	ID4010016	UNITED WATER IDAHO INC	WESTMORELAND	TETRACHLOROETHYLENE	6.5	UG/L	19-Aug-05	CWS	186,000
ID	ID4010016	UNITED WATER IDAHO INC	COUNTRY CLUB	TETRACHLOROETHYLENE	0.8	UG/L	14-Sep-05	CWS	186,000
ID	ID4010016	UNITED WATER IDAHO INC	COUNTRY CLUB	TETRACHLOROETHYLENE	0.9	UG/L	20-Jul-05	CWS	186,000
ID	ID4010016	UNITED WATER IDAHO INC	SWIFT #2 — ASR SUMMER WELL	TETRACHLOROETHYLENE	0.5	UG/L	10-Aug-05	CWS	186,000
ID	ID4010016	UNITED WATER IDAHO INC	CLIFFSIDE BACK UP WELL	TETRACHLOROETHYLENE	4.6	UG/L	30-Mar-05	CWS	186,000
ID	ID4010016	UNITED WATER IDAHO INC	COUNTRY CLUB	TETRACHLOROETHYLENE	1.2	UG/L	16-Jun-05	CWS	186,000
ID	ID4010016	UNITED WATER IDAHO INC	CHAMBERLAIN #2	TETRACHLOROETHYLENE	0.6	UG/L	01-Aug-00	CWS	186,000
ID	ID4010016	UNITED WATER IDAHO INC	CLIFFSIDE BACK UP WELL	TETRACHLOROETHYLENE	6.1	UG/L	01-Mar-05	CWS	186,000
ID	ID4010016	UNITED WATER IDAHO INC	CLIFFSIDE BACK UP WELL	TETRACHLOROETHYLENE	4.3	UG/L	12-Jan-05	CWS	186,000
ID	ID4010016	UNITED WATER IDAHO INC	CHAMBERLAIN #2	TETRACHLOROETHYLENE	0.5	UG/L	16-Aug-04	CWS	186,000
ID	ID4010016	UNITED WATER IDAHO INC	SWIFT #2 — ASR SUMMER WELL	TETRACHLOROETHYLENE	2.6	UG/L	23-Feb-01	CWS	186,000
ID	ID4010016	UNITED WATER IDAHO INC	WESTMORELAND	TETRACHLOROETHYLENE	3.4	UG/L	29-Sep-00	CWS	186,000
					3.12				
ID	ID4010016	UNITED WATER IDAHO INC	SWIFT #2 — ASR SUMMER WELL	TRICHLOROETHYLENE	0.5	UG/L	19-Aug-05	CWS	186,000
ID	ID4010016	UNITED WATER IDAHO INC	SWIFT #2 — ASR SUMMER WELL	TRICHLOROETHYLENE	0.5	UG/L	07-Aug-03	CWS	186,000
ID	ID4010016	UNITED WATER IDAHO INC	SWIFT #2 — ASR SUMMER WELL	TRICHLOROETHYLENE	0.5	UG/L	26-Sep-02	CWS	186,000
					0.5				

APPENDIX 7-10 Columbus Surface Water Data

State	PWSID	Water System Name	Analyte	Concentration	Unit	Sample Date	PWSType	Pop
GA	GA2150000	Columbus	BROMODICHLOROMETHANE	12	ug/L	19-Jun-01	CWS	185,000
GA	GA2150000	Columbus	BROMODICHLOROMETHANE	11	ug/L	22-Mar-04	CWS	185,000
GA	GA2150000	Columbus	BROMODICHLOROMETHANE	10	ug/L	19-May-03	CWS	185,000
GA	GA2150000	Columbus	BROMODICHLOROMETHANE	14	ug/L	18-Apr-00	CWS	185,000
GA	GA2150000	Columbus	BROMODICHLOROMETHANE	12	ug/L	29-May-02	CWS	185,000
GA	GA2150000	Columbus	BROMODICHLOROMETHANE	12	ug/L	31-May-05	CWS	185,000
GA	GA2150000	Columbus	CHLORODIBROMOMETHANE	3.2	ug/L	19-Jun-01	CWS	185,000
GA	GA2150000	Columbus	CHLORODIBROMOMETHANE	3.7	ug/L	22-Mar-04	CWS	185,000
GA	GA2150000	Columbus	CHLORODIBROMOMETHANE	1.6	ug/L	19-May-03	CWS	185,000
GA	GA2150000	Columbus	CHLORODIBROMOMETHANE	4.5	ug/L	18-Apr-00	CWS	185,000
GA	GA2150000	Columbus	CHLORODIBROMOMETHANE	4.8	ug/L	29-May-02	CWS	185,000
GA	GA2150000	Columbus	CHLORODIBROMOMETHANE	3	ug/L	31-May-05	CWS	185,000
				7.65				
GA	GA2150000	Columbus	CHLOROFORM	36	ug/L	19-Jun-01	CWS	185,000
GA	GA2150000	Columbus	CHLOROFORM	23	ug/L	22-Mar-04	CWS	185,000
GA	GA2150000	Columbus	CHLOROFORM	45	ug/L	19-May-03	CWS	185,000
GA	GA2150000	Columbus	CHLOROFORM	27	ug/L	18-Apr-00	CWS	185,000
GA	GA2150000	Columbus	CHLOROFORM	22	ug/L	29-May-02	CWS	185,000
GA	GA2150000	Columbus	CHLOROFORM	34	ug/L	31-May-05	CWS	185,000
				31.2				
GA	GA2150000	Columbus	DIBROMOACETIC ACID	2.5	ug/L	13-Feb-02	CWS	185,000
GA	GA2150000	Columbus	DIBROMOACETIC ACID	1.6	ug/L	13-Feb-02	CWS	185,000
GA	GA2150000	Columbus	DIBROMOACETIC ACID	2.6	ug/L	13-Feb-02	CWS	185,000
GA	GA2150000	Columbus	DIBROMOACETIC ACID	1.2	ug/L	05-Nov-03	CWS	185,000
GA	GA2150000	Columbus	DIBROMOACETIC ACID	1.5	ug/L	05-Nov-03	CWS	185,000
GA	GA2150000	Columbus	DIBROMOACETIC ACID	1.2	ug/L	05-Nov-03	CWS	185,000
GA	GA2150000	Columbus	DIBROMOACETIC ACID	1.3	ug/L	05-Nov-03	CWS	185,000
GA	GA2150000	Columbus	DIBROMOACETIC ACID	1.7	ug/L	06-Nov-02	CWS	185,000
GA	GA2150000	Columbus	DIBROMOACETIC ACID	1.4	ug/L	26-Jul-04	CWS	185,000
GA	GA2150000	Columbus	DIBROMOACETIC ACID	1.8	ug/L	26-Jul-04	CWS	185,000
GA	GA2150000	Columbus	DIBROMOACETIC ACID	1.9	ug/L	26-Jul-04	CWS	185,000
GA	GA2150000	Columbus	DIBROMOACETIC ACID	1.6	ug/L	26-Jul-04	CWS	185,000
GA	GA2150000	Columbus	DIBROMOACETIC ACID	1.1	ug/L	03-May-04	CWS	185,000
				1.65				

APPENDIX 7-10 *Continued*

State	PWSID	Water System Name	Analyte	Concentration	Unit	Sample Date	PWSType	Pop
GA	GA2150000	Columbus	DICHLOROACETIC ACID	15	ug/L	26-Nov-01	CWS	185,000
GA	GA2150000	Columbus	DICHLOROACETIC ACID	12	ug/L	26-Nov-01	CWS	185,000
GA	GA2150000	Columbus	DICHLOROACETIC ACID	13	ug/L	26-Nov-01	CWS	185,000
GA	GA2150000	Columbus	DICHLOROACETIC ACID	14	ug/L	26-Nov-01	CWS	185,000
GA	GA2150000	Columbus	DICHLOROACETIC ACID	17	ug/L	13-Feb-02	CWS	185,000
GA	GA2150000	Columbus	DICHLOROACETIC ACID	10	ug/L	13-Feb-02	CWS	185,000
GA	GA2150000	Columbus	DICHLOROACETIC ACID	12	ug/L	13-Feb-02	CWS	185,000
GA	GA2150000	Columbus	DICHLOROACETIC ACID	15	ug/L	13-Feb-02	CWS	185,000
GA	GA2150000	Columbus	DICHLOROACETIC ACID	9.4	ug/L	05-Nov-03	CWS	185,000
GA	GA2150000	Columbus	DICHLOROACETIC ACID	11	ug/L	05-Nov-03	CWS	185,000
GA	GA2150000	Columbus	DICHLOROACETIC ACID	14	ug/L	05-Nov-03	CWS	185,000
GA	GA2150000	Columbus	DICHLOROACETIC ACID	12	ug/L	05-Nov-03	CWS	185,000
GA	GA2150000	Columbus	DICHLOROACETIC ACID	38	ug/L	01-Aug-05	CWS	185,000
GA	GA2150000	Columbus	DICHLOROACETIC ACID	36	ug/L	01-Aug-05	CWS	185,000
GA	GA2150000	Columbus	DICHLOROACETIC ACID	36	ug/L	01-Aug-05	CWS	185,000
GA	GA2150000	Columbus	DICHLOROACETIC ACID	41	ug/L	01-Aug-05	CWS	185,000
GA	GA2150000	Columbus	DICHLOROACETIC ACID	18	ug/L	06-Nov-02	CWS	185,000
GA	GA2150000	Columbus	DICHLOROACETIC ACID	11	ug/L	06-Nov-02	CWS	185,000
GA	GA2150000	Columbus	DICHLOROACETIC ACID	11	ug/L	06-Nov-02	CWS	185,000
GA	GA2150000	Columbus	DICHLOROACETIC ACID	10	ug/L	06-Nov-02	CWS	185,000
GA	GA2150000	Columbus	DICHLOROACETIC ACID	9.5	ug/L	02-Nov-05	CWS	185,000
GA	GA2150000	Columbus	DICHLOROACETIC ACID	9.8	ug/L	02-Nov-05	CWS	185,000
GA	GA2150000	Columbus	DICHLOROACETIC ACID	13	ug/L	02-Nov-05	CWS	185,000
GA	GA2150000	Columbus	DICHLOROACETIC ACID	10	ug/L	02-Nov-05	CWS	185,000
GA	GA2150000	Columbus	DICHLOROACETIC ACID	16	ug/L	07-Feb-05	CWS	185,000
GA	GA2150000	Columbus	DICHLOROACETIC ACID	18	ug/L	07-Feb-05	CWS	185,000
GA	GA2150000	Columbus	DICHLOROACETIC ACID	18	ug/L	07-Feb-05	CWS	185,000
GA	GA2150000	Columbus	DICHLOROACETIC ACID	16	ug/L	07-Feb-05	CWS	185,000
GA	GA2150000	Columbus	DICHLOROACETIC ACID	35	ug/L	22-May-01	CWS	185,000
GA	GA2150000	Columbus	DICHLOROACETIC ACID	23	ug/L	22-May-01	CWS	185,000
GA	GA2150000	Columbus	DICHLOROACETIC ACID	22	ug/L	22-May-01	CWS	185,000
GA	GA2150000	Columbus	DICHLOROACETIC ACID	22	ug/L	22-May-01	CWS	185,000
GA	GA2150000	Columbus	DICHLOROACETIC ACID	31	ug/L	08-Nov-04	CWS	185,000

GA	GA2150000	Columbus	DICHLOROACETIC ACID	24	ug/L	08-Nov-04	CWS	185,000
GA	GA2150000	Columbus	DICHLOROACETIC ACID	28	ug/L	08-Nov-04	CWS	185,000
GA	GA2150000	Columbus	DICHLOROACETIC ACID	29	ug/L	08-Nov-04	CWS	185,000
GA	GA2150000	Columbus	DICHLOROACETIC ACID	23	ug/L	02-May-05	CWS	185,000
GA	GA2150000	Columbus	DICHLOROACETIC ACID	22	ug/L	02-May-05	CWS	185,000
GA	GA2150000	Columbus	DICHLOROACETIC ACID	24	ug/L	02-May-05	CWS	185,000
GA	GA2150000	Columbus	DICHLOROACETIC ACID	26	ug/L	02-May-05	CWS	185,000
GA	GA2150000	Columbus	DICHLOROACETIC ACID	39	ug/L	12-May-03	CWS	185,000
GA	GA2150000	Columbus	DICHLOROACETIC ACID	45	ug/L	12-May-03	CWS	185,000
GA	GA2150000	Columbus	DICHLOROACETIC ACID	34	ug/L	12-May-03	CWS	185,000
GA	GA2150000	Columbus	DICHLOROACETIC ACID	39	ug/L	12-May-03	CWS	185,000
GA	GA2150000	Columbus	DICHLOROACETIC ACID	20	ug/L	11-Feb-03	CWS	185,000
GA	GA2150000	Columbus	DICHLOROACETIC ACID	18	ug/L	11-Feb-03	CWS	185,000
GA	GA2150000	Columbus	DICHLOROACETIC ACID	19	ug/L	11-Feb-03	CWS	185,000
GA	GA2150000	Columbus	DICHLOROACETIC ACID	20	ug/L	11-Feb-03	CWS	185,000
GA	GA2150000	Columbus	DICHLOROACETIC ACID	25	ug/L	21-Jul-03	CWS	185,000
GA	GA2150000	Columbus	DICHLOROACETIC ACID	27	ug/L	21-Jul-03	CWS	185,000
GA	GA2150000	Columbus	DICHLOROACETIC ACID	28	ug/L	21-Jul-03	CWS	185,000
GA	GA2150000	Columbus	DICHLOROACETIC ACID	33	ug/L	21-Jul-03	CWS	185,000
GA	GA2150000	Columbus	DICHLOROACETIC ACID	12	ug/L	26-Jul-04	CWS	185,000
GA	GA2150000	Columbus	DICHLOROACETIC ACID	12	ug/L	26-Jul-04	CWS	185,000
GA	GA2150000	Columbus	DICHLOROACETIC ACID	13	ug/L	26-Jul-04	CWS	185,000
GA	GA2150000	Columbus	DICHLOROACETIC ACID	18	ug/L	26-Jul-04	CWS	185,000
GA	GA2150000	Columbus	DICHLOROACETIC ACID	13	ug/L	21-Feb-01	CWS	185,000
GA	GA2150000	Columbus	DICHLOROACETIC ACID	15	ug/L	21-Feb-01	CWS	185,000
GA	GA2150000	Columbus	DICHLOROACETIC ACID	17	ug/L	21-Feb-01	CWS	185,000
GA	GA2150000	Columbus	DICHLOROACETIC ACID	15	ug/L	21-Feb-01	CWS	185,000
GA	GA2150000	Columbus	DICHLOROACETIC ACID	23	ug/L	17-May-00	CWS	185,000
GA	GA2150000	Columbus	DICHLOROACETIC ACID	16	ug/L	17-May-00	CWS	185,000
GA	GA2150000	Columbus	DICHLOROACETIC ACID	18	ug/L	17-May-00	CWS	185,000
GA	GA2150000	Columbus	DICHLOROACETIC ACID	17	ug/L	17-May-00	CWS	185,000
GA	GA2150000	Columbus	DICHLOROACETIC ACID	21	ug/L	15-Aug-00	CWS	185,000
GA	GA2150000	Columbus	DICHLOROACETIC ACID	21	ug/L	15-Aug-00	CWS	185,000
GA	GA2150000	Columbus	DICHLOROACETIC ACID	18	ug/L	15-Aug-00	CWS	185,000
GA	GA2150000	Columbus	DICHLOROACETIC ACID	23	ug/L	15-Aug-00	CWS	185,000
GA	GA2150000	Columbus	DICHLOROACETIC ACID	25	ug/L	29-May-02	CWS	185,000

APPENDIX 7-10 *Continued*

State	PWSID	Water System Name	Analyte	Concentration	Unit	Sample Date	PWSType	Pop
GA	GA2150000	Columbus	DICHLOROACETIC ACID	16	ug/L	29-May-02	CWS	185,000
GA	GA2150000	Columbus	DICHLOROACETIC ACID	17	ug/L	29-May-02	CWS	185,000
GA	GA2150000	Columbus	DICHLOROACETIC ACID	13	ug/L	29-May-02	CWS	185,000
GA	GA2150000	Columbus	DICHLOROACETIC ACID	20	ug/L	27-Aug-01	CWS	185,000
GA	GA2150000	Columbus	DICHLOROACETIC ACID	17	ug/L	27-Aug-01	CWS	185,000
GA	GA2150000	Columbus	DICHLOROACETIC ACID	20	ug/L	27-Aug-01	CWS	185,000
GA	GA2150000	Columbus	DICHLOROACETIC ACID	23	ug/L	27-Aug-01	CWS	185,000
GA	GA2150000	Columbus	DICHLOROACETIC ACID	7.2	ug/L	10-Feb-04	CWS	185,000
GA	GA2150000	Columbus	DICHLOROACETIC ACID	12	ug/L	10-Feb-04	CWS	185,000
GA	GA2150000	Columbus	DICHLOROACETIC ACID	9.9	ug/L	10-Feb-04	CWS	185,000
GA	GA2150000	Columbus	DICHLOROACETIC ACID	9.5	ug/L	10-Feb-04	CWS	185,000
GA	GA2150000	Columbus	DICHLOROACETIC ACID	13	ug/L	03-May-04	CWS	185,000
GA	GA2150000	Columbus	DICHLOROACETIC ACID	11	ug/L	03-May-04	CWS	185,000
GA	GA2150000	Columbus	DICHLOROACETIC ACID	13	ug/L	03-May-04	CWS	185,000
GA	GA2150000	Columbus	DICHLOROACETIC ACID	17	ug/L	03-May-04	CWS	185,000
GA	GA2150000	Columbus	DICHLOROACETIC ACID	15	ug/L	29-Nov-00	CWS	185,000
GA	GA2150000	Columbus	DICHLOROACETIC ACID	12	ug/L	29-Nov-00	CWS	185,000
GA	GA2150000	Columbus	DICHLOROACETIC ACID	11	ug/L	29-Nov-00	CWS	185,000
GA	GA2150000	Columbus	DICHLOROACETIC ACID	11	ug/L	29-Nov-00	CWS	185,000
GA	GA2150000	Columbus	DICHLOROACETIC ACID	10	ug/L	13-Aug-02	CWS	185,000
GA	GA2150000	Columbus	DICHLOROACETIC ACID	12	ug/L	13-Aug-02	CWS	185,000
GA	GA2150000	Columbus	DICHLOROACETIC ACID	3.4	ug/L	13-Aug-02	CWS	185,000
GA	GA2150000	Columbus	DICHLOROACETIC ACID	11	ug/L	13-Aug-02	CWS	185,000
GA	GA2150000	Columbus	DICHLOROACETIC ACID	9.1	ug/L	14-Feb-00	CWS	185,000
GA	GA2150000	Columbus	DICHLOROACETIC ACID	9.3	ug/L	14-Feb-00	CWS	185,000
GA	GA2150000	Columbus	DICHLOROACETIC ACID	11	ug/L	14-Feb-00	CWS	185,000
GA	GA2150000	Columbus	DICHLOROACETIC ACID	15	ug/L	14-Feb-00	CWS	185,000
				18.4				
GA	GA2150000	Columbus	MONOCHLOROACETIC ACID	3.4	ug/L	01-Aug-05	CWS	185,000
GA	GA2150000	Columbus	MONOCHLOROACETIC ACID	3.1	ug/L	01-Aug-05	CWS	185,000
GA	GA2150000	Columbus	MONOCHLOROACETIC ACID	3.2	ug/L	01-Aug-05	CWS	185,000
GA	GA2150000	Columbus	MONOCHLOROACETIC ACID	3.8	ug/L	01-Aug-05	CWS	185,000
GA	GA2150000	Columbus	MONOCHLOROACETIC ACID	3.8	ug/L	12-May-03	CWS	185,000
GA	GA2150000	Columbus	MONOCHLOROACETIC ACID	4.4	ug/L	12-May-03	CWS	185,000

GA	GA2150000	Columbus	MONOCHLOROACETIC ACID	3.9	ug/L	12-May-03	CWS	185,000
GA	GA2150000	Columbus	MONOCHLOROACETIC ACID	4.3	ug/L	12-May-03	CWS	185,000
GA	GA2150000	Columbus	MONOCHLOROACETIC ACID	4.4	ug/L	11-Feb-03	CWS	185,000
GA	GA2150000	Columbus	MONOCHLOROACETIC ACID	4.6	ug/L	11-Feb-03	CWS	185,000
GA	GA2150000	Columbus	MONOCHLOROACETIC ACID	6.1	ug/L	11-Feb-03	CWS	185,000
GA	GA2150000	Columbus	MONOCHLOROACETIC ACID	6.1	ug/L	11-Feb-03	CWS	185,000
GA	GA2150000	Columbus	MONOCHLOROACETIC ACID	5.4	ug/L	10-Feb-04	CWS	185,000
				4.35				
GA	GA2150000	Columbus	TRICHLOROACETIC ACID	8	ug/L	26-Nov-01	CWS	185,000
GA	GA2150000	Columbus	TRICHLOROACETIC ACID	7.2	ug/L	26-Nov-01	CWS	185,000
GA	GA2150000	Columbus	TRICHLOROACETIC ACID	7.8	ug/L	26-Nov-01	CWS	185,000
GA	GA2150000	Columbus	TRICHLOROACETIC ACID	6.3	ug/L	26-Nov-01	CWS	185,000
GA	GA2150000	Columbus	TRICHLOROACETIC ACID	9.8	ug/L	13-Feb-02	CWS	185,000
GA	GA2150000	Columbus	TRICHLOROACETIC ACID	4.7	ug/L	13-Feb-02	CWS	185,000
GA	GA2150000	Columbus	TRICHLOROACETIC ACID	7.6	ug/L	13-Feb-02	CWS	185,000
GA	GA2150000	Columbus	TRICHLOROACETIC ACID	8.6	ug/L	13-Feb-02	CWS	185,000
GA	GA2150000	Columbus	TRICHLOROACETIC ACID	5.4	ug/L	05-Nov-03	CWS	185,000
GA	GA2150000	Columbus	TRICHLOROACETIC ACID	5.5	ug/L	05-Nov-03	CWS	185,000
GA	GA2150000	Columbus	TRICHLOROACETIC ACID	6.6	ug/L	05-Nov-03	CWS	185,000
GA	GA2150000	Columbus	TRICHLOROACETIC ACID	7.4	ug/L	05-Nov-03	CWS	185,000
GA	GA2150000	Columbus	TRICHLOROACETIC ACID	26	ug/L	01-Aug-05	CWS	185,000
GA	GA2150000	Columbus	TRICHLOROACETIC ACID	26	ug/L	01-Aug-05	CWS	185,000
GA	GA2150000	Columbus	TRICHLOROACETIC ACID	24	ug/L	01-Aug-05	CWS	185,000
GA	GA2150000	Columbus	TRICHLOROACETIC ACID	21	ug/L	01-Aug-05	CWS	185,000
GA	GA2150000	Columbus	TRICHLOROACETIC ACID	10	ug/L	06-Nov-02	CWS	185,000
GA	GA2150000	Columbus	TRICHLOROACETIC ACID	7.2	ug/L	06-Nov-02	CWS	185,000
GA	GA2150000	Columbus	TRICHLOROACETIC ACID	6.3	ug/L	06-Nov-02	CWS	185,000
GA	GA2150000	Columbus	TRICHLOROACETIC ACID	5.5	ug/L	06-Nov-02	CWS	185,000
GA	GA2150000	Columbus	TRICHLOROACETIC ACID	6	ug/L	02-Nov-05	CWS	185,000
GA	GA2150000	Columbus	TRICHLOROACETIC ACID	4.2	ug/L	02-Nov-05	CWS	185,000
GA	GA2150000	Columbus	TRICHLOROACETIC ACID	3.9	ug/L	02-Nov-05	CWS	185,000
GA	GA2150000	Columbus	TRICHLOROACETIC ACID	4.1	ug/L	02-Nov-05	CWS	185,000
GA	GA2150000	Columbus	TRICHLOROACETIC ACID	9.8	ug/L	07-Feb-05	CWS	185,000
GA	GA2150000	Columbus	TRICHLOROACETIC ACID	14	ug/L	07-Feb-05	CWS	185,000
GA	GA2150000	Columbus	TRICHLOROACETIC ACID	11	ug/L	07-Feb-05	CWS	185,000
GA	GA2150000	Columbus	TRICHLOROACETIC ACID	15	ug/L	07-Feb-05	CWS	185,000

APPENDIX 7-10 *Continued*

State	PWSID	Water System Name	Analyte	Concentration	Unit	Sample Date	PWSType	Pop
GA	GA2150000	Columbus	TRICHLOROACETIC ACID	9.8	ug/L	22-May-01	CWS	185,000
GA	GA2150000	Columbus	TRICHLOROACETIC ACID	12	ug/L	22-May-01	CWS	185,000
GA	GA2150000	Columbus	TRICHLOROACETIC ACID	22	ug/L	22-May-01	CWS	185,000
GA	GA2150000	Columbus	TRICHLOROACETIC ACID	11	ug/L	22-May-01	CWS	185,000
GA	GA2150000	Columbus	TRICHLOROACETIC ACID	22	ug/L	08-Nov-04	CWS	185,000
GA	GA2150000	Columbus	TRICHLOROACETIC ACID	22	ug/L	08-Nov-04	CWS	185,000
GA	GA2150000	Columbus	TRICHLOROACETIC ACID	20	ug/L	08-Nov-04	CWS	185,000
GA	GA2150000	Columbus	TRICHLOROACETIC ACID	21	ug/L	08-Nov-04	CWS	185,000
GA	GA2150000	Columbus	TRICHLOROACETIC ACID	13	ug/L	02-May-05	CWS	185,000
GA	GA2150000	Columbus	TRICHLOROACETIC ACID	13	ug/L	02-May-05	CWS	185,000
GA	GA2150000	Columbus	TRICHLOROACETIC ACID	13	ug/L	02-May-05	CWS	185,000
GA	GA2150000	Columbus	TRICHLOROACETIC ACID	13	ug/L	02-May-05	CWS	185,000
GA	GA2150000	Columbus	TRICHLOROACETIC ACID	17	ug/L	12-May-03	CWS	185,000
GA	GA2150000	Columbus	TRICHLOROACETIC ACID	19	ug/L	12-May-03	CWS	185,000
GA	GA2150000	Columbus	TRICHLOROACETIC ACID	25	ug/L	12-May-03	CWS	185,000
GA	GA2150000	Columbus	TRICHLOROACETIC ACID	22	ug/L	12-May-03	CWS	185,000
GA	GA2150000	Columbus	TRICHLOROACETIC ACID	16	ug/L	11-Feb-03	CWS	185,000
GA	GA2150000	Columbus	TRICHLOROACETIC ACID	17	ug/L	11-Feb-03	CWS	185,000
GA	GA2150000	Columbus	TRICHLOROACETIC ACID	15	ug/L	11-Feb-03	CWS	185,000
GA	GA2150000	Columbus	TRICHLOROACETIC ACID	14	ug/L	11-Feb-03	CWS	185,000
GA	GA2150000	Columbus	TRICHLOROACETIC ACID	16	ug/L	21-Jul-03	CWS	185,000
GA	GA2150000	Columbus	TRICHLOROACETIC ACID	14	ug/L	21-Jul-03	CWS	185,000
GA	GA2150000	Columbus	TRICHLOROACETIC ACID	14	ug/L	21-Jul-03	CWS	185,000
GA	GA2150000	Columbus	TRICHLOROACETIC ACID	14	ug/L	21-Jul-03	CWS	185,000
GA	GA2150000	Columbus	TRICHLOROACETIC ACID	7.4	ug/L	26-Jul-04	CWS	185,000
GA	GA2150000	Columbus	TRICHLOROACETIC ACID	5.9	ug/L	26-Jul-04	CWS	185,000
GA	GA2150000	Columbus	TRICHLOROACETIC ACID	5.3	ug/L	26-Jul-04	CWS	185,000
GA	GA2150000	Columbus	TRICHLOROACETIC ACID	5.2	ug/L	26-Jul-04	CWS	185,000
GA	GA2150000	Columbus	TRICHLOROACETIC ACID	8.1	ug/L	21-Feb-01	CWS	185,000
GA	GA2150000	Columbus	TRICHLOROACETIC ACID	5.7	ug/L	21-Feb-01	CWS	185,000
GA	GA2150000	Columbus	TRICHLOROACETIC ACID	5.4	ug/L	21-Feb-01	CWS	185,000
GA	GA2150000	Columbus	TRICHLOROACETIC ACID	5.6	ug/L	21-Feb-01	CWS	185,000
GA	GA2150000	Columbus	TRICHLOROACETIC ACID	7.3	ug/L	17-May-00	CWS	185,000
GA	GA2150000	Columbus	TRICHLOROACETIC ACID	12	ug/L	17-May-00	CWS	185,000

GA	GA2150000	Columbus	TRICHLOROACETIC ACID	11	ug/L	17-May-00	CWS	185,000
GA	GA2150000	Columbus	TRICHLOROACETIC ACID	8.4	ug/L	17-May-00	CWS	185,000
GA	GA2150000	Columbus	TRICHLOROACETIC ACID	19	ug/L	15-Aug-00	CWS	185,000
GA	GA2150000	Columbus	TRICHLOROACETIC ACID	22	ug/L	15-Aug-00	CWS	185,000
GA	GA2150000	Columbus	TRICHLOROACETIC ACID	19	ug/L	15-Aug-00	CWS	185,000
GA	GA2150000	Columbus	TRICHLOROACETIC ACID	20	ug/L	15-Aug-00	CWS	185,000
GA	GA2150000	Columbus	TRICHLOROACETIC ACID	11	ug/L	29-May-02	CWS	185,000
GA	GA2150000	Columbus	TRICHLOROACETIC ACID	9.6	ug/L	29-May-02	CWS	185,000
GA	GA2150000	Columbus	TRICHLOROACETIC ACID	15	ug/L	29-May-02	CWS	185,000
GA	GA2150000	Columbus	TRICHLOROACETIC ACID	8.7	ug/L	29-May-02	CWS	185,000
GA	GA2150000	Columbus	TRICHLOROACETIC ACID	12	ug/L	27-Aug-01	CWS	185,000
GA	GA2150000	Columbus	TRICHLOROACETIC ACID	12	ug/L	27-Aug-01	CWS	185,000
GA	GA2150000	Columbus	TRICHLOROACETIC ACID	12	ug/L	27-Aug-01	CWS	185,000
GA	GA2150000	Columbus	TRICHLOROACETIC ACID	9.6	ug/L	27-Aug-01	CWS	185,000
GA	GA2150000	Columbus	TRICHLOROACETIC ACID	6.8	ug/L	10-Feb-04	CWS	185,000
GA	GA2150000	Columbus	TRICHLOROACETIC ACID	4.6	ug/L	10-Feb-04	CWS	185,000
GA	GA2150000	Columbus	TRICHLOROACETIC ACID	6.9	ug/L	10-Feb-04	CWS	185,000
GA	GA2150000	Columbus	TRICHLOROACETIC ACID	9.7	ug/L	10-Feb-04	CWS	185,000
GA	GA2150000	Columbus	TRICHLOROACETIC ACID	7.7	ug/L	03-May-04	CWS	185,000
GA	GA2150000	Columbus	TRICHLOROACETIC ACID	4.7	ug/L	03-May-04	CWS	185,000
GA	GA2150000	Columbus	TRICHLOROACETIC ACID	5.5	ug/L	03-May-04	CWS	185,000
GA	GA2150000	Columbus	TRICHLOROACETIC ACID	6.6	ug/L	03-May-04	CWS	185,000
GA	GA2150000	Columbus	TRICHLOROACETIC ACID	6.2	ug/L	29-Nov-00	CWS	185,000
GA	GA2150000	Columbus	TRICHLOROACETIC ACID	6.7	ug/L	29-Nov-00	CWS	185,000
GA	GA2150000	Columbus	TRICHLOROACETIC ACID	6.9	ug/L	29-Nov-00	CWS	185,000
GA	GA2150000	Columbus	TRICHLOROACETIC ACID	7.1	ug/L	29-Nov-00	CWS	185,000
GA	GA2150000	Columbus	TRICHLOROACETIC ACID	9.9	ug/L	13-Aug-02	CWS	185,000
GA	GA2150000	Columbus	TRICHLOROACETIC ACID	7.2	ug/L	13-Aug-02	CWS	185,000
GA	GA2150000	Columbus	TRICHLOROACETIC ACID	6.4	ug/L	13-Aug-02	CWS	185,000
GA	GA2150000	Columbus	TRICHLOROACETIC ACID	6.1	ug/L	13-Aug-02	CWS	185,000
GA	GA2150000	Columbus	TRICHLOROACETIC ACID	9.1	ug/L	14-Feb-00	CWS	185,000
GA	GA2150000	Columbus	TRICHLOROACETIC ACID	4.7	ug/L	14-Feb-00	CWS	185,000
GA	GA2150000	Columbus	TRICHLOROACETIC ACID	5.5	ug/L	14-Feb-00	CWS	185,000
GA	GA2150000	Columbus	TRICHLOROACETIC ACID	5.1	ug/L	14-Feb-00	CWS	185,000
				11.1				

APPENDIX 7-11 Dayton Groundwater Data

State	PWSID	pws_name	SUBSTANCE_	LEVEL	UNIT	DATE	PWSType	Pop
OH	OH5703512	Dayton Public Water System	Atrazine	1.78	UG/L	26-Jun-00	CWS	166,000
OH	OH5703512	Dayton Public Water System	Atrazine	1.78	UG/L	26-Jun-00	CWS	166,000
OH	OH5703512	Dayton Public Water System	Atrazine	0.41	UG/L	14-Aug-00	CWS	166,000
OH	OH5703512	Dayton Public Water System	Atrazine	0.41	UG/L	14-Aug-00	CWS	166,000
OH	OH5703512	Dayton Public Water System	Atrazine	4.1	UG/L	30-May-01	CWS	166,000
OH	OH5703512	Dayton Public Water System	Atrazine	4.1	UG/L	30-May-01	CWS	166,000
OH	OH5703512	Dayton Public Water System	Atrazine	0.413	UG/L	25-Jun-01	CWS	166,000
OH	OH5703512	Dayton Public Water System	Atrazine	0.413	UG/L	25-Jun-01	CWS	166,000
OH	OH5703512	Dayton Public Water System	Atrazine	0.4	UG/L	25-Jun-04	CWS	166,000
OH	OH5703512	Dayton Public Water System	Atrazine	0.4	UG/L	25-Jun-04	CWS	166,000
OH	OH5703512	Dayton Public Water System	Atrazine	0.16	UG/L	25-Jun-01	CWS	166,000
OH	OH5703512	Dayton Public Water System	Atrazine	0.16	UG/L	25-Jun-01	CWS	166,000
				1.21				
OH	OH5703512	Dayton Public Water System	Bromodichloromethane					
			Average of 401 samples	2.17	UG/L		CWS	166,000
OH	OH5703512	Dayton Public Water System	Carbon tetrachloride	0.26	UG/L	17-Feb-00	CWS	166,000
OH	OH5703512	Dayton Public Water System	Carbon tetrachloride	0.26	UG/L	17-Feb-00	CWS	166,000
				0.26				
OH	OH5703512	Dayton Public Water System	Chloromethane	0.7	UG/L	21-May-04	CWS	166,000
OH	OH5703512	Dayton Public Water System	Chloromethane	0.7	UG/L	21-May-04	CWS	166,000
OH	OH5703512	Dayton Public Water System	Chloromethane	0.6	UG/L	07-Mar-05	CWS	166,000
OH	OH5703512	Dayton Public Water System	Chloromethane	0.6	UG/L	07-Mar-05	CWS	166,000
				0.65				
OH	OH5703512	Dayton Public Water System	Dibromoacetic acid					
			Average of 149 samples	2.25	UG/L		CWS	166,000
OH	OH5703512	Dayton Public Water System	Dibromochloromethane					
			Average of 371 samaples	2.22	UG/L		CWS	166,000
OH	OH5703512	Dayton Public Water System	Dichloroacetic acid					
			Average of 148 samples	2.27	UG/L		CWS	166,000
OH	OH5703512	Dayton Public Water System	cis-1,2-Dichloroethene					
			Average of 239 samples	0.45	UG/L		CWS	166,000

OH	OH5703512	Dayton Public Water System	Metolachlor	0.902	UG/L	30-May-01	CWS	166,000
OH	OH5703512	Dayton Public Water System	Metolachlor	0.902	UG/L	30-May-01	CWS	166,000
				0.9				
OH	OH5703512	Dayton Public Water System	Monobromoacetic acid	1	UG/L	20-Feb-03	CWS	166,000
OH	OH5703512	Dayton Public Water System	Monobromoacetic acid	1	UG/L	20-Feb-03	CWS	166,000
OH	OH5703512	Dayton Public Water System	Monochloroacetic acid	1.8	UG/L	17-Oct-02	CWS	166,000
OH	OH5703512	Dayton Public Water System	Monochloroacetic acid	1.8	UG/L	17-Oct-02	CWS	166,000
OH	OH5703512	Dayton Public Water System	Monochloroacetic acid	20	UG/L	07-Jul-03	CWS	166,000
OH	OH5703512	Dayton Public Water System	Monochloroacetic acid	20	UG/L	07-Jul-03	CWS	166,000
OH	OH5703512	Dayton Public Water System	Monochloroacetic acid	1.7	UG/L	17-Oct-02	CWS	166,000
OH	OH5703512	Dayton Public Water System	Monochloroacetic acid	1.7	UG/L	17-Oct-02	CWS	166,000
OH	OH5703512	Dayton Public Water System	Monochloroacetic acid	1.5	UG/L	17-Oct-02	CWS	166,000
OH	OH5703512	Dayton Public Water System	Monochloroacetic acid	1.5	UG/L	17-Oct-02	CWS	166,000
OH	OH5703512	Dayton Public Water System	Monochloroacetic acid	1.2	UG/L	17-Oct-02	CWS	166,000
OH	OH5703512	Dayton Public Water System	Monochloroacetic acid	1.2	UG/L	17-Oct-02	CWS	166,000
OH	OH5703512	Dayton Public Water System	Monochloroacetic acid	1.4	UG/L	17-Oct-02	CWS	166,000
OH	OH5703512	Dayton Public Water System	Monochloroacetic acid	1.4	UG/L	17-Oct-02	CWS	166,000
				4.09				
OH	OH5703512	Dayton Public Water System	Naphthalene	0.28	UG/L	09-Jan-01	CWS	166,000
OH	OH5703512	Dayton Public Water System	Naphthalene	0.28	UG/L	09-Jan-01	CWS	166,000
				0.28				
OH	OH5703512	Dayton Public Water System	Simazine	0.84	MG/L	30-May-01	CWS	166,000
OH	OH5703512	Dayton Public Water System	Simazine	0.84	MG/L	30-May-01	CWS	166,000
				0.84				
OH	OH5701315	Montgomery Co:Water System #1 PWS	Trichloroacetic acid	1.2	UG/L	26-Mar-02	CWS	53,816
OH	OH5701315	Montgomery Co:Water System #1 PWS	Trichloroacetic acid	1.2	UG/L	26-Mar-02	CWS	53,816
OH	OH5701315	Montgomery Co:Water System #1 PWS	Trichloroacetic acid	1.2	UG/L	26-Mar-02	CWS	53,816
OH	OH5701315	Montgomery Co:Water System #1 PWS	Trichloroacetic acid	1.2	UG/L	26-Mar-02	CWS	53,816
OH	OH5701315	Montgomery Co:Water System #1 PWS	Trichloroacetic acid	1.2	UG/L	20-Jun-02	CWS	53,816
OH	OH5701315	Montgomery Co:Water System #1 PWS	Trichloroacetic acid	1.2	UG/L	20-Jun-02	CWS	53,816
OH	OH5701315	Montgomery Co:Water System #1 PWS	Trichloroacetic acid	1.2	UG/L	20-Jun-02	CWS	53,816
OH	OH5701315	Montgomery Co:Water System #1 PWS	Trichloroacetic acid	1.2	UG/L	20-Jun-02	CWS	53,816
OH	OH5701315	Montgomery Co:Water System #1 PWS	Trichloroacetic acid	1	UG/L	20-Jun-02	CWS	53,816
OH	OH5701315	Montgomery Co:Water System #1 PWS	Trichloroacetic acid	1	UG/L	20-Jun-02	CWS	53,816
OH	OH5701315	Montgomery Co:Water System #1 PWS	Trichloroacetic acid	1	UG/L	20-Jun-02	CWS	53,816

APPENDIX 7-11 *Continued*

State	PWSID	pws_name	SUBSTANCE_	LEVEL	UNIT	DATE	PWSType	Pop
OH	OH5701315	Montgomery Co:Water System #1 PWS	Trichloroacetic acid	1	UG/L	20-Jun-02	CWS	53,816
OH	OH5703512	Dayton Public Water System	Trichloroacetic acid	1.2	UG/L	07-Jul-03	CWS	166,000
OH	OH5703512	Dayton Public Water System	Trichloroacetic acid	1.2	UG/L	07-Jul-03	CWS	166,000
OH	OH5703512	Dayton Public Water System	Trichloroacetic acid	3.3	UG/L	20-Feb-04	CWS	166,000
OH	OH5703512	Dayton Public Water System	Trichloroacetic acid	3.3	UG/L	20-Feb-04	CWS	166,000
OH	OH5703512	Dayton Public Water System	Trichloroacetic acid	1.29	UG/L	09-Jul-04	CWS	166,000
OH	OH5703512	Dayton Public Water System	Trichloroacetic acid	1.29	UG/L	09-Jul-04	CWS	166,000
OH	OH5703512	Dayton Public Water System	Trichloroacetic acid	2.29	UG/L	19-Nov-04	CWS	166,000
OH	OH5703512	Dayton Public Water System	Trichloroacetic acid	2.29	UG/L	19-Nov-04	CWS	166,000
OH	OH5703512	Dayton Public Water System	Trichloroacetic acid	1	UG/L	20-Feb-03	CWS	166,000
OH	OH5703512	Dayton Public Water System	Trichloroacetic acid	1	UG/L	20-Feb-03	CWS	166,000
OH	OH5703512	Dayton Public Water System	Trichloroacetic acid	2.1	UG/L	20-Feb-04	CWS	166,000
OH	OH5703512	Dayton Public Water System	Trichloroacetic acid	2.1	UG/L	20-Feb-04	CWS	166,000
OH	OH5703512	Dayton Public Water System	Trichloroacetic acid	1.5	UG/L	18-Jan-02	CWS	166,000
OH	OH5703512	Dayton Public Water System	Trichloroacetic acid	1.5	UG/L	18-Jan-02	CWS	166,000
OH	OH5703512	Dayton Public Water System	Trichloroacetic acid	1.2	UG/L	02-Aug-02	CWS	166,000
OH	OH5703512	Dayton Public Water System	Trichloroacetic acid	1.2	UG/L	02-Aug-02	CWS	166,000
OH	OH5703512	Dayton Public Water System	Trichloroacetic acid	1	UG/L	20-Feb-03	CWS	166,000
OH	OH5703512	Dayton Public Water System	Trichloroacetic acid	1	UG/L	20-Feb-03	CWS	166,000
OH	OH5703512	Dayton Public Water System	Trichloroacetic acid	1	UG/L	20-Feb-03	CWS	166,000
OH	OH5703512	Dayton Public Water System	Trichloroacetic acid	1	UG/L	20-Feb-03	CWS	166,000
				1.42				
OH	OH5703512	Dayton Public Water System	Trichloroethene	0.31	UG/L	14-Jan-02	CWS	166,000
OH	OH5703512	Dayton Public Water System	Trichloroethene	0.31	UG/L	14-Jan-02	CWS	166,000
OH	OH5703512	Dayton Public Water System	Trichloroethene	0.33	UG/L	21-Jan-03	CWS	166,000
OH	OH5703512	Dayton Public Water System	Trichloroethene	0.33	UG/L	21-Jan-03	CWS	166,000
				0.32				
OH	OH5700412	Cains Mobile Home Court	Trichlorofluoromethane	0.8	UG/L	14-Sep-00	CWS	70
OH	OH5700412	Cains Mobile Home Court	Trichlorofluoromethane	0.7	UG/L	18-Dec-03	CWS	70
				0.75				

APPENDIX 7-12 Dayton Surface Water Data

Expr1000	PWSID	pws_name	source	SUBSTANCE_	RESULT_VAL	RESULT_UNI	SAMPLE_COL	PWSType	PSource	Pop
OH	OH5701503	Montgomery Co:Water System #2 PWS	PSW	Dibromoacetic acid	2.5	UG/L	20-Jun-02	CWS	SWP	109,000
OH	OH5701503	Montgomery Co:Water System #2 PWS	PSW	Dibromoacetic acid	2	UG/L	19-Sep-02	CWS	SWP	109,000
OH	OH5701503	Montgomery Co:Water System #2 PWS	PSW	Dibromoacetic acid	2.4	UG/L	21-Mar-03	CWS	SWP	109,000
OH	OH5701503	Montgomery Co:Water System #2 PWS	PSW	Dibromoacetic acid	1.4	UG/L	13-Jun-03	CWS	SWP	109,000
OH	OH5701503	Montgomery Co:Water System #2 PWS	PSW	Dibromoacetic acid	2.69	UG/L	12-Sep-03	CWS	SWP	109,000
OH	OH5701503	Montgomery Co:Water System #2 PWS	PSW	Dibromoacetic acid	1.4	UG/L	11-Mar-04	CWS	SWP	109,000
OH	OH5701503	Montgomery Co:Water System #2 PWS	PSW	Dibromoacetic acid	1.86	UG/L	11-Jun-04	CWS	SWP	109,000
OH	OH5701503	Montgomery Co:Water System #2 PWS	PSW	Dibromoacetic acid	2.51	UG/L	10-Sep-04	CWS	SWP	109,000
OH	OH5701503	Montgomery Co:Water System #2 PWS	PSW	Dibromoacetic acid	2.2	UG/L	29-Dec-04	CWS	SWP	109,000
OH	OH5701503	Montgomery Co:Water System #2 PWS	PSW	Dibromoacetic acid	1.72	UG/L	28-Feb-05	CWS	SWP	109,000
OH	OH5701503	Montgomery Co:Water System #2 PWS	PSW	Dibromoacetic acid	1.9	UG/L	20-Jun-02	CWS	SWP	109,000
OH	OH5701503	Montgomery Co:Water System #2 PWS	PSW	Dibromoacetic acid	1.4	UG/L	19-Sep-02	CWS	SWP	109,000
OH	OH5701503	Montgomery Co:Water System #2 PWS	PSW	Dibromoacetic acid	1.8	UG/L	20-Jun-02	CWS	SWP	109,000
OH	OH5701503	Montgomery Co:Water System #2 PWS	PSW	Dibromoacetic acid	2.5	UG/L	19-Sep-02	CWS	SWP	109,000
OH	OH5701503	Montgomery Co:Water System #2 PWS	PSW	Dibromoacetic acid	1.8	UG/L	20-Jun-02	CWS	SWP	109,000
OH	OH5701503	Montgomery Co:Water System #2 PWS	PSW	Dibromoacetic acid	1.8	UG/L	19-Sep-02	CWS	SWP	109,000
OH	OH5701503	Montgomery Co:Water System #2 PWS	PSW	Dichloroacetic acid	2.4	UG/L	26-Mar-02	CWS	SWP	109,000
OH	OH5701503	Montgomery Co:Water System #2 PWS	PSW	Dichloroacetic acid	2.5	UG/L	20-Jun-02	CWS	SWP	109,000
OH	OH5701503	Montgomery Co:Water System #2 PWS	PSW	Dichloroacetic acid	1.3	UG/L	19-Sep-02	CWS	SWP	109,000
OH	OH5701503	Montgomery Co:Water System #2 PWS	PSW	Dichloroacetic acid	1.7	UG/L	21-Mar-03	CWS	SWP	109,000
OH	OH5701503	Montgomery Co:Water System #2 PWS	PSW	Dichloroacetic acid	1.4	UG/L	13-Jun-03	CWS	SWP	109,000
OH	OH5701503	Montgomery Co:Water System #2 PWS	PSW	Dichloroacetic acid	4.54	UG/L	12-Sep-03	CWS	SWP	109,000
OH	OH5701503	Montgomery Co:Water System #2 PWS	PSW	Dichloroacetic acid	1.39	UG/L	11-Mar-04	CWS	SWP	109,000
OH	OH5701503	Montgomery Co:Water System #2 PWS	PSW	Dichloroacetic acid	2.35	UG/L	11-Jun-04	CWS	SWP	109,000
OH	OH5701503	Montgomery Co:Water System #2 PWS	PSW	Dichloroacetic acid	1.05	UG/L	10-Sep-04	CWS	SWP	109,000
OH	OH5701503	Montgomery Co:Water System #2 PWS	PSW	Dichloroacetic acid	2	UG/L	29-Dec-04	CWS	SWP	109,000
OH	OH5701503	Montgomery Co:Water System #2 PWS	PSW	Dichloroacetic acid	1.7	UG/L	28-Feb-05	CWS	SWP	109,000
OH	OH5701503	Montgomery Co:Water System #2 PWS	PSW	Dichloroacetic acid	2.3	UG/L	20-Jun-02	CWS	SWP	109,000
OH	OH5701503	Montgomery Co:Water System #2 PWS	PSW	Dichloroacetic acid	1.6	UG/L	20-Jun-02	CWS	SWP	109,000
OH	OH5701503	Montgomery Co:Water System #2 PWS	PSW	Dichloroacetic acid	1.5	UG/L	19-Sep-02	CWS	SWP	109,000
OH	OH5701503	Montgomery Co:Water System #2 PWS	PSW	Dichloroacetic acid	2.1	UG/L	26-Mar-02	CWS	SWP	109,000
OH	OH5701503	Montgomery Co:Water System #2 PWS	PSW	Dichloroacetic acid	2.3	UG/L	20-Jun-02	CWS	SWP	109,000
					2					
OH	OH5701503	Montgomery Co:Water System #2 PWS	PSW	Monochloroacetic acid	4.4	UG/L	20-Jun-02	CWS	SWP	109,000
OH	OH5701503	Montgomery Co:Water System #2 PWS	PSW	Trichloroacetic acid	1.2	UG/L	20-Jun-02	CWS	SWP	109,000
OH	OH5701503	Montgomery Co:Water System #2 PWS	PSW	Trichloroacetic acid	1.3	UG/L	20-Jun-02	CWS	SWP	109,000
					1.25					

APPENDIX 7-13 Fresno Groundwater Data

State	PWSID	Water System Name	Source	Analyte	Level	Unit	Sample Date	PWSType	Pop
CA	CA1010007	FRESNO, CITY OF	WELL 243-RAW-INACTIVE	1,1,1-TRICHLOROETHANE	0.88	UG/L	6/26/2001 0:00	CWS	438,000
CA	CA1010001	BAKMAN WATER COMPANY	WELL 02	1,1-DICHLOROETHYLENE	0.31	UG/L	11/6/2002 0:00	CWS	8,500
CA	CA1010007	FRESNO, CITY OF	WELL 203-DESTROYED	1,1-DICHLOROETHYLENE	1.2	UG/L	6/11/1998 0:00	CWS	438,000
CA	CA1010007	FRESNO, CITY OF	WELL 203-DESTROYED	1,1-DICHLOROETHYLENE	1.2	UG/L	9/22/1998 0:00	CWS	438,000
CA	CA1010007	FRESNO, CITY OF	WELL 203-DESTROYED	1,1-DICHLOROETHYLENE	1.3	UG/L	4/24/1998 0:00	CWS	438,000
CA	CA1010007	FRESNO, CITY OF	WELL 203-DESTROYED	1,1-DICHLOROETHYLENE	1.4	UG/L	2/4/1998 0:00	CWS	438,000
CA	CA1010007	FRESNO, CITY OF	WELL 203-DESTROYED	1,1-DICHLOROETHYLENE	1.7	UG/L	1/9/1997 0:00	CWS	438,000
CA	CA1010007	FRESNO, CITY OF	WELL 203-DESTROYED	1,1-DICHLOROETHYLENE	2.3	UG/L	7/16/1997 0:00	CWS	438,000
CA	CA1010007	FRESNO, CITY OF	WELL 201-INF	1,1-DICHLOROETHYLENE	5.8	UG/L	3/21/2001 0:00	CWS	438,000
CA	CA1010007	FRESNO, CITY OF	WELL 201-INF	1,1-DICHLOROETHYLENE	6.6	UG/L	5/16/2001 0:00	CWS	438,000
CA	CA1010007	FRESNO, CITY OF	WELL 201-INF	1,1-DICHLOROETHYLENE	6.7	UG/L	5/11/2001 0:00	CWS	438,000
CA	CA1010007	FRESNO, CITY OF	WELL 201-INF	1,1-DICHLOROETHYLENE	7.8	UG/L	8/14/2003 0:00	CWS	438,000
CA	CA1010007	FRESNO, CITY OF	WELL 201-INF	1,1-DICHLOROETHYLENE	8.7	UG/L	10/6/2003 0:00	CWS	438,000
CA	CA1010007	FRESNO, CITY OF	WELL 201-INF	1,1-DICHLOROETHYLENE	9	UG/L	4/10/2001 0:00	CWS	438,000
CA	CA1010007	FRESNO, CITY OF	WELL 201-INF	1,1-DICHLOROETHYLENE	9.3	UG/L	4/11/2001 0:00	CWS	438,000
CA	CA1010007	FRESNO, CITY OF	WELL 201-INF	1,1-DICHLOROETHYLENE	13	UG/L	2/24/2004 0:00	CWS	438,000
CA	CA1010007	FRESNO, CITY OF	SITE 201-GAC	1,1-DICHLOROETHYLENE	15	UG/L	3/5/2004 0:00	CWS	438,000
CA	CA1010007	FRESNO, CITY OF	SITE 201-GAC	1,1-DICHLOROETHYLENE	16	UG/L	2/24/2004 0:00	CWS	438,000
					6.31				
CA	CA1010007	FRESNO, CITY OF	WELL 039A-RAW	1,2,3-TRICHLOROPROPANE	5.00E-03	UG/L	3/3/2005 0:00	CWS	438,000
CA	CA1010007	FRESNO, CITY OF	WELL 198-RAW	1,2,3-TRICHLOROPROPANE	5.00E-03	UG/L	3/17/2005 0:00	CWS	438,000
CA	CA1010007	FRESNO, CITY OF	WELL 021A-RAW	1,2,3-TRICHLOROPROPANE	6.00E-03	UG/L	3/27/2002 0:00	CWS	438,000
CA	CA1010007	FRESNO, CITY OF	WELL 220-RAW	1,2,3-TRICHLOROPROPANE	7.00E-03	UG/L	1/22/2002 0:00	CWS	438,000
CA	CA1010007	FRESNO, CITY OF	WELL 065-RAW	1,2,3-TRICHLOROPROPANE	7.00E-03	UG/L	1/29/2002 0:00	CWS	438,000
CA	CA1010007	FRESNO, CITY OF	WELL 018A-RAW	1,2,3-TRICHLOROPROPANE	7.00E-03	UG/L	2/26/2002 0:00	CWS	438,000
CA	CA1010007	FRESNO, CITY OF	WELL 085 INF	1,2,3-TRICHLOROPROPANE	8.00E-03	UG/L	3/12/2002 0:00	CWS	438,000
CA	CA1010007	FRESNO, CITY OF	WELL 014A-RAW	1,2,3-TRICHLOROPROPANE	8.00E-03	UG/L	2/20/2002 0:00	CWS	438,000
CA	CA1010007	FRESNO, CITY OF	WELL 219-RAW	1,2,3-TRICHLOROPROPANE	9.00E-03	UG/L	1/22/2002 0:00	CWS	438,000
CA	CA1010007	FRESNO, CITY OF	WELL 039A-RAW	1,2,3-TRICHLOROPROPANE	9.00E-03	UG/L	2/13/2002 0:00	CWS	438,000
CA	CA1010007	FRESNO, CITY OF	WELL 039A-RAW	1,2,3-TRICHLOROPROPANE	9.00E-03	UG/L	8/15/2001 0:00	CWS	438,000
CA	CA1010007	FRESNO, CITY OF	WELL 021A-RAW	1,2,3-TRICHLOROPROPANE	9.00E-03	UG/L	7/18/2001 0:00	CWS	438,000
CA	CA1010007	FRESNO, CITY OF	WELL 277 INF	1,2,3-TRICHLOROPROPANE	0.01	UG/L	3/8/2002 0:00	CWS	438,000
CA	CA1010007	FRESNO, CITY OF	WELL 277 INF	1,2,3-TRICHLOROPROPANE	0.01	UG/L	3/22/2005 0:00	CWS	438,000
CA	CA1010007	FRESNO, CITY OF	WELL 275 INF	1,2,3-TRICHLOROPROPANE	0.01	UG/L	3/8/2002 0:00	CWS	438,000
CA	CA1010007	FRESNO, CITY OF	WELL 275 INF	1,2,3-TRICHLOROPROPANE	0.01	UG/L	3/23/2005 0:00	CWS	438,000
CA	CA1010007	FRESNO, CITY OF	WELL 217 INF	1,2,3-TRICHLOROPROPANE	0.01	UG/L	3/30/2005 0:00	CWS	438,000
CA	CA1010007	FRESNO, CITY OF	WELL 203A-RAW	1,2,3-TRICHLOROPROPANE	0.01	UG/L	3/17/2005 0:00	CWS	438,000
CA	CA1010007	FRESNO, CITY OF	WELL 059-RAW	1,2,3-TRICHLOROPROPANE	0.01	UG/L	1/22/2002 0:00	CWS	438,000
CA	CA1010007	FRESNO, CITY OF	WELL 059-RAW	1,2,3-TRICHLOROPROPANE	0.01	UG/L	9/7/2001 0:00	CWS	438,000

CA	CA1010007	FRESNO, CITY OF	WELL 048-RAW	1,2,3-TRICHLOROPROPANE	0.01	UG/L	7/10/2001 0:00	CWS	438,000
CA	CA1010007	FRESNO, CITY OF	WELL 298-RAW	1,2,3-TRICHLOROPROPANE	0.01	UG/L	2/20/2002 0:00	CWS	438,000
CA	CA1010007	FRESNO, CITY OF	WELL 298-RAW	1,2,3-TRICHLOROPROPANE	0.01	UG/L	7/13/2001 0:00	CWS	438,000
CA	CA1010007	FRESNO, CITY OF	WELL 184 INF	1,2,3-TRICHLOROPROPANE	0.01	UG/L	3/22/2005 0:00	CWS	438,000
CA	CA1010007	FRESNO, CITY OF	WELL 289-2 INF	1,2,3-TRICHLOROPROPANE	0.01	UG/L	3/8/2002 0:00	CWS	438,000
CA	CA1010007	FRESNO, CITY OF	WELL 137 INF	1,2,3-TRICHLOROPROPANE	0.01	UG/L	3/12/2002 0:00	CWS	438,000
CA	CA1010007	FRESNO, CITY OF	WELL 102-RAW	1,2,3-TRICHLOROPROPANE	0.01	UG/L	3/22/2005 0:00	CWS	438,000
CA	CA1010007	FRESNO, CITY OF	WELL 164-1-RAW	1,2,3-TRICHLOROPROPANE	0.01	UG/L	3/23/2005 0:00	CWS	438,000
CA	CA1010007	FRESNO, CITY OF	WELL 164-1-RAW	1,2,3-TRICHLOROPROPANE	0.01	UG/L	9/14/2001 0:00	CWS	438,000
CA	CA1010007	FRESNO, CITY OF	WELL 240-RAW	1,2,3-TRICHLOROPROPANE	0.02	UG/L	1/23/2002 0:00	CWS	438,000
CA	CA1010007	FRESNO, CITY OF	WELL 240-RAW	1,2,3-TRICHLOROPROPANE	0.02	UG/L	8/30/2001 0:00	CWS	438,000
CA	CA1010007	FRESNO, CITY OF	WELL 059-RAW	1,2,3-TRICHLOROPROPANE	0.02	UG/L	3/15/2005 0:00	CWS	438,000
CA	CA1010007	FRESNO, CITY OF	WELL 048-RAW	1,2,3-TRICHLOROPROPANE	0.02	UG/L	3/17/2005 0:00	CWS	438,000
CA	CA1010007	FRESNO, CITY OF	WELL 165-2-RAW	1,2,3-TRICHLOROPROPANE	0.02	UG/L	1/22/2002 0:00	CWS	438,000
CA	CA1010007	FRESNO, CITY OF	WELL 231-RAW	1,2,3-TRICHLOROPROPANE	0.03	UG/L	1/12/2005 0:00	CWS	438,000
CA	CA1010007	FRESNO, CITY OF	WELL 032B (NEW SOURCE)-RAW	1,2,3-TRICHLOROPROPANE	0.03	UG/L	3/17/2005 0:00	CWS	438,000
CA	CA1010007	FRESNO, CITY OF	WELL 137 INF	1,2,3-TRICHLOROPROPANE	0.03	UG/L	3/30/2005 0:00	CWS	438,000
CA	CA1010007	FRESNO, CITY OF	WELL 231-RAW	1,2,3-TRICHLOROPROPANE	0.04	UG/L	3/15/2005 0:00	CWS	438,000
CA	CA1010007	FRESNO, CITY OF	WELL 240-RAW	1,2,3-TRICHLOROPROPANE	0.04	UG/L	3/16/2005 0:00	CWS	438,000
CA	CA1010007	FRESNO, CITY OF	WELL 231-RAW	1,2,3-TRICHLOROPROPANE	0.06	UG/L	1/23/2002 0:00	CWS	438,000
CA	CA1010007	FRESNO, CITY OF	WELL 231-RAW	1,2,3-TRICHLOROPROPANE	0.07	UG/L	1/31/2002 0:00	CWS	438,000
CA	CA1010007	FRESNO, CITY OF	WELL 165-2-RAW	1,2,3-TRICHLOROPROPANE	0.07	UG/L	3/10/2005 0:00	CWS	438,000
CA	CA1010007	FRESNO, CITY OF	WELL 231-RAW	1,2,3-TRICHLOROPROPANE	0.08	UG/L	8/30/2001 0:00	CWS	438,000
CA	CA1010007	FRESNO, CITY OF	WELL 135A INF	1,2,3-TRICHLOROPROPANE	0.08	UG/L	3/8/2002 0:00	CWS	438,000
CA	CA1010007	FRESNO, CITY OF	WELL 164-2 INF	1,2,3-TRICHLOROPROPANE	0.09	UG/L	3/23/2005 0:00	CWS	438,000
CA	CA1010007	FRESNO, CITY OF	WELL 135A INF	1,2,3-TRICHLOROPROPANE	0.09	UG/L	3/30/2005 0:00	CWS	438,000
CA	CA1010007	FRESNO, CITY OF	WELL 110 INF	1,2,3-TRICHLOROPROPANE	0.24	UG/L	3/7/2002 0:00	CWS	438,000
CA	CA1010007	FRESNO, CITY OF	WELL 070-INF	1,2,3-TRICHLOROPROPANE	0.25	UG/L	2/4/2005 0:00	CWS	438,000
CA	CA1010007	FRESNO, CITY OF	WELL 070-INF	1,2,3-TRICHLOROPROPANE	0.25	UG/L	4/18/2005 0:00	CWS	438,000
CA	CA1010007	FRESNO, CITY OF	WELL 070-INF	1,2,3-TRICHLOROPROPANE	0.27	UG/L	6/15/2005 0:00	CWS	438,000
CA	CA1010007	FRESNO, CITY OF	WELL 070-INF	1,2,3-TRICHLOROPROPANE	0.28	UG/L	1/12/2005 0:00	CWS	438,000
CA	CA1010007	FRESNO, CITY OF	WELL 063-RAW	1,2,3-TRICHLOROPROPANE	0.61	UG/L	1/22/2002 0:00	CWS	438,000
CA	CA1010007	FRESNO, CITY OF	WELL 063-RAW	1,2,3-TRICHLOROPROPANE	0.64	UG/L	9/7/2001 0:00	CWS	438,000
CA	CA1010007	FRESNO, CITY OF	WELL 063-RAW	1,2,3-TRICHLOROPROPANE	1	UG/L	9/27/2000 0:00	CWS	438,000
CA	CA1000004	BELMONT WATER CORP	WELL 1A-STANDBY	1,2,3-TRICHLOROPROPANE	0.39	UG/L	11/5/2002 0:00	CWS	100
					9.09E-02				
CA	CA1010007	FRESNO, CITY OF	SITE 184 (GAC)	1,2,3-TRICHLOROPROPANE	9.00E-03	UG/L	9/19/2001 0:00	CWS	438,000
CA	CA1010007	FRESNO, CITY OF	SITE 275 (GAC)	1,2,3-TRICHLOROPROPANE	0.01	UG/L	9/20/2001 0:00	CWS	438,000
CA	CA1010007	FRESNO, CITY OF	SITE 289 (GAC)	1,2,3-TRICHLOROPROPANE	0.01	UG/L	3/23/2005 0:00	CWS	438,000
CA	CA1010007	FRESNO, CITY OF	SITE 277 (GAC)	1,2,3-TRICHLOROPROPANE	0.01	UG/L	9/20/2001 0:00	CWS	438,000
CA	CA1010007	FRESNO, CITY OF	SITE 289 (GAC)	1,2,3-TRICHLOROPROPANE	0.02	UG/L	9/14/2001 0:00	CWS	438,000
CA	CA1010007	FRESNO, CITY OF	SITE 082 (GAC)	1,2,3-TRICHLOROPROPANE	0.02	UG/L	9/18/2001 0:00	CWS	438,000

APPENDIX 7-13 *Continued*

State	PWSID	Water System Name	Source	Analyte	Level	Unit	Sample Date	PWSType	Pop
CA	CA1010007	FRESNO, CITY OF	SITE 070 (GAC)	1,2,3-TRICHLOROPROPANE	0.07	UG/L	1/12/2005 0:00	CWS	438,000
CA	CA1010007	FRESNO, CITY OF	SITE 164 (GAC)	1,2,3-TRICHLOROPROPANE	0.08	UG/L	9/14/2001 0:00	CWS	438,000
CA	CA1010007	FRESNO, CITY OF	SITE 164 (GAC)	1,2,3-TRICHLOROPROPANE	0.1	UG/L	3/7/2002 0:00	CWS	438,000
CA	CA1010007	FRESNO, CITY OF	SITE 070 (GAC)	1,2,3-TRICHLOROPROPANE	0.11	UG/L	2/4/2005 0:00	CWS	438,000
CA	CA1010007	FRESNO, CITY OF	SITE 070 (GAC)	1,2,3-TRICHLOROPROPANE	0.12	UG/L	2/4/2005 0:00	CWS	438,000
CA	CA1010007	FRESNO, CITY OF	SITE 110 (GAC)	1,2,3-TRICHLOROPROPANE	0.13	UG/L	9/14/2001 0:00	CWS	438,000
CA	CA1010007	FRESNO, CITY OF	SITE 070 (GAC)	1,2,3-TRICHLOROPROPANE	0.16	UG/L	4/18/2005 0:00	CWS	438,000
CA	CA1010007	FRESNO, CITY OF	SITE 135 (GAC)	1,2,3-TRICHLOROPROPANE	0.18	UG/L	9/20/2001 0:00	CWS	438,000
CA	CA1010007	FRESNO, CITY OF	SITE 070 (GAC)	1,2,3-TRICHLOROPROPANE	0.2	UG/L	9/18/2001 0:00	CWS	438,000
CA	CA1010007	FRESNO, CITY OF	SITE 070 (GAC)	1,2,3-TRICHLOROPROPANE	0.26	UG/L	6/15/2005 0:00	CWS	438,000
CA	CA1010007	FRESNO, CITY OF	SITE 070 (GAC) VESSEL 3 EFFLUENT	1,2,3-TRICHLOROPROPANE	0.5	UG/L	5/18/2005 0:00	CWS	438,000
					1.17E-01				
CA	CA1010007	FRESNO, CITY OF	SITE 070 (PTA)-TREATED	1,2-DICHLOROPROPANE	0.6	UG/L	3/3/1999 0:00	CWS	438,000
CA	CA1010007	FRESNO, CITY OF	WELL 070-INF	1,2-DICHLOROPROPANE	0.5	UG/L	2/9/1999 0:00	CWS	438,000
CA	CA1010007	FRESNO, CITY OF	WELL 070-INF	1,2-DICHLOROPROPANE	0.6	UG/L	8/23/1999 0:00	CWS	438,000
CA	CA1010007	FRESNO, CITY OF	WELL 070-INF	1,2-DICHLOROPROPANE	0.6	UG/L	3/3/1999 0:00	CWS	438,000
CA	CA1010007	FRESNO, CITY OF	WELL 110 INF	1,2-DICHLOROPROPANE	0.61	UG/L	5/7/2001 0:00	CWS	438,000
CA	CA1010007	FRESNO, CITY OF	WELL 070-INF	1,2-DICHLOROPROPANE	0.65	UG/L	12/3/1999 0:00	CWS	438,000
CA	CA1010007	FRESNO, CITY OF	WELL 070-INF	1,2-DICHLOROPROPANE	0.67	UG/L	10/8/2002 0:00	CWS	438,000
CA	CA1010007	FRESNO, CITY OF	WELL 070-INF	1,2-DICHLOROPROPANE	0.7	UG/L	5/7/1999 0:00	CWS	438,000
CA	CA1010007	FRESNO, CITY OF	WELL 070-INF	1,2-DICHLOROPROPANE	0.7	UG/L	10/30/1998 0:00	CWS	438,000
CA	CA1010007	FRESNO, CITY OF	WELL 070-INF	1,2-DICHLOROPROPANE	0.71	UG/L	3/15/1999 0:00	CWS	438,000
CA	CA1010007	FRESNO, CITY OF	WELL 070-INF	1,2-DICHLOROPROPANE	0.8	UG/L	1/29/1999 0:00	CWS	438,000
CA	CA1010007	FRESNO, CITY OF	WELL 070-INF	1,2-DICHLOROPROPANE	0.8	UG/L	11/23/1999 0:00	CWS	438,000
CA	CA1010007	FRESNO, CITY OF	WELL 070-INF	1,2-DICHLOROPROPANE	0.85	UG/L	11/8/2001 0:00	CWS	438,000
CA	CA1010007	FRESNO, CITY OF	WELL 070-INF	1,2-DICHLOROPROPANE	0.86	UG/L	11/2/1999 0:00	CWS	438,000
CA	CA1010007	FRESNO, CITY OF	WELL 070-INF	1,2-DICHLOROPROPANE	0.87	UG/L	10/10/2003 0:00	CWS	438,000
CA	CA1010007	FRESNO, CITY OF	WELL 070-INF	1,2-DICHLOROPROPANE	0.89	UG/L	2/8/2002 0:00	CWS	438,000
CA	CA1010007	FRESNO, CITY OF	WELL 070-INF	1,2-DICHLOROPROPANE	0.9	UG/L	8/18/1999 0:00	CWS	438,000
CA	CA1010007	FRESNO, CITY OF	WELL 070-INF	1,2-DICHLOROPROPANE	0.9	UG/L	3/22/1999 0:00	CWS	438,000
CA	CA1010007	FRESNO, CITY OF	WELL 070-INF	1,2-DICHLOROPROPANE	0.9	UG/L	11/16/1999 0:00	CWS	438,000
CA	CA1010007	FRESNO, CITY OF	WELL 070-INF	1,2-DICHLOROPROPANE	0.95	UG/L	6/4/2002 0:00	CWS	438,000
CA	CA1010007	FRESNO, CITY OF	WELL 070-INF	1,2-DICHLOROPROPANE	0.95	UG/L	12/23/1999 0:00	CWS	438,000
CA	CA1010007	FRESNO, CITY OF	WELL 070-INF	1,2-DICHLOROPROPANE	0.97	UG/L	11/10/1999 0:00	CWS	438,000
CA	CA1010007	FRESNO, CITY OF	WELL 070-INF	1,2-DICHLOROPROPANE	0.98	UG/L	7/28/1999 0:00	CWS	438,000
CA	CA1010007	FRESNO, CITY OF	WELL 070-INF	1,2-DICHLOROPROPANE	1	UG/L	5/10/1999 0:00	CWS	438,000
CA	CA1010007	FRESNO, CITY OF	WELL 070-INF	1,2-DICHLOROPROPANE	1	UG/L	9/2/1999 0:00	CWS	438,000
CA	CA1010007	FRESNO, CITY OF	WELL 070-INF	1,2-DICHLOROPROPANE	1	UG/L	1/15/1997 0:00	CWS	438,000
CA	CA1010007	FRESNO, CITY OF	WELL 070-INF	1,2-DICHLOROPROPANE	1	UG/L	2/2/1999 0:00	CWS	438,000
CA	CA1010007	FRESNO, CITY OF	WELL 070-INF	1,2-DICHLOROPROPANE	1.1	UG/L	6/14/1999 0:00	CWS	438,000
CA	CA1010007	FRESNO, CITY OF	WELL 070-INF	1,2-DICHLOROPROPANE	1.1	UG/L	7/13/1999 0:00	CWS	438,000

CA	CA1010007	FRESNO, CITY OF	WELL 070-INF	1,2-DICHLOROPROPANE	1.1	UG/L	1/15/1997 0:00	CWS	438,000
CA	CA1010007	FRESNO, CITY OF	WELL 070-INF	1,2-DICHLOROPROPANE	1.2	UG/L	11/6/2002 0:00	CWS	438,000
CA	CA1010007	FRESNO, CITY OF	WELL 070-INF	1,2-DICHLOROPROPANE	1.2	UG/L	3/15/2005 0:00	CWS	438,000
CA	CA1010007	FRESNO, CITY OF	WELL 070-INF	1,2-DICHLOROPROPANE	1.2	UG/L	6/3/1999 0:00	CWS	438,000
CA	CA1010007	FRESNO, CITY OF	WELL 070-INF	1,2-DICHLOROPROPANE	1.2	UG/L	6/10/1999 0:00	CWS	438,000
CA	CA1010007	FRESNO, CITY OF	WELL 070-INF	1,2-DICHLOROPROPANE	1.2	UG/L	6/29/1999 0:00	CWS	438,000
CA	CA1010007	FRESNO, CITY OF	WELL 070-INF	1,2-DICHLOROPROPANE	1.2	UG/L	3/9/1999 0:00	CWS	438,000
CA	CA1010007	FRESNO, CITY OF	WELL 070-INF	1,2-DICHLOROPROPANE	1.2	UG/L	7/10/2000 0:00	CWS	438,000
CA	CA1010007	FRESNO, CITY OF	WELL 070-INF	1,2-DICHLOROPROPANE	1.3	UG/L	7/5/2002 0:00	CWS	438,000
CA	CA1010007	FRESNO, CITY OF	WELL 070-INF	1,2-DICHLOROPROPANE	1.3	UG/L	9/20/1999 0:00	CWS	438,000
CA	CA1010007	FRESNO, CITY OF	WELL 070-INF	1,2-DICHLOROPROPANE	1.3	UG/L	10/5/1999 0:00	CWS	438,000
CA	CA1010007	FRESNO, CITY OF	WELL 070-INF	1,2-DICHLOROPROPANE	1.3	UG/L	7/21/1999 0:00	CWS	438,000
CA	CA1010007	FRESNO, CITY OF	WELL 070-INF	1,2-DICHLOROPROPANE	1.3	UG/L	5/11/2001 0:00	CWS	438,000
CA	CA1010007	FRESNO, CITY OF	WELL 070-INF	1,2-DICHLOROPROPANE	1.3	UG/L	2/3/2000 0:00	CWS	438,000
CA	CA1010007	FRESNO, CITY OF	WELL 070-INF	1,2-DICHLOROPROPANE	1.3	UG/L	10/18/1999 0:00	CWS	438,000
CA	CA1010007	FRESNO, CITY OF	WELL 070-INF	1,2-DICHLOROPROPANE	1.4	UG/L	10/11/1999 0:00	CWS	438,000
CA	CA1010007	FRESNO, CITY OF	WELL 070-INF	1,2-DICHLOROPROPANE	1.4	UG/L	5/20/1999 0:00	CWS	438,000
CA	CA1010007	FRESNO, CITY OF	WELL 070-INF	1,2-DICHLOROPROPANE	1.4	UG/L	10/1/1999 0:00	CWS	438,000
CA	CA1010007	FRESNO, CITY OF	WELL 070-INF	1,2-DICHLOROPROPANE	1.4	UG/L	8/5/1999 0:00	CWS	438,000
CA	CA1010007	FRESNO, CITY OF	WELL 070-INF	1,2-DICHLOROPROPANE	1.5	UG/L	5/28/1999 0:00	CWS	438,000
CA	CA1010007	FRESNO, CITY OF	WELL 070-INF	1,2-DICHLOROPROPANE	1.6	UG/L	9/16/1999 0:00	CWS	438,000
CA	CA1010007	FRESNO, CITY OF	WELL 070-INF	1,2-DICHLOROPROPANE	1.7	UG/L	9/8/1999 0:00	CWS	438,000
CA	CA1010007	FRESNO, CITY OF	WELL 070-INF	1,2-DICHLOROPROPANE	1.9	UG/L	11/6/2003 0:00	CWS	438,000
CA	CA1010007	FRESNO, CITY OF	WELL 070-INF	1,2-DICHLOROPROPANE	1.9	UG/L	4/2/1999 0:00	CWS	438,000
CA	CA1010007	FRESNO, CITY OF	WELL 070-INF	1,2-DICHLOROPROPANE	2	UG/L	8/9/1999 0:00	CWS	438,000
CA	CA1010007	FRESNO, CITY OF	WELL 070-INF	1,2-DICHLOROPROPANE	2	UG/L	4/10/2000 0:00	CWS	438,000
CA	CA1010007	FRESNO, CITY OF	WELL 070-INF	1,2-DICHLOROPROPANE	2	UG/L	3/10/2000 0:00	CWS	438,000
CA	CA1010007	FRESNO, CITY OF	WELL 070-INF	1,2-DICHLOROPROPANE	2.1	UG/L	4/9/1999 0:00	CWS	438,000
CA	CA1010007	FRESNO, CITY OF	WELL 070-INF	1,2-DICHLOROPROPANE	2.2	UG/L	6/9/2000 0:00	CWS	438,000
CA	CA1010007	FRESNO, CITY OF	WELL 070-INF	1,2-DICHLOROPROPANE	2.2	UG/L	5/4/2000 0:00	CWS	438,000
CA	CA1010007	FRESNO, CITY OF	WELL 070-INF	1,2-DICHLOROPROPANE	2.4	UG/L	10/4/2001 0:00	CWS	438,000
CA	CA1010007	FRESNO, CITY OF	WELL 070-INF	1,2-DICHLOROPROPANE	2.7	UG/L	9/14/2001 0:00	CWS	438,000
CA	CA1010007	FRESNO, CITY OF	WELL 070-INF	1,2-DICHLOROPROPANE	2.8	UG/L	12/3/2001 0:00	CWS	438,000
CA	CA1010007	FRESNO, CITY OF	WELL 070-INF	1,2-DICHLOROPROPANE	2.9	UG/L	11/16/2004 0:00	CWS	438,000
					1.28				
CA	CA1010007	FRESNO, CITY OF	WELL 040-DESTROYED	1,3,5-TRIMETHYLBENZENE	0.9	UG/L	5/18/1998 0:00	CWS	438,000
CA	CA1010007	FRESNO, CITY OF	WELL 040A-RAW-INACTIVE	1,3,5-TRIMETHYLBENZENE	0.9	UG/L	5/18/1998 0:00	CWS	438,000
CA	CA1010007	FRESNO, CITY OF	WELL 040A-RAW-INACTIVE	1,2,4-TRIMETHYLBENZENE	2	UG/L	5/18/1998 0:00	CWS	438,000
					1				
CA	CA1010007	FRESNO, CITY OF	WELL 286 INF	2,2-DICHLOROPROPANE	0.53	UG/L	6/30/1999 0:00	CWS	438,000
CA	CA1010007	FRESNO, CITY OF	WELL 286 INF	2,2-DICHLOROPROPANE	0.61	UG/L	6/30/1999 0:00	CWS	438,000
CA	CA1010007	FRESNO, CITY OF	WELL 215-RAW-INACTIVE	2,2-DICHLOROPROPANE	9	UG/L	6/5/2001 0:00	CWS	438,000
CA	CA1010007	FRESNO, CITY OF	WELL 279 INF	2,2-DICHLOROPROPANE	80	UG/L	9/7/1999 0:00	CWS	438,000

APPENDIX 7-13 *Continued*

State	PWSID	Water System Name	Source	Analyte	Level	Unit	Sample Date	PWSType	Pop
CA	CA1010007	FRESNO, CITY OF	WELL 257	2,2-DICHLOROPROPANE	85	UG/L	9/7/1999 0:00	CWS	438,000
					35				
CA	CA1010007	FRESNO, CITY OF	WELL 159 INF	BROMOMETHANE	0.66	UG/L	7/22/2002 0:00	CWS	438,000
CA	CA1010007	FRESNO, CITY OF	SITE 070 (GAC) VESSEL 1 EFFLUENT	BROMOMETHANE	1.1	UG/L	4/5/2002 0:00	CWS	438,000
CA	CA1010007	FRESNO, CITY OF	SITE 070 (GAC)	BROMOMETHANE	1.1	UG/L	4/5/2002 0:00	CWS	438,000
					1				
CA	CA1010007	FRESNO, CITY OF	SITE 070 (GAC) VESSEL 2 EFFLUENT	CHLOROETHANE	0.52	UG/L	7/10/2000 0:00	CWS	438,000
CA	CA1010007	FRESNO, CITY OF	SITE 070 (GAC)	CHLOROETHANE	0.7	UG/L	3/3/1999 0:00	CWS	438,000
CA	CA1010007	FRESNO, CITY OF	SITE 279 (PTA)	CHLOROETHANE	0.7	UG/L	3/3/1999 0:00	CWS	438,000
CA	CA1010007	FRESNO, CITY OF	WELL 213A-RAW	CHLOROETHANE	0.61	UG/L	5/3/2000 0:00	CWS	438,000
					0.63				
CA	CA1010007	FRESNO, CITY OF	SITE 070 (GAC) VESSEL 1 EFFLUENT	CHLOROMETHANE	0.54	UG/L	4/5/2002 0:00	CWS	438,000
CA	CA1010007	FRESNO, CITY OF	SITE 283/286 (GAC)	CHLOROMETHANE	2	UG/L	7/3/2002 0:00	CWS	438,000
CA	CA1010007	FRESNO, CITY OF	SITE 283/286 (GAC)	CHLOROMETHANE	0.97	UG/L	8/6/2001 0:00	CWS	438,000
					1.17				
CA	CA1010007	FRESNO, CITY OF	WELL 065-RAW	CHLOROMETHANE	0.5	UG/L	4/20/1999 0:00	CWS	438,000
CA	CA1010007	FRESNO, CITY OF	WELL 070-INF	CHLOROMETHANE	0.5	UG/L	4/5/2002 0:00	CWS	438,000
CA	CA1010007	FRESNO, CITY OF	WELL 014A-RAW	CHLOROMETHANE	0.5	UG/L	12/15/1998 0:00	CWS	438,000
CA	CA1010007	FRESNO, CITY OF	WELL 070-INF	CHLOROMETHANE	0.6	UG/L	1/11/2002 0:00	CWS	438,000
CA	CA1010007	FRESNO, CITY OF	WELL 020-RAW	CHLOROMETHANE	0.62	UG/L	3/1/2002 0:00	CWS	438,000
CA	CA1010007	FRESNO, CITY OF	WELL 273	CHLOROMETHANE	0.74	UG/L	5/1/2002 0:00	CWS	438,000
CA	CA1010007	FRESNO, CITY OF	WELL 213A-RAW	CHLOROMETHANE	0.9	UG/L	12/15/1998 0:00	CWS	438,000
CA	CA1010007	FRESNO, CITY OF	WELL 213A-RAW	CHLOROMETHANE	0.98	UG/L	5/3/2000 0:00	CWS	438,000
CA	CA1010007	FRESNO, CITY OF	WELL 213A-RAW	CHLOROMETHANE	1	UG/L	3/1/2000 0:00	CWS	438,000
CA	CA1010007	FRESNO, CITY OF	WELL 215-RAW-INACTIVE	CHLOROMETHANE	1.4	UG/L	8/3/2001 0:00	CWS	438,000
CA	CA1010007	FRESNO, CITY OF	WELL 213A-RAW	CHLOROMETHANE	1.4	UG/L	10/7/1999 0:00	CWS	438,000
CA	CA1010007	FRESNO, CITY OF	WELL 215-RAW-INACTIVE	CHLOROMETHANE	1.6	UG/L	4/4/2002 0:00	CWS	438,000
CA	CA1010007	FRESNO, CITY OF	WELL 213A-RAW	CHLOROMETHANE	2	UG/L	4/4/2002 0:00	CWS	438,000
CA	CA1010007	FRESNO, CITY OF	WELL 213A-RAW	CHLOROMETHANE	2.4	UG/L	8/1/2001 0:00	CWS	438,000
CA	CA1010007	FRESNO, CITY OF	WELL 213A-RAW	CHLOROMETHANE	2.6	UG/L	7/10/2000 0:00	CWS	438,000
CA	CA1010007	FRESNO, CITY OF	WELL 213A-RAW	CHLOROMETHANE	3	UG/L	9/5/2000 0:00	CWS	438,000
CA	CA1010007	FRESNO, CITY OF	WELL 213A-RAW	CHLOROMETHANE	4.2	UG/L	5/1/2002 0:00	CWS	438,000
					1.46				
CA	CA1010007	FRESNO, CITY OF	SITE 283/286 (GAC)	CIS-1,2-DICHLOROETHYLENE	0.52	UG/L	10/11/2000 0:00	CWS	438,000
CA	CA1010007	FRESNO, CITY OF	SITE 283/286 (GAC)	CIS-1,2-DICHLOROETHYLENE	0.6	UG/L	10/11/2000 0:00	CWS	438,000
					0.56				

CA	CA1010007	FRESNO, CITY OF	84 SAMPLES	CIS-1,2-DICHLOROETHYLENE	1.78	UG/L	10/9/1997 0:00	CWS	438,000
					1.78				
CA	CA1010017	FRESNO CO WATERWORKS	WELL R8-RAW	DIAZINON	507	UG/L	10/27/2004 0:00	CWS	2,042
CA	CA1010007	FRESNO, CITY OF	1,004 SAMPLES-GAC	DIBROMOCHLOROPROPANE (DBCP)	0.08	UG/L	1/8/1999 0:00	CWS	438,000
CA	CA1010007	FRESNO, CITY OF	3,175 SAMPLES	DIBROMOCHLOROPROPANE (DBCP)	0.34	UG/L	8/5/2002 0:00	CWS	438,000
CA	CA1010007	FRESNO, CITY OF	SITE 283/286 (GAC)	DICHLORODIFLUOROMETHANE	0.53	UG/L	4/1/2004 0:00	CWS	438,000
CA	CA1010007	FRESNO, CITY OF	SITE 283/286 (GAC)	DICHLORODIFLUOROMETHANE	0.62	UG/L	6/17/2003 0:00	CWS	438,000
CA	CA1010007	FRESNO, CITY OF	SITE 283/286 (GAC)	DICHLORODIFLUOROMETHANE	0.77	UG/L	5/8/2003 0:00	CWS	438,000
CA	CA1010007	FRESNO, CITY OF	SITE 283/286 (GAC)	DICHLORODIFLUOROMETHANE	1.3	UG/L	4/12/2004 0:00	CWS	438,000
CA	CA1010007	FRESNO, CITY OF	SITE 283/286 (GAC)	DICHLORODIFLUOROMETHANE	1.3	UG/L	10/7/2003 0:00	CWS	438,000
CA	CA1010007	FRESNO, CITY OF	SITE 283/286 (GAC)	DICHLORODIFLUOROMETHANE	1.3	UG/L	12/2/2003 0:00	CWS	438,000
CA	CA1010007	FRESNO, CITY OF	SITE 283/286 (GAC)	DICHLORODIFLUOROMETHANE	1.3	UG/L	11/5/2002 0:00	CWS	438,000
CA	CA1010007	FRESNO, CITY OF	SITE 283/286 (GAC)	DICHLORODIFLUOROMETHANE	0.8	UG/L	4/19/2004 0:00	CWS	438,000
CA	CA1010007	FRESNO, CITY OF	SITE 283/286 (GAC)	DICHLORODIFLUOROMETHANE	0.88	UG/L	8/6/2003 0:00	CWS	438,000
CA	CA1010007	FRESNO, CITY OF	SITE 283/286 (GAC)	DICHLORODIFLUOROMETHANE	0.91	UG/L	11/7/2003 0:00	CWS	438,000
CA	CA1010007	FRESNO, CITY OF	SITE 283/286 (GAC)	DICHLORODIFLUOROMETHANE	0.96	UG/L	4/26/2004 0:00	CWS	438,000
CA	CA1010007	FRESNO, CITY OF	SITE 283/286 (GAC)	DICHLORODIFLUOROMETHANE	0.96	UG/L	5/27/2003 0:00	CWS	438,000
CA	CA1010007	FRESNO, CITY OF	SITE 283/286 (GAC)	DICHLORODIFLUOROMETHANE	0.99	UG/L	6/4/2003 0:00	CWS	438,000
CA	CA1010007	FRESNO, CITY OF	SITE 283/286 (GAC)	DICHLORODIFLUOROMETHANE	1	UG/L	8/14/2002 0:00	CWS	438,000
CA	CA1010007	FRESNO, CITY OF	SITE 283/286 (GAC)	DICHLORODIFLUOROMETHANE	1	UG/L	6/25/2003 0:00	CWS	438,000
CA	CA1010007	FRESNO, CITY OF	SITE 283/286 (GAC)	DICHLORODIFLUOROMETHANE	1.1	UG/L	5/12/2003 0:00	CWS	438,000
CA	CA1010007	FRESNO, CITY OF	SITE 283/286 (GAC)	DICHLORODIFLUOROMETHANE	1.2	UG/L	4/5/2004 0:00	CWS	438,000
CA	CA1010007	FRESNO, CITY OF	SITE 283/286 (GAC)	DICHLORODIFLUOROMETHANE	1.2	UG/L	10/8/2003 0:00	CWS	438,000
CA	CA1010007	FRESNO, CITY OF	SITE 283/286 (GAC)	DICHLORODIFLUOROMETHANE	1.4	UG/L	3/3/2004 0:00	CWS	438,000
CA	CA1010007	FRESNO, CITY OF	SITE 283/286 (GAC)	DICHLORODIFLUOROMETHANE	1.4	UG/L	2/3/2004 0:00	CWS	438,000
CA	CA1010007	FRESNO, CITY OF	SITE 283/286 (GAC)	DICHLORODIFLUOROMETHANE	1.5	UG/L	5/7/2004 0:00	CWS	438,000
CA	CA1010007	FRESNO, CITY OF	SITE 283/286 (GAC)	DICHLORODIFLUOROMETHANE	1.5	UG/L	5/12/2004 0:00	CWS	438,000
CA	CA1010007	FRESNO, CITY OF	SITE 283/286 (GAC)	DICHLORODIFLUOROMETHANE	1.5	UG/L	8/29/2001 0:00	CWS	438,000
CA	CA1010007	FRESNO, CITY OF	SITE 283/286 (GAC)	DICHLORODIFLUOROMETHANE	1.9	UG/L	7/3/2001 0:00	CWS	438,000
CA	CA1010007	FRESNO, CITY OF	SITE 283/286 (GAC)	DICHLORODIFLUOROMETHANE	1.9	UG/L	4/24/2003 0:00	CWS	438,000
CA	CA1010007	FRESNO, CITY OF	SITE 217 (GAC)	DICHLORODIFLUOROMETHANE	4.2	UG/L	3/12/2002 0:00	CWS	438,000
CA	CA1010007	FRESNO, CITY OF	SITE 283/286 (GAC)	DICHLORODIFLUOROMETHANE	2.3	UG/L	5/6/2003 0:00	CWS	438,000
CA	CA1010007	FRESNO, CITY OF	SITE 283/286 (GAC)	DICHLORODIFLUOROMETHANE	2.5	UG/L	4/29/2003 0:00	CWS	438,000
CA	CA1010007	FRESNO, CITY OF	SITE 217 (GAC)	DICHLORODIFLUOROMETHANE	4.8	UG/L	9/19/2001 0:00	CWS	438,000
					1.48				
CA	CA1010007	FRESNO, CITY OF	257 SAMPLES	DICHLORODIFLUOROMETHANE	9.8	UG/L	7/8/1997 0:00	CWS	438,000
CA	CA1010007	FRESNO, CITY OF	WELL 186 INF	DICHLOROMETHANE	0.52	UG/L	6/27/2001 0:00	CWS	438,000
CA	CA1010007	FRESNO, CITY OF	SITE 070 (GAC) VESSEL 1 EFFLUENT	DICHLOROMETHANE	0.66	UG/L	10/18/2001 0:00	CWS	438,000
CA	CA1010007	FRESNO, CITY OF	WELL 264	DICHLOROMETHANE	0.8	UG/L	7/2/1999 0:00	CWS	438,000
					0.66				

APPENDIX 7-13 *Continued*

State	PWSID	Water System Name	Source	Analyte	Level	Unit	Sample Date	PWSType	Pop
CA	CA1010339	California State University Fresno	WELL 06	DIETHYL PHTHALATE	1.4	UG/L	11/6/2002 0:00	CWS	20,000
CA	CA1010007	FRESNO, CITY OF	105 SAMPLES GAC	ETHYLENE DIBROMIDE (EDB)	0.04	UG/L	1/27/1998 0:00	CWS	438,000
CA	CA1010007	FRESNO, CITY OF	351 SAMPLES	ETHYLENE DIBROMIDE (EDB)	0.08	UG/L	4/3/1997 0:00	CWS	438,000
CA	CA1010007	FRESNO, CITY OF	WELL 019A-RAW	METHYL TERT BUTYL ETHER (MTBE)	5	UG/L	7/10/2001 0:00	CWS	438,000
CA	CA1010007	FRESNO, CITY OF	WELL 033A-RAW	MOLINATE	5.7	UG/L	2/27/2002 0:00	CWS	438,000
CA	CA1010007	FRESNO, CITY OF	WELL 211-RAW	MONOCHLOROBENZENE	0.52	UG/L	3/19/2001 0:00	CWS	438,000
CA	CA1010007	FRESNO, CITY OF	WELL 040-DESTROYED	N-BUTYLBENZENE	0.9	UG/L	5/18/1998 0:00	CWS	438,000
CA	CA1010007	FRESNO, CITY OF	WELL 040A-RAW-INACTIVE	N-BUTYLBENZENE	0.9	UG/L	5/18/1998 0:00	CWS	438,000
CA	CA1010007	FRESNO, CITY OF	WELL 287	STYRENE	0.52	UG/L	11/4/2002 0:00	CWS	438,000
CA	CA1010007	FRESNO, CITY OF	321 SAMPLES	TETRACHLOROETHYLENE	1.43	UG/L	12/1/2003 0:00	CWS	438,000
CA	CA1010007	FRESNO, CITY OF	WELL 221	THIOBENCARB	7.2	UG/L	4/2/1999 0:00	CWS	438,000
CA	CA1010013	HUME LAKE CHRISTIAN CAMPS, INC	FRENCH DRAIN-ABANDONED	TOLUENE	0.5	UG/L	2/16/1999 0:00	CWS	3,000
CA	CA1010013	HUME LAKE CHRISTIAN CAMPS, INC	WELL 02	TOLUENE	0.51	UG/L	2/16/1999 0:00	CWS	3,000
CA	CA1010007	FRESNO, CITY OF	WELL 264	TOLUENE	0.52	UG/L	5/1/2002 0:00	CWS	438,000
CA	CA1010007	FRESNO, CITY OF	WELL 264	TOLUENE	0.54	UG/L	7/3/2001 0:00	CWS	438,000
CA	CA1010007	FRESNO, CITY OF	WELL 284	TOLUENE	0.58	UG/L	5/1/2002 0:00	CWS	438,000
CA	CA1010007	FRESNO, CITY OF	WELL 265-INACTIVE	TOLUENE	0.6	UG/L	9/7/1999 0:00	CWS	438,000
CA	CA1010007	FRESNO, CITY OF	WELL 266	TOLUENE	0.61	UG/L	5/1/2002 0:00	CWS	438,000
CA	CA1010007	FRESNO, CITY OF	WELL 267	TOLUENE	0.63	UG/L	5/1/2002 0:00	CWS	438,000
CA	CA1010007	FRESNO, CITY OF	WELL 273	TOLUENE	0.68	UG/L	5/1/2002 0:00	CWS	438,000
CA	CA1010007	FRESNO, CITY OF	WELL 213A-RAW	TOLUENE	0.68	UG/L	5/1/2002 0:00	CWS	438,000
CA	CA1010007	FRESNO, CITY OF	WELL 287	TOLUENE	0.7	UG/L	5/1/2002 0:00	CWS	438,000
CA	CA1010007	FRESNO, CITY OF	WELL 258	TOLUENE	0.8	UG/L	5/1/2002 0:00	CWS	438,000
CA	CA1010007	FRESNO, CITY OF	SITE 070 (GAC) VESSEL 3 EFFLUENT	TOLUENE	0.9	UG/L	3/16/2004 0:00	CWS	438,000
CA	CA1010007	FRESNO, CITY OF	WELL 286 INF	TOLUENE	1	UG/L	12/5/2001 0:00	CWS	438,000
CA	CA1010007	FRESNO, CITY OF	WELL 117 (KVA WELL)-RAW	TOLUENE	3.9	UG/L	11/25/1997 0:00	CWS	438,000
					0.88				
CA	CA1009215	FELGER FARMS	TREATMENT PLANT-TREATED	TRICHLOROACETIC ACID (TCAA)	3.5	UG/L	7/13/2004 0:00	CWS	108
CA	CA1009035	PILIBOS BROTHERS RANCH	TREATMENT PLANT-TREATED	TRICHLOROACETIC ACID (TCAA)	5.6	UG/L	7/13/2004 0:00	CWS	64
CA	CA1009023	BRITZ/COLUSA	TREATMENT PLANT-TREATED	TRICHLOROACETIC ACID (TCAA)	6.8	UG/L	8/28/2003 0:00	CWS	170
					5.3				
CA	CA1010007	FRESNO, CITY OF	50 SAMPLES-GAC	TRICHLOROETHYLENE	2.06	UG/L	8/4/2004 0:00	CWS	438,000
CA	CA1010007	FRESNO, CITY OF	630 SAMPLES	TRICHLOROETHYLENE	15.2	UG/L	10/13/2004 0:00	CWS	438,000

CA	CA1010007	FRESNO, CITY OF	WELL 125-RAW	XYLENES (TOTAL)	0.5	UG/L	5/29/2001 0:00	CWS	438,000
CA	CA1010007	FRESNO, CITY OF	WELL 283 INF	XYLENES (TOTAL)	0.64	UG/L	5/2/2002 0:00	CWS	438,000
CA	CA1010007	FRESNO, CITY OF	WELL 257	XYLENES (TOTAL)	0.73	UG/L	5/1/2002 0:00	CWS	438,000
CA	CA1010007	FRESNO, CITY OF	WELL 279 INF	XYLENES (TOTAL)	0.78	UG/L	5/1/2002 0:00	CWS	438,000
CA	CA1010007	FRESNO, CITY OF	WELL 280	XYLENES (TOTAL)	0.79	UG/L	5/1/2002 0:00	CWS	438,000
CA	CA1010007	FRESNO, CITY OF	WELL 040-DESTROYED	XYLENES (TOTAL)	0.8	UG/L	5/18/1998 0:00	CWS	438,000
CA	CA1010007	FRESNO, CITY OF	WELL 040A-RAW-INACTIVE	XYLENES (TOTAL)	0.8	UG/L	5/18/1998 0:00	CWS	438,000
CA	CA1010007	FRESNO, CITY OF	WELL 264	XYLENES (TOTAL)	0.84	UG/L	5/1/2002 0:00	CWS	438,000
CA	CA1010007	FRESNO, CITY OF	WELL 284	XYLENES (TOTAL)	0.99	UG/L	5/1/2002 0:00	CWS	438,000
CA	CA1010007	FRESNO, CITY OF	WELL 266	XYLENES (TOTAL)	1	UG/L	5/1/2002 0:00	CWS	438,000
CA	CA1010007	FRESNO, CITY OF	WELL 267	XYLENES (TOTAL)	1.1	UG/L	5/1/2002 0:00	CWS	438,000
CA	CA1010007	FRESNO, CITY OF	WELL 213A-RAW	XYLENES (TOTAL)	1.1	UG/L	5/1/2002 0:00	CWS	438,000
CA	CA1010007	FRESNO, CITY OF	WELL 273	XYLENES (TOTAL)	1.2	UG/L	5/1/2002 0:00	CWS	438,000
CA	CA1010007	FRESNO, CITY OF	WELL 287	XYLENES (TOTAL)	1.2	UG/L	5/1/2002 0:00	CWS	438,000
CA	CA1010007	FRESNO, CITY OF	WELL 258	XYLENES (TOTAL)	1.5	UG/L	5/1/2002 0:00	CWS	438,000
CA	CA1010007	FRESNO, CITY OF	WELL 286 INF	XYLENES (TOTAL)	1.5	UG/L	12/5/2001 0:00	CWS	438,000
CA	CA1010007	FRESNO, CITY OF	WELL 280	M,P-XYLENE	0.56	UG/L	5/1/2002 0:00	CWS	438,000
CA	CA1010007	FRESNO, CITY OF	WELL 257	M,P-XYLENE	0.57	UG/L	5/1/2002 0:00	CWS	438,000
CA	CA1010007	FRESNO, CITY OF	WELL 279 INF	M,P-XYLENE	0.62	UG/L	5/1/2002 0:00	CWS	438,000
CA	CA1010007	FRESNO, CITY OF	WELL 264	M,P-XYLENE	0.63	UG/L	5/1/2002 0:00	CWS	438,000
CA	CA1010007	FRESNO, CITY OF	WELL 284	M,P-XYLENE	0.67	UG/L	5/1/2002 0:00	CWS	438,000
CA	CA1010007	FRESNO, CITY OF	WELL 266	M,P-XYLENE	0.75	UG/L	5/1/2002 0:00	CWS	438,000
CA	CA1010007	FRESNO, CITY OF	WELL 213A-RAW	M,P-XYLENE	0.77	UG/L	5/1/2002 0:00	CWS	438,000
CA	CA1010007	FRESNO, CITY OF	WELL 267	M,P-XYLENE	0.78	UG/L	5/1/2002 0:00	CWS	438,000
CA	CA1010007	FRESNO, CITY OF	WELL 273	M,P-XYLENE	0.89	UG/L	5/1/2002 0:00	CWS	438,000
CA	CA1010007	FRESNO, CITY OF	WELL 287	M,P-XYLENE	0.89	UG/L	5/1/2002 0:00	CWS	438,000
CA	CA1010007	FRESNO, CITY OF	WELL 258	M,P-XYLENE	1.1	UG/L	5/1/2002 0:00	CWS	438,000
CA	CA1010007	FRESNO, CITY OF	WELL 286 INF	M,P-XYLENE	1.1	UG/L	12/5/2001 0:00	CWS	438,000
					0.89				

APPENDIX 7-14 Fresno Surface Water Data

State	PWSID	Water System Name	Source	Analyte	Level	Unit	Sample Date	PWSType	Pop
CA	CA1009023	BRITZ/COLUSA	TREATMENT PLANT — TREATED	DIBROMOACETIC ACID (DBAA)	4.9	UG/L	8/28/2003 0:00	CWS	170
CA	CA1009035	PILIBOS BROTHERS RANCH	TREATMENT PLANT — TREATED	DIBROMOACETIC ACID (DBAA)	7.1	UG/L	7/13/2004 0:00	CWS	64
CA	CA1009215	FELGER FARMS	TREATMENT PLANT — TREATED	DIBROMOACETIC ACID (DBAA)	9.8	UG/L	7/13/2004 0:00	CWS	108
					7.26				
CA	CA1009023	BRITZ/COLUSA	TREATMENT PLANT — TREATED	DICHLOROACETIC ACID (DCAA)	8.9	UG/L	8/28/2003 0:00	CWS	170
CA	CA1009035	PILIBOS BROTHERS RANCH	TREATMENT PLANT — TREATED	DICHLOROACETIC ACID (DCAA)	7.8	UG/L	7/13/2004 0:00	CWS	64
CA	CA1009215	FELGER FARMS	TREATMENT PLANT — TREATED	DICHLOROACETIC ACID (DCAA)	6.4	UG/L	7/13/2004 0:00	CWS	108
					7.7				
CA	CA1000484	FCSA #34/BRIGHTON CREST	MILLERTON LAKE — RAW	DICHLOROMETHANE	2.8	UG/L	10/18/2001 0:00	CWS	139
CA	CA1000484	FCSA #34/BRIGHTON CREST	MILLERTON LAKE — RAW	DICHLOROMETHANE	2.8	UG/L	10/18/2001 0:00	CWS	139
					2.8				
CA	CA1009023	BRITZ/COLUSA	TREATMENT PLANT — TREATED	HALOACETIC ACIDS (5) (HAA5)	21	UG/L	8/28/2003 0:00	CWS	170
CA	CA1009035	PILIBOS BROTHERS RANCH	TREATMENT PLANT — TREATED	HALOACETIC ACIDS (5) (HAA5)	20	UG/L	7/13/2004 0:00	CWS	64
CA	CA1009215	FELGER FARMS	TREATMENT PLANT — TREATED	HALOACETIC ACIDS (5) (HAA5)	20	UG/L	7/13/2004 0:00	CWS	108
					20.3				

APPENDIX 7-15 Grand Rapids Groundwater Data

Expr1000	Federal PWS ID	NAME	point	analyteName	result	units	collectDate	PWSType	Pop
MI	MI0040337	TAMARAC VILLAGE	Treated Public Distribution System	BROMOCHLOROACETIC ACID	1	ug/L	13-Aug-04	CWS	872
MI	MI0040664	APPLE GROVE ESTATES	Treated Public Distribution System	CHLOROFORM	1.1	ug/L	03-Aug-04	CWS	125
MI	MI0040664	APPLE GROVE ESTATES	Pressure Tank/Plant Tap	CHLOROFORM	1.1	ug/L	06-Jul-05	CWS	125
MI	MI0040084	HICKORY HILLS	Treated Public Distribution System	CHLOROFORM	1.2	ug/L	26-Aug-04	CWS	708
MI	MI0040084	HICKORY HILLS	Treated Public Distribution System	CHLOROFORM	1.2	ug/L	26-Aug-04	CWS	708
MI	MI0040337	TAMARAC VILLAGE	Treated Public Distribution System	CHLOROFORM	5.4	ug/L	20-Aug-01	CWS	872
MI	MI0040337	TAMARAC VILLAGE	Treated Public Distribution System	CHLOROFORM	2.9	ug/L	20-Aug-01	CWS	872
MI	MI0040337	TAMARAC VILLAGE	Treated Public Distribution System	CHLOROFORM	15.1	ug/L	13-Aug-04	CWS	872
MI	MI0040337	TAMARAC VILLAGE	Treated Public Distribution System	CHLOROFORM	1.6	ug/L	13-Aug-04	CWS	872
MI	MI0040337	TAMARAC VILLAGE	Treated Public Distribution System	CHLOROFORM	22.6	ug/L	13-Aug-04	CWS	872
MI	MI0040337	TAMARAC VILLAGE	Treated Public Distribution System	CHLOROFORM	4.1	ug/L	23-Aug-04	CWS	872
MI	MI0040337	TAMARAC VILLAGE	Treated Public Distribution System	CHLOROFORM	0.7	ug/L	23-Aug-04	CWS	872
					5.18				
MI	MI0040337	TAMARAC VILLAGE	Treated Public Distribution System	DICHLOROACETIC ACID	12	ug/L	13-Aug-04	CWS	872
MI	MI0040337	TAMARAC VILLAGE	Treated Public Distribution System	DICHLOROACETIC ACID	8	ug/L	13-Aug-04	CWS	872
MI	MI0040337	TAMARAC VILLAGE	Treated Public Distribution System	DICHLOROACETIC ACID	19	ug/L	03-Aug-05	CWS	872
MI	MI0040337	TAMARAC VILLAGE	Treated Public Distribution System	DICHLOROACETIC ACID	11	ug/L	03-Aug-05	CWS	872
MI	MI0040337	TAMARAC VILLAGE	Treated Public Distribution System	DICHLOROACETIC ACID	9	ug/L	03-Aug-05	CWS	872
					11.8				
MI	MI0040426	WHITE'S MOBILE HOME PLAZA	Untreated Public Distribution System	DICHLOROBENZENE, 1,2-	0.5	ug/L	03-Jan-00	CWS	268
MI	MI0040426	WHITE'S MOBILE HOME PLAZA	Public System Well	DICHLOROBENZENE, 1,2-	0.9	ug/L	09-Oct-00	CWS	268
MI	MI0040426	WHITE'S MOBILE HOME PLAZA	Untreated Public Distribution System	DICHLOROBENZENE, 1,2-	0.9	ug/L	03-Jan-01	CWS	268
MI	MI0040426	WHITE'S MOBILE HOME PLAZA	Untreated Public Distribution System	DICHLOROBENZENE, 1,2-	0.8	ug/L	02-Apr-01	CWS	268
MI	MI0040426	WHITE'S MOBILE HOME PLAZA	Untreated Public Distribution System	DICHLOROBENZENE, 1,2-	0.8	ug/L	10-Jul-01	CWS	268
MI	MI0040426	WHITE'S MOBILE HOME PLAZA	Untreated Public Distribution System	DICHLOROBENZENE, 1,2-	0.8	ug/L	08-Jan-02	CWS	268
MI	MI0040426	WHITE'S MOBILE HOME PLAZA	Untreated Public Distribution System	DICHLOROBENZENE, 1,2-	0.8	ug/L	02-Apr-02	CWS	268
MI	MI0040426	WHITE'S MOBILE HOME PLAZA	Untreated Public Distribution System	DICHLOROBENZENE, 1,2-	0.8	ug/L	08-Jul-02	CWS	268
MI	MI0040426	WHITE'S MOBILE HOME PLAZA	Untreated Public Distribution System	DICHLOROBENZENE, 1,2-	0.9	ug/L	01-Oct-02	CWS	268
MI	MI0040426	WHITE'S MOBILE HOME PLAZA	Public System Well	DICHLOROBENZENE, 1,2-	0.5	ug/L	08-Apr-03	CWS	268
MI	MI0040426	WHITE'S MOBILE HOME PLAZA	Public System Well	DICHLOROBENZENE, 1,2-	0.7	ug/L	07-Jul-03	CWS	268
MI	MI0040426	WHITE'S MOBILE HOME PLAZA	Public System Well	DICHLOROBENZENE, 1,2-	0.8	ug/L	01-Oct-03	CWS	268
MI	MI0040426	WHITE'S MOBILE HOME PLAZA	Public System Well	DICHLOROBENZENE, 1,2-	0.9	ug/L	05-Apr-04	CWS	268
MI	MI0040426	WHITE'S MOBILE HOME PLAZA	Untreated Public Distribution System	DICHLOROBENZENE, 1,2-	0.7	ug/L	04-Oct-04	CWS	268
MI	MI0040426	WHITE'S MOBILE HOME PLAZA	Untreated Public Distribution System	DICHLOROBENZENE, 1,2-	1	ug/L	04-Apr-05	CWS	268
MI	MI0040426	WHITE'S MOBILE HOME PLAZA	Public System Well	DICHLOROBENZENE, 1,2-	0.6	ug/L	02-Jan-03	CWS	268
					0.78				

APPENDIX 7-15 *Continued*

Expr1000	Federal PWS ID	NAME	point	analyteName	result	units	collectDate	PWSType	Pop
MI	MI0040426	WHITE'S MOBILE HOME PLAZA	Public System Well	DICHLOROBENZENE, 1,3-	0.6	ug/L	09-Oct-00	CWS	268
MI	MI0040426	WHITE'S MOBILE HOME PLAZA	Untreated Public Distribution System	DICHLOROBENZENE, 1,3-	0.6	ug/L	03-Jan-01	CWS	268
MI	MI0040426	WHITE'S MOBILE HOME PLAZA	Untreated Public Distribution System	DICHLOROBENZENE, 1,3-	0.5	ug/L	02-Apr-01	CWS	268
MI	MI0040426	WHITE'S MOBILE HOME PLAZA	Untreated Public Distribution System	DICHLOROBENZENE, 1,3-	0.5	ug/L	10-Jul-01	CWS	268
MI	MI0040426	WHITE'S MOBILE HOME PLAZA	Untreated Public Distribution System	DICHLOROBENZENE, 1,3-	0.5	ug/L	08-Jan-02	CWS	268
MI	MI0040426	WHITE'S MOBILE HOME PLAZA	Untreated Public Distribution System	DICHLOROBENZENE, 1,3-	0.5	ug/L	02-Apr-02	CWS	268
MI	MI0040426	WHITE'S MOBILE HOME PLAZA	Untreated Public Distribution System	DICHLOROBENZENE, 1,3-	0.5	ug/L	08-Jul-02	CWS	268
MI	MI0040426	WHITE'S MOBILE HOME PLAZA	Untreated Public Distribution System	DICHLOROBENZENE, 1,3-	0.6	ug/L	01-Oct-02	CWS	268
MI	MI0040426	WHITE'S MOBILE HOME PLAZA	Public System Well	DICHLOROBENZENE, 1,3-	0.5	ug/L	07-Jul-03	CWS	268
MI	MI0040426	WHITE'S MOBILE HOME PLAZA	Public System Well	DICHLOROBENZENE, 1,3-	0.5	ug/L	01-Oct-03	CWS	268
MI	MI0040426	WHITE'S MOBILE HOME PLAZA	Public System Well	DICHLOROBENZENE, 1,3-	0.6	ug/L	05-Apr-04	CWS	268
MI	MI0040426	WHITE'S MOBILE HOME PLAZA	Untreated Public Distribution System	DICHLOROBENZENE, 1,3-	0.5	ug/L	04-Oct-04	CWS	268
MI	MI0040426	WHITE'S MOBILE HOME PLAZA	Untreated Public Distribution System	DICHLOROBENZENE, 1,3-	0.7	ug/L	04-Apr-05	CWS	268
					0.55				
MI	MI0040426	WHITE'S MOBILE HOME PLAZA	Public System Well	DICHLOROBENZENE, 1,4-	0.7	ug/L	09-Oct-00	CWS	268
MI	MI0040426	WHITE'S MOBILE HOME PLAZA	Untreated Public Distribution System	DICHLOROBENZENE, 1,4-	0.7	ug/L	03-Jan-01	CWS	268
MI	MI0040426	WHITE'S MOBILE HOME PLAZA	Untreated Public Distribution System	DICHLOROBENZENE, 1,4-	0.6	ug/L	02-Apr-01	CWS	268
MI	MI0040426	WHITE'S MOBILE HOME PLAZA	Untreated Public Distribution System	DICHLOROBENZENE, 1,4-	0.6	ug/L	10-Jul-01	CWS	268
MI	MI0040426	WHITE'S MOBILE HOME PLAZA	Untreated Public Distribution System	DICHLOROBENZENE, 1,4-	0.6	ug/L	08-Jan-02	CWS	268
MI	MI0040426	WHITE'S MOBILE HOME PLAZA	Untreated Public Distribution System	DICHLOROBENZENE, 1,4-	0.6	ug/L	02-Apr-02	CWS	268
MI	MI0040426	WHITE'S MOBILE HOME PLAZA	Untreated Public Distribution System	DICHLOROBENZENE, 1,4-	0.6	ug/L	08-Jul-02	CWS	268
MI	MI0040426	WHITE'S MOBILE HOME PLAZA	Untreated Public Distribution System	DICHLOROBENZENE, 1,4-	0.7	ug/L	01-Oct-02	CWS	268
MI	MI0040426	WHITE'S MOBILE HOME PLAZA	Public System Well	DICHLOROBENZENE, 1,4-	0.5	ug/L	08-Apr-03	CWS	268
MI	MI0040426	WHITE'S MOBILE HOME PLAZA	Public System Well	DICHLOROBENZENE, 1,4-	0.5	ug/L	07-Jul-03	CWS	268
MI	MI0040426	WHITE'S MOBILE HOME PLAZA	Public System Well	DICHLOROBENZENE, 1,4-	0.5	ug/L	01-Oct-03	CWS	268
MI	MI0040426	WHITE'S MOBILE HOME PLAZA	Public System Well	DICHLOROBENZENE, 1,4-	0.7	ug/L	05-Apr-04	CWS	268
MI	MI0040426	WHITE'S MOBILE HOME PLAZA	Untreated Public Distribution System	DICHLOROBENZENE, 1,4-	0.5	ug/L	04-Oct-04	CWS	268
MI	MI0040426	WHITE'S MOBILE HOME PLAZA	Untreated Public Distribution System	DICHLOROBENZENE, 1,4-	0.8	ug/L	04-Apr-05	CWS	268
					0.61				
MI	MI0040337	TAMARAC VILLAGE	Treated Public Distribution System	DICHLOROBROMOMETHANE	0.5	ug/L	20-Aug-01	CWS	872
MI	MI0040337	TAMARAC VILLAGE	Treated Public Distribution System	DICHLOROBROMOMETHANE	2.3	ug/L	13-Aug-04	CWS	872
MI	MI0040337	TAMARAC VILLAGE	Treated Public Distribution System	DICHLOROBROMOMETHANE	2.8	ug/L	13-Aug-04	CWS	872
MI	MI0040337	TAMARAC VILLAGE	Treated Public Distribution System	DICHLOROBROMOMETHANE	0.6	ug/L	23-Aug-04	CWS	872
					1.55				
MI	MI0040008	DOGWOOD MANOR MOBILE HOME PARK	Untreated Public Distribution System	TETRAHYDROFURAN	59	ug/L	21-May-01	CWS	190

APPENDIX 7-16 Grand Rapids Surface Water Data

Expr1000	Federal PWS ID	NAME	point	analyteName	result	units	collectDate	PWSType	Pop
MI	MI0002790	GRAND RAPIDS	Pressure Tank/Plant Tap	CHLORODIBROMOMETHANE	4	ug/L	29-Jun-00	CWS	200,000
MI	MI0002790	GRAND RAPIDS	Pressure Tank/Plant Tap	CHLORODIBROMOMETHANE	2	ug/L	14-Feb-01	CWS	200,000
MI	MI0002790	GRAND RAPIDS	Pressure Tank/Plant Tap	CHLORODIBROMOMETHANE	2.5	ug/L	29-May-01	CWS	200,000
MI	MI0002790	GRAND RAPIDS	Pressure Tank/Plant Tap	CHLORODIBROMOMETHANE	2.5	ug/L	29-Jun-01	CWS	200,000
MI	MI0002790	GRAND RAPIDS	Pressure Tank/Plant Tap	CHLORODIBROMOMETHANE	3.8	ug/L	30-Aug-01	CWS	200,000
MI	MI0002790	GRAND RAPIDS	Pressure Tank/Plant Tap	CHLORODIBROMOMETHANE	2.9	ug/L	26-Nov-01	CWS	200,000
MI	MI0002790	GRAND RAPIDS	Pressure Tank/Plant Tap	CHLORODIBROMOMETHANE	2.2	ug/L	13-Feb-02	CWS	200,000
MI	MI0002790	GRAND RAPIDS	Pressure Tank/Plant Tap	CHLORODIBROMOMETHANE	2.1	ug/L	19-Dec-02	CWS	200,000
MI	MI0002790	GRAND RAPIDS	Pressure Tank/Plant Tap	CHLORODIBROMOMETHANE	2	ug/L	27-Mar-03	CWS	200,000
MI	MI0002790	GRAND RAPIDS	Pressure Tank/Plant Tap	CHLORODIBROMOMETHANE	2.1	ug/L	02-Mar-04	CWS	200,000
MI	MI0002790	GRAND RAPIDS	Other	CHLORODIBROMOMETHANE	2	ug/L	08-Mar-05	CWS	200,000
					2.56				
MI	MI0002790	GRAND RAPIDS	Pressure Tank/Plant Tap	CHLOROFORM	12.7	ug/L	29-Jun-00	CWS	200,000
MI	MI0002790	GRAND RAPIDS	Pressure Tank/Plant Tap	CHLOROFORM	6	ug/L	14-Feb-01	CWS	200,000
MI	MI0002790	GRAND RAPIDS	Pressure Tank/Plant Tap	CHLOROFORM	7.9	ug/L	29-May-01	CWS	200,000
MI	MI0002790	GRAND RAPIDS	Pressure Tank/Plant Tap	CHLOROFORM	6.9	ug/L	29-Jun-01	CWS	200,000
MI	MI0002790	GRAND RAPIDS	Pressure Tank/Plant Tap	CHLOROFORM	20	ug/L	30-Aug-01	CWS	200,000
MI	MI0002790	GRAND RAPIDS	Pressure Tank/Plant Tap	CHLOROFORM	13.6	ug/L	26-Nov-01	CWS	200,000
MI	MI0002790	GRAND RAPIDS	Pressure Tank/Plant Tap	CHLOROFORM	6.3	ug/L	13-Feb-02	CWS	200,000
MI	MI0002790	GRAND RAPIDS	Pressure Tank/Plant Tap	CHLOROFORM	4.2	ug/L	19-Dec-02	CWS	200,000
MI	MI0002790	GRAND RAPIDS	Pressure Tank/Plant Tap	CHLOROFORM	2.3	ug/L	27-Mar-03	CWS	200,000
MI	MI0002790	GRAND RAPIDS	Pressure Tank/Plant Tap	CHLOROFORM	3	ug/L	02-Mar-04	CWS	200,000
MI	MI0002790	GRAND RAPIDS	Other	CHLOROFORM	10.4	ug/L	08-Mar-05	CWS	200,000
					8.48				
MI	MI0002790	GRAND RAPIDS	Pressure Tank/Plant Tap	DICHLOROBROMOMETHANE	8.1	ug/L	29-Jun-00	CWS	200,000
MI	MI0002790	GRAND RAPIDS	Pressure Tank/Plant Tap	DICHLOROBROMOMETHANE	4.8	ug/L	14-Feb-01	CWS	200,000
MI	MI0002790	GRAND RAPIDS	Pressure Tank/Plant Tap	DICHLOROBROMOMETHANE	5.3	ug/L	29-May-01	CWS	200,000
MI	MI0002790	GRAND RAPIDS	Pressure Tank/Plant Tap	DICHLOROBROMOMETHANE	5.9	ug/L	29-Jun-01	CWS	200,000
MI	MI0002790	GRAND RAPIDS	Pressure Tank/Plant Tap	DICHLOROBROMOMETHANE	10	ug/L	30-Aug-01	CWS	200,000
MI	MI0002790	GRAND RAPIDS	Pressure Tank/Plant Tap	DICHLOROBROMOMETHANE	8.7	ug/L	26-Nov-01	CWS	200,000
MI	MI0002790	GRAND RAPIDS	Pressure Tank/Plant Tap	DICHLOROBROMOMETHANE	5.5	ug/L	13-Feb-02	CWS	200,000
MI	MI0002790	GRAND RAPIDS	Pressure Tank/Plant Tap	DICHLOROBROMOMETHANE	4	ug/L	19-Dec-02	CWS	200,000
MI	MI0002790	GRAND RAPIDS	Pressure Tank/Plant Tap	DICHLOROBROMOMETHANE	2.9	ug/L	27-Mar-03	CWS	200,000
MI	MI0002790	GRAND RAPIDS	Pressure Tank/Plant Tap	DICHLOROBROMOMETHANE	3.6	ug/L	02-Mar-04	CWS	200,000
MI	MI0002790	GRAND RAPIDS	Other	DICHLOROBROMOMETHANE	6.3	ug/L	08-Mar-05	CWS	200,000
					5.92				

APPENDIX 7-17 Harrisburg Groundwater Data

Expr1000	PWSID	Expr1002	CMTNAME	RESULT	Expr1	SAMPDATE	PWSType	Pop
PA	PA7210054	center square water co	ATRAZINE (SOC)	0.2	ug/L	19-Jun-00	CWS	1613
PA	PA7220941	harrisburg christian sch	DICHLOROMETHANE	1.2	ug/L	31-Jul-02	NTNCWS	350
PA	PA7220070	mountaindale apartments	STYRENE	0.8	ug/L	03-Aug-01	CWS	30
PA	PA7220941	harrisburg christian sch	STYRENE	6.4	ug/L	31-Jul-02	NTNCWS	350
PA	PA7220941	harrisburg christian sch	STYRENE	1.1	ug/L	05-Nov-02	NTNCWS	350
PA	PA7220941	harrisburg christian sch	STYRENE	1	ug/L	14-Feb-03	NTNCWS	350
PA	PA7220941	harrisburg christian sch	STYRENE	1.3	ug/L	13-May-03	NTNCWS	350
PA	PA7220941	harrisburg christian sch	STYRENE	3.7	ug/L	21-Jul-03	NTNCWS	350
PA	PA7220941	harrisburg christian sch	STYRENE	1	ug/L	21-Oct-03	NTNCWS	350
PA	PA7220941	harrisburg christian sch	STYRENE	0.5	ug/L	11-Feb-04	NTNCWS	350
				1.98				
PA	PA7220941	harrisburg christian sch	TOLUENE	0.5	ug/L	11-May-01	NTNCWS	350
PA	PA7220941	harrisburg christian sch	TOLUENE	0.5	ug/L	11-May-01	NTNCWS	350
PA	PA7670061	newberry sys	TRICHLOROETHYLENE	0.6	ug/L	15-May-01	CWS	4499
PA	PA7670061	newberry sys	TRICHLOROETHYLENE	0.7	ug/L	15-Jul-03	CWS	4499
PA	PA7670061	newberry sys	TRICHLOROETHYLENE	0.6	ug/L	26-Feb-04	CWS	4499
PA	PA7670061	newberry sys	XYLENES (TOTAL)	0.73	ug/L	26-Jul-02	CWS	4499
PA	PA7670061	newberry sys	XYLENES (TOTAL)	0.5	ug/L	15-Oct-02	CWS	4499
				0.62				

APPENDIX 7-18 Harrisburg Surface Water Data

Expr1000	PWSID	Expr1002	CMTNAME	RESULT	Expr1	SAMPDATE	PWSType	Pop
PA	PA7220015	united water pennsylvania	ATRAZINE (SOC)	0.2	ug/L	19-Jun-00	CWS	86,500

APPENDIX 7-19 Los Angeles Water Data

State	PWSID	Water System Name	Source	Analyte	Concentration	Unit	Sample Date	PWSType	PSource	Pop
CA	CA1910067	LOS ANGELES-CITY, DEPT. OF WATER & POWER	297 SAMPLES — WELLS	1,1-DICHLOROETHANE	0.92	UG/L				
CA	CA1910067	LOS ANGELES-CITY, DEPT. OF WATER & POWER	278 SAMPLES — WELLS	1,2-DICHLOROETHANE	0.83	UG/L				
CA	CA1910067	LOS ANGELES-CITY, DEPT. OF WATER & POWER	991 SAMPLES — WELLS	1,1-DICHLOROETHYLENE	2.77	UG/L				
CA	CA1910067	LOS ANGELES-CITY, DEPT. OF WATER & POWER	732 SAMPLES — WELLS	CIS-1,2-DICHLOROETHYLENE	2.2	UG/L				
CA	CA1910067	LOS ANGELES-CITY, DEPT. OF WATER & POWER	805 SAMPLES — WELLS	DICHLORODIFLUOROMETHANE	2.72	UG/L				
CA	CA1910067	LOS ANGELES-CITY, DEPT. OF WATER & POWER	NORTH HOLLYWOOD AERATION TOWER OUTLET	DICHLOROMETHANE	0.87	UG/L	1/23/2002 0:00	CWS	SW	3,828,700
CA	CA1910067	LOS ANGELES-CITY, DEPT. OF WATER & POWER	NORTH HOLLYWOOD AERATION WELL 07	1,2-DICHLOROPROPANE	0.68	UG/L	1/19/2005 0:00	CWS	SW	3,828,700
CA	CA1910067	LOS ANGELES-CITY, DEPT. OF WATER & POWER	NORTH HOLLYWOOD AERATION WELL 05	1,2-DICHLOROPROPANE	0.65	UG/L	1/23/2002 0:00	CWS	SW	3,828,700
CA	CA1910067	LOS ANGELES-CITY, DEPT. OF WATER & POWER	NORTH HOLLYWOOD AERATION WELL 05	1,2-DICHLOROPROPANE	0.57	UG/L	3/18/2002 0:00	CWS	SW	3,828,700
CA	CA1910067	LOS ANGELES-CITY, DEPT. OF WATER & POWER	NORTH HOLLYWOOD AERATION WELL 05	1,2-DICHLOROPROPANE	0.55	UG/L	4/19/2002 0:00	CWS	SW	3,828,700
CA	CA1910067	LOS ANGELES-CITY, DEPT. OF WATER & POWER	NORTH HOLLYWOOD AERATION WELL 05	1,2-DICHLOROPROPANE	0.55	UG/L	2/21/2002 0:00	CWS	SW	3,828,700
CA	CA1910067	LOS ANGELES-CITY, DEPT. OF WATER & POWER	NORTH HOLLYWOOD AERATION WELL 05	1,2-DICHLOROPROPANE	0.83	UG/L	6/21/2001 0:00	CWS	SW	3,828,700
CA	CA1910067	LOS ANGELES-CITY, DEPT. OF WATER & POWER	NORTH HOLLYWOOD AERATION WELL 05	1,2-DICHLOROPROPANE	0.89	UG/L	7/11/2001 0:00	CWS	SW	3,828,700
CA	CA1910067	LOS ANGELES-CITY, DEPT. OF WATER & POWER	NORTH HOLLYWOOD AERATION WELL 05	1,2-DICHLOROPROPANE	0.69	UG/L	12/5/2001 0:00	CWS	SW	3,828,700
CA	CA1910067	LOS ANGELES-CITY, DEPT. OF WATER & POWER	NORTH HOLLYWOOD AERATION WELL 05	1,2-DICHLOROPROPANE	0.66	UG/L	8/28/2001 0:00	CWS	SW	3,828,700
CA	CA1910067	LOS ANGELES-CITY, DEPT. OF WATER & POWER	NORTH HOLLYWOOD AERATION WELL 05	1,2-DICHLOROPROPANE	0.67	UG/L	9/19/2001 0:00	CWS	SW	3,828,700
CA	CA1910067	LOS ANGELES-CITY, DEPT. OF WATER & POWER	NORTH HOLLYWOOD AERATION WELL 05	1,2-DICHLOROPROPANE	0.54	UG/L	4/17/2003 0:00	CWS	SW	3,828,700
CA	CA1910067	LOS ANGELES-CITY, DEPT. OF WATER & POWER	NORTH HOLLYWOOD AERATION WELL 05	1,2-DICHLOROPROPANE	0.5	UG/L	10/29/2003 0:00	CWS	SW	3,828,700
CA	CA1910067	LOS ANGELES-CITY, DEPT. OF WATER & POWER	NORTH HOLLYWOOD AERATION WELL 05	1,2-DICHLOROPROPANE	0.6	UG/L	1/28/2004 0:00	CWS	SW	3,828,700
CA	CA1910067	LOS ANGELES-CITY, DEPT. OF WATER & POWER	NORTH HOLLYWOOD AERATION WELL 05	1,2-DICHLOROPROPANE	0.53	UG/L	9/24/2002 0:00	CWS	SW	3,828,700
CA	CA1910067	LOS ANGELES-CITY, DEPT. OF WATER & POWER	NORTH HOLLYWOOD AERATION WELL 05	1,2-DICHLOROPROPANE	0.54	UG/L	10/17/2002 0:00	CWS	SW	3,828,700

APPENDIX 7-19 *Continued*

State	PWSID	Water System Name	Source	Analyte	Concentration	Unit	Sample Date	PWSType	PSource	Pop
CA	CA1910067	LOS ANGELES-CITY, DEPT. OF WATER & POWER	NORTH HOLLYWOOD AERATION WELL 05	1,2-DICHLOROPROPANE	0.55	UG/L	7/18/2002 0:00	CWS	SW	3,828,700
CA	CA1910067	LOS ANGELES-CITY, DEPT. OF WATER & POWER	NORTH HOLLYWOOD AERATION WELL 05	1,2-DICHLOROPROPANE	0.51	UG/L	8/27/2002 0:00	CWS	SW	3,828,700
CA	CA1910067	LOS ANGELES-CITY, DEPT. OF WATER & POWER	NORTH HOLLYWOOD AERATION WELL 05	1,2-DICHLOROPROPANE	0.6	UG/L	12/10/2002 0:00	CWS	SW	3,828,700
CA	CA1910067	LOS ANGELES-CITY, DEPT. OF WATER & POWER	NORTH HOLLYWOOD AERATION WELL 05	1,2-DICHLOROPROPANE	0.61	UG/L	1/21/2003 0:00	CWS	SW	3,828,700
CA	CA1910067	LOS ANGELES-CITY, DEPT. OF WATER & POWER	NORTH HOLLYWOOD AERATION WELL 05	1,2-DICHLOROPROPANE	0.62	UG/L	11/8/2002 0:00	CWS	SW	3,828,700
CA	CA1910067	LOS ANGELES-CITY, DEPT. OF WATER & POWER	NORTH HOLLYWOOD AERATION WELL 05	1,2-DICHLOROPROPANE	1.02	UG/L	7/13/1998 0:00	CWS	SW	3,828,700
CA	CA1910067	LOS ANGELES-CITY, DEPT. OF WATER & POWER	NORTH HOLLYWOOD AERATION WELL 05	1,2-DICHLOROPROPANE	0.6	UG/L	3/3/1998 0:00	CWS	SW	3,828,700
CA	CA1910067	LOS ANGELES-CITY, DEPT. OF WATER & POWER	NORTH HOLLYWOOD AERATION WELL 05	1,2-DICHLOROPROPANE	0.5	UG/L	12/11/1997 0:00	CWS	SW	3,828,700
CA	CA1910067	LOS ANGELES-CITY, DEPT. OF WATER & POWER	NORTH HOLLYWOOD AERATION WELL 05	1,2-DICHLOROPROPANE	1.13	UG/L	3/28/2001 0:00	CWS	SW	3,828,700
CA	CA1910067	LOS ANGELES-CITY, DEPT. OF WATER & POWER	NORTH HOLLYWOOD AERATION WELL 05	1,2-DICHLOROPROPANE	0.98	UG/L	4/25/2001 0:00	CWS	SW	3,828,700
CA	CA1910067	LOS ANGELES-CITY, DEPT. OF WATER & POWER	NORTH HOLLYWOOD AERATION WELL 05	1,2-DICHLOROPROPANE	1.01	UG/L	11/21/2000 0:00	CWS	SW	3,828,700
CA	CA1910067	LOS ANGELES-CITY, DEPT. OF WATER & POWER	NORTH HOLLYWOOD AERATION WELL 05	1,2-DICHLOROPROPANE	0.62	UG/L	12/4/1998 0:00	CWS	SW	3,828,700
CA	CA1910067	LOS ANGELES-CITY, DEPT. OF WATER & POWER	NORTH HOLLYWOOD AERATION WELL 05	1,2-DICHLOROPROPANE	0.52	UG/L	1/19/2005 0:00	CWS	SW	3,828,700
CA	CA1910067	LOS ANGELES-CITY, DEPT. OF WATER & POWER	NORTH HOLLYWOOD AERATION WELL 05	1,2-DICHLOROPROPANE	0.61	UG/L	4/26/2004 0:00	CWS	SW	3,828,700
CA	CA1910067	LOS ANGELES-CITY, DEPT. OF WATER & POWER	NORTH HOLLYWOOD AERATION WELL 04	1,2-DICHLOROPROPANE	1.02	UG/L	10/26/2000 0:00	CWS	SW	3,828,700
CA	CA1910067	LOS ANGELES-CITY, DEPT. OF WATER & POWER	NORTH HOLLYWOOD AERATION WELL 04	1,2-DICHLOROPROPANE	1	UG/L	8/16/2000 0:00	CWS	SW	3,828,700
CA	CA1910067	LOS ANGELES-CITY, DEPT. OF WATER & POWER	NORTH HOLLYWOOD AERATION WELL 04	1,2-DICHLOROPROPANE	0.6	UG/L	9/6/2000 0:00	CWS	SW	3,828,700
CA	CA1910067	LOS ANGELES-CITY, DEPT. OF WATER & POWER	NORTH HOLLYWOOD AERATION WELL 04	1,2-DICHLOROPROPANE	1	UG/L	8/7/2000 0:00	CWS	SW	3,828,700
CA	CA1910067	LOS ANGELES-CITY, DEPT. OF WATER & POWER	NORTH HOLLYWOOD AERATION WELL 02	1,2-DICHLOROPROPANE	0.55	UG/L	1/19/2005 0:00	CWS	SW	3,828,700
CA	CA1910067	LOS ANGELES-CITY, DEPT. OF WATER & POWER	NORTH HOLLYWOOD WELL 27 — INACTIVE	1,2-DICHLOROPROPANE	1.18 0.71	UG/L	9/23/1997 0:00	CWS	SW	3,828,700
CA	CA1910067	LOS ANGELES-CITY, DEPT. OF WATER & POWER	VERDUGO WELL 11	DIQUAT	4	UG/L	8/7/2003 0:00	CWS	SW	3,828,700

CA	CA1910067	LOS ANGELES-CITY, DEPT. OF WATER & POWER	MANHATTAN WELL 06-A	DIQUAT	4 44	UG/L	9/17/2003 0:00	CWS	SW	3,828,700
CA	CA1910067	LOS ANGELES-CITY, DEPT. OF WATER & POWER	NORTH HOLLYWOOD AERATION WELL 05	ETHYLBENZENE	1	UG/L	8/7/2000 0:00	CWS	SW	3,828,700
CA	CA1910067	LOS ANGELES-CITY, DEPT. OF WATER & POWER	NORTH HOLLYWOOD AERATION WELL 04	ETHYLBENZENE	1	UG/L	8/7/2000 0:00	CWS	SW	3,828,700
CA	CA1910067	LOS ANGELES-CITY, DEPT. OF WATER & POWER	RINALDI TOLUCA WELL 15	ETHYLBENZENE	0.6 0.87	UG/L	2/6/2004 0:00	CWS	SW	3,828,700
CA	CA1910067	LOS ANGELES-CITY, DEPT. OF WATER & POWER	NORTH HOLLYWOOD AERATION WELL 08	ETHYLENE DIBROMIDE (EDB)	0.02	UG/L	4/13/1998 0:00	CWS	SW	3,828,700
CA	CA1910067	LOS ANGELES-CITY, DEPT. OF WATER & POWER	NORTH HOLLYWOOD AERATION WELL 08	ETHYLENE DIBROMIDE (EDB)	0.02 0.02	UG/L	1/14/1998 0:00	CWS	SW	3,828,700
CA	CA1910067	LOS ANGELES-CITY, DEPT. OF WATER & POWER	MISSION WELL 07	INDENO (1,2,3-CD) PYRENE	0.05	UG/L	4/10/2003 0:00	CWS	SW	3,828,700
CA	CA1910067	LOS ANGELES-CITY, DEPT. OF WATER & POWER	TUJUNGA WELL 10	METHYL ISOBUTYL KETONE	0.61	UG/L	3/25/2000 0:00	CWS	SW	3,828,700
CA	CA1910067	LOS ANGELES-CITY, DEPT. OF WATER & POWER	NORTH HOLLYWOOD AERATION WELL 05	METHYL ISOBUTYL KETONE	5	UG/L	8/7/2000 0:00	CWS	SW	3,828,700
CA	CA1910067	LOS ANGELES-CITY, DEPT. OF WATER & POWER	POLLOCK WELL 06	METHYL ISOBUTYL KETONE	1 2.2	UG/L	3/2/2000 0:00	CWS	SW	3,828,700
CA	CA1910067	LOS ANGELES-CITY, DEPT. OF WATER & POWER	TUJUNGA WELL 05	METHYL TERT BUTYL ETHER (MTBE)	0.5	UG/L	3/11/1997 0:00	CWS	SW	3,828,700
CA	CA1910067	LOS ANGELES-CITY, DEPT. OF WATER & POWER	TUJUNGA WELL 04	METHYL TERT BUTYL ETHER (MTBE)	1.8	UG/L	3/11/1997 0:00	CWS	SW	3,828,700
CA	CA1910067	LOS ANGELES-CITY, DEPT. OF WATER & POWER	WHITNALL WELL 05	METHYL TERT BUTYL ETHER (MTBE)	3.09	UG/L	5/18/2000 0:00	CWS	SW	3,828,700
CA	CA1910067	LOS ANGELES-CITY, DEPT. OF WATER & POWER	WHITNALL WELL 05	METHYL TERT BUTYL ETHER (MTBE)	3.53	UG/L	4/29/2000 0:00	CWS	SW	3,828,700
CA	CA1910067	LOS ANGELES-CITY, DEPT. OF WATER & POWER	WHITNALL WELL 05	METHYL TERT BUTYL ETHER (MTBE)	5.74	UG/L	4/4/2000 0:00	CWS	SW	3,828,700
CA	CA1910067	LOS ANGELES-CITY, DEPT. OF WATER & POWER	VERDUGO WELL 02 — INACTIVE	METHYL TERT BUTYL ETHER (MTBE)	1	UG/L	5/5/1997 0:00	CWS	SW	3,828,700
CA	CA1910067	LOS ANGELES-CITY, DEPT. OF WATER & POWER	VERDUGO WELL 02 — INACTIVE	METHYL TERT BUTYL ETHER (MTBE)	2.3	UG/L	1/31/1997 0:00	CWS	SW	3,828,700
CA	CA1910067	LOS ANGELES-CITY, DEPT. OF WATER & POWER	VERDUGO WELL 02 — INACTIVE	METHYL TERT BUTYL ETHER (MTBE)	1	UG/L	3/21/1997 0:00	CWS	SW	3,828,700
CA	CA1910067	LOS ANGELES-CITY, DEPT. OF WATER & POWER	NORTH HOLLYWOOD WELL 17 — INACTIVE	METHYL TERT BUTYL ETHER (MTBE)	3.5 2.5	UG/L	2/4/1997 0:00	CWS	SW	3,828,700
CA	CA1910067	LOS ANGELES-CITY, DEPT. OF WATER & POWER	RINALDI TOLUCA WELL 12	N-NITROSODIMETHYLAMINE (NDMA)	4.00E-03	UG/L	12/12/2001 0:00	CWS	SW	3,828,700
CA	CA1910067	LOS ANGELES-CITY, DEPT. OF WATER & POWER	RIVER SUPPLY CONDUIT BCL — TREATED — 771 SAMPLES	TETRACHLOROETHYLENE	0.7	UG/L	6/14/2002 0:00	CWS	SW	3,828,700

APPENDIX 7-19 *Continued*

State	PWSID	Water System Name	Source	Analyte	Concentration	Unit	Sample Date	PWSType	PSource	Pop
CA			2535 SAMPLES — WELLS	TETRACHLOROETHYLENE	4.59	UG/L				
CA	CA1910067	LOS ANGELES-CITY, DEPT. OF WATER & POWER	RIVER SUPPLY CONDUIT BCL — TREATED	TOLUENE	0.6	UG/L	6/6/2001 0:00	CWS	SW	3,828,700
CA	CA1910067	LOS ANGELES-CITY, DEPT. OF WATER & POWER	NORTH HOLLYWOOD WELL 04	TOLUENE	4.52 2.56	UG/L	9/25/2002 0:00	CWS	SW	3,828,700
CA			302 SAMPLES — WELLS	1,1,2-TRICHLOROETHANE	1.54	UG/L				
CA	CA1910067	LOS ANGELES-CITY, DEPT. OF WATER & POWER	RIVER SUPPLY CONDUIT BCL — TREATED — 1252 SAMPLES	TRICHLOROETHYLENE	1.35	UG/L	4/23/2002 0:00	CWS	SW	3,828,700
CA			3723 SAMPLES — WELLS	TRICHLOROETHYLENE	10.7	UG/L				
CA	CA1910067	LOS ANGELES-CITY, DEPT. OF WATER & POWER	TUJUNGA WELL 12	1,1,2-TRICHLORO-1,2,2-TRIFLUOROETHANE	0.6	UG/L	3/25/2000 0:00	CWS	SW	3,828,700
CA	CA1910067	LOS ANGELES-CITY, DEPT. OF WATER & POWER	TUJUNGA WELL 12	1,1,2-TRICHLORO-1,2,2-TRIFLUOROETHANE	1.08	UG/L	12/14/2000 0:00	CWS	SW	3,828,700
CA	CA1910067	LOS ANGELES-CITY, DEPT. OF WATER & POWER	NORTH HOLLYWOOD AERATION TOWER INLET	1,1,2-TRICHLORO-1,2,2-TRIFLUOROETHANE	1	UG/L	3/16/2000 0:00	CWS	SW	3,828,700
CA	CA1910067	LOS ANGELES-CITY, DEPT. OF WATER & POWER	NORTH HOLLYWOOD AERATION TOWER INLET	1,1,2-TRICHLORO-1,2,2-TRIFLUOROETHANE	1.26	UG/L	2/17/2000 0:00	CWS	SW	3,828,700
CA	CA1910067	LOS ANGELES-CITY, DEPT. OF WATER & POWER	NORTH HOLLYWOOD AERATION TOWER INLET	1,1,2-TRICHLORO-1,2,2-TRIFLUOROETHANE	0.85	UG/L	7/18/2000 0:00	CWS	SW	3,828,700
CA	CA1910067	LOS ANGELES-CITY, DEPT. OF WATER & POWER	NORTH HOLLYWOOD AERATION TOWER INLET	1,1,2-TRICHLORO-1,2,2-TRIFLUOROETHANE	0.84	UG/L	5/10/2000 0:00	CWS	SW	3,828,700
CA	CA1910067	LOS ANGELES-CITY, DEPT. OF WATER & POWER	NORTH HOLLYWOOD AERATION TOWER INLET	1,1,2-TRICHLORO-1,2,2-TRIFLUOROETHANE	1.14	UG/L	1/25/2000 0:00	CWS	SW	3,828,700
CA	CA1910067	LOS ANGELES-CITY, DEPT. OF WATER & POWER	NORTH HOLLYWOOD AERATION TOWER INLET	1,1,2-TRICHLORO-1,2,2-TRIFLUOROETHANE	1	UG/L	9/5/2000 0:00	CWS	SW	3,828,700
CA	CA1910067	LOS ANGELES-CITY, DEPT. OF WATER & POWER	NORTH HOLLYWOOD AERATION TOWER INLET	1,1,2-TRICHLORO-1,2,2-TRIFLUOROETHANE	0.51	UG/L	4/25/2001 0:00	CWS	SW	3,828,700
CA	CA1910067	LOS ANGELES-CITY, DEPT. OF WATER & POWER	NORTH HOLLYWOOD AERATION WELL 06	1,1,2-TRICHLORO-1,2,2-TRIFLUOROETHANE	0.5	UG/L	8/11/1998 0:00	CWS	SW	3,828,700
CA	CA1910067	LOS ANGELES-CITY, DEPT. OF WATER & POWER	NORTH HOLLYWOOD AERATION WELL 06	1,1,2-TRICHLORO-1,2,2-TRIFLUOROETHANE	0.62	UG/L	8/11/1998 0:00	CWS	SW	3,828,700
CA	CA1910067	LOS ANGELES-CITY, DEPT. OF WATER & POWER	NORTH HOLLYWOOD AERATION WELL 04	1,1,2-TRICHLORO-1,2,2-TRIFLUOROETHANE	0.6	UG/L	12/11/1997 0:00	CWS	SW	3,828,700
CA	CA1910067	LOS ANGELES-CITY, DEPT. OF WATER & POWER	NORTH HOLLYWOOD AERATION WELL 04	1,1,2-TRICHLORO-1,2,2-TRIFLUOROETHANE	1.04	UG/L	10/15/1997 0:00	CWS	SW	3,828,700
CA	CA1910067	LOS ANGELES-CITY, DEPT. OF WATER & POWER	NORTH HOLLYWOOD AERATION WELL 01-INACTVE	1,1,2-TRICHLORO-1,2,2-TRIFLUOROETHANE	0.75	UG/L	5/11/1998 0:00	CWS	SW	3,828,700
CA	CA1910067	LOS ANGELES-CITY, DEPT. OF WATER & POWER	NORTH HOLLYWOOD AERATION WELL 01-INACTVE	1,1,2-TRICHLORO-1,2,2-TRIFLUOROETHANE	0.5	UG/L	12/11/1997 0:00	CWS	SW	3,828,700

CA	CA1910067	LOS ANGELES-CITY, DEPT. OF WATER & POWER	NORTH HOLLYWOOD AERATION WELL 01-INACTVE	1,1,2-TRICHLORO-1,2,2-TRIFLUOROETHANE	1.8	UG/L	11/12/1997 0:00	CWS	SW	3,828,700
CA	CA1910067	LOS ANGELES-CITY, DEPT. OF WATER & POWER	NORTH HOLLYWOOD AERATION WELL 08	1,1,2-TRICHLORO-1,2,2-TRIFLUOROETHANE	2.17	UG/L	11/21/2000 0:00	CWS	SW	3,828,700
CA	CA1910067	LOS ANGELES-CITY, DEPT. OF WATER & POWER	NORTH HOLLYWOOD AERATION WELL 08	1,1,2-TRICHLORO-1,2,2-TRIFLUOROETHANE	2.2	UG/L	4/13/1998 0:00	CWS	SW	3,828,700
CA	CA1910067	LOS ANGELES-CITY, DEPT. OF WATER & POWER	NORTH HOLLYWOOD AERATION WELL 08	1,1,2-TRICHLORO-1,2,2-TRIFLUOROETHANE	2.3	UG/L	2/11/1998 0:00	CWS	SW	3,828,700
CA	CA1910067	LOS ANGELES-CITY, DEPT. OF WATER & POWER	NORTH HOLLYWOOD AERATION WELL 08	1,1,2-TRICHLORO-1,2,2-TRIFLUOROETHANE	2.21	UG/L	3/3/1998 0:00	CWS	SW	3,828,700
CA	CA1910067	LOS ANGELES-CITY, DEPT. OF WATER & POWER	NORTH HOLLYWOOD AERATION WELL 08	1,1,2-TRICHLORO-1,2,2-TRIFLUOROETHANE	1.68	UG/L	5/11/1998 0:00	CWS	SW	3,828,700
CA	CA1910067	LOS ANGELES-CITY, DEPT. OF WATER & POWER	NORTH HOLLYWOOD AERATION WELL 08	1,1,2-TRICHLORO-1,2,2-TRIFLUOROETHANE	1.27	UG/L	7/13/1998 0:00	CWS	SW	3,828,700
CA	CA1910067	LOS ANGELES-CITY, DEPT. OF WATER & POWER	NORTH HOLLYWOOD AERATION WELL 08	1,1,2-TRICHLORO-1,2,2-TRIFLUOROETHANE	1.73	UG/L	8/11/1998 0:00	CWS	SW	3,828,700
CA	CA1910067	LOS ANGELES-CITY, DEPT. OF WATER & POWER	NORTH HOLLYWOOD AERATION WELL 08	1,1,2-TRICHLORO-1,2,2-TRIFLUOROETHANE	0.87	UG/L	6/17/1998 0:00	CWS	SW	3,828,700
CA	CA1910067	LOS ANGELES-CITY, DEPT. OF WATER & POWER	NORTH HOLLYWOOD AERATION WELL 08	1,1,2-TRICHLORO-1,2,2-TRIFLUOROETHANE	3.18	UG/L	10/15/1997 0:00	CWS	SW	3,828,700
CA	CA1910067	LOS ANGELES-CITY, DEPT. OF WATER & POWER	NORTH HOLLYWOOD AERATION WELL 08	1,1,2-TRICHLORO-1,2,2-TRIFLUOROETHANE	1.77	UG/L	11/12/1997 0:00	CWS	SW	3,828,700
CA	CA1910067	LOS ANGELES-CITY, DEPT. OF WATER & POWER	NORTH HOLLYWOOD AERATION WELL 08	1,1,2-TRICHLORO-1,2,2-TRIFLUOROETHANE	2.2	UG/L	12/11/1997 0:00	CWS	SW	3,828,700
CA	CA1910067	LOS ANGELES-CITY, DEPT. OF WATER & POWER	NORTH HOLLYWOOD AERATION WELL 08	1,1,2-TRICHLORO-1,2,2-TRIFLUOROETHANE	2.44	UG/L	4/13/2000 0:00	CWS	SW	3,828,700
CA	CA1910067	LOS ANGELES-CITY, DEPT. OF WATER & POWER	NORTH HOLLYWOOD AERATION WELL 08	1,1,2-TRICHLORO-1,2,2-TRIFLUOROETHANE	4	UG/L	2/17/2000 0:00	CWS	SW	3,828,700
CA	CA1910067	LOS ANGELES-CITY, DEPT. OF WATER & POWER	NORTH HOLLYWOOD AERATION WELL 08	1,1,2-TRICHLORO-1,2,2-TRIFLUOROETHANE	3	UG/L	7/25/2000 0:00	CWS	SW	3,828,700
CA	CA1910067	LOS ANGELES-CITY, DEPT. OF WATER & POWER	NORTH HOLLYWOOD AERATION WELL 08	1,1,2-TRICHLORO-1,2,2-TRIFLUOROETHANE	3.08	UG/L	9/5/2000 0:00	CWS	SW	3,828,700
CA	CA1910067	LOS ANGELES-CITY, DEPT. OF WATER & POWER	NORTH HOLLYWOOD AERATION WELL 08	1,1,2-TRICHLORO-1,2,2-TRIFLUOROETHANE	1.63	UG/L	11/10/1998 0:00	CWS	SW	3,828,700
CA	CA1910067	LOS ANGELES-CITY, DEPT. OF WATER & POWER	NORTH HOLLYWOOD AERATION WELL 08	1,1,2-TRICHLORO-1,2,2-TRIFLUOROETHANE	2.27	UG/L	12/4/1998 0:00	CWS	SW	3,828,700
CA	CA1910067	LOS ANGELES-CITY, DEPT. OF WATER & POWER	NORTH HOLLYWOOD AERATION WELL 08	1,1,2-TRICHLORO-1,2,2-TRIFLUOROETHANE	2.44	UG/L	9/22/1998 0:00	CWS	SW	3,828,700
CA	CA1910067	LOS ANGELES-CITY, DEPT. OF WATER & POWER	NORTH HOLLYWOOD AERATION WELL 08	1,1,2-TRICHLORO-1,2,2-TRIFLUOROETHANE	1.86	UG/L	10/21/1998 0:00	CWS	SW	3,828,700
CA	CA1910067	LOS ANGELES-CITY, DEPT. OF WATER & POWER	NORTH HOLLYWOOD AERATION WELL 08	1,1,2-TRICHLORO-1,2,2-TRIFLUOROETHANE	5	UG/L	1/25/2000 0:00	CWS	SW	3,828,700
CA	CA1910067	LOS ANGELES-CITY, DEPT. OF WATER & POWER	NORTH HOLLYWOOD AERATION WELL 08	1,1,2-TRICHLORO-1,2,2-TRIFLUOROETHANE	2.7 1.73	UG/L	4/14/1999 0:00	CWS	SW	3,828,700

APPENDIX 7-19 *Continued*

State	PWSID	Water System Name	Source	Analyte	Concentration	Unit	Sample Date	PWSType	PSource	Pop
CA	CA1910067	LOS ANGELES-CITY, DEPT. OF WATER & POWER	TUJUNGA WELL 07	TRICHLOROFLUOROMETHANE	5.29	UG/L	2/26/2004 0:00	CWS	SW	3,828,700
CA	CA1910067	LOS ANGELES-CITY, DEPT. OF WATER & POWER	TUJUNGA WELL 09	TRICHLOROFLUOROMETHANE	0.55	UG/L	1/22/2002 0:00	CWS	SW	3,828,700
CA	CA1910067	LOS ANGELES-CITY, DEPT. OF WATER & POWER	TUJUNGA WELL 09	TRICHLOROFLUOROMETHANE	1	UG/L	7/26/2000 0:00	CWS	SW	3,828,700
CA	CA1910067	LOS ANGELES-CITY, DEPT. OF WATER & POWER	TUJUNGA WELL 12	TRICHLOROFLUOROMETHANE	7.8	UG/L	9/10/2003 0:00	CWS	SW	3,828,700
CA	CA1910067	LOS ANGELES-CITY, DEPT. OF WATER & POWER	TUJUNGA WELL 12	TRICHLOROFLUOROMETHANE	8.58	UG/L	7/8/2003 0:00	CWS	SW	3,828,700
CA	CA1910067	LOS ANGELES-CITY, DEPT. OF WATER & POWER	TUJUNGA WELL 12	TRICHLOROFLUOROMETHANE	11	UG/L	3/4/2004 0:00	CWS	SW	3,828,700
CA	CA1910067	LOS ANGELES-CITY, DEPT. OF WATER & POWER	TUJUNGA WELL 12	TRICHLOROFLUOROMETHANE	9.6	UG/L	2/26/2004 0:00	CWS	SW	3,828,700
CA	CA1910067	LOS ANGELES-CITY, DEPT. OF WATER & POWER	TUJUNGA WELL 12	TRICHLOROFLUOROMETHANE	7.76	UG/L	10/22/2003 0:00	CWS	SW	3,828,700
CA	CA1910067	LOS ANGELES-CITY, DEPT. OF WATER & POWER	TUJUNGA WELL 12	TRICHLOROFLUOROMETHANE	15	UG/L	12/9/2003 0:00	CWS	SW	3,828,700
CA	CA1910067	LOS ANGELES-CITY, DEPT. OF WATER & POWER	TUJUNGA WELL 12	TRICHLOROFLUOROMETHANE	7.6	UG/L	2/18/2003 0:00	CWS	SW	3,828,700
CA	CA1910067	LOS ANGELES-CITY, DEPT. OF WATER & POWER	TUJUNGA WELL 12	TRICHLOROFLUOROMETHANE	5.82	UG/L	3/25/2003 0:00	CWS	SW	3,828,700
CA	CA1910067	LOS ANGELES-CITY, DEPT. OF WATER & POWER	TUJUNGA WELL 12	TRICHLOROFLUOROMETHANE	5.6	UG/L	3/17/2003 0:00	CWS	SW	3,828,700
CA	CA1910067	LOS ANGELES-CITY, DEPT. OF WATER & POWER	TUJUNGA WELL 12	TRICHLOROFLUOROMETHANE	5.55	UG/L	1/17/2003 0:00	CWS	SW	3,828,700
CA	CA1910067	LOS ANGELES-CITY, DEPT. OF WATER & POWER	TUJUNGA WELL 12	TRICHLOROFLUOROMETHANE	5.99	UG/L	12/9/2002 0:00	CWS	SW	3,828,700
CA	CA1910067	LOS ANGELES-CITY, DEPT. OF WATER & POWER	TUJUNGA WELL 12	TRICHLOROFLUOROMETHANE	6.17	UG/L	3/25/2003 0:00	CWS	SW	3,828,700
CA	CA1910067	LOS ANGELES-CITY, DEPT. OF WATER & POWER	TUJUNGA WELL 12	TRICHLOROFLUOROMETHANE	11.2	UG/L	6/19/2003 0:00	CWS	SW	3,828,700
CA	CA1910067	LOS ANGELES-CITY, DEPT. OF WATER & POWER	TUJUNGA WELL 12	TRICHLOROFLUOROMETHANE	8.42	UG/L	4/5/2004 0:00	CWS	SW	3,828,700
CA	CA1910067	LOS ANGELES-CITY, DEPT. OF WATER & POWER	TUJUNGA WELL 12	TRICHLOROFLUOROMETHANE	5.94	UG/L	4/23/2003 0:00	CWS	SW	3,828,700
CA	CA1910067	LOS ANGELES-CITY, DEPT. OF WATER & POWER	TUJUNGA WELL 12	TRICHLOROFLUOROMETHANE	6.12	UG/L	1/12/2005 0:00	CWS	SW	3,828,700
CA	CA1910067	LOS ANGELES-CITY, DEPT. OF WATER & POWER	TUJUNGA WELL 12	TRICHLOROFLUOROMETHANE	9.49	UG/L	12/7/2004 0:00	CWS	SW	3,828,700
CA	CA1910067	LOS ANGELES-CITY, DEPT. OF WATER & POWER	TUJUNGA WELL 12	TRICHLOROFLUOROMETHANE	9.86	UG/L	11/18/2004 0:00	CWS	SW	3,828,700

CA	CA1910067	LOS ANGELES-CITY, DEPT. OF WATER & POWER	TUJUNGA WELL 12	TRICHLOROFLUOROMETHANE	36.8	UG/L	3/22/2005 0:00	CWS	SW	3,828,700
CA	CA1910067	LOS ANGELES-CITY, DEPT. OF WATER & POWER	TUJUNGA WELL 12	TRICHLOROFLUOROMETHANE	26.1	UG/L	5/13/2005 0:00	CWS	SW	3,828,700
CA	CA1910067	LOS ANGELES-CITY, DEPT. OF WATER & POWER	TUJUNGA WELL 12	TRICHLOROFLUOROMETHANE	21.1	UG/L	4/8/2005 0:00	CWS	SW	3,828,700
CA	CA1910067	LOS ANGELES-CITY, DEPT. OF WATER & POWER	TUJUNGA WELL 12	TRICHLOROFLUOROMETHANE	8.94	UG/L	2/11/2005 0:00	CWS	SW	3,828,700
CA	CA1910067	LOS ANGELES-CITY, DEPT. OF WATER & POWER	TUJUNGA WELL 12	TRICHLOROFLUOROMETHANE	7.51	UG/L	6/8/2004 0:00	CWS	SW	3,828,700
CA	CA1910067	LOS ANGELES-CITY, DEPT. OF WATER & POWER	TUJUNGA WELL 12	TRICHLOROFLUOROMETHANE	7.98	UG/L	5/24/2004 0:00	CWS	SW	3,828,700
CA	CA1910067	LOS ANGELES-CITY, DEPT. OF WATER & POWER	TUJUNGA WELL 12	TRICHLOROFLUOROMETHANE	10	UG/L	1/20/2004 0:00	CWS	SW	3,828,700
CA	CA1910067	LOS ANGELES-CITY, DEPT OF WATER & POWER	TUJUNGA WELL 12	TRICHLOROFLUOROMETHANE	6.38	UG/L	11/25/2002 0:00	CWS	SW	3,828,700
CA	CA1910067	LOS ANGELES-CITY, DEPT. OF WATER & POWER	TUJUNGA WELL 12	TRICHLOROFLUOROMETHANE	5.33	UG/L	10/10/2002 0:00	CWS	SW	3,828,700
CA	CA1910067	LOS ANGELES-CITY, DEPT. OF WATER & POWER	TUJUNGA WELL 12	TRICHLOROFLUOROMETHANE	0.61	UG/L	12/14/2000 0:00	CWS	SW	3,828,700
CA	CA1910067	LOS ANGELES-CITY, DEPT. OF WATER & POWER	TUJUNGA WELL 12	TRICHLOROFLUOROMETHANE	1.31	UG/L	2/2/2001 0:00	CWS	SW	3,828,700
CA	CA1910067	LOS ANGELES-CITY, DEPT. OF WATER & POWER	TUJUNGA WELL 12	TRICHLOROFLUOROMETHANE	1.29	UG/L	1/9/2001 0:00	CWS	SW	3,828,700
CA	CA1910067	LOS ANGELES-CITY, DEPT. OF WATER & POWER	TUJUNGA WELL 11	TRICHLOROFLUOROMETHANE	10.1 8.76	UG/L	5/13/2005 0:00	CWS	SW	3,828,700
CA	CA1910067	LOS ANGELES-CITY, DEPT. OF WATER & POWER	NORTH HOLLYWOOD AERATION WELL 05	XYLENES (TOTAL)	1	UG/L	8/16/2000 0:00	CWS	SW	3,828,700
CA	CA1910067	LOS ANGELES-CITY, DEPT. OF WATER & POWER	NORTH HOLLYWOOD AERATION WELL 05	XYLENES (TOTAL)	6	UG/L	8/7/2000 0:00	CWS	SW	3,828,700
CA	CA1910067	LOS ANGELES-CITY, DEPT. OF WATER & POWER	NORTH HOLLYWOOD AERATION WELL 05	XYLENES (TOTAL)	1	UG/L	7/27/2000 0:00	CWS	SW	3,828,700
CA	CA1910067	LOS ANGELES-CITY, DEPT. OF WATER & POWER	NORTH HOLLYWOOD AERATION WELL 04	XYLENES (TOTAL)	79	UG/L	8/7/2000 0:00	CWS	SW	3,828,700
CA	CA1910067	LOS ANGELES-CITY, DEPT. OF WATER & POWER	NORTH HOLLYWOOD AERATION WELL 05	M,P-XYLENE	1	UG/L	8/16/2000 0:00	CWS	SW	3,828,700
CA	CA1910067	LOS ANGELES-CITY, DEPT. OF WATER & POWER	NORTH HOLLYWOOD AERATION WELL 05	M,P-XYLENE	1	UG/L	7/27/2000 0:00	CWS	SW	3,828,700
CA	CA1910067	LOS ANGELES-CITY, DEPT. OF WATER & POWER	NORTH HOLLYWOOD AERATION WELL 05	M,P-XYLENE	5	UG/L	8/7/2000 0:00	CWS	SW	3,828,700
CA	CA1910067	LOS ANGELES-CITY, DEPT. OF WATER & POWER	NORTH HOLLYWOOD AERATION WELL 04	M,P-XYLENE	1	UG/L	8/7/2000 0:00	CWS	SW	3,828,700
CA	CA1910067	LOS ANGELES-CITY, DEPT. OF WATER & POWER	NORTH HOLLYWOOD AERATION WELL 05	O-XYLENE	1	UG/L	8/7/2000 0:00	CWS	SW	3,828,700
CA	CA1910067	LOS ANGELES-CITY, DEPT. OF WATER & POWER	NORTH HOLLYWOOD AERATION WELL 04	O-XYLENE	78 17.4	UG/L	8/7/2000 0:00	CWS	SW	3,828,700

APPENDIX 7-20 Milwaukee Groundwater Data

State	PWSID	SYSTEM	SAMPLE SITE	ANALYTE	CONC	UOM	DATE	PWSType	PSource	Pop
WI	WI2670696	HILLTOP HIGHLANDS MHP 1	Entry Point	1,1,1-TRICHLOROETHANE	0.1	UG/L	05-May-97	CWS	GW	35
WI	WI2670706	HILLTOP HIGHLANDS MHP 3	Entry Point	1,1,1-TRICHLOROETHANE	0.16	UG/L	22-Mar-95	CWS	GW	35
WI	WI2670696	HILLTOP HIGHLANDS MHP 1	Entry Point	1,1,1-TRICHLOROETHANE	0.13	UG/L	29-Feb-96	CWS	GW	35
WI	WI2670696	HILLTOP HIGHLANDS MHP 1	Entry Point	1,1,1-TRICHLOROETHANE	0.2	UG/L	06-Apr-94	CWS	GW	35
WI	WI2670696	HILLTOP HIGHLANDS MHP 1	Entry Point	1,1,1-TRICHLOROETHANE	0.14	UG/L	18-Jul-95	CWS	GW	35
WI	WI2670696	HILLTOP HIGHLANDS MHP 1	Entry Point	1,1,1-TRICHLOROETHANE	0.15	UG/L	22-Mar-95	CWS	GW	35
WI	WI2670696	HILLTOP HIGHLANDS MHP 1	Entry Point	1,1,1-TRICHLOROETHANE	0.17	UG/L	22-Jun-95	CWS	GW	35
WI	WI1140130	MAJESTIC HILLS	Entry Point	1,1,1-TRICHLOROETHANE	0.067	UG/L	15-Apr-97	CWS	GW	550
WI	WI2301080	TAN OAK APTS	Entry Point	1,1,1-TRICHLOROETHANE	0.1	UG/L	13-Dec-94	CWS	GW	325
WI	WI1140130	MAJESTIC HILLS	Entry Point	1,1,1-TRICHLOROETHANE	0.067	UG/L	15-Apr-97	CWS	GW	550
					0.13					
WI	WI2670696	HILLTOP HIGHLANDS MHP 1	Entry Point	1,1-DICHLOROETHANE	0.2	UG/L	22-Jun-95	CWS	GW	35
WI	WI2670696	HILLTOP HIGHLANDS MHP 1	Entry Point	1,1-DICHLOROETHANE	0.2	UG/L	18-Jul-95	CWS	GW	35
WI	WI2670696	HILLTOP HIGHLANDS MHP 1	Entry Point	1,1-DICHLOROETHANE	0.4	UG/L	06-Apr-94	CWS	GW	35
WI	WI2670706	HILLTOP HIGHLANDS MHP 3	Entry Point	1,1-DICHLOROETHANE	0.22	UG/L	22-Mar-95	CWS	GW	35
WI	WI2670696	HILLTOP HIGHLANDS MHP 1	Entry Point	1,1-DICHLOROETHANE	0.2	UG/L	22-Mar-95	CWS	GW	35
WI	WI4700271	DAKOTA CAPITAL PARK	Entry Point	1,1-DICHLOROETHANE	0.2	UG/L	22-Mar-94	CWS	GW	102
WI	WI2670696	HILLTOP HIGHLANDS MHP 1	Entry Point	1,1-DICHLOROETHANE	0.19	UG/L	05-May-97	CWS	GW	35
WI	WI2670696	HILLTOP HIGHLANDS MHP 1	Entry Point	1,1-DICHLOROETHANE	0.22	UG/L	29-Feb-96	CWS	GW	35
					0.23					
WI	WI2670778	HILLTOP HIGHLANDS MHP 6	Entry Point	1,2-DICHLOROETHANE	0.4	UG/L	01-Jul-94	CWS	GW	28
WI	WI2670706	HILLTOP HIGHLANDS MHP 3	Entry Point	1,2-DICHLOROETHANE	0.4	UG/L	30-Jun-94	CWS	GW	35
WI	WI2670705	HILLTOP HIGHLANDS MHP 2	Entry Point	1,2-DICHLOROETHANE	0.4	UG/L	01-Jul-94	CWS	GW	35
WI	WI2670705	HILLTOP HIGHLANDS MHP 2	Entry Point	1,2-DICHLOROETHANE	0.4	UG/L	30-Jun-94	CWS	GW	35
WI	WI2670696	HILLTOP HIGHLANDS MHP 1	Entry Point	1,2-DICHLOROETHANE	0.3	UG/L	01-Jul-94	CWS	GW	35
WI	WI2670778	HILLTOP HIGHLANDS MHP 6	Entry Point	1,2-DICHLOROETHANE	0.4	UG/L	30-Jun-94	CWS	GW	28
WI	WI2670696	HILLTOP HIGHLANDS MHP 1	Entry Point	1,2-DICHLOROETHANE	0.3	UG/L	30-Jun-94	CWS	GW	35
WI	WI2670706	HILLTOP HIGHLANDS MHP 3	Entry Point	1,2-DICHLOROETHANE	0.4	UG/L	01-Jul-94	CWS	GW	35
WI	WI1140130	MAJESTIC HILLS	Entry Point	1,2-DICHLOROETHANE	0.88	UG/L	17-Apr-96	CWS	GW	550
WI	WI1140130	MAJESTIC HILLS	Entry Point	1,2-DICHLOROETHANE	0.4	UG/L	28-Feb-95	CWS	GW	550
					0.43					
WI	WI4700271	DAKOTA CAPITAL PARK	Distribution System	BROMODICHLOROMETHANE	0.855	UG/L	08-Sep-04	CWS	GW	102
WI	WI2670706	HILLTOP HIGHLANDS MHP 3	Entry Point	BROMODICHLOROMETHANE	0.55	UG/L	14-Jun-00	CWS	GW	35
WI	WI2670705	HILLTOP HIGHLANDS MHP 2	Entry Point	BROMODICHLOROMETHANE	2	UG/L	18-Jul-95	CWS	GW	35
WI	WI2670778	HILLTOP HIGHLANDS MHP 6	Entry Point	BROMODICHLOROMETHANE	0.12	UG/L	22-Mar-95	CWS	GW	28
WI	WI2670778	HILLTOP HIGHLANDS MHP 6	Entry Point	BROMODICHLOROMETHANE	1.1	UG/L	09-Dec-98	CWS	GW	28
WI	WI2670706	HILLTOP HIGHLANDS MHP 3	Entry Point	BROMODICHLOROMETHANE	0.61	UG/L	29-Feb-96	CWS	GW	35
WI	WI2670705	HILLTOP HIGHLANDS MHP 2	Entry Point	BROMODICHLOROMETHANE	0.14	UG/L	22-Jun-95	CWS	GW	35
WI	WI2670778	HILLTOP HIGHLANDS MHP 6	Entry Point	BROMODICHLOROMETHANE	0.6	UG/L	13-Sep-99	CWS	GW	28
WI	WI2671430	HILLTOP HIGHLANDS MHP 5	Entry Point	BROMODICHLOROMETHANE	0.6	UG/L	04-Apr-01	CWS	GW	90
WI	WI2410589	ROBERT WILLIAM PARK WATER ASSOCIATION	Distribution System	BROMODICHLOROMETHANE	0.19	UG/L	18-Nov-04	CWS	GW	1065

WI	WI2410589	ROBERT WILLIAM PARK WATER ASSOCIATION	Distribution System	BROMODICHLOROMETHANE	0.19	UG/L	18-Nov-04	CWS	GW	1065
WI	WI2410589	ROBERT WILLIAM PARK WATER ASSOCIATION	Entry Point	BROMODICHLOROMETHANE	1.5	UG/L	29-Oct-96	CWS	GW	1065
WI	WI2670696	HILLTOP HIGHLANDS MHP 1	Entry Point	BROMODICHLOROMETHANE	0.088	UG/L	29-Feb-96	CWS	GW	35
WI	WI2410118	BRIGGS AND STRATTON CORP	Entry Point	BROMODICHLOROMETHANE	0.11	UG/L	10-Nov-97	NTNCWS	GW	2500
WI	WI1141023	SENSIENT FLAVORS CORP	Distribution System	BROMODICHLOROMETHANE	0.17	UG/L	21-Jul-04	NTNCWS	GW	94
WI	WI1140130	MAJESTIC HILLS	Entry Point	BROMODICHLOROMETHANE	0.2	UG/L	28-Feb-95	CWS	GW	550
					0.56					
WI	WI2671430	HILLTOP HIGHLANDS MHP 5	Entry Point	BROMOFORM	0.5	UG/L	22-Jan-02	CWS	GW	90
WI	WI2460150	LAUREL LAKES 508	Entry Point	BROMOMETHANE	0.52	UG/L	19-Mar-96	CWS	GW	50
WI	WI2460137	LAUREL LAKES 608	Entry Point	BROMOMETHANE	0.52	UG/L	19-Mar-96	CWS	GW	80
WI	WI2460142	LINDEN LANE APTS 117	Entry Point	BROMOMETHANE	0.85	UG/L	19-Mar-96	CWS	GW	90
WI	WI2460154	LINDEN LANE APTS 141	Entry Point	BROMOMETHANE	0.29	UG/L	19-Mar-96	CWS	GW	75
WI	WI2670705	HILLTOP HIGHLANDS MHP 2	Entry Point	BROMOMETHANE	0.18	UG/L	18-Jul-95	CWS	GW	35
WI	WI2460158	HEIDEL RD APTS 152	Entry Point	BROMOMETHANE	0.5	UG/L	19-Mar-96	CWS	GW	40
WI	WI2460152	HEIDEL RD APTS 139	Entry Point	BROMOMETHANE	0.47	UG/L	19-Mar-96	CWS	GW	40
					0.48					
WI	WI2670705	HILLTOP HIGHLANDS MHP 2	Entry Point	CARBON TETRACHLORIDE	0.72	UG/L	18-Jul-95	CWS	GW	35
WI	WI2670696	HILLTOP HIGHLANDS MHP 1	Entry Point	CHLOROFORM	0.17	UG/L	29-Feb-96	CWS	GW	35
WI	WI4700271	DAKOTA CAPITAL PARK	Distribution System	CHLOROFORM	2.13	UG/L	08-Sep-04	CWS	GW	102
WI	WI2670705	HILLTOP HIGHLANDS MHP 2	Entry Point	CHLOROFORM	22	UG/L	18-Jul-95	CWS	GW	35
WI	WI4700271	DAKOTA CAPITAL PARK	Entry Point	CHLOROFORM	0.13	UG/L	04-Dec-00	CWS	GW	102
WI	WI2411769	GENERAL MITCHELL (IAP) ANG	Entry Point	CHLOROFORM	0.2	UG/L	13-Dec-95	NTNCWS	GW	25
WI	WI2410589	ROBERT WILLIAM PARK WATER ASSOCIATION	Distribution System	CHLOROFORM	0.18	UG/L	18-Nov-04	CWS	GW	1065
WI	WI2410589	ROBERT WILLIAM PARK WATER ASSOCIATION	Distribution System	CHLOROFORM	0.24	UG/L	18-Nov-04	CWS	GW	1065
WI	WI2410589	ROBERT WILLIAM PARK WATER ASSOCIATION	Entry Point	CHLOROFORM	4.4	UG/L	29-Oct-96	CWS	GW	1065
WI	WI4700271	DAKOTA CAPITAL PARK	Distribution System	CHLOROFORM	0.267	UG/L	08-Sep-04	CWS	GW	102
WI	WI2650116	PIONEER ESTATES OF DELAVAN	Entry Point	CHLOROFORM	0.75	UG/L	20-Dec-00	CWS	GW	240
WI	WI2650116	PIONEER ESTATES OF DELAVAN	Entry Point	CHLOROFORM	0.77	UG/L	23-Sep-02	CWS	GW	240
WI	WI2410118	BRIGGS AND STRATTON CORP	Entry Point	CHLOROFORM	0.37	UG/L	10-Nov-97	NTNCWS	GW	2500
WI	WI2670778	HILLTOP HIGHLANDS MHP 6	Entry Point	CHLOROFORM	0.6	UG/L	13-Sep-99	CWS	GW	28
WI	WI2670778	HILLTOP HIGHLANDS MHP 6	Entry Point	CHLOROFORM	0.4	UG/L	20-Dec-94	CWS	GW	28
WI	WI2670706	HILLTOP HIGHLANDS MHP 3	Entry Point	CHLOROFORM	0.38	UG/L	14-Jun-00	CWS	GW	35
WI	WI4700271	DAKOTA CAPITAL PARK	Entry Point	CHLOROFORM	0.083	UG/L	14-Aug-95	CWS	GW	102
WI	WI2670706	HILLTOP HIGHLANDS MHP 3	Entry Point	CHLOROFORM	0.44	UG/L	29-Feb-96	CWS	GW	35
WI	WI2670778	HILLTOP HIGHLANDS MHP 6	Entry Point	CHLOROFORM	0.24	UG/L	22-Mar-95	CWS	GW	28
WI	WI2670705	HILLTOP HIGHLANDS MHP 2	Entry Point	CHLOROFORM	0.19	UG/L	22-Jun-95	CWS	GW	35
WI	WI2671430	HILLTOP HIGHLANDS MHP 5	Entry Point	CHLOROFORM	1.1	UG/L	01-Mar-00	CWS	GW	90
WI	WI2301080	TAN OAK APTS	Entry Point	CHLOROFORM	1	UG/L	10-Mar-93	CWS	GW	325
WI	WI1110104	OAKWOOD TERRACE	Entry Point	CHLOROFORM	0.13	UG/L	27-Dec-96	CWS	GW	100
WI	WI1140130	MAJESTIC HILLS	Entry Point	CHLOROFORM	3.1	UG/L	28-Feb-95	CWS	GW	550
WI	WI1141023	SENSIENT FLAVORS CORP	Distribution System	CHLOROFORM	0.58	UG/L	21-Jul-04	NTNCWS	GW	94
					1.66					
WI	WI2670696	HILLTOP HIGHLANDS MHP 1	Entry Point	CHLOROMETHANE (METHYLCHLORIDE)	0.08	UG/L	18-Jul-95	CWS	GW	35
WI	WI2670778	HILLTOP HIGHLANDS MHP 6	Entry Point	CHLOROMETHANE (METHYLCHLORIDE)	0.092	UG/L	25-Sep-95	CWS	GW	28

APPENDIX 7-20 *Continued*

State	PWSID	SYSTEM	SAMPLE SITE	ANALYTE	CONC	UOM	DATE	PWSType	PSource	Pop
WI	WI2460550	U S POST OFFICE THIENSVILLE	Entry Point	CHLOROMETHANE (METHYLCHLORIDE)	0.16	UG/L	24-Jun-96	NTNCWS	GW	45
WI	WI2670705	HILLTOP HIGHLANDS MHP 2	Entry Point	CHLOROMETHANE (METHYLCHLORIDE)	1.9	UG/L	18-Jul-95	CWS	GW	35
WI	WI2410118	BRIGGS AND STRATTON CORP	Entry Point	CHLOROMETHANE (METHYLCHLORIDE)	0.21	UG/L	10-Nov-97	NTNCWS	GW	2500
WI	WI2670696	HILLTOP HIGHLANDS MHP 1	Entry Point	CHLOROMETHANE (METHYLCHLORIDE)	0.13	UG/L	29-Feb-96	CWS	GW	35
WI	WI2670705	HILLTOP HIGHLANDS MHP 2	Entry Point	CHLOROMETHANE (METHYLCHLORIDE)	0.09	UG/L	22-Jun-95	CWS	GW	35
WI	WI2670706	HILLTOP HIGHLANDS MHP 3	Entry Point	CHLOROMETHANE (METHYLCHLORIDE)	0.075	UG/L	25-Sep-95	CWS	GW	35
WI	WI1540126	ROCKVALE MHP	Entry Point	CHLOROMETHANE (METHYLCHLORIDE)	0.14	UG/L	15-Nov-95	CWS	GW	800
WI	WI1140130	MAJESTIC HILLS	Entry Point	CHLOROMETHANE (METHYLCHLORIDE)	0.2	UG/L	12-Dec-96	CWS	GW	550
WI	WI1140130	MAJESTIC HILLS	Entry Point	CHLOROMETHANE (METHYLCHLORIDE)	0.11	UG/L	17-Apr-96	CWS	GW	550
					0.29					
WI	WI2650116	PIONEER ESTATES OF DELAVAN	Well	DI(2-ETHYLHEXYL) PHTHALATE	12	UG/L	22-Jan-98	CWS	GW	240
WI	WI2411769	GENERAL MITCHELL (IAP) ANG	Entry Point	DI(2-ETHYLHEXYL) PHTHALATE	2	UG/L	16-Apr-01	NTNCWS	GW	25
WI	WI2410603	FRANKLIN MOBILE, L.L.C., #2	Entry Point	DI(2-ETHYLHEXYL) PHTHALATE	5.8	UG/L	08-Nov-00	CWS	GW	102
WI	WI2460152	HEIDEL RD APTS 139	Entry Point	DI(2-ETHYLHEXYL) PHTHALATE	1.7	UG/L	29-Aug-00	CWS	GW	40
WI	WI1280101	OAK RIDGE MOBILE COM 1 & 2	Entry Point	DI(2-ETHYLHEXYL) PHTHALATE	0.9	UG/L	17-May-00	CWS	GW	133
					4.48					
WI	WI2670778	HILLTOP HIGHLANDS MHP 6	Entry Point	DIBROMOCHLOROMETHANE	0.9	UG/L	13-Sep-99	CWS	GW	28
WI	WI2670778	HILLTOP HIGHLANDS MHP 6	Entry Point	DIBROMOCHLOROMETHANE	1.4	UG/L	09-Dec-98	CWS	GW	28
WI	WI2671430	HILLTOP HIGHLANDS MHP 5	Entry Point	DIBROMOCHLOROMETHANE	0.5	UG/L	04-Apr-01	CWS	GW	90
WI	WI2410589	ROBERT WILLIAM PARK WATER ASSOCIATION	Entry Point	DIBROMOCHLOROMETHANE	0.6	UG/L	29-Oct-96	CWS	GW	1065
WI	WI2410118	BRIGGS AND STRATTON CORP	Entry Point	DIBROMOCHLOROMETHANE	0.09	UG/L	10-Nov-97	NTNCWS	GW	2500
WI	WI2671430	HILLTOP HIGHLANDS MHP 5	Entry Point	DIBROMOCHLOROMETHANE	1.1	UG/L	01-Mar-00	CWS	GW	90
WI	WI2670706	HILLTOP HIGHLANDS MHP 3	Entry Point	DIBROMOCHLOROMETHANE	0.44	UG/L	29-Feb-96	CWS	GW	35
WI	WI4700271	DAKOTA CAPITAL PARK	Distribution System	DIBROMOCHLOROMETHANE	0.465	UG/L	08-Sep-04	CWS	GW	102
					0.69					
WI	WI4700271	DAKOTA CAPITAL PARK	Distribution System	DICHLOROACETIC ACID	3.85	UG/L	08-Sep-04	CWS	GW	102
WI	WI2410552	NORTH SHORE EAST WATER TRUST	Entry Point	DICHLOROMETHANE	2.1	UG/L	20-Jan-04	CWS	GW	600
WI	WI2460189	THIENSVILLE PROFESSIONAL PARK	Entry Point	DICHLOROMETHANE	0.31	UG/L	26-Feb-04	NTNCWS	GW	127
WI	WI2460189	THIENSVILLE PROFESSIONAL PARK	Entry Point	DICHLOROMETHANE	0.32	UG/L	26-Apr-05	NTNCWS	GW	127
WI	WI2670705	HILLTOP HIGHLANDS MHP 2	Entry Point	DICHLOROMETHANE	0.36	UG/L	18-Jul-95	CWS	GW	35
WI	WI2460113	PIONEER GRAFTON MOBILE HM PK	Entry Point	DICHLOROMETHANE	0.6	UG/L	03-Dec-01	CWS	GW	150
WI	WI2410118	BRIGGS AND STRATTON CORP	Entry Point	DICHLOROMETHANE	0.15	UG/L	10-Nov-97	NTNCWS	GW	2500
WI	WI2410552	NORTH SHORE EAST WATER TRUST	Entry Point	DICHLOROMETHANE	1.8	UG/L	25-Jun-03	CWS	GW	600
WI	WI2410552	NORTH SHORE EAST WATER TRUST	Entry Point	DICHLOROMETHANE	1.9	UG/L	25-Jun-03	CWS	GW	600
WI	WI1540126	ROCKVALE MHP	Entry Point	DICHLOROMETHANE	0.5	UG/L	05-Jun-01	CWS	GW	800
					0.83					
WI	WI2670778	HILLTOP HIGHLANDS MHP 6	Distribution System	HAA5	7.4	UG/L	23-Aug-04	CWS	GW	28
WI	WI2671430	HILLTOP HIGHLANDS MHP 5	Distribution System	HAA5	8.4	UG/L	23-Aug-04	CWS	GW	90
WI	WI2670696	HILLTOP HIGHLANDS MHP 1	Distribution System	HAA5	1	UG/L	23-Aug-04	CWS	GW	35

WI	WI4700271	DAKOTA CAPITAL PARK	Distribution System	HAA5	0.314	UG/L	08-Sep-04	CWS	GW	102
WI	WI4700271	DAKOTA CAPITAL PARK	Distribution System	HAA5	37.64	UG/L	08-Sep-04	CWS	GW	102
					10.4					
WI	WI2670778	HILLTOP HIGHLANDS MHP 6	Distribution System	MONOCHLOROACETIC ACID	7.4	UG/L	23-Aug-04	CWS	GW	28
WI	WI4700271	DAKOTA CAPITAL PARK	Distribution System	MONOCHLOROACETIC ACID	31.1	UG/L	08-Sep-04	CWS	GW	102
WI	WI2671430	HILLTOP HIGHLANDS MHP 5	Distribution System	MONOCHLOROACETIC ACID	8.4	UG/L	23-Aug-04	CWS	GW	90
					15.6					
WI	WI2670696	HILLTOP HIGHLANDS MHP 1	Entry Point	P-DICHLOROBENZENE	0.1	UG/L	06-Apr-94	CWS	GW	35
WI	WI2681944	REGENCY COURT 275	Entry Point	P-DICHLOROBENZENE	0.42	UG/L	10-Mar-05	NTNCWS	GW	105
WI	WI2681944	REGENCY COURT 275	Entry Point	P-DICHLOROBENZENE	0.18	UG/L	17-Oct-02	NTNCWS	GW	105
WI	WI2460137	LAUREL LAKES 608	Entry Point	P-DICHLOROBENZENE	0.3	UG/L	19-Mar-96	CWS	GW	80
WI	WI2670696	HILLTOP HIGHLANDS MHP 1	Entry Point	P-DICHLOROBENZENE	0.2	UG/L	01-Jul-94	CWS	GW	35
WI	WI2460158	HEIDEL RD APTS 152	Entry Point	P-DICHLOROBENZENE	0.27	UG/L	19-Mar-96	CWS	GW	40
WI	WI2670696	HILLTOP HIGHLANDS MHP 1	Entry Point	P-DICHLOROBENZENE	0.2	UG/L	30-Jun-94	CWS	GW	35
WI	WI2460150	LAUREL LAKES 508	Entry Point	P-DICHLOROBENZENE	0.24	UG/L	19-Mar-96	CWS	GW	50
WI	WI2460154	LINDEN LANE APTS 141	Entry Point	P-DICHLOROBENZENE	0.27	UG/L	19-Mar-96	CWS	GW	75
WI	WI2680825	ENTERPRISE BLUEMOUND	Entry Point	P-DICHLOROBENZENE	0.47	UG/L	02-Dec-04	NTNCWS	GW	150
WI	WI2460152	HEIDEL RD APTS 139	Entry Point	P-DICHLOROBENZENE	0.28	UG/L	19-Mar-96	CWS	GW	40
					0.27					
WI	WI2460189	THIENSVILLE PROFESSIONAL PARK	Entry Point	STYRENE	0.5	UG/L	13-Mar-95	NTNCWS	GW	127
WI	WI2460189	THIENSVILLE PROFESSIONAL PARK	Entry Point	STYRENE	0.3	UG/L	24-Apr-95	NTNCWS	GW	127
WI	WI1140130	MAJESTIC HILLS	Entry Point	STYRENE	0.2	UG/L	28-Feb-95	CWS	GW	550
					0.33					
WI	WI2670696	HILLTOP HIGHLANDS MHP 1	Entry Point	TETRACHLOROETHYLENE	0.086	UG/L	22-Jun-95	CWS	GW	35
WI	WI2670696	HILLTOP HIGHLANDS MHP 1	Entry Point	TETRACHLOROETHYLENE	0.086	UG/L	18-Jul-95	CWS	GW	35
WI	WI2670696	HILLTOP HIGHLANDS MHP 1	Entry Point	TETRACHLOROETHYLENE	0.075	UG/L	22-Mar-95	CWS	GW	35
WI	WI2670706	HILLTOP HIGHLANDS MHP 3	Entry Point	TETRACHLOROETHYLENE	0.077	UG/L	22-Mar-95	CWS	GW	35
					0.08					
WI	WI2460150	LAUREL LAKES 508	Entry Point	TOLUENE	0.3	UG/L	29-Aug-00	CWS	GW	50
WI	WI2680825	ENTERPRISE BLUEMOUND	Entry Point	TOLUENE	0.18	UG/L	31-Dec-97	NTNCWS	GW	150
WI	WI2460137	LAUREL LAKES 608	Entry Point	TOLUENE	0.24	UG/L	29-Aug-00	CWS	GW	80
WI	WI2460158	HEIDEL RD APTS 152	Entry Point	TOLUENE	0.47	UG/L	24-Apr-03	CWS	GW	40
WI	WI2410552	NORTH SHORE EAST WATER TRUST	Entry Point	TOLUENE	27	UG/L	20-Jan-04	CWS	GW	600
WI	WI2650116	PIONEER ESTATES OF DELAVAN	Entry Point	TOLUENE	9.8	UG/L	22-Jan-98	CWS	GW	240
WI	WI1280101	OAK RIDGE MOBILE COM 1 & 2	Entry Point	TOLUENE	0.24	UG/L	05-May-05	CWS	GW	133
WI	WI2301080	TAN OAK APTS	Entry Point	TOLUENE	0.7	UG/L	10-Mar-93	CWS	GW	325
WI	WI1280101	OAK RIDGE MOBILE COM 1 & 2	Well	TOLUENE	1.9	UG/L	28-Apr-04	CWS	GW	133
					4.54					
WI	WI2460550	U S POST OFFICE THIENSVILLE	Entry Point	TRIAZINE SCREEN	0.1	PPB	26-Jun-91	NTNCWS	GW	45
WI	WI2670696	HILLTOP HIGHLANDS MHP 1	Distribution System	TRICHLOROACETIC ACID	1	UG/L	23-Aug-04	CWS	GW	35
WI	WI4700271	DAKOTA CAPITAL PARK	Distribution System	TRICHLOROACETIC ACID	0.314	UG/L	08-Sep-04	CWS	GW	102
WI	WI4700271	DAKOTA CAPITAL PARK	Distribution System	TRICHLOROACETIC ACID	2.69	UG/L	08-Sep-04	CWS	GW	102
					1.33					

APPENDIX 7-20 *Continued*

State	PWSID	SYSTEM	SAMPLE SITE	ANALYTE	CONC	UOM	DATE	PWSType	PSource	Pop
WI	WI2670705	HILLTOP HIGHLANDS MHP 2	Entry Point	TTHM	24	UG/L	18-Jul-95	CWS	GW	35
WI	WI2670705	HILLTOP HIGHLANDS MHP 2	Entry Point	TTHM	0.33	UG/L	22-Jun-95	CWS	GW	35
WI	WI4700271	DAKOTA CAPITAL PARK	Distribution System	TTHM	0.267	UG/L	08-Sep-04	CWS	GW	102
WI	WI4700271	DAKOTA CAPITAL PARK	Entry Point	TTHM	0.13	UG/L	04-Dec-00	CWS	GW	102
WI	WI4700271	DAKOTA CAPITAL PARK	Distribution System	TTHM	3.45	UG/L	08-Sep-04	CWS	GW	102
WI	WI4700271	DAKOTA CAPITAL PARK	Entry Point	TTHM	0.083	UG/L	14-Aug-95	CWS	GW	102
WI	WI2671430	HILLTOP HIGHLANDS MHP 5	Entry Point	TTHM	0.5	UG/L	22-Jan-02	CWS	GW	90
WI	WI2671430	HILLTOP HIGHLANDS MHP 5	Entry Point	TTHM	1.1	UG/L	04-Apr-01	CWS	GW	90
WI	WI2670778	HILLTOP HIGHLANDS MHP 6	Entry Point	TTHM	2.5	UG/L	09-Dec-98	CWS	GW	28
WI	WI2670778	HILLTOP HIGHLANDS MHP 6	Entry Point	TTHM	2.1	UG/L	13-Sep-99	CWS	GW	28
WI	WI2671430	HILLTOP HIGHLANDS MHP 5	Entry Point	TTHM	2.2	UG/L	01-Mar-00	CWS	GW	90
WI	WI2670778	HILLTOP HIGHLANDS MHP 6	Entry Point	TTHM	0.36	UG/L	22-Mar-95	CWS	GW	28
WI	WI2410589	ROBERT WILLIAM PARK WATER ASSOCIATION	Distribution System	TTHM	0.43	UG/L	18-Nov-04	CWS	GW	1065
WI	WI2410589	ROBERT WILLIAM PARK WATER ASSOCIATION	Distribution System	TTHM	0.37	UG/L	18-Nov-04	CWS	GW	1065
WI	WI2410589	ROBERT WILLIAM PARK WATER ASSOCIATION	Entry Point	TTHM	6.5	UG/L	29-Oct-96	CWS	GW	1065
WI	WI2670706	HILLTOP HIGHLANDS MHP 3	Entry Point	TTHM	1.49	UG/L	29-Feb-96	CWS	GW	35
WI	WI2670706	HILLTOP HIGHLANDS MHP 3	Entry Point	TTHM	0.93	UG/L	14-Jun-00	CWS	GW	35
WI	WI2650116	PIONEER ESTATES OF DELAVAN	Entry Point	TTHM	0.77	UG/L	23-Sep-02	CWS	GW	240
WI	WI2650116	PIONEER ESTATES OF DELAVAN	Entry Point	TTHM	0.75	UG/L	20-Dec-00	CWS	GW	240
WI	WI2410118	BRIGGS AND STRATTON CORP	Entry Point	TTHM	0.57	UG/L	10-Nov-97	NTNCWS	GW	2500
WI	WI2411769	GENERAL MITCHELL (IAP) ANG	Entry Point	TTHM	0.2	UG/L	13-Dec-95	NTNCWS	GW	25
WI	WI2670696	HILLTOP HIGHLANDS MHP 1	Entry Point	TTHM	0.258	UG/L	29-Feb-96	CWS	GW	35
WI	WI1110104	OAKWOOD TERRACE	Entry Point	TTHM	0.13	UG/L	27-Dec-96	CWS	GW	100
WI	WI1141023	SENSIENT FLAVORS CORP	Distribution System	TTHM	0.75	UG/L	21-Jul-04	NTNCWS	GW	94
WI	WI2301080	TAN OAK APTS	Entry Point	TTHM	1	UG/L	10-Mar-93	CWS	GW	325
WI	WI1140130	MAJESTIC HILLS	Entry Point	TTHM	3.3	UG/L	28-Feb-95	CWS	GW	550
					2.03					
WI	WI2410552	NORTH SHORE EAST WATER TRUST	Entry Point	XYLENES, TOTAL	74	UG/L	20-Jan-04	CWS	GW	600
WI	WI2650116	PIONEER ESTATES OF DELAVAN	Well	XYLENES, TOTAL	0.64	UG/L	07-May-98	CWS	GW	240
					37.3					

APPENDIX 7-21 Milwaukee Suface Water Data

State	PWSID	SYSTEM	SAMPLE SITE	ANALYTE	CONC	UOM	DATE	PWSType	PSource	Pop
WI	WI2460112	WE ENERGIES 1	Entry Point	1,1,1-TRICHLOROETHANE	0.69	UG/L	16-Dec-94	CWS	SWP	5,470
WI	WI2410100	MILWAUKEE WATERWORKS	Entry Point	1,2-DICHLOROPROPANE	0.3	UG/L	26-Feb-96	CWS	SW	682,332
WI	WI2410100	MILWAUKEE WATERWORKS-461 SAMPLES	Distribution System	BROMODICHLOROMETHANE	4.82	UG/L	11-Feb-85	CWS	SW	682,332
WI	WI2413515	NORTH SHORE WATER COMMISSION	Distribution System	BROMOFORM	0.6	UG/L	08-Oct-03	CWS	SW	35,598
WI	WI2413515	NORTH SHORE WATER COMMISSION	Distribution System	BROMOFORM	0.2	UG/L	26-Jan-98	CWS	SW	35,598
WI	WI2413515	NORTH SHORE WATER COMMISSION	Distribution System	BROMOFORM	0.3	UG/L	29-Feb-00	CWS	SW	35,598
WI	WI2413515	NORTH SHORE WATER COMMISSION	Distribution System	BROMOFORM	0.36	UG/L	17-Jan-05	CWS	SW	35,598
WI	WI2413515	NORTH SHORE WATER COMMISSION	Distribution System	BROMOFORM	0.2	UG/L	26-May-98	CWS	SW	35,598
WI	WI2413515	NORTH SHORE WATER COMMISSION	Distribution System	BROMOFORM	0.35	UG/L	04-Apr-05	CWS	SW	35,598
WI	WI2413515	NORTH SHORE WATER COMMISSION	Distribution System	BROMOFORM	0.6	UG/L	03-Sep-03	CWS	SW	35,598
WI	WI2410100	MILWAUKEE WATERWORKS	Distribution System	BROMOFORM	1.2	UG/L	09-Feb-98	CWS	SW	682,332
WI	WI2410100	MILWAUKEE WATERWORKS	Entry Point	BROMOFORM	0.3	UG/L	08-Mar-93	CWS	SW	682,332
WI	WI2410100	MILWAUKEE WATERWORKS	Distribution System	BROMOFORM	1.5	UG/L	12-Oct-99	CWS	SW	682,332
WI	WI2413515	NORTH SHORE WATER COMMISSION	Distribution System	BROMOFORM	0.2	UG/L	06-Sep-95	CWS	SW	35,598
WI	WI2410100	MILWAUKEE WATERWORKS	Entry Point	BROMOFORM	0.1	UG/L	12-Feb-96	CWS	SW	682,332
WI	WI2413515	NORTH SHORE WATER COMMISSION	Distribution System	BROMOFORM	0.2	UG/L	27-Apr-95	CWS	SW	35,598
WI	WI2413515	NORTH SHORE WATER COMMISSION	Distribution System	BROMOFORM	0.2	UG/L	08-Feb-01	CWS	SW	35,598
WI	WI2410100	MILWAUKEE WATERWORKS	Distribution System	BROMOFORM	0.2	UG/L	07-Aug-95	CWS	SW	682,332
WI	WI2410100	MILWAUKEE WATERWORKS	Distribution System	BROMOFORM	0.2	UG/L	23-Oct-95	CWS	SW	682,332
WI	WI2410100	MILWAUKEE WATERWORKS	Distribution System	BROMOFORM	0.2	UG/L	24-Oct-96	CWS	SW	682,332
WI	WI2410100	MILWAUKEE WATERWORKS	Distribution System	BROMOFORM	0.2	UG/L	27-Feb-95	CWS	SW	682,332
WI	WI2410100	MILWAUKEE WATERWORKS	Entry Point	BROMOFORM	0.3	UG/L	14-Jun-99	CWS	SW	682,332
WI	WI2413515	NORTH SHORE WATER COMMISSION	Distribution System	BROMOFORM	0.4	UG/L	31-Jul-00	CWS	SW	35,598
WI	WI2413515	NORTH SHORE WATER COMMISSION	Distribution System	BROMOFORM	0.1	UG/L	14-May-97	CWS	SW	35,598
WI	WI2413515	NORTH SHORE WATER COMMISSION	Distribution System	BROMOFORM	0.6	UG/L	06-May-03	CWS	SW	35,598
WI	WI2410100	MILWAUKEE WATERWORKS	Entry Point	BROMOFORM	0.1	UG/L	09-Mar-98	CWS	SW	682,332
WI	WI2410100	MILWAUKEE WATERWORKS	Distribution System	BROMOFORM	1.2	UG/L	14-May-02	CWS	SW	682,332
WI	WI2410100	MILWAUKEE WATERWORKS	Distribution System	BROMOFORM	0.2	UG/L	03-Jun-93	CWS	SW	682,332
WI	WI2410100	MILWAUKEE WATERWORKS	Distribution System	BROMOFORM	1	UG/L	17-Apr-00	CWS	SW	682,332
WI	WI2410100	MILWAUKEE WATERWORKS	Distribution System	BROMOFORM	0.1	UG/L	26-Apr-94	CWS	SW	682,332
WI	WI2410100	MILWAUKEE WATERWORKS	Distribution System	BROMOFORM	0.2	UG/L	21-Feb-95	CWS	SW	682,332
WI	WI2413515	NORTH SHORE WATER COMMISSION	Distribution System	BROMOFORM	0.1	UG/L	29-Jul-96	CWS	SW	35,598
WI	WI2410100	MILWAUKEE WATERWORKS	Distribution System	BROMOFORM	0.2	UG/L	08-Nov-94	CWS	SW	682,332
WI	WI2413515	NORTH SHORE WATER COMMISSION	Distribution System	BROMOFORM	0.6	UG/L	08-Oct-03	CWS	SW	35,598
WI	WI2413515	NORTH SHORE WATER COMMISSION	Distribution System	BROMOFORM	0.3	UG/L	14-Oct-97	CWS	SW	35,598
WI	WI2413515	NORTH SHORE WATER COMMISSION	Distribution System	BROMOFORM	0.6	UG/L	08-Oct-03	CWS	SW	35,598
WI	WI2413515	NORTH SHORE WATER COMMISSION	Distribution System	BROMOFORM	0.1	UG/L	04-Feb-97	CWS	SW	35,598
WI	WI2413515	NORTH SHORE WATER COMMISSION	Distribution System	BROMOFORM	0.2	UG/L	03-Nov-95	CWS	SW	35,598
WI	WI2413515	NORTH SHORE WATER COMMISSION	Distribution System	BROMOFORM	0.5	UG/L	22-Nov-99	CWS	SW	35,598
WI	WI2413515	NORTH SHORE WATER COMMISSION	Distribution System	BROMOFORM	0.5	UG/L	11-May-99	CWS	SW	35,598

APPENDIX 7-21 *Continued*

State	PWSID	SYSTEM	SAMPLE SITE	ANALYTE	CONC	UOM	DATE	PWSType	PSource	Pop
WI	WI2413515	NORTH SHORE WATER COMMISSION	Distribution System	BROMOFORM	0.2	UG/L	06-Dec-94	CWS	SW	35,598
WI	WI2413515	NORTH SHORE WATER COMMISSION	Distribution System	BROMOFORM	0.2	UG/L	04-Apr-96	CWS	SW	35,598
WI	WI2413515	NORTH SHORE WATER COMMISSION	Distribution System	BROMOFORM	0.6	UG/L	03-Sep-03	CWS	SW	35,598
WI	WI2413515	NORTH SHORE WATER COMMISSION	Distribution System	BROMOFORM	0.6	UG/L	03-Sep-03	CWS	SW	35,598
WI	WI2413515	NORTH SHORE WATER COMMISSION	Distribution System	BROMOFORM	0.2	UG/L	22-Jan-96	CWS	SW	35,598
WI	WI2413515	NORTH SHORE WATER COMMISSION	Distribution System	BROMOFORM	0.5	UG/L	06-May-03	CWS	SW	35,598
WI	WI2413515	NORTH SHORE WATER COMMISSION	Distribution System	BROMOFORM	0.6	UG/L	08-Oct-03	CWS	SW	35,598
WI	WI2413515	NORTH SHORE WATER COMMISSION	Distribution System	BROMOFORM	0.7	UG/L	03-Sep-03	CWS	SW	35,598
WI	WI2410100	MILWAUKEE WATERWORKS	Distribution System	BROMOFORM	0.4	UG/L	14-Nov-94	CWS	SW	682,332
WI	WI2413515	NORTH SHORE WATER COMMISSION	Distribution System	BROMOFORM	0.5	UG/L	06-May-03	CWS	SW	35,598
WI	WI2413515	NORTH SHORE WATER COMMISSION	Distribution System	BROMOFORM	0.2	UG/L	22-Oct-96	CWS	SW	35,598
WI	WI2410100	MILWAUKEE WATERWORKS	Distribution System	BROMOFORM	0.1	UG/L	18-Jul-94	CWS	SW	682,332
WI	WI2413515	NORTH SHORE WATER COMMISSION	Distribution System	BROMOFORM	0.3	UG/L	08-May-00	CWS	SW	35,598
WI	WI2413515	NORTH SHORE WATER COMMISSION	Distribution System	BROMOFORM	0.2	UG/L	21-Jul-97	CWS	SW	35,598
WI	WI2413515	NORTH SHORE WATER COMMISSION	Entry Point	BROMOFORM	0.2	UG/L	27-Apr-95	CWS	SW	35,598
WI	WI2413515	NORTH SHORE WATER COMMISSION	Distribution System	BROMOFORM	0.5	UG/L	18-Nov-02	CWS	SW	35,598
WI	WI2413515	NORTH SHORE WATER COMMISSION	Entry Point	BROMOFORM	0.2	UG/L	03-Mar-99	CWS	SW	35,598
WI	WI2410100	MILWAUKEE WATERWORKS	Entry Point	BROMOFORM	0.1	UG/L	26-Feb-96	CWS	SW	682,332
WI	WI2410100	MILWAUKEE WATERWORKS	Distribution System	BROMOFORM	0.1	UG/L	19-Apr-95	CWS	SW	682,332
WI	WI2410100	MILWAUKEE WATERWORKS	Distribution System	BROMOFORM	1.3	UG/L	14-May-02	CWS	SW	682,332
WI	WI2410100	MILWAUKEE WATERWORKS	Distribution System	BROMOFORM	1.3	UG/L	09-Feb-98	CWS	SW	682,332
WI	WI2413515	NORTH SHORE WATER COMMISSION	Entry Point	BROMOFORM	0.2	UG/L	02-Mar-99	CWS	SW	35,598
WI	WI2410100	MILWAUKEE WATERWORKS	Distribution System	BROMOFORM	1.1	UG/L	14-May-02	CWS	SW	682,332
WI	WI2413515	NORTH SHORE WATER COMMISSION	Distribution System	BROMOFORM	0.4	UG/L	04-Aug-98	CWS	SW	35,598
WI	WI2413515	NORTH SHORE WATER COMMISSION	Entry Point	BROMOFORM	0.2	UG/L	25-May-94	CWS	SW	35,598
WI	WI2413515	NORTH SHORE WATER COMMISSION	Entry Point	BROMOFORM	0.3	UG/L	22-Jan-96	CWS	SW	35,598
WI	WI2413515	NORTH SHORE WATER COMMISSION	Entry Point	BROMOFORM	0.3	UG/L	17-Jan-05	CWS	SW	35,598
WI	WI2410100	MILWAUKEE WATERWORKS	Entry Point	BROMOFORM	0.1	UG/L	09-Mar-98	CWS	SW	682,332
WI	WI2410100	MILWAUKEE WATERWORKS	Entry Point	BROMOFORM	0.3	UG/L	08-Mar-93	CWS	SW	682,332
WI	WI2413515	NORTH SHORE WATER COMMISSION	Distribution System	BROMOFORM	0.4	UG/L	19-Oct-98	CWS	SW	35,598
WI	WI2413515	NORTH SHORE WATER COMMISSION	Distribution System	BROMOFORM	0.4	UG/L	31-Oct-00	CWS	SW	35,598
WI	WI2410100	MILWAUKEE WATERWORKS	Distribution System	BROMOFORM	0.1	UG/L	27-Apr-93	CWS	SW	682,332
WI	WI2410100	MILWAUKEE WATERWORKS	Entry Point	BROMOFORM	0.2	UG/L	06-Mar-95	CWS	SW	682,332
WI	WI2410100	MILWAUKEE WATERWORKS	Distribution System	BROMOFORM	0.2	UG/L	12-Feb-96	CWS	SW	682,332
WI	WI2410100	MILWAUKEE WATERWORKS	Distribution System	BROMOFORM	0.2	UG/L	24-Jan-94	CWS	SW	682,332
WI	WI2410100	MILWAUKEE WATERWORKS	Distribution System	BROMOFORM	0.1	UG/L	14-Aug-96	CWS	SW	682,332
WI	WI2410100	MILWAUKEE WATERWORKS	Distribution System	BROMOFORM	0.4	UG/L	27-Aug-96	CWS	SW	682,332
WI	WI2410100	MILWAUKEE WATERWORKS	Entry Point	BROMOFORM	0.2	UG/L	15-Jun-99	CWS	SW	682,332
WI	WI2410100	MILWAUKEE WATERWORKS	Distribution System	BROMOFORM	9	UG/L	13-Nov-00	CWS	SW	682,332
WI	WI2410100	MILWAUKEE WATERWORKS	Distribution System	BROMOFORM	1	UG/L	17-Apr-00	CWS	SW	682,332

WI	WI2410100	MILWAUKEE WATERWORKS	Distribution System	BROMOFORM	0.2	UG/L	06-Nov-95	CWS	SW	682,332
WI	WI2410100	MILWAUKEE WATERWORKS	Distribution System	BROMOFORM	1.3	UG/L	12-Nov-02	CWS	SW	682,332
WI	WI2410100	MILWAUKEE WATERWORKS	Distribution System	BROMOFORM	0.2	UG/L	18-Nov-93	CWS	SW	682,332
WI	WI2410100	MILWAUKEE WATERWORKS	Distribution System	BROMOFORM	0.1	UG/L	15-Aug-95	CWS	SW	682,332
WI	WI2410100	MILWAUKEE WATERWORKS	Distribution System	BROMOFORM	0.1	UG/L	17-Jun-96	CWS	SW	682,332
WI	WI2410100	MILWAUKEE WATERWORKS	Distribution System	BROMOFORM	0.1	UG/L	09-Feb-94	CWS	SW	682,332
WI	WI2410100	MILWAUKEE WATERWORKS	Distribution System	BROMOFORM	1.1	UG/L	14-May-02	CWS	SW	682,332
WI	WI2413515	NORTH SHORE WATER COMMISSION	Distribution System	BROMOFORM	0.3	UG/L	21-Feb-95	CWS	SW	35,598
WI	WI2410100	MILWAUKEE WATERWORKS	Distribution System	BROMOFORM	0.2	UG/L	21-Oct-96	CWS	SW	682,332
					0.49					
WI	WI2410100	MILWAUKEE WATERWORKS	Entry Point	CARBON TETRACHLORIDE	0.1	UG/L	12-Feb-96	CWS	SW	682,332
WI	WI2410100	MILWAUKEE WATERWORKS-428 SAMPLES	Entry Point	CHLOROFORM	7.82	UG/L	14-Feb-05	CWS	SW	682,332
WI	WI2460112	WE ENERGIES 1	Entry Point	CHLOROMETHANE (METHYLCHLORIDE)	0.15	UG/L	31-Jul-96	CWS	SWP	5,470
WI	WI2410100	MILWAUKEE WATERWORKS	Entry Point	CHLOROMETHANE (METHYLCHLORIDE)	1.1	UG/L	19-Aug-02	CWS	SW	682,332
					0.63					
WI	WI2413515	NORTH SHORE WATER COMMISSION	Distribution System	DIBROMOACETIC ACID	1.1	UG/L	12-Mar-02	CWS	SW	35,598
WI	WI2413515	NORTH SHORE WATER COMMISSION	Distribution System	DIBROMOACETIC ACID	1.1	UG/L	12-Mar-02	CWS	SW	35,598
WI	WI2413515	NORTH SHORE WATER COMMISSION	Distribution System	DIBROMOACETIC ACID	1	UG/L	12-Mar-02	CWS	SW	35,598
					1.05					
WI	WI2410100	MILWAUKEE WATERWORKS-400 SAMPLES	Distribution System	DIBROMOCHLOROMETHANE	2.44	UG/L	19-Feb-02	CWS	SW	682,332
WI	WI2413515	NORTH SHORE WATER COMMISSION	Distribution System	DICHLOROACETIC ACID	1.9	UG/L	08-Oct-03	CWS	SW	35,598
WI	WI2413515	NORTH SHORE WATER COMMISSION	Distribution System	DICHLOROACETIC ACID	2.2	UG/L	08-Oct-03	CWS	SW	35,598
WI	WI2413515	NORTH SHORE WATER COMMISSION	Distribution System	DICHLOROACETIC ACID	4.1	UG/L	13-Oct-04	CWS	SW	35,598
WI	WI2413515	NORTH SHORE WATER COMMISSION	Distribution System	DICHLOROACETIC ACID	1.6	UG/L	03-Sep-03	CWS	SW	35,598
WI	WI2413515	NORTH SHORE WATER COMMISSION	Distribution System	DICHLOROACETIC ACID	3.2	UG/L	19-Jan-04	CWS	SW	35,598
WI	WI2413515	NORTH SHORE WATER COMMISSION	Distribution System	DICHLOROACETIC ACID	1.4	UG/L	03-Sep-03	CWS	SW	35,598
WI	WI2413515	NORTH SHORE WATER COMMISSION	Distribution System	DICHLOROACETIC ACID	1.8	UG/L	28-Jan-03	CWS	SW	35,598
WI	WI2413515	NORTH SHORE WATER COMMISSION	Distribution System	DICHLOROACETIC ACID	2.4	UG/L	28-Jan-03	CWS	SW	35,598
WI	WI2413515	NORTH SHORE WATER COMMISSION	Distribution System	DICHLOROACETIC ACID	4.7	UG/L	05-Apr-04	CWS	SW	35,598
WI	WI2413515	NORTH SHORE WATER COMMISSION	Distribution System	DICHLOROACETIC ACID	1.4	UG/L	08-Oct-03	CWS	SW	35,598
WI	WI2413515	NORTH SHORE WATER COMMISSION	Distribution System	DICHLOROACETIC ACID	1.6	UG/L	03-Sep-03	CWS	SW	35,598
WI	WI2413515	NORTH SHORE WATER COMMISSION	Distribution System	DICHLOROACETIC ACID	5.2	UG/L	17-Jan-05	CWS	SW	35,598
WI	WI2413515	NORTH SHORE WATER COMMISSION	Distribution System	DICHLOROACETIC ACID	7.4	UG/L	18-Apr-02	CWS	SW	35,598
WI	WI2413515	NORTH SHORE WATER COMMISSION	Distribution System	DICHLOROACETIC ACID	7.3	UG/L	04-Apr-05	CWS	SW	35,598
WI	WI2413515	NORTH SHORE WATER COMMISSION	Distribution System	DICHLOROACETIC ACID	2.6	UG/L	24-Jul-02	CWS	SW	35,598
WI	WI2413515	NORTH SHORE WATER COMMISSION	Distribution System	DICHLOROACETIC ACID	2.2	UG/L	28-Jan-03	CWS	SW	35,598
WI	WI2413515	NORTH SHORE WATER COMMISSION	Distribution System	DICHLOROACETIC ACID	1	UG/L	08-Oct-03	CWS	SW	35,598
WI	WI2413515	NORTH SHORE WATER COMMISSION	Distribution System	DICHLOROACETIC ACID	6.9	UG/L	18-Apr-02	CWS	SW	35,598
WI	WI2413515	NORTH SHORE WATER COMMISSION	Distribution System	DICHLOROACETIC ACID	6.3	UG/L	08-Jul-04	CWS	SW	35,598

APPENDIX 7-21 *Continued*

State	PWSID	SYSTEM	SAMPLE SITE	ANALYTE	CONC	UOM	DATE	PWSType	PSource	Pop
WI	WI2413515	NORTH SHORE WATER COMMISSION	Distribution System	DICHLOROACETIC ACID	4.1	UG/L	06-May-03	CWS	SW	35,598
WI	WI2413515	NORTH SHORE WATER COMMISSION	Distribution System	DICHLOROACETIC ACID	2	UG/L	12-Mar-02	CWS	SW	35,598
WI	WI2413515	NORTH SHORE WATER COMMISSION	Distribution System	DICHLOROACETIC ACID	1.3	UG/L	28-Jan-03	CWS	SW	35,598
WI	WI2413515	NORTH SHORE WATER COMMISSION	Distribution System	DICHLOROACETIC ACID	5.1	UG/L	06-May-03	CWS	SW	35,598
WI	WI2413515	NORTH SHORE WATER COMMISSION	Distribution System	DICHLOROACETIC ACID	3.1	UG/L	24-Jul-02	CWS	SW	35,598
WI	WI2413515	NORTH SHORE WATER COMMISSION	Distribution System	DICHLOROACETIC ACID	7.5	UG/L	18-Apr-02	CWS	SW	35,598
WI	WI2413515	NORTH SHORE WATER COMMISSION	Distribution System	DICHLOROACETIC ACID	2.7	UG/L	06-May-03	CWS	SW	35,598
WI	WI2413515	NORTH SHORE WATER COMMISSION	Distribution System	DICHLOROACETIC ACID	2.4	UG/L	12-Mar-02	CWS	SW	35,598
WI	WI2413515	NORTH SHORE WATER COMMISSION	Distribution System	DICHLOROACETIC ACID	2.5	UG/L	12-Mar-02	CWS	SW	35,598
WI	WI2413515	NORTH SHORE WATER COMMISSION	Distribution System	DICHLOROACETIC ACID	2.8	UG/L	24-Jul-02	CWS	SW	35,598
WI	WI2413515	NORTH SHORE WATER COMMISSION	Distribution System	DICHLOROACETIC ACID	4	UG/L	03-Sep-03	CWS	SW	35,598
WI	WI2413515	NORTH SHORE WATER COMMISSION	Distribution System	DICHLOROACETIC ACID	6.5	UG/L	24-Jul-02	CWS	SW	35,598
WI	WI2413515	NORTH SHORE WATER COMMISSION	Distribution System	DICHLOROACETIC ACID	3.3	UG/L	18-Nov-02	CWS	SW	35,598
WI	WI2413515	NORTH SHORE WATER COMMISSION	Distribution System	DICHLOROACETIC ACID	3.5	UG/L	18-Apr-02	CWS	SW	35,598
WI	WI2413515	NORTH SHORE WATER COMMISSION	Distribution System	DICHLOROACETIC ACID	3.4	UG/L	18-Nov-02	CWS	SW	35,598
WI	WI2413515	NORTH SHORE WATER COMMISSION	Distribution System	DICHLOROACETIC ACID	6.3	UG/L	06-May-03	CWS	SW	35,598
WI	WI2413515	NORTH SHORE WATER COMMISSION	Distribution System	DICHLOROACETIC ACID	4.1	UG/L	12-Mar-02	CWS	SW	35,598
WI	WI2410100	MILWAUKEE WATERWORKS	Distribution System	DICHLOROACETIC ACID	1.4	UG/L	15-Feb-05	CWS	SW	682,332
WI	WI2410100	MILWAUKEE WATERWORKS	Distribution System	DICHLOROACETIC ACID	1	UG/L	11-May-04	CWS	SW	682,332
WI	WI2410100	MILWAUKEE WATERWORKS	Distribution System	DICHLOROACETIC ACID	1	UG/L	11-May-04	CWS	SW	682,332
WI	WI2410100	MILWAUKEE WATERWORKS	Distribution System	DICHLOROACETIC ACID	1.3	UG/L	17-Aug-04	CWS	SW	682,332
WI	WI2410100	MILWAUKEE WATERWORKS	Distribution System	DICHLOROACETIC ACID	1.1	UG/L	19-Aug-03	CWS	SW	682,332
WI	WI2410100	MILWAUKEE WATERWORKS	Distribution System	DICHLOROACETIC ACID	1.6	UG/L	19-Aug-03	CWS	SW	682,332
WI	WI2410100	MILWAUKEE WATERWORKS	Distribution System	DICHLOROACETIC ACID	1	UG/L	17-Feb-04	CWS	SW	682,332
WI	WI2410100	MILWAUKEE WATERWORKS	Distribution System	DICHLOROACETIC ACID	1.5	UG/L	17-Aug-04	CWS	SW	682,332
WI	WI2410100	MILWAUKEE WATERWORKS	Distribution System	DICHLOROACETIC ACID	1	UG/L	11-May-04	CWS	SW	682,332
WI	WI2410100	MILWAUKEE WATERWORKS	Distribution System	DICHLOROACETIC ACID	1.2	UG/L	20-Aug-02	CWS	SW	682,332
WI	WI2410100	MILWAUKEE WATERWORKS	Distribution System	DICHLOROACETIC ACID	1.2	UG/L	20-Aug-02	CWS	SW	682,332
WI	WI2410100	MILWAUKEE WATERWORKS	Distribution System	DICHLOROACETIC ACID	1.6	UG/L	19-Aug-03	CWS	SW	682,332
WI	WI2410100	MILWAUKEE WATERWORKS	Distribution System	DICHLOROACETIC ACID	1.5	UG/L	18-Feb-03	CWS	SW	682,332
WI	WI2410100	MILWAUKEE WATERWORKS	Distribution System	DICHLOROACETIC ACID	1.7	UG/L	19-Aug-03	CWS	SW	682,332
WI	WI2410100	MILWAUKEE WATERWORKS	Distribution System	DICHLOROACETIC ACID	1.6	UG/L	19-Aug-03	CWS	SW	682,332
WI	WI2410100	MILWAUKEE WATERWORKS	Distribution System	DICHLOROACETIC ACID	1	UG/L	11-May-04	CWS	SW	682,332
WI	WI2410100	MILWAUKEE WATERWORKS	Distribution System	DICHLOROACETIC ACID	1.1	UG/L	17-Aug-04	CWS	SW	682,332
WI	WI2410100	MILWAUKEE WATERWORKS	Distribution System	DICHLOROACETIC ACID	1.7	UG/L	17-Aug-04	CWS	SW	682,332
WI	WI2410100	MILWAUKEE WATERWORKS	Distribution System	DICHLOROACETIC ACID	1.4	UG/L	09-Nov-04	CWS	SW	682,332
WI	WI2410100	MILWAUKEE WATERWORKS	Distribution System	DICHLOROACETIC ACID	2	UG/L	17-Aug-04	CWS	SW	682,332
WI	WI2410100	MILWAUKEE WATERWORKS	Distribution System	DICHLOROACETIC ACID	1.6	UG/L	15-Feb-05	CWS	SW	682,332
WI	WI2410100	MILWAUKEE WATERWORKS	Distribution System	DICHLOROACETIC ACID	1	UG/L	20-Aug-02	CWS	SW	682,332
WI	WI2410100	MILWAUKEE WATERWORKS	Distribution System	DICHLOROACETIC ACID	1.7	UG/L	15-Feb-05	CWS	SW	682,332

WI	WI2410100	MILWAUKEE WATERWORKS	Distribution System	DICHLOROACETIC ACID	1.6	UG/L	15-Feb-05	CWS	SW	682,332
WI	WI2410100	MILWAUKEE WATERWORKS	Distribution System	DICHLOROACETIC ACID	1.5	UG/L	15-Feb-05	CWS	SW	682,332
WI	WI2410100	MILWAUKEE WATERWORKS	Distribution System	DICHLOROACETIC ACID	1.7	UG/L	15-Feb-05	CWS	SW	682,332
WI	WI2410100	MILWAUKEE WATERWORKS	Distribution System	DICHLOROACETIC ACID	1.3	UG/L	17-Aug-04	CWS	SW	682,332
WI	WI2410100	MILWAUKEE WATERWORKS	Distribution System	DICHLOROACETIC ACID	1.6	UG/L	15-Feb-05	CWS	SW	682,332
WI	WI2410100	MILWAUKEE WATERWORKS	Distribution System	DICHLOROACETIC ACID	2.1	UG/L	17-Aug-04	CWS	SW	682,332
WI	WI2410100	MILWAUKEE WATERWORKS	Distribution System	DICHLOROACETIC ACID	1.2	UG/L	09-Nov-04	CWS	SW	682,332
WI	WI2410100	MILWAUKEE WATERWORKS	Distribution System	DICHLOROACETIC ACID	1.3	UG/L	09-Nov-04	CWS	SW	682,332
WI	WI2410100	MILWAUKEE WATERWORKS	Distribution System	DICHLOROACETIC ACID	1.2	UG/L	09-Nov-04	CWS	SW	682,332
WI	WI2410100	MILWAUKEE WATERWORKS	Distribution System	DICHLOROACETIC ACID	1.4	UG/L	09-Nov-04	CWS	SW	682,332
WI	WI2410100	MILWAUKEE WATERWORKS	Distribution System	DICHLOROACETIC ACID	1.3	UG/L	09-Nov-04	CWS	SW	682,332
WI	WI2410100	MILWAUKEE WATERWORKS	Distribution System	DICHLOROACETIC ACID	1.8	UG/L	18-Aug-04	CWS	SW	682,332
WI	WI2410100	MILWAUKEE WATERWORKS	Distribution System	DICHLOROACETIC ACID	1.5	UG/L	15-Feb-05	CWS	SW	682,332
					2.51					
WI	WI2460112	WE ENERGIES 1	Entry Point	DICHLOROMETHANE	1.14	UG/L	16-Dec-94	CWS	SWP	5,470
WI	WI2410100	MILWAUKEE WATERWORKS	Well	DICHLOROMETHANE	0.9	UG/L	17-Aug-89	CWS	SW	682,332
					1.02					
WI	WI2460112	WE ENERGIES 1	Entry Point	ETHYLBENZENE	0.4	UG/L	13-Jun-95	CWS	SWP	5,470
WI	WI2410100	MILWAUKEE WATERWORKS	Distribution System	HAA5	1.2	UG/L	09-Nov-04	CWS	SW	682,332
WI	WI2410100	MILWAUKEE WATERWORKS	Distribution System	HAA5	1.7	UG/L	15-Feb-05	CWS	SW	682,332
WI	WI2410100	MILWAUKEE WATERWORKS	Distribution System	HAA5	1.6	UG/L	15-Feb-05	CWS	SW	682,332
WI	WI2410100	MILWAUKEE WATERWORKS	Distribution System	HAA5	1.5	UG/L	15-Feb-05	CWS	SW	682,332
WI	WI2410100	MILWAUKEE WATERWORKS	Distribution System	HAA5	1.3	UG/L	09-Nov-04	CWS	SW	682,332
WI	WI2410100	MILWAUKEE WATERWORKS	Distribution System	HAA5	1.6	UG/L	15-Feb-05	CWS	SW	682,332
WI	WI2410100	MILWAUKEE WATERWORKS	Distribution System	HAA5	1.5	UG/L	15-Feb-05	CWS	SW	682,332
WI	WI2410100	MILWAUKEE WATERWORKS	Distribution System	HAA5	1.7	UG/L	15-Feb-05	CWS	SW	682,332
WI	WI2410100	MILWAUKEE WATERWORKS	Distribution System	HAA5	1.4	UG/L	15-Feb-05	CWS	SW	682,332
WI	WI2410100	MILWAUKEE WATERWORKS	Distribution System	HAA5	1.6	UG/L	15-Feb-05	CWS	SW	682,332
WI	WI2410100	MILWAUKEE WATERWORKS	Distribution System	HAA5	1.2	UG/L	09-Nov-04	CWS	SW	682,332
WI	WI2410100	MILWAUKEE WATERWORKS	Distribution System	HAA5	1.6	UG/L	19-Aug-03	CWS	SW	682,332
WI	WI2410100	MILWAUKEE WATERWORKS	Distribution System	HAA5	3.3	UG/L	19-Feb-02	CWS	SW	682,332
WI	WI2410100	MILWAUKEE WATERWORKS	Distribution System	HAA5	1	UG/L	11-May-04	CWS	SW	682,332
WI	WI2410100	MILWAUKEE WATERWORKS	Distribution System	HAA5	3	UG/L	19-Aug-03	CWS	SW	682,332
WI	WI2410100	MILWAUKEE WATERWORKS	Distribution System	HAA5	1	UG/L	11-May-04	CWS	SW	682,332
WI	WI2410100	MILWAUKEE WATERWORKS	Distribution System	HAA5	2.6	UG/L	19-Aug-03	CWS	SW	682,332
WI	WI2410100	MILWAUKEE WATERWORKS	Distribution System	HAA5	1	UG/L	17-Feb-04	CWS	SW	682,332
WI	WI2410100	MILWAUKEE WATERWORKS	Distribution System	HAA5	1	UG/L	20-Aug-02	CWS	SW	682,332
WI	WI2410100	MILWAUKEE WATERWORKS	Distribution System	HAA5	1	UG/L	11-May-04	CWS	SW	682,332
WI	WI2410100	MILWAUKEE WATERWORKS	Distribution System	HAA5	2.3	UG/L	17-Aug-04	CWS	SW	682,332
WI	WI2410100	MILWAUKEE WATERWORKS	Distribution System	HAA5	1.2	UG/L	20-Aug-02	CWS	SW	682,332
WI	WI2410100	MILWAUKEE WATERWORKS	Distribution System	HAA5	1.5	UG/L	18-Feb-03	CWS	SW	682,332
WI	WI2410100	MILWAUKEE WATERWORKS	Distribution System	HAA5	1.4	UG/L	09-Nov-04	CWS	SW	682,332
WI	WI2410100	MILWAUKEE WATERWORKS	Distribution System	HAA5	1	UG/L	11-May-04	CWS	SW	682,332

APPENDIX 7-21 *Continued*

State	PWSID	SYSTEM	SAMPLE SITE	ANALYTE	CONC	UOM	DATE	PWSType	PSource	Pop
WI	WI2410100	MILWAUKEE WATERWORKS	Distribution System	HAA5	1.2	UG/L	20-Aug-02	CWS	SW	682,332
WI	WI2410100	MILWAUKEE WATERWORKS	Distribution System	HAA5	3.1	UG/L	19-Aug-03	CWS	SW	682,332
WI	WI2410100	MILWAUKEE WATERWORKS	Distribution System	HAA5	2.9	UG/L	19-Aug-03	CWS	SW	682,332
WI	WI2410100	MILWAUKEE WATERWORKS	Distribution System	HAA5	1.7	UG/L	17-Aug-04	CWS	SW	682,332
WI	WI2410100	MILWAUKEE WATERWORKS	Distribution System	HAA5	3	UG/L	17-Aug-04	CWS	SW	682,332
WI	WI2410100	MILWAUKEE WATERWORKS	Distribution System	HAA5	1.5	UG/L	17-Aug-04	CWS	SW	682,332
WI	WI2410100	MILWAUKEE WATERWORKS	Distribution System	HAA5	3.1	UG/L	17-Aug-04	CWS	SW	682,332
WI	WI2410100	MILWAUKEE WATERWORKS	Distribution System	HAA5	2.4	UG/L	17-Aug-04	CWS	SW	682,332
WI	WI2410100	MILWAUKEE WATERWORKS	Distribution System	HAA5	1.8	UG/L	18-Aug-04	CWS	SW	682,332
WI	WI2410100	MILWAUKEE WATERWORKS	Distribution System	HAA5	1.4	UG/L	09-Nov-04	CWS	SW	682,332
WI	WI2410100	MILWAUKEE WATERWORKS	Distribution System	HAA5	1.3	UG/L	09-Nov-04	CWS	SW	682,332
WI	WI2413515	NORTH SHORE WATER COMMISSION	Distribution System	HAA5	3.5	UG/L	08-Oct-03	CWS	SW	35,598
WI	WI2413515	NORTH SHORE WATER COMMISSION	Distribution System	HAA5	7.1	UG/L	03-Sep-03	CWS	SW	35,598
WI	WI2413515	NORTH SHORE WATER COMMISSION	Distribution System	HAA5	5.8	UG/L	19-Jan-04	CWS	SW	35,598
WI	WI2413515	NORTH SHORE WATER COMMISSION	Distribution System	HAA5	16.6	UG/L	18-Apr-02	CWS	SW	35,598
WI	WI2413515	NORTH SHORE WATER COMMISSION	Distribution System	HAA5	6.8	UG/L	06-May-03	CWS	SW	35,598
WI	WI2413515	NORTH SHORE WATER COMMISSION	Distribution System	HAA5	6.3	UG/L	03-Sep-03	CWS	SW	35,598
WI	WI2413515	NORTH SHORE WATER COMMISSION	Distribution System	HAA5	10.7	UG/L	18-Apr-02	CWS	SW	35,598
WI	WI2413515	NORTH SHORE WATER COMMISSION	Distribution System	HAA5	4	UG/L	28-Jan-03	CWS	SW	35,598
WI	WI2413515	NORTH SHORE WATER COMMISSION	Distribution System	HAA5	7	UG/L	24-Jul-02	CWS	SW	35,598
WI	WI2413515	NORTH SHORE WATER COMMISSION	Distribution System	HAA5	3.6	UG/L	18-Nov-02	CWS	SW	35,598
WI	WI2413515	NORTH SHORE WATER COMMISSION	Distribution System	HAA5	6	UG/L	18-Nov-02	CWS	SW	35,598
WI	WI2413515	NORTH SHORE WATER COMMISSION	Distribution System	HAA5	4.7	UG/L	28-Jan-03	CWS	SW	35,598
WI	WI2413515	NORTH SHORE WATER COMMISSION	Distribution System	HAA5	5.3	UG/L	18-Nov-02	CWS	SW	35,598
WI	WI2413515	NORTH SHORE WATER COMMISSION	Distribution System	HAA5	5.4	UG/L	08-Oct-03	CWS	SW	35,598
WI	WI2413515	NORTH SHORE WATER COMMISSION	Distribution System	HAA5	3.5	UG/L	08-Oct-03	CWS	SW	35,598
WI	WI2413515	NORTH SHORE WATER COMMISSION	Distribution System	HAA5	13.3	UG/L	08-Jul-04	CWS	SW	35,598
WI	WI2410100	MILWAUKEE WATERWORKS	Distribution System	HAA5	1.1	UG/L	17-Aug-04	CWS	SW	682,332
WI	WI2413515	NORTH SHORE WATER COMMISSION	Distribution System	HAA5	6.1	UG/L	12-Mar-02	CWS	SW	35,598
WI	WI2413515	NORTH SHORE WATER COMMISSION	Distribution System	HAA5	8.1	UG/L	24-Jul-02	CWS	SW	35,598
WI	WI2413515	NORTH SHORE WATER COMMISSION	Distribution System	HAA5	7.3	UG/L	06-May-03	CWS	SW	35,598
WI	WI2413515	NORTH SHORE WATER COMMISSION	Distribution System	HAA5	11.7	UG/L	06-May-03	CWS	SW	35,598
WI	WI2413515	NORTH SHORE WATER COMMISSION	Distribution System	HAA5	6.8	UG/L	24-Jul-02	CWS	SW	35,598
WI	WI2413515	NORTH SHORE WATER COMMISSION	Distribution System	HAA5	5.1	UG/L	12-Mar-02	CWS	SW	35,598
WI	WI2413515	NORTH SHORE WATER COMMISSION	Distribution System	HAA5	19	UG/L	18-Apr-02	CWS	SW	35,598
WI	WI2413515	NORTH SHORE WATER COMMISSION	Distribution System	HAA5	2.9	UG/L	28-Jan-03	CWS	SW	35,598
WI	WI2413515	NORTH SHORE WATER COMMISSION	Distribution System	HAA5	7.3	UG/L	03-Sep-03	CWS	SW	35,598
WI	WI2413515	NORTH SHORE WATER COMMISSION	Distribution System	HAA5	9.2	UG/L	06-May-03	CWS	SW	35,598
WI	WI2413515	NORTH SHORE WATER COMMISSION	Distribution System	HAA5	6.4	UG/L	03-Sep-03	CWS	SW	35,598
WI	WI2413515	NORTH SHORE WATER COMMISSION	Distribution System	HAA5	5.6	UG/L	12-Mar-02	CWS	SW	35,598

WI	WI2413515	NORTH SHORE WATER COMMISSION	Distribution System	HAA5	3	UG/L	28-Jan-03	CWS	SW	35,598
WI	WI2413515	NORTH SHORE WATER COMMISSION	Distribution System	HAA5	14.5	UG/L	18-Apr-02	CWS	SW	35,598
WI	WI2413515	NORTH SHORE WATER COMMISSION	Distribution System	HAA5	5.3	UG/L	08-Oct-03	CWS	SW	35,598
WI	WI2413515	NORTH SHORE WATER COMMISSION	Distribution System	HAA5	10.5	UG/L	24-Jul-02	CWS	SW	35,598
WI	WI2413515	NORTH SHORE WATER COMMISSION	Distribution System	HAA5	8.61	UG/L	17-Jan-05	CWS	SW	35,598
WI	WI2413515	NORTH SHORE WATER COMMISSION	Distribution System	HAA5	7.7	UG/L	13-Oct-04	CWS	SW	35,598
WI	WI2413515	NORTH SHORE WATER COMMISSION	Distribution System	HAA5	10.7	UG/L	04-Apr-05	CWS	SW	35,598
WI	WI2413515	NORTH SHORE WATER COMMISSION	Distribution System	HAA5	8.2	UG/L	05-Apr-04	CWS	SW	35,598
WI	WI2413515	NORTH SHORE WATER COMMISSION	Distribution System	HAA5	5.7	UG/L	18-Nov-02	CWS	SW	35,598
WI	WI2413515	NORTH SHORE WATER COMMISSION	Distribution System	HAA5	8.4	UG/L	12-Mar-02	CWS	SW	35,598
					4.69					
WI	WI2413515	NORTH SHORE WATER COMMISSION	Distribution System	MONOBROMOACETIC ACID	0.91	UG/L	17-Jan-05	CWS	SW	35,598
WI	WI2413515	NORTH SHORE WATER COMMISSION	Distribution System	MONOCHLOROACETIC ACID	2.3	UG/L	18-Apr-02	CWS	SW	35,598
WI	WI2410100	MILWAUKEE WATERWORKS	Distribution System	MONOCHLOROACETIC ACID	3.3	UG/L	19-Feb-02	CWS	SW	682,332
WI	WI2413515	NORTH SHORE WATER COMMISSION	Distribution System	MONOCHLOROACETIC ACID	3.1	UG/L	18-Nov-02	CWS	SW	35,598
WI	WI2413515	NORTH SHORE WATER COMMISSION	Distribution System	MONOCHLOROACETIC ACID	5.1	UG/L	18-Apr-02	CWS	SW	35,598
					2.34					
WI	WI2460112	WE ENERGIES 1	Entry Point	TETRACHLOROETHYLENE	0.3	UG/L	13-Jun-95	CWS	SWP	5,470
WI	WI2410100	MILWAUKEE WATERWORKS	Entry Point	TOLUENE	0.5	UG/L	08-Mar-93	CWS	SW	682,332
WI	WI2460112	WE ENERGIES 1	Entry Point	TOLUENE	0.6	UG/L	30-Mar-95	CWS	SWP	5,470
WI	WI2460112	WE ENERGIES 1	Entry Point	TOLUENE	2.2	UG/L	13-Jun-95	CWS	SWP	5,470
					1.1					
WI	WI2410100	MILWAUKEE WATERWORKS-469 SAMPLES	Distribution System	TTHM	14.1	UG/L	29-Oct-90	CWS	SW	682,332
WI	WI2460112	WE ENERGIES 1	Entry Point	XYLENES, TOTAL	1.3	UG/L	13-Jun-95	CWS	SWP	5,470
WI	WI2410168	GLENDALE WATERWORKS	Well	XYLENES, TOTAL	0.756	UG/L	19-Sep-89	CWS	SWP	14,088
					1.03					

APPENDIX 7-22 New York Groundwater Data

State	PWSID	SYSTEM	SAMPLE SITE	ANALYTE	CONC	UOM	DATE	PWSType	PSource	Pop
NY	NY3920885	MOUNT EBO WS		METHYL TERT BUTYL ETHER (MTBE)	1.7	UG/L	14-Jan-02	CWS	GW	1500
NY	NY3921970	VERIZON		METHYL TERT BUTYL ETHER (MTBE)	1.1	UG/L	11-Mar-05	TNCWS	GW	25

APPENDIX 7-23 Oklahoma City Groundwater Data

State	PWSID	SYSTEM	SAMPLE SITE	ANALYTE	CONC	UOM	DATE	PWSType	Pop
OK	OK2005501	NICHOLS HILLS	WELL 1B	BENZENE	1.2	UG/L	19-Apr-05	CWS	4056
OK	OK2005524	APPLEWOOD MHP	WELL 1	BROMODICHLOROMETHANE	3.8	UG/L	24-Aug-04	CWS	62
OK	OK2005524	APPLEWOOD MHP	WELL 1	CHLORODIBROMOMETHANE	5	UG/L	24-Aug-04	CWS	62
OK	OK4005570	INDEPENDENT WATER CORP	WELL 1	CHLORODIBROMOMETHANE	0.5 2.75	UG/L	04-Apr-01	CWS	125
OK	OK2005501	NICHOLS HILLS	WELL 7	CHLOROFORM	0.5	UG/L	15-Oct-03	CWS	4056
OK	OK2005524	APPLEWOOD MHP	WELL 1	CHLOROFORM	9.9	UG/L	24-Aug-04	CWS	62
OK	OK4005570	INDEPENDENT WATER CORP	WELL 1	CHLOROFORM	0.7	UG/L	10-Oct-00	CWS	125
OK	OK4005570	INDEPENDENT WATER CORP	WELL 1	CHLOROFORM	0.73	UG/L	04-Apr-01	CWS	125
OK	OK4005570	INDEPENDENT WATER CORP	WELL 1	CHLOROFORM	0.63 2.49	UG/L	15-Jun-01	CWS	125
OK	OK2005501	NICHOLS HILLS	WELL 1B	TOLUENE	0.7	UG/L	19-Apr-05	CWS	4056

APPENDIX 7-24 Oklahoma City Surface Water Data

State	PWSID	SYSTEM	SAMPLE SITE	ANALYTE	CONC	UOM	DATE	PWSType	Pop
OK	OK1020502	OKLAHOMA CITY OVERHOLSER	OKC OVERHOLSER WTP	BROMODICHLOROMETHANE	17.2	UG/L	28-Jul-00	CWS	46,000
OK	OK1020502	OKLAHOMA CITY OVERHOLSER	OKC OVERHOLSER WTP	BROMODICHLOROMETHANE	38	UG/L	23-Aug-01	CWS	46,000
OK	OK1020502	OKLAHOMA CITY OVERHOLSER	OKC OVERHOLSER WTP	BROMODICHLOROMETHANE	36	UG/L	23-Jul-02	CWS	46,000
OK	OK1020502	OKLAHOMA CITY OVERHOLSER	OKC OVERHOLSER WTP	BROMODICHLOROMETHANE	24	UG/L	23-Jul-03	CWS	46,000
OK	OK1020502	OKLAHOMA CITY OVERHOLSER	OKC OVERHOLSER WTP	BROMODICHLOROMETHANE	27	UG/L	18-Jun-04	CWS	46,000
OK	OK1020802	OKLAHOMA CITY DRAPER	OKC DRAPER WTP	BROMODICHLOROMETHANE	10.4	UG/L	16-Apr-02	CWS	276,000
OK	OK1020902	OKLAHOMA CITY HEFNER	OKC HEFNER WTP	BROMODICHLOROMETHANE	0.5	UG/L	13-Dec-02	CWS	276,000
					21.9				
OK	OK1020502	OKLAHOMA CITY OVERHOLSER	OKC OVERHOLSER WTP	CHLORODIBROMOMETHANE	4.28	UG/L	28-Jul-00	CWS	46,000
OK	OK1020502	OKLAHOMA CITY OVERHOLSER	OKC OVERHOLSER WTP	CHLORODIBROMOMETHANE	28.3	UG/L	23-Aug-01	CWS	46,000
OK	OK1020502	OKLAHOMA CITY OVERHOLSER	OKC OVERHOLSER WTP	CHLORODIBROMOMETHANE	25	UG/L	23-Jul-02	CWS	46,000
OK	OK1020502	OKLAHOMA CITY OVERHOLSER	OKC OVERHOLSER WTP	CHLORODIBROMOMETHANE	13.7	UG/L	23-Jul-03	CWS	46,000
OK	OK1020502	OKLAHOMA CITY OVERHOLSER	OKC OVERHOLSER WTP	CHLORODIBROMOMETHANE	14.5	UG/L	18-Jun-04	CWS	46,000
OK	OK1020802	OKLAHOMA CITY DRAPER	OKC DRAPER WTP	CHLORODIBROMOMETHANE	1.5	UG/L	16-Apr-02	CWS	276,000
					14.5				
OK	OK1020502	OKLAHOMA CITY OVERHOLSER	OKC OVERHOLSER WTP	CHLOROFORM	40	UG/L	28-Jul-00	CWS	46,000
OK	OK1020502	OKLAHOMA CITY OVERHOLSER	OKC OVERHOLSER WTP	CHLOROFORM	80	UG/L	23-Aug-01	CWS	46,000
OK	OK1020502	OKLAHOMA CITY OVERHOLSER	OKC OVERHOLSER WTP	CHLOROFORM	84	UG/L	23-Jul-02	CWS	46,000
OK	OK1020502	OKLAHOMA CITY OVERHOLSER	OKC OVERHOLSER WTP	CHLOROFORM	34	UG/L	23-Jul-03	CWS	46,000
OK	OK1020502	OKLAHOMA CITY OVERHOLSER	OKC OVERHOLSER WTP	CHLOROFORM	51	UG/L	18-Jun-04	CWS	46,000
OK	OK1020802	OKLAHOMA CITY DRAPER	OKC DRAPER WTP	CHLOROFORM	29.1	UG/L	16-Apr-02	CWS	276,000
OK	OK1020902	OKLAHOMA CITY HEFNER	OKC HEFNER WTP	CHLOROFORM	1	UG/L	13-Dec-02	CWS	276,000
					45.6				

APPENDIX 7-25 Orlando Groundwater Data

State	PWSID	NAME	Expr1	CODE_DESC	RESULTS	UNIT_OF_MEASURE	SAMPLE_DATE	PWSType	Pop
FL	FL3484250	ROBERT'S BUSINESS CENTER		CIS-1 2-DICHLOROETHENE	0.38	UG/L	14-Dec-04	NTNCWS	25
FL	FL3484132	OCUD/EASTERN WATER SYSTEM (3 WPS)		DI(2-ETHYLHEXYL)PHTHALATE	0.49	UG/L	02-Aug-04	CWS	138,674

APPENDIX 7-26 Phoenix Groundwater Data

State	PWSID	Water System Name	Analyte	Level	Unit	Sample Date	PWSType	Pop
AZ	AZ0407487	UNIVERSAL PROPULSION INC	1,1-DICHLOROETHYLENE	3.3	ug/L	13-May-05	NTNCWS	285
AZ	AZ0407695	AZ AMERICAN WTR CO — AGUA FRIA	ETHYLBENZENE	0.7	ug/L	02-May-05	CWS	34,000
AZ	AZ0407695	AZ AMERICAN WTR CO — AGUA FRIA	ETHYLBENZENE	0.7	ug/L	02-May-05	CWS	34,000
AZ	AZ0407695	AZ AMERICAN WTR CO — AGUA FRIA	ETHYLBENZENE	0.7	ug/L	02-May-05	CWS	34,000
AZ	AZ0407695	AZ AMERICAN WTR CO — AGUA FRIA	ETHYLBENZENE	0.7	ug/L	02-May-05	CWS	34,000
AZ	AZ0407695	AZ AMERICAN WTR CO — AGUA FRIA	ETHYLBENZENE	0.7	ug/L	02-May-05	CWS	34,000
AZ	AZ0407695	AZ AMERICAN WTR CO — AGUA FRIA	ETHYLBENZENE	0.7	ug/L	02-May-05	CWS	34,000
AZ	AZ0407695	AZ AMERICAN WTR CO — AGUA FRIA	ETHYLBENZENE	0.7	ug/L	02-May-05	CWS	34,000
AZ	AZ0407695	AZ AMERICAN WTR CO — AGUA FRIA	ETHYLBENZENE	0.7	ug/L	02-May-05	CWS	34,000
AZ	AZ0407695	AZ AMERICAN WTR CO — AGUA FRIA	ETHYLBENZENE	0.7	ug/L	02-May-05	CWS	34,000
AZ	AZ0407695	AZ AMERICAN WTR CO — AGUA FRIA	ETHYLBENZENE	0.7	ug/L	02-May-05	CWS	34,000
AZ	AZ0407695	AZ AMERICAN WTR CO — AGUA FRIA	ETHYLBENZENE	0.7	ug/L	02-May-05	CWS	34,000
AZ	AZ0407695	AZ AMERICAN WTR CO — AGUA FRIA	ETHYLBENZENE	0.7	ug/L	02-May-05	CWS	34,000
AZ	AZ0407695	AZ AMERICAN WTR CO — AGUA FRIA	ETHYLBENZENE	0.7	ug/L	02-May-05	CWS	34,000
AZ	AZ0407695	AZ AMERICAN WTR CO — AGUA FRIA	ETHYLBENZENE	0.7	ug/L	02-May-05	CWS	34,000
AZ	AZ0407695	AZ AMERICAN WTR CO — AGUA FRIA	ETHYLBENZENE	0.7	ug/L	02-May-05	CWS	34,000
AZ	AZ0407695	AZ AMERICAN WTR CO — AGUA FRIA	ETHYLBENZENE	0.7	ug/L	02-May-05	CWS	34,000
AZ	AZ0407695	AZ AMERICAN WTR CO — AGUA FRIA	ETHYLBENZENE	0.7	ug/L	02-May-05	CWS	34,000
AZ	AZ0407695	AZ AMERICAN WTR CO — AGUA FRIA	ETHYLBENZENE	0.7	ug/L	02-May-05	CWS	34,000
AZ	AZ0407695	AZ AMERICAN WTR CO — AGUA FRIA	ETHYLBENZENE	0.7	ug/L	02-May-05	CWS	34,000
AZ	AZ0407695	AZ AMERICAN WTR CO — AGUA FRIA	ETHYLBENZENE	0.7	ug/L	02-May-05	CWS	34,000
AZ	AZ0407511	BEARDSLEY WC #3	ETHYLBENZENE	2.7	ug/L	08-Sep-04	CWS	60
AZ	AZ0407511	BEARDSLEY WC #3	ETHYLBENZENE	1.7	ug/L	02-Jun-05	CWS	60
				1.7				
AZ	AZ0407740	REDHAWK POTABLE WS	TOLUENE	1.2	ug/L	08-Apr-05	NTNCWS	25
AZ	AZ0407511	BEARDSLEY WC #3	1,2,4-TRICHLOROBENZENE	0.94	ug/L	02-Jun-05	CWS	60
AZ	AZ0408015	HAVASU WATER	XYLENES	0.5	ug/L	14-Feb-05	CWS	3,440
AZ	AZ0408015	HAVASU WATER	XYLENES	0.5	ug/L	14-Feb-05	CWS	3,440
AZ	AZ0408015	HAVASU WATER	XYLENES	0.5	ug/L	14-Feb-05	CWS	3,440
AZ	AZ0408015	HAVASU WATER	XYLENES	0.5	ug/L	14-Feb-05	CWS	3,440
AZ	AZ0408015	HAVASU WATER	XYLENES	0.5	ug/L	14-Feb-05	CWS	3,440
AZ	AZ0408015	HAVASU WATER	XYLENES	0.5	ug/L	14-Feb-05	CWS	3,440
AZ	AZ0408015	HAVASU WATER	XYLENES	0.5	ug/L	14-Feb-05	CWS	3,440
AZ	AZ0408015	HAVASU WATER	XYLENES	0.5	ug/L	14-Feb-05	CWS	3,440
AZ	AZ0408015	HAVASU WATER	XYLENES	0.5	ug/L	14-Feb-05	CWS	3,440
AZ	AZ0408015	HAVASU WATER	XYLENES	0.9	ug/L	14-Feb-05	CWS	3,440
AZ	AZ0408015	HAVASU WATER	XYLENES	0.9	ug/L	14-Feb-05	CWS	3,440
AZ	AZ0408015	HAVASU WATER	XYLENES	0.9	ug/L	14-Feb-05	CWS	3,440
AZ	AZ0408015	HAVASU WATER	XYLENES	0.9	ug/L	14-Feb-05	CWS	3,440
AZ	AZ0408015	HAVASU WATER	XYLENES	0.9	ug/L	14-Feb-05	CWS	3,440
AZ	AZ0408015	HAVASU WATER	XYLENES	0.9	ug/L	14-Feb-05	CWS	3,440
AZ	AZ0408015	HAVASU WATER	XYLENES	0.9	ug/L	14-Feb-05	CWS	3,440

AZ	AZ0408015	HAVASU WATER	XYLENES	0.9	ug/L	14-Feb-05	CWS	3,440
AZ	AZ0408015	HAVASU WATER	XYLENES	0.9	ug/L	14-Feb-05	CWS	3,440
AZ	AZ0407740	REDHAWK POTABLE WS	XYLENES	4.4	ug/L	08-Apr-05	NTNCWS	25
AZ	AZ0407695	AZ AMERICAN WTR CO — AGUA FRIA	XYLENES	1.3	ug/L	15-Feb-05	CWS	34,000
AZ	AZ0407695	AZ AMERICAN WTR CO — AGUA FRIA	XYLENES	1.3	ug/L	15-Feb-05	CWS	34,000
AZ	AZ0407695	AZ AMERICAN WTR CO — AGUA FRIA	XYLENES	1.3	ug/L	15-Feb-05	CWS	34,000
AZ	AZ0407695	AZ AMERICAN WTR CO — AGUA FRIA	XYLENES	1.3	ug/L	15-Feb-05	CWS	34,000
AZ	AZ0407695	AZ AMERICAN WTR CO — AGUA FRIA	XYLENES	1.3	ug/L	15-Feb-05	CWS	34,000
AZ	AZ0407695	AZ AMERICAN WTR CO — AGUA FRIA	XYLENES	1.3	ug/L	15-Feb-05	CWS	34,000
AZ	AZ0407695	AZ AMERICAN WTR CO — AGUA FRIA	XYLENES	1.3	ug/L	15-Feb-05	CWS	34,000
AZ	AZ0407695	AZ AMERICAN WTR CO — AGUA FRIA	XYLENES	1.3	ug/L	15-Feb-05	CWS	34,000
AZ	AZ0407695	AZ AMERICAN WTR CO — AGUA FRIA	XYLENES	1.3	ug/L	15-Feb-05	CWS	34,000
AZ	AZ0407695	AZ AMERICAN WTR CO — AGUA FRIA	XYLENES	1.3	ug/L	15-Feb-05	CWS	34,000
AZ	AZ0407695	AZ AMERICAN WTR CO — AGUA FRIA	XYLENES	1.3	ug/L	15-Feb-05	CWS	34,000
AZ	AZ0407695	AZ AMERICAN WTR CO — AGUA FRIA	XYLENES	1.3	ug/L	15-Feb-05	CWS	34,000
AZ	AZ0407695	AZ AMERICAN WTR CO — AGUA FRIA	XYLENES	1.3	ug/L	15-Feb-05	CWS	34,000
AZ	AZ0407695	AZ AMERICAN WTR CO — AGUA FRIA	XYLENES	1.3	ug/L	15-Feb-05	CWS	34,000
AZ	AZ0407695	AZ AMERICAN WTR CO — AGUA FRIA	XYLENES	1.3	ug/L	15-Feb-05	CWS	34,000
AZ	AZ0407695	AZ AMERICAN WTR CO — AGUA FRIA	XYLENES	1.3	ug/L	15-Feb-05	CWS	34,000
AZ	AZ0407695	AZ AMERICAN WTR CO — AGUA FRIA	XYLENES	1.3	ug/L	15-Feb-05	CWS	34,000
AZ	AZ0407695	AZ AMERICAN WTR CO — AGUA FRIA	XYLENES	1.3	ug/L	15-Feb-05	CWS	34,000
AZ	AZ0407695	AZ AMERICAN WTR CO — AGUA FRIA	XYLENES	1.3	ug/L	15-Feb-05	CWS	34,000
AZ	AZ0407695	AZ AMERICAN WTR CO — AGUA FRIA	XYLENES	1.3	ug/L	15-Feb-05	CWS	34,000
AZ	AZ0407695	AZ AMERICAN WTR CO — AGUA FRIA	XYLENES	6	ug/L	02-May-05	CWS	34,000
AZ	AZ0407695	AZ AMERICAN WTR CO — AGUA FRIA	XYLENES	6	ug/L	02-May-05	CWS	34,000
AZ	AZ0407695	AZ AMERICAN WTR CO — AGUA FRIA	XYLENES	6	ug/L	02-May-05	CWS	34,000
AZ	AZ0407695	AZ AMERICAN WTR CO — AGUA FRIA	XYLENES	6	ug/L	02-May-05	CWS	34,000
AZ	AZ0407695	AZ AMERICAN WTR CO — AGUA FRIA	XYLENES	6	ug/L	02-May-05	CWS	34,000
AZ	AZ0407695	AZ AMERICAN WTR CO — AGUA FRIA	XYLENES	6	ug/L	02-May-05	CWS	34,000
AZ	AZ0407695	AZ AMERICAN WTR CO — AGUA FRIA	XYLENES	6	ug/L	02-May-05	CWS	34,000
AZ	AZ0407695	AZ AMERICAN WTR CO — AGUA FRIA	XYLENES	6	ug/L	02-May-05	CWS	34,000
AZ	AZ0407695	AZ AMERICAN WTR CO — AGUA FRIA	XYLENES	6	ug/L	02-May-05	CWS	34,000
AZ	AZ0407695	AZ AMERICAN WTR CO — AGUA FRIA	XYLENES	6	ug/L	02-May-05	CWS	34,000
AZ	AZ0407695	AZ AMERICAN WTR CO — AGUA FRIA	XYLENES	6	ug/L	02-May-05	CWS	34,000
AZ	AZ0407695	AZ AMERICAN WTR CO — AGUA FRIA	XYLENES	6	ug/L	02-May-05	CWS	34,000
AZ	AZ0407695	AZ AMERICAN WTR CO — AGUA FRIA	XYLENES	6	ug/L	02-May-05	CWS	34,000
AZ	AZ0407695	AZ AMERICAN WTR CO — AGUA FRIA	XYLENES	6	ug/L	02-May-05	CWS	34,000
AZ	AZ0407695	AZ AMERICAN WTR CO — AGUA FRIA	XYLENES	6	ug/L	02-May-05	CWS	34,000
AZ	AZ0407695	AZ AMERICAN WTR CO — AGUA FRIA	XYLENES	6	ug/L	02-May-05	CWS	34,000
AZ	AZ0407695	AZ AMERICAN WTR CO — AGUA FRIA	XYLENES	6	ug/L	02-May-05	CWS	34,000
AZ	AZ0407695	AZ AMERICAN WTR CO — AGUA FRIA	XYLENES	6	ug/L	02-May-05	CWS	34,000
AZ	AZ0407695	AZ AMERICAN WTR CO — AGUA FRIA	XYLENES	6	ug/L	02-May-05	CWS	34,000
AZ	AZ0407695	AZ AMERICAN WTR CO — AGUA FRIA	XYLENES	6	ug/L	02-May-05	CWS	34,000
AZ	AZ0407511	BEARDSLEY WC #3	XYLENES	10.6	ug/L	08-Sep-04	CWS	60
AZ	AZ0407511	BEARDSLEY WC #3	XYLENES	2	ug/L	16-Feb-05	CWS	60
AZ	AZ0407511	BEARDSLEY WC #3	XYLENES	8.4	ug/L	02-Jun-05	CWS	60
				2.96				

APPENDIX 7-27 Phoenix Surface Water Data

State	PWSID	Water System Name	Analyte	Concentration	Unit	Sample Date	PWSType	Pop
AZ	AZ0407504	ANTHEM WATER SYSTEM	2,4-D	0.2	ug/L	28-Feb-05	CWS	11,000
AZ	AZ0407504	ANTHEM WATER SYSTEM	2,4-D	0.2	ug/L	28-Feb-05	CWS	11,000
AZ	AZ0407504	ANTHEM WATER SYSTEM	2,4-D	0.2	ug/L	28-Feb-05	CWS	11,000
AZ	AZ0407504	ANTHEM WATER SYSTEM	2,4-D	0.2	ug/L	28-Feb-05	CWS	11,000
AZ	AZ0407504	ANTHEM WATER SYSTEM	2,4-D	0.2	ug/L	28-Feb-05	CWS	11,000
				0.2				
AZ	AZ0407504	ANTHEM WATER SYSTEM	DI(2-ETHYLHEXYL) PHTHALATE	1.1	ug/L	28-Feb-05	CWS	11,000
AZ	AZ0407504	ANTHEM WATER SYSTEM	DI(2-ETHYLHEXYL) PHTHALATE	1.1	ug/L	28-Feb-05	CWS	11,000
AZ	AZ0407504	ANTHEM WATER SYSTEM	DI(2-ETHYLHEXYL) PHTHALATE	1.1	ug/L	28-Feb-05	CWS	11,000
AZ	AZ0407504	ANTHEM WATER SYSTEM	DI(2-ETHYLHEXYL) PHTHALATE	1.1	ug/L	28-Feb-05	CWS	11,000
AZ	AZ0407504	ANTHEM WATER SYSTEM	DI(2-ETHYLHEXYL) PHTHALATE	1.1	ug/L	28-Feb-05	CWS	11,000
				1.1				

APPENDIX 7-28 Portland Water Data

State	PWS1	PWSName	Source	ContamName	Results	Unit	SampDate	PWSType	PSource	Pop
OR	OR4100984	MOBILE VILLAGE INC	EP FOR WELL	CIS-1,2-DICHLOROETHYLENE	0.7	ug/L	19-Mar-02	CWS	GW	100
OR	OR4100984	MOBILE VILLAGE INC	EP FOR WELL	CIS-1,2-DICHLOROETHYLENE	0.9	ug/L	28-May-02	CWS	GW	100
OR	OR4100984	MOBILE VILLAGE INC	EP FOR WELL	CIS-1,2-DICHLOROETHYLENE	0.6	ug/L	24-Sep-02	CWS	GW	100
OR	OR4100984	MOBILE VILLAGE INC	EP FOR WELL	CIS-1,2-DICHLOROETHYLENE	0.5	ug/L	22-Apr-03	CWS	GW	100
OR	OR4100984	MOBILE VILLAGE INC	EP FOR WELL	CIS-1,2-DICHLOROETHYLENE	0.6	ug/L	26-Nov-02	CWS	GW	100
OR	OR4100984	MOBILE VILLAGE INC	WELL	CIS-1,2-DICHLOROETHYLENE	0.5	ug/L	23-Oct-01	CWS	GW	100
					0.63					
OR	OR4100984	MOBILE VILLAGE INC	EP FOR WELL	TETRACHLOROETHYLENE	5.8	ug/L	19-Mar-02	CWS	GW	100
OR	OR4100984	MOBILE VILLAGE INC	EP FOR WELL	TETRACHLOROETHYLENE	6.7	ug/L	28-May-02	CWS	GW	100
OR	OR4100984	MOBILE VILLAGE INC	EP FOR WELL	TETRACHLOROETHYLENE	5.4	ug/L	29-Jul-03	CWS	GW	100
OR	OR4100984	MOBILE VILLAGE INC	EP FOR WELL	TETRACHLOROETHYLENE	4.4	ug/L	24-Sep-02	CWS	GW	100
OR	OR4100984	MOBILE VILLAGE INC	EP FOR WELL	TETRACHLOROETHYLENE	4	ug/L	26-Nov-02	CWS	GW	100
OR	OR4100984	MOBILE VILLAGE INC	EP FOR WELL	TETRACHLOROETHYLENE	6.4	ug/L	22-Apr-03	CWS	GW	100
OR	OR4100984	MOBILE VILLAGE INC	WELL	TETRACHLOROETHYLENE	4.6	ug/L	26-Jun-01	CWS	GW	100
OR	OR4100984	MOBILE VILLAGE INC	WELL	TETRACHLOROETHYLENE	2.6	ug/L	23-Oct-01	CWS	GW	100
OR	OR4100657	PORTLAND BUREAU OF WATER WORKS	PARKROSE WELLS	TETRACHLOROETHYLENE	1.5	ug/L	28-Mar-00	CWS	SW	456,000
OR	OR4100657	PORTLAND BUREAU OF WATER WORKS	WELL #19	TETRACHLOROETHYLENE	0.5	ug/L	27-Mar-00	CWS	SW	456,000
					4.19					
OR	OR4193957	PGE FARADAY DAM	WELL	P-DICHLOROBENZENE	0.8	ug/L	11-Sep-01	NTNCWS	GW	30
OR	OR4100657	PORTLAND BUREAU OF WATER WORKS	WELL #35	DIQUAT	0.4	ug/L	09-Nov-00	CWS	SW	456,000

APPENDIX 7-29 Reno Groundwater Data

State	NUMBER0	SYSNAME	ANLYTNAME	LEVEL	UNIT	COLLECTION_END_DT	PWSType	PSource	Pop
NV	NV0000901	ARROWCREEK	BROMODICHLOROMETHANE	0.7	UG/L	27-Jul-04	CWS	GW	1230
NV	NV0000191	ARROWHEAD MHP	BROMODICHLOROMETHANE	0.6	UG/L	09-Dec-04	CWS	GW	65
NV	NV0001085	DESERT SPRINGS UTILITY COMPANY	BROMODICHLOROMETHANE	3.3	UG/L	23-Jul-04	CWS	GW	5780
NV	NV0001085	DESERT SPRINGS UTILITY COMPANY	BROMODICHLOROMETHANE	0.72	UG/L	23-Jul-04	CWS	GW	5780
NV	NV0000832	DOUBLE DIAMOND	BROMODICHLOROMETHANE	2.6	UG/L	14-Jul-04	CWS	GW	4144
NV	NV0000832	DOUBLE DIAMOND	BROMODICHLOROMETHANE	2.3	UG/L	07-Jun-05	CWS	GW	4144
NV	NV0000832	DOUBLE DIAMOND	BROMODICHLOROMETHANE	1.2	UG/L	22-Jun-04	CWS	GW	4144
NV	NV0000832	DOUBLE DIAMOND	BROMODICHLOROMETHANE	1.8	UG/L	17-Mar-05	CWS	GW	4144
NV	NV0000832	DOUBLE DIAMOND	BROMODICHLOROMETHANE	1.4	UG/L	24-Mar-05	CWS	GW	4144
NV	NV0000832	DOUBLE DIAMOND	BROMODICHLOROMETHANE	4	UG/L	18-Oct-04	CWS	GW	4144
NV	NV0000832	DOUBLE DIAMOND	BROMODICHLOROMETHANE	0.64	UG/L	14-Jul-04	CWS	GW	4144
NV	NV0000832	DOUBLE DIAMOND	BROMODICHLOROMETHANE	2.2	UG/L	22-Jun-04	CWS	GW	4144
NV	NV0000832	DOUBLE DIAMOND	BROMODICHLOROMETHANE	5.1	UG/L	09-Mar-04	CWS	GW	4144
NV	NV0000832	DOUBLE DIAMOND	BROMODICHLOROMETHANE	5.6	UG/L	09-Mar-04	CWS	GW	4144
NV	NV0000832	DOUBLE DIAMOND	BROMODICHLOROMETHANE	4.1	UG/L	09-Mar-04	CWS	GW	4144
NV	NV0000832	DOUBLE DIAMOND	BROMODICHLOROMETHANE	0.71	UG/L	14-Jul-04	CWS	GW	4144
NV	NV0000832	DOUBLE DIAMOND	BROMODICHLOROMETHANE	4.6	UG/L	18-Oct-04	CWS	GW	4144
NV	NV0000832	DOUBLE DIAMOND	BROMODICHLOROMETHANE	1.5	UG/L	17-Mar-05	CWS	GW	4144
NV	NV0000832	DOUBLE DIAMOND	BROMODICHLOROMETHANE	1.4	UG/L	22-Jun-04	CWS	GW	4144
NV	NV0000832	DOUBLE DIAMOND	BROMODICHLOROMETHANE	3.2	UG/L	07-Jun-05	CWS	GW	4144
NV	NV0000832	DOUBLE DIAMOND	BROMODICHLOROMETHANE	2.1	UG/L	18-Oct-04	CWS	GW	4144
NV	NV0000832	DOUBLE DIAMOND	BROMODICHLOROMETHANE	4.5	UG/L	07-Jun-05	CWS	GW	4144
NV	NV0000195	FOUR SEASONS PARK	BROMODICHLOROMETHANE	0.0023	MG/L	31-Mar-03	CWS	GW	100
NV	NV0000198	HIDDEN VALLEY	BROMODICHLOROMETHANE	1.5	UG/L	13-Oct-04	CWS	GW	1522
NV	NV0000198	HIDDEN VALLEY	BROMODICHLOROMETHANE	6.1	UG/L	21-Jun-04	CWS	GW	1522
NV	NV0000198	HIDDEN VALLEY	BROMODICHLOROMETHANE	7.2	UG/L	14-Jul-04	CWS	GW	1522
NV	NV0000198	HIDDEN VALLEY	BROMODICHLOROMETHANE	4.1	UG/L	13-Oct-04	CWS	GW	1522
NV	NV0000198	HIDDEN VALLEY	BROMODICHLOROMETHANE	4.6	UG/L	07-Jun-05	CWS	GW	1522
NV	NV0000198	HIDDEN VALLEY	BROMODICHLOROMETHANE	1.1	UG/L	17-Mar-05	CWS	GW	1522
NV	NV0000198	HIDDEN VALLEY	BROMODICHLOROMETHANE	1.3	UG/L	17-Mar-05	CWS	GW	1522
NV	NV0000198	HIDDEN VALLEY	BROMODICHLOROMETHANE	5.4	UG/L	07-Jun-05	CWS	GW	1522
NV	NV0000198	HIDDEN VALLEY	BROMODICHLOROMETHANE	4.8	UG/L	21-Jun-04	CWS	GW	1522
NV	NV0000198	HIDDEN VALLEY	BROMODICHLOROMETHANE	5.1	UG/L	14-Jul-04	CWS	GW	1522
NV	NV0000199	KE TA MOBILE HOME PARK	BROMODICHLOROMETHANE	0.58	UG/L	02-Sep-04	CWS	GW	35
NV	NV0000202	LEMMON VALLEY WATER COMPANY	BROMODICHLOROMETHANE	1.2	UG/L	23-Jul-04	CWS	GW	2928
NV	NV0000732	MOUNT ROSE BOWL HOMEOWNERS ASSOCIATION	BROMODICHLOROMETHANE	2.5	UG/L	26-Oct-04	CWS	GW	35
NV	NV0000732	MOUNT ROSE BOWL HOMEOWNERS ASSOCIATION	BROMODICHLOROMETHANE	0.0017	MG/L	25-Oct-04	CWS	GW	35
NV	NV0000869	NATCHEZ GYM	BROMODICHLOROMETHANE	0.54	UG/L	11-Feb-00	NTNCWS	GW	175
NV	NV0000869	NATCHEZ GYM	BROMODICHLOROMETHANE	0.6	UG/L	23-Aug-04	NTNCWS	GW	175
NV	NV0002526	OLD WASHOE ESTATES WCUD	BROMODICHLOROMETHANE	1	UG/L	23-Jul-04	CWS	GW	112
NV	NV0000206	PIONEER HILLS MHP	BROMODICHLOROMETHANE	0.7	UG/L	16-Aug-04	CWS	GW	427

APPENDIX 7-29 *Continued*

State	NUMBER0	SYSNAME	ANLYTNAME	LEVEL	UNIT	COLLECTION_END_DT	PWSType	PSource	Pop
NV	NV0000754	RIVER BEND MHP	BROMODICHLOROMETHANE	4.5	UG/L	31-Aug-04	CWS	GW	90
NV	NV0003001	RIVERBELLE PROPERTIES LLC CEDARS	BROMODICHLOROMETHANE	0.61	UG/L	30-Sep-04	TNCWS	GW	40
NV	NV0000244	RIVERBELLE PROPERTIES LLC MHP	BROMODICHLOROMETHANE	4.9	UG/L	30-Sep-04	CWS	GW	245
NV	NV0000244	RIVERBELLE PROPERTIES LLC MHP	BROMODICHLOROMETHANE	1.5	UG/L	19-Aug-04	CWS	GW	245
NV	NV0000767	ROSEMOUNT WATER CO	BROMODICHLOROMETHANE	0.0016	MG/L	15-Jul-02	CWS	GW	54
NV	NV0000767	ROSEMOUNT WATER CO	BROMODICHLOROMETHANE	1	UG/L	18-Aug-04	CWS	GW	54
NV	NV0004082	SPRING CREEK WATER COMPANY	BROMODICHLOROMETHANE	7	UG/L	16-Jul-04	CWS	GW	1885
NV	NV0004082	SPRING CREEK WATER COMPANY	BROMODICHLOROMETHANE	12	UG/L	14-Mar-05	CWS	GW	1885
NV	NV0004082	SPRING CREEK WATER COMPANY	BROMODICHLOROMETHANE	6.9	UG/L	16-Jul-04	CWS	GW	1885
NV	NV0004082	SPRING CREEK WATER COMPANY	BROMODICHLOROMETHANE	8.2	UG/L	07-Jun-05	CWS	GW	1885
NV	NV0004082	SPRING CREEK WATER COMPANY	BROMODICHLOROMETHANE	14	UG/L	14-Mar-05	CWS	GW	1885
NV	NV0004082	SPRING CREEK WATER COMPANY	BROMODICHLOROMETHANE	14	UG/L	13-Oct-04	CWS	GW	1885
NV	NV0004082	SPRING CREEK WATER COMPANY	BROMODICHLOROMETHANE	7	UG/L	21-Jun-04	CWS	GW	1885
NV	NV0004082	SPRING CREEK WATER COMPANY	BROMODICHLOROMETHANE	12	UG/L	13-Oct-04	CWS	GW	1885
NV	NV0004082	SPRING CREEK WATER COMPANY	BROMODICHLOROMETHANE	5.9	UG/L	21-Jun-04	CWS	GW	1885
NV	NV0004082	SPRING CREEK WATER COMPANY	BROMODICHLOROMETHANE	8	UG/L	07-Jun-05	CWS	GW	1885
NV	NV0000801	STAMPMILL	BROMODICHLOROMETHANE	1	UG/L	23-Jul-04	CWS	GW	101
NV	NV0002525	SUNRISE ESTATES WATER SYSTEM	BROMODICHLOROMETHANE	0.74	UG/L	27-Jul-04	CWS	GW	86
NV	NV0000802	TIMBERLINE ESTATES	BROMODICHLOROMETHANE	1.4	UG/L	23-Jul-04	CWS	GW	55
NV	NV0000978	TRUCKEE CANYON WATER SYSTEM	BROMODICHLOROMETHANE	3.6	UG/L	14-Aug-04	TNCWS	GW	30
NV	NV0000207	UTILITIES INC OF NEVADA	BROMODICHLOROMETHANE	0.51	UG/L	21-Sep-04	CWS	GW	7805
NV	NV0000781	WESTERNER MOTEL	BROMODICHLOROMETHANE	0.011	UG/L	05-Apr-00	TNCWS	GW	25
				3.45					
NV	NV0001085	DESERT SPRINGS UTILITY COMPANY	CHLORODIBROMOMETHANE	8.8	UG/L	23-Jul-04	CWS	GW	5780
NV	NV0001085	DESERT SPRINGS UTILITY COMPANY	CHLORODIBROMOMETHANE	3.3	UG/L	23-Jul-04	CWS	GW	5780
NV	NV0000832	DOUBLE DIAMOND	CHLORODIBROMOMETHANE	1.2	UG/L	14-Jul-04	CWS	GW	4144
NV	NV0000832	DOUBLE DIAMOND	CHLORODIBROMOMETHANE	1.2	UG/L	24-Mar-05	CWS	GW	4144
NV	NV0000832	DOUBLE DIAMOND	CHLORODIBROMOMETHANE	0.77	UG/L	07-Jun-05	CWS	GW	4144
NV	NV0000832	DOUBLE DIAMOND	CHLORODIBROMOMETHANE	1.3	UG/L	17-Mar-05	CWS	GW	4144
NV	NV0000832	DOUBLE DIAMOND	CHLORODIBROMOMETHANE	2.1	UG/L	09-Mar-04	CWS	GW	4144
NV	NV0000832	DOUBLE DIAMOND	CHLORODIBROMOMETHANE	2.3	UG/L	09-Mar-04	CWS	GW	4144
NV	NV0000832	DOUBLE DIAMOND	CHLORODIBROMOMETHANE	1.7	UG/L	09-Mar-04	CWS	GW	4144
NV	NV0000832	DOUBLE DIAMOND	CHLORODIBROMOMETHANE	1.5	UG/L	18-Oct-04	CWS	GW	4144
NV	NV0000832	DOUBLE DIAMOND	CHLORODIBROMOMETHANE	0.93	UG/L	17-Mar-05	CWS	GW	4144
NV	NV0000832	DOUBLE DIAMOND	CHLORODIBROMOMETHANE	0.59	UG/L	22-Jun-04	CWS	GW	4144
NV	NV0000832	DOUBLE DIAMOND	CHLORODIBROMOMETHANE	1.2	UG/L	07-Jun-05	CWS	GW	4144
NV	NV0000832	DOUBLE DIAMOND	CHLORODIBROMOMETHANE	1.1	UG/L	18-Oct-04	CWS	GW	4144
NV	NV0000832	DOUBLE DIAMOND	CHLORODIBROMOMETHANE	1.6	UG/L	07-Jun-05	CWS	GW	4144
NV	NV0000832	DOUBLE DIAMOND	CHLORODIBROMOMETHANE	0.54	UG/L	22-Jun-04	CWS	GW	4144
NV	NV0000832	DOUBLE DIAMOND	CHLORODIBROMOMETHANE	1.9	UG/L	18-Oct-04	CWS	GW	4144

NV	NV0000198	HIDDEN VALLEY	CHLORODIBROMOMETHANE	1.6	UG/L	13-Oct-04	CWS	GW	1522
NV	NV0000198	HIDDEN VALLEY	CHLORODIBROMOMETHANE	0.8	UG/L	09-Mar-04	CWS	GW	1522
NV	NV0000198	HIDDEN VALLEY	CHLORODIBROMOMETHANE	4.5	UG/L	14-Jul-04	CWS	GW	1522
NV	NV0000198	HIDDEN VALLEY	CHLORODIBROMOMETHANE	3.9	UG/L	13-Oct-04	CWS	GW	1522
NV	NV0000198	HIDDEN VALLEY	CHLORODIBROMOMETHANE	2	UG/L	07-Jun-05	CWS	GW	1522
NV	NV0000198	HIDDEN VALLEY	CHLORODIBROMOMETHANE	3.9	UG/L	21-Jun-04	CWS	GW	1522
NV	NV0000198	HIDDEN VALLEY	CHLORODIBROMOMETHANE	1.3	UG/L	17-Mar-05	CWS	GW	1522
NV	NV0000198	HIDDEN VALLEY	CHLORODIBROMOMETHANE	1.3	UG/L	17-Mar-05	CWS	GW	1522
NV	NV0000198	HIDDEN VALLEY	CHLORODIBROMOMETHANE	2.4	UG/L	07-Jun-05	CWS	GW	1522
NV	NV0000198	HIDDEN VALLEY	CHLORODIBROMOMETHANE	2.7	UG/L	14-Jul-04	CWS	GW	1522
NV	NV0000198	HIDDEN VALLEY	CHLORODIBROMOMETHANE	3.1	UG/L	21-Jun-04	CWS	GW	1522
NV	NV0000202	LEMMON VALLEY WATER COMPANY	CHLORODIBROMOMETHANE	1.9	UG/L	23-Jul-04	CWS	GW	2928
NV	NV0000732	MOUNT ROSE BOWL HOMEOWNERS ASSOCIATION	CHLORODIBROMOMETHANE	0.0006	MG/L	25-Oct-04	CWS	GW	35
NV	NV0000869	NATCHEZ GYM	CHLORODIBROMOMETHANE	0.84	UG/L	11-Feb-00	NTNCWS	GW	175
NV	NV0000869	NATCHEZ GYM	CHLORODIBROMOMETHANE	1.6	UG/L	23-Aug-04	NTNCWS	GW	175
NV	NV0002526	OLD WASHOE ESTATES WCUD	CHLORODIBROMOMETHANE	2.2	UG/L	23-Jul-04	CWS	GW	112
NV	NV0000206	PIONEER HILLS MHP	CHLORODIBROMOMETHANE	0.96	UG/L	26-Jan-00	CWS	GW	427
NV	NV0000206	PIONEER HILLS MHP	CHLORODIBROMOMETHANE	0.9	UG/L	16-Aug-04	CWS	GW	427
NV	NV0000244	RIVERBELLE PROPERTIES LLC MHP	CHLORODIBROMOMETHANE	2.2	UG/L	30-Sep-04	CWS	GW	245
NV	NV0000244	RIVERBELLE PROPERTIES LLC MHP	CHLORODIBROMOMETHANE	0.7	UG/L	19-Aug-04	CWS	GW	245
NV	NV0000754	RIVER BEND MHP	CHLORODIBROMOMETHANE	1.1	UG/L	31-Aug-04	CWS	GW	90
NV	NV0001086	SKY RANCH WATER SERVICE CORPORATION	CHLORODIBROMOMETHANE	0.51	UG/L	21-Sep-04	CWS	GW	2030
NV	NV0004082	SPRING CREEK WATER COMPANY	CHLORODIBROMOMETHANE	1.9	UG/L	16-Jul-04	CWS	GW	1885
NV	NV0004082	SPRING CREEK WATER COMPANY	CHLORODIBROMOMETHANE	8.3	UG/L	13-Oct-04	CWS	GW	1885
NV	NV0004082	SPRING CREEK WATER COMPANY	CHLORODIBROMOMETHANE	1.2	UG/L	16-Jul-04	CWS	GW	1885
NV	NV0004082	SPRING CREEK WATER COMPANY	CHLORODIBROMOMETHANE	1.4	UG/L	07-Jun-05	CWS	GW	1885
NV	NV0004082	SPRING CREEK WATER COMPANY	CHLORODIBROMOMETHANE	7	UG/L	14-Mar-05	CWS	GW	1885
NV	NV0004082	SPRING CREEK WATER COMPANY	CHLORODIBROMOMETHANE	12	UG/L	13-Oct-04	CWS	GW	1885
NV	NV0004082	SPRING CREEK WATER COMPANY	CHLORODIBROMOMETHANE	0.57	UG/L	21-Jun-04	CWS	GW	1885
NV	NV0004082	SPRING CREEK WATER COMPANY	CHLORODIBROMOMETHANE	0.6	UG/L	21-Jun-04	CWS	GW	1885
NV	NV0004082	SPRING CREEK WATER COMPANY	CHLORODIBROMOMETHANE	8.1	UG/L	14-Mar-05	CWS	GW	1885
NV	NV0004082	SPRING CREEK WATER COMPANY	CHLORODIBROMOMETHANE	1.3	UG/L	07-Jun-05	CWS	GW	1885
NV	NV0004082	SPRING CREEK WATER COMPANY	CHLORODIBROMOMETHANE	1.4	UG/L	09-Mar-04	CWS	GW	1885
NV	NV0000801	STAMPMILL	CHLORODIBROMOMETHANE	2.2	UG/L	23-Jul-04	CWS	GW	101
NV	NV0000802	TIMBERLINE ESTATES	CHLORODIBROMOMETHANE	0.55	UG/L	23-Jul-04	CWS	GW	55
NV	NV0000978	TRUCKEE CANYON WATER SYSTEM	CHLORODIBROMOMETHANE	3.6	UG/L	14-Aug-04	TNCWS	GW	30
NV	NV0000207	UTILITIES INC OF NEVADA	CHLORODIBROMOMETHANE	0.73	UG/L	21-Sep-04	CWS	GW	7805
				2.28					

APPENDIX 7-30 Reno Surface Water Data

State	NUMBER0	SYSNAME	ANLYTNAME	LEVEL	UNIT	DATE	PWSType	PSource	Pop
NV	NV0000729	HORIZON HILLS GID WCUD	BROMODICHLOROMETHANE	8.5	UG/L	09-Mar-04	CWS	SWP	374
NV	NV0000729	HORIZON HILLS GID WCUD	BROMODICHLOROMETHANE	6.3	UG/L	13-Oct-04	CWS	SWP	374
NV	NV0000729	HORIZON HILLS GID WCUD	BROMODICHLOROMETHANE	7.7	UG/L	07-Jun-05	CWS	SWP	374
NV	NV0000729	HORIZON HILLS GID WCUD	BROMODICHLOROMETHANE	5.6	UG/L	16-Jul-04	CWS	SWP	374
NV	NV0000729	HORIZON HILLS GID WCUD	BROMODICHLOROMETHANE	6.2	UG/L	21-Jun-04	CWS	SWP	374
NV	NV0000729	HORIZON HILLS GID WCUD	BROMODICHLOROMETHANE	13	UG/L	09-Mar-05	CWS	SWP	374
NV	NV0000190	TRUCKEE MEADOWS WATER AUTHORITY	BROMODICHLOROMETHANE	1.6	UG/L	06-Jun-01	CWS	SW	212,760
NV	NV0000190	TRUCKEE MEADOWS WATER AUTHORITY	BROMODICHLOROMETHANE	12	UG/L	26-Mar-01	CWS	SW	212,760
NV	NV0000190	TRUCKEE MEADOWS WATER AUTHORITY	BROMODICHLOROMETHANE	0.61	UG/L	05-Jun-01	CWS	SW	212,760
NV	NV0000190	TRUCKEE MEADOWS WATER AUTHORITY	BROMODICHLOROMETHANE	0.85	UG/L	15-Oct-01	CWS	SW	212,760
NV	NV0000190	TRUCKEE MEADOWS WATER AUTHORITY	BROMODICHLOROMETHANE	0.85	UG/L	15-Oct-01	CWS	SW	212,760
NV	NV0000190	TRUCKEE MEADOWS WATER AUTHORITY	BROMODICHLOROMETHANE	8.14	UG/L	14-Feb-05	CWS	SW	212,760
NV	NV0000190	TRUCKEE MEADOWS WATER AUTHORITY	BROMODICHLOROMETHANE	13.6	UG/L	15-Feb-05	CWS	SW	212,760
NV	NV0000190	TRUCKEE MEADOWS WATER AUTHORITY	BROMODICHLOROMETHANE	6.57	UG/L	04-May-04	CWS	SW	212,760
NV	NV0000190	TRUCKEE MEADOWS WATER AUTHORITY	BROMODICHLOROMETHANE	11.3	UG/L	07-Dec-04	CWS	SW	212,760
NV	NV0000190	TRUCKEE MEADOWS WATER AUTHORITY	BROMODICHLOROMETHANE	5.87	UG/L	11-Aug-04	CWS	SW	212,760
NV	NV0000190	TRUCKEE MEADOWS WATER AUTHORITY	BROMODICHLOROMETHANE	10.9	UG/L	07-Dec-04	CWS	SW	212,760
NV	NV0000190	TRUCKEE MEADOWS WATER AUTHORITY	BROMODICHLOROMETHANE	2.43	UG/L	04-May-04	CWS	SW	212,760
NV	NV0000190	TRUCKEE MEADOWS WATER AUTHORITY	BROMODICHLOROMETHANE	5.68	UG/L	03-Feb-04	CWS	SW	212,760
NV	NV0000190	TRUCKEE MEADOWS WATER AUTHORITY	BROMODICHLOROMETHANE	8.22	UG/L	09-Aug-04	CWS	SW	212,760
NV	NV0000190	TRUCKEE MEADOWS WATER AUTHORITY	BROMODICHLOROMETHANE	9.18	UG/L	07-Dec-04	CWS	SW	212,760
NV	NV0000190	TRUCKEE MEADOWS WATER AUTHORITY	BROMODICHLOROMETHANE	9.81	UG/L	14-Feb-05	CWS	SW	212,760
NV	NV0000190	TRUCKEE MEADOWS WATER AUTHORITY	BROMODICHLOROMETHANE	6.02	UG/L	03-Feb-04	CWS	SW	212,760
NV	NV0000190	TRUCKEE MEADOWS WATER AUTHORITY	BROMODICHLOROMETHANE	10.5	UG/L	07-Dec-04	CWS	SW	212,760
NV	NV0000190	TRUCKEE MEADOWS WATER AUTHORITY	BROMODICHLOROMETHANE	11.2	UG/L	14-Feb-05	CWS	SW	212,760
NV	NV0000190	TRUCKEE MEADOWS WATER AUTHORITY	BROMODICHLOROMETHANE	5.15	UG/L	05-May-04	CWS	SW	212,760
NV	NV0000190	TRUCKEE MEADOWS WATER AUTHORITY	BROMODICHLOROMETHANE	7.12	UG/L	09-Aug-04	CWS	SW	212,760
NV	NV0000190	TRUCKEE MEADOWS WATER AUTHORITY	BROMODICHLOROMETHANE	6.08	UG/L	09-Aug-04	CWS	SW	212,760
NV	NV0000190	TRUCKEE MEADOWS WATER AUTHORITY	BROMODICHLOROMETHANE	4.3	UG/L	09-Aug-04	CWS	SW	212,760
NV	NV0000190	TRUCKEE MEADOWS WATER AUTHORITY	BROMODICHLOROMETHANE	5.36	UG/L	04-May-04	CWS	SW	212,760
NV	NV0000190	TRUCKEE MEADOWS WATER AUTHORITY	BROMODICHLOROMETHANE	9.17	UG/L	15-Feb-05	CWS	SW	212,760
NV	NV0000190	TRUCKEE MEADOWS WATER AUTHORITY	BROMODICHLOROMETHANE	8.59	UG/L	08-Dec-04	CWS	SW	212,760
NV	NV0000190	TRUCKEE MEADOWS WATER AUTHORITY	BROMODICHLOROMETHANE	4.09	UG/L	04-May-04	CWS	SW	212,760
NV	NV0000190	TRUCKEE MEADOWS WATER AUTHORITY	BROMODICHLOROMETHANE	4.02	UG/L	04-Feb-04	CWS	SW	212,760
NV	NV0000190	TRUCKEE MEADOWS WATER AUTHORITY	BROMODICHLOROMETHANE	5.04	UG/L	09-Aug-04	CWS	SW	212,760
NV	NV0000190	TRUCKEE MEADOWS WATER AUTHORITY	BROMODICHLOROMETHANE	4.76	UG/L	09-Aug-04	CWS	SW	212,760
NV	NV0000190	TRUCKEE MEADOWS WATER AUTHORITY	BROMODICHLOROMETHANE	5.56	UG/L	04-Feb-04	CWS	SW	212,760
NV	NV0000190	TRUCKEE MEADOWS WATER AUTHORITY	BROMODICHLOROMETHANE	9.54	UG/L	08-Dec-04	CWS	SW	212,760
NV	NV0000190	TRUCKEE MEADOWS WATER AUTHORITY	BROMODICHLOROMETHANE	5.83	UG/L	04-May-04	CWS	SW	212,760
NV	NV0000190	TRUCKEE MEADOWS WATER AUTHORITY	BROMODICHLOROMETHANE	6.04	UG/L	02-Feb-04	CWS	SW	212,760
NV	NV0000190	TRUCKEE MEADOWS WATER AUTHORITY	BROMODICHLOROMETHANE	10.7	UG/L	15-Feb-05	CWS	SW	212,760

NV	NV0000190	TRUCKEE MEADOWS WATER AUTHORITY	BROMODICHLOROMETHANE	6.39	UG/L	11-Aug-04	CWS	SW	212,760
NV	NV0000190	TRUCKEE MEADOWS WATER AUTHORITY	BROMODICHLOROMETHANE	6.14	UG/L	04-May-04	CWS	SW	212,760
NV	NV0000190	TRUCKEE MEADOWS WATER AUTHORITY	BROMODICHLOROMETHANE	6.67	UG/L	07-Dec-04	CWS	SW	212,760
NV	NV0000190	TRUCKEE MEADOWS WATER AUTHORITY	BROMODICHLOROMETHANE	11.8	UG/L	15-Feb-05	CWS	SW	212,760
NV	NV0000190	TRUCKEE MEADOWS WATER AUTHORITY	BROMODICHLOROMETHANE	8.05	UG/L	02-Feb-04	CWS	SW	212,760
NV	NV0000190	TRUCKEE MEADOWS WATER AUTHORITY	BROMODICHLOROMETHANE	7.49	UG/L	11-Aug-04	CWS	SW	212,760
NV	NV0000190	TRUCKEE MEADOWS WATER AUTHORITY	BROMODICHLOROMETHANE	5.76	UG/L	02-Feb-04	CWS	SW	212,760
NV	NV0000190	TRUCKEE MEADOWS WATER AUTHORITY	BROMODICHLOROMETHANE	6.61	UG/L	04-May-04	CWS	SW	212,760
NV	NV0000190	TRUCKEE MEADOWS WATER AUTHORITY	BROMODICHLOROMETHANE	6.89	UG/L	11-Aug-04	CWS	SW	212,760
NV	NV0000190	TRUCKEE MEADOWS WATER AUTHORITY	BROMODICHLOROMETHANE	13.4	UG/L	07-Dec-04	CWS	SW	212,760
NV	NV0000190	TRUCKEE MEADOWS WATER AUTHORITY	BROMODICHLOROMETHANE	14	UG/L	15-Feb-05	CWS	SW	212,760
NV	NV0000190	TRUCKEE MEADOWS WATER AUTHORITY	BROMODICHLOROMETHANE	6.13	UG/L	05-May-04	CWS	SW	212,760
NV	NV0000190	TRUCKEE MEADOWS WATER AUTHORITY	BROMODICHLOROMETHANE	6.9	UG/L	07-Dec-04	CWS	SW	212,760
NV	NV0000190	TRUCKEE MEADOWS WATER AUTHORITY	BROMODICHLOROMETHANE	2.77	UG/L	05-May-04	CWS	SW	212,760
NV	NV0000190	TRUCKEE MEADOWS WATER AUTHORITY	BROMODICHLOROMETHANE	5.9	UG/L	11-Aug-04	CWS	SW	212,760
NV	NV0000190	TRUCKEE MEADOWS WATER AUTHORITY	BROMODICHLOROMETHANE	8.3	UG/L	15-Feb-05	CWS	SW	212,760
NV	NV0000190	TRUCKEE MEADOWS WATER AUTHORITY	BROMODICHLOROMETHANE	3.49	UG/L	02-Feb-04	CWS	SW	212,760
NV	NV0000190	TRUCKEE MEADOWS WATER AUTHORITY	BROMODICHLOROMETHANE	10.4	UG/L	15-Feb-05	CWS	SW	212,760
NV	NV0000190	TRUCKEE MEADOWS WATER AUTHORITY	BROMODICHLOROMETHANE	10.9	UG/L	08-Dec-04	CWS	SW	212,760
NV	NV0000190	TRUCKEE MEADOWS WATER AUTHORITY	BROMODICHLOROMETHANE	6.37	UG/L	09-Aug-04	CWS	SW	212,760
NV	NV0000190	TRUCKEE MEADOWS WATER AUTHORITY	BROMODICHLOROMETHANE	4	UG/L	09-Aug-04	CWS	SW	212,760
NV	NV0000190	TRUCKEE MEADOWS WATER AUTHORITY	BROMODICHLOROMETHANE	4.71	UG/L	05-May-04	CWS	SW	212,760
NV	NV0000190	TRUCKEE MEADOWS WATER AUTHORITY	BROMODICHLOROMETHANE	4.7	UG/L	04-May-04	CWS	SW	212,760
NV	NV0000190	TRUCKEE MEADOWS WATER AUTHORITY	BROMODICHLOROMETHANE	11.6	UG/L	15-Feb-05	CWS	SW	212,760
NV	NV0000190	TRUCKEE MEADOWS WATER AUTHORITY	BROMODICHLOROMETHANE	5.46	UG/L	04-Feb-04	CWS	SW	212,760
NV	NV0000190	TRUCKEE MEADOWS WATER AUTHORITY	BROMODICHLOROMETHANE	7.66	UG/L	08-Dec-04	CWS	SW	212,760
NV	NV0000190	TRUCKEE MEADOWS WATER AUTHORITY	BROMODICHLOROMETHANE	6.2	UG/L	07-Dec-04	CWS	SW	212,760
NV	NV0000190	TRUCKEE MEADOWS WATER AUTHORITY	BROMODICHLOROMETHANE	4.05	UG/L	05-May-04	CWS	SW	212,760
NV	NV0000190	TRUCKEE MEADOWS WATER AUTHORITY	BROMODICHLOROMETHANE	4.78	UG/L	03-Feb-04	CWS	SW	212,760
NV	NV0000190	TRUCKEE MEADOWS WATER AUTHORITY	BROMODICHLOROMETHANE	5.44	UG/L	11-Aug-04	CWS	SW	212,760
NV	NV0000190	TRUCKEE MEADOWS WATER AUTHORITY	BROMODICHLOROMETHANE	4.38	UG/L	02-Feb-04	CWS	SW	212,760
NV	NV0000190	TRUCKEE MEADOWS WATER AUTHORITY	BROMODICHLOROMETHANE	9.68	UG/L	07-Dec-04	CWS	SW	212,760
NV	NV0000190	TRUCKEE MEADOWS WATER AUTHORITY	BROMODICHLOROMETHANE	5.55	UG/L	04-May-04	CWS	SW	212,760
NV	NV0000190	TRUCKEE MEADOWS WATER AUTHORITY	BROMODICHLOROMETHANE	11.6	UG/L	15-Feb-05	CWS	SW	212,760
NV	NV0000190	TRUCKEE MEADOWS WATER AUTHORITY	BROMODICHLOROMETHANE	7.69	UG/L	11-Aug-04	CWS	SW	212,760
NV	NV0000190	TRUCKEE MEADOWS WATER AUTHORITY	BROMODICHLOROMETHANE	5.02	UG/L	09-Aug-04	CWS	SW	212,760
NV	NV0000190	TRUCKEE MEADOWS WATER AUTHORITY	BROMODICHLOROMETHANE	7.67	UG/L	08-Dec-04	CWS	SW	212,760
NV	NV0000190	TRUCKEE MEADOWS WATER AUTHORITY	BROMODICHLOROMETHANE	7.8	UG/L	15-Feb-05	CWS	SW	212,760
NV	NV0000190	TRUCKEE MEADOWS WATER AUTHORITY	BROMODICHLOROMETHANE	3.68	UG/L	04-Feb-04	CWS	SW	212,760
NV	NV0000190	TRUCKEE MEADOWS WATER AUTHORITY	BROMODICHLOROMETHANE	2.86	UG/L	05-May-04	CWS	SW	212,760
NV	NV0000190	TRUCKEE MEADOWS WATER AUTHORITY	BROMODICHLOROMETHANE	5.3	UG/L	04-Feb-04	CWS	SW	212,760
NV	NV0000190	TRUCKEE MEADOWS WATER AUTHORITY	BROMODICHLOROMETHANE	10.5	UG/L	15-Feb-05	CWS	SW	212,760
NV	NV0000190	TRUCKEE MEADOWS WATER AUTHORITY	BROMODICHLOROMETHANE	3.64	UG/L	04-May-04	CWS	SW	212,760
NV	NV0000190	TRUCKEE MEADOWS WATER AUTHORITY	BROMODICHLOROMETHANE	7.22	UG/L	09-Aug-04	CWS	SW	212,760
NV	NV0000190	TRUCKEE MEADOWS WATER AUTHORITY	BROMODICHLOROMETHANE	9.8	UG/L	08-Dec-04	CWS	SW	212,760

APPENDIX 7-30 *Continued*

State	NUMBER0	SYSNAME	ANLYTNAME	LEVEL	UNIT	DATE	PWSType	PSource	Pop
NV	NV0000190	TRUCKEE MEADOWS WATER AUTHORITY	BROMODICHLOROMETHANE	3.2	UG/L	04-Feb-04	CWS	SW	212,760
NV	NV0000190	TRUCKEE MEADOWS WATER AUTHORITY	BROMODICHLOROMETHANE	6.57	UG/L	15-Feb-05	CWS	SW	212,760
NV	NV0000190	TRUCKEE MEADOWS WATER AUTHORITY	BROMODICHLOROMETHANE	3.26	UG/L	05-May-04	CWS	SW	212,760
NV	NV0000190	TRUCKEE MEADOWS WATER AUTHORITY	BROMODICHLOROMETHANE	6.55	UG/L	08-Dec-04	CWS	SW	212,760
NV	NV0000190	TRUCKEE MEADOWS WATER AUTHORITY	BROMODICHLOROMETHANE	4.06	UG/L	09-Aug-04	CWS	SW	212,760
NV	NV0000190	TRUCKEE MEADOWS WATER AUTHORITY	BROMODICHLOROMETHANE	8.35	UG/L	08-Dec-04	CWS	SW	212,760
NV	NV0000190	TRUCKEE MEADOWS WATER AUTHORITY	BROMODICHLOROMETHANE	5.31	UG/L	09-Aug-04	CWS	SW	212,760
NV	NV0000190	TRUCKEE MEADOWS WATER AUTHORITY	BROMODICHLOROMETHANE	8.57	UG/L	14-Feb-05	CWS	SW	212,760
NV	NV0000190	TRUCKEE MEADOWS WATER AUTHORITY	BROMODICHLOROMETHANE	4.79	UG/L	02-Feb-04	CWS	SW	212,760
NV	NV0000190	TRUCKEE MEADOWS WATER AUTHORITY	BROMODICHLOROMETHANE	5.37	UG/L	11-Aug-04	CWS	SW	212,760
NV	NV0000190	TRUCKEE MEADOWS WATER AUTHORITY	BROMODICHLOROMETHANE	5.42	UG/L	07-Dec-04	CWS	SW	212,760
NV	NV0000190	TRUCKEE MEADOWS WATER AUTHORITY	BROMODICHLOROMETHANE	4.3	UG/L	05-May-04	CWS	SW	212,760
NV	NV0000190	TRUCKEE MEADOWS WATER AUTHORITY	BROMODICHLOROMETHANE	13.4	UG/L	15-Feb-05	CWS	SW	212,760
NV	NV0000190	TRUCKEE MEADOWS WATER AUTHORITY	BROMODICHLOROMETHANE	10.1	UG/L	08-Dec-04	CWS	SW	212,760
NV	NV0000190	TRUCKEE MEADOWS WATER AUTHORITY	BROMODICHLOROMETHANE	10.3	UG/L	09-Aug-04	CWS	SW	212,760
NV	NV0000190	TRUCKEE MEADOWS WATER AUTHORITY	BROMODICHLOROMETHANE	4.11	UG/L	05-May-04	CWS	SW	212,760
NV	NV0000190	TRUCKEE MEADOWS WATER AUTHORITY	BROMODICHLOROMETHANE	10	UG/L	14-Feb-05	CWS	SW	212,760
NV	NV0000190	TRUCKEE MEADOWS WATER AUTHORITY	BROMODICHLOROMETHANE	5.01	UG/L	04-Feb-04	CWS	SW	212,760
NV	NV0000190	TRUCKEE MEADOWS WATER AUTHORITY	BROMODICHLOROMETHANE	9.9	UG/L	15-Feb-05	CWS	SW	212,760
NV	NV0000190	TRUCKEE MEADOWS WATER AUTHORITY	BROMODICHLOROMETHANE	8.33	UG/L	14-Feb-05	CWS	SW	212,760
NV	NV0000190	TRUCKEE MEADOWS WATER AUTHORITY	BROMODICHLOROMETHANE	4.97	UG/L	03-Feb-04	CWS	SW	212,760
NV	NV0000190	TRUCKEE MEADOWS WATER AUTHORITY	BROMODICHLOROMETHANE	7.33	UG/L	07-Dec-04	CWS	SW	212,760
NV	NV0000190	TRUCKEE MEADOWS WATER AUTHORITY	BROMODICHLOROMETHANE	4.15	UG/L	03-Feb-04	CWS	SW	212,760
NV	NV0000190	TRUCKEE MEADOWS WATER AUTHORITY	BROMODICHLOROMETHANE	4.79	UG/L	05-May-04	CWS	SW	212,760
NV	NV0000190	TRUCKEE MEADOWS WATER AUTHORITY	BROMODICHLOROMETHANE	9.21	UG/L	07-Dec-04	CWS	SW	212,760
				6.91					
NV	NV0000729	HORIZON HILLS GID WCUD	CHLORODIBROMOMETHANE	1.2	UG/L	09-Mar-04	CWS	SWP	374
NV	NV0000729	HORIZON HILLS GID WCUD	CHLORODIBROMOMETHANE	2.8	UG/L	09-Mar-05	CWS	SWP	374
NV	NV0000729	HORIZON HILLS GID WCUD	CHLORODIBROMOMETHANE	0.7	UG/L	21-Jun-04	CWS	SWP	374
NV	NV0000729	HORIZON HILLS GID WCUD	CHLORODIBROMOMETHANE	0.88	UG/L	16-Jul-04	CWS	SWP	374
NV	NV0000729	HORIZON HILLS GID WCUD	CHLORODIBROMOMETHANE	1.4	UG/L	13-Oct-04	CWS	SWP	374
NV	NV0000729	HORIZON HILLS GID WCUD	CHLORODIBROMOMETHANE	1.1	UG/L	07-Jun-05	CWS	SWP	374
NV	NV0000190	TRUCKEE MEADOWS WATER AUTHORITY	CHLORODIBROMOMETHANE	10	UG/L	26-Mar-01	CWS	SW	212,760
NV	NV0000190	TRUCKEE MEADOWS WATER AUTHORITY	CHLORODIBROMOMETHANE	1.92	UG/L	14-Feb-05	CWS	SW	212,760
NV	NV0000190	TRUCKEE MEADOWS WATER AUTHORITY	CHLORODIBROMOMETHANE	0.94	UG/L	09-Aug-04	CWS	SW	212,760
NV	NV0000190	TRUCKEE MEADOWS WATER AUTHORITY	CHLORODIBROMOMETHANE	0.55	UG/L	04-May-04	CWS	SW	212,760
NV	NV0000190	TRUCKEE MEADOWS WATER AUTHORITY	CHLORODIBROMOMETHANE	2.08	UG/L	15-Feb-05	CWS	SW	212,760
NV	NV0000190	TRUCKEE MEADOWS WATER AUTHORITY	CHLORODIBROMOMETHANE	2.18	UG/L	08-Dec-04	CWS	SW	212,760
NV	NV0000190	TRUCKEE MEADOWS WATER AUTHORITY	CHLORODIBROMOMETHANE	0.7	UG/L	04-Feb-04	CWS	SW	212,760
NV	NV0000190	TRUCKEE MEADOWS WATER AUTHORITY	CHLORODIBROMOMETHANE	0.63	UG/L	09-Aug-04	CWS	SW	212,760
NV	NV0000190	TRUCKEE MEADOWS WATER AUTHORITY	CHLORODIBROMOMETHANE	0.51	UG/L	09-Aug-04	CWS	SW	212,760
NV	NV0000190	TRUCKEE MEADOWS WATER AUTHORITY	CHLORODIBROMOMETHANE	0.89	UG/L	04-Feb-04	CWS	SW	212,760
NV	NV0000190	TRUCKEE MEADOWS WATER AUTHORITY	CHLORODIBROMOMETHANE	2.03	UG/L	08-Dec-04	CWS	SW	212,760

NV	NV0000190	TRUCKEE MEADOWS WATER AUTHORITY	CHLORODIBROMOMETHANE	0.56	UG/L	04-May-04	CWS	SW	212,760
NV	NV0000190	TRUCKEE MEADOWS WATER AUTHORITY	CHLORODIBROMOMETHANE	2.33	UG/L	15-Feb-05	CWS	SW	212,760
NV	NV0000190	TRUCKEE MEADOWS WATER AUTHORITY	CHLORODIBROMOMETHANE	4.7	UG/L	02-Feb-04	CWS	SW	212,760
NV	NV0000190	TRUCKEE MEADOWS WATER AUTHORITY	CHLORODIBROMOMETHANE	4.35	UG/L	15-Feb-05	CWS	SW	212,760
NV	NV0000190	TRUCKEE MEADOWS WATER AUTHORITY	CHLORODIBROMOMETHANE	1.68	UG/L	11-Aug-04	CWS	SW	212,760
NV	NV0000190	TRUCKEE MEADOWS WATER AUTHORITY	CHLORODIBROMOMETHANE	1.73	UG/L	04-May-04	CWS	SW	212,760
NV	NV0000190	TRUCKEE MEADOWS WATER AUTHORITY	CHLORODIBROMOMETHANE	4.92	UG/L	07-Dec-04	CWS	SW	212,760
NV	NV0000190	TRUCKEE MEADOWS WATER AUTHORITY	CHLORODIBROMOMETHANE	6.57	UG/L	02-Feb-04	CWS	SW	212,760
NV	NV0000190	TRUCKEE MEADOWS WATER AUTHORITY	CHLORODIBROMOMETHANE	1.5	UG/L	11-Aug-04	CWS	SW	212,760
NV	NV0000190	TRUCKEE MEADOWS WATER AUTHORITY	CHLORODIBROMOMETHANE	4.3	UG/L	02-Feb-04	CWS	SW	212,760
NV	NV0000190	TRUCKEE MEADOWS WATER AUTHORITY	CHLORODIBROMOMETHANE	2.87	UG/L	04-May-04	CWS	SW	212,760
NV	NV0000190	TRUCKEE MEADOWS WATER AUTHORITY	CHLORODIBROMOMETHANE	1.55	UG/L	11-Aug-04	CWS	SW	212,760
NV	NV0000190	TRUCKEE MEADOWS WATER AUTHORITY	CHLORODIBROMOMETHANE	5.78	UG/L	07-Dec-04	CWS	SW	212,760
NV	NV0000190	TRUCKEE MEADOWS WATER AUTHORITY	CHLORODIBROMOMETHANE	5.37	UG/L	15-Feb-05	CWS	SW	212,760
NV	NV0000190	TRUCKEE MEADOWS WATER AUTHORITY	CHLORODIBROMOMETHANE	0.63	UG/L	05-May-04	CWS	SW	212,760
NV	NV0000190	TRUCKEE MEADOWS WATER AUTHORITY	CHLORODIBROMOMETHANE	1.22	UG/L	09-Aug-04	CWS	SW	212,760
NV	NV0000190	TRUCKEE MEADOWS WATER AUTHORITY	CHLORODIBROMOMETHANE	3.4	UG/L	07-Dec-04	CWS	SW	212,760
NV	NV0000190	TRUCKEE MEADOWS WATER AUTHORITY	CHLORODIBROMOMETHANE	1.46	UG/L	11-Aug-04	CWS	SW	212,760
NV	NV0000190	TRUCKEE MEADOWS WATER AUTHORITY	CHLORODIBROMOMETHANE	2.8	UG/L	15-Feb-05	CWS	SW	212,760
NV	NV0000190	TRUCKEE MEADOWS WATER AUTHORITY	CHLORODIBROMOMETHANE	2.29	UG/L	02-Feb-04	CWS	SW	212,760
NV	NV0000190	TRUCKEE MEADOWS WATER AUTHORITY	CHLORODIBROMOMETHANE	2.44	UG/L	15-Feb-05	CWS	SW	212,760
NV	NV0000190	TRUCKEE MEADOWS WATER AUTHORITY	CHLORODIBROMOMETHANE	3.73	UG/L	08-Dec-04	CWS	SW	212,760
NV	NV0000190	TRUCKEE MEADOWS WATER AUTHORITY	CHLORODIBROMOMETHANE	0.93	UG/L	09-Aug-04	CWS	SW	212,760
NV	NV0000190	TRUCKEE MEADOWS WATER AUTHORITY	CHLORODIBROMOMETHANE	0.78	UG/L	09-Aug-04	CWS	SW	212,760
NV	NV0000190	TRUCKEE MEADOWS WATER AUTHORITY	CHLORODIBROMOMETHANE	0.5	UG/L	04-May-04	CWS	SW	212,760
NV	NV0000190	TRUCKEE MEADOWS WATER AUTHORITY	CHLORODIBROMOMETHANE	4.59	UG/L	15-Feb-05	CWS	SW	212,760
NV	NV0000190	TRUCKEE MEADOWS WATER AUTHORITY	CHLORODIBROMOMETHANE	0.9	UG/L	04-Feb-04	CWS	SW	212,760
NV	NV0000190	TRUCKEE MEADOWS WATER AUTHORITY	CHLORODIBROMOMETHANE	1.71	UG/L	08-Dec-04	CWS	SW	212,760
NV	NV0000190	TRUCKEE MEADOWS WATER AUTHORITY	CHLORODIBROMOMETHANE	4.22	UG/L	07-Dec-04	CWS	SW	212,760
NV	NV0000190	TRUCKEE MEADOWS WATER AUTHORITY	CHLORODIBROMOMETHANE	1.97	UG/L	05-May-04	CWS	SW	212,760
NV	NV0000190	TRUCKEE MEADOWS WATER AUTHORITY	CHLORODIBROMOMETHANE	3.15	UG/L	03-Feb-04	CWS	SW	212,760
NV	NV0000190	TRUCKEE MEADOWS WATER AUTHORITY	CHLORODIBROMOMETHANE	1.05	UG/L	11-Aug-04	CWS	SW	212,760
NV	NV0000190	TRUCKEE MEADOWS WATER AUTHORITY	CHLORODIBROMOMETHANE	4.82	UG/L	15-Feb-05	CWS	SW	212,760
NV	NV0000190	TRUCKEE MEADOWS WATER AUTHORITY	CHLORODIBROMOMETHANE	3.86	UG/L	02-Feb-04	CWS	SW	212,760
NV	NV0000190	TRUCKEE MEADOWS WATER AUTHORITY	CHLORODIBROMOMETHANE	6.99	UG/L	07-Dec-04	CWS	SW	212,760
NV	NV0000190	TRUCKEE MEADOWS WATER AUTHORITY	CHLORODIBROMOMETHANE	0.93	UG/L	04-May-04	CWS	SW	212,760
NV	NV0000190	TRUCKEE MEADOWS WATER AUTHORITY	CHLORODIBROMOMETHANE	4.84	UG/L	15-Feb-05	CWS	SW	212,760
NV	NV0000190	TRUCKEE MEADOWS WATER AUTHORITY	CHLORODIBROMOMETHANE	3.22	UG/L	11-Aug-04	CWS	SW	212,760
NV	NV0000190	TRUCKEE MEADOWS WATER AUTHORITY	CHLORODIBROMOMETHANE	0.52	UG/L	09-Aug-04	CWS	SW	212,760
NV	NV0000190	TRUCKEE MEADOWS WATER AUTHORITY	CHLORODIBROMOMETHANE	1.89	UG/L	08-Dec-04	CWS	SW	212,760
NV	NV0000190	TRUCKEE MEADOWS WATER AUTHORITY	CHLORODIBROMOMETHANE	1.5	UG/L	15-Feb-05	CWS	SW	212,760
NV	NV0000190	TRUCKEE MEADOWS WATER AUTHORITY	CHLORODIBROMOMETHANE	1.11	UG/L	04-Feb-04	CWS	SW	212,760
NV	NV0000190	TRUCKEE MEADOWS WATER AUTHORITY	CHLORODIBROMOMETHANE	3.31	UG/L	15-Feb-05	CWS	SW	212,760
NV	NV0000190	TRUCKEE MEADOWS WATER AUTHORITY	CHLORODIBROMOMETHANE	0.64	UG/L	04-May-04	CWS	SW	212,760
NV	NV0000190	TRUCKEE MEADOWS WATER AUTHORITY	CHLORODIBROMOMETHANE	1.24	UG/L	09-Aug-04	CWS	SW	212,760

APPENDIX 7-30 *Continued*

State	NUMBER0	SYSNAME	ANLYTNAME	LEVEL	UNIT	DATE	PWSType	PSource	Pop
NV	NV0000190	TRUCKEE MEADOWS WATER AUTHORITY	CHLORODIBROMOMETHANE	3.43	UG/L	08-Dec-04	CWS	SW	212,760
NV	NV0000190	TRUCKEE MEADOWS WATER AUTHORITY	CHLORODIBROMOMETHANE	0.52	UG/L	04-Feb-04	CWS	SW	212,760
NV	NV0000190	TRUCKEE MEADOWS WATER AUTHORITY	CHLORODIBROMOMETHANE	1.2	UG/L	15-Feb-05	CWS	SW	212,760
NV	NV0000190	TRUCKEE MEADOWS WATER AUTHORITY	CHLORODIBROMOMETHANE	0.91	UG/L	05-May-04	CWS	SW	212,760
NV	NV0000190	TRUCKEE MEADOWS WATER AUTHORITY	CHLORODIBROMOMETHANE	1.12	UG/L	08-Dec-04	CWS	SW	212,760
NV	NV0000190	TRUCKEE MEADOWS WATER AUTHORITY	CHLORODIBROMOMETHANE	2.07	UG/L	08-Dec-04	CWS	SW	212,760
NV	NV0000190	TRUCKEE MEADOWS WATER AUTHORITY	CHLORODIBROMOMETHANE	0.87	UG/L	09-Aug-04	CWS	SW	212,760
NV	NV0000190	TRUCKEE MEADOWS WATER AUTHORITY	CHLORODIBROMOMETHANE	1.93	UG/L	14-Feb-05	CWS	SW	212,760
NV	NV0000190	TRUCKEE MEADOWS WATER AUTHORITY	CHLORODIBROMOMETHANE	3.65	UG/L	02-Feb-04	CWS	SW	212,760
NV	NV0000190	TRUCKEE MEADOWS WATER AUTHORITY	CHLORODIBROMOMETHANE	1.19	UG/L	11-Aug-04	CWS	SW	212,760
NV	NV0000190	TRUCKEE MEADOWS WATER AUTHORITY	CHLORODIBROMOMETHANE	2.52	UG/L	15-Feb-05	CWS	SW	212,760
NV	NV0000190	TRUCKEE MEADOWS WATER AUTHORITY	CHLORODIBROMOMETHANE	5.53	UG/L	15-Feb-05	CWS	SW	212,760
NV	NV0000190	TRUCKEE MEADOWS WATER AUTHORITY	CHLORODIBROMOMETHANE	1.84	UG/L	04-May-04	CWS	SW	212,760
NV	NV0000190	TRUCKEE MEADOWS WATER AUTHORITY	CHLORODIBROMOMETHANE	6.41	UG/L	03-Feb-04	CWS	SW	212,760
NV	NV0000190	TRUCKEE MEADOWS WATER AUTHORITY	CHLORODIBROMOMETHANE	6.26	UG/L	07-Dec-04	CWS	SW	212,760
NV	NV0000190	TRUCKEE MEADOWS WATER AUTHORITY	CHLORODIBROMOMETHANE	1.5	UG/L	11-Aug-04	CWS	SW	212,760
NV	NV0000190	TRUCKEE MEADOWS WATER AUTHORITY	CHLORODIBROMOMETHANE	6.44	UG/L	07-Dec-04	CWS	SW	212,760
NV	NV0000190	TRUCKEE MEADOWS WATER AUTHORITY	CHLORODIBROMOMETHANE	1.45	UG/L	04-May-04	CWS	SW	212,760
NV	NV0000190	TRUCKEE MEADOWS WATER AUTHORITY	CHLORODIBROMOMETHANE	4.55	UG/L	15-Feb-05	CWS	SW	212,760
NV	NV0000190	TRUCKEE MEADOWS WATER AUTHORITY	CHLORODIBROMOMETHANE	4.23	UG/L	07-Dec-04	CWS	SW	212,760
NV	NV0000190	TRUCKEE MEADOWS WATER AUTHORITY	CHLORODIBROMOMETHANE	1.88	UG/L	05-May-04	CWS	SW	212,760
NV	NV0000190	TRUCKEE MEADOWS WATER AUTHORITY	CHLORODIBROMOMETHANE	5	UG/L	15-Feb-05	CWS	SW	212,760
NV	NV0000190	TRUCKEE MEADOWS WATER AUTHORITY	CHLORODIBROMOMETHANE	5.65	UG/L	08-Dec-04	CWS	SW	212,760
NV	NV0000190	TRUCKEE MEADOWS WATER AUTHORITY	CHLORODIBROMOMETHANE	7.19	UG/L	09-Aug-04	CWS	SW	212,760
NV	NV0000190	TRUCKEE MEADOWS WATER AUTHORITY	CHLORODIBROMOMETHANE	4.36	UG/L	14-Feb-05	CWS	SW	212,760
NV	NV0000190	TRUCKEE MEADOWS WATER AUTHORITY	CHLORODIBROMOMETHANE	2.3	UG/L	08-Dec-04	CWS	SW	212,760
NV	NV0000190	TRUCKEE MEADOWS WATER AUTHORITY	CHLORODIBROMOMETHANE	0.84	UG/L	04-Feb-04	CWS	SW	212,760
NV	NV0000190	TRUCKEE MEADOWS WATER AUTHORITY	CHLORODIBROMOMETHANE	2	UG/L	07-Dec-04	CWS	SW	212,760
NV	NV0000190	TRUCKEE MEADOWS WATER AUTHORITY	CHLORODIBROMOMETHANE	2.25	UG/L	14-Feb-05	CWS	SW	212,760
NV	NV0000190	TRUCKEE MEADOWS WATER AUTHORITY	CHLORODIBROMOMETHANE	1	UG/L	03-Feb-04	CWS	SW	212,760
NV	NV0000190	TRUCKEE MEADOWS WATER AUTHORITY	CHLORODIBROMOMETHANE	2.59	UG/L	07-Dec-04	CWS	SW	212,760
NV	NV0000190	TRUCKEE MEADOWS WATER AUTHORITY	CHLORODIBROMOMETHANE	2.49	UG/L	14-Feb-05	CWS	SW	212,760
NV	NV0000190	TRUCKEE MEADOWS WATER AUTHORITY	CHLORODIBROMOMETHANE	0.5	UG/L	09-Aug-04	CWS	SW	212,760
NV	NV0000190	TRUCKEE MEADOWS WATER AUTHORITY	CHLORODIBROMOMETHANE	0.94	UG/L	09-Aug-04	CWS	SW	212,760
NV	NV0000190	TRUCKEE MEADOWS WATER AUTHORITY	CHLORODIBROMOMETHANE	1.89	UG/L	14-Feb-05	CWS	SW	212,760
NV	NV0000190	TRUCKEE MEADOWS WATER AUTHORITY	CHLORODIBROMOMETHANE	0.57	UG/L	03-Feb-04	CWS	SW	212,760
NV	NV0000190	TRUCKEE MEADOWS WATER AUTHORITY	CHLORODIBROMOMETHANE	0.89	UG/L	03-Feb-04	CWS	SW	212,760
NV	NV0000190	TRUCKEE MEADOWS WATER AUTHORITY	CHLORODIBROMOMETHANE	1.69	UG/L	07-Dec-04	CWS	SW	212,760
NV	NV0000190	TRUCKEE MEADOWS WATER AUTHORITY	CHLORODIBROMOMETHANE	0.72	UG/L	03-Feb-04	CWS	SW	212,760
NV	NV0000190	TRUCKEE MEADOWS WATER AUTHORITY	CHLORODIBROMOMETHANE	2.17	UG/L	07-Dec-04	CWS	SW	212,760
				2.51					

APPENDIX 7-31 San Antonio Groundwater Data

State	PWS_ID	Source	CONSTITUENT	LEVEL_NO	Unit	COLL_DATE	PWSType	PSource	Pop
TX	0150120	bmwd leon springs mobile villa	ACETONE	24.9	ug/L	17-Jan-02	CWS	GW	741
TX	0150524	southwest isd elm creek	ACETONE	17	ug/L	30-Nov-00	NTNCWS	GW	700
TX	0150524	southwest isd elm creek	ACETONE	16.9	ug/L	30-Aug-01	NTNCWS	GW	700
TX	0150524	southwest isd elm creek	ACETONE	69.38	ug/L	25-May-05	NTNCWS	GW	700
TX	0150190	southwest isd main	ACETONE	24.8	ug/L	11-Apr-02	NTNCWS	GW	5,500
TX	0150018	san antonio water system	ACETONE	23	ug/L	29-Jan-03	CWS	GW	1,219,113
				29.3					
TX	0150534	bmwd woods of fair oaks	ATRAZINE	0.59	ug/L	18-Mar-04	CWS	GW	354
TX	0150524	southwest isd elm creek	Benzene, 2-chloro-1,3-dimethyl-	0.9	ug/L	25-May-05	NTNCWS	GW	700
TX	0150524	southwest isd elm creek	Benzene, 2-chloro-1,3-dimethyl-	0.9	ug/L	25-May-05	NTNCWS	GW	700
TX	0150524	southwest isd elm creek	Benzene, 4-bromo-1,2-dimethyl-	4.6	ug/L	25-May-05	NTNCWS	GW	700
TX	0150524	southwest isd elm creek	Benzene, 4-bromo-1,2-dimethyl-	4.6	ug/L	25-May-05	NTNCWS	GW	700
				2.75					
TX		143 SAMPLES	BROMODICHLOROMETHANE	4.52	ug/L				
TX	0150309	southwest research inst main comp	2-BUTANONE (MEK)	12	ug/L	21-Feb-00	NTNCWS	GW	2,400
TX	0150178	city of leon valley	2-BUTANONE	6.1	ug/L	11-Jul-01	CWS	GW	7,656
				6					
TX	0150430	bmwd country oaks estates	BUTYLBENZYLPHTHALATE	26.4	ug/L	23-Jan-03	CWS	GW	387
TX	0150524	southwest isd elm creek	Carbon Disulfide	2.25	ug/L	25-May-05	NTNCWS	GW	700
TX	0150524	southwest isd elm creek	Carbon Tetrachloride	0.59	ug/L	25-May-05	NTNCWS	GW	700
TX	0100017	bandera river ranch 1	CHLOROBENZENE	0.7	ug/L	08-May-01	CWS	GW	312
TX		244 SAMPLES	DIBROMOCHLOROMETHANE	4.23	ug/L				
TX	0150054	bmwd hill country	DIBROMOMETHANE	2.1	ug/L	15-Oct-01	CWS	GW	22,830
TX	0150010	city of kirby	1,4-DICHLOROBENZENE	0.5	ug/L	21-Mar-02	CWS	GW	8,623
TX	0150018	san antonio water system	DICHLOROBROMOMETHANE	1.3	ug/L	03-Nov-04	CWS	GW	1,219,113
TX	0150018	san antonio water system	DICHLOROBROMOMETHANE	2.9	ug/L	04-Nov-04	CWS	GW	1,219,113
				2.1					
TX	0150310	southwest research inst west comp	DICHLOROMETHANE	0.7	ug/L	21-Feb-00	NTNCWS	GW	200
TX	0150310	southwest research inst west comp	DICHLOROMETHANE	2.6	ug/L	30-May-00	NTNCWS	GW	200
TX	0150310	southwest research inst west comp	DICHLOROMETHANE	0.7	ug/L	09-Jul-02	NTNCWS	GW	200
TX	0150310	southwest research inst west comp	DICHLOROMETHANE	7.6	ug/L	17-Oct-02	NTNCWS	GW	200
TX	0150526	bmwd village green	DICHLOROMETHANE	0.6	ug/L	27-Jan-03	CWS	GW	207
TX	0150309	southwest research inst main comp	DICHLOROMETHANE	1.4	ug/L	17-Oct-02	NTNCWS	GW	2,400
TX	0150309	southwest research inst main comp	DICHLOROMETHANE	1.5	ug/L	17-Oct-02	NTNCWS	GW	2,400
				2.16					
TX	0460166	bmwd oakland estates	DI-(2-ETHYLHEXYL)ADIPATE	2.59	ug/L	18-Mar-04	CWS	GW	450
TX	0150018	san antonio water system	DI-(2-ETHYLHEXYL) PHTHALATE	4.19	ug/L	22-Jan-03	CWS	GW	1,219,113
TX	0150309	southwest research inst main comp	DIETHYLPHTHALATE	2.2	ug/L	03-Jun-04	NTNCWS	GW	2,400
TX	0150054	bmwd hill country	ETHYLBENZENE	1.3	ug/L	16-Apr-01	CWS	GW	22,830
TX	0150270	bmwd timberwood park	ETHYLBENZENE	1	ug/L	25-Oct-00	CWS	GW	5,475
TX	0150270	bmwd timberwood park	ETHYLBENZENE	2.3	ug/L	03-Jul-01	CWS	GW	5,475
TX	0150270	bmwd timberwood park	ETHYLBENZENE	0.6	ug/L	14-May-02	CWS	GW	5,475
TX	0150270	bmwd timberwood park	ETHYLBENZENE	1.6	ug/L	08-Jun-05	CWS	GW	5,475
TX	0150524	southwest isd elm creek	ETHYLBENZENE	1.1	ug/L	30-Nov-00	NTNCWS	GW	700
TX	0150524	southwest isd elm creek	ETHYLBENZENE	4.4	ug/L	30-Aug-01	NTNCWS	GW	700

APPENDIX 7-31 *Continued*

State	PWS_ID	Source	CONSTITUENT	LEVEL_NO	Unit	COLL_DATE	PWSType	PSource	Pop
TX	0150524	southwest isd elm creek	ETHYLBENZENE	1.39	ug/L	25-May-05	NTNCWS	GW	700
TX	0460027	city of garden ridge	ETHYLBENZENE	0.6	ug/L	22-Oct-02	CWS	GW	2,397
TX	0460027	city of garden ridge	ETHYLBENZENE	1.4	ug/L	17-Jul-03	CWS	GW	2,397
TX	0460027	city of garden ridge	ETHYLBENZENE	0.5	ug/L	26-Apr-04	CWS	GW	2,397
TX	0150068	city of live oak	ETHYLBENZENE	0.8	ug/L	12-Mar-02	CWS	GW	8,886
TX	0150190	southwest isd main	ETHYLBENZENE	1.7	ug/L	11-Apr-02	NTNCWS	GW	5,500
TX	0150018	san antonio water system	ETHYLBENZENE	1.1	ug/L	31-Aug-00	CWS	GW	1,219,113
TX	0150018	san antonio water system	ETHYLBENZENE	2.1	ug/L	04-Nov-04	CWS	GW	1,219,113
				1.46					
TX	0150120	bmwd leon springs mobile villa	FLUORENE	0.13	ug/L	25-Oct-00	CWS	GW	741
TX	0150524	southwest isd elm creek	HEXACHLOROBUTADIENE	1.3	ug/L	30-Nov-00	NTNCWS	GW	700
TX	0150309	southwest research inst main comp	METHYL METHACRYLATE	4.4	ug/L	21-Feb-00	NTNCWS	GW	2,400
TX	0150270	bmwd timberwood park	4-METHYL-2-PENTANONE (MIBK)	9.4	ug/L	25-Oct-00	CWS	GW	5,475
TX	0150270	bmwd timberwood park	4-METHYL-2-PENTANONE (MIBK)	21.9	ug/L	03-Jul-01	CWS	GW	5,475
TX	0460027	city of garden ridge	4-METHYL-2-PENTANONE (MIBK)	2.8	ug/L	22-Oct-02	CWS	GW	2,397
TX	0460027	city of garden ridge	4-METHYL-2-PENTANONE (MIBK)	7.9	ug/L	17-Jul-03	CWS	GW	2,397
TX	0460027	city of garden ridge	4-METHYL-2-PENTANONE (MIBK)	4.4	ug/L	26-Apr-04	CWS	GW	2,397
TX	0150018	san antonio water system	4-METHYL-2-PENTANONE (MIBK)	9.2	ug/L	31-Aug-00	CWS	GW	1,219,113
TX	0150524	southwest isd elm creek	4-METHYL-2-PENTANONE (MIBK)	4.4	ug/L	30-Nov-00	NTNCWS	GW	700
TX	0150524	southwest isd elm creek	4-METHYL-2-PENTANONE (MIBK)	12.3	ug/L	30-Aug-01	NTNCWS	GW	700
				9					
TX	0150120	bmwd leon springs mobile villa	PHENANTHRENE	0.16	ug/L	25-Oct-00	CWS	GW	741
TX	0150018	san antonio water system	STYRENE	0.6	ug/L	21-Feb-02	CWS	GW	1,219,113
TX	0150054	bmwd hill country	TETRACHLOROETHENE	0.6	ug/L	16-Apr-01	CWS	GW	22830
TX	0150054	bmwd hill country	TETRACHLOROETHENE	0.7	ug/L	25-Jun-02	CWS	GW	22830
TX	0150120	bmwd leon springs mobile villa	TETRACHLOROETHENE	1.6	ug/L	25-Oct-00	CWS	GW	741
TX	0150120	bmwd leon springs mobile villa	TETRACHLOROETHENE	1	ug/L	25-Oct-00	CWS	GW	741
TX	0150120	bmwd leon springs mobile villa	TETRACHLOROETHENE	1.2	ug/L	17-Jan-02	CWS	GW	741
TX	0150120	bmwd leon springs mobile villa	TETRACHLOROETHENE	2.5	ug/L	17-Jan-02	CWS	GW	741
TX	0150120	bmwd leon springs mobile villa	TETRACHLOROETHENE	2.1	ug/L	17-Jan-02	CWS	GW	741
TX	0150120	bmwd leon springs mobile villa	TETRACHLOROETHENE	0.6	ug/L	10-Sep-02	CWS	GW	741
TX	0150120	bmwd leon springs mobile villa	TETRACHLOROETHENE	0.6	ug/L	07-Oct-02	CWS	GW	741
TX	0150018	san antonio water system	TETRACHLOROETHENE	0.6	ug/L	14-Feb-00	CWS	GW	1,219,113
TX	0150018	san antonio water system	TETRACHLOROETHENE	0.7	ug/L	14-Feb-00	CWS	GW	1,219,113
TX	0150018	san antonio water system	TETRACHLOROETHENE	0.5	ug/L	31-Aug-00	CWS	GW	1,219,113
TX	0150018	san antonio water system	TETRACHLOROETHENE	0.8	ug/L	31-Aug-00	CWS	GW	1,219,113
TX	0150018	san antonio water system	TETRACHLOROETHENE	0.5	ug/L	12-Mar-01	CWS	GW	1,219,113
TX	0150018	san antonio water system	TETRACHLOROETHENE	1.2	ug/L	12-Mar-01	CWS	GW	1,219,113
TX	0150018	san antonio water system	TETRACHLOROETHENE	0.7	ug/L	13-Mar-01	CWS	GW	1,219,113
TX	0150018	san antonio water system	TETRACHLOROETHENE	0.6	ug/L	18-Sep-01	CWS	GW	1,219,113
TX	0150018	san antonio water system	TETRACHLOROETHENE	1.4	ug/L	18-Sep-01	CWS	GW	1,219,113
TX	0150018	san antonio water system	TETRACHLOROETHENE	0.9	ug/L	21-Feb-02	CWS	GW	1,219,113
TX	0150018	san antonio water system	TETRACHLOROETHENE	1.1	ug/L	24-Jul-02	CWS	GW	1,219,113
TX	0150018	san antonio water system	TETRACHLOROETHENE	0.6	ug/L	15-Oct-02	CWS	GW	1,219,113
TX	0150018	san antonio water system	TETRACHLOROETHENE	0.9	ug/L	21-Jan-03	CWS	GW	1,219,113
TX	0150018	san antonio water system	TETRACHLOROETHENE	0.6	ug/L	04-Aug-03	CWS	GW	1,219,113
TX	0150018	san antonio water system	TETRACHLOROETHENE	0.6	ug/L	03-Feb-04	CWS	GW	1,219,113
				0.94					

TX	0150018	san antonio water system	TETRAHYDROFURAN	39	ug/L	29-Jan-03	CWS	GW	1,219,113
TX	0150310	southwest research inst west comp	TETRAHYDROFURAN	3.5	ug/L	09-Apr-03	NTNCWS	GW	200
TX	0100017	bandera river ranch 1	TETRAHYDROFURAN	4.6	ug/L	08-May-01	CWS	GW	312
TX	0150120	bmwd leon springs mobile villa	TETRAHYDROFURAN	71.2	ug/L	17-Jan-02	CWS	GW	741
TX	0150092	city of shavano park	TETRAHYDROFURAN	5.4	ug/L	11-Jul-02	CWS	GW	2,040
TX	0150521	ram store 24 retail center	TETRAHYDROFURAN	5.8	ug/L	30-May-00	TNCWS	GW	100
				21.6					
TX	0940086	sunshine nut co	TOLUENE	4.8	ug/L	28-Aug-01	NTNCWS	GW	120
TX	0150120	bmwd leon springs mobile villa	TRICHLOROETHENE	0.6	ug/L	25-Oct-00	CWS	GW	741
TX	0460013	bmwd bulverde hills	1,2,4-TRIMETHYLBENZENE	1.6	ug/L	15-May-02	CWS	GW	840
TX	0150106	bexar county wcid 10 windcrest	M&P-XYLENE	1.3	ug/L	20-Mar-02	CWS	GW	5,105
TX	0150054	bmwd hill country	M&P-XYLENE	5	ug/L	16-Apr-01	CWS	GW	22,830
TX	0150054	bmwd hill country	O-XYLENE	3.2	ug/L	16-Apr-01	CWS	GW	22,830
TX	0150054	bmwd hill country	O-XYLENE	0.5	ug/L	14-May-02	CWS	GW	22,830
TX	0150524	southwest isd elm creek	M&P-XYLENE	3.3	ug/L	30-Nov-00	NTNCWS	GW	700
TX	0150524	southwest isd elm creek	M&P-XYLENE	10.5	ug/L	30-Aug-01	NTNCWS	GW	700
TX	0150524	southwest isd elm creek	M&P-XYLENE	2.95	ug/L	25-May-05	NTNCWS	GW	700
TX	0150524	southwest isd elm creek	O-XYLENE	2.6	ug/L	30-Nov-00	NTNCWS	GW	700
TX	0150524	southwest isd elm creek	O-XYLENE	8.5	ug/L	30-Aug-01	NTNCWS	GW	700
TX	0150524	southwest isd elm creek	O-XYLENE	5.05	ug/L	25-May-05	NTNCWS	GW	700
TX	0150524	southwest isd elm creek	Xylenes, Total	8	ug/L	25-May-05	NTNCWS	GW	700
TX	0150190	southwest isd main	M&P-XYLENE	3.6	ug/L	11-Apr-02	NTNCWS	GW	5,500
TX	0150190	southwest isd main	O-XYLENE	3.8	ug/L	11-Apr-02	NTNCWS	GW	5,500
TX	0150270	bmwd timberwood park	M&P-XYLENE	3.5	ug/L	25-Oct-00	CWS	GW	5,475
TX	0150270	bmwd timberwood park	M&P-XYLENE	13.5	ug/L	03-Jul-01	CWS	GW	5,475
TX	0150270	bmwd timberwood park	M&P-XYLENE	1.8	ug/L	14-May-02	CWS	GW	5,475
TX	0150270	bmwd timberwood park	M&P-XYLENE	6.57	ug/L	08-Jun-05	CWS	GW	5,475
TX	0150270	bmwd timberwood park	O-XYLENE	2.1	ug/L	25-Oct-00	CWS	GW	5,475
TX	0150270	bmwd timberwood park	O-XYLENE	6.7	ug/L	03-Jul-01	CWS	GW	5,475
TX	0150270	bmwd timberwood park	O-XYLENE	1	ug/L	14-May-02	CWS	GW	5,475
TX	0150270	bmwd timberwood park	O-XYLENE	0.5	ug/L	14-May-02	CWS	GW	5,475
TX	0150270	bmwd timberwood park	O-XYLENE	2.97	ug/L	08-Jun-05	CWS	GW	5,475
TX	0460027	city of garden ridge	M&P-XYLENE	2.4	ug/L	22-Oct-02	CWS	GW	2,397
TX	0460027	city of garden ridge	M&P-XYLENE	5.9	ug/L	17-Jul-03	CWS	GW	2,397
TX	0460027	city of garden ridge	M&P-XYLENE	2.2	ug/L	26-Apr-04	CWS	GW	2,397
TX	0460027	city of garden ridge	O-XYLENE	0.8	ug/L	22-Oct-02	CWS	GW	2,397
TX	0460027	city of garden ridge	O-XYLENE	2.4	ug/L	17-Jul-03	CWS	GW	2,397
TX	0460027	city of garden ridge	O-XYLENE	1	ug/L	26-Apr-04	CWS	GW	2,397
TX	0150068	city of live oak	M&P-XYLENE	1.7	ug/L	12-Mar-02	CWS	GW	8,886
TX	0150068	city of live oak	O-XYLENE	1	ug/L	12-Mar-02	CWS	GW	8,886
TX	0150270	bmwd timberwood park	Xylenes, Total	9.54	ug/L	08-Jun-05	CWS	GW	5,475
TX	0150018	san antonio water system	M&P-XYLENE	6.2	ug/L	31-Aug-00	CWS	GW	1,219,113
TX	0150018	san antonio water system	M&P-XYLENE	1.1	ug/L	13-Mar-01	CWS	GW	1,219,113
TX	0150018	san antonio water system	M&P-XYLENE	1.5	ug/L	17-Sep-01	CWS	GW	1,219,113
TX	0150018	san antonio water system	M&P-XYLENE	5.3	ug/L	04-Nov-04	CWS	GW	1,219,113
TX	0150018	san antonio water system	O-XYLENE	3.4	ug/L	31-Aug-00	CWS	GW	1,219,113
TX	0150018	san antonio water system	O-XYLENE	0.5	ug/L	12-Mar-01	CWS	GW	1,219,113
TX	0150018	san antonio water system	O-XYLENE	0.6	ug/L	13-Mar-01	CWS	GW	1,219,113
TX	0150018	san antonio water system	O-XYLENE	1.1	ug/L	17-Sep-01	CWS	GW	1,219,113
TX	0150018	san antonio water system	O-XYLENE	0.5	ug/L	19-Jun-02	CWS	GW	1,219,113
TX	0150018	san antonio water system	O-XYLENE	2.6	ug/L	04-Nov-04	CWS	GW	1,219,113
				3.58					

APPENDIX 7-32 San Antonio Surface Water Data

State	PWS_ID	Source	CONSTITUENT	LEVEL_NO	Unit	COLL_DATE	PWSType	PSource	Pop
TX	0150249	bmwd southside	ATRAZINE	0.25	ug/L	14-May-01	CWS	SW	116,406
TX	0150249	bmwd southside	ATRAZINE	0.27	ug/L	14-May-01	CWS	SW	116,406
TX	0150249	bmwd southside	ATRAZINE	0.27	ug/L	14-May-01	CWS	SW	116,406
TX	0150249	bmwd southside	ATRAZINE	0.54	ug/L	11-Apr-02	CWS	SW	116,406
				0.33					
TX	0150249	bmwd southside	BROMODICHLOROMETHANE	30	ug/L	07-Feb-01	CWS	SW	116,406
TX	0150249	bmwd southside	BROMODICHLOROMETHANE	11	ug/L	14-May-01	CWS	SW	116,406
TX	0150249	bmwd southside	BROMODICHLOROMETHANE	1.7	ug/L	14-May-01	CWS	SW	116,406
TX	0150249	bmwd southside	BROMODICHLOROMETHANE	11	ug/L	14-May-01	CWS	SW	116,406
TX	0150249	bmwd southside	BROMODICHLOROMETHANE	0.7	ug/L	14-May-01	CWS	SW	116,406
TX	0150249	bmwd southside	BROMODICHLOROMETHANE	11.9	ug/L	31-Jul-01	CWS	SW	116,406
TX	0150249	bmwd southside	BROMODICHLOROMETHANE	8.6	ug/L	06-Nov-01	CWS	SW	116,406
TX	0150249	bmwd southside	BROMODICHLOROMETHANE	1.3	ug/L	06-Nov-01	CWS	SW	116,406
TX	0150249	bmwd southside	BROMODICHLOROMETHANE	7.6	ug/L	06-Nov-01	CWS	SW	116,406
TX	0150249	bmwd southside	BROMODICHLOROMETHANE	2.3	ug/L	06-Nov-01	CWS	SW	116,406
TX	0150249	bmwd southside	BROMODICHLOROMETHANE	5.8	ug/L	06-Nov-01	CWS	SW	116,406
TX	0150249	bmwd southside	BROMODICHLOROMETHANE	8.7	ug/L	15-Jan-02	CWS	SW	116,406
TX	0150249	bmwd southside	BROMODICHLOROMETHANE	8.3	ug/L	15-Jan-02	CWS	SW	116,406
TX	0150249	bmwd southside	BROMODICHLOROMETHANE	9	ug/L	15-Jan-02	CWS	SW	116,406
TX	0150249	bmwd southside	BROMODICHLOROMETHANE	7.9	ug/L	15-Jan-02	CWS	SW	116,406
TX	0150249	bmwd southside	BROMODICHLOROMETHANE	15.4	ug/L	21-Aug-02	CWS	SW	116,406
TX	0150249	bmwd southside	BROMODICHLOROMETHANE	5.3	ug/L	23-Jan-03	CWS	SW	116,406
TX	0150249	bmwd southside	BROMODICHLOROMETHANE	12.47	ug/L	03-Aug-04	CWS	SW	116,406
TX	0150249	bmwd southside	BROMODICHLOROMETHANE	1.73	ug/L	09-Mar-05	CWS	SW	116,406
				8.46					

TX	0150249	bmwd southside	DIBROMOCHLOROMETHANE	0.5	ug/L	01-Feb-00	CWS	SW	116,406
TX	0150249	bmwd southside	DIBROMOCHLOROMETHANE	0.7	ug/L	01-Feb-00	CWS	SW	116,406
TX	0150249	bmwd southside	DIBROMOCHLOROMETHANE	13	ug/L	07-Feb-01	CWS	SW	116,406
TX	0150249	bmwd southside	DIBROMOCHLOROMETHANE	13	ug/L	14-May-01	CWS	SW	116,406
TX	0150249	bmwd southside	DIBROMOCHLOROMETHANE	2.2	ug/L	14-May-01	CWS	SW	116,406
TX	0150249	bmwd southside	DIBROMOCHLOROMETHANE	11	ug/L	14-May-01	CWS	SW	116,406
TX	0150249	bmwd southside	DIBROMOCHLOROMETHANE	1.4	ug/L	14-May-01	CWS	SW	116,406
TX	0150249	bmwd southside	DIBROMOCHLOROMETHANE	0.7	ug/L	15-May-01	CWS	SW	116,406
TX	0150249	bmwd southside	DIBROMOCHLOROMETHANE	19	ug/L	31-Jul-01	CWS	SW	116,406
TX	0150249	bmwd southside	DIBROMOCHLOROMETHANE	0.8	ug/L	06-Nov-01	CWS	SW	116,406
TX	0150249	bmwd southside	DIBROMOCHLOROMETHANE	12	ug/L	06-Nov-01	CWS	SW	116,406
TX	0150249	bmwd southside	DIBROMOCHLOROMETHANE	1.3	ug/L	06-Nov-01	CWS	SW	116,406
TX	0150249	bmwd southside	DIBROMOCHLOROMETHANE	10	ug/L	06-Nov-01	CWS	SW	116,406
TX	0150249	bmwd southside	DIBROMOCHLOROMETHANE	4.7	ug/L	06-Nov-01	CWS	SW	116,406
TX	0150249	bmwd southside	DIBROMOCHLOROMETHANE	7.5	ug/L	06-Nov-01	CWS	SW	116,406
TX	0150249	bmwd southside	DIBROMOCHLOROMETHANE	2.1	ug/L	15-Jan-02	CWS	SW	116,406
TX	0150249	bmwd southside	DIBROMOCHLOROMETHANE	7.4	ug/L	15-Jan-02	CWS	SW	116,406
TX	0150249	bmwd southside	DIBROMOCHLOROMETHANE	7.2	ug/L	15-Jan-02	CWS	SW	116,406
TX	0150249	bmwd southside	DIBROMOCHLOROMETHANE	7.5	ug/L	15-Jan-02	CWS	SW	116,406
TX	0150249	bmwd southside	DIBROMOCHLOROMETHANE	1.5	ug/L	15-Jan-02	CWS	SW	116,406
TX	0150249	bmwd southside	DIBROMOCHLOROMETHANE	7.2	ug/L	15-Jan-02	CWS	SW	116,406
TX	0150249	bmwd southside	DIBROMOCHLOROMETHANE	19.2	ug/L	21-Aug-02	CWS	SW	116,406
TX	0150249	bmwd southside	DIBROMOCHLOROMETHANE	9.8	ug/L	23-Jan-03	CWS	SW	116,406
TX	0150249	bmwd southside	DIBROMOCHLOROMETHANE	13.38	ug/L	03-Aug-04	CWS	SW	116,406
TX	0150249	bmwd southside	DIBROMOCHLOROMETHANE	1.8	ug/L	09-Mar-05	CWS	SW	116,406
				7					
TX	0150249	bmwd southside	M&P-XYLENE	1	ug/L	14-May-01	CWS	SW	116,406
TX	0150249	bmwd southside	O-XYLENE	0.5	ug/L	06-Nov-01	CWS	SW	116,406
				0.5					

APPENDIX 7-33 Savannah Groundwater Data

State	PWS_ID	Source	CONSTITUENT	LEVEL_NO	Unit	COLL_DATE	PWSType	PSource	Pop
GA	GA3050009	spring grove subdivision	BROMOFORM	0.85	ug/L	20-Sep-04	CWS	GW	68
GA	GA2290005	oak ridge est/hacklebarney s/d	BROMOFORM	0.87	ug/L	13-Jun-01	CWS	GW	419
GA	GA2290005	oak ridge est/hacklebarney s/d	BROMOFORM	1.2	ug/L	13-Jun-01	CWS	GW	419
GA	GA2290005	oak ridge est/hacklebarney s/d	BROMOFORM	0.63	ug/L	29-May-02	CWS	GW	419
GA	GA2290005	oak ridge est/hacklebarney s/d	BROMOFORM	0.70	ug/L	27-May-03	CWS	GW	419
GA	GA2290005	oak ridge est/hacklebarney s/d	BROMOFORM	0.89	ug/L	12-May-04	CWS	GW	419
GA	GA2290005	oak ridge est/hacklebarney s/d	BROMOFORM	1.1	ug/L	23-Feb-00	CWS	GW	419
GA	GA1910069	sutherland bluff plantation	BROMOFORM	0.60	ug/L	22-Aug-01	CWS	GW	84
GA	GA1910069	sutherland bluff plantation	BROMOFORM	2.4	ug/L	02-Jun-05	CWS	GW	84
GA	GA1910069	sutherland bluff plantation	BROMOFORM	0.72	ug/L	02-Jun-05	CWS	GW	84
GA	GA1030125	savannah electric-springfield opr. ctr.	BROMOFORM	0.69	ug/L	28-Oct-03	NTNCWS	GW	40
GA	GA1030125	savannah electric-springfield opr. ctr.	BROMOFORM	0.60	ug/L	01-May-01	NTNCWS	GW	40
GA	GA1030125	savannah electric-springfield opr. ctr.	BROMOFORM	0.88	ug/L	23-Jul-01	NTNCWS	GW	40
GA	GA1030125	savannah electric-springfield opr. ctr.	BROMOFORM	1.1	ug/L	03-Sep-03	NTNCWS	GW	40
GA	GA1030125	savannah electric-springfield opr. ctr.	BROMOFORM	0.73	ug/L	19-Feb-01	NTNCWS	GW	40
GA	GA1030125	savannah electric-springfield opr. ctr.	BROMOFORM	0.79	ug/L	19-Jan-04	NTNCWS	GW	40
GA	GA1030119	huntington station	BROMOFORM	3.9	ug/L	18-Aug-05	CWS	GW	122
GA	GA1030119	huntington station	BROMOFORM	1.0	ug/L	16-Feb-05	CWS	GW	122
GA	GA1030119	huntington station	BROMOFORM	3.3	ug/L	25-Apr-05	CWS	GW	122
GA	GA1030109	hunters mill subdivision	BROMOFORM	0.50	ug/L	09-Sep-02	CWS	GW	26
GA	GA1030109	hunters mill subdivision	BROMOFORM	0.72	ug/L	20-Mar-00	CWS	GW	26
GA	GA1030095	pine hill subdivision	BROMOFORM	0.53	ug/L	23-Jul-01	CWS	GW	676
GA	GA1030095	pine hill subdivision	BROMOFORM	0.73	ug/L	01-May-03	CWS	GW	676
GA	GA1030093	saddlebrook subdivision	BROMOFORM	1.5	ug/L	17-Nov-04	CWS	GW	283
GA	GA1030077	goshen villa subdivision	BROMOFORM	2.1	ug/L	15-Apr-03	CWS	GW	486
GA	GA1030077	goshen villa subdivision	BROMOFORM	2.1	ug/L	15-Apr-03	CWS	GW	486
GA	GA1030012	westwood heights subdivision	BROMOFORM	1.4	ug/L	30-Mar-05	CWS	GW	1092
GA	GA0510275	dean forest rd.tp-north	BROMOFORM	0.51	ug/L	04-Feb-03	CWS	GW	39
GA	GA0510056	white oaks subdivision	BROMOFORM	0.65	ug/L	27-Jun-00	CWS	GW	360
GA	GA0510139	heathcote farms subdivision	BROMOFORM	0.67	ug/L	27-Jun-00	CWS	GW	68
GA	GA0510130	candler general hospital	BROMOFORM	0.57	ug/L	08-Oct-02	CWS	GW	1400
GA	GA0290085	marsh harbor	BROMOFORM	0.53	ug/L	07-Oct-02	CWS	GW	55
GA	GA0310188	olney station	BROMOFORM	0.50	ug/L	09-Sep-02	CWS	GW	240
GA	GA0290080	buckhead development	BROMOFORM	1.6	ug/L	28-Feb-05	CWS	GW	605
GA	GA0510021	cottonvale estates	BROMOFORM	0.51	ug/L	12-Jan-00	CWS	GW	260
GA	GA0510022	dyches mobile estates	BROMOFORM	0.51	ug/L	11-Jan-00	CWS	GW	210
GA	GA0510024	estill hammock/spanish hammock	BROMOFORM	0.52	ug/L	26-Jan-00	CWS	GW	317
GA	GA0510034	larchmont utilities	BROMOFORM	4.8	ug/L	23-Jun-05	CWS	GW	6515
GA	GA0510034	larchmont utilities	BROMOFORM	2.5	ug/L	23-Jun-05	CWS	GW	6515
GA	GA0510034	larchmont utilities	BROMOFORM	8.6	ug/L	23-Jun-05	CWS	GW	6515
GA	GA0510050	southwinds community	BROMOFORM	0.50	ug/L	26-Jan-00	CWS	GW	450
GA	GA0250006	whispering pines subdivision	BROMOFORM	0.59	ug/L	20-Aug-02	CWS	GW	182
GA	GA0250006	whispering pines subdivision	BROMOFORM	0.87	ug/L	05-Oct-05	CWS	GW	182
GA	GA0290005	bryan acres subdivision	BROMOFORM	0.53	ug/L	09-Sep-02	CWS	GW	140
GA	GA0290076	oxford subdivision	BROMOFORM	0.51	ug/L	03-Sep-02	CWS	GW	309
GA	GA0290073	strathy hall two subdivision	BROMOFORM	0.67	ug/L	21-Feb-02	CWS	GW	325
				1.28					

GA		164 SAMPLES	BROMODICHLOROMETHANE	1.46	ug/L				
GA		122 SAMPLES	CHLOROFORM	1.53	ug/L				
GA		186 SAMPLES	CHLORODIBROMOMETHANE	1.47	ug/L				
GA	GA1830001	cedar brook mobile home park	DICHLOROACETIC ACID	1.5	ug/L	01-Sep-04	CWS	GW	55
GA	GA0510075	uga-skidaway oceanography inst	DICHLOROACETIC ACID	2.4	ug/L	27-Sep-05	NTNCWS	GW	108
GA	GA0510075	uga-skidaway oceanography inst	DICHLOROACETIC ACID	2.3	ug/L	16-Aug-04	NTNCWS	GW	108
GA	GA0510022	dyches mobile estates	DICHLOROACETIC ACID	1.3	ug/L	30-Sep-05	CWS	GW	210
				2.07					
GA	GA1030109	hunters mill subdivision	DICHLOROMETHANE	0.72	ug/L	09-Sep-02	CWS	GW	26
GA	GA1030095	pine hill subdivision	DICHLOROMETHANE	0.58	ug/L	01-May-03	CWS	GW	676
GA	GA0510056	white oaks subdivision	DICHLOROMETHANE	0.77	ug/L	09-Sep-02	CWS	GW	360
GA	GA0510008	savannah christian prep. school	DICHLOROMETHANE	0.76	ug/L	08-Oct-02	NTNCWS	GW	1150
GA	GA0290006	bryan woods estates	DICHLOROMETHANE	1.8	ug/L	09-Sep-02	CWS	GW	100
GA	GA0290005	bryan acres subdivision	DICHLOROMETHANE	2.0	ug/L	09-Sep-02	CWS	GW	140
GA	GA0290004	so atl-black creek farms subdivision	DICHLOROMETHANE	1.6	ug/L	09-Sep-02	CWS	GW	302
				1.18					
GA	GA0510034	larchmont utilities	ETHYLBENZENE	1.2	ug/L	23-Jun-05	CWS	GW	6515
GA	GA1030116	hampton creek subdivision	METHYL ETHYL KETONE (MEK)	2.4	ug/L	08-Oct-02	CWS	GW	26
GA	GA0510130	candler general hospital	MONOCHLOROACETIC ACID	3.2	ug/L	03-Sep-03	CWS	GW	1400
GA	GA0510092	derenne plaza condo	MONOCHLOROACETIC ACID	2.8	ug/L	03-Sep-03	CWS	GW	86
GA	GA0510046	river oaks subdivision	MONOCHLOROACETIC ACID	6.1	ug/L	03-Sep-03	CWS	GW	455
GA	GA0510014	azalea mobile home plaza	MONOCHLOROACETIC ACID	3.2	ug/L	03-Sep-03	CWS	GW	475
GA	GA0510008	savannah christian prep. school	MONOCHLOROACETIC ACID	3.6	ug/L	03-Sep-03	NTNCWS	GW	1150
				3.78					
GA	GA1030125	savannah electric-springfield opr. ctr.	TOLUENE	0.77	ug/L	23-Jul-01	NTNCWS	GW	40
GA	GA1030119	huntington station	TOLUENE	4.4	ug/L	18-Aug-05	CWS	GW	122
GA	GA1030119	huntington station	TOLUENE	0.89	ug/L	16-Feb-05	CWS	GW	122
GA	GA1030119	huntington station	TOLUENE	0.95	ug/L	25-Apr-05	CWS	GW	122
GA	GA1030116	hampton creek subdivision	TOLUENE	0.56	ug/L	23-Sep-02	CWS	GW	26
GA	GA1030116	hampton creek subdivision	TOLUENE	0.80	ug/L	03-Feb-03	CWS	GW	26
GA	GA0290080	buckhead development	TOLUENE	3.9	ug/L	28-Feb-05	CWS	GW	605
GA	GA0290080	buckhead development	TOLUENE	1.8	ug/L	23-Mar-05	CWS	GW	605
GA	GA0510275	dean forest rd.tp-north	TOLUENE	0.54	ug/L	05-Nov-01	CWS	GW	39
GA	GA0510275	dean forest rd.tp-north	TOLUENE	1.2	ug/L	12-Feb-02	CWS	GW	39
GA	GA0510255	airport offices & indust. park	TOLUENE	2.1	ug/L	12-Feb-02	NTNCWS	GW	80
				1.63					
GA	GA1910050	belle bluff estates	TRICHLOROACETIC ACID	1.6	ug/L	13-Sep-04	CWS	GW	73
GA	GA1830001	cedar brook mobile home park	TRICHLOROACETIC ACID	2.3	ug/L	01-Sep-04	CWS	GW	55
				1.95					
GA	GA0510034	larchmont utilities	1,2,4-TRIMETHYLBENZENE	3.7	ug/L	23-Jun-05	CWS	GW	6515
GA	GA0510034	larchmont utilities	1,3,5-TRIMETHYLBENZENE	0.65	ug/L	23-Jun-05	CWS	GW	6515
				2.18					
GA	GA1030116	hampton creek subdivision	XYLENES	0.74	ug/L	30-Jul-03	CWS	GW	26
GA	GA1030093	saddlebrook subdivision	XYLENES	1.3	ug/L	14-Mar-05	CWS	GW	283
GA	GA0510034	larchmont utilities	XYLENES	16	ug/L	23-Jun-05	CWS	GW	6515
GA	GA0510008	savannah christian prep. school	XYLENES	2.6	ug/L	08-Oct-02	NTNCWS	GW	1150
				5.16					

APPENDIX 7-34 Sioux Falls Surface Water Data

State	PWSID	SYSTEM	SAMPLE SITE	ANALYTE	CONC	DATE	PWSType	PSource	Pop
SD	SD4600294	Sioux Falls	TREATMENT PLANT	2,4-D	0.6	22-May-00	CWS	SW	128,000

APPENDIX 7-35 Springfield Surface Water Data

State	PWS_ID	Source	CONSTITUENT	LEVEL_NO	Unit	COLL_DATE	PWSType	PSource	Pop
IL	IL1671200	springfield	ATRAZINE	0.5	UG/L	07-Mar-02	CWS	SW	130,000
IL	IL1671200	springfield	ATRAZINE	1.5	UG/L	16-May-02	CWS	SW	130,000
IL	IL1671200	springfield	ATRAZINE	0.39	UG/L	04-Sep-02	CWS	SW	130,000
IL	IL1671200	springfield	ATRAZINE	0.39	UG/L	30-Oct-02	CWS	SW	130,000
IL	IL1671200	springfield	ATRAZINE	0.36	UG/L	03-Mar-03	CWS	SW	130,000
IL	IL1671200	springfield	ATRAZINE	0.39	UG/L	07-May-03	CWS	SW	130,000
IL	IL1671200	springfield	ATRAZINE	1.1	UG/L	02-Sep-03	CWS	SW	130,000
IL	IL1671200	springfield	ATRAZINE	0.78	UG/L	15-Dec-03	CWS	SW	130,000
IL	IL1671200	springfield	ATRAZINE	0.46	UG/L	09-Aug-04	CWS	SW	130,000
IL	IL1671200	springfield	ATRAZINE	0.53	UG/L	02-Nov-04	CWS	SW	130,000
IL	IL1671200	springfield	ATRAZINE	0.51	UG/L	18-Feb-04	CWS	SW	130,000
				0.63					
IL	IL1671200	springfield	SIMAZINE	1.2	UG/L	01-Feb-05	CWS	SW	130,000

APPENDIX 7-36 Tulsa Surface Water Data

State	PWSID	SYSTEM	SAMPLE SITE	ANALYTE	CONC	UOM	DATE	PWSType	PSource	Pop
OK	OK1020418	TULSA MOHAWK	TULSA MOHAWK WTP	BROMODICHLOROMETHANE	7.1	UG/L	22-May-02	CWS	SW	197,000
OK	OK1021307	TULSA AB JEWELL	TULSA A.B. JEWELL WT	BROMODICHLOROMETHANE	10.2	UG/L	18-Oct-02	CWS	SW	403,000
OK	OK1021616	SPAVINAW	SPAVINAW WTP	BROMODICHLOROMETHANE	0.8	UG/L	02-Aug-00	CWS	SW	563
OK	OK1021616	SPAVINAW	SPAVINAW WTP	BROMODICHLOROMETHANE	0.6	UG/L	02-Oct-00	CWS	SW	563
OK	OK1021616	SPAVINAW	SPAVINAW WTP	BROMODICHLOROMETHANE	0.8	UG/L	28-Nov-01	CWS	SW	563
OK	OK1021616	SPAVINAW	SPAVINAW WTP	BROMODICHLOROMETHANE	9.3	UG/L	05-Mar-02	CWS	SW	563
OK	OK1021616	SPAVINAW	SPAVINAW WTP	BROMODICHLOROMETHANE	22	UG/L	17-May-02	CWS	SW	563
OK	OK1021616	SPAVINAW	SPAVINAW WTP	BROMODICHLOROMETHANE	1.6	UG/L	08-Aug-02	CWS	SW	563
OK	OK1021616	SPAVINAW	SPAVINAW WTP	BROMODICHLOROMETHANE	8.3	UG/L	12-Feb-03	CWS	SW	563
					6.74					
OK	OK1020418	TULSA MOHAWK	TULSA MOHAWK WTP	CHLORODIBROMOMETHANE	4.1	UG/L	22-May-02	CWS	SW	197,000
OK	OK1021307	TULSA AB JEWELL	TULSA A.B. JEWELL WT	CHLORODIBROMOMETHANE	7.6	UG/L	18-Oct-02	CWS	SW	403,000
OK	OK1021616	SPAVINAW	SPAVINAW WTP	CHLORODIBROMOMETHANE	0.1	UG/L	02-Aug-00	CWS	SW	563
OK	OK1021616	SPAVINAW	SPAVINAW WTP	CHLORODIBROMOMETHANE	4.6	UG/L	05-Mar-02	CWS	SW	563
OK	OK1021616	SPAVINAW	SPAVINAW WTP	CHLORODIBROMOMETHANE	8.2	UG/L	17-May-02	CWS	SW	563
OK	OK1021616	SPAVINAW	SPAVINAW WTP	CHLORODIBROMOMETHANE	2.3	UG/L	12-Feb-03	CWS	SW	563
					4.48					
OK	OK1020418	TULSA MOHAWK	TULSA MOHAWK WTP	CHLOROFORM	7.4	UG/L	22-May-02	CWS	SW	197,000
OK	OK1021307	TULSA AB JEWELL	TULSA A.B. JEWELL WT	CHLOROFORM	7.6	UG/L	18-Oct-02	CWS	SW	403,000
OK	OK1021616	SPAVINAW	SPAVINAW WTP	CHLOROFORM	2.69	UG/L	03-May-00	CWS	SW	563
OK	OK1021616	SPAVINAW	SPAVINAW WTP	CHLOROFORM	2.9	UG/L	02-Aug-00	CWS	SW	563
OK	OK1021616	SPAVINAW	SPAVINAW WTP	CHLOROFORM	2	UG/L	02-Oct-00	CWS	SW	563
OK	OK1021616	SPAVINAW	SPAVINAW WTP	CHLOROFORM	0.7	UG/L	22-Aug-01	CWS	SW	563
OK	OK1021616	SPAVINAW	SPAVINAW WTP	CHLOROFORM	4	UG/L	28-Nov-01	CWS	SW	563
OK	OK1021616	SPAVINAW	SPAVINAW WTP	CHLOROFORM	15.6	UG/L	05-Mar-02	CWS	SW	563
OK	OK1021616	SPAVINAW	SPAVINAW WTP	CHLOROFORM	59	UG/L	17-May-02	CWS	SW	563
OK	OK1021616	SPAVINAW	SPAVINAW WTP	CHLOROFORM	7.8	UG/L	08-Aug-02	CWS	SW	563
OK	OK1021616	SPAVINAW	SPAVINAW WTP	CHLOROFORM	22.3	UG/L	12-Feb-03	CWS	SW	563
OK	OK1021616	SPAVINAW	SPAVINAW WTP	CHLOROFORM	1.5	UG/L	03-Feb-04	CWS	SW	563
					11.1					
OK	OK1021616	SPAVINAW	SPAVINAW WTP	XYLENES	1.24	UG/L	03-May-00	CWS	SW	563
OK	OK1021616	SPAVINAW	SPAVINAW WTP	XYLENES	2.4	UG/L	02-Aug-00	CWS	SW	563
					1.82					

APPENDIX 7-37 West Palm Beach Groundwater Data

State	PWSID	NAME	ANALYTE	RESULTS	UNIT_OF_MEASURE	SAMPLE_DATE	PWSType	PSource	Pop
FL	FL4504664	JUPITER FARMS COMMU. ELEM. SCHL.	DALAPON	2.6	UG/L	15-Nov-04	NTNCWS	GW	825

APPENDIX 7-38 West Palm Beach Surface Water

State	PWSID	NAME	ANALYTE	RESULTS	UNIT_OF_MEASURE	SAMPLE_DATE	PWSType	PSource	Pop
FL	FL4501559	WEST PALM BEACH CITY OF	ATRAZINE	0.21	UG/L	16-Aug-04	CWS	SW	87,769
FL	FL4501559	WEST PALM BEACH CITY OF	ATRAZINE	0.15	UG/L	07-Jun-04	CWS	SW	87,769
				0.18					

APPENDIX 7-39 Worcester Groundwater Data

State	PWS_ID	Source	ANALYTE	LEVEL_NO	Unit	COLL_DATE	PWSType	PSource	Pop
MA	MA2054010	CHARLTON CHINESE TAKEOUT	METHYL TERTIARY BUTYL ETHER	2.8	ug/L	19-Feb-03	TNCWS	GW	35

APPENDIX 7-40 Worcester Surface Water Data

State	PWS_ID	Source	POE	Constituent	Level_NO	Unit
MA	MA2348000	DPW, Water	Filtration	Chloroform	23	ug/L
MA	MA2348000	DPW, Water	Filtration	Chloroform	19	ug/L
MA	MA2348000	DPW, Water	Filtration	Chloroform	15	ug/L
MA	MA2348000	DPW, Water	Filtration	Chloroform	14	ug/L
MA	MA2348000	DPW, Water	Filtration	Chloroform	15	ug/L
					17.2	
MA	MA2348000	DPW, Water	Filtration	Bromodichloromethane	4.7	ug/L
MA	MA2348000	DPW, Water	Filtration	Bromodichloromethane	5.6	ug/L
MA	MA2348000	DPW, Water	Filtration	Bromodichloromethane	5.6	ug/L
MA	MA2348000	DPW, Water	Filtration	Bromodichloromethane	3.9	ug/L
MA	MA2348000	DPW, Water	Filtration	Bromodichloromethane	4.1	ug/L
					4.78	
MA	MA2348000	DPW, Water	Filtration	Chlorodibromomethane	0.8	ug/L
MA	MA2348000	DPW, Water	Filtration	Chlorodibromomethane	0.9	ug/L
MA	MA2348000	DPW, Water	Filtration	Chlorodibromomethane	0.67	ug/L
MA	MA2348000	DPW, Water	Filtration	Chlorodibromomethane	0.6	ug/L
MA	MA2348000	DPW, Water	Filtration	Chlorodibromomethane	0.5	ug/L
					0.7	

APPENDIX 8-1 EPA-Regulated Air Toxic List: Synthetic Organic Chemicals

Acetaldehyde	1,4-Dichlorobenzene(p)
Acetamide	3,3-Dichlorobenzidene
Acetonitrile	Dichloroethyl ether
Acetophenone	(Bis(2-chloroethyl)ether)
2-Acetylaminofluorene	1,3-Dichloropropene
Acrolein	Dichlorvos
Acrylamide	Diethanolamine
Acrylic acid	N,N-Diethyl aniline
Acrylonitrile	(N,N-Dimethylaniline)
Allyl chloride	Diethyl sulfate
4-Aminobiphenyl	3, 3-Dimethoxybenzidine
Aniline	Dimethyl aminoazobenzene
o-Anisidine	3,3′-Dimethyl benzidine
Benzene (including benzene from gasoline)	Dimethyl carbamoyl chloride
Benzidine	Dimethyl formamide
Benzotrichloride	1,1-Dimethyl hydrazine
Benzyl chloride	Dimethyl phthalate
Biphenyl	Dimethyl sulfate
Bis(2-ethylhexyl)phthalate	4,6-Dinitro-o-cresol, and salts
Bis(chloromethyl)ether	2,4-Dinitrophenol
Bromoform	2,4-Dinitrotoluene
1,3-Butadiene	1,4-Dioxane (1,4-diethyleneoxide)
Calcium cyanamide	1,2-Diphenylhydrazine
Caprolactam	Epichlorohydrin (I-Chloro-2,3-epoxypropane)
Captan	1,2-Epoxybutane
Carbaryl	Ethyl acrylate
Carbon disulfide	Ethyl benzene
Carbon tetrachloride	Ethyl carbamate (urethane)
Carbonyl sulfide	Ethyl chloride (chloroethane)
Catechol	Ethylene dibromide (dibromoethane)
Chloramben	Ethylene dichloride (1,2-dichloroethane)
Chlordane	Ethylene glycol
Chlorine	Ethylene imine (aziridine)
Chloroacetic acid	Ethylene oxide
2-Chloroacetophenone	Ethylene thiourea
Chlorobenzene	Ethylidene dichloride (1,1-dichloroethane)
Chlorobenzilate	Formaldehyde
Chloroform	Heptachlor
Chloromethyl ether	Hexachlorobenzene
Chloroprene	Hexachlorobutadiene
Cresols/cresylic acid (isomers and mixture)	Hexachlorocyclopentadiene
o-Cresol	Hexachloroethane
m-Cresol	Hexamethylene-1,6-diisocyanate
p-Cresol	Hexamethylphosphoramide
Cumene	Hexane hydrazine
2,4-D, salts and esters	Hydrochloric acid
DDE	Hydrogen fluoride (hydrofluoric acid)
Diazomethane	Hydrogen sulfide
Dibenzofurans	Hydroquinone
1,2-Dibromo-3-chloropropane	Isophorone
Dibutylphthalate	Lindane (all isomers)

APPENDIX 8-1 *Continued*

Maleic anhydride	1,3-Propane sultone
Methanol	beta-Propiolactone
Methoxychlor	Propionaldehyde
Methyl bromide (bromomethane)	Propoxur (baygon)
Methyl chloride (chloromethane)	Propylene dichloride (1,2-dichloropropane)
Methyl chloroform (1,1,1-trichloroethane)	Propylene oxide
Methyl ethyl ketone (2-butanone)	1,2-Propylenimine (2-methyl aziridine)
Methyl hydrazine	Quinoline
Methyl iodide (lodomethane)	Quinone
Methyl isobutyl ketone (hexone)	Styrene
Methyl isocyanate	Styrene oxide
Methyl methacrylate	2,3,7,8-Tetrachlorodibenzo-p-dioxin
Methyl tert butyl ether	1,1,2,2-Tetrachloroethane
4,4-Methylene bis(2-chloroaniline)	Tetrachloroethylene (perchloroethylene)
Methylene chloride (dichloromethane)	Toluene
Methylene diphenyl diisocyanate (MDI)	2,4-Toluene diamine
4,4-Methylenedianiline	2,4-Toluene diisocyanate
Naphthalene	o-Toluidine
Nitrobenzene	Toxaphene (chlorinated camphene)
4-Nitrobiphenyl	1,2,4-Trichlorobenzene
4-Nitrophenol	1,1,2-Trichloroethane
2-Nitropropane	Trichloroethylene
N-Nitroso-N-methylurea	2,4,5-Trichlorophenol
N-Nitrosodimethylamine	2,4,6-Trichlorophenol
N-Nitrosomorpholine	Triethylamine
Parathion	Trifluralin
Pentachloronitrobenzene (quintobenzene)	2,2,4-Trimethylpentane
Pentachlorophenol	Vinyl acetate
Phenol	Vinyl bromide
p-Phenylenediamine	Vinyl chloride
Phosgene	Vinylidene chloride (1,1-dichloroethylene)
Phosphine	Xylenes (isomers and mixture)
Phosphorus	o-Xylenes
Phthalic anhydride	m-Xylenes
Polychlorinated biphenyls (aroclors)	p-Xylenes

APPENDIX 8-2 Total Toxic Air Emissions — Bakersfield, CA

	Pounds/Year						
Chemical	**1996**	**1997**	**1998**	**1999**	**2000**	**2001**	**Total**
1,2,4-Trimethylbenzene	1,655	2,500	365	376	1,196	1,553	7,645
Benzene	4,969	4,840	2,369	2,879	2,010	2,760	19,827
Carbon disulfide	0	0	0	255	255	255	765
Carbonyl sulfide	0	0	0	505	505	505	1,515
Chlorobenzene	0	0	0	0	0	250	250
Creosote	1,337	1,029	0	0	0	0	2,366
Cyclohexane	2,800	3,650	1,000	1,700	2,000	1,819	12,969
Dioxin and dioxin-like compounds	0	0	0	0	6.8	0	6.8
Ethylene	2,550	1,950	1,000	1,158	1,000	1,000	8,658
Ethylbenzene	1,536	1,435	1,153	0	1,010	1,500	6,634
Methanol	29,250	32,250	20,616	19,317	20,151	18,786	140,370
Methyl tert butyl ether	7,050	9,150	6,910	4,000	2,700	3,600	33,410
N-hexane	5,400	6,755	2,610	2,505	2,915	2,850	23,035
N-methyl-2-pyrrolidone	87	500	500	500	87	83	1,757
Naphthalene	0	0	0	0	10	10	20
Propylene	7,150	5,150	1,950	1,750	1,750	1,750	19,500
Polycyclic aromatic compounds	0	0	0	0	13	141	154
Styrene	11,110	15,283	223,852	23,100	21,110	14,110	308,565
Toluene	7,097	6,319	4,087	3,410	3,847	4,464	29,224
Xylene (mixed isomers)	8,922	9,243	6,179	5,967	7,501	7,986	45,798
Total	90,913	100,054	272,591	67,422	68,066.8	63,422	662,468.8

APPENDIX 8-3 Total Toxic Air Emissions — Baton Rouge, LA

	Pounds/Year						
Chemical	**1996**	**1997**	**1998**	**1999**	**2000**	**2001**	**Total**
1,1-Dichloro-1-fluoroethane	9,100	9,125	0	266	0	0	18,491
1,2,4-Trimethylbenzene	12,482	17,649	440	0	262	185	31,018
2-Nitrophenol	13	14	17	15	17	15	91
1,1,1-Trichloroethane	4,750	4,505	6,805	14,135	9,968	7,156	47,319
1,1,1,2-Tetrachloroethane	5	7		75	25	25	137
1,1,2,2-Tetrachloro-1-fluoroethane	0	0	10	10	25	25	70
1,1,2,2-Trichloroethane	0	0	8	0	0	0	8
1,1,2,2-Tetrachloroethane	0	0	0	6	6	17	29
1,1,2-Trichloroethane	22	25	15	13	14	79	168
1,2-Dibromoethane	1,938	5,165	0	0	1,493	0	8,596
1,2-Dichloroethane	44,688	41,921	0	107	28,153	0	114,869
1,2-Dichloro-1,1,2-trifluoroethane	69,000	65,575	101,118	107,479	108,513	16,455	468,140
1,2-Dichloro-1,1-difluoroethane	0	0	10	10	25	25	70
1,2-Dichloroethane	0	0	29,676	25,372		38,044	93,092
1,3-Butadiene	52,938	51,947	43,349	52,579	52,548	58,194	311,555
1,2-Dibromoethane	0	0	4,073	2,560		1,373	8,006
1,4-Dioxane	1	0	0	0	0	0	1
1,1-Dichloro-1-fluoroethane	0	0	73,738	73,743	73,775	105,513	326,769
1-chloro-1,1,2,2-tetrafluoroethane	44,370	0	0	0	0	0	44,370
1-Chloro-1,1-difluoroethane	3,750	4,630	14,940	15,849	21,816	15,657	76,642
1,2,4-Trimethylbenzene	0	0	25,471	21,426	28,159	32,651	107,707
2,2-Dichloro-1,1,1-trifluoroethane	33,770	17,905	17,591	14,874	17,481	16,455	118,076
2-Chloro-1,1,1,2-tetrafluoroethane	56,710	525	6,383	3,628	0	2,696	69,942
4,4′-Isopropylidenediphenol	130	1	0	159	60	89	439
4,6-Dinitro-o-cresol	42	49	62	50	63	51	317
Acetaldehyde	11,041	4,541	6,154	2,862	739	1,012	26,349
Acrylic acid	0	0	0	0	0	784	784
Acetonitrile	10,901	7,000	10,900	11,533	9,707	8,000	58,041
Acrylonitrile	3,612	6,004	5,002	3,802	4,642	3,185	26,247
Allyl chloride	0	2	2	0	3	0	7
Anthracene	1,043	1,117	788	743	4	87	3,782
Benzene	167,329	94,765	67,633	64,059	75,972	81,947	551,705
Biphenyl	692	345	0	0	0	0	1,037
Butyl acrylate	2,660	2,074	1,681	829	818	544	8,606

APPENDIX 8-3 *Continued*

Chemical	Pounds/Year						
	1996	1997	1998	1999	2000	2001	Total
Butyraldehyde	0	1	0	0	1	0	2
Carbon disulfide	73	5	200	497	1,383	2,261	4,419
Carbon tetrachloride	505	5,083	615	686	741	708	8,338
Carbonyl sulfide	0	0	3,901	4,363	7,400	12,580	28,244
Catechol	1,242	2,128	1,179	1,129	1,998	792	8,468
Certain glycol ethers	3,043	2,405	3,149	366	2,574	2,438	13,975
Chloroform	81,183	48,125	7,240	8,171	6,726	6,400	157,845
Chlorodifluoromethane	624,000	572,250	1,071,880	939,415	804,782	606,878	4,619,205
Chlorobenzene	0	1	72	1	1	0	75
Chloroethane	21	1,902	94	124	678	15	2,834
Chloromethane	96,261	87,296	87,466	98,364	100,933	108,522	578,842
Chloroprene	1	2	4	2	3	3	15
Chlorotrifluoromethane	9,350	16,900	14,700	10,463	16,000	15,227	82,640
Cresol (mixed isomers)	619	5,343	2,007	1,842	1,747	1,826	13,384
Cumene	4,412	6,030	7,270	5,841	3,817	3,984	31,354
Cyclohexane	26,532	35,722	50,525	53,619	47,905	35,894	250,197
Diaminotoluene (mixed isomers)	0	0	0	0	0	1	1
Diphenylamine	0	2	2	2	3	2	11
Dimethylamine	0	20	0	0	0	0	20
Dichlorobromine	0	9,176	0	0	0	0	9,176
Dichloromethane	18,326	0	646	630	466	0	20,068
Dichlorodifluoromethane	35,540	14,595	26,302	36,728	39,998	26,581	179,744
Dichloorfluoromethane	0	0	0	0	3,109	2,346	5,455
Dichlorotetrafluoroethane (CFC-114)	30,860	67,400	44,340	33,625	45,745	44,114	266,084
Dicyclopentadiene	4,100	4,000	5,100	5,000	8,100	4,100	30,400
Dioxin and dioxin-like compounds	0	0	0	0	63.127	60.04	123.167
Ethyl acrylate	0	0	0	73	3	0	76
Epichlorohydrin	2	0	0	0	0	0	2
Ethylene	964,026	921,349	897,298	807,164	632,798	542,697	4,765,332
Ethylbenzene	52,000	43,086	46,583	32,956	34,164	35,283	244,072
Ethylene glycol	638	5,824	6,006	6,882	6,528	1,308	27,186
Ethylene dichloride	8	12		20	21	30	91
Ethylidene dichloride	0	0	24	0	0	0	24
Formaldehyde	11,753	11,589	27,109	23,340	30,306	16,166	120,263

Hexachlorobenzene	0	0	211	111	0	0	322
Hydroquinone	1,267	1,313	1,172	1,051	1,176	1,364	7,343
Isobutylaldehyde	0	7	0	0	0	0	7
Malic anhydride	15,551	25,362	41,471	38,035	17,601	49,801	187,821
Methanol	165,539	280,288	324,355	371,438	376,860	230,855	1,749,335
Methyl acrylate	1590	2,540	1,490	2,600	2,400	1,990	12,610
Methyl ethyl keytone	240,008	182,298	216,411	161,724	165,667	161,767	1,127,875
Methyl methacrylate	0	0	0	0	80	0	80
Methyl isobutyl keytone	100,154	69,508	80,613	51,824	61,757	89,005	452,861
Methyl tert butyl ether	262,166	157,286	85,312	76,970	87,592	111,271	780,597
Monochloropentafluoroethane	5,260	5,340	13,522	13,255	13,603	13,555	64,535
N,N-dimethylformamide	12,000	14,008	14,025	13,000	13,025	17,000	83,058
N-butyl alcohol	0	1	930	670	2	670	2,273
N-hexane	880,962	965,155	1,024,860	972,642	1,046,221	866,591	5,756,431
N-methyl-2-pyrrolidone	73,241	74,724	74,702	70,002	150,000	50,000	492,669
Naphthalene	11,317	17,251	28,737	22,423	19,632	21,139	120,499
Nitrobenzene	0	0	3,833	1,313	0	0	5,146
O-toluidine	0	0	0	2	0	0	2
O-xylene	810	1,980	3,800	1,520	2,640	1,671	12,421
Permethrin	500	500	0	10	10	10	1,030
Phenanthrene	23	26	24	18	1	2	94
Phenol	3,726	7,175	3,247	3,956	9,951	25,205	53,260
Phthalic anhydride	35,801	17,802	23,603	19,501	13,701	24,501	134,909
Polycyclic aromatic compounds	0	0	0	0	7,189	5,068	12,257
Propionaldehyde	0	0	0	1	0	0	1
Propylene	151,480	142,732	157,656	119,522	179,213	120,530	871,133
Pyridine	0	0	0	0	48	0	48
Sec-butyl alcohol	18,029	16,805	18,105	23,965	21,007	21,000	118,911
Styrene	212,846	205,166	211,529	185,470	182,047	155,109	1,152,167
Tert-butyl alcohol	730	442	678	398	355	505	3,108
Tetrachloroethylene (PCE)	7,932	3,361	4,119	3,549	3,905	3,792	26,658
Toluene	214,972	156,351	183,385	135,021	147,976	166,446	1,004,151
Toluene diisocyanate	0	0	1	3	0	0	4
Trichlorofluoromethane	41,100	37,070	21,040	21,040	21,067	21,040	162,357
Trichloroethylene (TCE)	14	18	148	256	122	830	1,388
Triethylamine	0	0	1	1	0	0	2
Vinyl acetate	432,011	467,291	544,001	491,114	460,001	382,000	2,776,418
Vinyl chloride	7,389	8,199	12,803	7,257	6,200	8,290	50,138
Xylene (mixed isomers)	143,230	122,812	150,411	125,894	147,223	147,352	836,922
Total	5,535,130	5,177,948	5,965,773	5,423,043	5,390,553	4,558,839	32,051,286

APPENDIX 8-4 Total Toxic Air Emissions — Binghamton, NY

	Pounds/Year						
Chemical	**1996**	**1997**	**1998**	**1999**	**2000**	**2001**	**Total**
Certain glycol ethers		0	0	10	10	0	20
Diethanolamine	10	0	0	10	10	10	40
Total	10	0	0	20	20	10	60

APPENDIX 8-5 Total Toxic Air Emissions — Birmingham, AL

	Pounds/Year						
Chemical	**1996**	**1997**	**1998**	**1999**	**2000**	**2001**	**Total**
1,2,4-Trimethylbenzene	21,071	25,764	15,121	13,317	19,015	16,444	110,732
1,2-Dichloropropane	0	0	0	0	0	4	4
2,4-Dimethylphenol	0	0	0	0	0	296	296
4,4-Isopropylidenediphenol		500	500	500	500	750	2,750
Acetonitrile	0	0	0	0	0	24	24
Anthracene	412	945	927	519	531	388	3,722
Benzene	26,412	53,640	46,933	93,579	94,571	63,419	378,554
Benzol peroxide	0	0	0	0	0	250	250
Benzo(G,H,I) perylene	0	0	0	0	128	107	235
Biphenyl	110	110	120	141	75	51	607
Carbonyl sulfide	217,262	365,000	356,300	551,519	525,010	246,720	2,261,811
Certain glycol ethers	118,277	163,892	126,842	75,392	62,257	40,399	587,059
Chlorobenzene	2,044	0	0	414	420	380	3,258
Creosote	15,200	8,000	8,400	10,571	1,667	1,666	45,504
Cresol (mixed isomers)	900	371	386	402	678	535	3,272
Cyclohexane	117,079	124,148	108,141	92,279	97,365	81,844	620,856
Dibenzofuran	1,035	1,009	1,103	1,356	353	1,006	5,862
Dimethylamine	0	0	301,027	0	0	0	301,027
Dimethyl Phthalate	0	0	0	402	0	0	402
Dioxin and dioxin-like compounds	0	0	0	0	2.48	2.55	5.03
Diisocyanates	3,879	0	0	0	0	0	3,879
Dichloromethane	261,470	347,000	0	16,855	15,338	4,075	644,738

APPENDIX 8-5 *Continued*

	Pounds/Year						
Chemical	**1996**	**1997**	**1998**	**1999**	**2000**	**2001**	**Total**
Ethyl acrylate	0	0	0	0	0	8,756	8,756
Ethylene	18,223	43,542	13,617	86,251	76,853	38,254	276,740
Ethylbenzene	11,860	1,069	11,159	15,073	13,016	2,917	55,094
Formaldehyde	10,800	0	0	0	0	0	10,800
Isopropyl alcohol	38,250	52,587	73,313	104,406	0	0	268,556
Methanol	26,433	37,427	19,254	14,995	14,625	3,687	116,421
Methyl ethyl keytone	172,124	120,864	111,888	118,702	142,505	71,352	737,435
Methyl isobutyl keytone	48,690	52,557	28,966	47,140	49,021	26,092	252,466
Methyl tert butyl ether	1,200	900	1,418	3,269	2,738	2,410	11,935
N-butyl alcohol	7,265	16,354	6,174	6,383	18,468	11,701	66,345
N-hexane	65,044	82,765	62,607	55,580	54,881	3,803	324,680
Naphthalene	40,540	42,474	7,735	10,961	6,259	4,829	112,798
O-xylene	5,121	5,407	30	608	3,763	765	15,694
Phenanthrene	93	67	205	391	382	401	1,539
Phenol	65,264	62,522	73,450	67,823	67,030	53,652	389,741
Polycyclic aromatic compounds	141	146	190	291	567	487	1,822
Propylene	0	0	1,220	2,120	11,525	4,846	19,711
Quinoline	112	102	107	128	102	83	634
Sec-butyl alcohol	21,611	25,741	23,167	0	0	0	70,519
Styrene	2,136	2,152	2,768	17,346	16,994	15,691	57,087
Tetrachloroethylene (PCE)	0	0	25	30	0	0	55
Toluene	133,356	208,739	324,172	153,448	180,527	229,521	1,229,763
Trichloroethylene (TCE)	20,081	14,701	15,824	23,122	29,585	18,571	121,884
Xylene (mixed isomers)	409,045	241,594	330,062	378,448	407,544	378,973	2,145,666
Total	1,882,540	2,102,089	2,073,151	1,963,761	1,914,295	1,335,152	11,270,988

APPENDIX 8-6 Total Toxic Air Emissions — Boise, ID

	Pounds/Year						
Chemical	**1996**	**1997**	**1998**	**1999**	**2000**	**2001**	**Total**
1,2,4-Trimethylbenzene	0	0	155	144	118	117	534
Benzene	0	0	5,976	5,214	3,534	12,706	27,430
Certain glycol ethers	250	250	255	255	255	255	1,520
Ethylbenzene	0	0	576	506	357	1,083	2,522
Ethylene glycol	0	5	260	255	260	265	1,045
Methyl ethyl keytone	16,400	10,700	19,300	16,700	16,200	14,000	93,300
Methyl tert butyl ether	0	0	907	764	629	636	2,936
N-butyl alcohol	0	0	10,780	0	0	0	10,780
N-hexane	0	0	0	9,580	6,640	21,260	37,480
N-methyl-2-pyrrolidone	1,800	1,005	1,010	1,010	1,010	786	6,621
Styrene	58	63,343	189,566	193,880	206,560	117,460	770,867
Toluene	0	0	8,616	7,874	5,299	15,840	37,629
Toluene-2,6-diisocyanate	0	0	0	1,738	0	0	1,738
Xylene (mixed isomers)	0	0	20,155	4,346	2,572	5,712	32,785
Total	18,508	75,303	257,556	242,266	243,434	190,120	1,027,187

APPENDIX 8-7 Total Toxic Air Emissions — Charleston, SC

	Pounds/Year						
Chemical	**1996**	**1997**	**1998**	**1999**	**2000**	**2001**	**Total**
1,2,4-Trimethylbenzene	1,700	3,100	6,596	3,880	6,260	3,682	25,218
1,2-Dichloroethane	39,618	7,380	9,541	27,661	43,317	36,394	163,911
1,2-Dibromoethane	252	0	0	0	0	0	252
Acetaldehyde	99,115	91,145	91,705	100,690	64,357	49,208	496,220
Acrylic acid	0	149	162	215	215	181	922
Benzene	117,000	107,000	136,558	90,352	51,129	17,233	519,272
Benzo(g,h,i) perylene	0	0	0	0	7	2	9
Bromomethane	315,400	296,542	214,273	186,812	42,301	20,617	1,075,945
Biphenyl	6,537	2,913	0	0	0	0	9,450
Certain glycol ethers	14,484	17,350	10	11,705	5,136	9,093	57,778
Chloroethane	0	0	0	0	2,035	7,413	9,448
Chloromethane	219,074	212,770	107,180	195,214	37,704	29,308	801,250
Cumene	0	0	4	0	0	11	15
Diethanolamine	2	0	0	665	0	0	667
Dimethylamine	265	255	335		1,706	796	3,357
Dioxin and dioxin-like compounds	0	0	0	0	1	1	2
Epichlorohydrin	2,354	1,484	0	0	0	0	3,838
Ethylene oxide	780	4,530	2,075	2,972	3,500	0	13,857
Ethylbenzene	0	0	452	252	190	13,131	14,025
Ethylene glycol	0	0	11	12	6	11	40
Formaldehyde	71,640	65,382	55,241	62,762	47,729	29,061	331,815

Maleic anhydride	1,141	550	408	557	467	379	3,502
M-cresol	10	10	10	8	8	4	50
Methanol	2,882,597	2,441,569	2,108,707	2,155,325	2,252,828	1,919,043	13,760,069
Methyl acrylate	600	500	622	477	477	456	3,132
Methyl ethyl keytone	47,306	38,934	34,495	36,956	21,142	60,077	238,910
Methyl iodide	45,005	52,705	40,705	46,300	53,200	40,200	278,115
Methylene bromide	0	0	0	348	124	120	592
Methyl isobutyl keytone	20,056	23,217	21,632	17,278	0	0	82,183
Methyl tert butyl ether	0	0	0	1,707	0	0	1,707
N-butyl alcohol	1,590	1,585	2,406	2,141	2,350	2,256	12,328
N-hexane	42,515	45,755	41,183	45,268	55,185	43,081	272,987
Naphthalene	255	0	0	0	0	0	255
P-xylene	330,800	216,000	78,600	72,400	55,800	54,900	808,500
P-cresol	10	10	10	8	8	4	50
Phenol	10	4,085	4,222	3,656	4,641	1	16,615
Propylene oxide	8,686	11,302	7,770	9,001	11,620	10,372	58,751
Polycyclic aromatic compounds	0	0	4	10	474	245	733
Styrene	3,100	4,000	48,000	76,400	98,942	95,633	326,075
Toluene	50,577	49,626	45,583	66,715	56,142	13,986	282,629
Trichloroethylene (TCE)	33,589	20,552	27,696	5,573	17,086	0	104,496
Xylene (mixed isomers)	38,378	3,256	27,811	22,421	16,152	56,573	164,591
Total	4,394,446	3,723,656	3,114,007	3,245,741	2,952,239	2,513,472	19,943,561

APPENDIX 8-8 Total Toxic Air Emissions — Columbus, GA

	Pounds/Year						
Chemical	**1996**	**1997**	**1998**	**1999**	**2000**	**2001**	**Total**
Acrylic acid	0	2	3	0	7	0	12
1,1-Dichloro-1-fluoroethane	13,124	0	0	0	0	0	13,124
1,2,4-Trimethylbenzene	14	9	70	67	83	80	323
2-Ethoxyethanol	0	0	0	0	0	593	593
2-Methoxyethanol	18,500	7,142	8,346	4,511	3,693	5,510	47,702
2-Mercaptobenzothiazole	54,700	0	0	0	0	0	54,700
1,2-Dichloropropane	0	0	10	0	0	0	10
Acrylamide	0	0	0	10	27	10	47
Acetophenone	0	0	0	0	12	0	12
Benzene	0	54,414	59,109	57,927	33,210	0	204,660
Butyl acetate	244	243	240	242	244	243	1456
Acrylic acid	2	0	0	0	0	0	2
Certain glycol ethers	38,652	14,202	0	0	43	27	52,924
Dimethylamine	38	0	20	500	235	281	1,074
Diisocyanates	1,018	1,008	1,000	500	450	399	4,375
Dichloromethane	110,500	500	11,000	0	0	0	122,000
Epichlorohydrin	70	477	89	500	33	28	1,197
Ethyl acrylate	0	0	0	0	2	2	4

Ethylbenzene	41	0	62	56	66	83	308
Ethylene glycol	17,500	5,000	507	507	8	14	23,536
Methanol	19,280		26,807	4,914	1,862	1,779	54,642
Methyl acrylate	0	19,141	0	0	0	0	19,141
Methyl ethyl keytone	208,111	154,839	5,868	5,461	10,910	16,771	401,960
Methyl methacrylate	291	266	293	290	297	266	1,703
Methyl isobutyl keytone	24,721	18,745	338	255	47	1,915	46,021
Methyl tert butyl ether	0	0	543	0	0	0	543
NnN-dimethylformamide	0	0	0	0	0	678	678
N-butyl alcohol	1,300	522	0	465	609	62	2,958
N-methyl-2-pyrrolidone	6,500	9,200	2,500	0	13,380	0	31,580
Naphthalene	27	24	0	24	27	0	102
Phenol	0	0	0	0	290	0	290
Polycylic aromatic compounds	0	0	0	0	4	0	4
Propylene oxide	0	0	0	0	0	115	115
Styrene	0	0	1	0	116	0	117
Toluene	0	0	0	0	138	137	275
Trichloroethylene (TCE)	11,140	13,564	5,390	22,108	28,387	29,619	110,208
Triethylamine	803	594	658	645	593	500	3,793
Vinyl acetate	342	341	312	305	0	0	1,300
Xylene (mixed isomers)	37,199	26,232	43,517	24,747	25,201	19,147	176,043
Total	564,117	326,465	166,683	124,289	120,028	78,273	1,379,532

APPENDIX 8-9 Total Toxic Air Emissions — Dayton, OH

	Pounds/Year						
Chemical	**1996**	**1997**	**1998**	**1999**	**2000**	**2001**	**Total**
1,1-Dichloro-1-fluoroethane	10,000	0	0	0	0	0	10,000
1,2,4-Trimethylbenzene	370	380	629	691	1,433	4,567	8,070
1,1,1-Trichloroethane	1,072	0	0	0	0	0	1,072
2-Nitropropane	11,000	0	0	0	0	0	11,000
2-Methoxyethanol	0	0	2	2	2	0	6
Acretonitrile	0	0	138	0	0	0	138
Acrylamide	28	0	0	0	0	0	28
Acrylic acid	17	0	0	0	0	0	17
Benzene	260	250	1,469	1,348	611	1,770	5,708
Butyl acrylate	488	0	0	0	0	0	488
Butyraldehyde	470	0	0	0	0	0	470
Certain glycol ethers	319,997	163,250	160,956	74,986	62,119	44,795	826,103
Chloroform	37,000	0	35,000	38,000	39,000	40,000	189,000
Cumene	0	0	3	14	3	0	20
Cyclohexane	24,526	134	1,841	1,523	84	7	28,115
Di(2-ethylhexyl) phthalate	17	22	18	10	10	9	86
Diethanolamine	223	221	228	0	0	0	672
Dichloromethane	13,515	13,775	2,823	6,660	3,031	536	40,340
Dioxin and dioxin-like compounds	0	0	0	0	0	1.4	1.4
Ethylbenzene	4,633	4,240	5,657	6,231	5,842	6,153	32,756

Ethylene glycol	33	34	80	440	425	401	1,413
Formaldehyde	422	0	0	0	0	0	422
Hydroquinone	3	0	0	0	0	0	3
Methanol	355,081	39,933	19,600	16,574	19,995	9,238	460,421
Methyl ethyl keytone	30,828	38,811	43,513	28,071	10,948	17,942	170,113
Methyl isobutyl keytone	170,932	52,017	41,226	11,906	9,168	5,313	290,562
Methyl methacrylate	267	0	0	0	0	0	267
Methyl tert butyl ether	0	0	500	0	0	0	500
N-butyl alcohol	30,274	26,003	13,157	18,195	16,200	27,085	130,914
N-hexane	119,891	125,653	116,346	122,322	1,922	1,430	487,564
N-methyl-2-pyrrolidone	0	0	1,600	11,955	2,493	356	16,404
Naphthalene	0	0	0	0	0	182	182
Phenol	59,900	10,885	6,405	4,705	8,285	2,981	93,161
Polychlorinated alkanes	45	16	0	34	0	0	95
Polycyclic aromatic compounds	0	0	0	0	2	3	5
Pyridine	0	0	10	50	33	0	93
Sec-butyl alcohol	55,500	46,700	17,910	12,640	0	0	132,750
Styrene	0	0	0	0	250	250	500
Tetrachloroethylene (PCE)	7,213	8,000	0	0	0	7,960	23,173
Thiram	0	11	0	0	0	0	11
Toluene	18,695	13,268	27,734	32,320	23,615	26,388	142,020
Trichloroethylene (TCE)	8,900	10,900	43,900	38,800	27,351	16,800	146,651
Xylene (mixed isomers)	62,721	55,014	38,393	60,298	37,836	33,725	287,987
Total	1,344,321	609,517	579,138	487,775	270,658	247,892.4	3,539,301

APPENDIX 8-10 Total Toxic Air Emissions — Denver, CO

	Pounds/Year						
Chemical	**1996**	**1997**	**1998**	**1999**	**2000**	**2001**	**Total**
1,1-Dichloro-1-fluoroethane	85,076	100,410	40,332	35,812	23,535	24,066	309,231
1,2,4-Trimethylbenzene	0	0	0	0	0	2,166	2,166
1,1,1-Trichloroethane	7,210	23,614	17,369	15,010	15,268	0	78,471
1,1,2,2-Tetrachloroethane	0	10	10	0	0	0	20
1,2,4-Trimethylbenzene	0	1,225	2,960	3,127	2,555	0	9,867
1,3-Butadiene	1,019	0	0	65	35	61	1,180
Benzene	15,421	13,263	11,046	8,719	7,261	7,793	63,503
Benzo(g,h,i) perylene	0	0	0	0	28	26	54
Certain glycol ethers	35,772	14,350	3,304	41,507	15	21	94,969
Chlorodifluoromethane	5	5	5	0	0	0	15
Chloroform	500	500	500	500	416	455	2,871
Creosote	6,690	5,803	10,249	8,571	6,604	6,600	44,517
Cumene	1,181	658	1,027	314	349	268	3,797
Cyclohexane	6,847	3,628	2,425	5,260	9,400	10,598	38,158
Di(2-ethylhexyl) phthalate	19	25	5	5	5	0	59
Diisocyanates	6,582	8,917	7,605	2,005	1,712	1,679	28,500
Dichloromethane	13,659	586	10,906	15,010	16,536	1,127	57,824
Dichlorodifluoromethane	5	5	5	0	0	0	15
Dioxin and dioxin-like compounds	0	0	0	0	0.595	0.494	1.089
Epichlorohydrin	0	0	0	3,467	0	0	3,467
Ethylene oxide	7,600	7,500	7,600	0	3,040	2,102	27,842

Ethylbenzene	4,718	1,156	2,091	3,448	4,476	4,054	19,943
Ethylene	10,368	4,983	3,690	2,703	3,704	2,707	28,155
Ethylene glycol	731	755	790	790	772	774	4,612
Formaldehyde	152	0	0	0	0	0	152
Isopropyl alcohol	0	0	0	675	0	0	675
Methanol	37,229	26,206	25,031	14,710	7,621	10,166	120,963
Methyl ethyl keytone	29,325	23,136	28,859	0	13,303	750	95,373
Methylmethacrylate	0	0	0	20,401	0	0	20,401
Methyl isobutyl keytone	4,025	1,365	502	0	0	11,097	16,989
Methyl tert butyl ether	0	0	12,855	11,820	1,850	2,825	29,350
Monochloropentafluoroethane	5	0	0	0	0	0	5
N-butyl alcohol	1,085	753	1,335	1,439	930	1,505	7,047
N-hexane	31,823	32,000	33,471	33,199	23,567	23,445	177,505
N-methyl-2-pyrrolidone	0	17,935	12,087	0	0	1	30,023
Naphthalene	116	309	308	0	16	22	771
Propylene	91,343	75,933	86,940	24,902	16,169	0	295,287
Propylene oxide	0	0	0	0	0	23,209	23,209
Polycyclic aromatic compounds	319	310	281	329	323	302	1,864
Styrene	6,878	12,627	16,777	15,781	12,803	34,854	99,720
Tert-butyl alcohol	0	0	0	10	0	0	10
Tetrachloroethylene (PCE)	388	0	0	388	0	0	776
Toluene	166,106	114,103	106,039	92,085	79,824	73,296	631,453
Trichloroethylene (TCE)	78	0	0	0	0	0	78
Xylene (mixed isomers)	66,048	47,325	20,607	13,936	18,967	17,684	184,567
Total	638,323	539,395	467,011	375,988	271,084.6	263,653.5	2,555,455

APPENDIX 8-11 Total Toxic Air Emissions — Fresno, CA

	Pounds/Year						
Chemical	**1996**	**1997**	**1998**	**1999**	**2000**	**2001**	**Total**
1,1-Dichloro-1-fluoroethane	8,550	10,833	13,380	9,827	7,368	5,345	55,303
Chlorodifluoromethane	10,647	10,776	13,031	9,221	6,723	5,285	55,683
Diethanolamine	5	0	0	0	0	0	5
Diisocyanates	115	0	0	5	0	0	120
Dichloromethane	0	0	50	0	0	0	50
Maneb	0	0	0	0	0	6	6
Malathion	250	0	0	0	0	0	250
Methyl ethyl keytone	0	0	11,500	0	0	0	11,500
Napthalene	0	0	0	0	0	4	4
Styrene	43,556	39,676	167,801	111,309	56,000	37,455	455,797
Tetrachloroethylene (PCE)	0	0	0	250	0	0	250
Thiobencarb	500	500	250	0	0	0	1,250
Trifluralin	0	0	0	0	4	2	6
Total	63,623	61,785	206,012	130,612	70,095	48,097	580,224

APPENDIX 8-12 Total Toxic Air Emissions — Grand Rapids, MI

	Pounds/Year						
Chemical	**1996**	**1997**	**1998**	**1999**	**2000**	**2001**	**Total**
1,2,4-Trimethylbenzene	287,039	266,445	193,358	167,853	200,451	59,021	1,174,167
1,1,1-Trichloroethane	60,401	51,235	0	0	0	0	111,636
2-Ethoxyethanol	60	0	0	0	0	0	60
Certain glycol ethers	137,381	107,255	164,564	134,021	175,149	54,155	772,525
Cumene	11,862	0	0	0	0	0	11,862
Cyclohexane	0	0	0	100	0	0	100
Diisocyanates	440	442	734	456	371	258	2,701
Dichloromethane	17,863	30,834	31,582	26,204	23,038	11,805	141,326
Dimethipin	0	0	0	0	250	250	500
Ethylene oxide	250	250	250	250	250	250	1,500
Ethylbenzene	363,693	268,606	174,322	80,873	120,533	67,862	1,075,889
Ethylene glycol	2,791	4,457	2	10	22	1,148	8,430
Formaldehyde	4,293	15,897	149,926	2,663	2,883	3,399	179,061
Isopropyl alcohol	21,349	11,681	10,259	11,891	1,424	0	56,604
Maleic anhydride	0	0	0	0	0	91	91
Methanol	16,539	24,470	95,068	90,081	77,308	41,530	344,996
Methyl ethyl keytone	355,719	247,476	215,788	177,423	130,540	80,886	1,207,832
Methyl isobutyl keytone	180,838	69,520	117,142	131,675	143,177	31,420	673,772
Methyl tert butyl ether	0	0	499	0	0	0	499
N-butyl alcohol	3,526	49,286	74,772	138,911	54,879	16,792	338,166
N-hexane	22,011	28,688	30,709	21,688	18,723	7,215	129,034
N-methyl-2-pyrrolidone	18,107	6,118	4,590	4,606	10,234	6,926	50,581
Naphthalene	0	0	0	0	1,949	1,917	3,866
Phenol	750	2,000	850	250	250	250	4,350
Phthalic anhydride	0	1,576	1,919	2,570	2,570	2,260	10,895
Polycyclic aromatic compounds	0	0	0	0	3	0	3
Sec-butyl alcohol	0	0	12,612	0	0	0	12,612
Styrene	476	1,704	0	983	95	4,069	7,327
Toluene	684,479	710,915	673,151	554,538	334,713	67,188	3,024,984
Toluene-2,4-diisocyanate	1	0	0	0	1	0	2
Triethylamine	0	0	0	0	86	0	86
Trichloroethylene (TCE)	198,373	162,466	122,835	75,607	55,685	41,191	656,157
Xylene (mixed isomers)	1,023,229	1,008,257	1,349,041	572,256	651,939	229,180	4,833,902
Total	3,411,470	3,069,578	3,423,973	2,194,909	2,006,523	729,063	14,835,516

APPENDIX 8-13 Total Toxic Air Emissions — Harrisburg, PA

Chemical	Pounds/Year 1996	1997	1998	1999	2000	2001	Total
1,2,4-Trimethylbenzene	0	0	0	0	0	75	75
Benzene	0	0	845	829	782	608	3,064
Certain glycol ethers	59,710	30,181	74,800	0	0	0	164,691
Cumene	0	0	5	6	6	0	17
Ethylbenzene	0	0	163	141	135	126	565
N-hexane	0	0	1,196	952	870	1,006	4,024
Toluene	0	0	1,276	1,173	1,097	1,022	4,568
Xylene (mixed isomers)	0	0	740	646	611	540	2,537
Total	59,710	30,181	79,025	3,747	3,501	3,377	179,541

APPENDIX 8-14 Total Toxic Air Emissions — Jackson, MS

Chemical	Pounds/Year						
	1996	1997	1998	1999	2000	2001	Total
1,1-Dichloro-1-fluoroethane	11,481	42,665	9,440	6,132	0	0	69,718
1,2,4-Trimethylbenzene	2,165	7,286	1,122	58,358	23,050	35,215	127,196
Certain glycol ethers	4,397	11,829	6,311	204,976	60,420	88,741	376,674
Chlorofluoromethane	2,237	15,666	1,874	1,223	0	0	21,000
Di(2-ethylhexl) phthalate	5	5	0	0	0	0	10
Diisocyanates	136	115	104	68	0	0	423
Dimethylphthalate	0	775	402	892	2,044	3,451	7,564
Ethylbenzene	8,481	511	643	31,096	1,382	3,142	45,255
Ethylene glycol	0	0	2	3	4	0	9
Hydramethylon	10	20	0	10	10	10	60
M-xylene	0	4,000	2,500	0	0	0	6,500
Methanol	40,100	0	330	21	90	149	40,690
Methyl ethyl keytone	22,702	9,794	11,028	38,933	7,561	9,948	99,966
Methyl isobutyl keytone	43	39	440	48	197	0	767
N-butyl alcohol	531	1,463	1,858	15,995	27,360	31,108	78,315
Naphthalene	27,518	1,915	1,058	9,546	7,853	10,527	58,417
Polycyclic aromatic compounds	0	0	0	0	579	650	1,229
Sec-butyl alcohol	0	0	0	0	13,300	8,080	21,380
Styrene	3,500	3,500	9,350	39,280	35,440	34,160	125,230
Tetrachloroethylene (PCE)	32,332	28,050	9,134	0	0	0	69,516
Toluene	46,307	3,557	4,791	4,902	5,548	7,728	72,833
Vinyl acetate	0	0	0	0	0	21,450	21,450
Triethylamine	0	0	0	0	6,660	0	6,660
Xylene (mixed isomers)	99,841	12,776	10,309	130,188	19,140	12,978	285,232
Total	301,786	143,966	70,696	541,671	210,638	267,337	1,536,094

APPENDIX 8-15 Total Toxic Air Emissions — Lafayette, LA

	Pounds/Year						
Chemical	**1996**	**1997**	**1998**	**1999**	**2000**	**2001**	**Total**
Certain glycol ethers	0	800	0	0	0	0	800
Sec-butyl alcohol	143	0	0	0	0	0	143
Toluene	571	850	0	0	0	0	1421
Xylene (mixed isomers)	1339	630	223	0	0	0	2192
Total	2053	2280	223	0	0	0	4556

APPENDIX 8-16 Total Toxic Air Emissions — Los Angeles, CA

	Pounds/Year						
Chemical	**1996**	**1997**	**1998**	**1999**	**2000**	**2001**	**Total**
1,1-Dichloro-1-fluoroethane	10,400	0	9,300	10,300	0	0	30,000
1,1,1-Trichloroethane	103,180	45,569	0	8,936	10,094	7,617	175,396
1,2-Dichloroethane	0	0	0	0	625	1,154	1,779
1,2,4-Trichlorobenzene	0	2,576	0	0	430	0	3,006
1,2,4-Trimethylbenzene	750	250	1,262	1,446	1,008	556	5,272
2,4-D	0	0	0	73	531	99	703
4,4-Isopropylidenediphenol	2	2	0	0	0	0	4
4,4-Methylenedianiline	21	22	78	23	27	31	202
Butyl acrylate	500	500	500	500	500	1,000	3,500
Certain glycol ethers	34,150	40,341	31,505	34,052	23,921	0	163,969
Chlorodifluoromethane	0	0	0	0	0	26,119	26,119
Cyclohexane	0	0	0	0	395	416	811
Di(2-ethylhexyl) phthalate	2,100	1,905	2,005	2,113	2,005	1,705	11,833
Dichloromethane	57,117	0	2,140	500	553	250	60,560
Dicyclopentadiene	36	35	40	262	119	94	586
Dimethylamine	656	303	270	560	909	1,211	3,909
Dimethyl sulfate	0	0	0	0	35	52	87
Diisocyanates	0	0	13	0	0	0	13
Dimethyl phthalate	0	0	0	0	1,870	1,352	3,222
Diethanolamine	0	0	400	410	0	0	810
Dioxin and dioxin-like compounds	0	0	0	0	0.8122	0.7118	1.524
Ethylbenzene	830	674	668	715	715	926	4,528
Ethylene glycol	1,147	2,837	3,749	3,739	3,645	3,978	19,095
Formaldehyde	429	266	406	512	191	0	1,804
Hexachlorobenzene	5	5	2	2	2	2	18
Malic anhydride	25	25	0	33	32	36	151
Mecoprop	0	0	0	33	0	76	109

APPENDIX 8-16 *Continued*

	Pounds/Year						
Chemical	**1996**	**1997**	**1998**	**1999**	**2000**	**2001**	**Total**
Methanol	94,703	116,114	164,231	34,262	62,083	44,021	515,414
Methyl ethyl keytone	35,633	38,425	9,909	12,801	1,445	2,467	100,680
Methyl methacrylate	544	500	500	500	500	500	3,044
Methyl isobutyl keytone	53,129	50,091	20,691	8,240	10,322	5,635	148,108
N-butyl alcohol	122,325	15,233	17,934	3,874	2,714	2,942	165,022
N-methyl-2-pyrrolidone	0	1,237	5	5	0	0	1,247
N,n-dimethylformamide	39	38	79	79	113	72	420
N-hexane	88	4,009	5	6	0	60	4,168
Naphthalene	0	0	6	501	1	0	508
O-xylene	0	113	58	36	44	51	302
Phenol	91	95	35	36	38	64	359
Phthalic anhydride		19	36	297	551	289	1,192
Quintozene	500	402	400	402	402	330	2,436
Sec-butyl alcohol	0	5	5	260	0	254	524
Styrene	20,727	16,714	14,549	9,992	9,190	44,754	115,926
Tert-butyl alcohol	35	0	0	0	0	0	35
Tetrachloroethylene (PCE)	70,224	33,800	26,791	10,524	0	5	141,344
Thiourea	0	0	0	0	82	1	83
Toluene	131,115	108,564	61,525	49,823	66,901	49,307	467,235
Toluene diisocyanate	95	98	0	154	163	143	653
Toluene-2,4-diisocyanate	0	0	5	0	0	4,446	4,451
Trichloroethylene (TCE)	0	0	0	1,055	3,633	3,500	8,188
Vinyl acetate	1,000	2,702	2,517	2,682	2,483	2,015	13,399
Xylene (mixed isomers)	53,870	81,691	23,932	28,362	30,379	13,938	232,172
Total	795,466	565,160	395,551	228,100	238,651.8	221,468.7	2,444,398

APPENDIX 8-17 Total Toxic Air Emissions — Milwaukee, WI

Chemical	Pounds/Year						
	1996	1997	1998	1999	2000	2001	Total
1,1-Dichloro-1-fluoroethane	11,975	0	213	0	0	0	12,188
1,2,4-Trimethylbenzene	40,046	32,504	24,950	23,810	800	764	122,874
2,Mercaptobenzothiazole	5	0	0	0	0	0	5
1,1,1-Trichloroethane	94	0	0	0	0	0	94
1,2-Dichloropropane	73,676	63,278	0	0	0	0	136,954
Acetaldehyde	12,061	52,600	86,400	87,400	111,600	111,694	461,755
Acetonitrile	0	0	0		5	5	10
Benzene	0	0	714	253	4,755	330	6,052
Certain glycol ethers	411,638	350,459	409,979	382,771	222,958	179,290	1,957,095
Chloroform	595	500	500	1,000	1,000	1,000	4,595
Chloromethane	8,184	20,134	0	0	17,832	6,320	52,470
Cyclohexane	2,737	0	505	0	0	1,200	4,442
Cumene hydroperoxide	0	479	0	0	0	0	479
Di(2-ethylhexl) phthalate	0	0	0	0	0		0
Diisocyanates	29	22	0	10	5	5	71
Dichloromethane	21,449	33,468	5,746	2,005	23,550	25,705	111,923
Dimethylamine	0	469	2,008	884	866	0	4,227
Epichlorohydrin	518	547	556	579	510	526	3,236
Ethylbenzene	583	621	791	456	467	362	3,280
Ethylene glycol	262	250	250	0	250	250	1,262
Methanol	7,975	9,644	17,717	2,997	4,562	26,442	69,337
Methyl ethyl keytone	24,136	33,221	27,989	18,242	15,419	7,961	126,968

APPENDIX 8-17 *Continued*

	Pounds/Year						
Chemical	**1996**	**1997**	**1998**	**1999**	**2000**	**2001**	**Total**
Methylmethacrylate	250	0	0	0	0	0	250
Methyl isobutyl keytone	18,708	15,727	12,757	12,696	2,311	2,555	64,754
Methyl tert-butyl ether	91,624	1,000	500	1,275	2,710	755	97,864
N-butyl alcohol	61,793	60,126	137,690	117,459	60,901	38,616	476,585
N-hexane	505	4,200	1,337	7,430	7,014	380	20,866
N,N-Dimethylformamide	0	10	0	0	10	0	20
N-methyl-2-pyrrolidone	13,193	18,495	16,347	20,510	19,669	21,011	109,225
Phenol	500	872	0	0	0	0	1,372
Phthalic anhydride	5	0	0	0	0	0	5
Pyridine	0				255	0	255
Propylene	1,058	958	974	555	779	0	4,324
Polycyclic aromatic compounds	0	0	0	0	3	1	4
Styrene	89,500	64,632	69,403	47,361	44,303	47,688	362,887
Tetrachloroethylene (PCE)	3,700	0	5	250	250	250	4,455
Toluene	14,604	13,077	73,976	79,875	76,726	66,812	325,070
Trichloroethylene (TCE)	62,376	38,256	36,411	1,820	5,901	2,010	146,774
Triethylamine	750	1,051	1,005	0	0	255	3,061
Vinyl acetate	0			4,470	1,540	0	6,010
Xylene (mixed isomers)	40,302	221,242	192,756	154,889	34,778	50,485	694,452
Total	1,014,831	1,037,842	1,121,479	968,997	661,729	592,672	5,397,550

APPENDIX 8-18 Total Toxic Air Emissions — New York, NY

	Pounds/Year						
Chemical	**1996**	**1997**	**1998**	**1999**	**2000**	**2001**	**Total**
Benzo(g,h,i) perylene	0	0	0	0	1	1	2
Dioxin and dioxin-like compounds	0	0	0	0	0.41	0.4885	0.8985
Polycylic aromatic compounds	0	0	0	0	0	5	5
Total	0	0	0	0	1.41	6.4885	7.8985

APPENDIX 8-19 Total Toxic Air Emissions — Ogden, UT

	Pounds/Year						
Chemical	**1996**	**1997**	**1998**	**1999**	**2000**	**2001**	**Total**
1,2,4-Trimethylbenzene	1,559	2,942	2,446	2,652	6,233	2,268	18,100
Certain glycol ethers			577	8			585
Diethanolamine			4	2			6
Dichloromethane			114	134			248
Ethylbenzene						450	450
Ethylene glycol			35	33		27	95
Methanol			830		1,010	1,301	3,141
Methyl ethyl keytone		10,971	26	42	11,994	11,474	34,507
Methyl isobutyl keytone	1,000	1,000	11,750	17,000	4,300	7,300	42,350
N-butyl alcohol	1,471	3,308	2,878	2,847	5,083	1,810	17,397
Naphthalene	936						936
Phosgene		5,400	5,000	5,200	4,800	5,400	25,800
Phenol	5,400						5,400
Propylene					250	750	1,000
Tetrachloroethylene (PCE)			10,973				10,973
Toluene	20,893	22,032		96			43,021
Trichloroethylene (TCE)	10,443	12,960	11,800	10,792		16,080	62,075
Xylene (mixed isomers)	3,344	2,092	13,016	13,515	9,562	2,200	43,729
Total	45,046	60,705	59,449	52,321	43,232	49,060	309,813

APPENDIX 8-20 Total Toxic Air Emissions — Oklahoma City, OK

Chemical	Pounds/Year						
	1996	1997	1998	1999	2000	2001	Total
1,2,4-Trimethylbenzene	83,290	141,100	122,900	132,300	87,100	5,960	572,650
1,1,1-Trichloroethane						9,200	9,200
Benzene	296		1,000	500	500		2,296
Benzo(g,h,i) perylene					1		1
Carbon disulfide	390	250	250	250	38,150	750	40,040
Certain glycol ethers	14,436	118,283	92,864	39,750	508	3,562	269,403
Di(2-ethylhexyl) phthalate		1,300	5,800	4,760	3,750	3,750	19,360
Dichloromethane	22,025	17,244	420	400	250	250	40,589
Ethylbenzene	58,413	66,890	97,761	109,987	102,200	12,520	447,771
Ethylene glycol	1,510	510	1,502	1,502	1,538	501	7,063
Methanol	13,154	22,200	33,869	35,141	38,049	8,833	151,246
Methyl ethyl keytone	72,688	72,208	21,825	46,504	13,662	33,831	260,718
Methyl isobutyl keytone	47,943	47,160	61,415	78,501	85,100	12,310	332,429
Methyl methacrylate						1,070	1,070
Methyl tert butyl ether			500	1,000	500		2,000
N-butyl alcohol	124,901	121,400	123,300	101,700	102,600	17,560	591,461
N-hexane	410,156	253,387	203,772	32,570	248,785	172,576	1,321,246
N-methyl-2-pyrrolidone					14,360		14,360
Polycyclic aromatic compounds					7	4	11
Propylene	208	241	87				536
Styrene	4,921	6,000	13,374	4,522	8,308	9,225	46,350
Tetrachloroethylene (PCE)					82	64	146
Toluene	52,695	46,198	52,473		81,403	26,097	258,866
Trichloroethylene (TCE)				82,212			82,212
Vinyl chloride	5,300	3,700	5,300	2,150	2,400	1,760	20,610
Xylene (mixed isomers)	408,809	411,874	489,813	571,539	587,793	109,061	2,578,889
Total	1,321,135	1,329,945	1,328,225	1,245,288	1,417,046	428,884	7,070,523

APPENDIX 8-21 Total Toxic Air Emissions — Orlando, FL

	Pounds/Year						
Chemical	**1996**	**1997**	**1998**	**1999**	**2000**	**2001**	**Total**
1,2,4-Trimethylbenzene			42	500	500	500	1,542
Biphenyl	3,669	2,758	2,586	69	4,136	4,462	17,680
Catechol					226	127	353
Certain glycol ethers	26,001	32,016	22,970	2,047	10,082	3,242	96,358
Dichloromethane	577,225	523,278	173,583	608,951	591,582	353,319	2,827,938
Diisocyanates	255	455	195	185	1,199	1,248	3,537
Dioxin and dioxin-like compounds						3.8	3.8
Ethylbenzene			117	255	255	255	882
Ethylene glycol	9,049	5,006	10,737	2,059	2,140	2,046	31,037
Methanol			907	2,000	2,375	2,103	7,385
Methyl ethyl keytone	9,000	10,310	1,013	2,511	1,350	1,000	25,184
Methyl isobutyl keytone	2,389	1,079	1,726	255	1,280	1,328	8,057
Methyl methacrylate	4,960	8,416	2,775	3,691	19,233	13,841	52,916
N-butyl alcohol				255	255		510
N-hexane			631	1,205	755	755	3,346
N-methyl-2-pyrrolidone		54	68	75	497	279	973
Polycyclic aromatic compounds					1	3	4
Sodium dimethyldithiocarbamate	2	2	2	2			8
Styrene	332,974	292,166	581,150	558,086	498,372	389,962	2,652,710
Toluene diisocyanate (mixed isomers)	209	850	218	313	302	272	2,164
Toluene	6,330	3,030	21,131	18,176	3,426	2,310	54,403
Vinyl acetate			269	288	148	199	904
Xylene (mixed isomers)			11,939	7,620	8,495	8,740	36,794
Total	972,063	879,420	832,059	1,208,543	1,146,609	7,859,94.8	5,824,689

APPENDIX 8-22 Total Toxic Air Emissions — Peoria, IL

Chemical	Pounds/Year						
	1996	1997	1998	1999	2000	2001	Total
Benzene				2,569		790	3,359
Acetaldehyde				474,961	545,975	238,000	1,258,936
Acrolein				35,364	40,648	22,074	98,086
Certain glycol ethers	42,364	25,076	24,100	20,500	21,900	15,600	149,540
Diisocyanates				5			5
Dioxin and dioxin-like compounds					2.77	2.65	5.42
Ethylene glycol		310	5	1	1	1	318
Methanol	80	66,450	86,707	540,367	548,245	352,756	1,594,605
Methyl ethyl keytone	15,691			52,704	60,578	32,339	161,312
Methyl isobutyl keytone		20,894	500	500		21	21,915
Propylene	1,041	1,041	890	1,300			4,272
Toluene	10,700	12,350	8,856	5,076			36,982
Vinyl acetate					28,646	15,400	44,046
Xylene (mixed isomers)	46,171	46,180	37,000	22,780	23,800	17,200	193,131
Total	116,047	172,301	158,058	1,156,127	1,269,796	694,183.7	3,566,512

APPENDIX 8-23 Total Toxic Air Emissions — Phoenix, AZ

Chemical	Pounds/Year						
	1996	1997	1998	1999	2000	2001	Total
1,1-Dichloro-1-fluoroethane	750	10	1,156	1,177	1,779	1,841	6,713
1,2,4-Trimethylbenzene	50	11,700	1,602	1,675	1,738	1,565	18,330
2-Mercaptobenzothiazole		22,000					22,000
2-Methoxyethanol			17,852	18,847	16,446	4,358	57,503
1,1,1-Trichloroethane	72,996						72,996
4,4-Isopropyllindenediphenol						5,000	5,000
Benzene	180		2,536	2,812	1,965	2,199	9,692
Certain glycol ethers	81,651	76,607	109,851	156,836	183,259	117,689	725,893
Creosote							
Chlorodifluoromethane	31,000	39,100	48,000	51,500	40,020	48,000	257,620
Cyclohexane			739	594	1,061	690	3,084
Di(2-ethylhexyl) phthalate	52,329						52,329
Dibutylphthalate		10	215	192	115		532
Dioxin and dioxin-like compounds					0.19		0.19
Diisocyanates	10	1,249	210	201	201	201	2,072
Dichloromethane		54,030	48,458	50,354	40,653	33,374	226,869
Ethylbenzene	1,406	579,673	1,824	1,558	1,408	2,839	588,708
Ethylene glycol	200	2,435	872	10,980	309	70	14,866
Formaldehyde	5,173	5,334	3,309	3,801	7,716	7,935	33,268
Methanol	96,008	86,918	35,895	12,314	11,632	11,323	254,090
Methyl acrylate			269	231			500

APPENDIX 8-23 *Continued*

	Pounds/Year						
Chemical	**1996**	**1997**	**1998**	**1999**	**2000**	**2001**	**Total**
Methyl ethyl keytone	100,771	49,660	54,088	43,077	31,218	32,546	311,360
Methyl methacrylate			8,869	6,031	28,824	23,834	67,558
Methyl isobutyl keytone			10	6	8	6	30
Methyl tert butyl ether	917		27,847	23,762	12,430	14,638	79,594
N-butyl alcohol	36,869	37,026	41,627	37,738	45,120	57,901	256,281
N-hexane	335,485	299,486	271,562	178,468	164,318	153,255	1,402,574
N-methyl-2-pyrrolidone	600	1,647	1,481	1,880	2,258	1,872	9,738
Naphthalene	10		90				100
Phenol						1,305	1,305
Polycyclic aromatic compounds					4	1	5
Propylene			3,229	2,914	3,000	4,571	13,714
Sec-butyl alcohol			8	6	8	4	26
Sodium dimethyldithiocarbamate	250	390	582	133	69	326	1,750
Styrene	108,902	292,513	394,225	421,670	431,044	318,833	1,967,187
Tert-butyl alcohol						1,417	1,417
Tetrachloroethylene (PCE)	15,650	35,774	12,790	21,319	36,146	30,888	152,567
Toluene	66,285	78,135	59,045	50,102	47,547	46,892	348,006
Toluene-2,6-diisocyanate	255						255
Trichloroethylene (TCE)	25,154	25,628	11,905	10,275		10,750	83,712
Xylene (mixed isomers)	26,189	13,824	19,442	35,856	17,208	18,001	130,520
Total	1,059,090	1,713,149	1,179,588	1,146,309	1,127,504	954,124	7,179,764

APPENDIX 8-24 Total Toxic Air Emissions — Portland, OR

Chemical	Pounds/Year 1996	1997	1998	1999	2000	2001	Total
1,1-Dichloro-1-fluoroethane	531	567		20,700			21,798
1,2,4-Trimethylbenzene	1,058	37,847	62,838	82,867	32,079	10,180	226,869
4,4-Isopropylidenediphenol					2,797	2,774	5,571
Anthracene		51	205	4			260
Benzene	6,534	2,197	9,877	11,730	14,274	23,206	67,818
Butyl acrylate	107	108	109		112	86	522
Certain glycol ethers	248,082	211,096	159,328	128,508	101,861	55,981	904,856
Cresol (mixed isomers)		563	563	559	1,100	1,100	3,885
Creosote			27				27
Cumene	273		22	204			499
Cyclohexane	1,501	1,730	2,953	4,634	3,900	3,616	18,334
Dibenzofuran			2,741				2,741
Dicofol	250						250
Dioxin and dioxin-like compounds					0.078		0.078
Diisocyanates	1	1				5	7
Dichloromethane	187,605	158,615	41,216	1,819	1,309	1,712	392,276
Epichlorohydrin	4,762	2,408	3,904	3,915	3,916	3,905	22,810
Ethylene acrylate	2						2
Ethylbenzene	10,000	25,539	49,179	41,257	25,515	11,126	162,616
Ethylene glycol	3,000	3,421	3,177	3,192	3,047	3,246	19,083
Formaldehyde	5,766	5,798	5,476	5,750	5,950	3,520	32,260
M-xylene		53,237	6,071	6,071	2,155		67,534
Methanol	353,075	250,120	191,955	246,336	252,588	274,328	1,568,402

APPENDIX 8-24 *Continued*

	Pounds/Year						
Chemical	**1996**	**1997**	**1998**	**1999**	**2000**	**2001**	**Total**
Methyl ethyl keytone	127,750	54,782	77,869	71,546	18,177	14,644	364,768
Methyl methacrylate	115	115	115		130	97	572
Methyl isobutyl keytone	354	3,139	145	4,181	772	467	9,058
Methyl tert butyl ether	17,668		2,287	346		374	20,675
N-butyl alcohol	42,080	81,082	53,007	72,103	34,252	7,804	290,328
N-hexane	28,631	9,273	22,885	26,194	32,776	22,733	142,492
N,N-dimethylformamide					170	200	370
N-methyl-2-pyrrolidone	96,521			2,617			99,138
Naphthalene	87		22,225		73	490	22,875
Phenol	10,765	28,069	16,262	22,567	19,411	15,184	112,258
Phthalic anhydride		1	1		1		3
Polycyclic aromatic compounds			1,778		6	6	1,790
Propylene					4,000		4,000
Sec-butyl alcohol	26,577	27,777		42,396	9,577		106,327
Styrene	33,853	37,493	47,721	68,598	73,490	41,910	303,065
Tetrachloroethylene (PCE)			9				9
Toluene	316,916	235,601	210,703	207,564	88,852	53,781	1,113,417
Toluene diisocyanate		36	34	32	26	17	145
Trichloroethylene (TCE)	5	10,741	9,761	30	8,041	3,499	32,077
Triethylamine			3,341	2,881	3,952	4,587	14,761
Vinyl acetate	37		75		64	71	247
Xylene (mixed isomers)	280,637	225,892	283,137	235,541	175,754	109,853	1,310,814
Total	1,804,543	1,467,299	1,290,996	1,314,142	920,127.1	670,502	7,467,609

APPENDIX 8-25 Total Toxic Air Emissions — Providence, RI

	Pounds/Year						
Chemical	**1996**	**1997**	**1998**	**1999**	**2000**	**2001**	**Total**
1,1-Dichloro-1-fluoroethane	2,455						2,455
1,2,4-Trimethylbenzene			102	85	760	848	1,795
1,1,1-Trichloroethane	10,518						10,518
Benzene			2,319	1,909	2,363	3,670	10,261
Certain glycol ethers	29,651	22,807	25,815	28,977	17,561	13,426	138,237
Chlorobenzene	15	16	18	4			53
Chlorohexane				160	709		869
Cyclohexane			151				151
Dibutylphthalate		1					1
Diethanolamine	1	1			250		252
Dichloromethane	1,384	2,295	584	101	184	117	4,665
Ethylbenzene		26	520	445	997	1,347	3,335
Ethylene glycol	1	1	2	10		10	24
Formaldehyde	1	2	2	1	1	1	8
Methanol	171	108	166	144	138	94	821
Methyl ethyl keytone	158,914	168,620	200,998	215,956	175,671	121,120	1,041,279
Methyl isobutyl keytone		12,000		25,502			37,502
Methyl tert butyl ether			34,187		28,292	32,035	94,514
N-butyl alcohol				1	6	5	12
N-hexane			4,341	3,256	3,832	4,249	15,678
N,N-dimethylformamide		1	3	5	3		12
Naphthalene			500		5		505
Polycyclic aromatic compounds					7	8	15
Styrene	7	5	8	8	7	4	39
Tetrachloroethylene (PCE)			400	500	500	500	1,900
Toluene	647,565	438,851	51,756	65,830	86,815	47,175	1,337,992
Trichloroethylene (TCE)	14,250						14,250
Triethylamine						40	40
Xylene (mixed isomers)	83,991	56,253	2,233	8,820	10,732	4,145	166,174
Total	948,924	700,987	324,105	351,714	328,833	228,794	2,883,357

APPENDIX 8-26 Total Toxic Air Emissions — Pueblo, CO

	Pounds/Year						
Chemical	**1996**	**1997**	**1998**	**1999**	**2000**	**2001**	**Total**
1,3-Butadiene	0	0	72	72	60	64	268
Benzene	0	0	3,579	3,579	2,491	3,122	12,771
Chlorofluoromethane	53,000	25,000	2,500	2,500	2,900	8,400	94,300
Creosote	0	0	0	22	0	0	22
Diisocyanates	0	0	0	29	27	0	56
Dioxin and dioxin-like compounds	0	0	0	0	0.34	0.34	0.68
Dichloromethane	0	0	2,247	1,615	1,538	0	5,400
Ethylene	0	0	790	0	676	702	2,168
Ethylene glycol	0	0	0	1	2	3	6
Methanol	0	0	673	1,000	731	518	2,922
Naphthalene	0	0	1,688	1,688	1,682	1,685	6,743
Polycylic aromatic compounds	0	0	0	0	13	11	24
Propylene	0	0	255	255	252	253	1,015
Toluene	12,000	12,000	284	284	282	0	24,850
Xylene (mixed isomers)	10,900	8,000	9,100	4,700	0	0	32,700
Total	75,900	45,000	21,188	15,745	10,654.34	14,758.34	183,245.7

APPENDIX 8-27 Total Toxic Air Emissions — Reno, NV

	Pounds/Year						
Chemical	**1996**	**1997**	**1998**	**1999**	**2000**	**2001**	**Total**
1,1-Dichloro-1-fluoroethane	10,756	9,719	10,171	0	0	0	30,646
1,2,4-Trimethylbenzene	0	8,191	7,058	0	0	0	15,249
Dichloromethane	0	0	0	7,160	8,984	0	16,144
Certain glycol ethers	14,148	11,346	3,300	0	0	0	28,794
Diisocyanates	0	0	30	0	0	0	30
Dimethylamine	38	50	0	0	0	0	88
Ethylbenzene	5	7	0	0	1,188	0	1,200
Toluene	299,324	619,855	555,859	471,286	319,721	235,337	2,501,382
Xylene (mixed isomers)	17	29	12	14	13	12	97
Total	324,288	649,197	576,430	478,460	329,906	235,349	2,593,630

APPENDIX 8-28 Total Toxic Air Emissions — San Antonio, TX

	Pounds/Year						
Chemical	**1996**	**1997**	**1998**	**1999**	**2000**	**2001**	**Total**
1,1-Dichloro-1-fluoroethane	0	0	0	0	0	2,772	2,772
1,1,1-Trichloroethane	0	0	1,376	0	0	0	1,376
1,2,4-Trimethylbenzene	150	415	1,794	1,745	1,781	1,283	7,168
Benzene	310	805	3,810	3,437	3,626	3,966	15,954
Certain glycol ethers	13,850	12,900	7,700		255	250	34,955
Chlorodifluoromethane	2,746	85,569	3	35,152	37,480	76,560	237,510
Cumene	0	0	505	265	500	10	1,280
Cyclohexane	0	0	1,255	1,000	1,000	755	4,010
Dichloromethane	1,800	1,000	27,455	26,593	33,000	34,200	124,048
Diisocyanates	0	0	2	2	3	3	10
Diethanolamine	0	0	0	10	10	10	30
Dioxin and dioxin-like compounds	0	0	0	0	7.3432	0.1288	7.472
Ethylene	0	0	0	0	0	2,957	2,957
Ethylbenzene	550	12,148	9,425	7,499	9,829	13	39,464
Ethylene glycol	0	143	242	195	12	0	592
Methanol	0	0	400	18,329	0	0	18,729
Methyl ethyl keytone	84,173	129,559	21,413	4,703	1,000	255	241,103
Methyl isobutyl keytone	41,443	59,500	413	755	5,309	755	108,175
Methyl tert butyl ether	0	0	0	3,531	50	104	3,685
N-butyl alcohol	9,315	26,591	32,732	22,275	25,106	18,440	134,459
N-methyl-2-pyrrolidone	840	2,500	510	480	250	0	4,580
N-hexane	0	0	8,325	5,329	3,929	5,507	23,090
Naphthalene	0	0	255	260	255	0	770
Polycyclic aromatic compounds	0	0	0	0	60	39	99
Styrene	0	0	18,900	31,968	20,399	15,950	87,217
Tetrachloroethylene (PCE)	88,000	29,500	10,451	13,000	0	0	140,951
Toluene	249,387	278,627	186,937	116,380	73,339	51,916	956,586
Trichloroethylene (TCE)	157,368	151,694	712	24,599	26,614	18,112	379,099
Xylene (mixed isomers)	88,285	123,611	93,526	40,626	43,216	36,894	426,158
Total	738,217	914,562	428,141	358,133	287,030.3	270,751.1	2,996,834

APPENDIX 8-29 Total Toxic Air Emissions — Savannah, GA

	Pounds/Year						
Chemical	**1996**	**1997**	**1998**	**1999**	**2000**	**2001**	**Total**
1,1-Dichloro-1-fluoroethane	4,050	4,200	9,619	8,728	9,280	10,954	46,831
1,2,4-Trimethylbenzene	31	0	0	0	0	0	31
1,1,1-Trichloroethane	75,000	32,000	39,000	0	0	0	146,000
Acetaldehyde	204,619	235,611	210,185	172,840	154,090	153,808	1,131,153
Benzene	1,262	0	604	0	0	0	1,866
Benzo(G,H,I) perylene	0	0	0	0	69	2	71
Biphenyl	4,950	0	0	0	0	0	4,950
Carbonyl sulfide	270,000	590,001	490,001	490,001	490,001	280,001	2,610,005
Certain glycol ethers	41,766	41,895	39,858	44,615	40,186	18,393	226,713
Chlorodifluoromethane	10,017	13,400	9,619	12,565	13,279	6,000	64,880
Creosote	2,320	2,100	2,130	2,300	3,290	2,500	14,640
Cresol (mixed isomers)	2	0	0	0	0	0	2
Cumene	5	0	0	0	0	0	5
Cyclohexane	4,241	0	580	0	405	363	5,589
Dibutyl phthalate	10,700	0	0	0	0	0	10,700
Diisocyanates	255	31	19	35	20	9	369
Dioxin and dioxin-like compounds	0	0	0	0	4.431	2.469	6.9
Epichlorohydrin	3,722	5,591	6,437	10,560	7,331	8,919	42,560
Ethylbenzene	82	14,300	0	0	0	0	14,382
Formaldehyde	3,447	32,520	30,084	31,315	27,287	21,357	146,010
Formic acid	0	22,650	0	0	0	0	22,650
M-xylene	0	0	0	0	0	47	47
Malic anhydride	441	410	460	410	410	505	2,636
Methanol	1,849,750	2,065,310	1,915,894	1,328,852	1,701,219	1,601,000	10,462,025
Methyl ethyl keytone	90,030	138,220	74,269	19,850	21,590	24,310	368,269
Methyl isobutyl keytone	14,600	0	0	0	0	0	14,600
N-hexane	8,526	5,962	10,838	6,001	7,547	6,900	45,774
Pentachlorophenol	4	0	0	3	0	13	20
Phenol	5,989	6,417	9,031	5,793	3,588	2,707	33,525
Polycyclic aromatic compounds	0	0	0	0	5,119	193	5,312
Styrene	23,917	16,707	19,642	21,270	24,948	12,583	119,067
Titanium tetrachloride	530	6,100	6,980	6,810	7,800	6,900	35,120
Toluene	274,912	331,320	308,047	290,965	294,885	282,785	1,782,914
Trichloroethylene (TCE)	0	0	14,800	0	0	0	14,800
Xylene (mixed isomers)	62,801	84,830	132,019	146,404	70,916	7,704	504,674
Total	2,967,969	3,649,575	3,330,116	2,599,317	2,883,264	2,447,955	17,878,197

APPENDIX 8-30 Total Toxic Air Emissions — Sioux Falls, SD

	Pounds/Year						
Chemical	**1996**	**1997**	**1998**	**1999**	**2000**	**2001**	**Total**
1,2,4-Trimethylbenzene	11,535	16,862	15,740	0	0	0	44,137
Certain glycol ethers	0	0	0	3,120	1,480	0	4,600
Ethylbenzene	14,733	13,394	12,201	0	0	0	40,328
Methyl ethyl keytone	35,251	19,352	0	0	0	0	54,603
Methyl isobutyl keytone	36,409	0	0	0	0	0	36,409
Polycyclic aromatic compounds	0	0	0	0	5	0	5
Styrene	48,000	48,000	146,375	122,344	128,600	41,473	534,792
Toluene	27,476	0	0	0	17,967	25,994	71,437
Xylene (mixed isomers)	173,379	205,733	199,171	147,597	117,031	144,655	987,566
Total	346,783	303,341	373,487	273,061	265,083	212,122	1,773,877

APPENDIX 8-31 Total Toxic Air Emissions — Springfield, MO

	Pounds/Year						
Chemical	**1996**	**1997**	**1998**	**1999**	**2000**	**2001**	**Total**
1,2,4-Trimethylbenzene	43,510	46,607	35,963	31,071	16,158	0	173,309
Butyl acetate	1,204	0	0	0	7	10	1,221
Certain glycol ethers	26,056	20,592	14,375	9,853	10,086	1,000	81,962
Chloroethane	0	0	0	0	0	7,929	7,929
Chloromethane	26,640	21,260	5,863	11,034	15,053	0	79,850
Creosote	7,367	5,525	7,600	3,600	3,000	2,900	29,992
Cyclohexane	19,913	21,371	0	8,000	14,300	13,210	76,794
Di(2-ethylexyl) phthalate	73	62	0	40	42	30	247
Dimethylamine	71,001	84,759	0	52,207	0	0	207,967
Diisocyanates	2,729	1,474	715	3,769	3,425	2,625	14,737
Dioxin and dioxin-like compounds	0	0	0	0	0.81	0.77	1.58
Dichloromethane	0	0	39,373	0	38,708	25,460	103,541
Ethylbenzene	37,195	38,150	5,997	5,664	0	0	87,006
Ethylene glycol	0	0	0	0	2	4	6
Formaldehyde	1,000	1,000	1,000	1,000	250	0	4,250
Methanol	10,601	11,150	1,000	15,381	9,468	3,276	50,876
Methyl ethyl keytone	78,511	127,373	0	98,586	34,900	42,210	381,580
Methyl isobutyl keytone	22,914	24,905	0	900	5,400	5,210	59,329
N-butyl alcohol	6,947	5,568	5,287	5,119	0	0	22,921
N-hexane	8,589	8,424	0	3,624	10,800	10,430	41,867
Polycylic aromatic compounds	0	0	0	0	8	4	12
Styrene	4,229	4,646	4,574	4,771	7,598	7,888	33,706
Tetrabromobisphenol A	0	0	0	0	16	20	36
Toluene	324,558	371,245	44,240	139,110	137,280	129,892	1,146,325
Toluene diisocyanate (mixed isomers)	0	138	0	80	80	70	368
Xylene (mixed isomers)	240,545	229,666	52,788	81,659	67,410	16,110	688,178
Total	933,582	1,023,915	218,775	475,468	373,991.8	268,278.8	3,294,011

APPENDIX 8-32 Total Toxic Air Emissions — Tacoma, WA

	Pounds/Year						
Chemical	**1996**	**1997**	**1998**	**1999**	**2000**	**2001**	**Total**
1,1,1-Trichloroethane	0	0	0	0	0	2,526	2,526
1,2-Dibromoethane	0	0	0	0	0	2,692	2,692
Benzene	9,120	9,042	2,872	2,138	2,121	3,301	28,594
Benzo(g,h,i) perylene	0	0	0	0	70	1	71
Acetaldehyde	36,004	38,904	45,004	45,604	30,704	31,100	227,320
Acetonitrile	0	0	0	0	0	6,424	6,424
Carbonyl sulfide	178,000	182,000	167,000	201,184	128,661	55,941	912,786
Certain glycol ethers	1,400	1,600	1,900	1,740	3,124	1,950	11,714
Chloroform	25,000	29,700	37,100	34,800	0	3,527	130,127
Chloromethane	0	0	0	0	538	0	538
Cresol (mixed isomers)	34,000	38,400	32,700	30,400	0	1,428	136,928
Creosote	0	255	1,000	1,805	1,947	955	5,962
Cyclohexane	0	0	1,528	1,147	1,607	6,802	11,084
Dichlorodifluoromethane	27,163	34,000	22,030	21,548	11,000	12,000	127,741
Diisocyanates	5	5	5	5	5	5	30
Dioxin and dioxin-like compounds	0	0	0	0	10.31	23.5754	33.8854
Dichloromethane	24,421	17,303	16,193	16,082	17,167	18,122	109,288
Ethylbenzene	18,400	14,598	11,290	10,320	821	4,049	59,478
Ethylene glycol	11,756	12,153	3,212	3,421	2,980	66,731	100,253
Formaldehyde	2,897	19,696	21,498	20,581	21,109	25,207	110,988

APPENDIX 8-32 *Continued*

	Pounds/Year						
Chemical	**1996**	**1997**	**1998**	**1999**	**2000**	**2001**	**Total**
Methanol	791,186	949,271	814,037	597,860	528,399	505,285	4,186,038
Methyl ethyl keytone	18,429	61,102	51,951	34,544	34,485	66,351	266,862
Methyl isobutyl keytone	17,935	13,252	20	29	11,123	11,725	54,084
Methyl tert butyl ether	0	0	0	0	0	2,694	2,694
N-butyl alcohol	0	228	0	0	4,540	3,272	8,040
N,N-dimethylformamide	0	0	0	0	0	4,575	4,575
N-hexane	0	0	3,759	2,985	6,972	12,003	25,719
N-methyl-2-pyrrolidone	1,554	19,000	5,999	5,816	11,747	17,782	61,898
Naphthalene	0	0	56	43	128	121	348
Phenol	22,528	19,836	20,970	24,667	29,200	31,033	148,234
Polycylic aromatic compounds	5,070	5,170	4,760	3,420	5,390	26	23,836
Pyridine	0	0	0	0	0	4,907	4,907
Styrene	13,036	20,588	47,709	62,413	78,726	16,489	238,961
Tetrachloroethylene (PCE)	0	21,392	26,105	10,365	0	732	58,594
Toluene	105,410	114,875	43,880	42,071	75,741	72,148	454,125
Trichloroethylene (TCE)	0	0	0	0	0	2,753	2,753
Xylene (mixed isomers)	155,133	115,500	79,147	73,584	71,189	73,020	567,573
Total	1,498,447	1,737,870	1,461,725	1 248,572	1,079,504	1,067,701	8,093,819

APPENDIX 8-33 Total Toxic Air Emissions — Tulsa, OK

Chemical	Pounds/Year						
	1996	1997	1998	1999	2000	2001	Total
1,2,4-Trichlorobenzene	0	0	0	1,460	0	5,050	6,510
1,2,4-Trimethylbenzene	17,491	17,870	54,991	1,000	53,485	6,811	151,648
1,1,1-Trichloroethane	10,000	11,280	0	0	0	0	21,280
1,3-Butadiene	920	494	487	134	218	75	2,328
4,4-Isopropylidenediphenol	0	250	0	0	0	0	250
Acetophenone	255	255	5	7	7	0	529
Benzene	22,034	20,645	30,506	19,219	34,968	20,210	147,582
Benzo(g,h,i) perylene	0	0	0	17	0	3	20
Carbonyl sulfide	0	677	648	563	874	413	3,175
Certain glycol ethers	1,645	2,025	2,054	2,105	2,136	1,479	11,444
Chlorodifluoromethane	36,780	29,530	50,750	24,708	17,000	18,863	177,631
Cumene	210	0	0	0	0	0	210
Cyclohexane	20,446	19,688	23,449	8,549	25,499	7,495	105,126
Dichloromethane	12,308	12,308	11,508	0	755	0	36,879
Diethanolamine	3,724	6,766	6,806	9,357	4,497	9,333	40,483
Dioxin and dioxin-like compounds	0	0	0	1.602	0	0.702	2.304
Ethylene	14,255	6,696	7,319	8,047	5,672	5,330	47,319
Ethylbenzene	21,670	21,850	16,214	6,570	15,334	5,811	87,449
Ethylene glycol	11,743	1,510	1,437	1,710	2,221	400	19,021
Formaldehyde	10	10	5	2	2	0	29

APPENDIX 8-33 *Continued*

	Pounds/Year						
Chemical	**1996**	**1997**	**1998**	**1999**	**2000**	**2001**	**Total**
Hydrazine	10	0	0	0	0	0	10
Methanol	5,513	5,263	5,285	3,111	2,041	1,448	22,661
Methyl ethyl keytone	235,149	176,740	236,296	159,041	232,664	94,355	1,134,245
Methyl methacrylate	1,450	8,488	750	750	750	750	12,938
Methyl isobutyl keytone	1,473	1,500	1,661	2,132	1,164	26,785	34,715
Methyl tert butyl ether	17,660	26,940	10	0	0	8,800	53,410
N-butyl alcohol	0	0	0	500	0	504	1,004
N-hexane	77,678	70,199	86,795	25,443	96,174	42,103	398,392
Naphthalene	18,813	18,592	2,891	2,491	3,349	900	47,036
Phenol	737	121	124	5	115	250	1,352
Polycyclic aromatic compounds	0	51	50	562	0	118	781
Propargyl alcohol	10	10	10	0	0	0	30
Propylene	86,323	137,217	84,208	27,004	74,942	16,521	426,215
Styrene	22,605	17,983	19,703	17,304	15,256	21,255	114,106
Tetrachloroethylene (PCE)	82,066	91,488	42,014	38,880	19,451	10,438	284,337
Toluene	174,008	225,655	265,315	134,009	198,747	86,103	1,083,837
Trichloroethylene (TCE)	0	13,449	12,900	18,570	18,896	0	63,815
Xylene (mixed isomers)	162,178	126,869	127,569	78,866	99,589	40,777	635,848
Total	1,059,164	1,072,419	1,091,760	592,117.6	925,806	432,380.7	5,173,647

APPENDIX 8-34 Total Toxic Air Emissions — Washington, D.C.

	Pounds/Year						
Chemical	**1996**	**1997**	**1998**	**1999**	**2000**	**2001**	**Total**
Dioxin and dioxin-like compounds	0	0	0	0	0.12	0.1	0.22
Total	0	0	0	0	0.12	0.1	0.22

APPENDIX 8-35 Total Toxic Air Emissions — West Palm Beach, FL

	Pounds/Year						
Chemical	**1996**	**1997**	**1998**	**1999**	**2000**	**2001**	**Total**
1,1,1-Trichloroethane	8,415	14,820	0	0	0	0	23,235
1,2-Butylene oxide	0	0	0	0	0	10	10
Benzene	500	0	500	500	0	0	1,500
Dioxin and dioxin-like compounds	0	0	0	0	1.57	2.05	3.62
Ethylbenzene	255	0	255	255	0	0	765
Ethylene glycol	0	5,353	6,268	7,465	7,465	0	26,551
Methyl ethyl keytone	1,000	1,000	1,000	1,000	1,000	1,000	6,000
N-hexane	500	0	500	500	0	0	1,500
Polycyclic aromatic compounds	0	0	1	1	187	2	191
Styrene	3,294	3,394	4,119	4,127	4,560	21,887	41,381
Toluene	500	0	500	500	0	0	1,500
Trichloroethylene (TCE)	13,999	0	0	0	0	0	13,999
Xylene (mixed isomers)	255	0	255	255	0	0	765
Total	28,718	24,567	13,398	14,603	13,213.57	22,901.05	117,400.6

APPENDIX 8-36 Total Toxic Air Emissions — Worcester, MA

	Pounds/Year						
Chemical	**1996**	**1997**	**1998**	**1999**	**2000**	**2001**	**Total**
Benzo(g,h,i) perylene	0	0	0	0	16	0	16
Certain glycol ethers	9,157	10,183	10,706	9,788	9,524	9,193	58,551
Ethylene glycol	1,500	7,740	6,548	6,704	6,420	5,865	34,777
Formaldehyde	250	0	0	0	0	0	250
Methanol	40,983	50,164	67,620	64,288	48,578	10,559	282,192
Methyl ethyl keytone	47,213	51,282	46,997	54,544	45,902	11,575	257,513
Methyl isobutyl keytone	0	0	695	748	589	0	2,032
Phenol	7,600	8,966	8,024	5,223	4,417	6,481	40,711
Polycylic aromatic compounds	0	0	0	0	1,507	32	1,539
Toluene	18,062	23,727	20,655	21,277	15,085	3,474	102,280
Trichloroethylene (TCE)	22,026	20,169	16,170	18,099	17,662	0	94,126
Xylene (mixed isomers)	1,752	2,403	3,030	1,490	2,776	1,102	12,553
Total	148,543	174,634	180,445	182,161	152,476	48,281	886,540

APPENDIX 9-1 Proposition 65 Safe Harbor Levels for No Significant Risk—Carcinogens (2005)

Carcinogen	ug/day
A-alpha-C (2-amino-9H-pyrido[2,3-b]indole)	2
Acetaldehyde	90 (inhalation)
Acetamide	10
2-Acetylaminofluorene	0.2
Acrylamide	0.2
Acrylonitrile	0.7
Actinomycin D	0.00008
AF-2 [2-(2-furyl)-3(5-nitro-2-furyl)acrylamide]	3
Aldrin	0.04
2-Aminoanthraquinone	20
o-Aminoazotoluene	0.2
4-Aminobiphenyl	0.03
3-Amino-9-ethylcarbazole hydrochloride	9
1-Amino-2-methylanthraquinone	5
2-Amino-5-(5-nitro-2-furyl)-1,3,4-thiadiazole	0.04
Amitrole	0.7
Aniline	100
o-Anisidine	5
o-Anisidine hydrochloride	7
Aramite	20
Arsenic	0.06 (inhalation)
	10 (except inhalation)
Asbestos	100 fibers/day (inhalation)
Auramine	0.8
Azaserine	0.06
Azathioprine	0.4
Azobenzene	6
Benz [a] anthracene	0.033 (oral)
Benzene	6.4 (oral)
	13 (inhalation)
Benzidine	0.001
Benzo[b]fluoranthene	0.096 (oral)
Benzo[j]fluoranthene	0.11 (oral)
Benzofuran	1.1
Benzo[a]pyrene	0.06
Benzyl chloride	4
Benzyl violet 4B	30
Beryllium	0.1
Beryllium oxide	0.1
Beryllium sulfate	0.0002
Bis(2-chloroethyl)ether	0.3
Bis(chloromethyl)ether	0.02
Bromodichloromethane	5
Bromoform	64
1,3-Butadiene	0.4
Butylated hydroxyanisole	4000
beta-Butyrolactone	0
Cadmium	0.05 (inhalation)

APPENDIX 9-1 *Continued*

Carcinogen	ug/day
Captafol	5
Captan	300
Carbazole	4.1
Carbon tetrachloride	5
N-Carboxymethyl-N-nitrosourea	0.70
Chlorambucil	0.002
Chlordane	0.5
Chlordecone (kepone)	0.04
Chlorendic acid	8
Chlorinated paraffins	8
Chloroethane (ethyl chloride)	150
Chloroform	20 (oral)
	40 (inhalation)
Chloromethyl ether (technical grade)	0.3
3-Chloro-2-methylpropene	5
4-Chloro-ortho-phenylenediamine	40
Chlorothalonil	200
p-Chloro-ortho-toluidine	3
p-Chloro-o-toluidine, hydrochloride	3.3
Chlorozotocin	0.003
Chromium (hexavalent)	0.001 (inhalation)
Chrysene	0.35 (oral)
C.I. Basic Red 9 monohydrochloride	3
Cinnamyl anthranilate	200
Coke oven emissions	0.3
p-Cresidine	5
Cupferron	3
Cyclophosphamide (anhydrous)	1
Cyclophosphamide (hydrated)	1
D&C Red No.	9100
Dacarbazine	0.01
Daminozide	40
Dantron (chrysazin; 1,8-dihydroxyanthraquinone)	9
DDT, DDE, DDD (in combination)	2
DDVP (Dichlorvos)	2
2,4-Diaminoanisole	30
2,4-Diaminoanisole sulfate	50
4,4′-Diaminodiphenyl ether (4,4′-oxydianiline)	5
2,4-Diaminotoluene	0.2
Dibenz[a,h]anthracene	0.2
7H-Dibenzo[c,g]carbazole	0.0030 (oral)
Dibenzo[a,h]pyrene	0.0054 (oral)
Dibenzo[a,i]pyrene	0.0050 (oral)
1,2-Dibromo-3-chloropropane	0.1
p-Dichorobenzene	20
3,3′-Dichlorobenzidine	0.6
1,1-Dichloroethane	100
1,2-Dichloroethane (ethylene dichloride)	10
Dichloromethane (methylene chloride)	200 (inhalation)
	50

APPENDIX 9-1 *Continued*

Carcinogen	ug/day
1,2-Dichloropropane	9.7
Dieldrin	0.04
Di(2-ethylhexyl)phthalate (DEHP)	310
Diethylstilbesterol	0.002
Diglycidyl resorcinol ether (DGRE)	0.4
Dihydrosafrole	20
3,3′-Dimethoxybenzidine (o-Dianisidine)	0.15
3,3′-Dimethoxybenzidine dihydrochloride	0.19
4-Dimethylaminoazobenzene	0.2
trans-2-[(Dimethylamino)methylimino]-5 [2-(5-nitro-2-furyl)vinyl]-1,3,4-oxadiazole	2
7,12-Dimethylbenz(a)anthracene	0.003
3,3′-Dimethylbenzidine (o-Tolidine)	0.044
3,3′-Dimethylbenzidine dihydrochloride	0.059
Dimethylcarbamoyl chloride	0.05
1,2-Dimethylhydrazine	0.001
Dimethylvinyl chloride	20
2,4-Dinitrotoluene	2
1,4-Dioxane	30
Direct Black 38 (technical grade)	0.09
Direct Blue 6 (technical grade)	0.09
Direct Brown 95 (technical grade)	0.1
Disperse Blue 1	200
Epichlorohydrin	9
Estradiol 17b	0.02
Ethyl-4,4′-dichlorobenzilate (chlorobenzilate)	7
Ethylene dibromide	0.2 (oral) 3 (inhalation)
Ethylene oxide	2
Ethylene thiourea	20
Ethyleneimine	0.01
Folpet	200
Formaldehyde (gas)	40
2-(2-Formylhydrazino)-4-(5-nitro-2-furyl)thiazole	0.3
Furmecyclox	20
GIu-P-1 (2-Amino-6-methyldipyrido[l,2-a:3′,2′-d]imidazole)	0.1
Glu-P-2 (2-Aminodipyrido[1,2-a:3′,2′-d]-imidazole)	0.5
Gyrornitrin (acetaldehyde methylformylhydrazone)	0.07
HC Blue 1	10
Heptachlor	0.2
Heptachlorepoxide	0.08
Hexachlorobenzene	0.4
Hexachlorocyclohexane	
alpha isomer	0.3
beta isomer	0.5
gamma isomer	0.6
technical grade	0.2
Hexachlorodibenzodioxin	0.0002
Hexachloroethane	20
Hydrazine	0.04

Carcinogen	ug/day
Hydrazine sulfate	0.2
Hydrazobenzene (1,2-diphenylhydrazine)	0.8
IQ (2-amino-3-methylimidazo[4,5-f]quinoline)	0.5
Isobutyl nitrite	7.4
Lasiocarpine	0.09
Lead	15 (oral)
Lead acetate	23 (oral)
Lead phosphate	58 (oral)
Lead subacetate	41 (oral)
Me-A-alpha-C (2-amino-3-methyl-9H-pyrido[2,3-b]indole)	0.6
MeIQ (2-amino-3,4-dimethylimidazo-[4,5 f]quinoline)	0.46
MeIQx (2-amino-3,8-dimethylimidazo-[4,5 f]quinoxaline)	0.41
Melphalan	0.005
2-Methylaziridine (propyleneimine)	0.028
Methyl carbamate	160
3-Methylcholanthrene	0.03
5-Methylchrysene	0.0084 (oral)
4,4′-Methylene bis(2-chloroaniline)	0.5
4,4′-Methylene bis(N,N-dimethyl)benzeneamine	20
4,4′-Methylene bis(2-methylaniline)	0.8
4,4′-Methylenedianiline	0.4
4,4′-Methylenedianiline dihydrochloride	0.6
Methylhydrazine	0.058 (oral) 0.090 (inhalation)
Methylhydrazine sulfate	0.18
Methyl methane sulfonate	7
2-Methyl-l-nitroanthraquinone	0.2
N-Methyl-N′-nitro-N-nitrosoguanidine	0.08
Methylthiouracil	2
Michler's ketone	0.8
Mirex	0.04
Mitomycin C	0.00009
Monocrotaline	0.07
5-(Morpholinomethyl)-3-[(5-nitrofurfurylidene)-amino]-2-oxazolidinone	0.18
MX (3-chloro-4-(dichloromethyl)-5-hydroxy-2(5M-furanone)	0.11
Nalidixic acid	28
Naphthalene	5.8
2-Naphthylamine	0.4
Nickel refinery dust	0.8
Nickel subsulfide	0.4
Nitrilotriacetic acid	100
Nitrilotriacetic acid, trisodium salt monohydrate	70
5-Nitroacenaphthene	6
5-Nitro-o-anisidine	10
Nitrofen (technical grade)	9
Nitrofurazone	0.5
1-[(5-Nitrofurfurylidene)-amino]-2-imidazolidinone	0.4
N-[4-(5-Nitro-2-furyl)-2-thiazolyl] acetamide	0.5
N-Nitrosodi-n-butylamine	0.06

APPENDIX 9-1 *Continued*

Carcinogen	ug/day
N-Nitrosodiethanolamine	0.3
N-Nitrosodiethylamine	0.02
N-Nitrosodimethylamine	0.04
p-Nitrosodiphenylamine	30
N-Nitrosodiphenylamine	80
N-Nitrosodi-n-propylamine	0.1
N-Nitroso-N-ethylurea	0.03
4-(N-Nitrosomethylamino)-1-(3-pyridyl)-i-butanone	0.014
N-Nitrosomethylethylamine	0.03
N-Nitroso-N-methylurea	0.006
N-Nitroso-N-methylurethane	0.006
N-Nitrosomorpholine	0.1
N-Nitrosonornicotine	0.5
N-Nitrosopiperidine	0.07
N-Nitrosopyrrolidine	0.3
Pentachlorophenol	40
Phenacetin	300
Phenazopyridine	4
Phenazopyridine hydrochloride	5
Phenesterin	0.005
Phenobarbital	2
Phenoxybenzamine	0.2
Phenoxybenzamine hydrochloride	0.3
o-Phenylenediamine	26
o-Phenylenediamine dihydrochloride	44
Phenyl glycidyl ether	5.0
Phenylhydrazine	1.0
Phenylhydrazine hydrochloride	1.4
o-Phenylphenate, sodium	200
Polybrominated biphenyls	0.02
Polychlorinated biphenyls	0.09
Polygeenan	1200
Ponceau MX	200
Ponceau 3R	40
Potassium bromate	1
Procarbazine	0.05
Procarbazine hydrochloride	0.06
1,3-Propane sultone	0.3
beta-Propiolactone	0.05
Propylthiouracil	0.7
Reserpine	0.06
Safrole	3
Sterigmatocystin	0.02
Streptozotocin	0.006
Styrene oxide	4
Sulfallate	4
2,3,7,8-Tetrachlorodibenzo p-dioxin	0.000005
1,1,2,2-Tetrachloroethane	3
Tetrachloroethylene	14

APPENDIX 9-1 *Continued*

Carcinogen	ug/day
Tetranitromethane	0.059
Thioacetamide	0.1
4,4′-Thiodianiline	0.05
Thiourea	10
Toluene diisocyanate	20
ortho-Toluidine	4
ortho-Toluidine hydrochloride	5
Toxaphene	0.6
Trichloroethylene	50 (oral) 80 (inhalation)
2,4,6-Trichlorophenol	10
Trimethyl phosphate	24
Tris(1-aziridinyl)phosphine sulfide (thiotepa)	0.06
Tris(2,3-dibromopropyl)phosphate	0.3
Trp-P-1 (tryptophan-P-1)	0.03
Trp-P-2 (tryptophan-P-2)	0.2
Urethane (ethyl carbamate)	0.7
Vinyl chloride	3
Vinyl trichloride (1,1,2-trichloroethane)	10
2,6-Xylidine	110

APPENDIX 10-1 Organic Chemicals Monitored by the CDC

Polycyclic Aromatic Hydrocarbons
1-Hydroxybenz[a]anthracene
3-Hydroxybenz[a]anthracene
9-Hydroxybenz[a]anthracene
1-Hydroxybenzo[c]phenanthrene
2-Hydroxybenzo[c]phenanthrene
3-Hydroxybenzo[c]phenanthrene
1-Hydroxychrysene
2-Hydroxychrysene
3-Hydroxychrysene
4-Hydroxychrysene
6-Hydroxychrysene
3-Hydroxyfluoranthene
2-Hydroxyfluorene
3-Hydroxyfluorene
9-Hydroxyfluorene
1-Hydroxyphenanthrene
2-Hydroxyphenanthrene
3-Hydroxyphenanthrene
4-Hydroxyphenanthrene
9-Hydroxyphenanthrene
1-Hydroxypyrene
1-Hydroxynaphthalene
2-Hydroxynaphthalene
Polychlorinated Dibenzo-p-Dioxins, Polychlorinated Dibenzofurans, and Polychlorinated Biphenyls
1,2,3,4,6,7,8,9-Octachlorodibenzo p-dioxin (OCDD)
1,2,3,4,6,7,8-Heptachlorodibenzo p-dioxin (HpCDD)
1,2,3,4,7,8-Hexachlorodibenzo p-dioxin (HxCDD)
1,2,3,6,7,8-Hexachlorodibenzo p-dioxin (HxCDD)
1,2,3,7,8,9-Hexachlorodibenzo p-dioxin (HxCDD)
1,2,3,7,8-Pentachlorodibenzo-p-dioxin (PeCDD)
2,3,7,8-Tetrachlorodibenzo-p-dioxin (TCDD)
1,2,3,4,6,7,8,9-Octachlorodibenzofuran (OCDF)
1,2,3,4,6,7,8-Heptachlorodibenzofuran (HpCDF)
1,2,3,4,7,8,9-Heptachlorodibenzofuran (HpCDF)
1,2,3,4,7,8-Hexachlorodibenzofuran (HxCDF)
1,2,3,6,7,8-Hexachlorodibenzofuran (HxCDF)
1,2,3,7,8,9-Hexachlorodibenzofuran (HxCDF)
1,2,3,7,8-Pentachlorodibenzofuran (PeCDF)
2,3,4,6,7,8-Hexachlorodibenzofuran (HxCDF)
2,3,4,7,8-Pentachlorodibenzofuran (PeCDF)
2,3,7,8-Tetrachlorodibenzofuran (TCDF)
2,4,4′-Trichlorobiphenyl (PCB 28)
2,3′,4,4′-Tetrachlorobiphenyl (PCB 66)
2,4,4′,5-Tetrachlorobiphenyl (PCB 74)
3,4,4′,5-Tetrachlorobiphenyl (PCB 81)
2,3,3′,4,4′-Pentachlorobiphenyl (PCB 105)
2,3′,4,4′,5-Pentachlorobiphenyl (PCB 118)
3,3′,4,4′,5-Pentachlorobiphenyl (PCB 126)
2,3,3′,4,4′,5-Hexachlorobiphenyl (PCB 156)

Polychlorinated Dibenzo-p-Dioxins, Polychlorinated Dibenzofurans, and Polychlorinated Biphenyls

2,3,3′,4,4′,5′-Hexachlorobiphenyl (PCB 157)
2,3′,4,4′,5,5′-Hexachlorobiphenyl (PCB 167)
3,3′,4,4′,5,5′-Hexachlorobiphenyl (PCB 169)
2,3,3′,4,4′,5,5′-Heptachlorobiphenyl (PCB 189)
2,2′,5,5′-Tetrachlorobiphenyl (PCB 52)
2,2′,3,4,5′-Pentachlorobiphenyl (PCB 87)
2,2′,4,4′,5-Pentachlorobiphenyl (PCB 99)
2,2′,4,5,5′-Pentachlorobiphenyl (PCB 101)
2,3,3′,4′,6-Pentachlorobiphenyl (PCB 110)
2,2′,3,3′,4,4′-Hexachlorobiphenyl (PCB 128)
2,2′,3,4,4′,5′ and 2,3,3′,4,4′,6-Hexachlorobiphenyl (PCB 138 and 158)
2,2′,3,4′,5,5′-Hexachlorobiphenyl (PCB 146)
2,2′,3,4′,5′,6-Hexachlorobiphenyl (PCB 149)
2,2′,3,5,5′,6-1-Hexachlorobiphenyl (PCB 151)
2,2′,4,4′,5,5′-Hexachlorobiphenyl (PCB 153)
2,2′,3,3′,4,4′,5-Heptachlorobiphenyl (PCB 170)
2,2′,3,3′,4,5,5′-Heptachlorobiphenyl (PCB 172)
2,2′,3,3′,4,5′,6′-Heptachlorobiphenyl (PCB 177)
2,2′,3,3′,5,5′,6-Heptachlorobiphenyl (PCB 178)
2,2′,3,4,4′,5,5′-Heptachlorobiphenyl (PCB 180)
2,2′,3,4,4′,5′,6-Heptachlorobiphenyl (PCB 183)
2,2′,3,4′,5,5′,6-Heptachlorobiphenyl (PCB 187)
2,2′,3,3′,4,4′,5,5′-Octachlorobiphenyl (PCB 194)
2,2′,3,3′,4,4′,5,6-Octachlorobiphenyl (PCB 195)
2,2′,3,3′,4,4′,5,6′ and 2,2′,3,4,4′,5,5′,6-Octachlorobiphenyl (PCB 196 and 203)
2,2′,3,3′,4,5,5′,6′-Octachlorobiphenyl (PCB 199)
2,2′,3,3′,4,4′,5,5′,6-Nonachlorobiphenyl (PCB 206)

Phthalates

Mono-methyl phthalate
Mono-ethyl phthalate
Mono-n-butyl phthalate
Mono-isobutyl phthalate
Mono-benzyl phthalate
Mono-cyclohexyl phthalate
Mono-2-ethylhexyl phthalate
Mono-(2-ethyl-5-hydroxyhexyl) phthalate
Mono-(2-ethyl-5-oxohexyl) phthalate
Mono-3-carboxypropyl phthalate
Mono-n-octyl phthalate
Mono-isononyl phthalate

Organochlorine Pesticides

Hexachlorobenzene
Hexachlorocyclohexane
beta Hexachlorocyclohexane
gamma Hexachlorocyclohexane
Pentachlorophenol
2,4,5-Trichlorophenol
2,4,6-Trichlorophenol

Organochlorine Pesticides
Dichlorodiphenyltrichloroethane
p,p′-DDT
p,p′-DDE
n,p′-DDT
Chlordane
Oxychlordane
trans-Nonachlor
Heptachlor
Heptachlor epoxide
Mirex
Aldrin
Dieldrin
Ealdrin
Organophosphate Pesticides
Dimethylphosphate
Dimethylthiophosphate
Dimethyldithiophosphate
Diethylphosphate
Diethyldithiophosphate
Malathion dicarboxylic acid
para-Nitrophenol
3,5,6-Trichloro-2-pyridinol
2-Isopropyl-4-methyl-6-hydroxypyrimidine
2-(Diethylamino)-6-methylpyrimidin-4-ouone
3-Chloro-7-hydroxy-4-methyl-2H-chromen-2-one/ol
Herbicides
2,4,5-Trichlorophenoxyacetic acid
2,4-Dichlorophenoxyacetic acid
2,4-Dichlorophenol
Atrazine mercapturate
Alachlor mercapturate
Acetochlor mercapturate
Metolachlor mercapturate
Pyrethroids Pesticides
4-Fluoro-3-phenoxybenzoic acid
cis-3-(2,2-Dichlorovinyl)-2,2-dimethylcyclopropane carboxylic acid trans-3-(2,2-Dichlorovinyl)-2,2-dimethylcyclopropane carboxylic acid
cis-3-(2,2-Dibromovinyl)-2,2-dimethylcyclopropane carboxylic acid
3-Phenoxybenzoic acid
Other Pesticides
N,N-Diethyl-3-methylbenzamide
ortho-Phenylphenol
2,5-Dichlorophenol
Carbamate Insecticides
2-Isopropoxyphenol
Carbofuranphenol

GLOSSARY

- AA — Aplastic anemia
- ADD — Attention deficit disorder
- ADHD — Attention deficit hyperactivity disorder
- AL — Adult-onset leukemias
- ALL — Acute lymphocytic leukemia is the most common type of leukemia in young children.
- AML — Acute myeloid leukemia (AML) occurs in both adults and children. This type of leukemia is sometimes called acute nonlymphocytic leukemia (ANLL).
- CAA — Clean Air Act of 1955, reauthorized in 1963 and 1982
- CAAA — Clean Air Act Amendments of 1966, 1970, 1982, and 1990
- CERCLA — Comprehensive Environmental Response, Compensation, and Liability Act
- CL — Childhood leukemias
- CLL — Chronic lymphocytic leukemia (CLL) most often affects adults over the age of 55. It sometimes occurs in younger adults, but it almost never affects children.
- CML — Chronic myeloid leukemia (CML) occurs mainly in adults. A very small number of children also develop this disease.
- CNS — Central nervous system
- CPSA — Consumer Product Safety Act
- CWA — Clean Water Act
- DBCP — Dibromochloropropane
- DDT — 4,4′-(2,2,2-trichloroethane-1,1-cliyl)bis(chlorobenzene)
- ESA — Endangered Species Act
- FDA — Food and Drug Administration
- FEPCA — Federal Environmental Pesticide Control Act

- FIFRA — Federal Hazardous Substances Labeling Act
- FQPA — Food Quality Protection Act amended the FDCA and the FIFRA
- Halothane (2-Bromo-2-chloro-1,1,1-trifluoroethane) is a general anaesthetic delivered via inhalation that is still used in veterinary medicine and outside the United States because its is relatively inexpensive.
- HCB — Hexachlorobenzene
- HCH — Hexachlorocyclohexane
- HCSA — Hazard Communication Standard Act
- Hepatotoxicity — Toxic to liver
- IARC — International Agency for Research on Cancer
- IQ — Intelligence quotient
- kg — Kilogram
- LOEL — Lowest observed effect level
- ug/day — Micrograms per day
- MM — Multiple myeloma
- MPRSA — Marine Protection, Research, and Sanctuaries Act
- NEPA — National Environmental Protection Act
- NIOSH — National Institute of Occupational Safety Health
- NHL — Non-hodgkin's lymphomas
- NOEL — The "no observable effect level" is determined by identifying the level of exposure that has been shown to not pose any harm to humans or laboratory animals. Prop 65 then requires this "no observable effect level" to be divided by 1000 in order to provide an ample margin of safety. Businesses subject to Proposition 65 are required to provide a warning if they cause exposures to chemicals listed as causing birth defects or reproductive harm that exceeds 1/1000th of the "no observable effect level."
- NRDC — Natural Resource Defense Council
- NSRL — No significant risk level. The level of exposure that would result in not more than one excess case of cancer in 100,000 individuals exposed to the chemical over a 70-year lifetime. In other words, a person exposed to the chemical at the "no significant risk level" for 70 years would not have more than a "one in 100,000" chance of developing cancer as a result of that exposure.
- OR — Odds ratio
- OSHA — Occupational Safety Health Act
- OSHA — Occupational Safety Health Administration
- PBDEs — Polybrominated diphenyl ethers
- PCB — Polychlorinated biphenyls
- PCE — Perchloroethylene
- PHS — Public Health Service
- PPM — Parts per million
- PPPA — Poison Prevention Packaging Act

- PRGs — Preliminary Remediation Goals
- Prop 65 — Safe Drinking Water and Toxic Enforcement Act or Proposition 65 or Prop 65. OEHHA — Office of Environmental Health and Hazard Assessment
- PWA — Ports and Waterways Safety Act
- RBC — Red blood cells (also called RBCs or erythrocytes) carry oxygen from the lungs to the body's tissues and take carbon dioxide from the tissues back to the lungs. The red blood cells give blood its color.
- RCRA — Resource Conservation and Recovery Act
- RfD — reference dose
- SARA — Superfund Amendment and Reauthorization Act
- SDWA — Safe Drinking Water Act
- SMCRA — Surface Mining Control and Reclamation Act
- SMR — standard mortality ratio (SMR)
- TCE — trichloroethylene
- Trihalolmethanes — disifection byproducts created in drinking water after chlorination of the raw water.
- TSCA — Toxic Substances Control Act
- WBC — White blood cells (also called WBCs or leukocytes) help the body fight infections and other diseases.
- USPHS — U. S. Public Health Service
- USEPA — United States Environmental Protection Agency
- PFCs — perfluorochemicals
- PCDD — polychlorinated dibenzodioxins
- PCDF — polychlorinated dibenzofurans
- PBDD — polybrominated dibenzodioxins
- PBDF — polybrominated dibenzofurans
- PAHs — polyaromatic hydrocarbons
- EWG — Environmental Working Group
- TSCA — Toxic Substances Control Act of 1976
- COC — Chemical of concern
- AL — alimentary system (liver or
- DEV — development of a fetus
- END — endocrine system
- EYE — ocular system
- HEM — hematopoietic system
- IMM — the immune system
- NS — nervous system
- REP — reproductive system
- RESP — respiratory system
- SKIN — the dermal system
- NOEC — no-observable effect concentration
- EDC — Endocrine-disrupting compounds

- NMDRC — nonmonotonic dose-response curve
- MCL — maximum contaminant level
- Safe Harbor Level — consist of no significant risk levels for chemicals listed as causing cancer and maximum allowable dose levels for chemicals listed as causing birth defects or other reproductive harm.

INDEX